Longman Mathematics for IGCSE Book 2

D A Turner, I A Potts,
W R J Waite, B V Hony

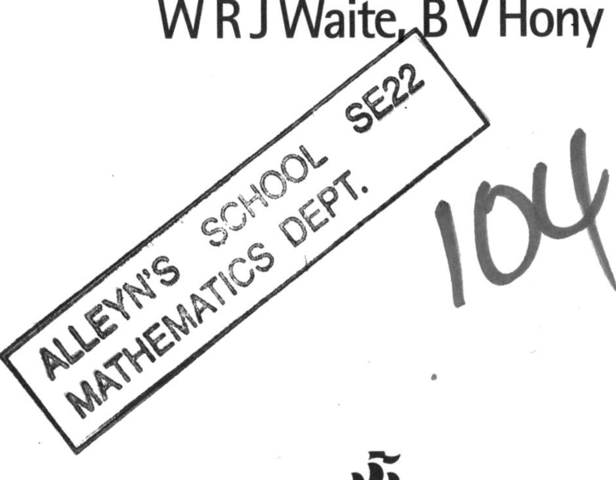

PEARSON
Longman

Pearson Education Limited
Edinburgh Gate
Harlow
Essex
CM20 2JE
England

First published 2006
Third impression 2006
ISBN-13: 978-1-4058-0212-3
ISBN-10:1-4058-0212-X

Prepared for publication by Tech-Set Ltd.

Printed by CPI Bath

Acknowledgements

We are grateful to Edexcel, NEAB and OCR for permission to reproduce copyright examination questions. Edexcel, NEAB and OCR can accept no responsibility whatsoever for the accuracy or method of working in the answers, where given.

We are grateful to the following for permission to reproduce photographs:

Actionplus: p.170, p.294; **Alamy:** p.2 (Shout Pictures), p.377 (John Terence Turner); **Empics:** p.460; **Mary Evans Picture Library:** p.16, p.282; **Getty Images:** p.454; **Frank Lane Picture Library:** p.83, p.301; **Rex Features:** p.291, p.455; **Superstock:** p.454.

Every effort has been made to trace the copyright holders and we apologise in advance for any unintentional omissions. We would be pleased to insert the appropriate acknowledgement in any subsequent edition of this publication.

Picture research by Kevin Brown

Note to teachers: Even numbered answers can be found at www.longman.co.uk/igcse/ .

Contents

* Each section of Unit 5 starts with a summary of all topics.
The revision exercises then test *all* topics of the IGCSE syllabus followed.

Course Structure

Unit 4

Unit 5 (Revision)

Preface

This two-book series is written for students following the IGCSE Higher tier specification for the Edexcel examination board. It comprises a Student's Book for each year of the course.

The course has been structured to enable these two books to be used in a sequential manner, both in the classroom and by students working on their own.

Each book contains five units of work. Each unit contains five sections in the topic areas: Number, Algebra, Graphs and Sequences, Shape and Space, and Handling Data.

In each unit, there are concise explanations and worked examples, plus numerous exercises that will help students to build up confidence.

Paired questions, with answers to the odd-numbered questions at the end of the Student's Book, allow students to check their answers and monitor their own progress. More difficult questions, to stretch the more able student, appear at the end of some exercises, and are identified by blue question numbers.

Parallel exercises are provided, allowing students to consolidate basic principles before being challenged with more difficult questions.

- **Non-starred exercises** are designed for students working towards IGCSE grades B/C.
- **Starred exercises** are designed to challenge students working towards IGCSE grades A/A*.

Real data has been used where possible and ICT links are identified by the icon  to give opportunities for students to investigate topics further using these skills. Both within the sections and at the end of each unit, challenges and investigations encourage students to think for themselves.

- **Challenges** provide questions applying the basic principles in unusual situations.
- **Investigations** prepare students for more advanced work.
  Students can refer to the 'How to Tackle Investigations' section at the end of Book 1 before embarking on any of these investigations.

Consolidation is a recurring theme throughout the course and general skills are reinforced at the end of each unit.

- **Pairs of parallel revision exercises** appear at the end of each of the five sections within each unit.
- **Numeracy practice exercises** provide opportunities for students to flex their basic arithmetic and algebraic skills.
- **Fact finders** test numerical comprehension of real data. Students use the information supplied to answer thought-provoking questions. The shaded questions are more challenging.
- **Summaries** précis the major points of each unit.
- **Examination practice papers** test students' understanding of material and terminate each of the units in Book 1 and the first four units of Book 2.

Revision is vital to the success of every student and has been covered in Unit 5 of Book 2:

- Summaries of all IGCSE topics plus ten Revision exercises
- Four examination papers covering the entire syllabus.

Number 1

Inverse proportion

An example of **direct proportion** is: when one quantity is *multiplied* by two, the other quantity is also *multiplied* by two.

For example, if 2 kg of apples cost $3, then 4 kg of the same type of apple costs $6.

This relationship produces a straight line graph through the origin.

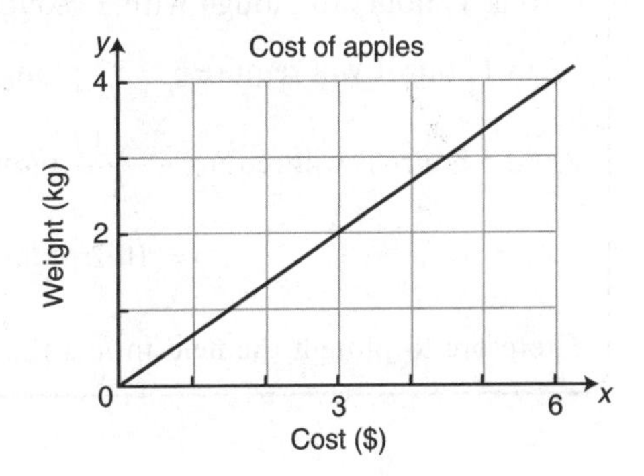

However, with **inverse proportion**, when one quantity is *multiplied* by two, the other quantity is *divided* by two.

For example, if one machine produces 100 lamp shades in one hour, then it will take two machines half an hour to produce the same number of identical lamp shades.

This relationship produces a **hyperbola**.

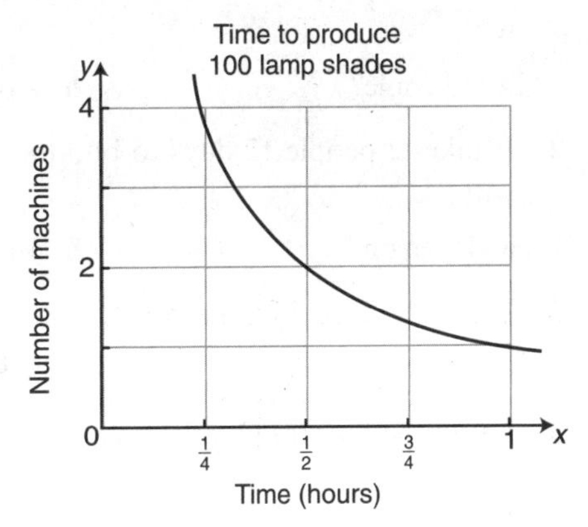

Activity 1

For a distance of 160 km, a train travels at a constant speed of 80 km/hour for 2 hours. Clearly, if the speed were halved, the same journey would take twice as long. This is an example of **inverse proportion**. This can be written as:

It takes 2 hours to travel 160 km at 80 km/hour

∴ It takes 4 hours to travel 160 km at 40 km/hour

♦ Copy and complete this table for the train journey over x km:

Time (hours)	1	2	4	6	7	8
Speed (km/hour)		80	40			

To find the speed for 6 and 7 hours, look at the product of the speed and time for the other entries.

♦ Plot the points on a suitable graph of speed against time. Join the points with a *smooth curve*. Comment on the shape of the curve.

Example 1

Last year, a farmer used 3 ploughs to plough a field and it took 17 hours. This year the same job must be done in less than 5 hours. How many ploughs will be needed?

It took 17 hours to plough with 3 ploughs

$\therefore$ in 1 hour it will require 3×17 ploughs

$\therefore$ in 5 hours it will require $\frac{3 \times 17}{5}$ ploughs

$= 10.2$ ploughs

Therefore to plough the field in *less* than 5 hours the farmer will need to use 11 ploughs.

Exercise 1

1 It takes 1 person 8 days to dig a trench. To dig a similar trench, how long would it take with

a 2 people? **b** 4 people? **c** 3 people?

2 It takes 2 people 12 days to build a wall. To build a similar wall, how long would it take with:

a 1 person? **b** 4 people? **c** 5 people?

3 Using the facts in Question 1, how many people are required to dig a similar trench in

a 1 day? **b** half a day? **c** a quarter of a day?

4 Using the facts in Question 2, how many people are required to build a similar wall in

a 2 days? **b** 3 days? **c** 4 days?

5 It has been estimated that it took 4000 men 30 years to build the largest pyramid, at Giza, in Egypt, over four and a half thousand years ago. How long would it have taken with

a 2000 men? **b** 8000 men? **c** 100 men?

6 Using the facts in Question 5, how many men would have been required to build it in

a 3 years? **b** 6 years? **c** 300 years?

7 Over a given distance, a train travels at a constant speed of 120 km/hour for 3 hours.

a Over the same distance, how long will it take a train travelling at 30 km/hour?

b Find the speed of another train that takes 5 hours to travel the same distance.

8 A car's average fuel consumption is 40 km/litre (km per litre) and, over a certain distance, it uses 4 litres of fuel.

a Over the same distance, another car uses 5 litres. Find its fuel consumption in km/litre.

b The fuel consumption of another car is 8 km/litre. How much fuel did this car use to travel the same distance?

Exercise 1★

1 In an exam room, the total power of all the light bulbs has to be 3000 watts.

a Copy and complete this table.

Number of light bulbs (N)	Power of each bulb (P)
6	
	600
2	
	100
N	P

b Write down the relationship between N and P.

2 The construction of the Channel Tunnel created about 100 000 man-years of employment in the UK. In theory, this meant that it could have been built by 20 000 men in 5 years.

a Copy and complete this table.

Number of years (N)	Number of men (M)
1	
	50 000
4	
	10 000
N	M

b Write down the relationship between N and M.

3 In a scientific experiment three different substances, A, B and C, *each of the same mass*, are used. Substance A has a density of 4 g/cm^3 and a volume of 3 cm^3.

a Find the density of 1 cm^3 of substance B.

b Find the volume of substance C with a density of 8 g/cm^3.

4 A water tank is filled in 8 minutes with a rate of flow of 40 litres/minute.

a How long will it take to fill at 32 litres/minute?

b What rate of flow will fill it in 25 minutes?

5 Using the facts for Question 2, copy and complete this table.

Number of men	Number of tunnels	Time in years
100 000	4	
100 000		2
20 000	8	
	2	0.5

6 One mosquito can produce 9 million young in 500 hours. Copy and complete this table.

Number of mosquitoes	Number of young	Time
1	18 000	
1		1 second
	9 million	1 hour
500		500 hours

7 One cow belches 200 g of methane in a day. Copy and complete these statements.

a 1 cow belches 1 kg of methane in ... days.

b ... cows belch 1000 kg of methane in 5 days.

c 3100 million cows belch ... tonnes of methane in 365 days.

(It is estimated that there are about 3100 million cows in the world.)

8 On average, an urban household in Europe produces 2.5 kg of waste per day.

a Calculate, in kilograms, correct to 2 significant figures, the amount of waste produced by each household in 1 year.

b Calculate, in tonnes, correct to 2 significant figures, the amount of waste produced by a small town of 5000 households in 1 year.

c Calculate, in tonnes, correct to 2 significant figures, the amount of waste produced by a city of 200 000 households in 1 year.

9 During the University Boat Race, a commentator said, 'Every member of the crew does the equivalent amount of work of someone who lifts a 25 kg sack of potatoes, from the floor to shoulder height, 36 times a minute for 18 minutes'.

a What total mass is 'lifted' by one crew member during the 18-minute race?
Give your answer to 2 significant figures.

b A lorry is to be loaded with sacks of potatoes each of mass 25 kg.
Work out, correct to 1 significant figure, how long it should take

(i) 1 crew member to load 4 tonnes

(ii) 8 crew members to load 4 tonnes

(iii) 4 crew members to load 4 tonnes

(iv) 8 crew members to load 1 tonne

Recurring decimals

Remember

- All fractions can be written as decimals which either *terminate* or produce a set of *recurring* digits.
- Fractions that produce terminating decimals have, in their simplest form, denominators with only 2 or 5 as factors. This is because 2 and 5 are the only factors of 10 (*decimal* system).
- The dot notation is used to indicate which digits recur.
  For example $0.2323\ldots = 0.\dot{2}\dot{3}$, $0.056056\ldots = 0.\dot{0}5\dot{6}$.

Example 2

Change $0.\dot{5}$ to a fraction.

$x = 0.5555555\ldots$	(Multiply both sides by 10)
$10 \times x = 5.5555555\ldots$	(Subtract the top from the bottom)
$9 \times x = 5$	(Divide both sides by 9)
$x = \frac{5}{9}$	

Example 3

Change $0.\dot{7}\dot{9}$ to a fraction.

$x = 0.797979\ldots$	(Multiply both sides by 100)
$100 \times x = 79.797979\ldots$	(Subtract the top from the bottom)
$99 \times x = 79$	(Divide both sides by 99)
$x = \frac{79}{99}$	

Example 4

Change $0.\dot{1}2\dot{3}$ to a fraction.

$x = 0.123123\ldots$	(Multiply both sides by 1000)
$1000 \times x = 123.123123\ldots$	(Subtract the top from the bottom)
$999 \times x = 123$	(Divide both sides by 999)
$x = \frac{123}{999}$	

Key Point

To change a simple recurring decimal to a fraction

No. of repeating digits	First, multiply by	Last, divide by
1	10	9
2	100	99
3	1000	999

Exercise 2

For Questions 1–6, change the fraction to a terminating decimal.

1 $\frac{3}{8}$ **2** $\frac{1}{20}$ **3** $\frac{2}{25}$ **4** $\frac{3}{16}$ **5** $\frac{9}{32}$ **6** $\frac{11}{80}$

For Questions 7–14, change the fraction to a recurring decimal, writing your answers using the dot notation.

7 $\frac{2}{9}$ **8** $\frac{5}{9}$ **9** $\frac{2}{11}$ **10** $\frac{4}{11}$

11 $\frac{4}{15}$ **12** $\frac{8}{15}$ **13** $\frac{7}{18}$ **14** $\frac{17}{18}$

For Questions 15–18, *without* doing any calculation, write down the fractions that produce terminating decimals.

15 $\frac{5}{11}, \frac{9}{16}, \frac{2}{3}, \frac{5}{6}, \frac{2}{15}$

16 $\frac{5}{7}, \frac{4}{33}, \frac{5}{32}, \frac{7}{30}, \frac{3}{8}$

17 $\frac{2}{19}, \frac{3}{20}, \frac{5}{48}, \frac{5}{64}, \frac{13}{22}$

18 $\frac{3}{40}, \frac{5}{17}, \frac{7}{80}, \frac{9}{25}, \frac{9}{24}$

For Questions 19–30, change each of these recurring decimals to a fraction in its simplest form.

19 $0.\dot{3}$ **20** $0.\dot{4}$ **21** $0.\dot{5}$ **22** $0.\dot{6}$ **23** $0.\dot{7}$ **24** $0.\dot{9}$

25 $0.0\dot{7}$ **26** $0.0\dot{1}$ **27** $0.0\dot{3}$ **28** $0.0\dot{2}$ **29** $0.0\dot{5}$ **30** $0.0\dot{6}$

Exercise 2★

For Questions 1–8, change each fraction to a recurring decimal, writing your answers using the dot notation.

1 $\frac{7}{15}$ **2** $\frac{11}{18}$ **3** $\frac{7}{150}$ **4** $\frac{11}{180}$

5 $2\frac{10}{33}$ **6** $4\frac{4}{33}$ **7** $\frac{149}{495}$ **8** $\frac{139}{495}$

For Questions 9–12, *without* doing any calculation, write down the fractions that produce terminating decimals.

9 $\frac{7}{17}, \frac{11}{16}, \frac{2}{3}, \frac{7}{40}, \frac{3}{15}$

10 $\frac{2}{7}, \frac{7}{40}, \frac{5}{32}, \frac{1}{30}, \frac{7}{8}$

11 $\frac{3}{17}, \frac{19}{20}, \frac{3}{25}, \frac{5}{64}, \frac{13}{24}$

12 $\frac{11}{125}, \frac{5.5}{128}, \frac{7}{81}, \frac{9}{512}, \frac{9}{48}$

For Questions 13–24, change each of these recurring decimals to a fraction in its simplest form.

13 $0.\dot{2}\dot{4}$ **14** $0.\dot{3}\dot{8}$ **15** $0.\dot{3}\dot{0}$ **16** $0.\dot{9}\dot{3}$ **17** $9.0\dot{1}\dot{9}$ **18** $8.0\dot{2}\dot{9}$

19 $0.0\dot{2}\dot{7}$ **20** $0.0\dot{3}\dot{6}$ **21** $0.\dot{4}1\dot{2}$ **22** $0.\dot{1}0\dot{1}$ **23** $0.\dot{3}8\dot{4}$ **24** $0.\dot{4}7\dot{4}$

For Questions 25–28, change each recurring decimal to a fraction.

25 $0.1\dot{2}$ **26** $0.8\dot{6}$ **27** $0.0\dot{5}\dot{6}$ **28** $0.1\dot{5}\dot{6}$

For Questions 29–30, write each answer as a recurring decimal.

29 $0.7\dot{3} \times 0.\dot{0}\dot{5}$ **30** $0.0\dot{7} \times 0.\dot{2}14285\dot{7}$

Exercise 3 (Revision)

1 A rounders field can be cut with 2 mowers in 3 hours.

a How long would it take to cut the field with 3 mowers?

b How many mowers would be required to cut the field in 1 hour?

2 A car's average fuel consumption is 60 km/litre and, over a certain distance, it uses 3 litres of fuel.

a Over the same distance, another car uses 12 litres. Find its fuel consumption in km/litre.

b The fuel consumption of another car is 25 km/litre. How much fuel did this car use to travel the same distance?

3 Write these fractions as terminating decimals.

a $\frac{1}{5}$ **b** $\frac{1}{8}$ **c** $\frac{1}{20}$

4 Change these fractions to recurring decimals, writing your answers using the dot notation.

a $\frac{1}{6}$ **b** $\frac{4}{9}$ **c** $\frac{3}{7}$

5 Change these recurring decimals to fractions.

a $0.\dot{2}$ **b** $0.0\dot{7}$ **c** $0.\dot{2}\dot{3}$

Exercise 3★ (Revision)

1 The average fuel consumption of Mrs Singh's car is 15 km/litre. On a particular journey, she used 48 litres of fuel.

a If she had used 45 litres, find her fuel consumption.

b If the fuel consumption had been 18 km/litre, how much fuel would she have used?

2 One honey bee has to travel 75 000 km to produce 500 g of honey.
Copy and complete these statements.

a ... honey bees would each have to travel 3000 km to produce a total of 500 g.

b 25 honey bees would each have to travel 60 km to produce a total of ... g.

c 25 honey bees would each have to travel ... km to produce a total of 1 tonne.

3 Given that $\frac{1}{13} = 0.\dot{0}7692\dot{3}$, write $\frac{2}{13}$ as a recurring decimal.

4 Which of these produce terminating decimals? $\frac{11}{25}$, $\frac{5}{12}$, $\frac{7}{256}$, $\frac{9}{500}$, $\frac{9}{15}$

5 Write these recurring decimals as fractions in their simplest form.

a $0.\dot{2}$ **b** $0.0\dot{1}$ **c** $0.\dot{6}\dot{7}$ **d** $3.0\dot{4}\dot{5}$

6 Change $\frac{11}{90}$ to a recurring decimal.

Proportion

If two quantities are related to each other, given enough information, it is possible to write a formula describing this relationship.

Activity 2

Copy and complete this table to show which paired items are directly related.

Variables	Related? Y/N
Area of a circle (A) and its radius (r)	Y
Circumference of a circle (C) and its diameter (d)	
Volume of water in a tank (V) and its weight (w)	
Distance travelled (D) at constant speed and time taken (t)	
Number of pages in a book (N) and its thickness (t)	
Mathematical ability (M) and a person's height (h)	
Wave height in the sea (W) and wind speed (s)	
Grill temperature (T) and time to toast bread (t)	

Direct proportion

Linear relationships

When water is poured into an empty fish tank, each litre poured in increases the depth by a fixed amount.

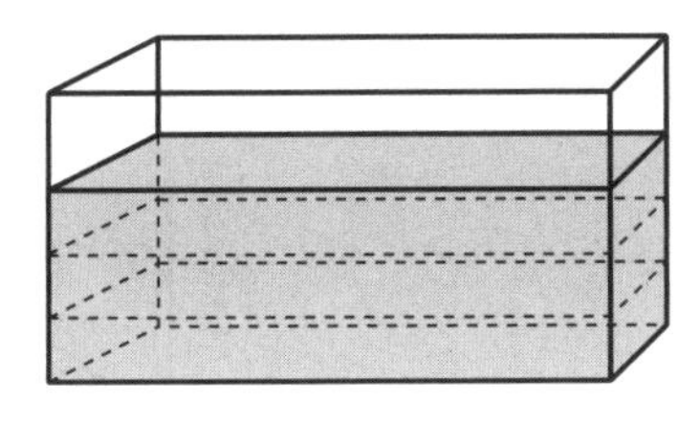

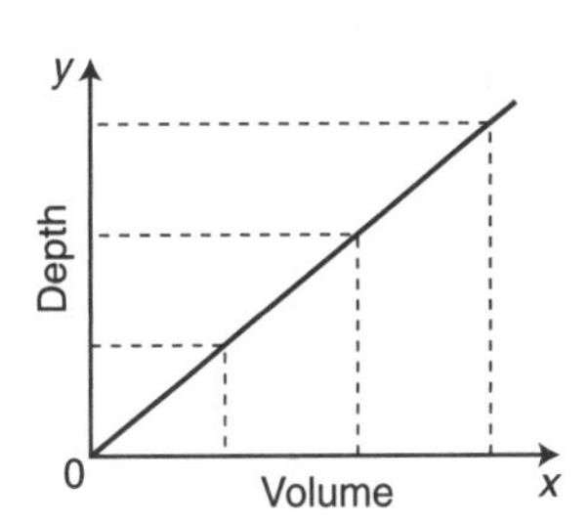

A graph of depth, y, against volume, x, is a straight line through the origin, showing a linear relationship.

In this case, y is **directly proportional** to x. If y is doubled, so is x. If y is halved, so is x, etc.

This relationship can be expressed in *any* of these ways.

- y is directly proportional to x.
- y varies directly with x.
- y varies as x.

All these statements mean the same.

In symbols, direct proportion relationships can be written as $y \propto x$. The $\propto$ sign can then be replaced by '$= k$' to give $y = kx$, where k is the **constant of proportionality**. The graph of $y = kx$ is the equation of a straight line through the origin, with slope k.

Key Point

y is directly proportional to x
is written as $y \propto x$
and this means $y = kx$, for some fixed value, k.

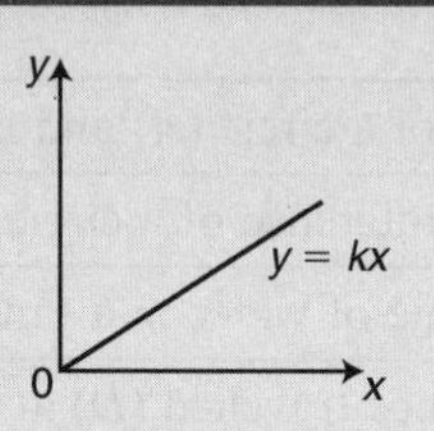

Example 1

The extension, y cm, of a spring is directly proportional to the mass, x kg, hanging from it.
If $y = 12$ cm when $x = 3$ kg, find

a the formula for y in terms of x

b the extension y cm when a 7 kg mass is attached

c the mass x kg that produces a 20 cm extension

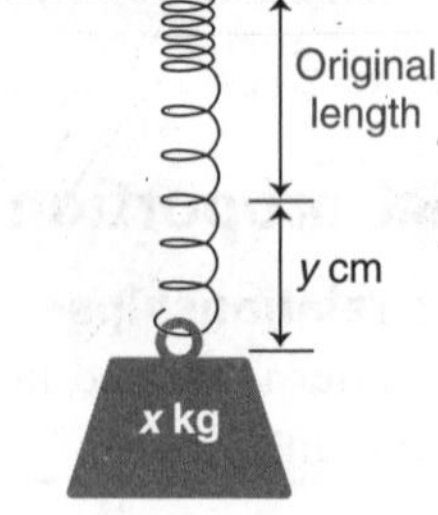
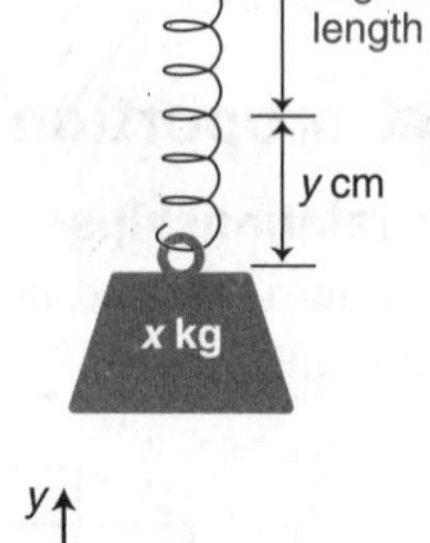

a y is proportional to x, so $y \propto x$ $\quad y = kx$

$y = 12$ when $x = 3$ $\quad 12 = k \times 3$

$k = 4$

The formula is therefore $y = 4x$.

b When $x = 7$ $\quad y = 4 \times 7$

The extension produced from a 7 kg mass is 28 cm.

c When $y = 20$ cm $\quad 20 = 4x$

$x = 5$

The extension produced from a 5 kg mass is 20 cm.

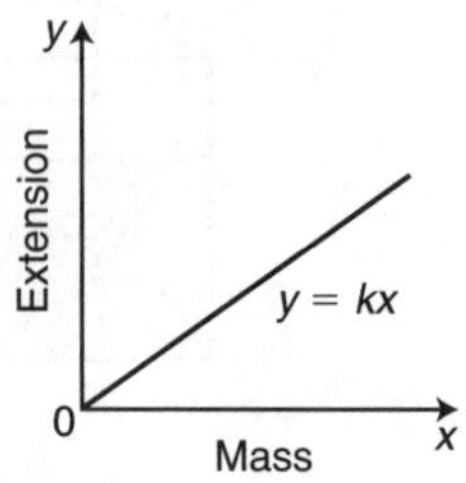

Exercise 4

1 y is directly proportional to x. If $y = 10$ when $x = 2$, find

a the formula for y in terms of x

b y when $x = 6$

c x when $y = 25$

2 d is directly proportional to t. If $d = 100$ when $t = 25$, find

a the formula for d in terms of t

b d when $t = 15$

c t when $d = 180$

3 An elastic string's extension y cm varies as the mass x kg that hangs from it. The string extends 4 cm when a 2 kg mass is attached.

a Find the formula for y in terms of x.

b Find y when $x = 5$.

c Find x when $y = 15$.

4 A bungee jumping rope's extension e m varies as the mass M kg of the person attached to it. If $e = 4$ m when $M = 80$ kg, find

a the formula for e in terms of M

b the extension for a person with a mass of 100 kg

c the mass of a person when the extension is 6 m

5 An ice-cream seller discovers that, on any particular day, the number of sales (I) is directly proportional to the temperature (t °C). 1500 sales are made when the temperature is 20 °C. How many sales might be expected on a day with a temperature forecast of 26 °C?

6 The number of people in a swimming pool (N) varies as the daily temperature (t °C). 175 people swim when the temperature is 25 °C. The pool's capacity is 200 people. Will people have to queue and wait if the temperature reaches 30 °C?

Exercise 4★

1 The speed of a stone, v m/s, falling off a cliff is directly proportional to the time, t seconds, after release. Its speed is 4.9 m/s after 0.5 s.

a Find the formula for v in terms of t.

b What is the speed after 5 s?

c At what time is the speed 24.5 m/s?

2 The cost, c cents, of a tin of salmon varies directly with its mass, m g. The cost of a 450 g tin is 150 cents.

a Find the formula for c in terms of m.

b How much does a 750 g tin cost?

c What is the mass of a tin costing $2?

3 The distance a honey bee travels, d km, is directly proportional to the mass of honey, m g, it produces. A bee travels 150 000 km to produce 1 kg of honey.

a Find the formula for d in terms of m.

b What distance is travelled by a bee to produce 10 g of honey?

c What mass of honey is produced by a bee travelling once around the world, a distance of 40 000 km?

4 The mass of sugar, m g, used in making oat-meal cookies varies directly as the number of cookies, n. 3.25 kg are used to make 500 cookies.

a Find the formula for m in terms of n.

b What mass of sugar is needed for 150 cookies?

c How many cookies can be made using 10 kg of sugar?

5 The height of a tree, h m, varies directly with its age, y years. A 9 m tree is 6 years old.

a Find the formula for h in terms of y.

b What height is a tree that is 6 months old?

c What is the age of a tree that is 50 cm tall?

6 The yearly profit (\$$P$ million) made by an Internet company, 'Line-On', is directly proportional to the annual amount spent (\$$x$ million) on its advertising on TV, radio, newspapers and the Internet. Its profit in one year from TV advertising alone amounts to \$6 million at a cost of \$1.5 million.

a Copy and complete this table.

Advertising medium	\$$x$ million	\$$P$ million
TV	1.5	6
Radio	0.5	
Newspapers		5
Internet	0.1	

b How much total profit was made by 'Line-On' in this particular year?

c What was the total amount spent by the company on advertising itself in this year?

d Express x as a percentage of P. Comment on your answer.

Nonlinear relationships

Water is poured into an empty inverted cone. Each litre poured in will result in a different depth increase.

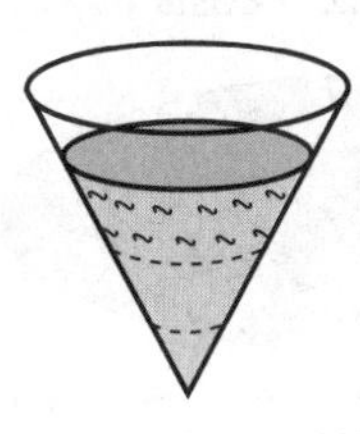

A graph of volume, y, against depth, x, will illustrate a direct **nonlinear relationship**.

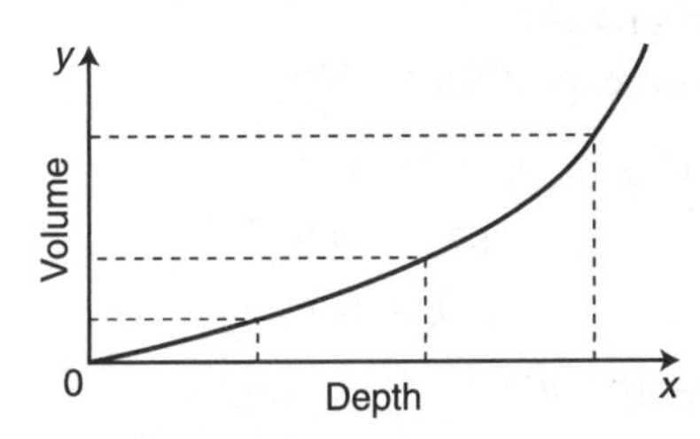

This relationship can be expressed in *either* of these ways.

- y is directly proportional to x cubed.
- y varies as x cubed.

Both these statements mean the same.

In symbols, this relationship is written as $y \propto x^3$. The $\propto$ sign can then be replaced by '$= k$' to give $y = kx^3$, where k is the constant of proportionality.

Key Point

'y varies as x squared' is written as $y \propto x^2$ and this means $y = kx^2$, for some fixed value, k.

Example 2

Express these relationships as equations with constants of proportionality.

a y is directly proportional to x squared. $\quad y \propto x^2 \Rightarrow y = kx^2$

b m varies directly with the cube of n. $\quad m \propto n^3 \Rightarrow m = kn^3$

c s is directly proportional to the square root of t. $\quad s \propto \sqrt{t} \Rightarrow s = k\sqrt{t}$

d v squared varies as the cube of w. $\quad v^2 \propto w^3 \Rightarrow v^2 = kw^3$

Example 3

The cost of Luciano's take-away pizzas (C cents) is directly proportional to the square of the diameter (d cm) of the pizza. A 30 cm pizza costs 675 cents.

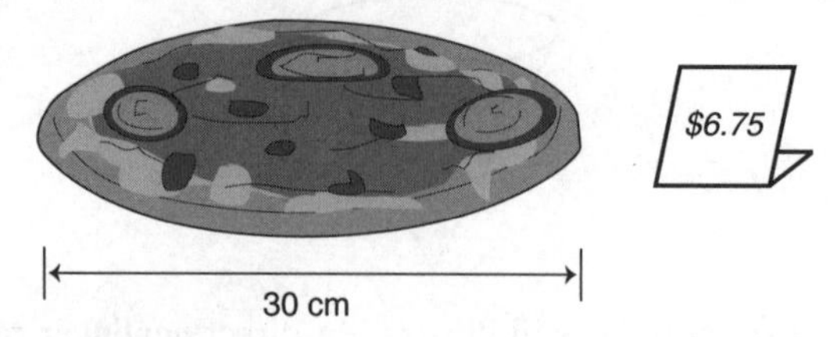

a What is the price of a 20 cm pizza?
b What size of pizza should you expect for $4.50?

a C is proportional to d^2, so $C \propto d^2$ $\quad C = kd^2$

$C = 675$ when $d = 30$ $\quad 675 = k(30)^2$

$k = 0.75$

The formula is therefore $C = 0.75d^2$.

When $d = 20$ $\quad C = 0.75(20)^2$

$C = 300$

The cost of a 20 cm pizza is $3.

b When $C = 450$ $\quad 450 = 0.75d^2$

$d^2 = 600$

$d = \sqrt{600} = 24.5$ (3 s.f.)

A $4.50 pizza should be 24.5 cm in diameter.

Exercise 5

1 y is directly proportional to the square of x. If $y = 100$ when $x = 5$, find
a the formula for y in terms of x **b** y when $x = 6$
c x when $y = 64$

2 p varies directly as the square of q. If $p = 72$ when $q = 6$, find
a the formula for p in terms of q **b** p when $q = 3$
c q when $p = 98$

3 v is directly proportional to the cube of w. If $v = 16$ when $w = 2$, find
a the formula for v in terms of w **b** v when $w = 3$
c w when $v = 128$

4 m varies directly as the square root of n. If $m = 10$ when $n = 1$, find
a the formula for m in terms of n **b** m when $n = 4$
c n when $m = 50$

5 The distance fallen by a parachutist, y m, is directly proportional to the square of the time taken, t secs. If 20 m are fallen in 2 s, find

a the formula expressing y in terms of t

b the distance fallen through in 3 s

c the time taken to fall 100 m

6 'Espirit' perfume is available in bottles of different volumes of similar shapes. The price, \$$P$, is directly proportional to the cube of the bottle height, h cm. A 10 cm high bottle is \$50. Find

a the formula for P in terms of h

b the price of a 12 cm high bottle

c the height of a bottle of 'Espirit' costing \$25.60

Exercise 5★

1 If f is directly proportional to g^2, copy and complete this table.

g	2	4	
f	12		108

2 If m is directly proportional to n^3, copy and complete this table.

n	1		5
m	4	32	

3 The resistance to motion, R newtons, of the 'Storm' racing car is directly proportional to the square of its speed, s km/hour. When the car travels at 160 km/hour it experiences a 500 newton resistance.

a Find the formula for R in terms of s.

b What is the car's speed when it experiences a resistance of 250 newtons?

4 The height of Giants, H metres, is directly proportional to the cube root of their age, y years. An 8-year-old Giant is 3 m tall.

a Find the formula for H in terms of y.

b What age is a 12 m tall Giant?

5 The surface area of a sphere is directly proportional to the square of its radius. A sphere of radius 10 cm must be increased to a radius of x cm if its surface area is to be doubled. Find x.

6 The mass of spherical cannon balls is directly proportional to the cube of their diameter. A cannon ball of diameter 100 cm must be decreased to a diameter of y cm if its mass is to be halved. Find y.

Activity 3

The German astronomer Kepler (1571–1630) devised three astronomical laws.
Kepler's third law gives the relationship between the orbital period, t days, of a planet around the Sun, and its mean distance, d km, from the Sun.

In simple terms, this law states that t^2 is directly proportional to d^3.

- Find a formula relating t and d, given that the Earth is 150 million km from the Sun.
- Copy and complete this table.

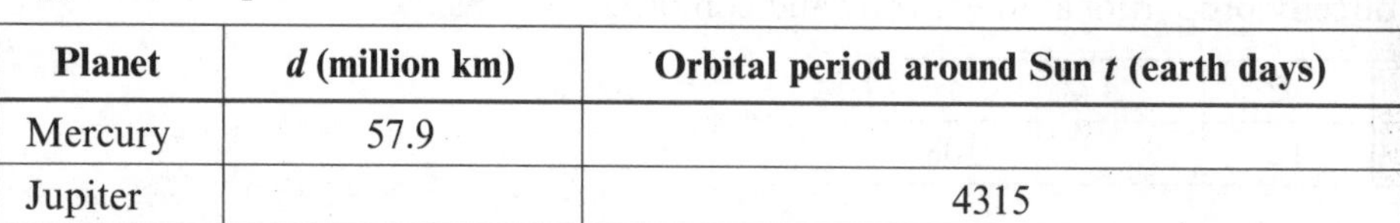

Planet	d (million km)	Orbital period around Sun t (earth days)
Mercury	57.9	
Jupiter		4315

- Try to find the values of d and t for other planets in the Solar System, and see if they fit the same relationship.

Inverse proportion

The temperature of a cup of coffee decreases as time increases.

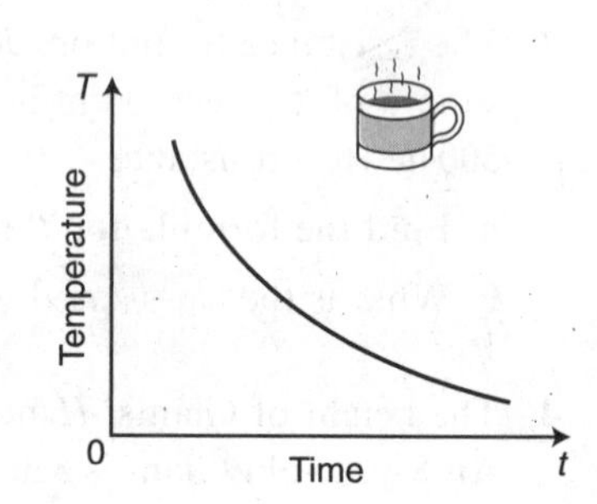

A graph of temperature (T) against time (t) shows an **inverse relationship**.

This can be expressed as: 'T is inversely proportional to t.'
In symbols, this is written as

$$T \propto \frac{1}{t}$$

The $\propto$ sign can then be replaced by '$= k$', so

$$T = \frac{k}{t}$$

where k is the constant of proportionality.

Key Point

If y is inversely proportional to x, the graph of y plotted against x looks like this.

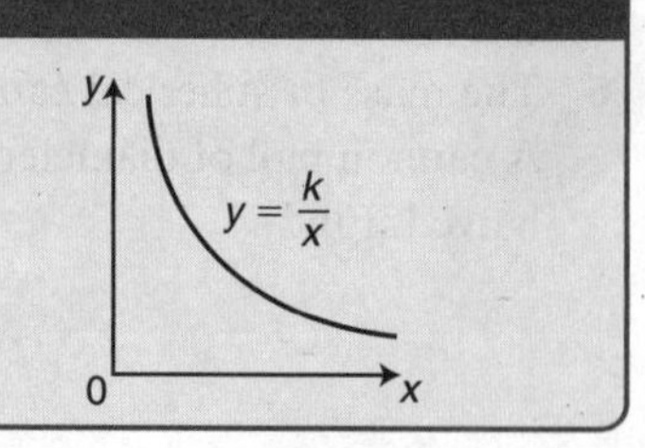

Example 4

Express these equations as relationships with constants of proportionality.

a y is inversely proportional to x squared. $y \propto \frac{1}{x^2} \Rightarrow y = \frac{k}{x^2}$

b m varies inversely as the cube of n. $m \propto \frac{1}{n^3} \Rightarrow m = \frac{k}{n^3}$

c s is inversely proportional to the square root of t. $s \propto \frac{1}{\sqrt{t}} \Rightarrow s = \frac{k}{\sqrt{t}}$

d v squared varies inversely as the cube of w. $v^2 \propto \frac{1}{w^3} \Rightarrow v^2 = \frac{k}{w^3}$

Example 5

Sound intensity, I dB (decibels), is inversely proportional to the square of the distance, d m, from the source. At a music festival, it is 110 dB, 3 m away from a loudspeaker.

a Find the formula relating I and d.

I is inversely proportional to d^2 $\quad I = \frac{k}{d^2}$

$I = 110$ when $d = 3$ $\quad 110 = \frac{k}{3^2}$

$k = 990$

The formula is therefore $I = \frac{990}{d^2}$.

b Calculate the sound intensity 2 m away from the speaker.

When $d = 2$ $\quad I = \frac{990}{2^2} = 247.5$

The sound intensity is 247.5 dB, 2 m away (enough to cause deafness).

c At what distance away from the speakers is the sound intensity 50 dB?

When $I = 50$ $\quad 50 = \frac{990}{d^2}$

$d^2 = 19.8$

$d = 4.45$ (3 s.f.)

The sound intensity is 50 dB, 4.45 m away from the speakers.

Exercise 6

1 y is inversely proportional to x. If $y = 4$ when $x = 3$, find

a the formula for y in terms of x **b** y when $x = 2$ **c** x when $y = 3$

2 d varies inversely with t. If $d = 10$ when $t = 25$, find

a the formula for d in terms of t **b** d when $t = 2$ **c** t when $d = 50$

3 m varies inversely with the square of n. If $m = 4$ when $n = 3$, find

a the formula for m in terms of n

b m when $n = 2$ **c** n when $m = 1$

4 V varies inversely with the cube of w. If $V = 12.5$ when $w = 2$, find

a the formula for V in terms of w

b V when $w = 1$ **c** w when $V = 0.8$

5 Light intensity, I candle-power, from a lighthouse is inversely proportional to the square of the distance, d m, of an object from this light source. If $I = 10^5$ when $d = 2$ m, find

a the formula for I in terms of d

b the light intensity at 2 km

6 The life-expectancy, L days, of a cockroach varies inversely with the square of the density, d people/m^2, of the human population near its habitat. If $L = 100$ when $d = 0.05$, find

a the formula for L in terms of d

b the life-expectancy of a cockroach in an area where the human population density is 0.1 people/m^2

Exercise 6★

1 If a is inversely proportional to b^2, copy and complete this table.

b	2	5	
a	50		2

2 A scientist gathers this data.

t	1	4		10
r	20		4	2

a Which of these relationships describes the collected data?

$$r \propto \frac{1}{\sqrt{t}} \qquad r \propto \frac{1}{t} \qquad r \propto \frac{1}{t^2}$$

b Copy and complete the table.

3 The electrical resistance, R ohm, of a fixed length of wire is inversely proportional to the square of its radius, r mm. If $R = 0.5$ when $r = 2$, find

a the formula for R in terms of r

b the resistance of a wire of 3 mm radius

4 The cost of Mrs Janus's electricity bill, \$$C$, varies inversely with the average temperature, t °C, over the period of the bill. If the bill is \$200 when the temperature is 25 °C, find

a the formula expressing C in terms of t

b the bill when the temperature is 18 °C

c the temperature generating a bill of \$400

5 The number of people shopping at Tang's Cornershop per day, N, varies inversely with the square root of the average outside temperature, t °C.

a Copy and complete this table.

Day	N	t
Mon	400	25
Tues		20
Wed	500	

b The remainder of the week (Thurs to Sat) has a hot spell with a constant daily average temperature of 30 °C. What is the average number of people per day who shop at Tang's for that week? (The shop is closed on Sundays.)

6 The time for a pendulum to swing, T s, is inversely proportional to the square root of the acceleration due to gravity, g m/s^2. On Earth $g = 9.8$, but on the Moon $g = 1.9$. Find the time of swing on the Moon of a pendulum whose time taken to swing on Earth is 2 s.

Activity 4

This graph shows an inverse relationship between the body mass, M kg, of mammals and their average heart pulse, P beats/min.

- ♦ Use the graph to complete this table.

	P (beats/min)	M (kg)
Hare		3
Dog	135	
Man		70
Horse	65	

- ♦ An unproven theory in biology states that the hearts of all mammals beat the same number of beats in an average life-span.
  - ▸ If man lives on average for 75 years, calculate the total number of heart beats in an average life-span.
  - ▸ Test out this theory by calculating the expected life-span of the creatures in the table above.

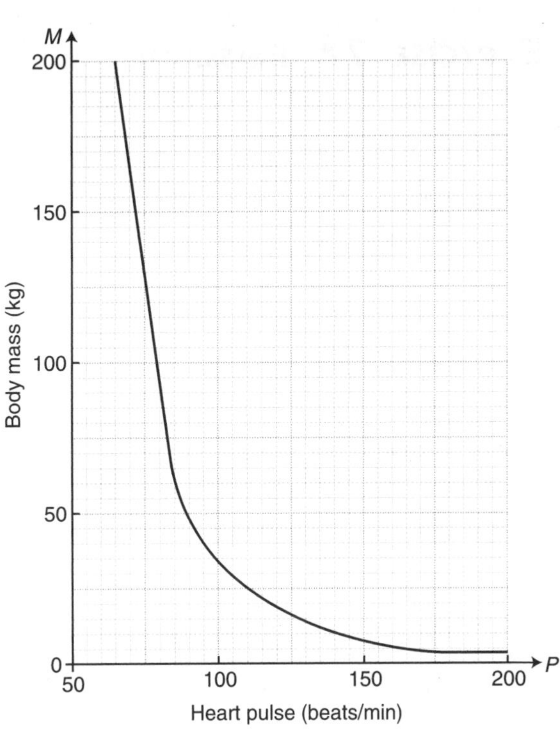

Exercise 7 (Revision)

1 y is directly proportional to x. If $y = 12$ when $x = 2$, find

a the formula for y in terms of x **b** y when $x = 7$ **c** x when $y = 66$

2 p varies as the square of q. If $p = 20$ when $q = 2$, find

a the formula for p in terms of q **b** p when $q = 10$ **c** q when $p = 605$

3 The cost, \$$c$, of laying floor tiles is directly proportional to the square of the area, $a\,\text{m}^2$, to be covered. If a $40\,\text{m}^2$ kitchen floor costs \$1200 to tile, find

a the formula for c in terms of a

b the cost of tiling a floor of area $30\,\text{m}^2$

c the area of floor covered by these tiles costing \$600

4 The time taken, t hours, to make a set of 20 curtains is inversely proportional to the number of people, n, who work on them. One person would take 80 hours to finish the task. Copy and complete this table.

n	1	2	4	
t	80			10

Exercise 7★ (Revision)

1 y squared varies as z cubed. If $y = 20$ when $z = 2$, find

a the formula relating y to z

b y when $z = 4$

c z when $y = 100$

2 m is inversely proportional to the square root of n.
If $m = 2.5 \times 10^7$ when $n = 1.25 \times 10^{-7}$, find

a the formula for m in terms of n

b m when $n = 7.5 \times 10^{-4}$

c n when m is one million

3 The frequency of radio waves, f MHz, varies inversely as their wavelengths, μ metres. If Radio 1 has $f = 99$ and $\mu = 3$, what is the wavelength of the BBC World Service on 198 kHz?

4 If y is inversely proportional to the nth power of x, copy and complete this table.

x	0.25	1	4	25
y		10	5	

Find a formula for y in terms of x.

Graphs 1

Cubic graphs $y = ax^3 + bx^2 + cx + d$

In cubic curves the highest power of x is x^3.

These curves have distinctive shapes and can be used to model real-life situations.

Example 1

Draw the graph of $y = 2x^3 + 2x^2 - 4x$ for $-3 \leqslant x \leqslant 2$ by compiling a suitable table.

x	-3	-2	-1	0	$\frac{1}{2}$	1	2
$2x^3$	-54	-16	-2	0	$\frac{1}{4}$	2	16
$2x^2$	18	8	2	0	$\frac{1}{2}$	2	8
$-4x$	12	8	4	0	-2	-4	-8
y	-24	0	4	0	$-1\frac{1}{4}$	0	16

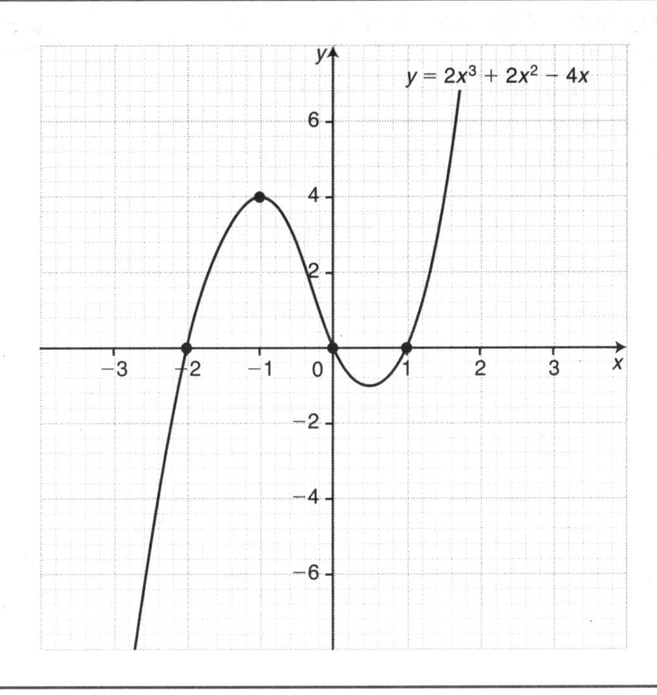

Remember

Cubic graphs have distinctive shapes that depend on the value of a.

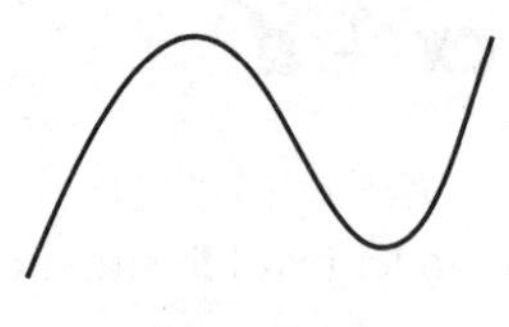

a is positive

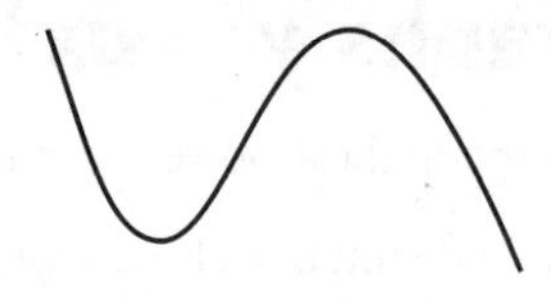

a is negative

Exercise 8

Draw the graphs of these equations between the stated x values after compiling a suitable table.

1 $y = x^3 + 2 \quad -3 \leqslant x \leqslant 3$

2 $y = x^3 - 2 \quad -3 \leqslant x \leqslant 3$

3 $y = x^3 + 3x \quad -3 \leqslant x \leqslant 3$

4 $y = x^3 - 3x \quad -3 \leqslant x \leqslant 3$

5 $y = x^3 + x^2 - 2x \quad -3 \leqslant x \leqslant 3$

6 $y = x^3 - 3x^2 + x \quad -2 \leqslant x \leqslant 3$

7 This water tank has dimensions as shown, in metres.

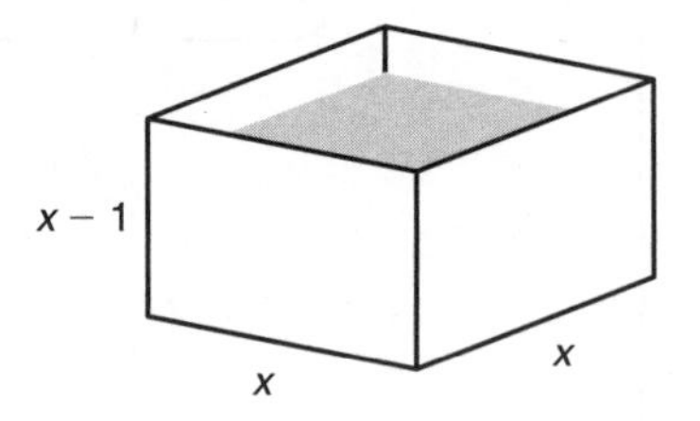

a Show that the volume of the tank, $V\,\text{m}^3$, is given by the formula $V = x^3 - x^2$.

b Draw the graph of V against x for $2 \leqslant x \leqslant 5$ by first constructing a suitable table of values.

c Use your graph to estimate the volume of a tank for which the base area is $16\,\text{m}^2$.

d What are the dimensions of a tank of volume $75\,\text{m}^3$?

8 The cross-section of a hilly region can be drawn as the graph of $y = x^3 - 8x^2 + 16x + 8$ for $0 \leqslant x \leqslant 5$, where x is measured in kilometres and y is the height above sea level in metres.

a Draw the cross-section by first constructing a suitable table for $0 \leqslant x \leqslant 5$.

b The peak is called Triblik and at the base of the valley is Vim Tarn. Mark these two features on your cross-section, and estimate the height of Triblik above Vim Tarn.

Exercise 8★

Draw the graphs of these equations between the stated x values after compiling a suitable table of values.

1 $y = 2x^3 - x^2 + x - 3 \quad -3 \leqslant x \leqslant 3$

2 $y = 2x^3 - 2x^2 - 24x \quad -4 \leqslant x \leqslant 4$

3 $y = -2x^3 + 3x^2 + 4x \quad -3 \leqslant x \leqslant 3$

4 $y = -x^3 - 2x^2 + 11x + 12 \quad -4 \leqslant x \leqslant 4$

5 A bullet is fired and its velocity, v metres per second, t seconds later is $v = 27t - t^3$, where $0 \leqslant t \leqslant 5$.

a Draw the graph of v against t after first compiling a table for the given values of t.

b Use your graph to estimate the greatest velocity of the bullet, and the time at which this occurs.

c For how long does the bullet travel faster than $30\,\text{m/s}$?

6 A toy is made that comprises a cylinder of diameter $2x$ cm and height x cm upon which is fixed a right circular cone of base radius x centimetres and height 6 cm.

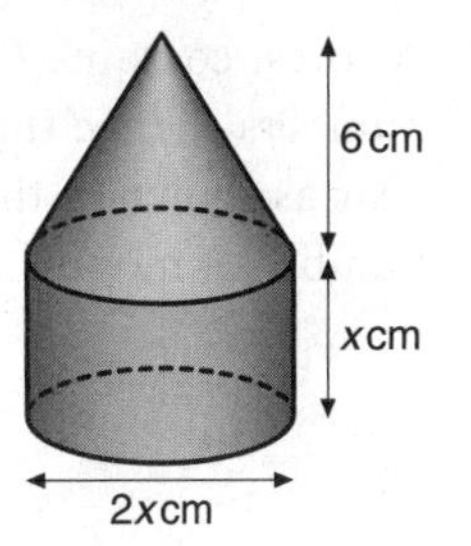

a Volume of a right circular cone $= \frac{1}{3} \times$ base area $\times$ height.
Show that the total volume V, in cubic centimetres, of the toy is given by $V = \pi x^2(x + 2)$.

b Draw the graph of V against x for $0 \leqslant x \leqslant 5$ after compiling a suitable table.

c Use your graph to find the volume of the toy of diameter 7 cm.

d What is the curved surface area of the cylinder if the total volume of the toy is 300 cm^3?

7 A closed cylindrical can of height h cm and radius r cm is made from a thin sheet of metal. The *total* surface area is $100\pi\,\text{cm}^2$.

a Show that $h = \dfrac{50}{r} - r$.
Hence show that the volume of the can, $V\,\text{cm}^3$, is given by $V = 50\pi r - \pi r^3$.

b Draw the graph of V against r for $0 \leqslant r \leqslant 7$ by first compiling a suitable table of values.

c Use the graph to estimate the greatest possible volume of the can.

d What is the diameter and height of the can of maximum volume?

8 An open box is made from a thin square metal sheet measuring 10 cm by 10 cm. Four squares of side x centimetres are cut away, and the remaining sides are folded upwards to make the box of depth x centimetres.

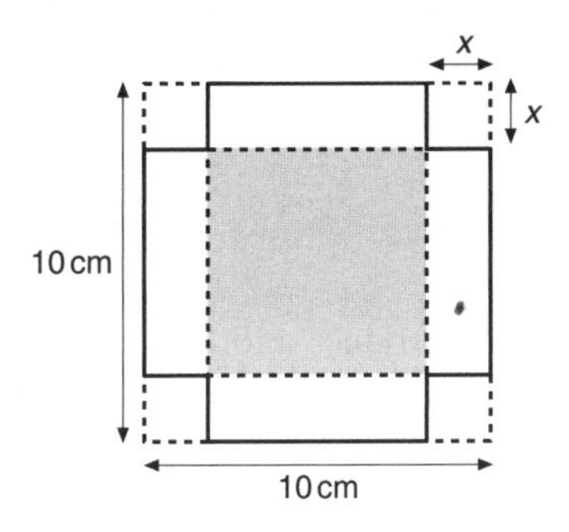

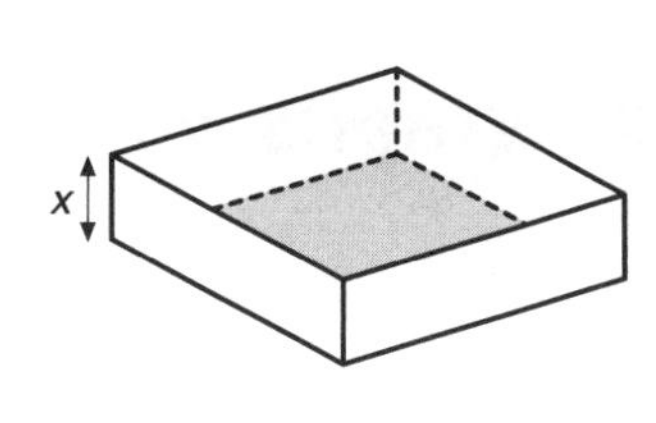

a Show that the side length of the box is $(10 - 2x)$ cm.

b Show that the volume V in cm^3 of the box is given by the formula $V = 100x - 40x^2 + 4x^3$, for $0 \leqslant x \leqslant 5$.

c Draw the graph of V against x by first constructing a table of suitable values.

d Use your graph to estimate the maximum volume of the box, and state its dimensions.

Activity 5

A forest contains F foxes and R rabbits. Their numbers change throughout the course of a given year as shown in the graph of F against R. t is the number of months after 1 January.

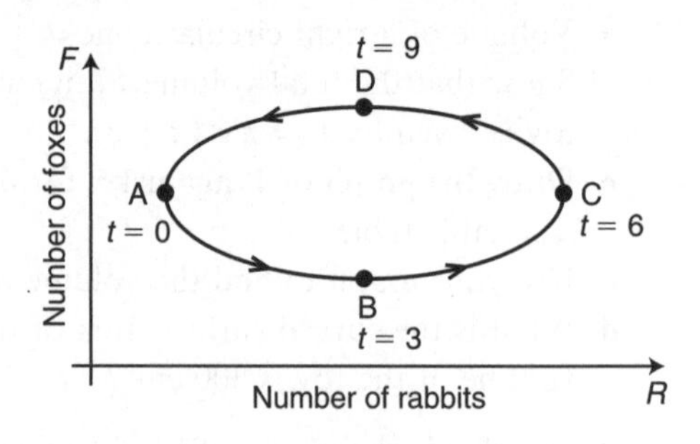

- Copy and complete this table.

Year interval	Fox numbers	Rabbit numbers	Reason
Jan–Mar (A–B)	Decreasing	Increasing	Fewer foxes to eat rabbits
Apr–June (B–C)			
Jul–Sep (C–D)			
Oct–Dec (D–A)			

Sketch two graphs of F against t and R against t for the interval $0 \leqslant t \leqslant 12$, placing the horizontal axes as shown for comparison.

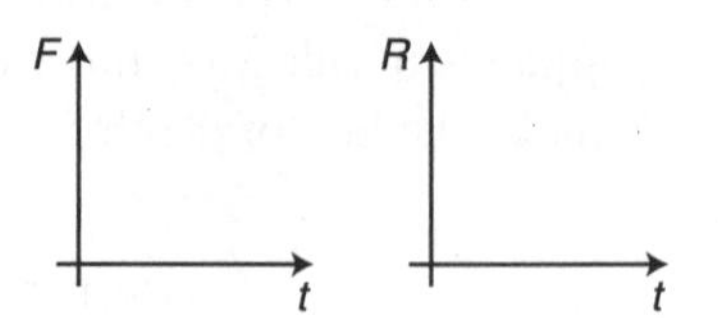

Reciprocal graphs $y = \frac{a}{x}$

Reciprocal graphs have x as the denominator, and they produce another type of curve called a **hyperbola**.

Activity 6

- Copy and complete these tables of values for the equations.

$y = \frac{3}{x}$

x	−3	−2	−1	0	1	2	3
y	−1				3		

$y = -\frac{3}{x}$

x	−3	−2	−1	0	1	2	3
y	1				−3		

Why are there no values for y when $x = 0$?

- Draw one set of x and y axes, and use them to plot the graphs of both equations, labelling each curve.

Remember

The graph of $y = \frac{a}{x}$ is a **hyperbola**.

The curve approaches, but never touches, the axes. Such lines are called **asymptotes**.

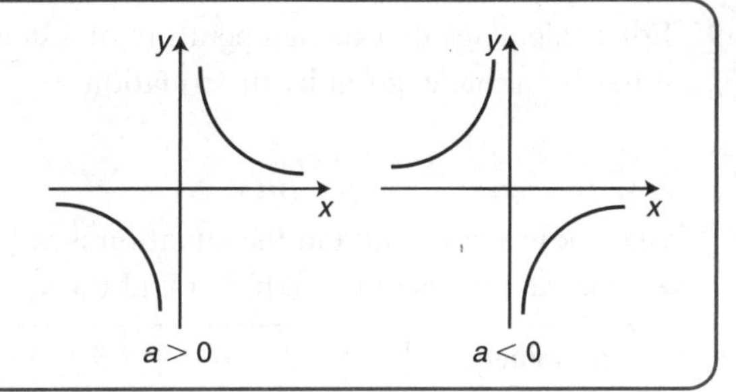

Other reciprocal graphs which involve division by an x term include, for example, $y = 1 + \frac{4}{x}$, $y = 2x - \frac{1}{x}$ and $y = \frac{12}{x - 1}$.

Exercise 9

Draw the graphs between the stated x values after compiling a suitable table.

1 $y = \frac{4}{x}$ $\quad -4 \leqslant x \leqslant 4$

2 $y = -\frac{4}{x}$ $\quad -4 \leqslant x \leqslant 4$

3 $y = \frac{10}{x}$ $\quad -5 \leqslant x \leqslant 5$

4 $y = -\frac{8}{x}$ $\quad -4 \leqslant x \leqslant 4$

5 An insect colony decreases after the spread of a virus. Its population y after t months is given by the equation

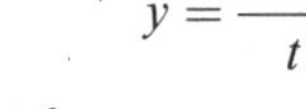

$$y = \frac{2000}{t}$$

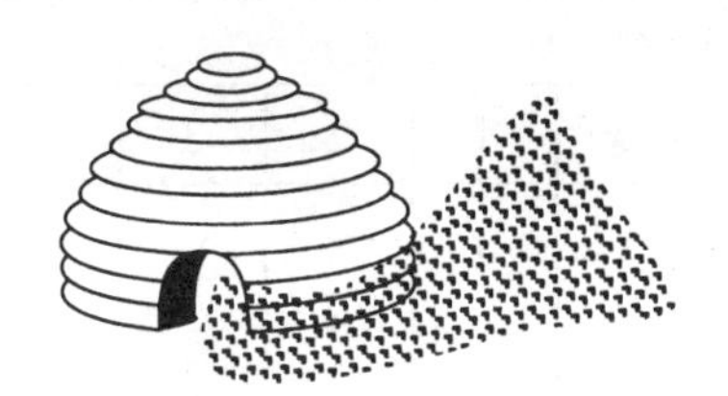

valid for $1 \leqslant t \leqslant 6$.

a Copy and complete this table.

***t* (months)**	1	2	3	4	5	6
y	2000					

b Draw a graph of y against t.

c Use your graph to estimate when the population decreases by 70% from its size after 1 month.

d How long does it take for the population to decrease from 1500 to 500?

6 A water tank springs a leak. The volume v, in m^3, at time t hours after the leak occurs is given by the equation

$$v = \frac{1000}{t}$$

valid for $1 \leqslant t \leqslant 20$.

a Copy and complete this table.

***t* (hours)**	1	5		15	
***v* (m^3)**			100		50

b Draw the graph of v against t.

c Use your graph to estimate when the volume of water is reduced by $750\,m^3$ from its value after 1 hour.

d How much water has been lost between 8 hours and 16 hours?

7 Edna calculates that the temperature of a cup of tea is t in degrees Celsius, m minutes after it has been made, given by the equation

$$t = \frac{k}{m}$$

where k is a constant and the equation is valid for $5 \leqslant m \leqslant 10$.

a Use the figures in this table to find the value of k, and hence copy and complete the table.

***m* (minutes)**	5	6	7	8	9	10
***t* (°C)**						40

b Draw the graph of t against m.

c Use your graph to estimate the temperature of a cup of tea after 450 s.

d After how long is the temperature of the cup of tea 60 °C?

e Edna is fussy, and she only drinks tea at a temperature of between 50 °C and 75 °C. Use your graph to find between what times Edna will drink her cup of tea.

Example 2

Draw the graph of $y = \frac{10}{x} + x - 5$ for $1 \leqslant x \leqslant 6$.

Construct a table and plot a graph from it.

x	1	2	3	4	5	6
$\frac{10}{x}$	10	5	$3\frac{1}{3}$	$2\frac{1}{2}$	2	$1\frac{2}{3}$
x	1	2	3	4	5	6
-5	-5	-5	-5	-5	-5	-5
y	6	2	$1\frac{1}{3}$	$1\frac{1}{2}$	2	$2\frac{2}{3}$

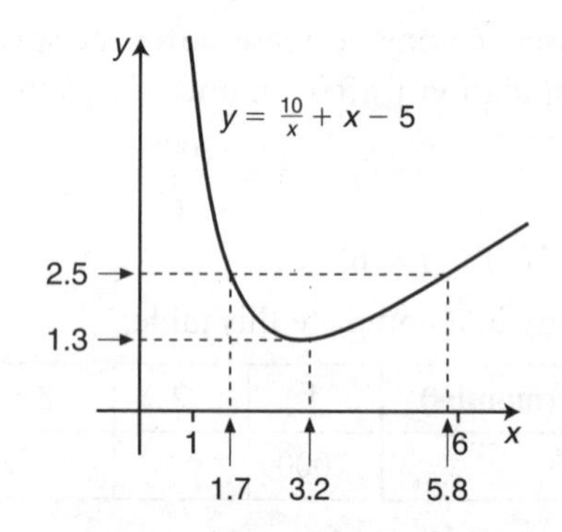

Use the graph to estimate the smallest value of y, and the value of x where this occurs.

The graph shows that the minimum value of $y \simeq 1.3$ when $x \simeq 3.2$.

For what values of x is $y = 2.5$? When $y = 2.5$, $x \simeq 1.7$ or 5.8.

Exercise 9★

Draw these graphs between the stated x values after compiling a suitable table.

1 $y = 1 + \frac{4}{x}$ $\quad -4 \leqslant x \leqslant 4$

2 $y = \frac{6}{x-2}$ $\quad -3 \leqslant x \leqslant 6$

3 $y = x^2 + \frac{2}{x}$ $\quad -4 \leqslant x \leqslant 4$

4 $y = \frac{8}{x} + x - 4$ $\quad 1 \leqslant x \leqslant 6$

5 Jacqui is training for an athletics match at school. She experiments with various angles of projection x for putting the shot, and finds that for $30° \leqslant x \leqslant 60°$ the horizontal distance d, in metres, is given by

$$d = 100 - x - \frac{2000}{x}$$

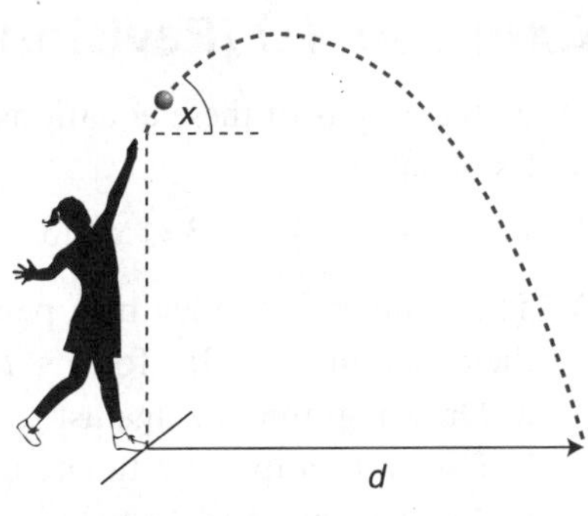

a Copy and complete this table.
Then draw a graph of d against x for $30° \leqslant x \leqslant 60°$.

b Use your graph to estimate the greatest distance obtained by Jacqui, and the angle necessary to achieve it.

c What values can x take if Jacqui is to put the shot at least 9 m?

x (°)	30	35	40	45	50	55	60
100			100				
$-x$			-40				
$-\frac{2000}{x}$			-50				
d (m)			10				

6 A snowboard company called Zoom hires out x hundred boards per week. The amount received R and the costs C, both measured in \$1000s, are given by

$$R = \frac{6x}{x+1} \quad \text{and} \quad C = x + 1$$

both valid for $0 \leqslant x \leqslant 5$.

a Compile a table for $0 \leqslant x \leqslant 5$ and then draw graphs of R against x and C against x on a single set of axes.

b Zoom's profit P in \$1000s per week is given by $P = R - C$. Use your graphs to estimate how many boards must be loaned out per week for Zoom to make a profit.

c What is the greatest weekly profit that Zoom can make, and the number of boards that must be hired out per week for this to be achieved?

7 The Dimox paint factory wants to store $50\,\text{m}^3$ of paint in a closed cylindrical tank. To reduce costs, it wants to use the minimum possible surface area (including the top and bottom).

a If the total surface area of the tank is $A\,\text{m}^2$, and the radius is r m, show that the height h m of the tank is given by

$$h = \frac{50}{\pi r^2}$$

and use this to show that

$$A = 2\pi r^2 + \frac{100}{r}$$

b Construct a suitable table of values for $1 \leqslant r \leqslant 5$ and draw a graph of A against r.

c Use your graph to estimate the value of r that produces the smallest surface area, and this value of A.

Exercise 10 (Revision)

Draw the graphs of these equations between the stated x values by first compiling suitable tables of values.

1 $y = x^3 + x - 3 \quad -3 \leqslant x \leqslant 3$

2 $y = x^3 + x^2 + 3 \quad -3 \leqslant x \leqslant 3$

3 The distance s m fallen by a pebble from a clifftop t seconds after the pebble falls is given by the equation $s = 4.9t^2$, for $0 \leqslant t \leqslant 4$.

a Draw a graph of s against t.

b Use your graph to estimate the distance fallen by the pebble after 2.5 s.

c At what time has the pebble fallen 50 m?

4 The profit P in \$ millions earned by Pixel Internet after t years is given by the equation $P = t^3 - 6t - 6$, for $0 \leqslant t \leqslant 6$.

a Draw a graph of P against t in the given range by first compiling a suitable table.

b Use your graph to estimate when Pixel Internet first made a profit.

c How long will it take for a profit of \$100 000 000 to be made?

5 Nick goes on a strict diet, which claims that his weight w kg after t weeks between weeks 30 and 40 will be given by

$$w = \frac{k}{t}$$

where k is a constant whole number.

a Use the data in this table to find the value of k, and hence copy and complete the table, giving your answers correct to 2 significant figures.

***t* (weeks)**	30	32	34	36	38	40
***w* (kg)**						70

b Draw a graph of w against t.

c Use your graph to estimate when Nick's weight should reach 80 kg.

d Why is there only a limited range of t values for which the given equation works?

6 The temperature of Kim's cup of coffee, t °C, m minutes after it has been poured into her mug, is given by the equation

$$t = \frac{425}{m} \quad \text{valid for } 5 \leqslant m \leqslant 10$$

a Copy and complete this table and use it to draw the graph of t against m for $5 \leqslant m \leqslant 10$.

m	5	6	7	8	9	10
t	85			53.1		

b Kim likes her coffee between the temperatures of 50 °C and 70 °C.
Use your graph to find at what times Kim should drink her coffee.

Exercise 10★ (Revision)

Draw the graphs of these equations between the stated x values by first compiling tables of values.

1 $y = 2x^3 - x^2 - 3x \quad -3 \leqslant x \leqslant 3$

2 $y = 3x(x + 2)^2 - 5 \quad -3 \leqslant x \leqslant 2$

3 The equation for the flight path of a golfer's shot is $y = 0.2x - 0.001x^2$, for $0 \leqslant x \leqslant 200$, where y m is the ball's height, and x m is the horizontal distance moved by the ball.

  a Draw a graph of y against x by first compiling a suitable table of values between the stated x values.

  b Use your graph to estimate the maximum height of the ball.

  c Between what distances is the ball at least 5 m above the ground?

4 The flow $Q\,\text{m}^3/\text{s}$ of a small river t hours after midnight is monitored after a storm, and is given by the equation $Q = t^3 - 8t^2 + 14t + 10$, for $0 \leqslant t \leqslant 5$.

  a Draw a graph of Q against t by first constructing a suitable table between the stated t values.

  b Use your graph to estimate the maximum flow and the time when this occurs.

  c Between what times does the river flood, if this occurs when the flow exceeds $10\,\text{m}^3/\text{s}$?

5 A pig farmer wants to enclose $600\,\text{m}^2$ for a rectangular pig pen. Three of the sides will consist of fencing, while the remaining side will consist of an existing stone wall. Two sides of the fence will be of length x metres.

  a Write the length of the third side in terms of x.

  b Show that the total length L in metres of the fence is given by

$$L = 2x + \frac{600}{x}$$

  c Construct a suitable table of values for $5 \leqslant x \leqslant 40$. Use this table to draw a graph of L against x.

  d Use your graph to estimate the minimum fence length possible, and the value of x for which it occurs.

  e What values can x take, if the fence length is not to exceed 75 m?

6 The graph of $y = x^3 - 5x^2 + ax + b$, where a and b are constants, passes through the points P(0, 5) and Q(1, 6).

  a Show that the values of a and b are both 5 by making sensible substitutions using the co-ordinates of P and Q.

  b Copy and complete this table for $-1 \leqslant x \leqslant 4$.

x	-1	0	1	2	3	4
y		5	6			9

  c Draw the graph of $y = x^3 - 5x^2 + 5x + 5$.

  d The curve represents a cross-section of a hillside. The top of the curve (R) represents the top of a hill and the bottom of the curve (S) represents the bottom of a valley. State the co-ordinates of R and S.

Congruence

If two figures are the same size and shape (one can be placed exactly on top of the other) they are **congruent**.

For example, an isosceles triangle ABC can be cut from a folded piece of paper.

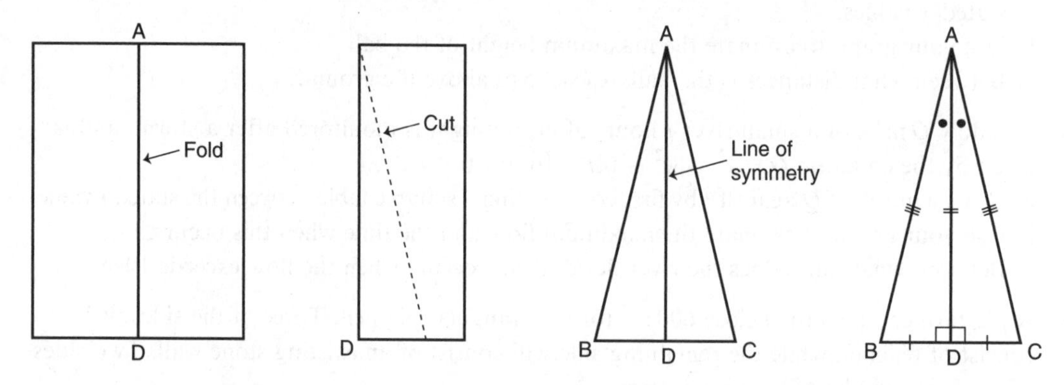

Because of the fold (line of symmetry), triangle ABD is exactly the same size and shape as triangle ACD. The triangles are congruent. Reversing this argument, $\triangle ABD \equiv \triangle ACD$ *because* of the line of symmetry.

Activity 7

- Use a pair of compasses and a protractor to construct the two triangles T_1 and T_2 below.

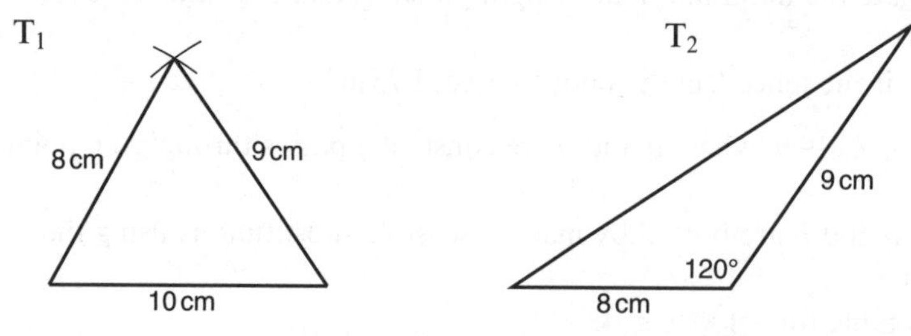

All triangles drawn from triangle T_1 should be exactly the same size and shape because three sides are given (SSS). This is one condition for congruency. In triangle T_2, two sides and the included angle are given (SAS), which is another condition for congruency.

- Triangle ABC is to be drawn where $\angle CAB = 30°$ and $AB = 10\,cm$.
  Copy the figure on the right and investigate all possible lengths of BC.
  Comment on your results.

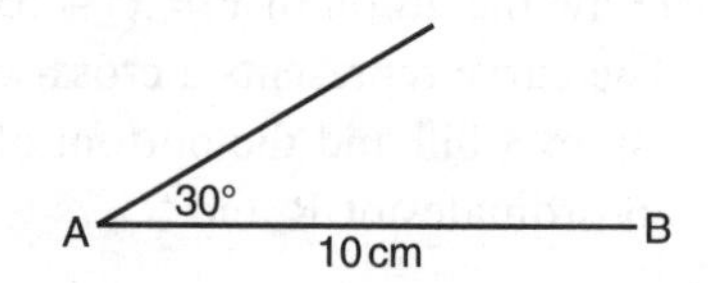

Key Point

For triangles to be **congruent** (the same size and shape), the corresponding sides and angles must be: **SSS**, **SAS**, **AAS** or **ASA** (in that order) or **RHS** (right angle, hypotenuse, side).

Example 1

Why are triangles ABC and XYZ congruent?

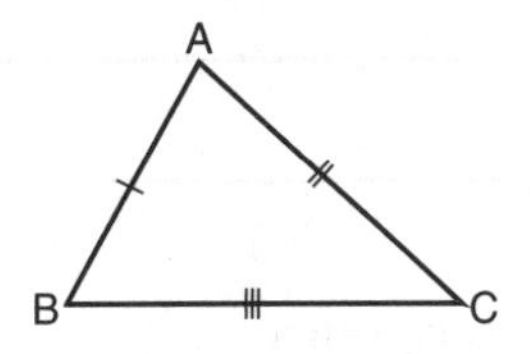

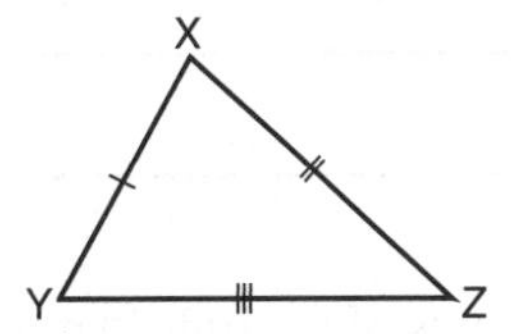

AB = XY
AC = XZ
BC = YZ

Triangles ABC and XYZ are congruent because the lengths of their corresponding sides are equal (SSS).

Remember

When trying to prove congruence, or calculate a length or angle using congruence:

- Only use the given facts.
- Draw a reasonably accurate and neat diagram to show all known facts.
- Give a reason for each statement in the proof.

Example 2

Triangles ABC and XYZ are congruent (SAS). Write down three deductions that follow from this fact.

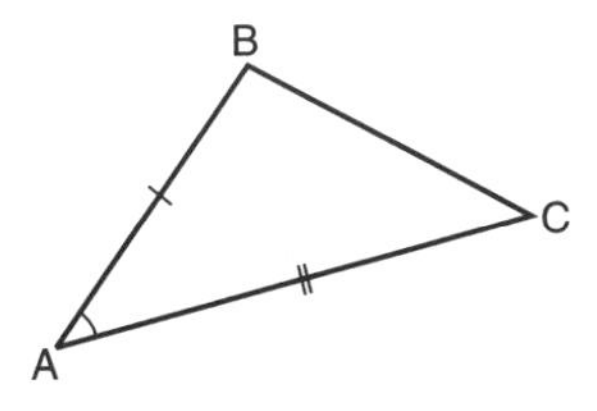

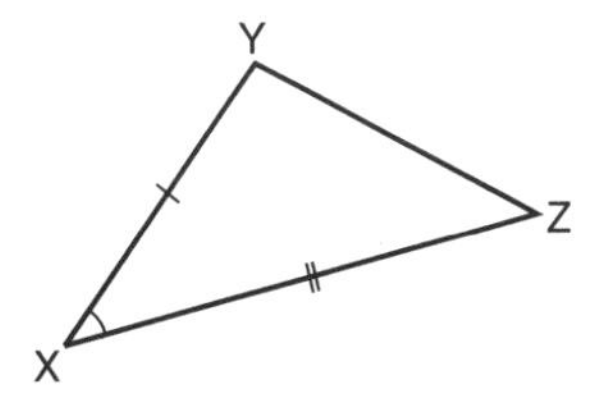

$\angle ACB = \angle XZY$, BC = YZ, $\angle ABC = \angle XYZ$

Example 3

Triangles ABC and XYZ are congruent (RHS). Write down three deductions that follow from this fact.

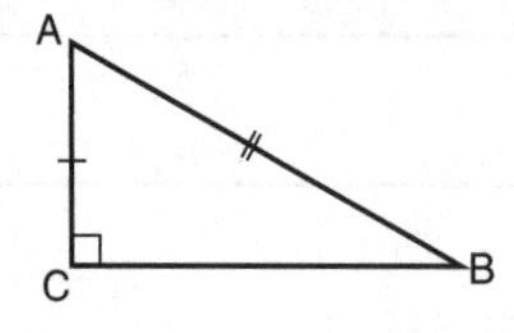

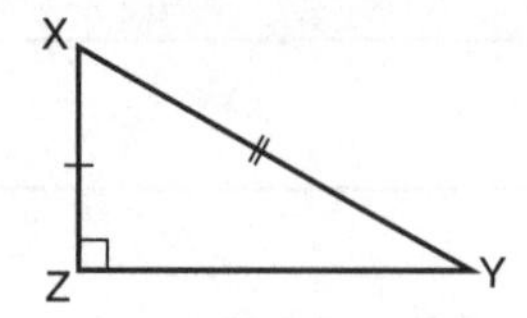

$\angle CAB = \angle ZXY$, CB = ZY, $\angle ABC = \angle XYZ$

Example 4

Prove that the diagonals of a parallelogram bisect each other.

Draw a parallelogram.

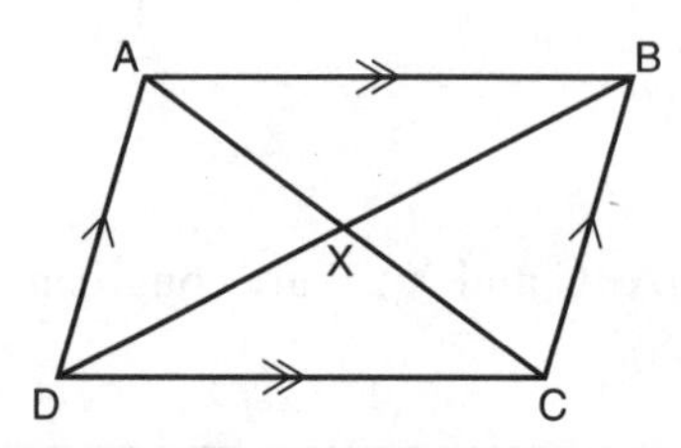

(To prove that triangles ABX and CDX are congruent, show that one pair of corresponding sides are equal and two corresponding angles are equal.)

State the triangles that are to be proved to be congruent. — In triangles ABX and CDX

State the three conditions for congruency and give the reason in brackets in each case.

1 AB = CD (Opp. sides of //gram)
2 $\angle BAX = \angle DCX$ (Alternate angles)
3 $\angle AXB = \angle CXD$ (Vertically opposite)

Write the conclusion.

$\therefore \triangle ABX \equiv \triangle CDX$ (AAS)
$\therefore$ AX = CX

Exercise 11

For Questions 1–6, state whether pairs of triangles are congruent and give the condition, e.g. SSS, SAS.

1

6 cm, 78°, 5 cm

6 cm, 43°, 78°, 5 cm

2

5 cm, 20°, 3 cm

30°, 3 cm, 5 cm

3

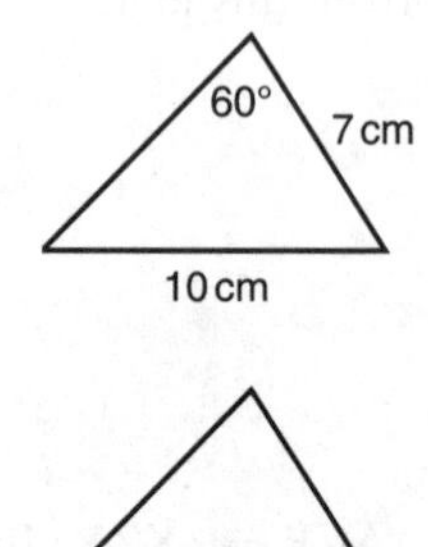

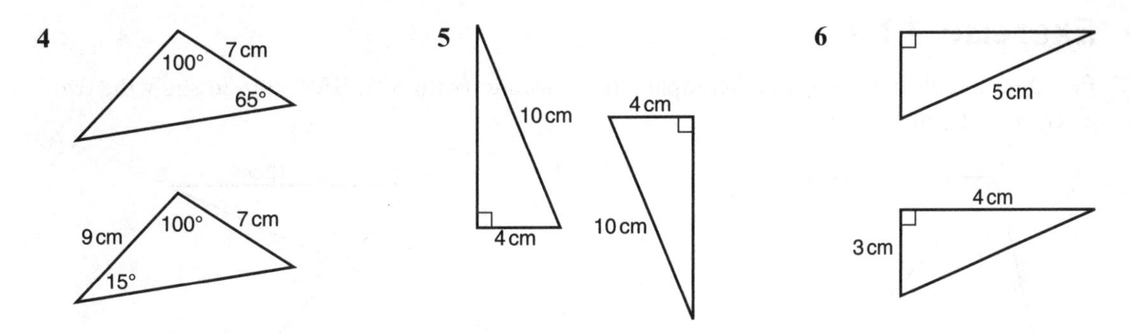

In Questions 7 and 8, the triangles ABC and XYZ are congruent. Write down three deductions about the sides and angles that follow from this fact.

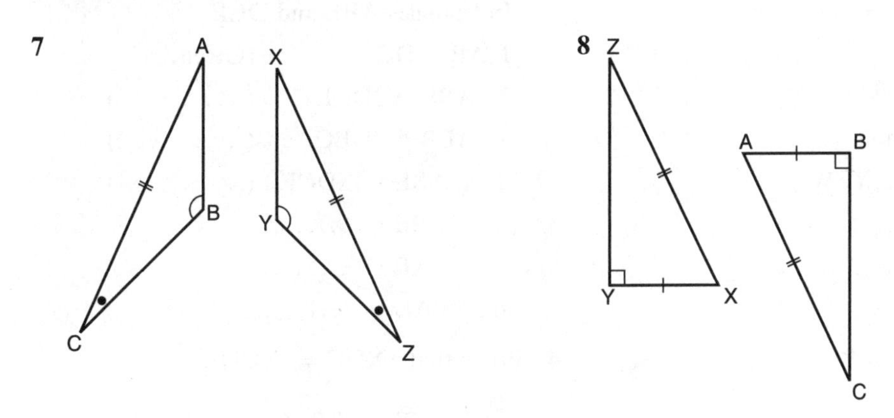

For Questions 9 and 10, copy and complete the workings (with SSS, SAS, etc.) to show the two given triangles are congruent.

9

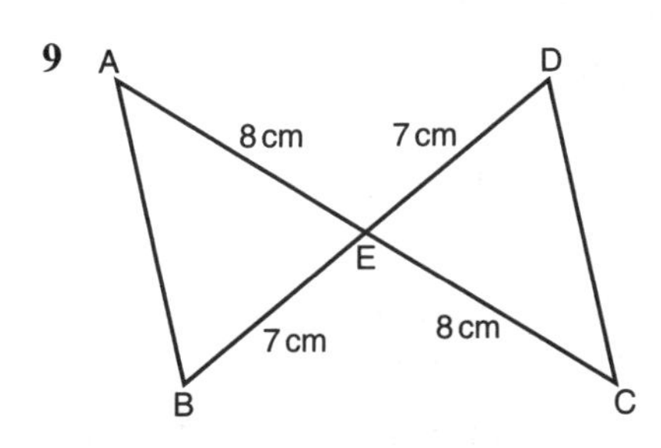

In triangles ABE and CDE

1 BE = DE (Given)

2 AE = CE (...........)

3 ∠AEB = ∠CED (...........)

∴ △ABE ≡ △CDE (...........)

∴ ∠EDC = ...

∠ECD = ...

and AB = ...

10

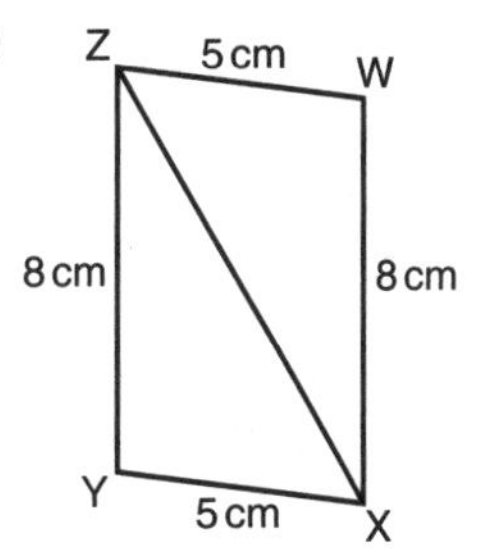

In triangles WXZ and YZX

1 WZ = YX (...........)

2 WX = YZ (...........)

3 Line XZ is common

∴ △WXZ ≡ △YZX (...........)

∴ ∠ZWX = ...

∠WZX = ...

and ∠XZY = ...

Exercise 11★

For Questions 1 and 2, copy and complete the workings (with SSS, SAS, etc.) to show the two given triangles are congruent.

1

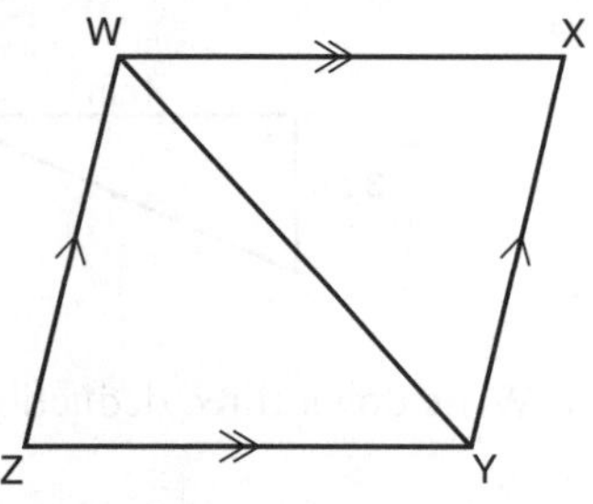

In triangles WXY and YZW

1 WX = YZ (...........)

2 WZ = YX (...........)

3 WY is common

$\therefore$ $\triangle$WXY $\equiv$ $\triangle$YZW (...........)

$\therefore$ $\angle$WXY = ...

$\angle$XWY = ...

and $\angle$XYW = ...

2

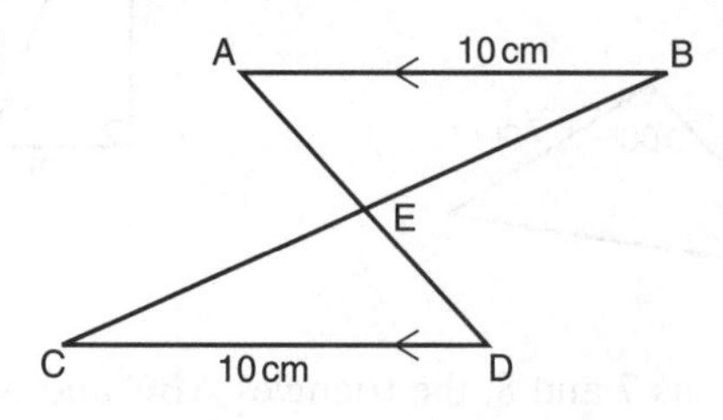

In triangles ABE and DCE

1 AB = DC (Given)

2 $\angle$ABE = $\angle$DCE (...........)

3 $\angle$AEB = $\angle$DEC (...........)

$\therefore$ $\triangle$ABE $\equiv$ $\triangle$DCE (...........)

$\therefore$ BE = ...

AE = ...

and $\angle$BAE = ...

3 Prove that AD = BC.

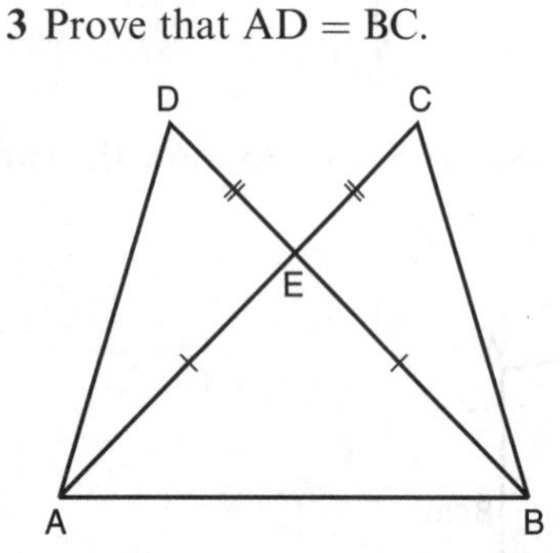

4 Prove that $\angle$XDC = $\angle$XCD.

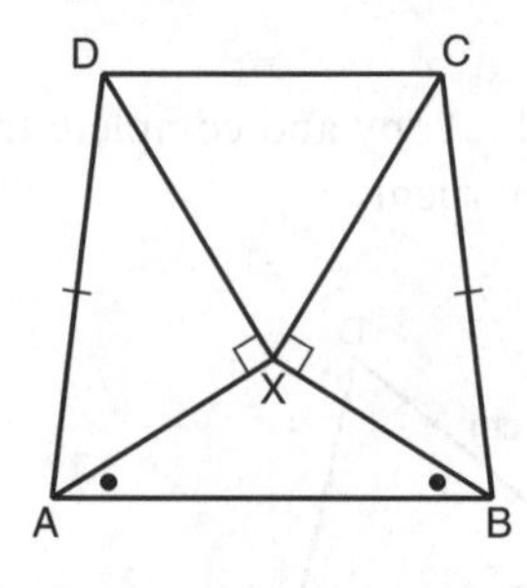

5 P is the mid-point of the side YZ of triangle XYZ. PA is the perpendicular from P to XY and PB is the perpendicular from P to XZ. If AP = BP prove that XY = XZ.

6 Triangle XYZ is isosceles where XY = XZ. A is a point on XY and B is a point on XZ such that AX = BX. Prove that AZ = BY.

7 The construction shows the angle bisector of $\angle$ABC where BM = BN and MO = NO. Use congruent triangles to prove that BX bisects $\angle$ABC.

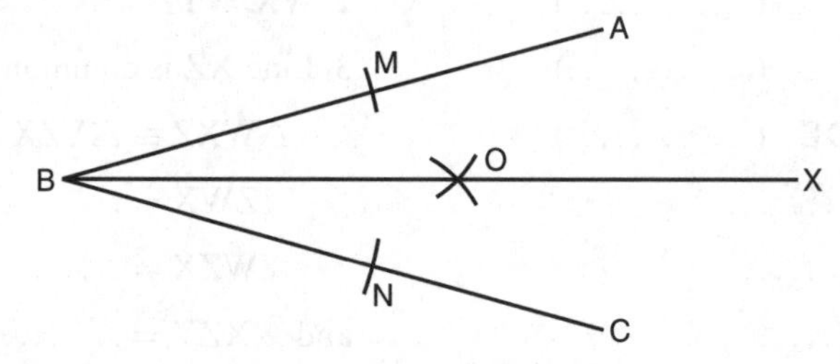

8 The construction shows the perpendicular from P on to the line AB where PL = PM and LX = MX. Use congruent triangles to prove that ∠PYB = 90°.

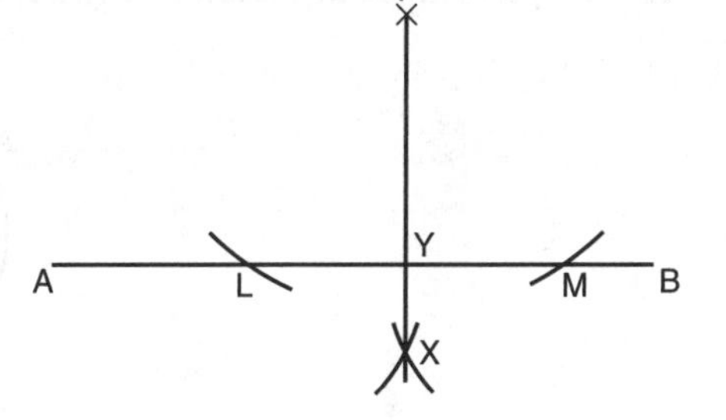

9 ABCD is a square, and the points E, F, G and H shown are such that EB = FC = GD = HA. Prove that EFGH is also a square.

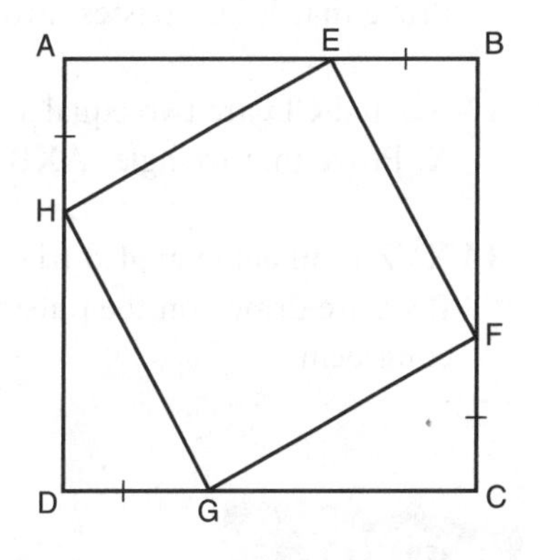

10 D is the mid-point of BC in the triangle ABC. BX and CY are perpendiculars. Prove that CY = BX.

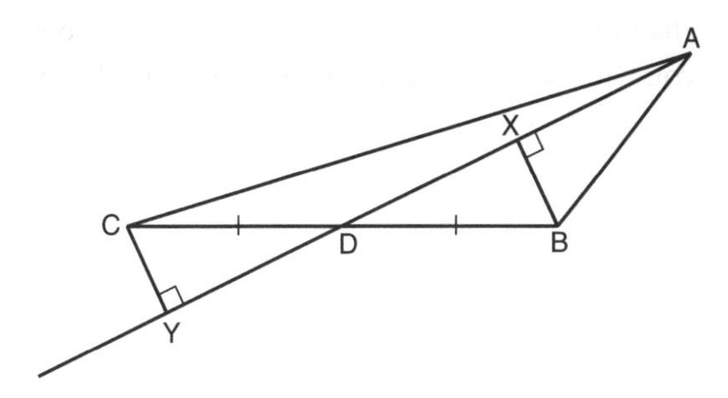

11 PX and PY are tangents to the circle, centre O. Prove that PX = PY.

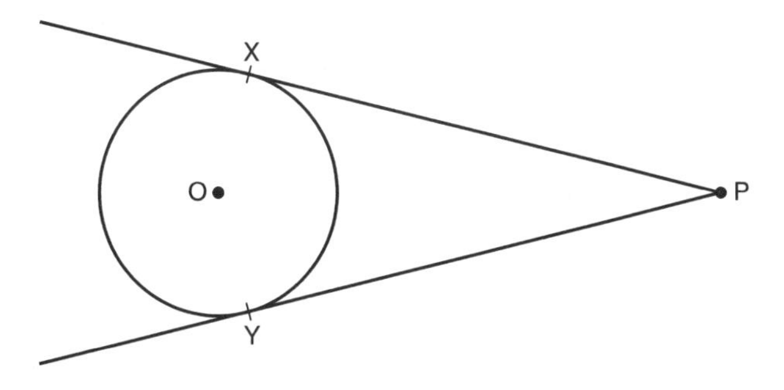

12 The perpendicular bisector of the chord AB passes through X.

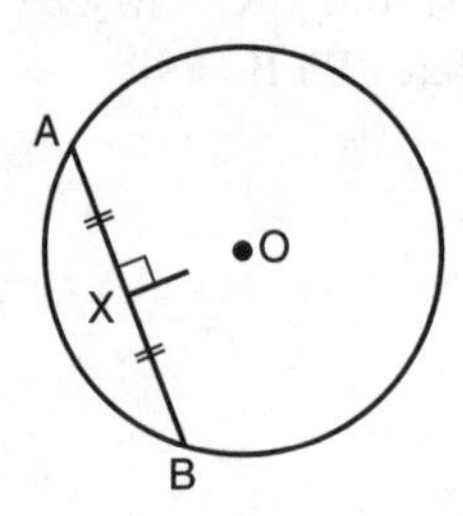

Prove that it also passes through the centre of the circle O.

13 AB and CD are two equal straight lines. The perpendicular bisectors of AC and BD meet at X. Prove that triangles AXB and CXD are congruent.

14 XYZ is an acute-angled triangle. On the sides XY and XZ equilateral triangles AXY and BXZ are drawn on the outside of triangle XYZ. Prove that triangles AXZ and YXB are congruent.

Activity 8

Practical surveying with a Silva compass

A Silva compass is used to measure a magnetic bearing: that is, the angle between magnetic North and a given direction. For their own safety, all hill walkers should be able to use such a compass. In this surveying exercise, you will use a Silva compass to work out the length of a hedge from a distance.

How to use a Silva compass

- Point the compass in the required direction.
- Rotate the 'compass housing' so that the 'black arrow' lines up with the magnetic needle.
- Read off the bearing.

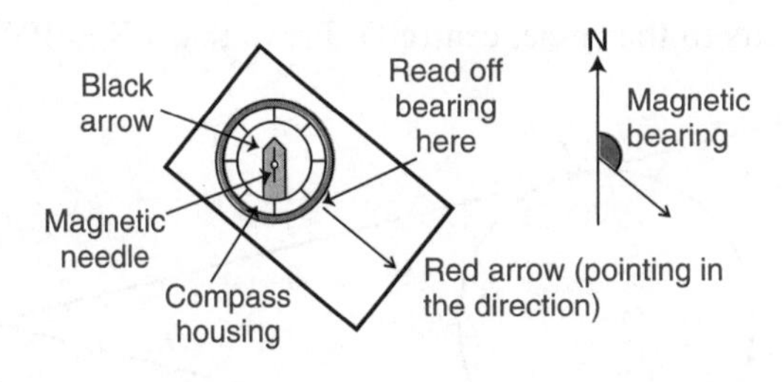

Finding the length of a hedge XY by scale drawing

A 'base line' AB is marked with two posts 50 metres apart and positioned so that BA is pointing to magnetic North. (Your teacher will set this up.)

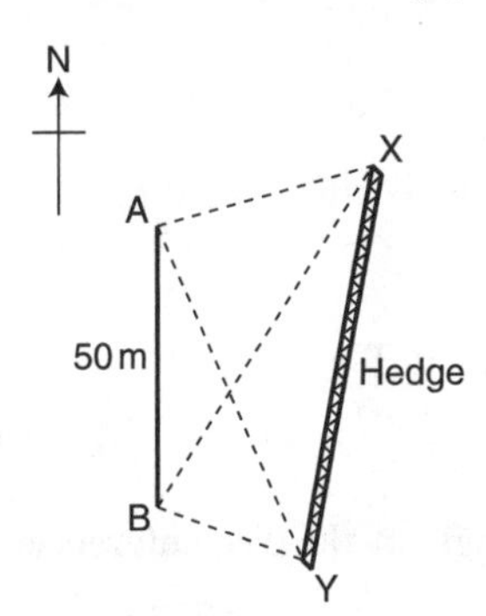

- Stand at A and measure the magnetic bearing of X and of Y. Record your results.
- Stand at B and measure the magnetic bearing of X and of Y. Record your results.
- Make a neat scale drawing and work out the length of the hedge XY.
- With a tape measure, measure the actual length of the hedge and calculate the percentage error in your answer by scale drawing.

Repeat, but this time the 50-metre long 'base line' AB is positioned so that BA is *not* pointing to magnetic North. In this case you must also measure the magnetic bearing of A from B.

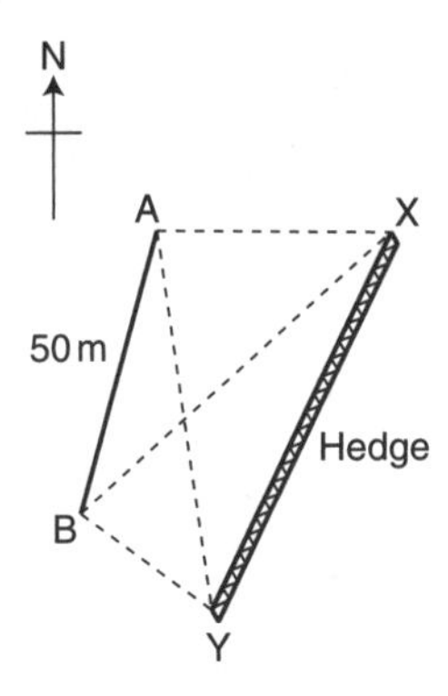

Investigate

Investigate the circumstances under which

a the length of the 'base line' and

b its position relative to the hedge

affect the accuracy of the method of calculating the length of a hedge in Activity 8.

Circles

Remember

Radii, tangent, chord

- AB is a chord
- The letter O will always indicate the centre of a circle
- $\triangle$OAB is isosceles
- XY is a tangent to the circle at T
- $\angle$YTO = 90°

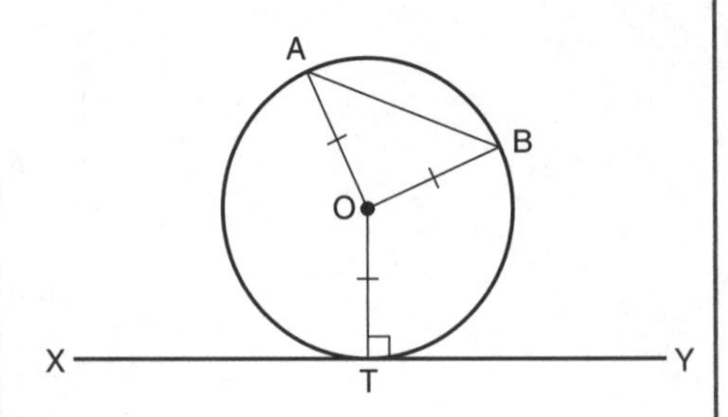

Angle at the centre is twice the angle at the circumference

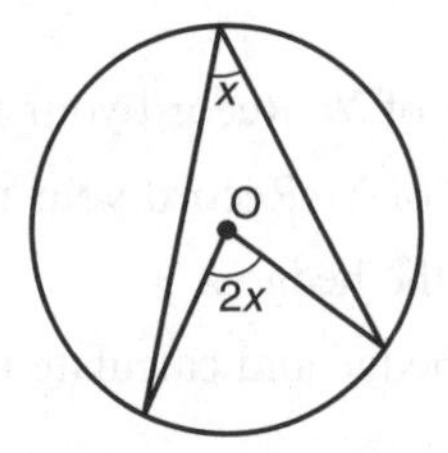

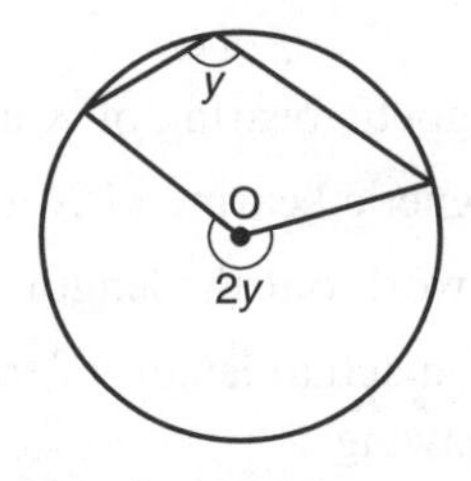

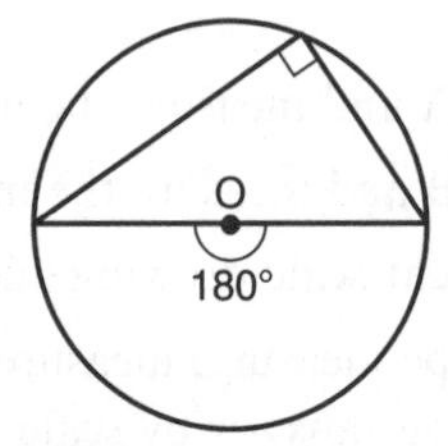

Opposite angles of a cyclic quadrilateral sum to 180°

$a° + b° = 180°$
$x° + y° = 180°$

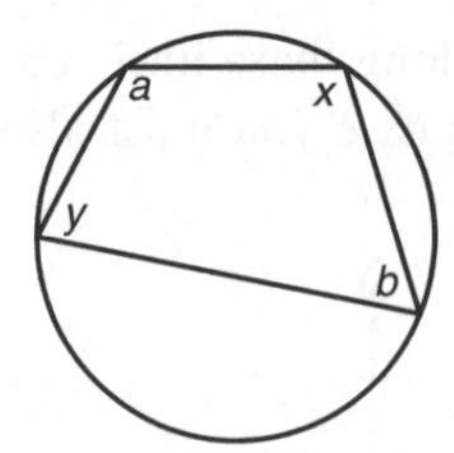

Angles in the same segment are equal

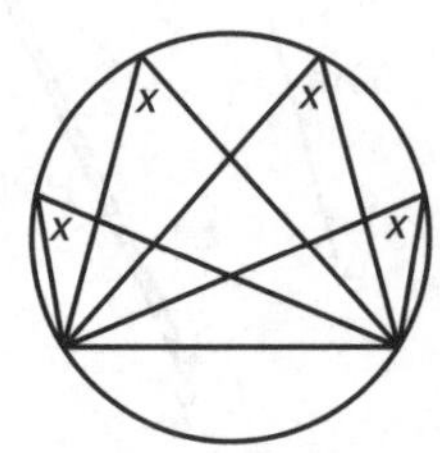

Activity 9

Use the diagram to prove that the angle at the centre is twice the angle at the circumference, and use this to prove that

- angles in the same segment are equal
- opposite angles of a cyclic quadrilateral sum to 180°.

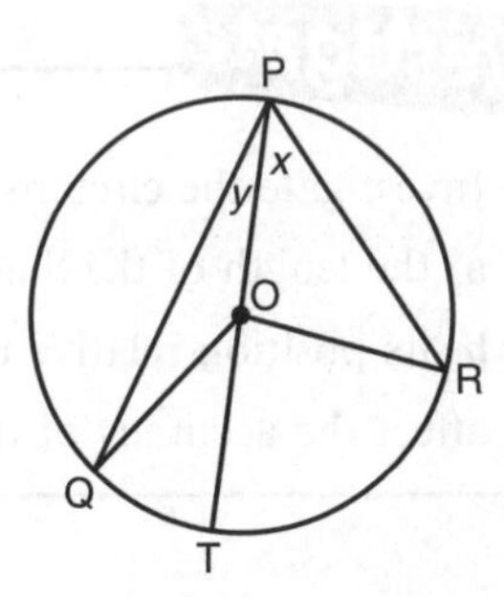

Remember

When trying to find angles or lengths in circles:

- Use the basic facts of geometry (as summarised on pages 38 and 397).
- A figure is **cyclic** if a circle can be drawn through its vertices. The vertices are **concyclic** points.
- Always draw a neat diagram, and include all the facts. Use a pair of compasses to draw all circles.
- Give a reason, in brackets, after each statement.

Example 5

Find ∠PNM.

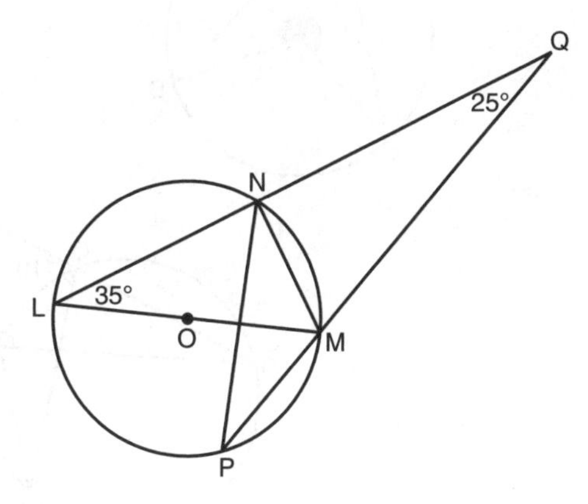

∠NPM = 35° (Angles in the same segment)

∠LNM = 90° (Angle at centre is 180°)

∠MNQ = 90° (△LMN is right-angled)

∴ ∠PNM = 30° (Angle sum of triangle PNQ)

Example 6

Prove that XY meets OZ at right angles.

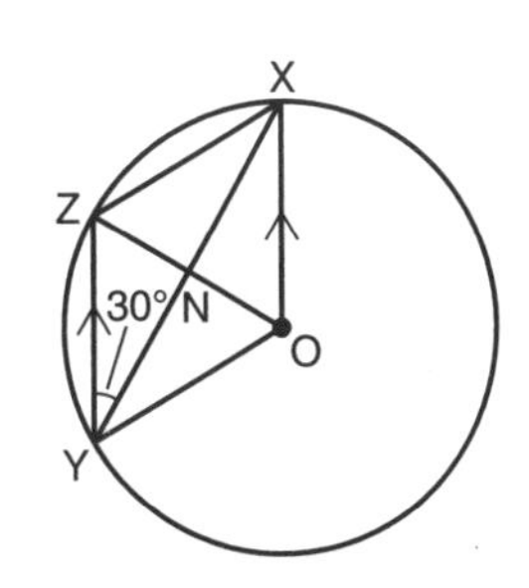

∠OXY = 30° (Alternate to ∠ZYX)

∴ ∠OYX = 30° (△OXY is isosceles)

∠YZO = 60° (△YZO is isosceles)

∴ ∠ZNY = 90° (Angle sum of triangle ZNY)

Exercise 12

For Questions 1–10, find the coloured angles, fully explaining your reasoning.

1

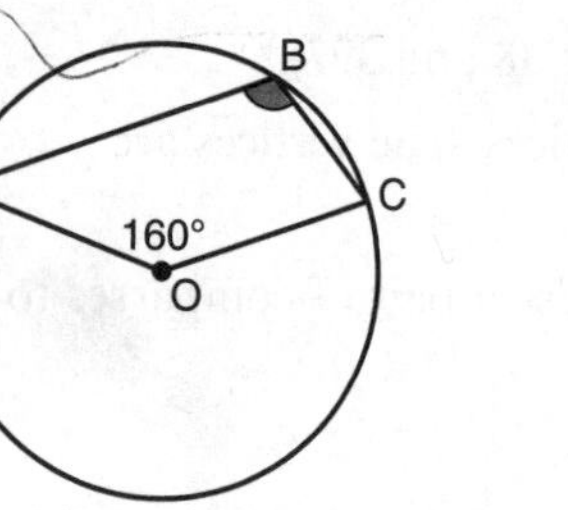

2

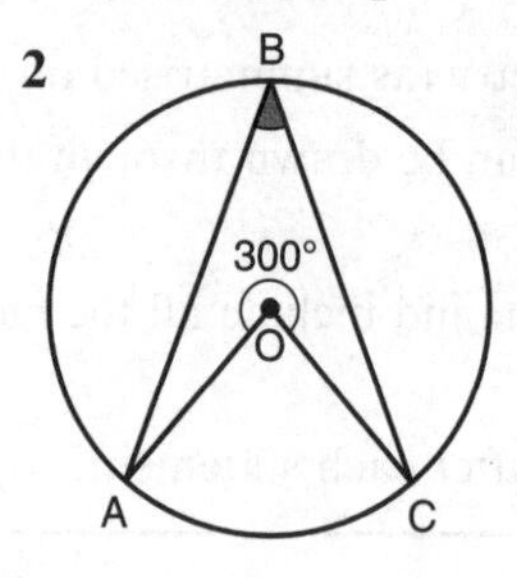

3

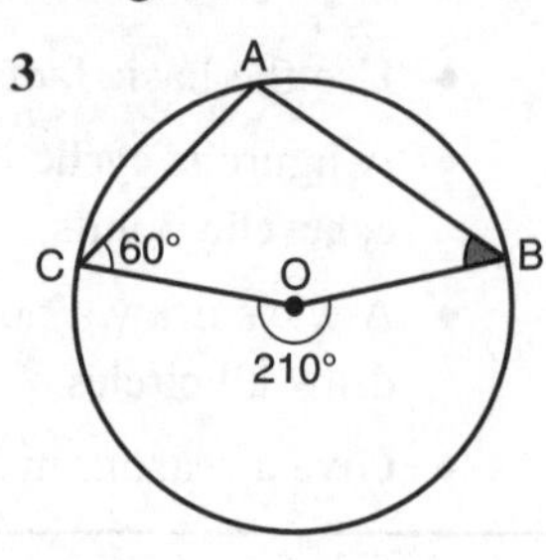

4

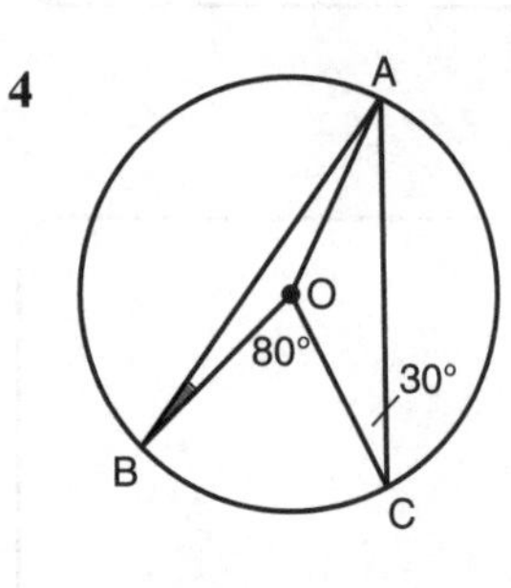

5

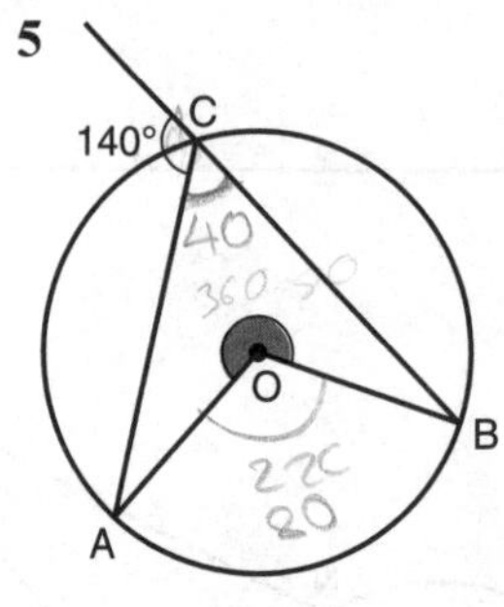

6

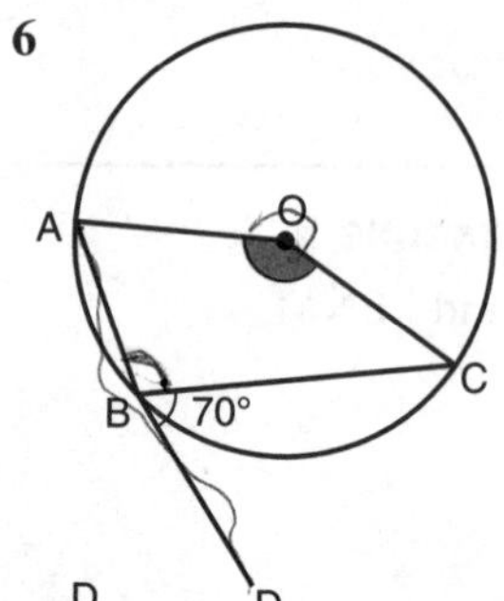

7

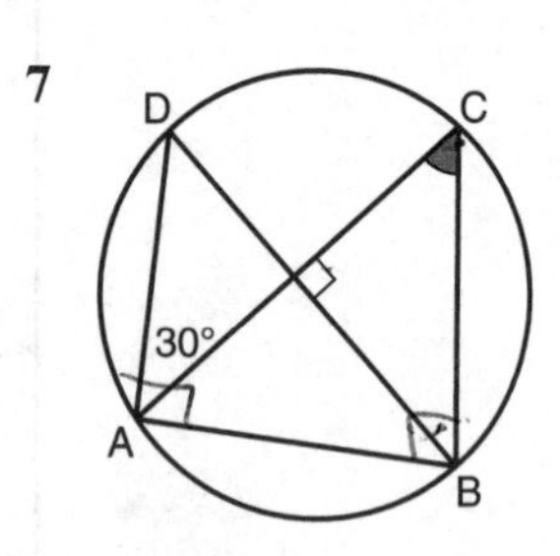

8

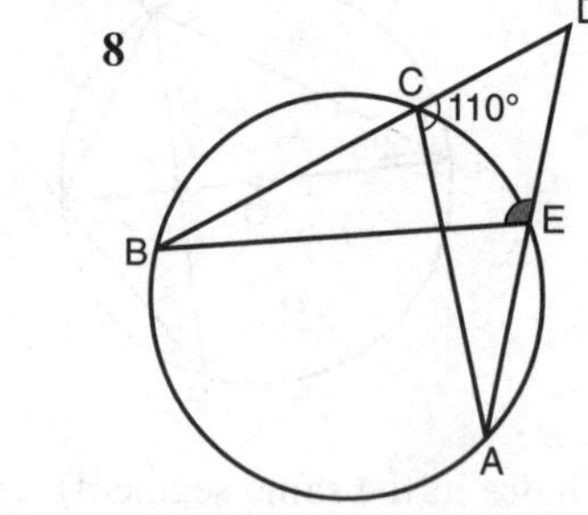

9

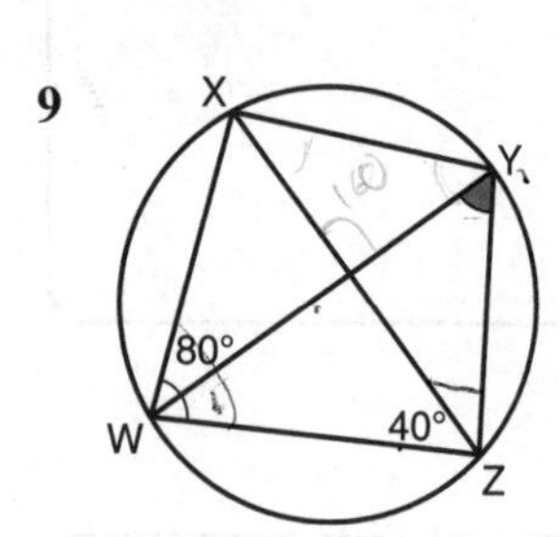

10

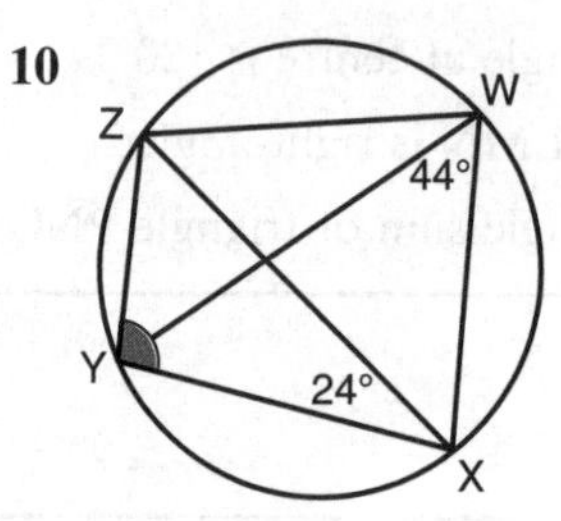

For Questions 11 and 12, prove that the points ACBD are concyclic.

11

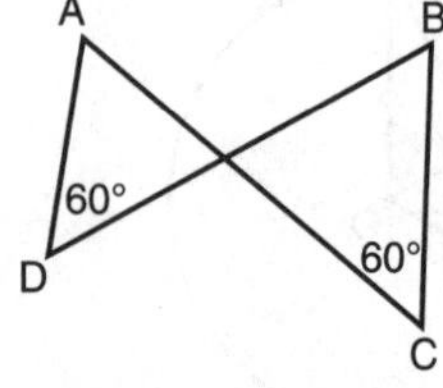

12

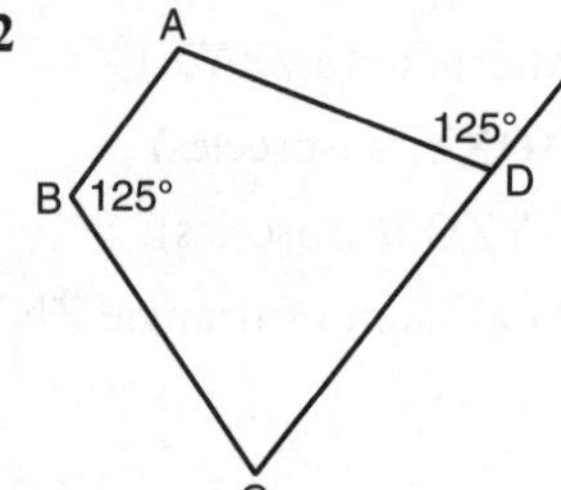

13 LK and MN are two equal chords of a circle. X and Y are their mid-points.
By considering the shape of LKMN, prove that XY makes equal angles with LK and MN.

14 Two circles of different radii and with centres P and Q intersect at X and Y.
By considering the shape of XPYQ, prove that XY meets PQ at right angles.

Exercise 12★

For Questions 1–8, find the coloured angles, fully explaining your reasoning.

1

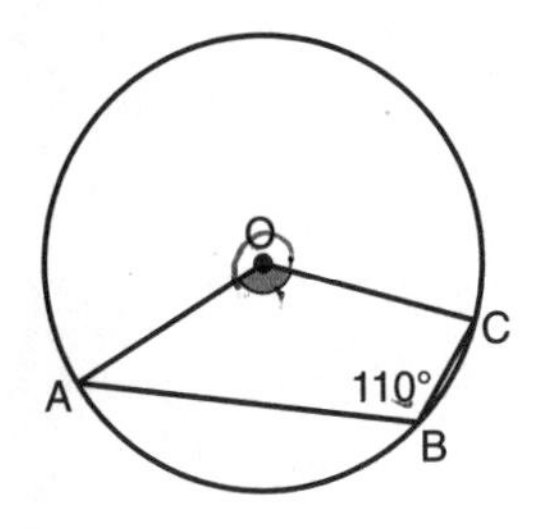

2

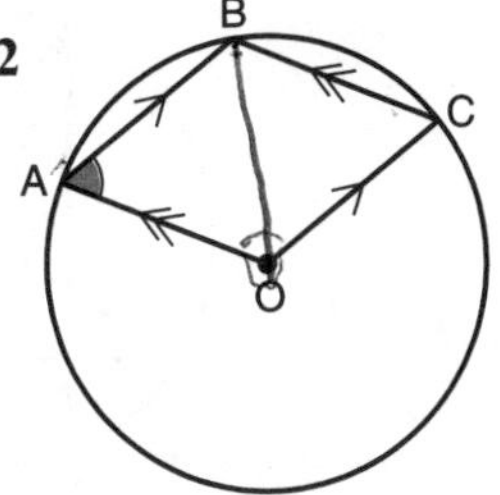

3

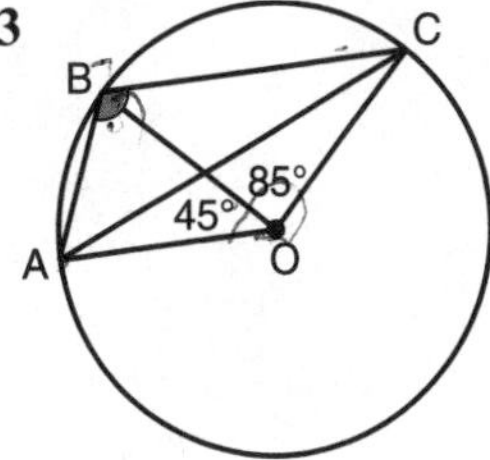

4

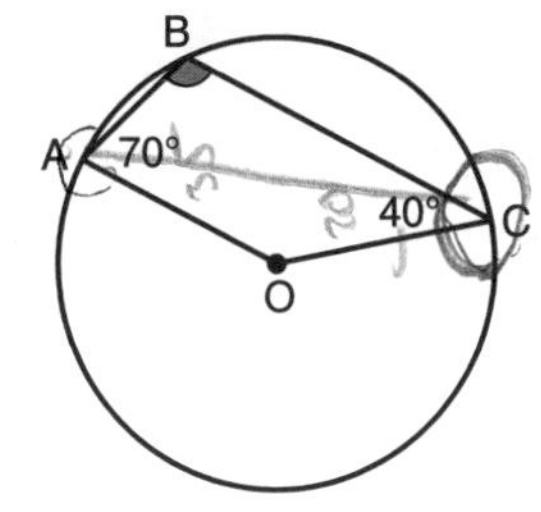

5

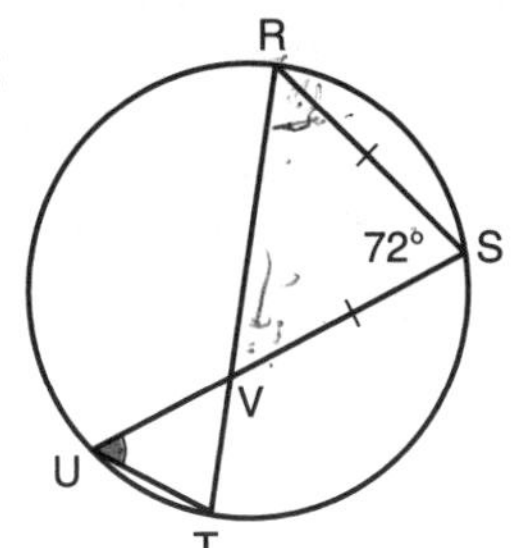

6

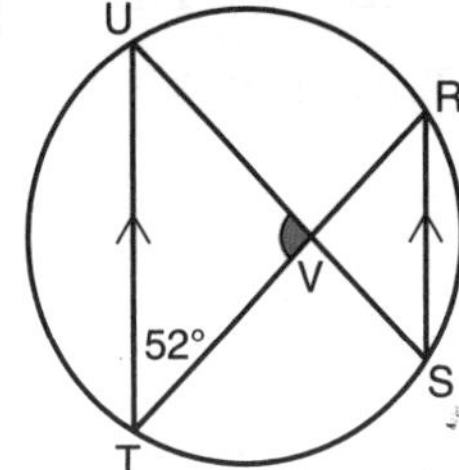

7

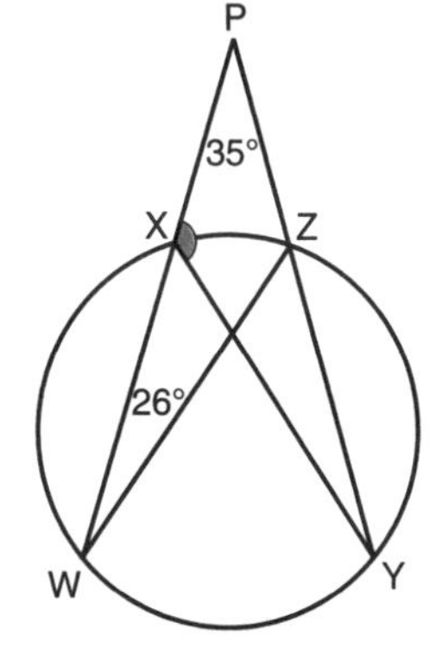

8

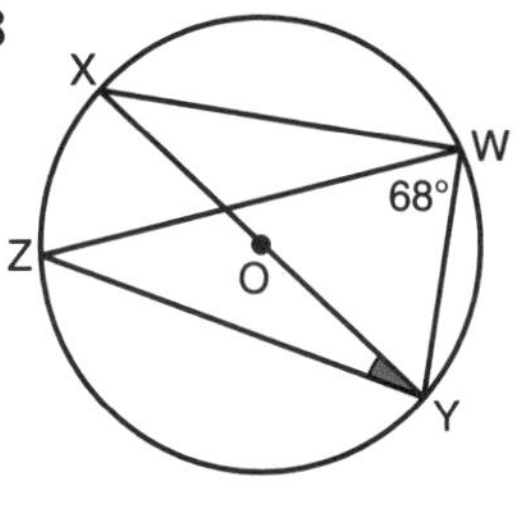

9 Find, in terms of x, $\angle$AOX.

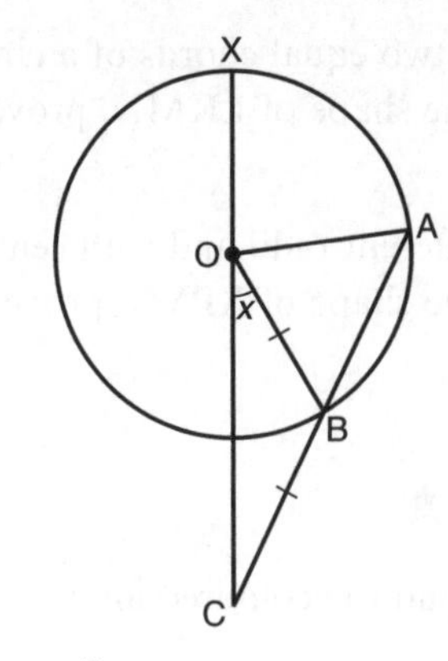

10 Find, in terms of x, $\angle$BCD.

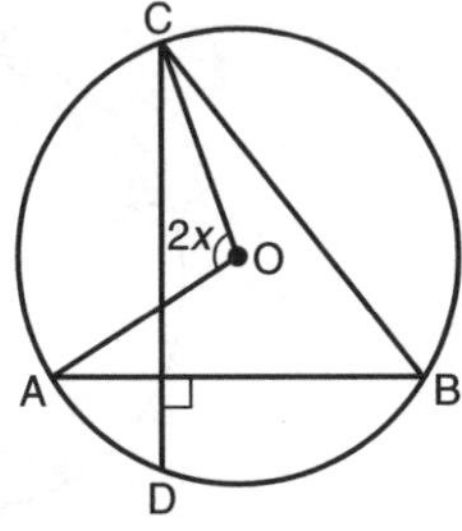

11 ABCD is a quadrilateral in which AB = AD and BD = CD.
Let $\angle$DBA = $x°$ and $\angle$DBC = $2x°$. Prove that A, B, C and D are concyclic.

12 ABCDEF is a hexagon inscribed in a circle. By joining AD, prove that
$\angle$ABC + $\angle$CDE + $\angle$EFA = 360°.

13 Prove that $\angle$CEA = $\angle$BDA.

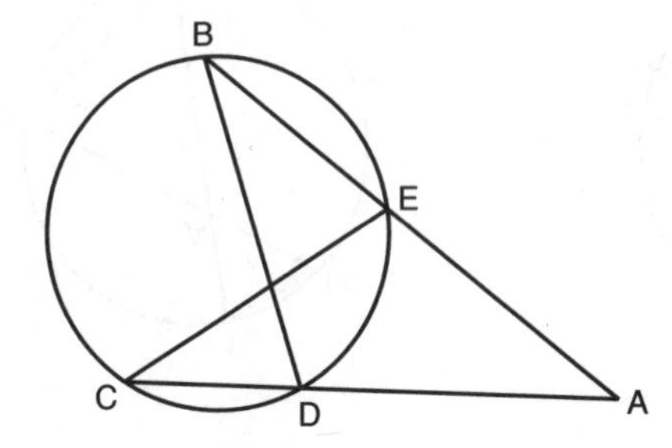

14 In the figure, OX is the diameter of the smaller circle, which cuts XY at A. Prove that AX = AY.

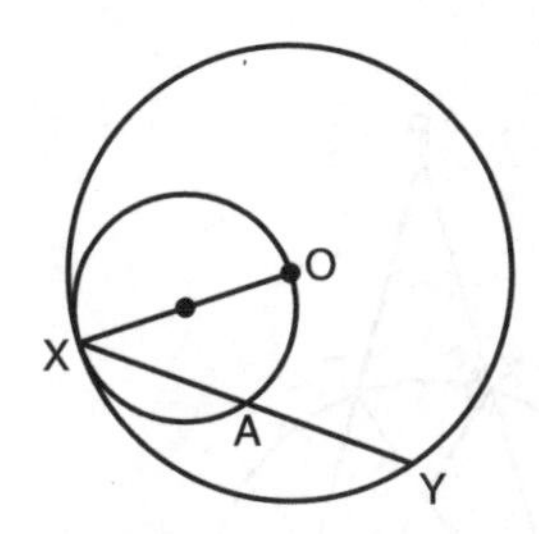

15 WXYZ is a cyclic quadrilateral. The sides XY and WZ produced meet at Q.
The sides XW and YZ produced meet at P. $\angle$WPZ = 30° and $\angle$YQZ = 20°.
Find the angles of the quadrilateral.

16 PQ and PR are any two chords of a circle, centre O. The diameter, perpendicular to PQ, cuts PR at X. Prove that the points Q, O, X and R are concyclic.

Angles in the alternate segment

Activity 10

Copy and complete the table one row at a time by calculating the sizes of the angles on each diagram.

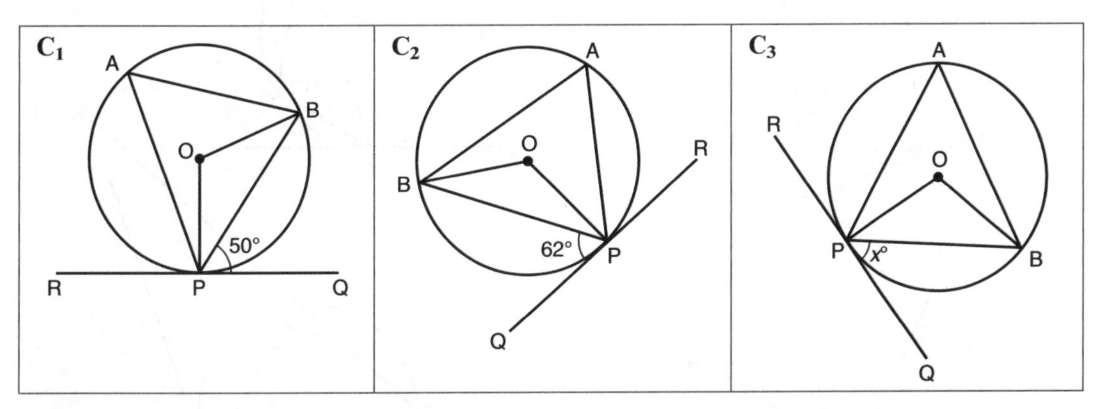

	$Q\hat{P}B$	$O\hat{P}B$	$O\hat{B}P$	$B\hat{O}P$	$B\hat{A}P$
C_1	50°				
C_2	62°				
C_3	x°				

Add reasons for your answers in circle C_3 to form a proof of the Alternate Segment Theorem.

Key Point

The angle between chord and tangent is equal to the angle in the alternate segment.

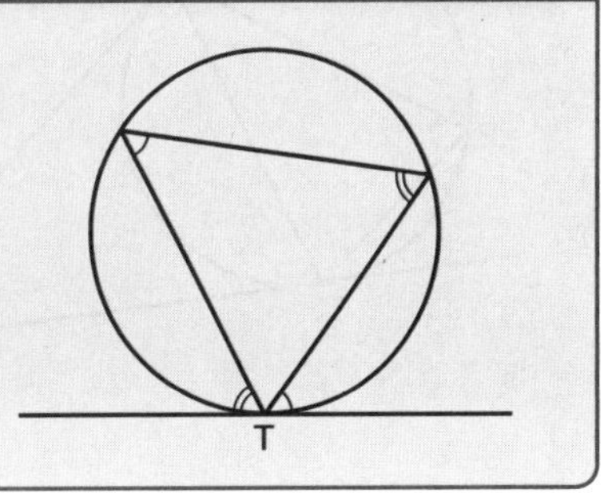

Exercise 13

In this exercise, O indicates the centre of a circle and T indicates a tangent to the circle.

For Questions 1–8, find the coloured angles, fully explaining your reasons.

1

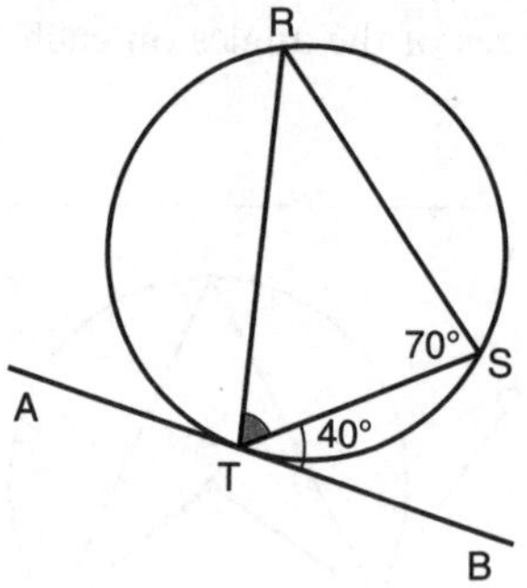

R, 70°, S, A, T, 40°, B

2

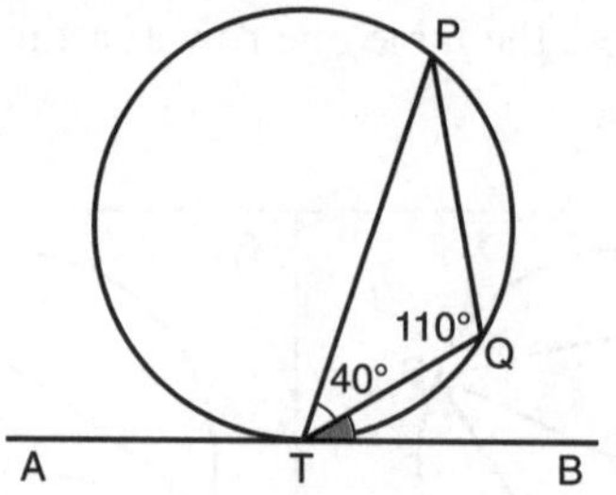

P, 110°, Q, 40°, A, T, B

3

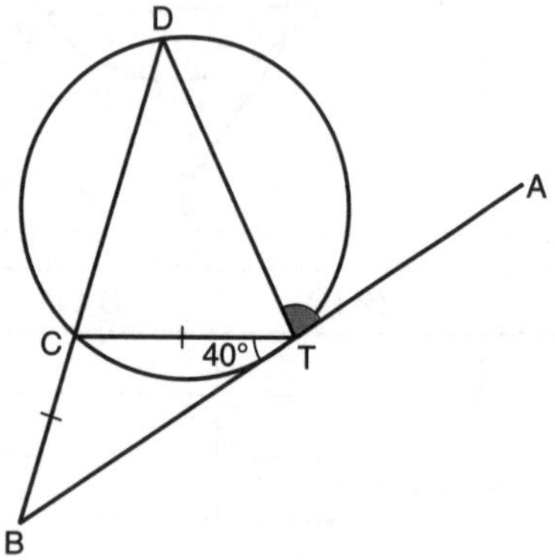

D, A, C, 40°, T, B

4

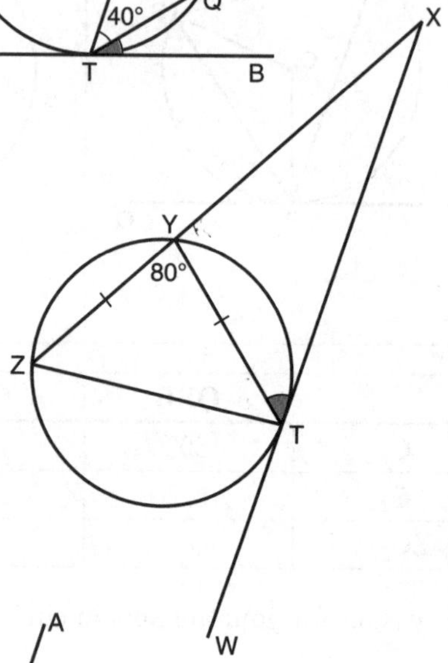

X, Y, 80°, Z, T, W

5

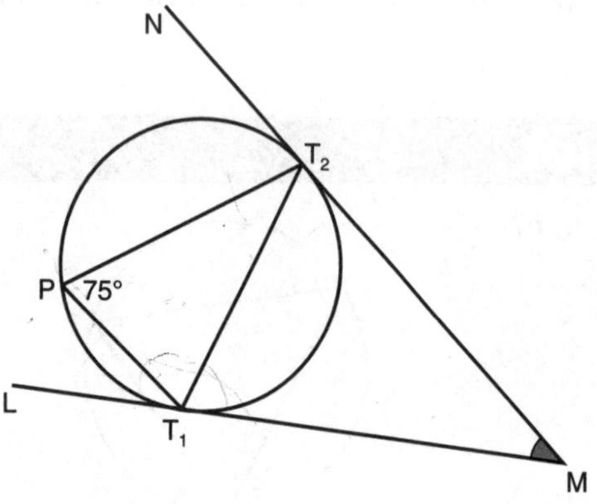

N, T_2, P, 75°, L, T_1, M

6

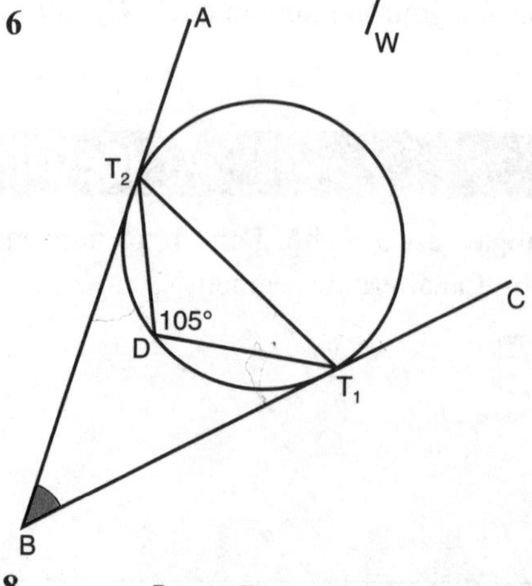

A, T_2, 105°, D, C, T_1, B

7

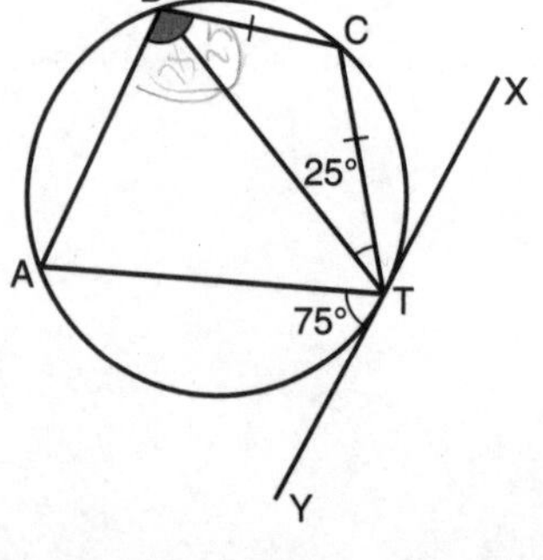

B, C, X, 25°, A, T, 75°, Y

8

D, C, E, 50°, 60°, A, T, B

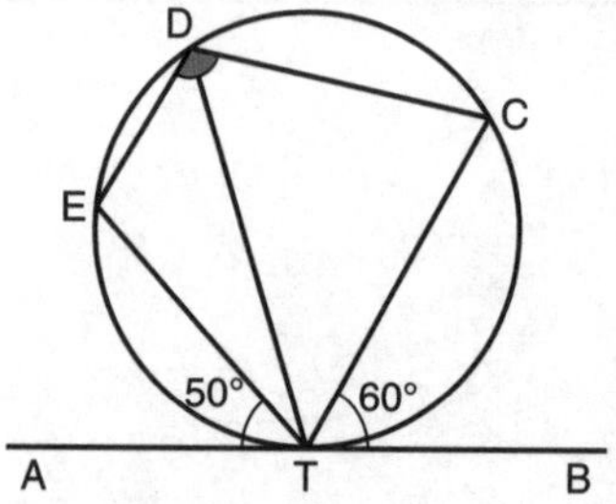

9 Find

a ∠OTX **b** ∠TOB **c** ∠OBT **d** ∠ATY

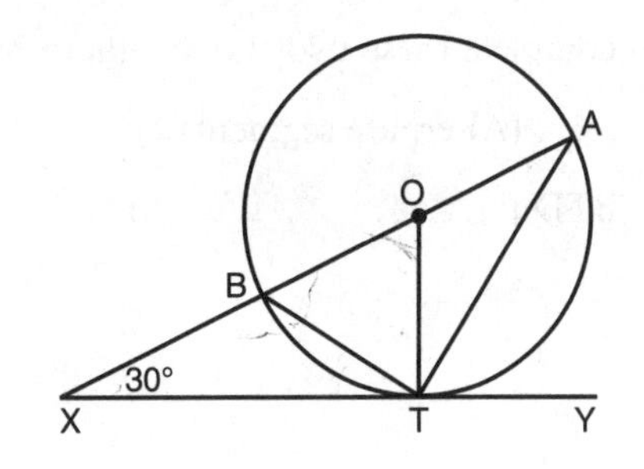

10 Find

a ∠OTB **b** ∠OTC **c** ∠OCT **d** ∠DTA

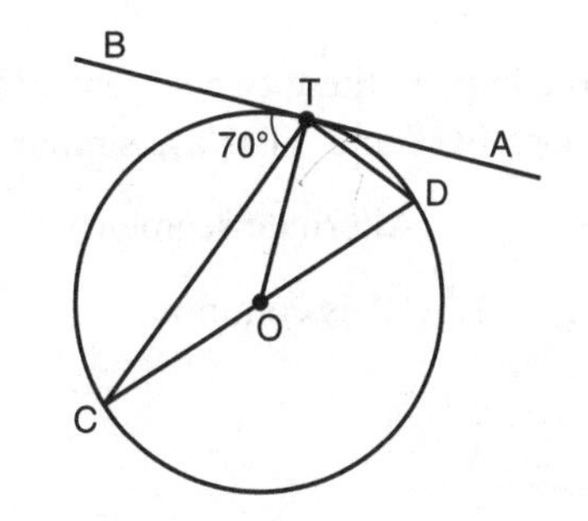

11 Copy and complete these two statements to prove ∠NPT = ∠PLT.

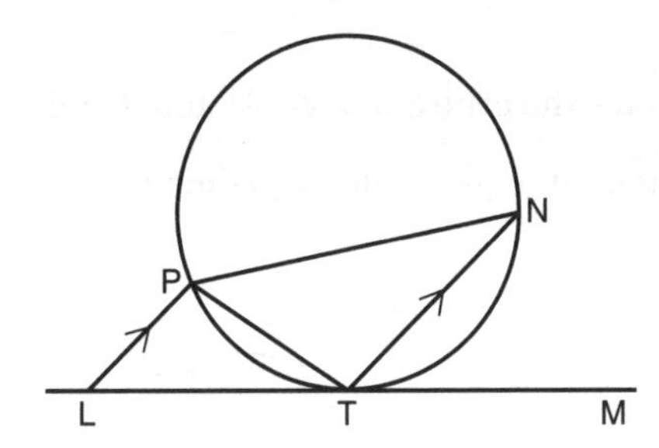

∠NTM = ... (Alternate segment)

∠PLT = ... (Corresponding angles)

12 Copy and complete these two statements to prove ∠ATF = ∠BAF.

∠ATF = ∠FDT (. . . . .)

∠FDT = ∠BAF (. . . . .)

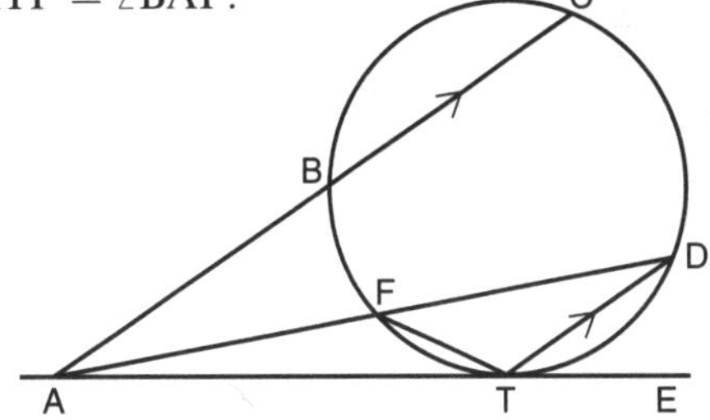

13 Copy and complete these two statements to prove ∠ATC = ∠BTD.

∠ATC = ... (Alternate segment)

∠ABT = ∠BTD (. . . . .)

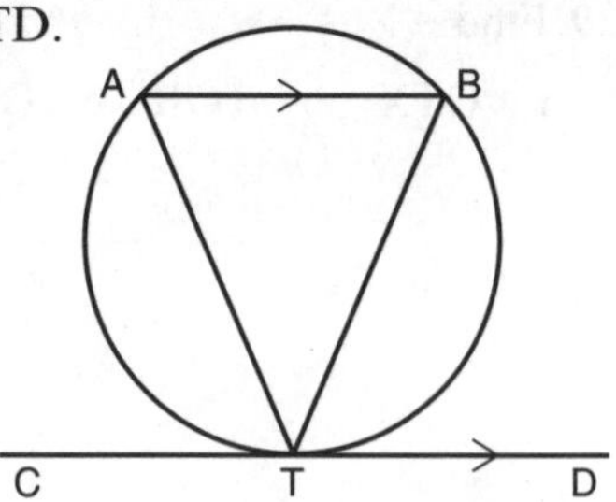

14 Copy and complete these two statements to prove that triangles BCT and TBD are similar.

∠DCT = ... ° (Alternate segment)

. . . . is common.

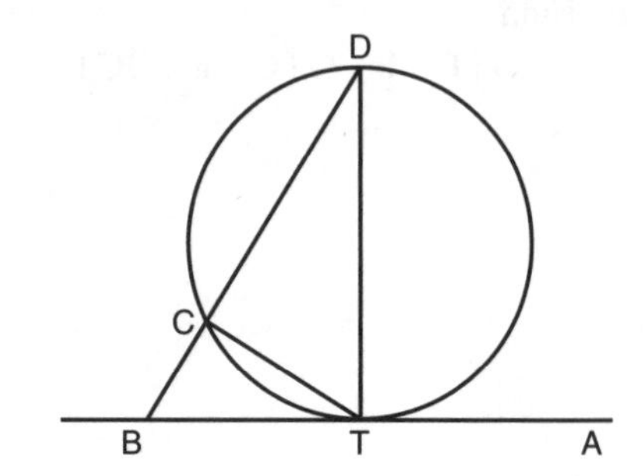

Exercise 13★

On all given diagrams, O indicates the centre of a circle and T indicates a tangent to the circle.

For Questions 1–4, find the coloured angles, fully explaining your reasons.

1

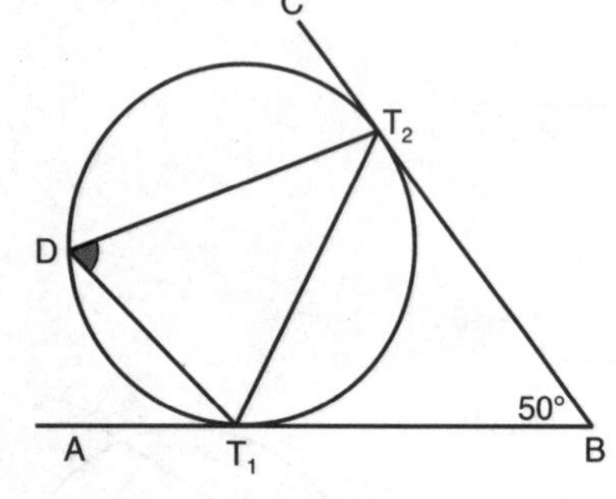

2

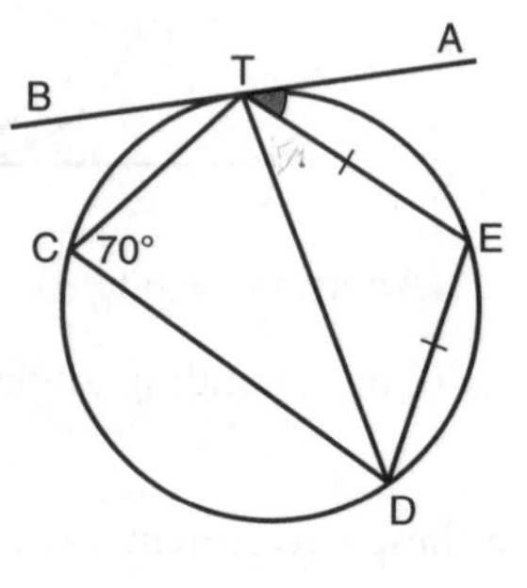

3

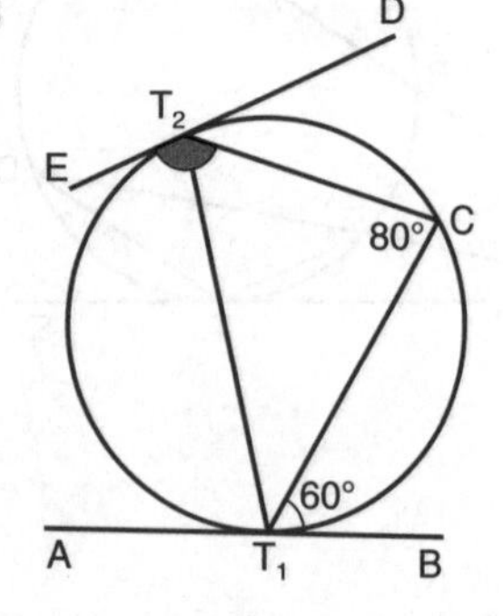

4

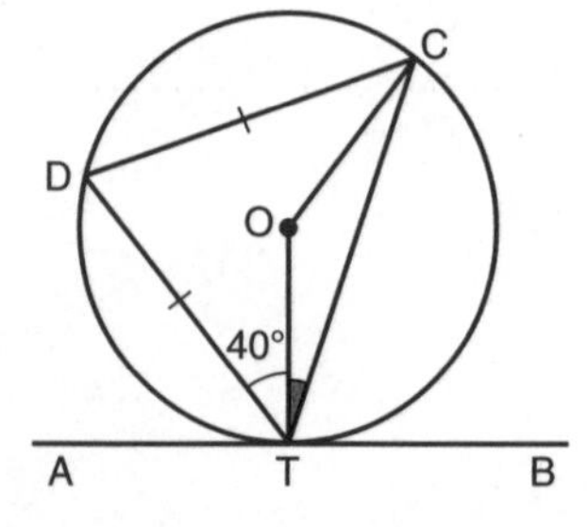

5 Prove that AB is the diameter.

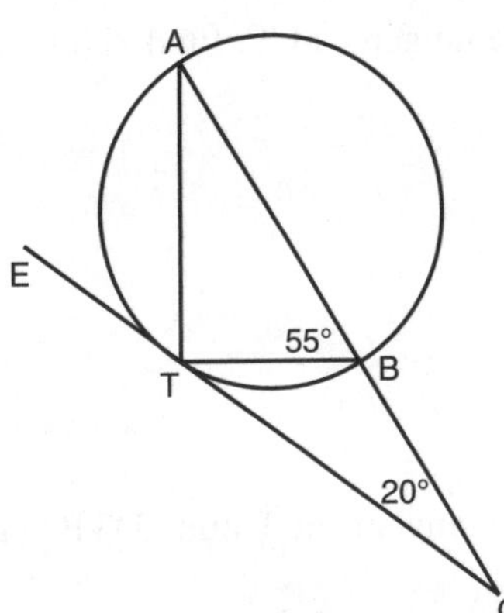

6 Given that $\angle BCT = \angle TCD$, prove that $\angle TBC = 90°$.

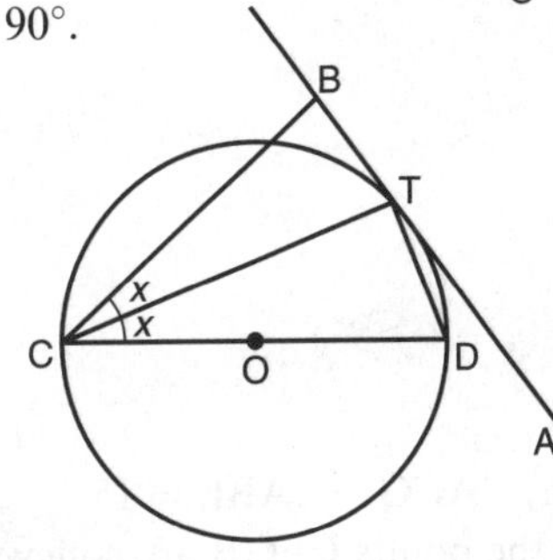

7 The inscribed circle of XYZ touches XY at A, YZ at B and XZ at C. If $\angle ZXY = 68°$ and $\angle ZYX = 44°$, find

a $\angle ABC$ **b** $\angle ACB$

8 PRY and PQX are tangents to the circle RSQ. If $\angle SRY = 154°$ and $\angle SQX = 136°$, find

a $\angle RSQ$ **b** $\angle RPQ$

9 Find

a $\angle DAE$

b $\angle BED$

c Prove that triangle ACD is isosceles.

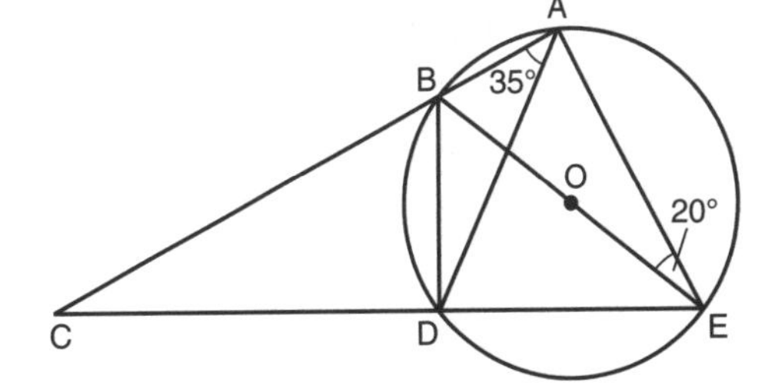

10 Find

a $\angle DTA$

b $\angle BCT$

c Prove that triangles BCT and BTD are similar.

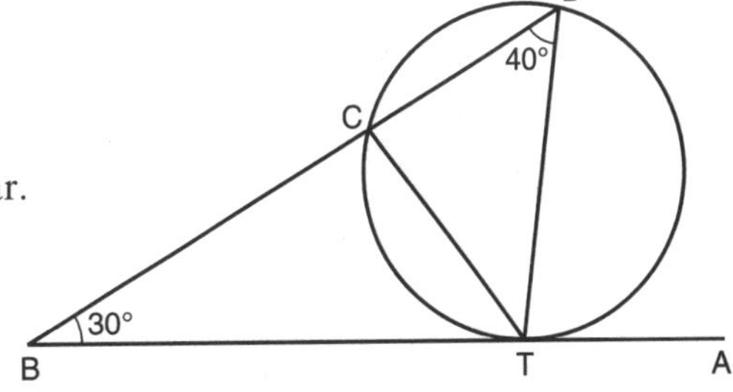

11 CD and AB are tangents at T. Find ∠ETF.

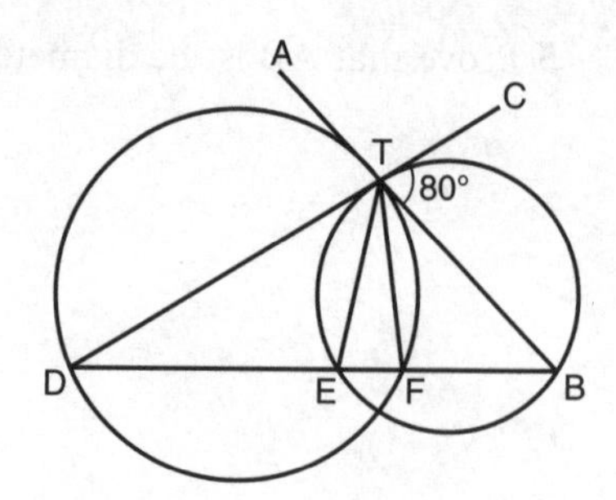

12 AB and CD are tangents at T and ∠DTB is less than 90°.
Find ∠DEB.

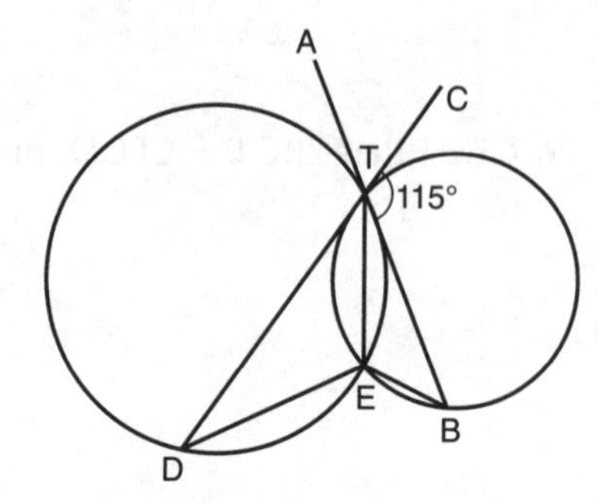

13 **a** Explain why ∠ACG = ∠ABF = 15°.
b Prove that the points CFGB are concyclic.

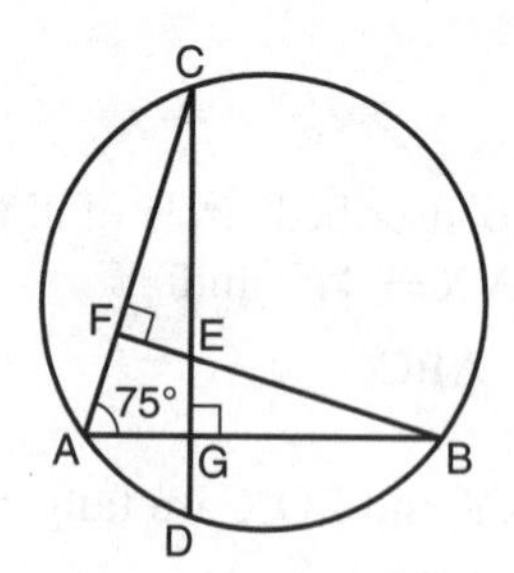

14 AB and DE are tangents to the circle, centre O.

a Write down all the angles equal to
(i) 70° (ii) 20°

b Prove that B is the mid-point of DT_2.

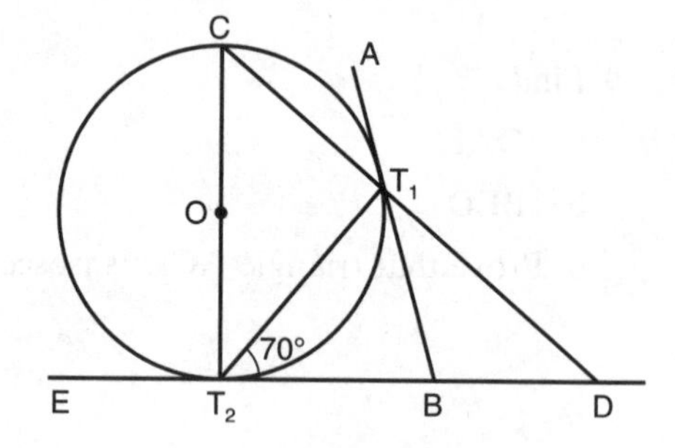

15 **a** Giving reasons, find, in terms of x, the angles EOC and CAE. Use your answers to show that triangle ABE is isosceles.

b If BE = CE prove that BE will be the tangent to the larger circle at E.

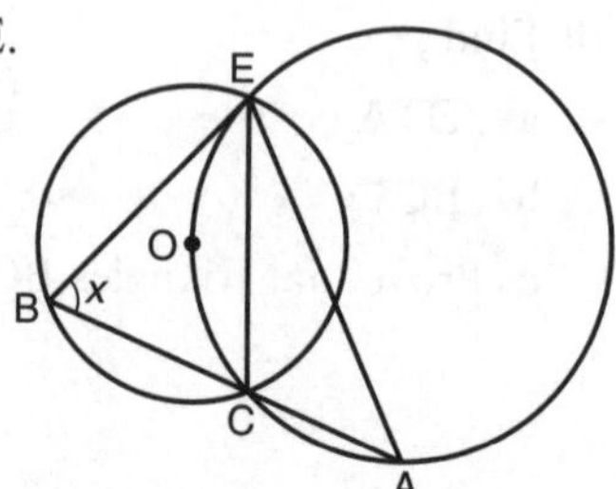

Intersecting chords and tangents

Two chords intersecting inside a circle

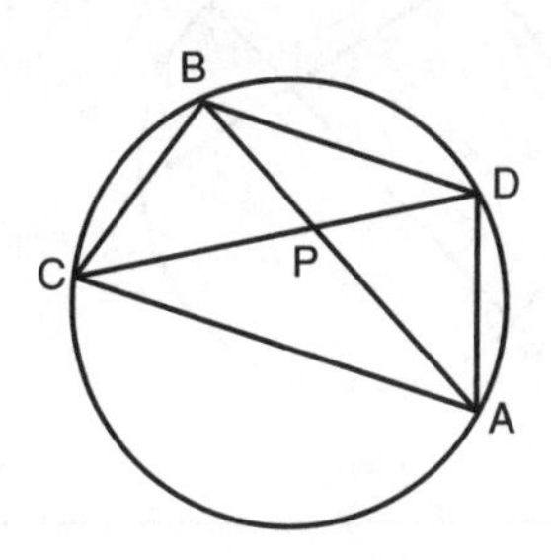

$AP \times PB = CP \times PD$

$A\hat{P}D = B\hat{P}C$ (Vertically opposite angles)

$C\hat{D}A = C\hat{B}A$ (Angles in same segment; chord AC)

(And $B\hat{A}D = B\hat{C}D$)

$\therefore$ APD and CPB are similar.

$\therefore \dfrac{AP}{CP} = \dfrac{PD}{PB}$

$\therefore AP \times PB = CP \times PD$

Two chords intersecting outside a circle

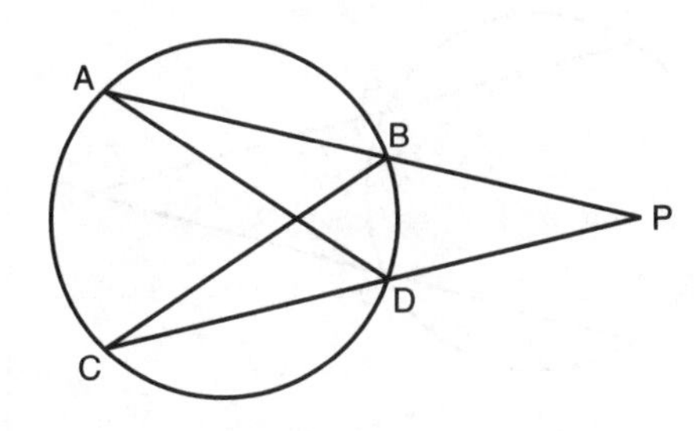

$AP \times PB = CP \times PD$

$B\hat{A}D = B\hat{C}D$ (Angles in same segment; chord BD)

$\hat{P}$ is common to $\triangle$s APD and CPB

(And $A\hat{D}P = C\hat{B}P$)

$\therefore$ $\triangle$s APD and CPB are similar.

$\therefore \dfrac{AP}{CP} = \dfrac{PD}{PB}$

$\therefore AP \times PB = CP \times PD$

When one chord becomes a tangent

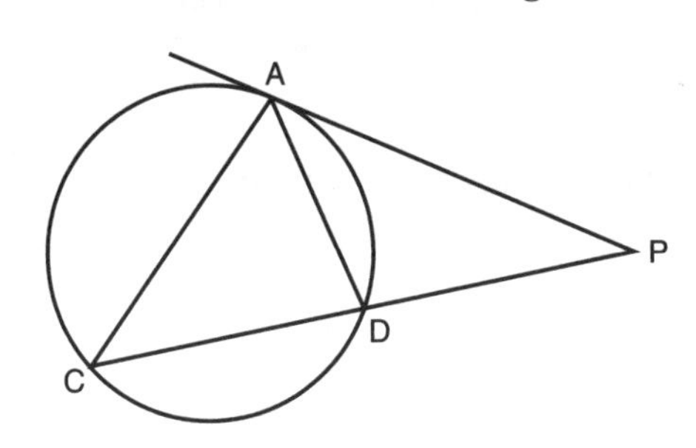

$AP^2 = CP \times PD$

A and B 'become' the same point; so

$AP \times PB \rightarrow AP \times PA = AP^2$

Or,

$\triangle$s APD and CPA are similar.

When both chords become tangents

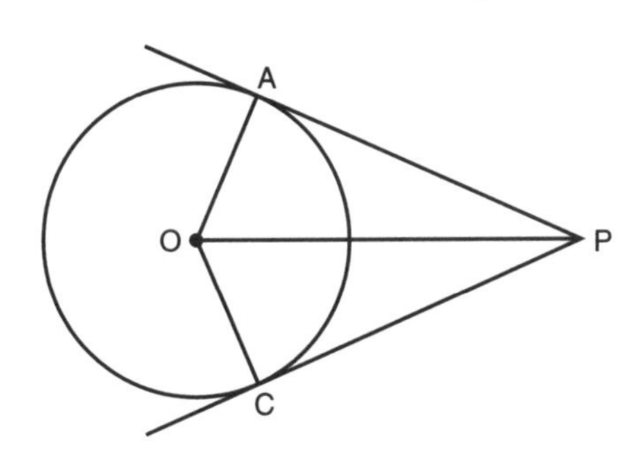

$AP = CP$

Now, C and D 'become' the same point; so

$CP \times PD \rightarrow CP \times PC = CP^2$

$AP^2 = CP^2$

Or,

$\triangle$s APO and CPO are congruent.

Example 7

AP = 12 cm, PD = 8 cm and CP = 9 cm. Find BP.

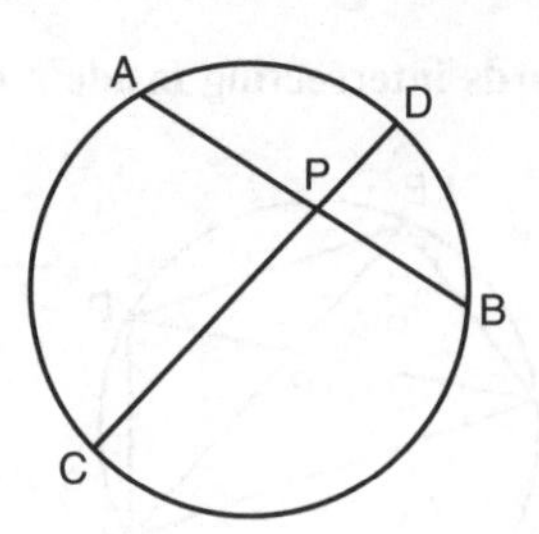

Let BP = x.
Now, AP × PB = CP × PD

$\therefore \quad 12 \times x = 9 \times 8$

$\therefore \quad x = \frac{9 \times 8}{12}$

$\therefore \quad x = 6\,\text{cm}$

Example 8

CD = 7 cm, DP = 5 cm and BP = 6 cm. Find AB.

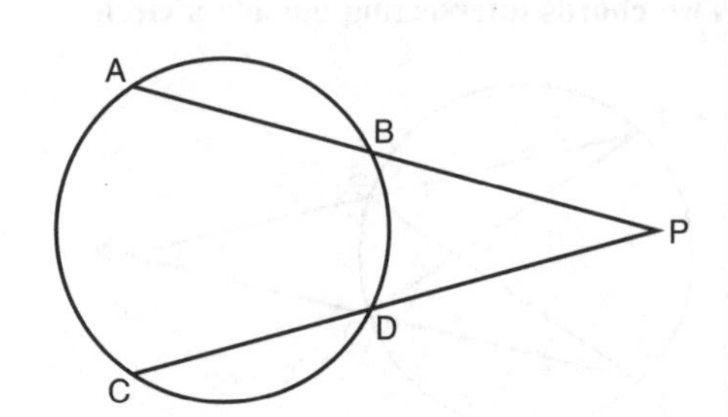

Let AB = x.
Now, AP × PB = CP × PD

AP = AB + BP = $x + 6$
CP = CD + DP = 12

$\therefore \quad (x + 6) \times 6 = 12 \times 5$

$\therefore \quad x + 6 = 10$

$\therefore \quad x = 4\,\text{cm}$

Exercise 14

1 AP = 15 cm, PB = 6 cm and DP = 4 cm. Find CP.

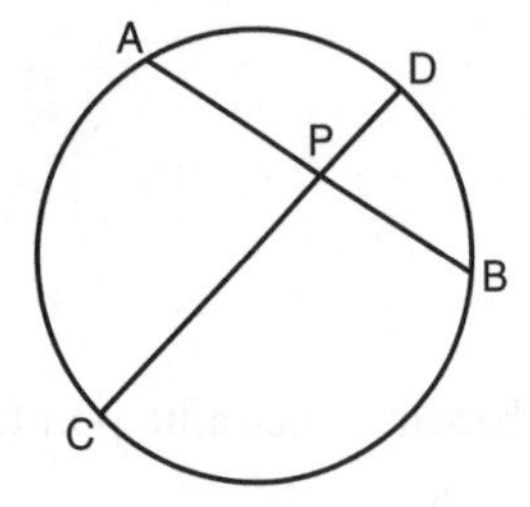

2 CP = 20 cm, PB = 7 cm and AP = 15 cm. Find DP.

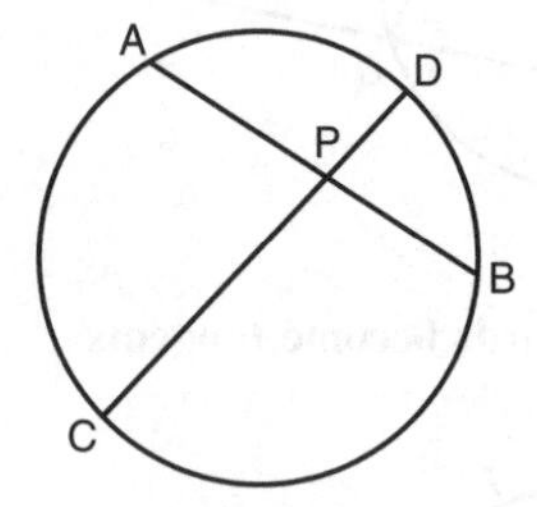

3 AB = 10 cm, PB = 3 cm and DP = 2 cm. Find CP.

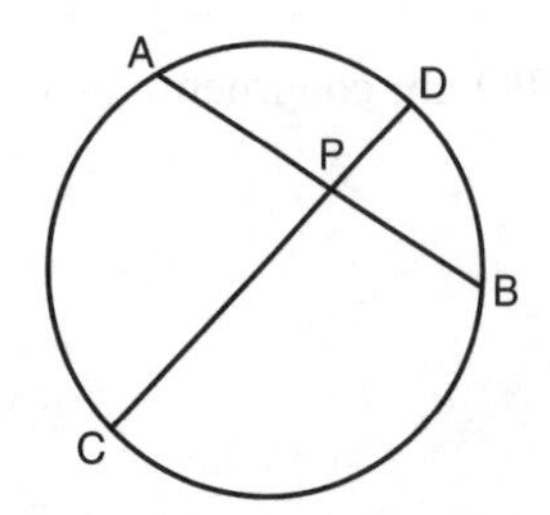

4 DP = 4 cm, PB = 6 cm and AB = 15 cm. Find CP.

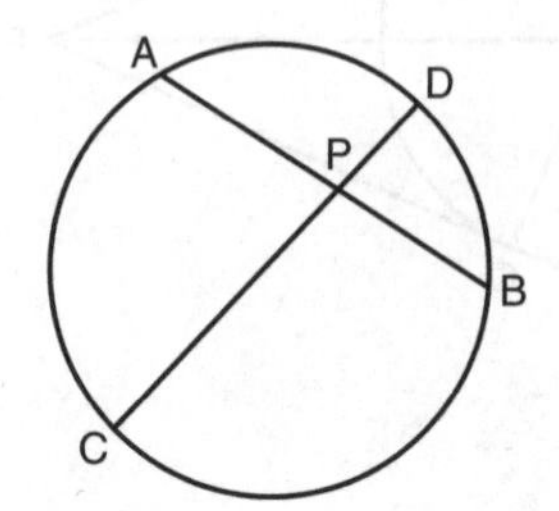

5 AP = 15 cm, BP = 6 cm and DP = 5 cm. Find CP.

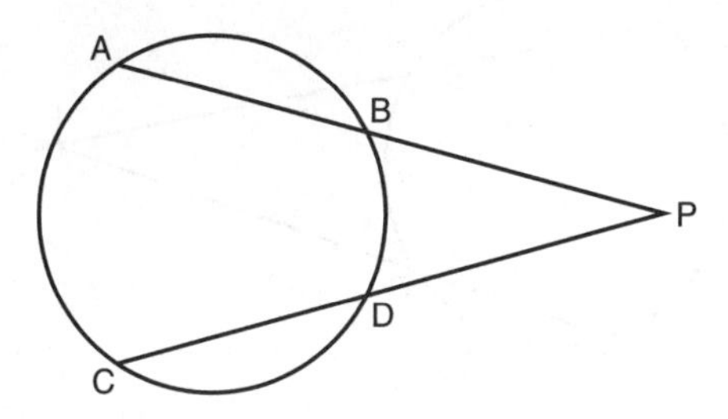

6 AB = 8 cm, BP = 12 cm and DP = 10 cm. Find CD.

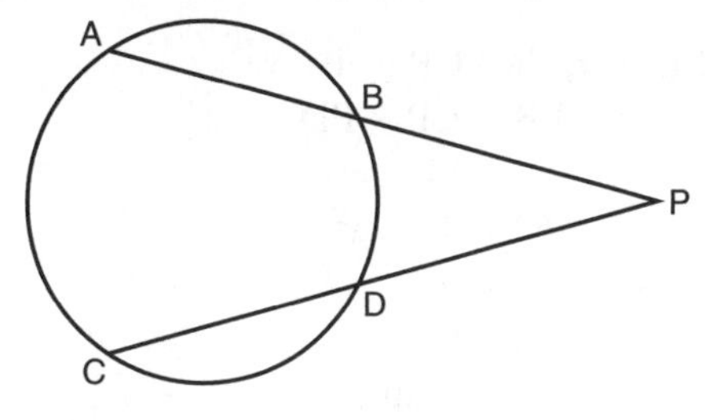

7 EX = 8 cm, HX = 6 cm and GH = 10 cm. Find EF.

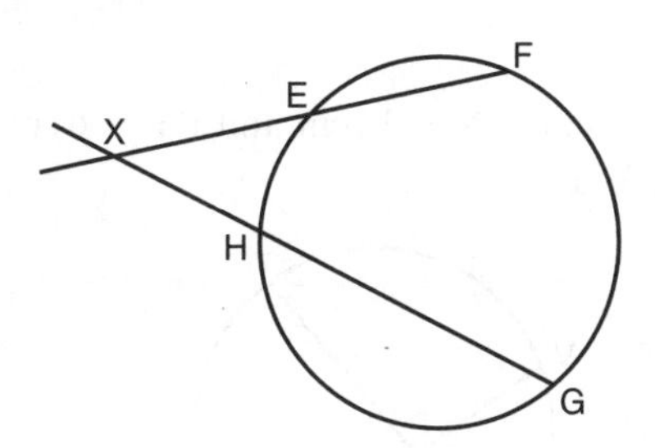

8 QR = 5 cm, PQ = 7 cm and RS = 4 cm. Find ST.

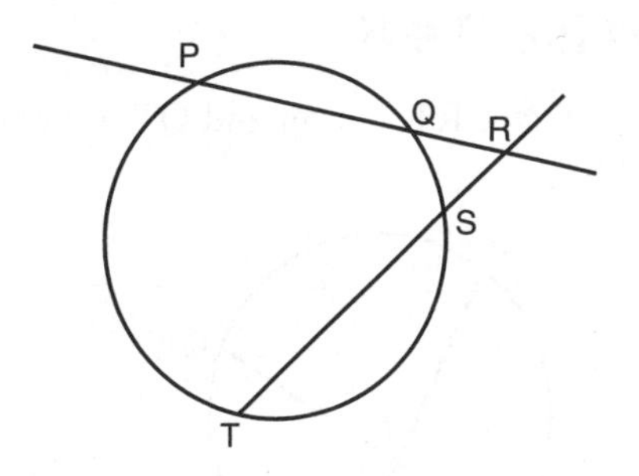

9 PT is a tangent to the circle. PS = 9 cm and SR = 7 cm. Calculate the length PT.

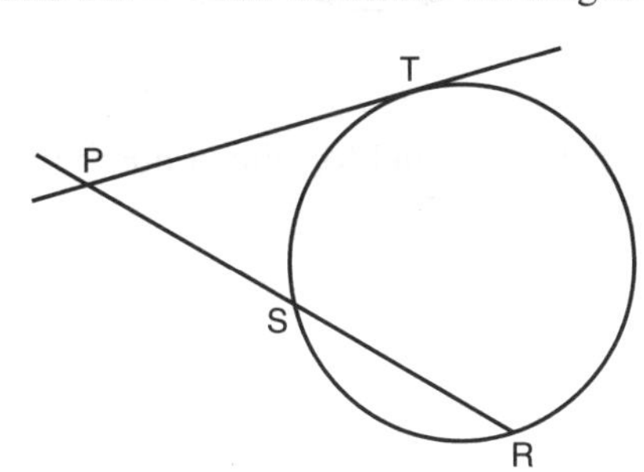

10 PQ is a tangent to the circle. PQ = 20 cm and QR = 16 cm. Calculate the length SR.

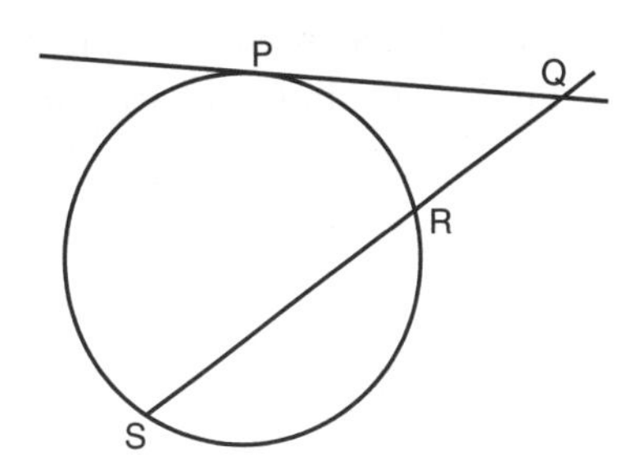

Example 9

AP = 8 cm, PB = 5 cm and CD = 14 cm. Find PD.

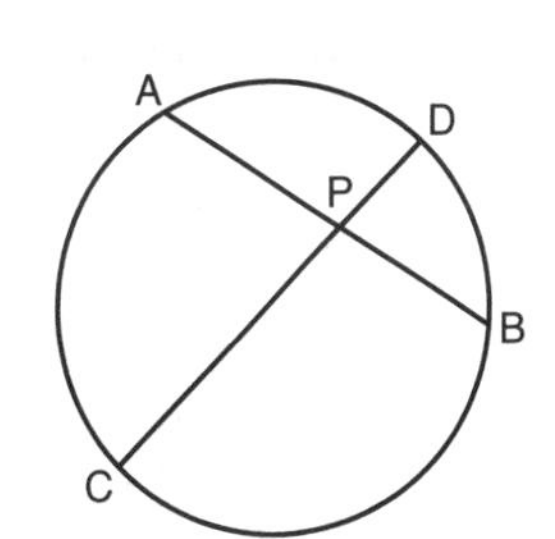

Let PD = x, then CP = $14 - x$.

Now, AP × PB = CP × PD

$\therefore \quad 8 \times 5 = (14 - x)x$

$40 = 14x - x^2$

$x^2 - 14x + 40 = 0$

$(x - 10)(x - 4) = 0$

$\therefore \quad x = 4$ or 10 cm

Example 10

AB = 7 cm, PB = 5 cm and CD = 4 cm. Find PD.

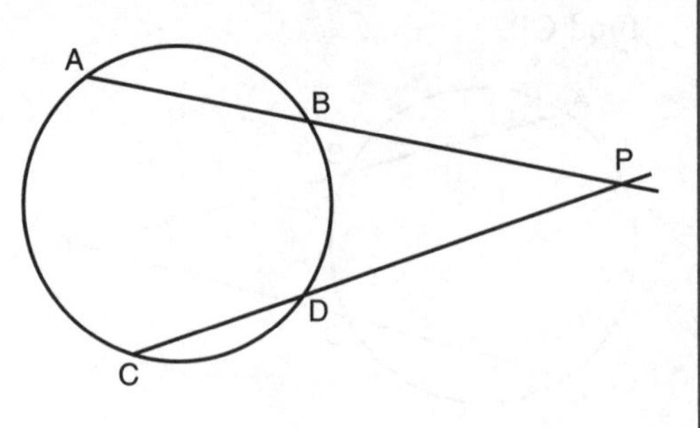

Let PD = x, then CP = $4 + x$.

Now, AP × PB = CP × PD

$\therefore \quad 12 \times 5 = (4 + x)x$

$60 = 4x + x^2$

$x^2 + 4x - 60 = 0$

$(x - 6)(x + 10) = 0$

$\therefore \quad x = 6\,\text{cm}$

Exercise 14★

1 PR = 10 cm, RT = 4 cm and QT = 8 cm. Find ST.

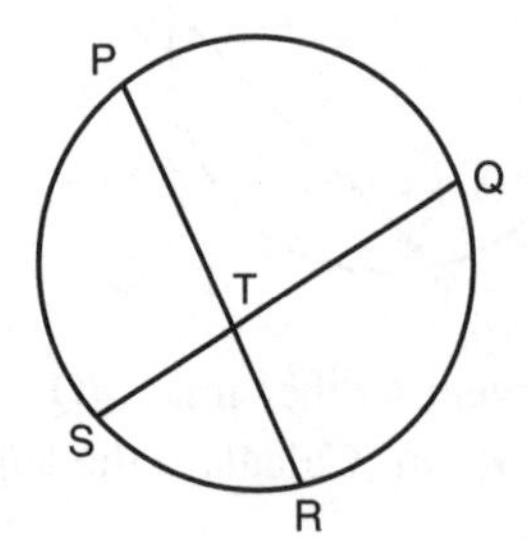

2 VP = 7 cm, VX = 16 cm and PY = 6 cm. Find WY.

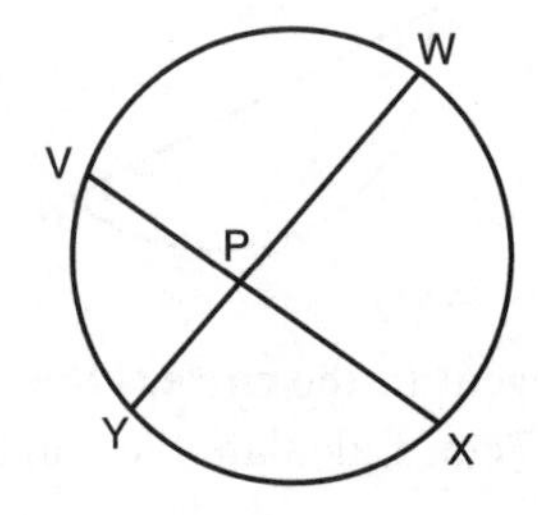

3 PR = 23 cm, PT = 8 cm, SQ = 22 cm and QT = x. Form a quadratic equation and find x.

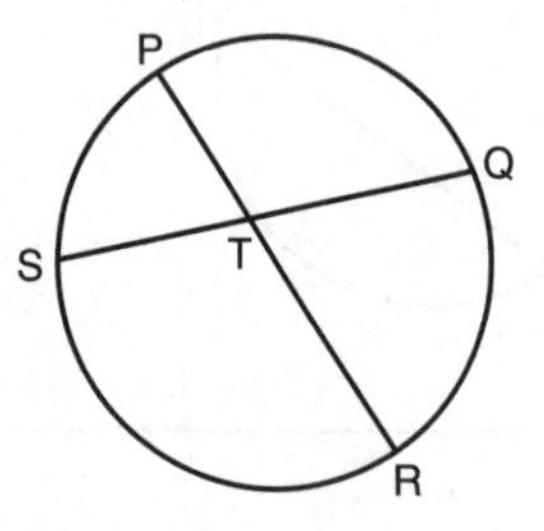

4 WP = YP = 10 cm, XZ = 5 × PZ. By putting PZ = x, and forming a quadratic equation, find x.

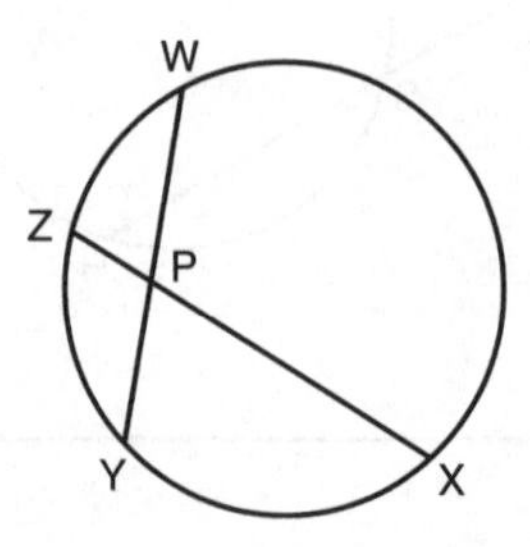

5 BC = 9 cm, AB = 7 cm and DE = 10 cm. Find CD.

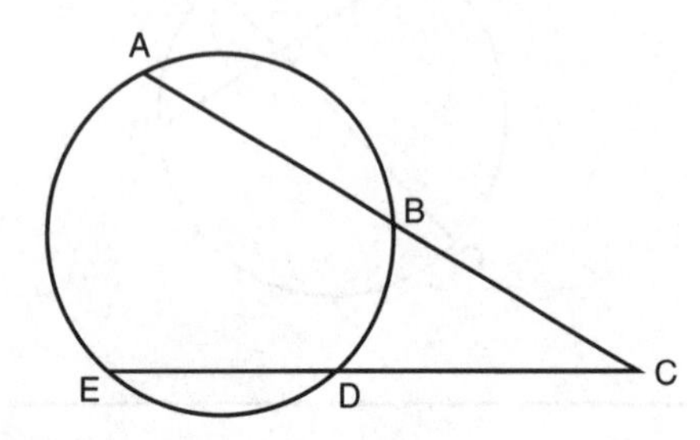

6 EF = 1 cm, IH = 8 cm and HG = 12 cm. Calculate FG.

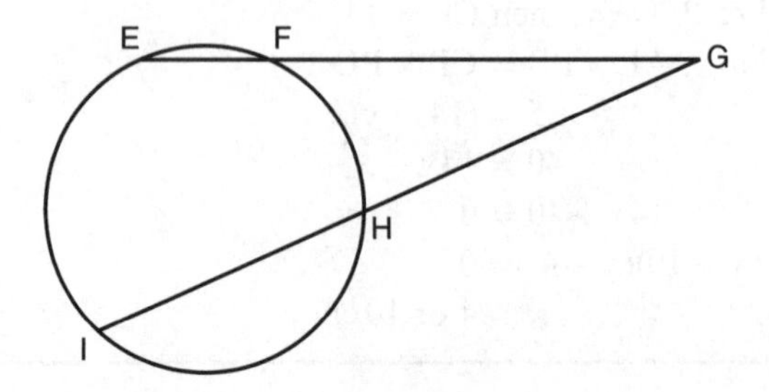

7 AC = 30 cm and AB = 14 cm.
Calculate CE when DE = 4 cm.

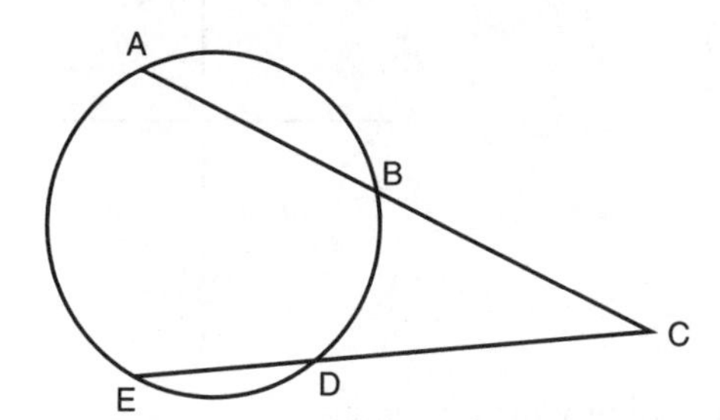

8 GT is a tangent to the circle. GT = 6 cm, IH = 3.5 cm and GH = x. Find x.

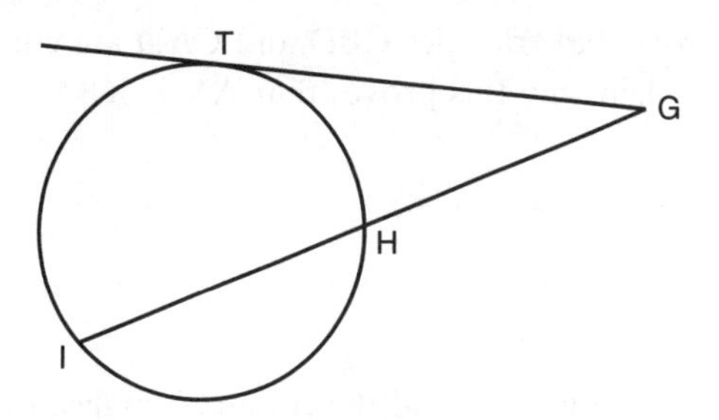

9 Let SQ = x, RQ = y and the radius = r.

a Show that the Intersecting Chord Theorem and Pythagoras' Theorem in △OSQ give the same equation in x, y and r.

b Calculate x when $y = 8$ and $r = 5$.

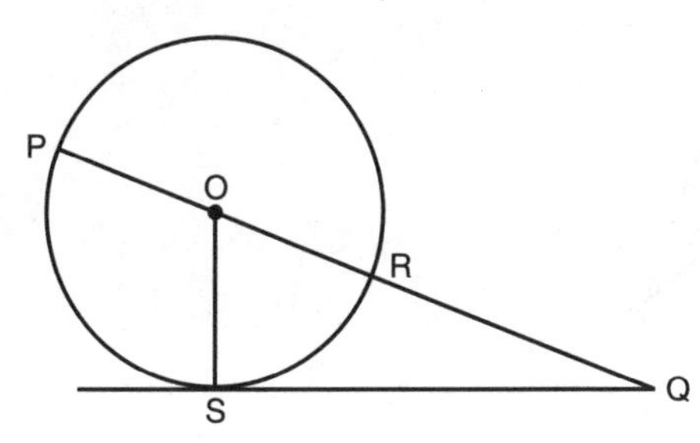

10 PQ is a tangent to the circle. PS passes through the centre of the circle at O. SR = 64 mm and RQ = 36 mm.

a Calculate the length PQ.

b Thus calculate the radius of the circle.

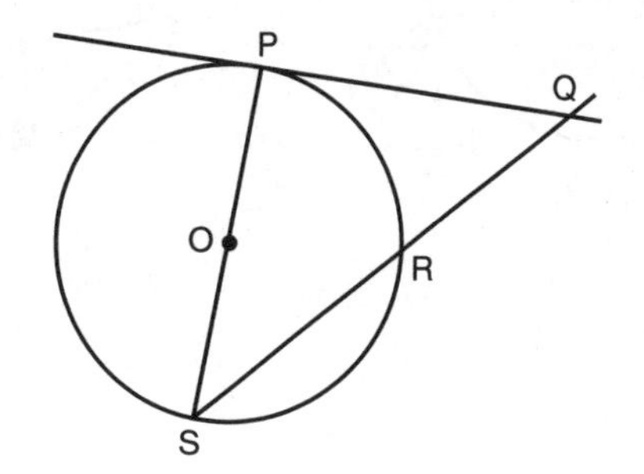

Exercise 15 (Revision)

In this exercise, O is the centre of a circle and T indicates a tangent to the circle.

1 Copy and complete the working to show that triangles ABE and CDE are congruent.

In triangles ABE and CDE

1 DE = ... (Given)

2 ∠BAE = ... (Alternate)

3 ∠AEB = ∠CED (............)

∴ △ABE ≡ △CDE (............)

∴ ∠ABE = ...

AB = ...

AE = ...

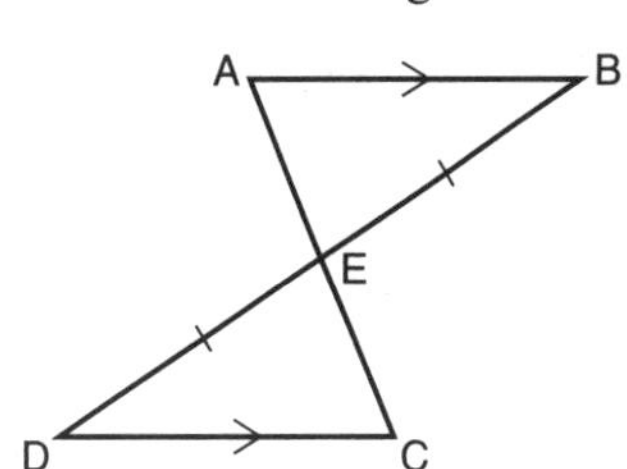

2 The construction shows the perpendicular bisector of the line AB.

Prove that triangles CBD and CAD are congruent.
Explain why this proves that AX = XB.

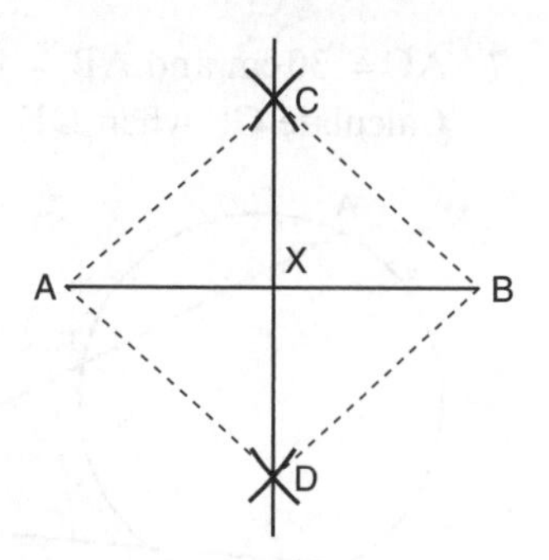

For Questions 3–6, find the coloured angles, fully explaining your reasoning.

3

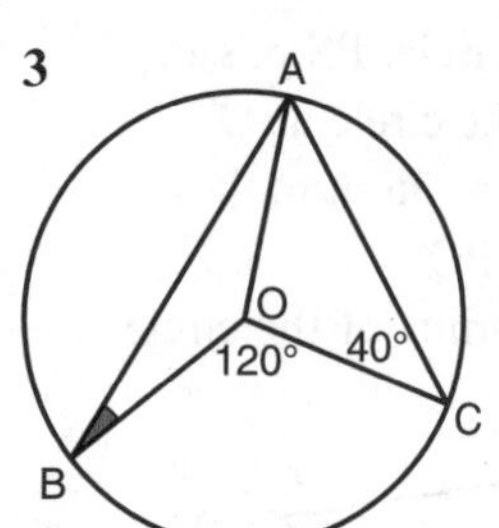

4

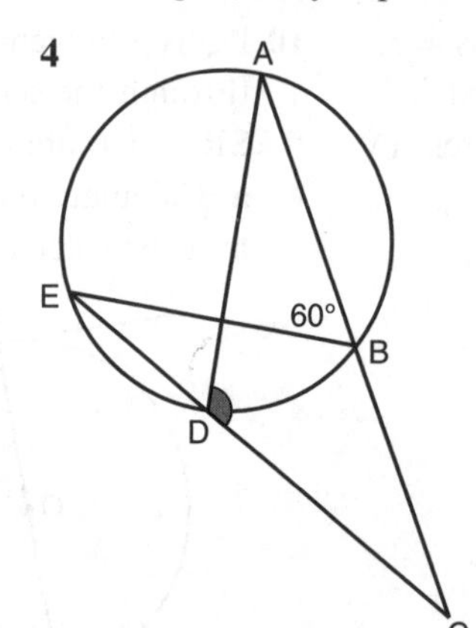

5

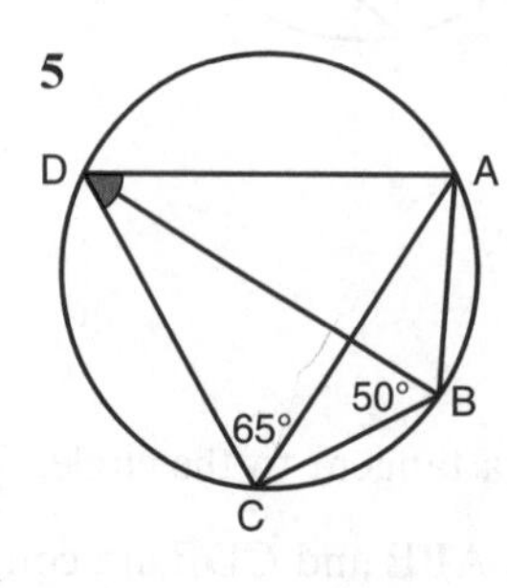

6

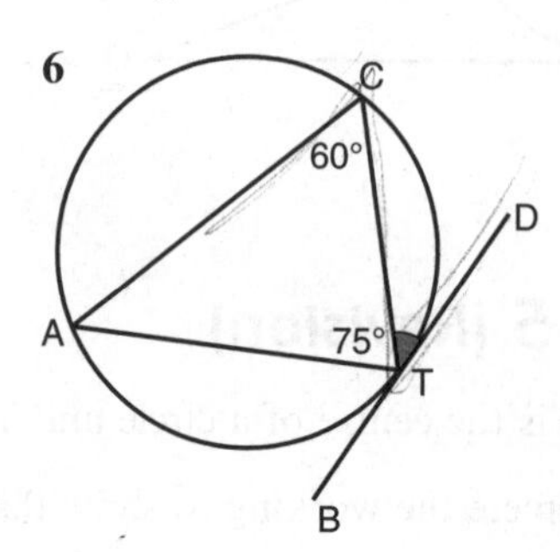

7 AP = 9 cm, AD = 15 cm and BP = 4 cm.
Find BC.

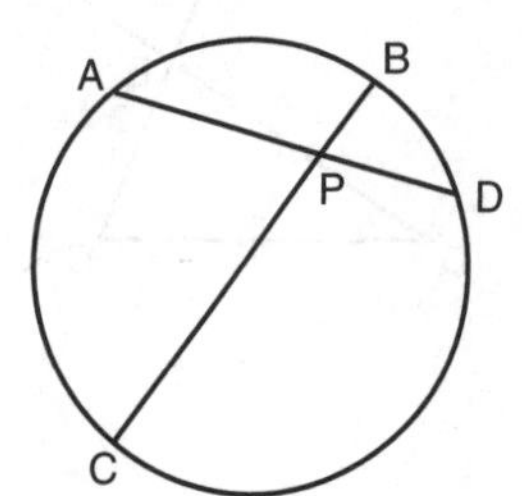

8 EF = 5 cm, EG = 14 cm and GH = 8 cm.
Calculate HI.

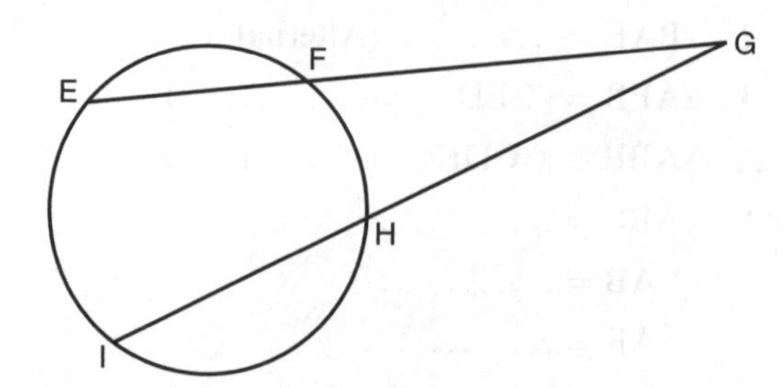

9 Find

a $\angle DCT$

b $\angle TCB$

c $\angle CTB$

d $\angle TDC$

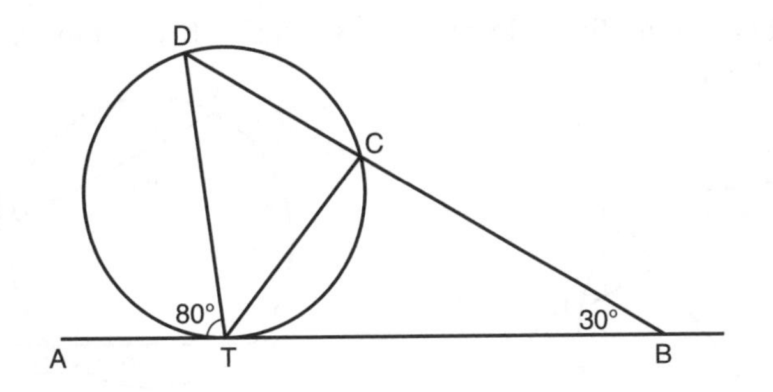

Exercise 15★ (Revision)

1 The construction shows the perpendicular through W on AB. Prove that triangles ZXW and ZYW are congruent.

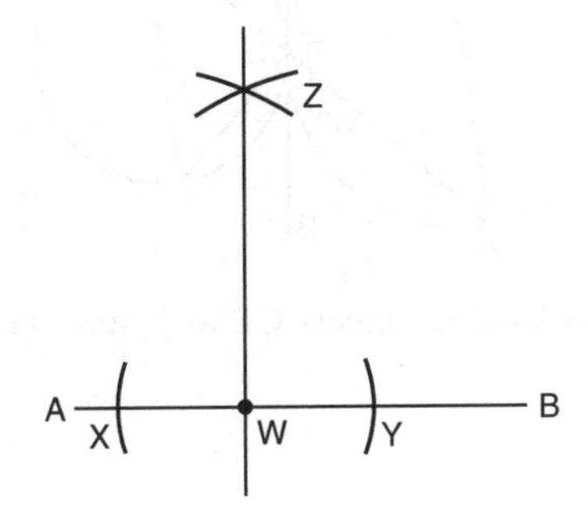

2 In the figure let $\angle ABO = x$.

a Show that $\angle EAO = \angle BOC = 45° + x$.

b Use congruent triangles to prove that $EA = OC$.

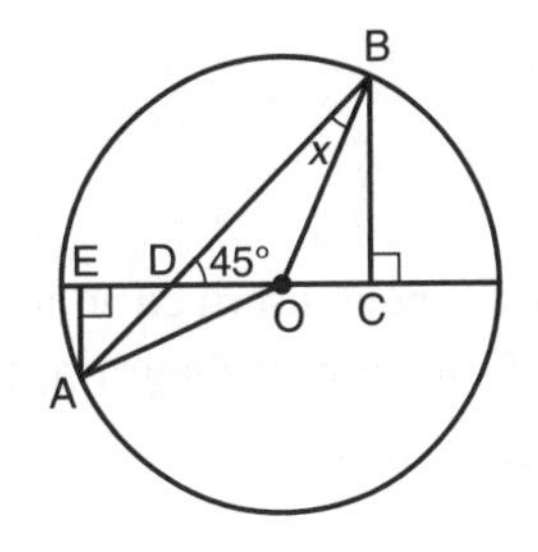

3 Find

a $\angle ETD$

b $\angle TEB$

c Prove that EC is the diameter of the circle.

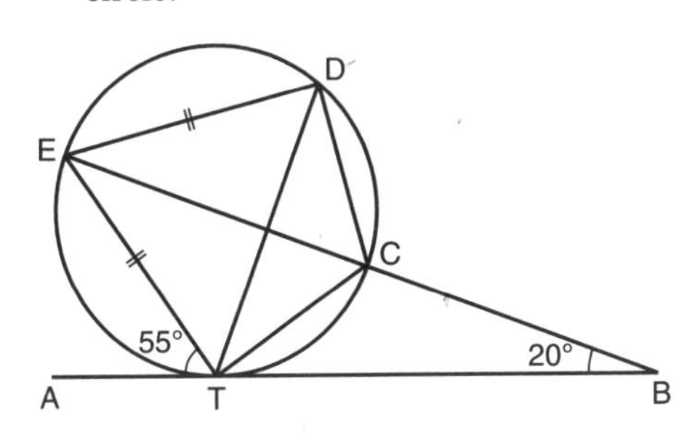

4 PQ is a tangent to the circle.

a Form an equation in x.

b Solve the equation to find length RQ.

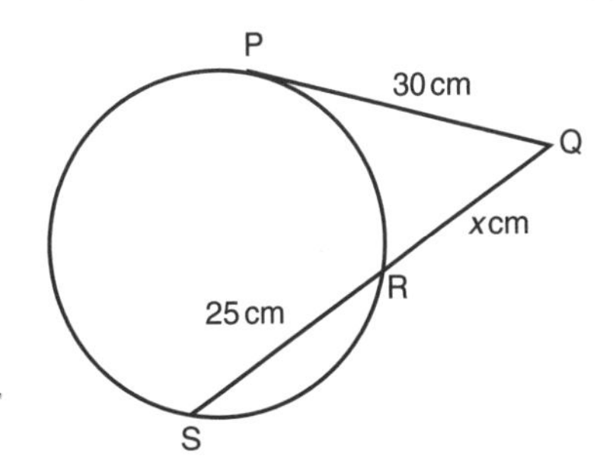

5 AP = 9 cm, AD = 13 cm, BC = 15 cm, PB = x and CP = y.
Find x and y.

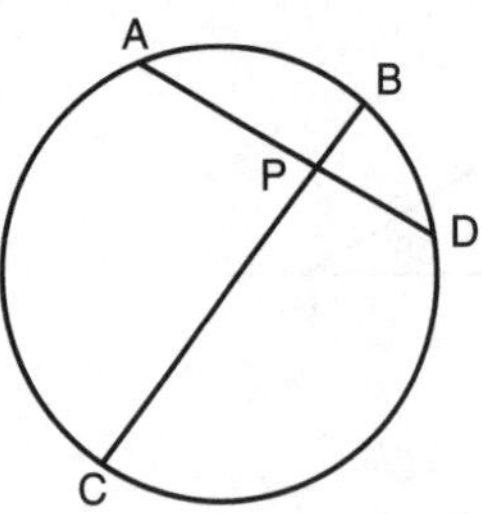

6 AC = 10 cm. Find x and y.

7 AB is the common tangent.

a Find ∠XZT.

b Find ∠WVT.

c Prove that XZ is parallel to WV.

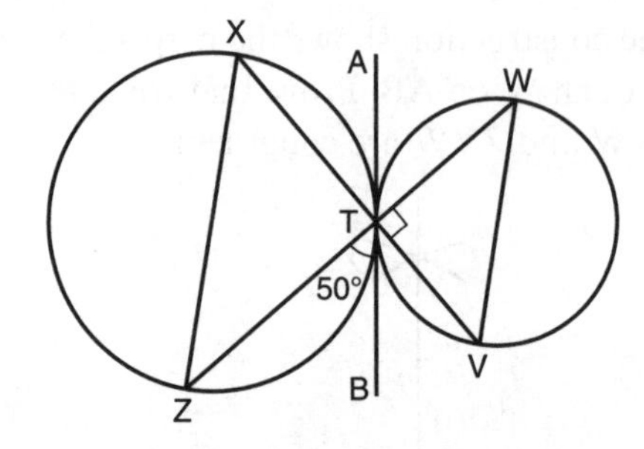

8 ABC is any triangle inscribed in a circle. The bisector of angle A meets CB at X and the circle at Y. Draw a neat diagram, then prove that

a triangle BCY is isosceles

b ∠ABY = ∠BXY

Sets 1

Remember

A **set** is a collection of objects, described by a list or a rule. $A = \{1, 3, 5\}$

Each object is an **element** or **member** of the set. $1 \in A, 2 \notin A$

Sets are **equal** if they have exactly the same elements. $B = \{5, 3, 1\}, B = A$

The **number of elements** of set A is given by $n(A)$. $n(A) = 3$

The **empty set** is the set with no members. $\{\}$ or $\emptyset$

The **universal set** contains all the elements being discussed in a particular problem. $\mathscr{E}$

$\mathscr{E}$

B is a **subset** of A if every member of B is a member of A. $B \subset A$

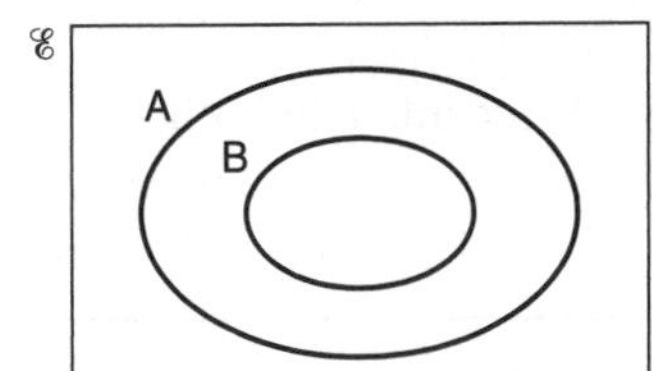

The **complement** of set A is the set of all elements not in A. A'

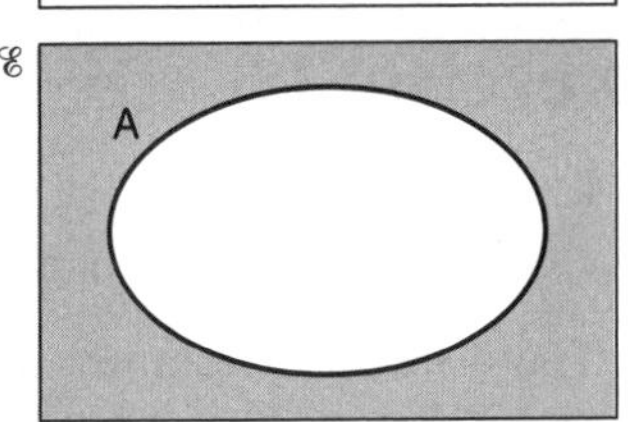

The **intersection** of A and B is the set of elements which are in both A and B. $A \cap B$

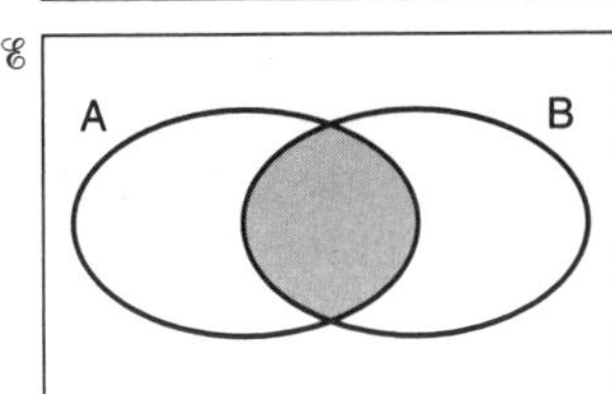

The **union** of A and B is the set of elements which are in A or B or both. $A \cup B$

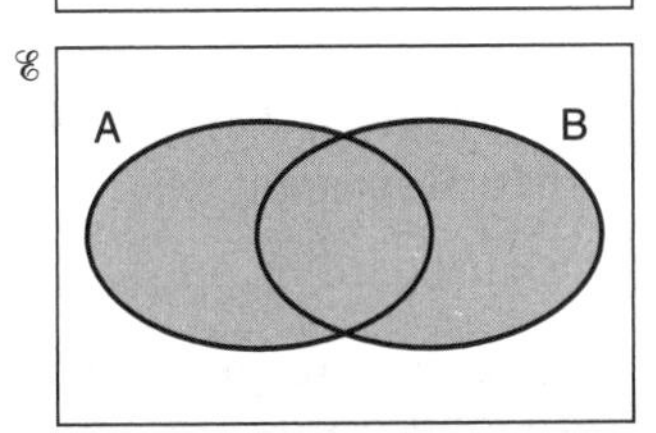

Problems involving sets

Entering the information from a problem into a Venn diagram usually means the numbers in the sets can be worked out. Sometimes it is easier to use some algebra as well.

Example 1

In a class of 23 students, 15 like coffee, 13 like tea and 4 students don't like either drink. How many like

a tea only **b** coffee only **c** both drinks?

Enter the information into a Venn diagram in stages. Let C be the set of coffee drinkers and T the set of tea drinkers. Let x be the number of students who like both. The 4 students who don't like either drink can be put in, along with x for the students who like both.

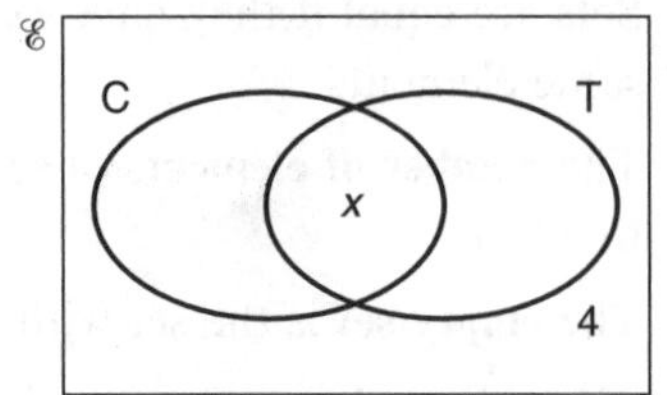

Now the other information can be added.

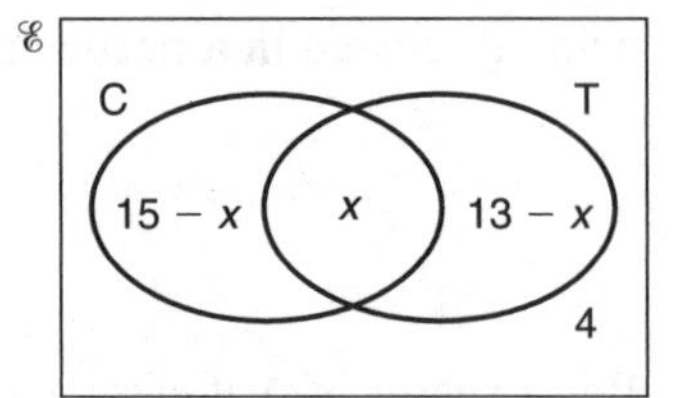

The number who like coffee, tea or both is $23 - 4 = 19$.
This means $n(C \cup T) = 19$.
So $(15 - x) + x + (13 - x) = 19$
$\Rightarrow 28 - x = 19$
$\Rightarrow x = 9$
So 9 students like both.
The number liking tea only is $13 - x = 4$.
The number liking coffee only is $15 - x = 6$.

Exercise 16

1 In a class of 40 students, 18 had watched 'Next Door' last night, 23 had watched 'Westenders' and 7 had watched both programmes. How many students did not watch either programme?

2 There are 182 spectators at a football match. 79 are wearing a hat, 62 are wearing a scarf and 27 are wearing a hat but not a scarf. How many are wearing neither a hat nor a scarf?

3 In one form in a school, 13 students are studying media, 12 students are studying sociology, 8 students are studying both and 5 students are doing neither subject. How many students are there in the form?

4 In a town, 9 shops rent out videos, 12 shops rent out DVDs, 7 shops rent out both and 27 shops rent out neither videos nor DVDs. How many shops are there in the town?

5 A youth group has 31 members. 15 like skateboarding, 13 like roller-skating and 8 don't like either. How many like

a skateboarding only **b** roller-skating only **c** both?

6 52 students are going on a skiing trip. 28 have skied before, 30 have snowboarded before, while 12 have done neither. How many have done both sports before?

Exercise 16★

1 A social club has 40 members. 18 like singing, 7 like both singing and dancing, while 6 like neither. How many like dancing?

2 At Tom's party there were both pizzas and burgers to eat. Some people had one of each. The number who had a burger only was seven more than the number who had both. Twice as many people ate a pizza only as the number who had both. The number who ate neither was the same as the number who had both. If there were 57 people at the party, how many people ate both?

3 In the end-of-year exams, 68 students took Mathematics, 72 took Physics and 77 took Chemistry. 44 took Mathematics and Physics, 55 took Physics and Chemistry, 50 took Mathematics and Chemistry, while 32 took all three subjects. Draw a Venn diagram to represent this information and hence calculate how many students took these three exams.

4 A group of 40 teenagers have all seen the film 'Parry Hotter'. 22 have seen it on DVD, 23 have seen it on video and 17 have seen it at the cinema. 12 have seen it on DVD and video, 6 have seen it on video and at the cinema, and 7 have seen it on DVD and at the cinema. How many have seen it on DVD, on video and at the cinema? (Hint – in the Venn diagram let x be the number who have seen it on DVD, on video and at the cinema.)

5 In a form of 25 students, 19 have scientific calculators and 14 have graphic calculators. If x students have both and y students have neither, what are the largest and smallest possible values of x and y?

6 It is claimed that 75% of teenagers can ride a bike and 65% can swim. What can be said about the percentage who do both?

Identifying sets by shading

Sometimes it can be difficult to find the intersection or union of sets in a Venn diagram. If one set is shaded in one direction and the other set in another direction, then the intersection is given wherever there is cross shading; the union is given by any shading at all.

Example 2

Show on a Venn diagram

a $A' \cap B$ **b** $A' \cup B$

The diagrams show first the sets A and B, then the set A′ shaded one way, then the set B shaded the other way.

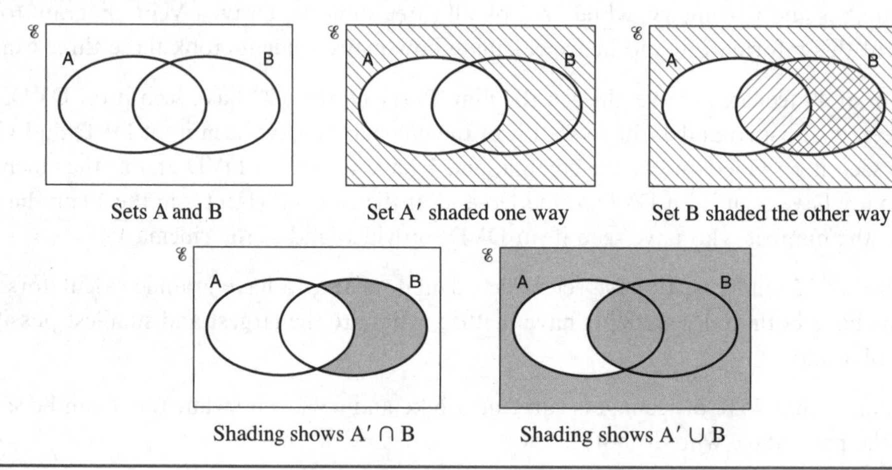

Sets A and B — Set A′ shaded one way — Set B shaded the other way

Shading shows $A' \cap B$ — Shading shows $A' \cup B$

Exercise 17

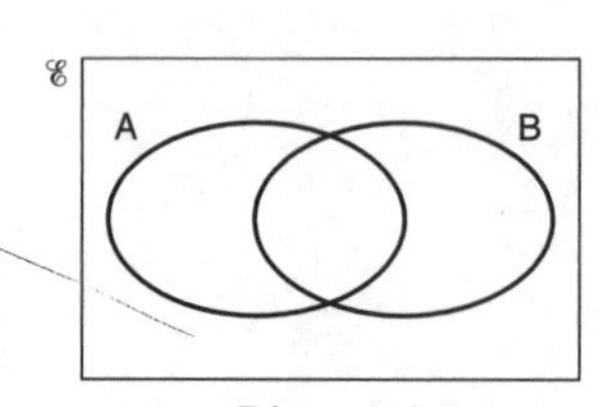

Diagram 1

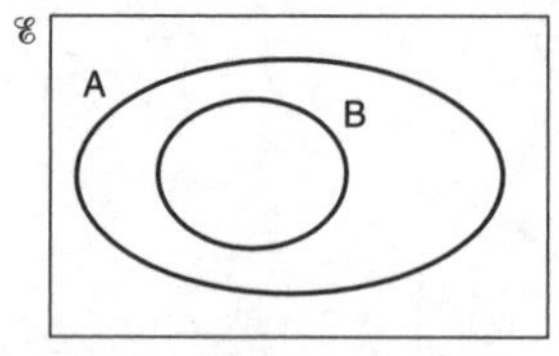

Diagram 2

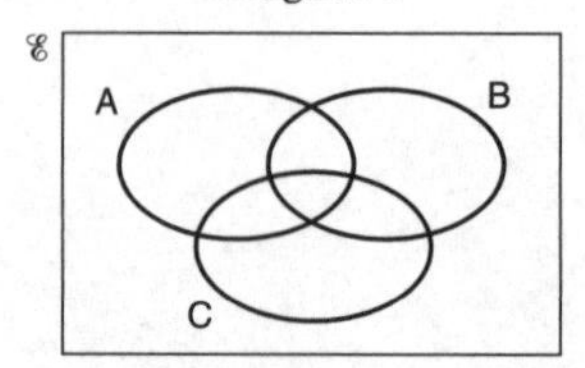

Diagram 3

1 On copies of diagram 1 shade the following sets:
- **a** $A \cap B'$
- **b** $A \cup B'$
- **c** $A' \cap B'$
- **d** $A' \cup B'$

2 On copies of diagram 2 shade the following sets:
- **a** $A \cap B'$
- **b** $A \cup B'$
- **c** $A' \cap B'$
- **d** $A' \cup B'$

3 On copies of diagram 3 shade the following sets:
- **a** $A \cap B \cap C$
- **b** $A' \cup (B \cap C)$

4 On copies of diagram 3 shade the following sets:
- **a** $(A \cup B') \cap C$
- **b** $A \cup B \cup C'$

5 Describe the shaded sets using set notation:

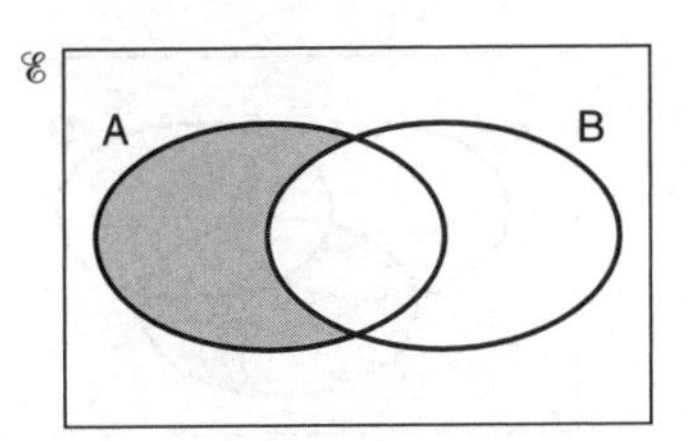

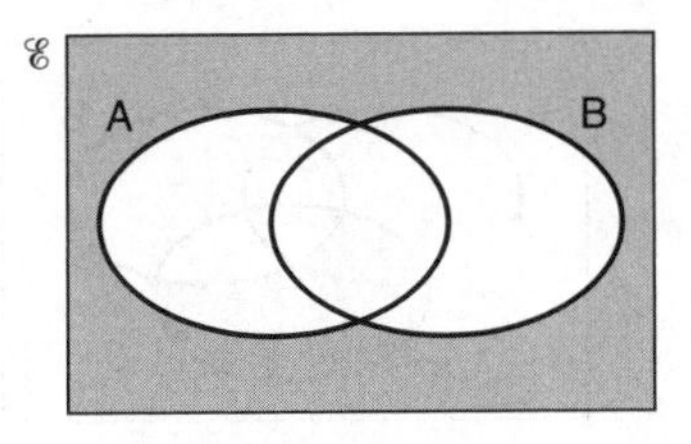

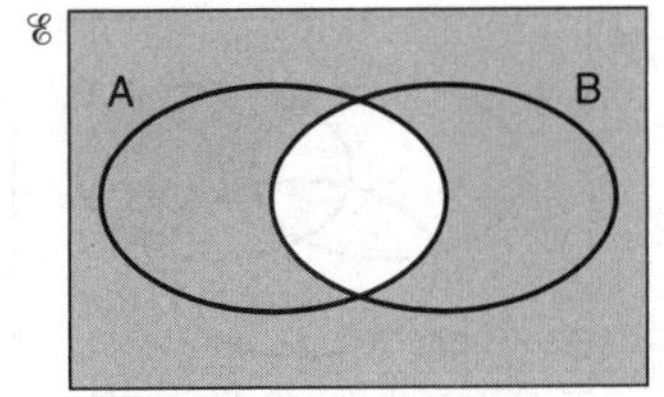

6 Describe the shaded sets using set notation:

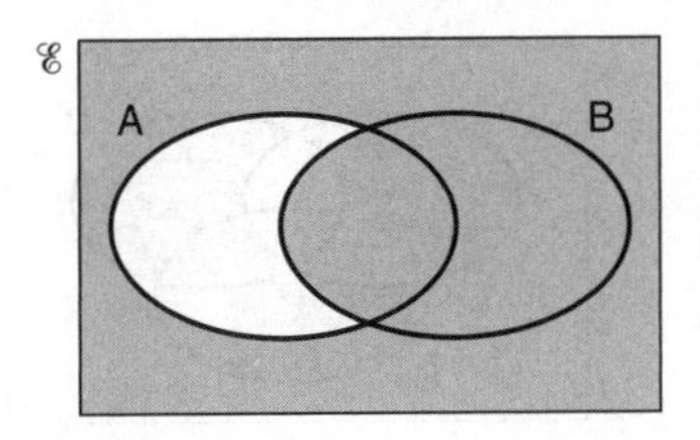

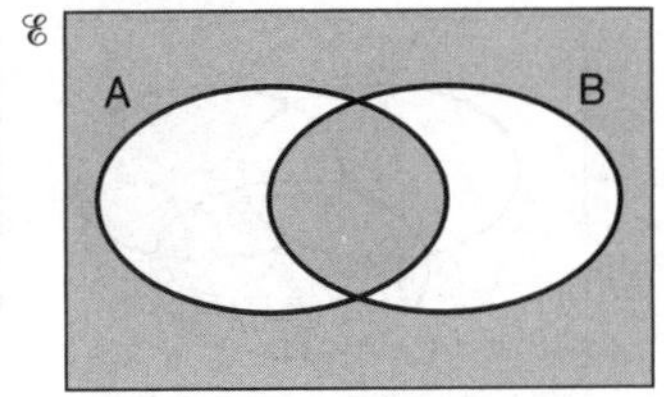

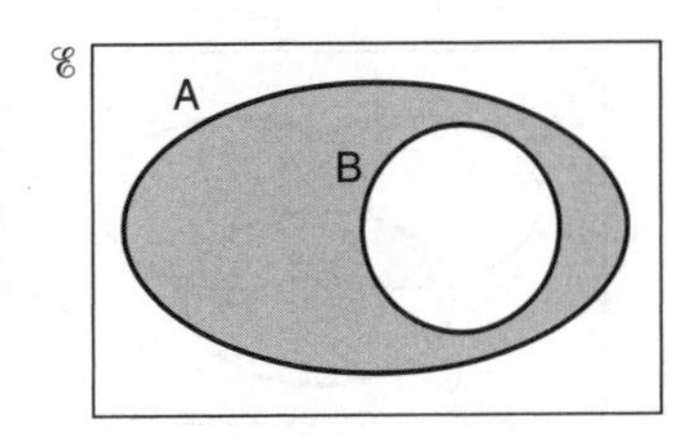

Exercise 17★

1 On copies of diagram 1 shade the following sets:

a $(A' \cap B)'$

b $(A \cup B')'$

c $(A \cap B')'$

d $(A \cup B)'$

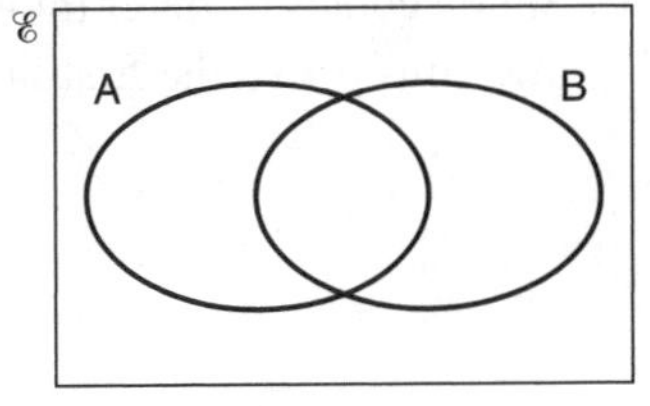

Diagram 1

2 On copies of diagram 2 shade the following sets:

a $(A \cap B')'$

b $(A \cup B')'$

c $(A \cap B)'$

d $(A' \cup B)'$

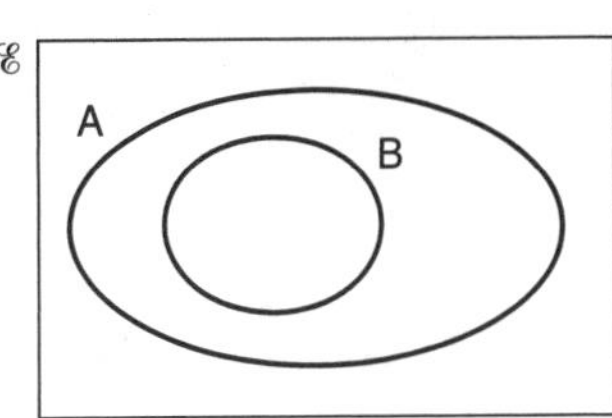

Diagram 2

3 On copies of diagram 3 shade the following sets:

a $A \cap B \cap C$

b $(A \cap B \cap C)'$

c $(A \cap B') \cup C$

d $(A \cup B)' \cap C$

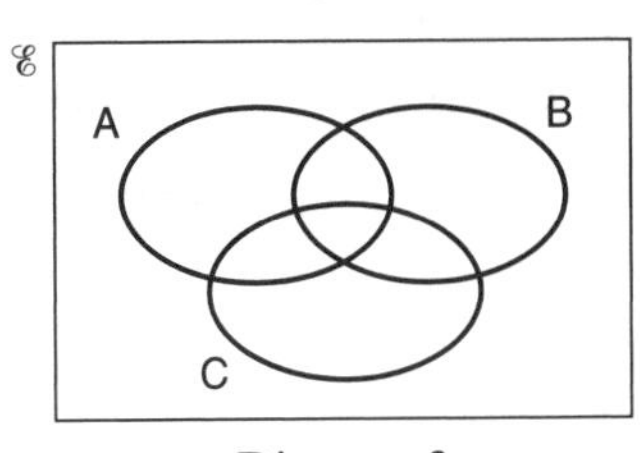

Diagram 3

4 On copies of diagram 3 shade the following sets:

a $A \cup (B' \cap C)$

b $(A \cap B) \cup C'$

c $A \cup (B \cap C)$

d $(A \cup B) \cap C$

5 Describe the shaded sets using set notation:

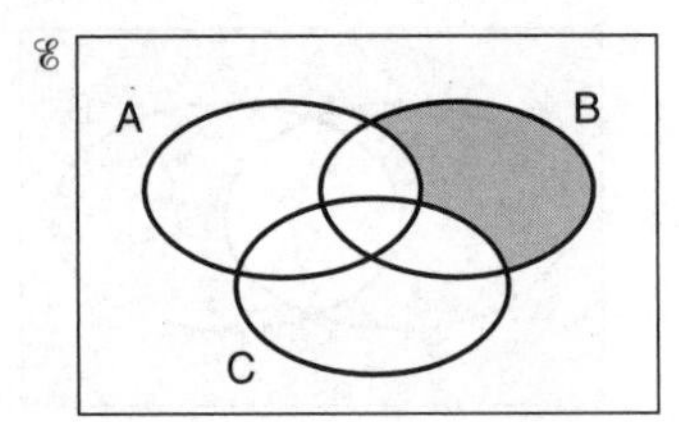

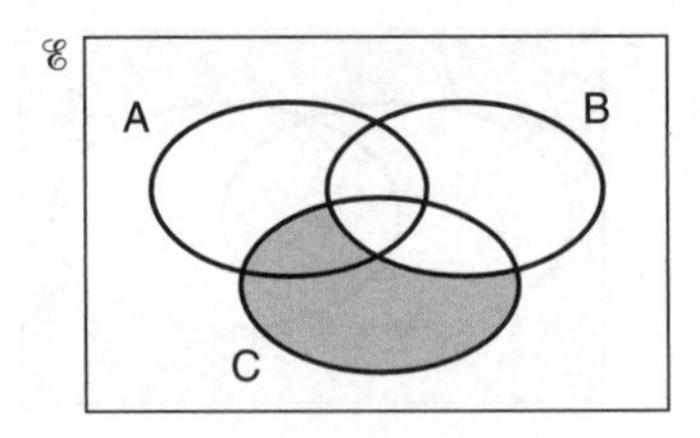

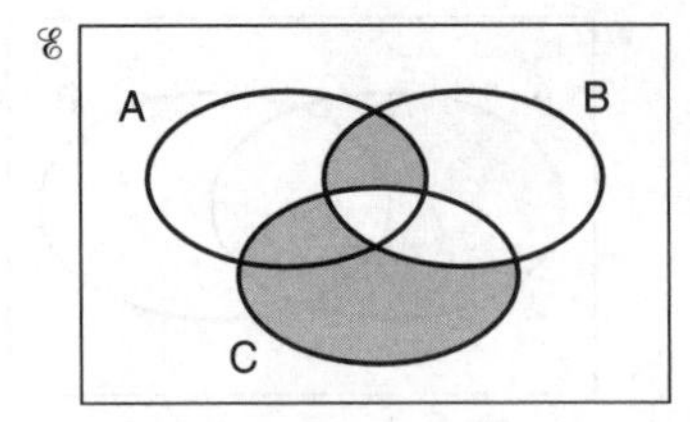

6 Describe the shaded sets using set notation:

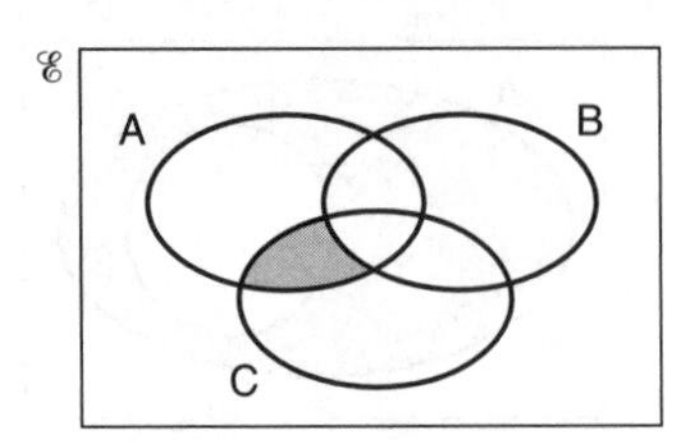

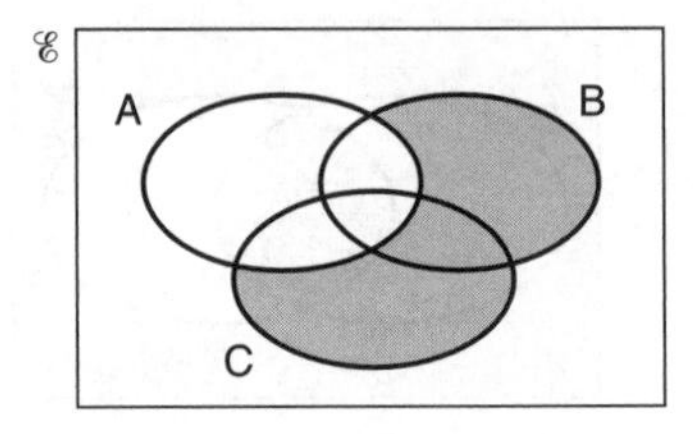

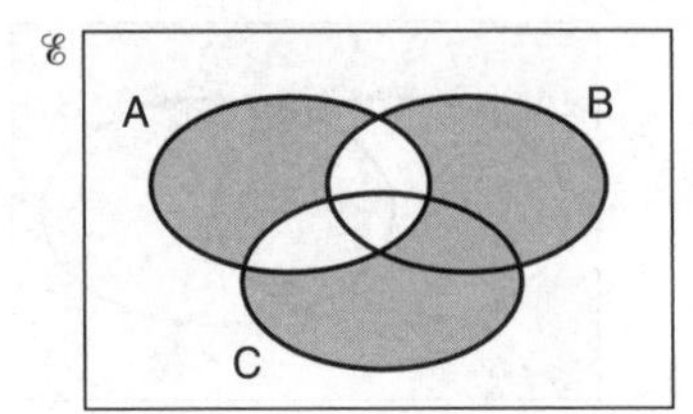

7 This question is about **De Morgan's Laws**.
These state that $(A \cup B)' = A' \cap B'$ and $(A \cap B)' = A' \cup B'$.

Shade copies of diagram 4 to show the following sets: $A \cup B$, $(A \cup B)'$, A', B' and $A' \cap B'$ and thus prove the first law. Use a similar method to prove the second law.

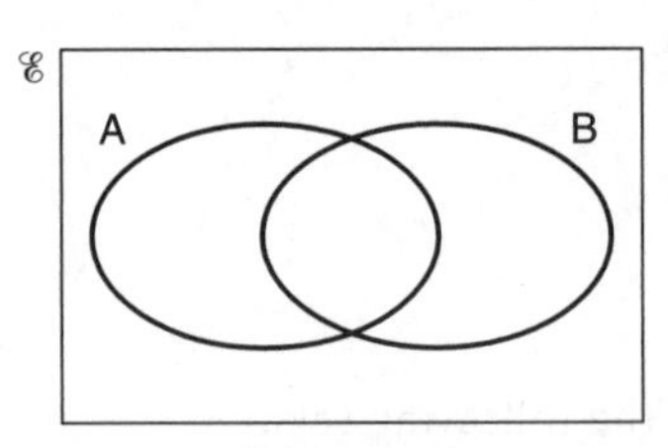

Diagram 4

Set-builder notation

Sets can be described using **set-builder notation**:
$A = \{x \text{ such that } x > 2\}$ means 'A is the set of all x such that x is greater than 2'.
Rather than write '**such that**', the notation $A = \{x : x > 2\}$ is used.

$B = \{x : x > 2, x \text{ is a positive integer}\}$ means the set of positive integers x such that x is greater than 2. This means $B = \{3, 4, 5, 6, \ldots\}$.

Certain sets of numbers are used so frequently that they are given special symbols.

$\mathbb{N}$ is the set of natural numbers or positive integers $\{1, 2, 3, 4, \ldots\}$.
$\mathbb{Z}$ is the set of integers $\{\ldots, -2, -1, 0, 1, 2, \ldots\}$.
$\mathbb{Q}$ is the set of rational numbers.
$\mathbb{R}$ is the set of real numbers.

Example 3

The set $B = \{3, 4, 5, 6, \ldots\}$ is written as $B = \{x : x > 2, x \in \mathbb{N}\}$.
The set $A = \{x : x > 2, x \in \mathbb{R}\}$ is the set of all real numbers greater than two.
Note that $3.2 \in A$ but $3.2 \notin B$, so $A \neq B$.
The set $\{x : x \text{ is even}, x \in \mathbb{N}\}$ is the set $\{2, 4, 6, \ldots\}$.
The set $\{x : x = 3y, y \in \mathbb{N}\}$ is the set $\{3, 6, 9, 12, \ldots\}$.

Exercise 18

1 List the following sets:

a $\{x : x \text{ is a weekday beginning with T}\}$
b $\{z : z \text{ is a colour in traffic lights}\}$
c $\{x : x < 7, x \in \mathbb{N}\}$
d $\{x : -2 < x < 7, x \in \mathbb{Z}\}$

2 List the following sets:

a $\{x : x \text{ is a continent}\}$
b $\{y : y \text{ is a Mathematics teacher in your school}\}$
c $\{x : x \leqslant 5, x \in \mathbb{N}\}$
d $\{x : -4 < x \leqslant 2, x \in \mathbb{Z}\}$

3 Express in set-builder notation the set of natural numbers which are

a less than 7
b greater than 4
c between 2 and 11 inclusive
d between −3 and 3
e odd
f prime

4 Express in set-builder notation the set of natural numbers which are

a greater than −3
b less than or equal to 9
c between 5 and 19
d between −4 and 31 inclusive
e multiples of 5
f factors of 48

Exercise 18★

1 $A = \{x : x \leqslant 6, x \in \mathbb{N}\}$, $B = \{x : x = 2y, y \in A\}$, $C = \{1, 3, 5, 7, 9, 11\}$
List the following sets:

a B
b $\{x : x = 2y + 1, y \in C\}$
c $A \cap B$
d Give a rule to describe $B \cup C$.

2 $A = \{x : -2 \leqslant x \leqslant 2, x \in \mathbb{Z}\}$, $B = \{x : x = y^2, y \in A\}$, $C = \{x : x = 2^y, y \in A\}$
List the following sets:

a B
b C
c $A \cap B \cap C$
d $\{(x, y) : x = y, x \in A, y \in C\}$

3 List the sets

a $\{x : 2^x = -1, x \in \mathbb{R}\}$
b $\{2^{-x} : 0 \leqslant x < 5, x \in \mathbb{Z}\}$
c $\{x : x^2 + x - 6 = 0, x \in \mathbb{N}\}$
d $\{x : x^2 + x - 6 = 0, x \in \mathbb{Z}\}$

4 List the sets

a $\{x : x^2 + 1 = 0, x \in \mathbb{R}\}$

b $\{2^x : 0 \leqslant x \leqslant 5, x \in \mathbb{Z}\}$

c $\{x : x^2 + 2x - 6 = 0, x \in \mathbb{Q}\}$

d $\{x : x^2 + 2x - 6 = 0, x \in \mathbb{R}\}$

5 Show the sets $\mathbb{N}$, $\mathbb{Z}$, $\mathbb{Q}$, $\mathbb{R}$ in a Venn diagram.

Exercise 19 (Revision)

1 $n(\text{A}) = 20$, $n(\text{A} \cap \text{B}) = 7$ and $n(\text{A}' \cap \text{B}) = 10$.

a Draw a Venn diagram to show this information.

b Find $n(\text{B})$.

c Find $n(\text{A} \cup \text{B})$.

2 In a class of 20 students, 16 drink tea, 12 drink coffee and 2 students drink neither drink.

a How many drink tea only?

b How many drink coffee only?

c How many drink both drinks?

3 In a town in Belgium, all the inhabitants speak either French or Flemish. 69% speak French and 48% speak Flemish.

a What percentage speak both languages?

b What percentage speak French only?

c What percentage speak Flemish only?

4 On copies of the diagram, shade the following sets:

a $\text{A}' \cap \text{B}$

b $(\text{A} \cup \text{B})'$

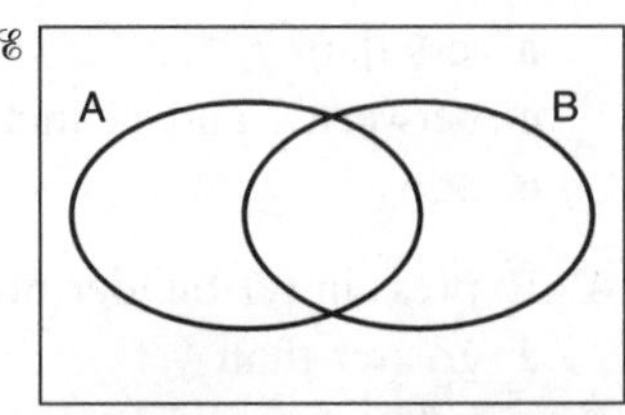

5 Describe the shaded set using set notation.

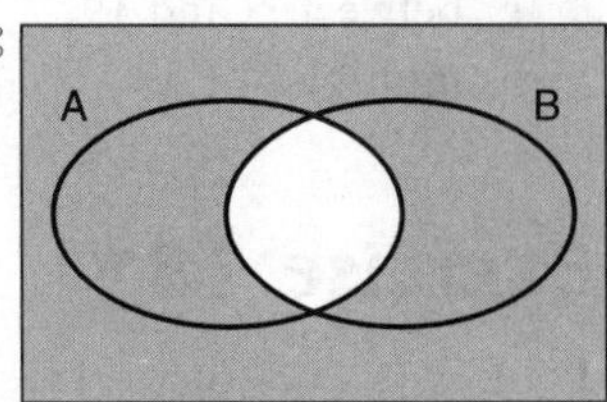

6 List the following sets:

a $\{x : -3 < x < 4, x \in \mathbb{Z}\}$

b $\{x : x < 5, x \in \mathbb{N}\}$

c $\{x : -6 < x < -1, x \in \mathbb{N}\}$

7 Express in set-builder notation the set of natural numbers which are

a even

b factors of 24

c between −1 and 4 inclusive

Exercise 19★ (Revision)

1 $n(A \cap B \cap C) = 2$, $n(A \cup B \cup C)' = 5$, $n(A) = n(B) = n(C) = 15$ and $n(A \cap B) = n(A \cap C) = n(B \cap C) = 6$. How many are in the universal set?

2 A youth club has 140 members. 80 listen to pop music, 40 to rock music, 75 to heavy metal and 2 members don't listen to any music. 15 members listen to pop and rock only, 12 to rock and heavy metal only, while 10 listen to pop and heavy metal only. How many listen to all three types of music?

3 In a group of 50 students at a summer school, 15 play tennis, 20 play cricket, 20 swim and 7 students do nothing. 3 students play tennis and cricket, 6 students play cricket and swim, while 5 students play tennis and swim. How many do all three sports?

4 On copies of the diagram, shade the following sets:

a $(A' \cap B) \cup C$

b $A \cap (B \cup C)$

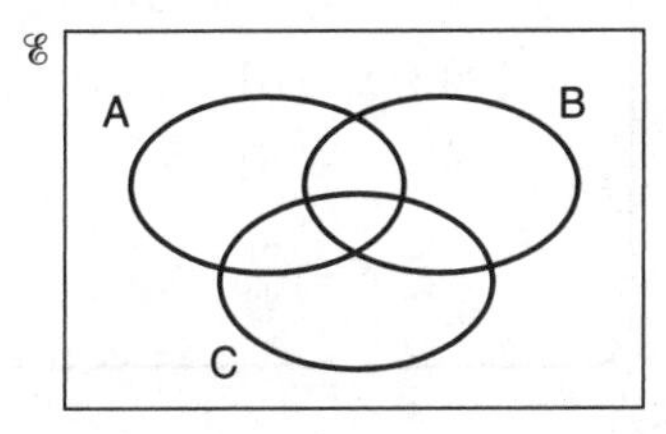

5 Describe the shaded sets using set notation.

a

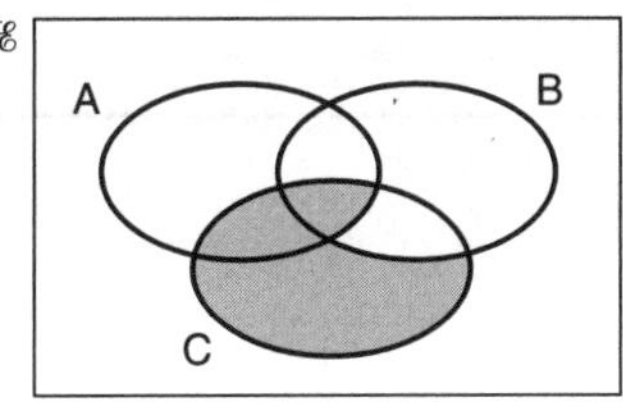

b

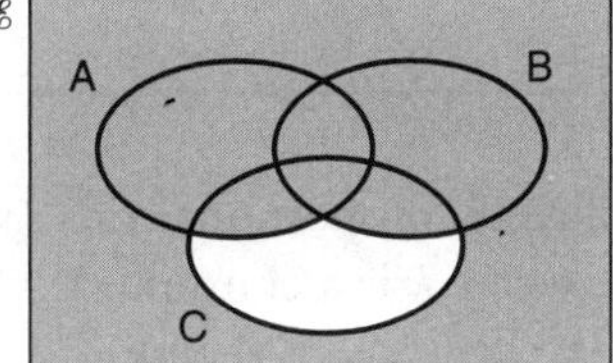

6 List the sets

a $\{x : x^2 - 1 = 0, x \in \mathbb{R}\}$

b $\{x : x^2 + 4x = 0, x \in \mathbb{Q}\}$

c $\{x : x^2 + 2x + 2 = 0, x \in \mathbb{R}\}$

7 Express in set-builder notation the set of natural numbers which are

a greater than -5

b between 4 and 12

c multiples of 3

Summary 1

Number

Inverse proportion

It takes 2 hours to cut a school playing field with 3 mowers.

In 1 hour it would take 6 mowers ($\div$ LHS by 2 and $\times$ RHS by 2).

In 1.5 hours it would take 4 mowers ($\times$ LHS by 1.5 and $\div$ RHS by 1.5).

To change a recurring decimal to a fraction

$x = 0.616161\ldots$	(Multiply both sides by 100)
$100 \times x = 61.616161\ldots$	(Subtract the top from the bottom)
$99 \times x = 61$	(Divide both sides by 99)
$x = \frac{61}{99}$	

Algebra

- If y is directly proportional to x ($y \propto x$)
  $y = kx$, where k is a constant.
  If $y = 10$ when $x = 5$, then $k = 2$.
  $y = 2x$
  So if $x = 10$, $y = 20$.

- If y is directly proportional to x^2 ($y \propto x^2$)
  $y = kx^2$, where k is a constant.
  If $y = 100$ when $x = 5$, then $k = 4$.
  $y = 4x^2$
  So if $x = 10$, $y = 400$.

- If y is inversely proportional to x^3 $\left(y \propto \frac{1}{x^3}\right)$
  $y = \frac{k}{x^3}$, where k is a constant.
  If $y = 10$ when $x = 2$, then $k = 80$.
  $y = \frac{80}{x^3}$
  So if $x = 0.1$, $y = 80\,000$.

Graphs

Cubic graphs $y = ax^3 + bx^2 + cx + d$

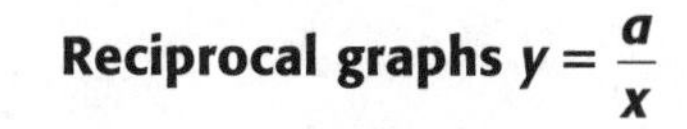

Reciprocal graphs $y = \frac{a}{x}$

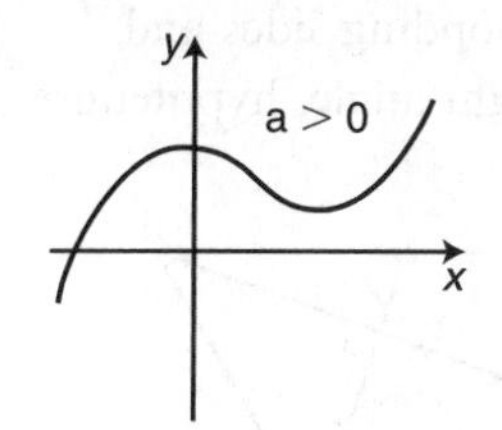

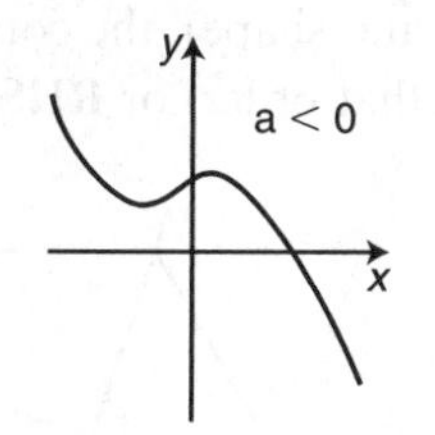

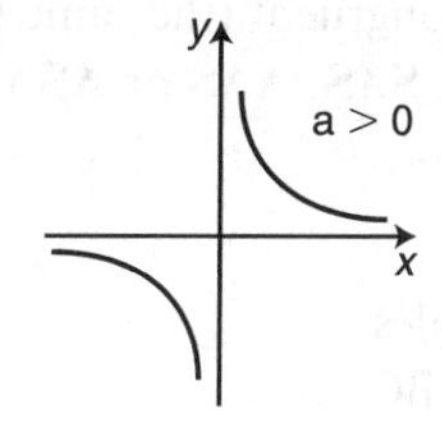

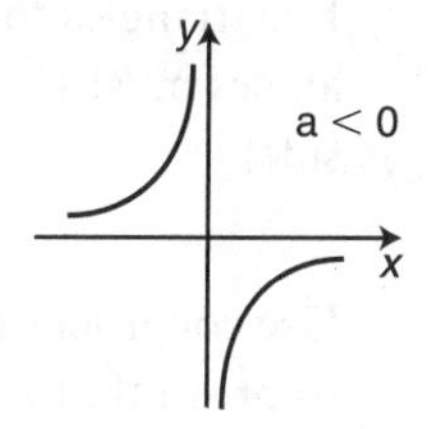

Graphs of type $y = Ax + \frac{B}{x^2} + \frac{C}{x^3}$ (A, B and C are constants)

To draw this graph for $-3 \leqslant x \leqslant 3$ compile a suitable table of values, plot the points and draw a smooth curve.

x	-3	-2	-1	0	1	2	3
Ax	...	...	...	...	...	...	...
$\frac{B}{x^2}$	...	...	...	...	...	...	...
$\frac{C}{x^3}$	...	...	...	...	...	...	...
y	...	...	...	...	...	...	...

Shape and space

Congruence

For triangles to be congruent (the same size and shape), the corresponding sides and angles must be: **SSS**, **SAS**, **AAS** or **ASA** (in that order) or **RHS** (right angle, hypotenuse, side).

Use congruent triangles to prove that AD = BC.

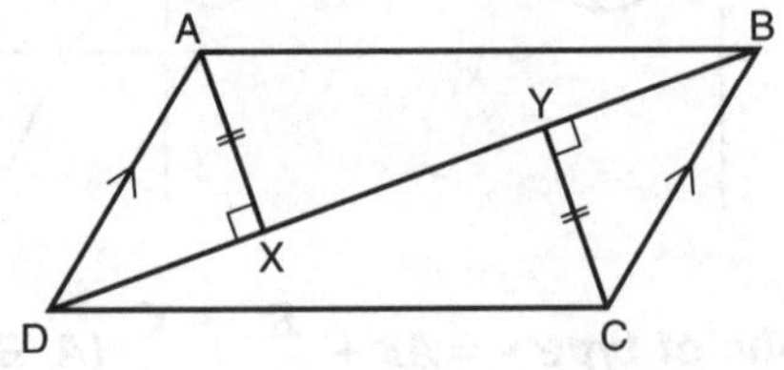

- State the triangles.
- State the three conditions for congruency and give reasons in brackets.
- Write the conclusion.

In triangles BYC and DXA

1	AX = CY	(Given)
2	∠DBC = ∠BDA	(Alternate angles)
3	∠BYC = ∠DXA	(Given)

∴ △BYC ≡ △DXA (SAA)

∴ AD = BC

Circle properties

Angle at the centre is twice the angle at the circumference

Opposite angles of a cyclic quadrilateral sum to $180° \Rightarrow w + y = 180°$, $x + z = 180°$

Angles in the same segment are equal

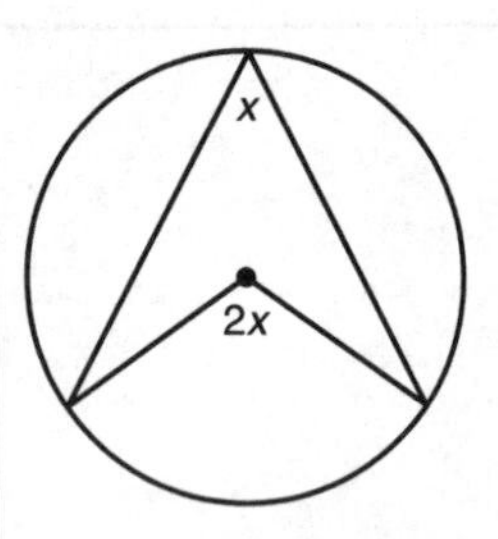

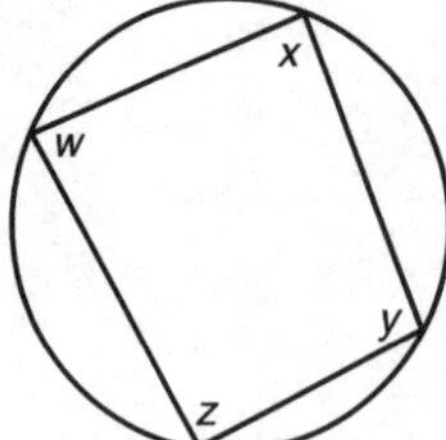

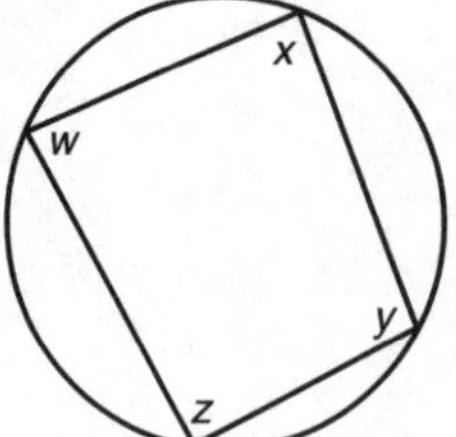

Alternate Segment Theorem

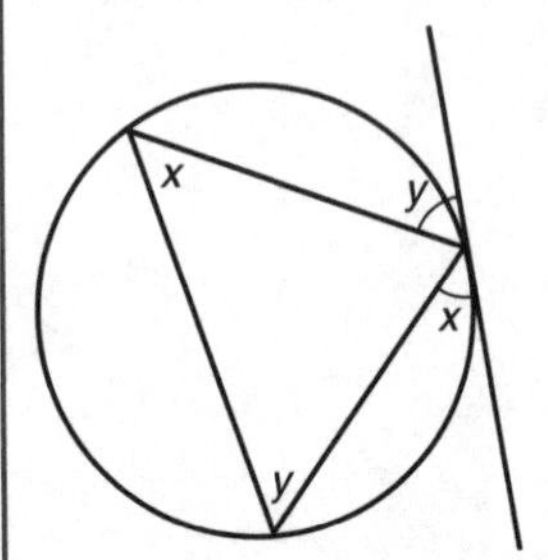

Intersecting chords

AP × PB = CP × PD

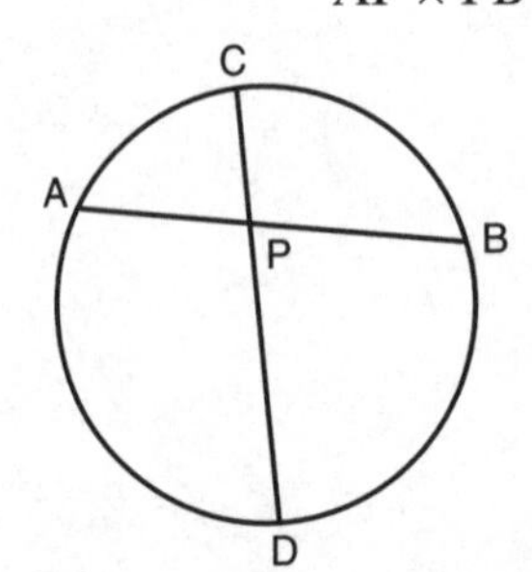

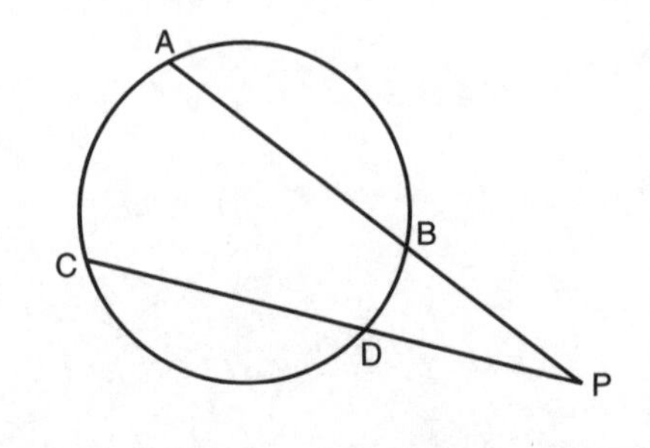

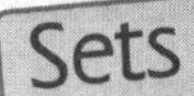

A **set** is a collection of objects, described by a list or a rule. $A = \{1, 3, 5\}$

Each object is an **element** or **member** of the set. $1 \in A, 2 \notin A$

Sets are **equal** if they have exactly the same elements. $B = \{5, 3, 1\}, B = A$

The **number of elements** of set A is given by $n(A)$. $n(A) = 3$

The **empty set** is the set with no members. $\{\}$ or $\emptyset$

The **universal set** contains all the elements being discussed in a particular problem. $\mathscr{E}$

B is a **subset** of A if every member of B is a member of A. $B \subset A$

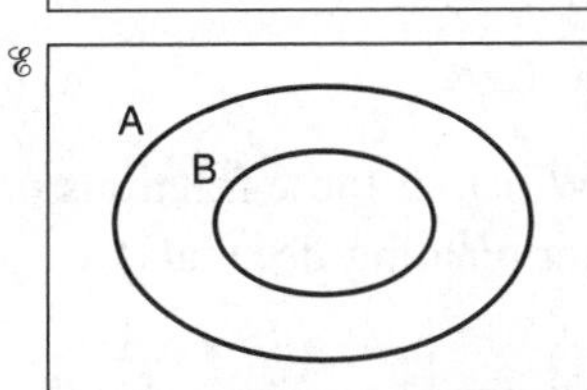

The **complement** of set A is the set of all elements not in A. A'

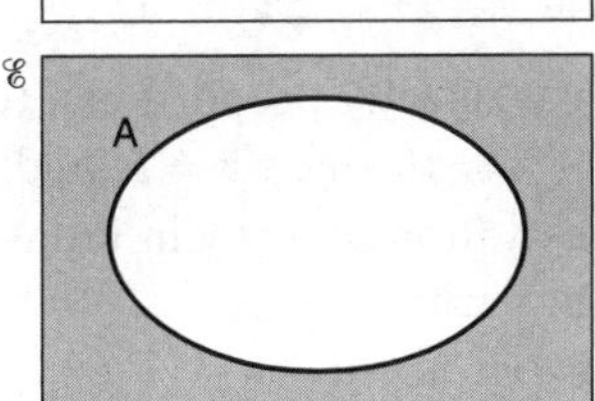

The **intersection** of A and B is the set of elements which are in both A and B. $A \cap B$

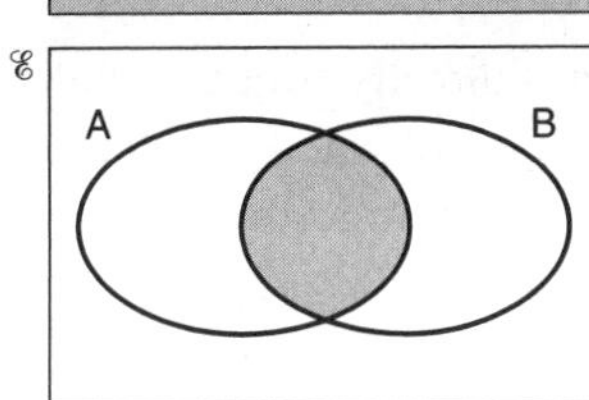

The **union** of A and B is the set of elements which are in A or B or both. $A \cup B$

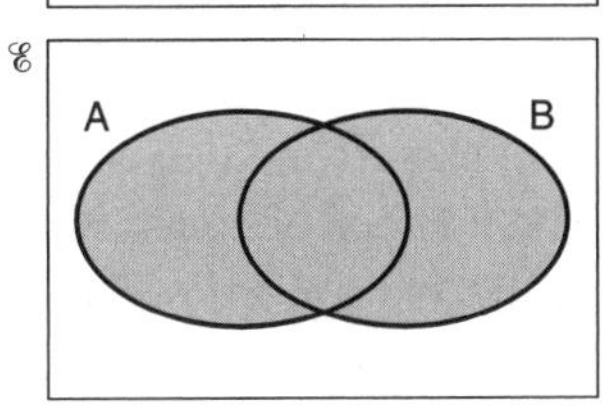

In **set-builder notation**, $A = \{x : x > 2, x \in \mathbb{R}\}$ means A is the set of all x such that x is greater than 2, where x is a real number.

$\mathbb{N}$ is the set of natural numbers or positive integers $\{1, 2, 3, 4, \ldots\}$.

$\mathbb{Z}$ is the set of integers $\{\ldots, -2, -1, 0, 1, 2, \ldots\}$.

$\mathbb{Q}$ is the set of rational numbers.

$\mathbb{R}$ is the set of real numbers.

Examination practice 1

1 The average fuel consumption of a Boeing 757 is 0.4 mpg (miles per gallon) and in flying from London to Nice it uses 2400 gallons.

a On the same route another plane uses 800 gallons. Find its fuel consumption.

b The fuel consumption of another plane is 0.5 mpg. How much fuel would it use on the same route?

2 Change these recurring decimals to fractions.

a $0.\dot{5}\dot{8}$ **b** $0.\dot{5}6\dot{7}$

3 Which of these fractions produce terminating decimals?

$\frac{9}{40}$ $\frac{7}{15}$ $\frac{7}{8}$ $\frac{11}{12}$ $\frac{7}{128}$

4 y is directly proportional to x.
If $x = 10$ when $y = 5$, find:

a a formula for y in terms of x

b y when $x = 5$

c x when $y = \frac{1}{2}$

5 p is directly proportional to q squared.
If $q = 10$ when $p = 20$, find:

a a formula for p in terms of q

b p when $q = 20$

c q when $p = 20$

6 A machine produces coins, of a fixed thickness, from a given volume of metal. The number of coins, N, produced is inversely proportional to the square of the diameter, d.

a 4000 coins are made of diameter 1.5 cm. Find the value of the constant of proportionality, k.

b Find the formula for N in terms of d.

c Find the number of coins that can be produced of diameter 2 cm.

d If 1000 coins are produced, find their diameter.

7 x varies directly as the cube root of y.
When $y = 27$, $x = 12$.

a Find a formula connecting x and y.

b Find the value of y when $x = 32$.

8 Dave launches a model aeroplane. The height, h m, of the plane t seconds after launch is given by

$$h = \frac{t^2}{4} - t + 1.5$$

valid for $0 \leqslant t \leqslant 5$.

a Make a table of values giving h for $0 \leqslant t \leqslant 5$ and draw a graph of h against t.

b Use your graph to find the height above the ground at the launch.

c What is the minimum height above the ground and the time at which this occurs?

d Between what times is the plane less than one metre above the ground?

9 **a** Draw the graph of
$y = x^3 + 3x^2 - x - 3$ for $-5 \leqslant x \leqslant 3$.

b State the x values where this graph cuts the x-axis.

10 The petrol consumption (y kilometres per litre) of a car is related to its speed (x kilometres per hour) by the formula

$$y = 70 - \frac{x}{3} - \frac{2400}{x}$$

a Copy and complete the table of values.

x	60	70	80	90	100	110	120	130
$\frac{x}{3}$		23.3	26.7		33.3	36.7		43.3
$\frac{2400}{x}$		34.3		26.7		21.8		18.5
y		12.4				11.5		8.2

b Draw the graph of y against x. Draw the x-axis for $60 \leqslant x \leqslant 130$ using a scale of 2 cm to 10 km/h and the y-axis for $7 \leqslant y \leqslant 14$ using a scale of 4 cm to 1 km per litre.

c Use your graph to estimate the petrol consumption at 115 km/h.

d Use your graph to estimate the most economical speed at which to travel to conserve petrol.

11 TPK is a tangent to the circle.
TSQ is a straight line.
PQ = QR
∠QPK = 50°
∠STP = 26°

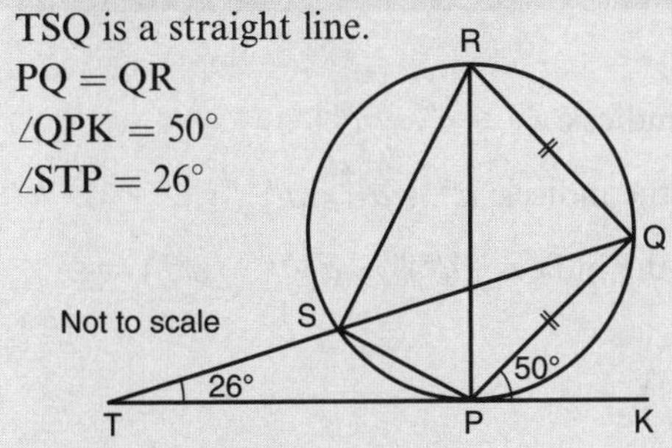

Calculate the size of

a angle PQR

b angle QRS

NEAB

12

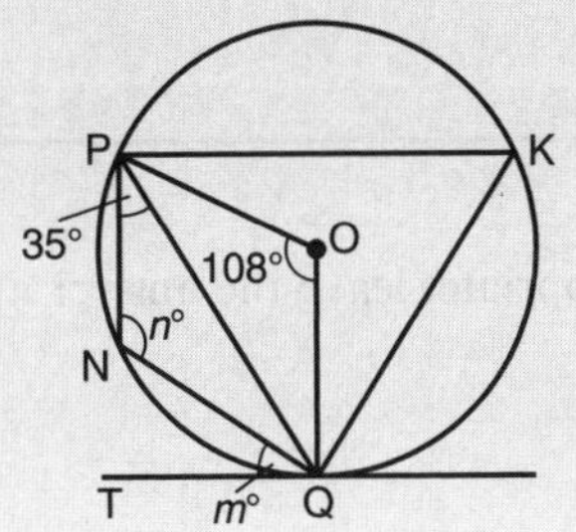

This diagram is NOT drawn accurately.

O is the centre of the circle. P, K, Q, N are points on the circumference.

QT is the tangent to the circle at Q.
Angle POQ = 108°
Angle NPQ = 35°

a Calculate the values of m and n.

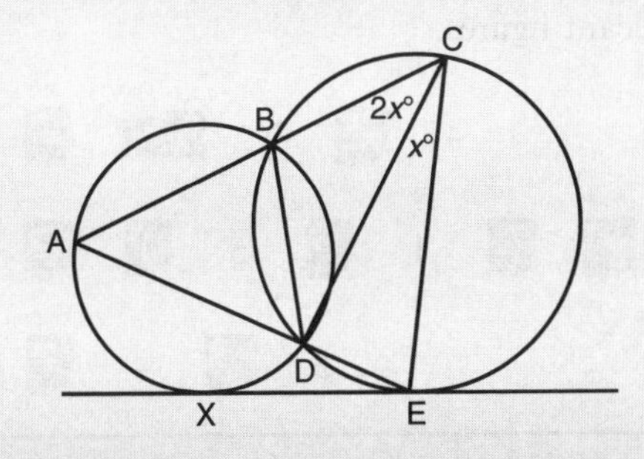

ABC and ADE are straight lines. CE is a diameter.

Angle DCE = $x°$ and angle BCD = $2x°$.

b Find, in terms of x, the size of the angle
(i) ABD (ii) DBE (iii) BAD

c Explain why BE × AC = AE × CD

LONDON

13

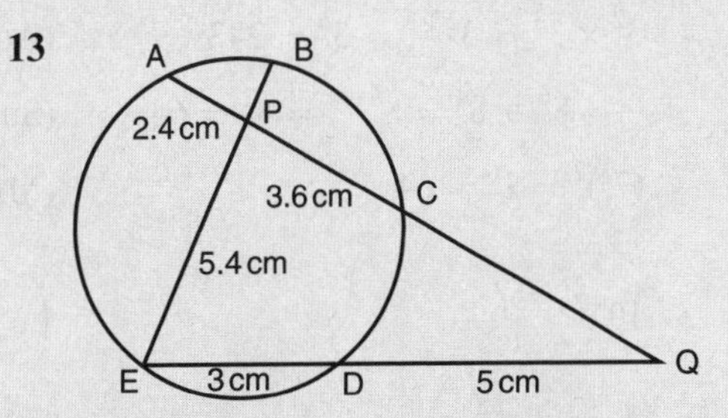

AP = 2.4 cm, CP = 3.6 cm, EP = 5.4 cm, DE = 3 cm and DQ = 5 cm.

a Work out the length BP.

b Calculate the length CQ.

14 A = {2, 4, 6, 8} B = {3, 4, 5, 6}

a List the members of the set
(i) $A \cap B$ (ii) $A \cup B$

b Explain clearly the meaning of
(i) $4 \in B$ (ii) $n(B) = 4$

15 In a class of 24 students, 22 have a mobile phone and 10 have a portable DVD player. One student has neither.

a Draw a Venn diagram to show this information.

b How many students have
(i) a mobile phone only?
(ii) a portable DVD player only
(iii) both a mobile phone and a portable DVD player?

Number 2

Negative and fractional indices

Key Points

- Rules of indices

  $3^2 \times 3^3 = 3^{2+3} = 3^5 = 243$ (Add the indices: $a^m \times a^n = a^{m+n}$)

  $\frac{8^6}{8^4} = 8^6 \div 8^4 = 8^{6-4} = 8^2 = 64$ (Subtract the indices: $a^m \div a^n = a^{m-n}$)

  $(2^3)^2 = 2^6 = 64$ (Multiply the indices: $(a^m)^n = a^{m \times n} = a^{mn}$)

- $10^{-2} = \frac{1}{10^2}$ $\left(a^{-n} = \frac{1}{a^n}\right)$

- $3^{\frac{1}{2}} \times 3^{\frac{1}{2}} = 3^1 \quad \therefore \quad 3^{\frac{1}{2}} = \sqrt{3}$ $\left(a^{\frac{1}{2}} = \sqrt{a} \text{ and similarly } a^{\frac{1}{3}} = \sqrt[3]{a}\right)$

Single negative and fractional indices

Example 1

Without using a calculator, evaluate these and, where appropriate, leave the answer as a fraction.

a $4^{-3} = \frac{1}{4^3} = \frac{1}{64}$

b $125^{\frac{1}{3}} = \sqrt[3]{125} = 5$

c $3^{-3} \times 3^2 = 3^{(-3+2)} = 3^{-1} = \frac{1}{3}$

d $6^{-4} \div 6^{-2} = 6^{(-4--2)} = 6^{-2} = \frac{1}{6^2} = \frac{1}{36}$

e $(3^{-1})^2 = 3^{(-1 \times 2)} = 3^{-2} = \frac{1}{3^2} = \frac{1}{9}$

Example 2

Use a calculator to work out these, correct to 3 significant figures.

a $4^{-3} = 0.015625 = 0.0156$

4  3  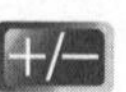 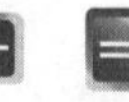

b $60^{\frac{1}{5}} = 2.27$

60  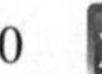  1 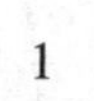  5  

c $(5^{-2})^{-2} = 5^{(-2 \times -2)} = 5^4 = 625$

5  4 

Example 3

Simplify these.

a $a^{\frac{1}{2}} \times a^{\frac{1}{2}} \times a = a^{\left(\frac{1}{2}+\frac{1}{2}+1\right)} = a^2$

b $3a^2 \times 2a^{-3} = 6 \times a^{2+(-3)} = 6a^{-1} = \frac{6}{a}$

c $(6a^{-2}) \div (2a^2) = 3a^{(-2-2)} = 3a^{-4} = \frac{3}{a^4}$

Alternatively: $\frac{6}{a^2} \div 2a^2 = \frac{6}{a^2} \times \frac{1}{2a^2} = \frac{3}{a^4}$

Exercise 20

Without using a calculator, evaluate these and, where appropriate, leave the answer as a fraction.

1 3^{-2} **2** 4^{-2} **3** 10^{-1} **4** 10^{-2}

5 2^{-3} **6** 5^{-2} **7** 8^{-1} **8** 12^{-2}

9 $81^{\frac{1}{2}}$ **10** $64^{\frac{1}{2}}$ **11** 4^{-3} **12** 3^{-3}

13 $27^{\frac{1}{3}}$ **14** $8^{\frac{1}{3}}$ **15** $2^2 \times 2^{-1}$ **16** $2^3 \times 2^{-1}$

17 $2^{-1} \div 2^3$ **18** $2^{-2} \div 2^3$ **19** $(3^{-1})^2$ **20** $(2^2)^{-2}$

21 $2^{-2} \times 2^{-2}$ **22** $3^{-1} \times 3^{-2}$ **23** $2^{-2} \div 2^{-1}$ **24** $3^{-3} \div 3^{-2}$

25 $(2^{-2})^{-2}$ **26** $(3^{-3})^{-2}$

Use your calculator to work out these, correct to 3 significant figures.

27 5^4 **28** 4^3 **29** $23^{\frac{1}{3}}$ **30** $24^{\frac{1}{2}}$

31 $6^2 \times 6^{-3}$ **32** $7^2 \times 7^{-4}$ **33** $3^{-1} \times 3^{-3}$ **34** $4^{-2} \times 4^{-1}$

35 $6^2 \div 6^{-4}$ **36** $5^2 \div 5^{-3}$ **37** $3^{-1} \div 3^2$ **38** $2^{-2} \div 2^3$

39 $(3^2)^{-1}$ **40** $(4^{-1})^3$ **41** $(8^{-2})^{-1}$ **42** $(9^{-2})^{-2}$

Simplify these.

43 $a^2 \times a^{-1}$ **44** $b^4 \times b^{-2}$ **45** $c^{-1} \div c^2$ **46** $d^3 \div d^{-2}$

47 $e^2 \times e^3 \times e^{-4}$ **48** $f^3 \times f^2 \times f^{-4}$ **49** $(a^2)^{-1}$ **50** $(g^{-1})^2$

Exercise 20★

Without using a calculator, evaluate these and, where appropriate, leave the answer as a fraction.

1 2^{-6} **2** 3^{-4} **3** $125^{\frac{1}{3}}$ **4** $81^{\frac{1}{4}}$ **5** 6^0

6 3.8^0 **7** $2^{10} \times 2^3 \times 2^{-4}$ **8** $3^4 \times 3^{-3} \div 3^2$ **9** $200 \times (4^2)^{-1}$ **10** $900 \times (3^{-1})^2$

11 $\dfrac{4^{-2}}{4^{-3}}$ 12 $\dfrac{6^{-2}}{6^{-4}}$ 13 0.1×0.1^{-2} 14 0.2×0.2^{-3} 15 $\left(\frac{1}{2}\right)^{-3}$

16 $\left(\frac{2}{3}\right)^{-2}$ 17 $\left(\frac{1}{4}\right)^{\frac{1}{2}}$ 18 $\left(2\frac{1}{4}\right)^{\frac{1}{2}}$ 19 $(-6)^{-2}$ 20 $\left(-\frac{1}{3}\right)^{-3}$

Use your calculator to work out these, correct to 3 significant figures.

21 1.4^{-3} 22 1.02^{-6} 23 $362^{\frac{1}{9}}$

24 $14.6^{\frac{1}{4}}$ 25 $3^{-3} \times 3^{-2} \times 3^{-1}$ 26 $4^{-5} \times 4^{-4} \div 4^{-5}$

Simplify these.

27 $a^{-2} \times a^2 \div a^{-2}$ 28 $b^4 \times b^{-3} \div b^{-2}$ 29 $2(c^2)^{-2}$

30 $(2c^{-1})^2$ 31 $a^{-2} + a^{-2} + a^{-2}$ 32 $b^{-3} + 3b^{-3} - 2b^{-3}$

33 $3a^{-2} \times 4a$ 34 $4b^2 \times 2b^{-3}$ 35 $a^{\frac{1}{2}} \times a^{\frac{1}{2}}$

36 $b^{\frac{1}{4}} \div b^{\frac{1}{4}}$ 37 $(c^{-2})^{\frac{1}{2}}$ 38 $(d^{\frac{1}{3}})^{-3}$

Solve these for x.

39 $9^{\frac{1}{x}} = 3$ 40 $8^{\frac{1}{x}} = 2$ 41 $81^{\frac{1}{2}} = 3^x$

42 $8^{\frac{1}{3}} = 2^x$ 43 $2^x = \frac{1}{8}$ 44 $3^x = \frac{1}{9}$

45 Find k if $x^k = \sqrt[3]{x} \div \dfrac{1}{x^2}$ **46** Find k if $x^{k-1} = \dfrac{(x^2)^{-2}}{x^3}$

Simplify these.

47 $1 \div a^3$ **48** $1 \div a^{-2}$ **49** $b^{-2} \div \dfrac{1}{b^3}$

50 $c \div \dfrac{1}{c^{-3}}$ **51** $a^{-2} + \dfrac{1}{a^2}$ **52** $4a^{-2} - \dfrac{4}{a^2}$

53 $\dfrac{3}{a^2} + 2a^{-2}$ **54** $(2m)^{-2} \times 2m^2$ **55** $(3a)^{-2} \div \dfrac{1}{3a^2}$

56 $8b^4 \times 2b \times (2b)^{-2}$ **57** $(-3a)^3 \div (3a^{-3})$ **58** $(8d^{-3}) \div (2d^{-1})$

59 $\left(\dfrac{a}{b}\right)^{-1}$

60 Find a and b if $2^{2a} = 64$ and $10^b = 0.001$

Combined negative and fractional indices

$16^{\frac{3}{4}}$ can be written as $\left(16^{\frac{1}{4}}\right)^3 = (\sqrt[4]{16})^3 = 2^3 = 8$

This shows that $16^{-\frac{3}{4}} = \dfrac{1}{16^{\frac{3}{4}}} = \dfrac{1}{2^3} = \dfrac{1}{8}$

Exercise 21

Without using a calculator, evaluate these and, where appropriate, leave the answer as a fraction.

1 $9^{-\frac{1}{2}}$ **2** $16^{-\frac{1}{2}}$ **3** $100^{-\frac{1}{2}}$ **4** $25^{-\frac{1}{2}}$

5 $81^{-\frac{1}{2}}$ **6** $125^{-\frac{1}{3}}$ **7** $64^{-\frac{1}{3}}$ **8** $4^{\frac{3}{2}}$

9 $9^{\frac{3}{2}}$ **10** $4^{-\frac{3}{2}}$ **11** $9^{-\frac{3}{2}}$

Use your calculator to work out these, correct to 3 significant figures.

12 $6^{-\frac{1}{2}}$ **13** $20^{-\frac{1}{2}}$ **14** $100^{-\frac{1}{4}}$ **15** $50^{-\frac{1}{4}}$

16 $5^{\frac{3}{2}}$ **17** $11^{\frac{3}{2}}$ **18** $3^{-\frac{3}{2}}$ **19** $7^{-\frac{3}{2}}$

Simplify these.

20 $a^{2\frac{1}{2}} \times a^{-\frac{1}{2}}$ **21** $b^{-\frac{1}{3}} \times b^{-\frac{1}{3}} \times b^{-\frac{1}{3}}$ **22** $c^{2\frac{1}{2}} \div c^{-\frac{1}{2}}$

23 $d^{\frac{2}{3}} \div d^{-\frac{1}{3}}$ **24** $(e^{\frac{2}{3}})^3$ **25** $(f^2)^{-\frac{1}{2}}$

Exercise 21★

Without using a calculator, evaluate these and, where appropriate, leave the answer as a fraction.

1 $36^{-\frac{1}{2}}$ **2** $144^{-\frac{1}{2}}$ **3** $1 \div 216^{-\frac{1}{3}}$ **4** $1 \div 1000^{-\frac{1}{3}}$

5 $8^{\frac{4}{3}}$ **6** $1^{\frac{3}{2}}$ **7** $8^{-\frac{4}{3}}$ **8** $1^{-\frac{3}{2}}$

Use your calculator to work out these, correct to 3 significant figures.

9 $0.6^{-\frac{1}{2}}$ **10** $0.3^{-\frac{1}{2}}$ **11** $1.4^{-\frac{1}{3}}$ **12** $3.7^{-\frac{1}{3}}$

13 $2.01^{\frac{3}{2}}$ **14** $0.909^{\frac{3}{2}}$ **15** $2.01^{-\frac{3}{2}}$ **16** $0.909^{-\frac{3}{2}}$

Simplify these.

17 $a^{\frac{1}{2}} \times a^{-2\frac{1}{2}}$ **18** $b^{-\frac{1}{2}} \times b^{-2\frac{1}{2}}$ **19** $c^{-2\frac{1}{3}} \div c^{-\frac{1}{3}}$ **20** $d^{1\frac{1}{5}} \div d^{-\frac{4}{5}}$

21 $(e^{-\frac{1}{2}})^{-2}$ **22** $(f^{-\frac{1}{3}})^{-6}$ **23** $(27^2)^{\frac{1}{3}}$ **24** $(16^3)^{\frac{1}{4}}$

25 $216^{-\frac{2}{3}}$ **26** $1000^{-\frac{2}{3}}$

Use your calculator to work out these, correct to 3 significant figures.

27 $1 \div \dfrac{1}{1.02^{-2.5}}$ **28** $\left(1 \div \dfrac{1}{1.02^{-2.5}}\right)^{-2.5}$

Find the values of x and y in these equations.

29 $(a^x b^y)^{\frac{1}{6}} = a^{\frac{1}{2}} \times b^{-\frac{1}{3}}$

30 $\sqrt[5]{a^4 b^{-3}} = a^x b^y$

Exercise 22 (Revision)

Work out these and, where appropriate, leave the answer as a fraction.

1 2^{-4} **2** $100^{\frac{1}{2}}$ **3** $64^{\frac{1}{3}}$ **4** $64^{\frac{1}{2}}$

5 $25^{\frac{1}{2}}$ **6** $125^{\frac{1}{3}}$ **7** 8^{-2} **8** $16^{-\frac{1}{2}}$

9 $121^{-\frac{1}{2}}$ **10** $3^3 \times 3^{-1}$ **11** $3^{-1} \div 3^3$ **12** $(3^{-3})^{-1}$

13 $36^{-\frac{1}{2}}$ **14** $9^{\frac{3}{2}}$

Simplify these.

15 $a^3 \times a^{-1}$ **16** $a^3 \div a^{-1}$ **17** $(d^{-1})^2$

18 $b^{\frac{1}{2}} \times b^{-2}$ **19** $b^{\frac{1}{2}} \div b^{-2}$ **20** $\left(c^{-\frac{1}{2}}\right)^{-2}$

Exercise 22★ (Revision)

Work out these and, where appropriate, leave the answer as a fraction.

1 2^{-5} **2** $216^{\frac{1}{3}}$ **3** 9.7^0 **4** $27 \times (3^{-1})^2$

5 $2^8 \times 2^{-2} \times 2^{-4}$ **6** $49^{-\frac{1}{2}}$ **7** $\left(\frac{1}{2}\right)^{-3} \div 2^{-3}$ **8** $16^{\frac{3}{4}}$

9 $\left(\frac{1}{27}\right)^{\frac{2}{3}}$ **10** $(125)^{-\frac{4}{3}}$ **11** $(0.125)^{-\frac{2}{3}}$ **12** $\left(4^{\frac{1}{3}}\right)^{-1\frac{1}{2}}$

Simplify these.

13 $3c^3 \times c^{-2}$ **14** $b^{-2} + 4b^{-2}$ **15** $a^3 \times a^{-2} \div a^{-1}$ **16** $3 \times (c^{-1})^2$

17 $2\left(a^{\frac{1}{3}}\right)^{-3}$ **18** $d^{\frac{1}{2}} \times d^{\frac{2}{3}}$ **19** $(-3a)^3 \div (3a^{-3})$ **20** $(2b^2)^{-1} \div (-2b)^{-2}$

Algebra 2

Solving quadratic equations

Quadratic equations can be written as $ax^2 + bx + c = 0$ where a, b and c are constants.

Solving quadratic equations by factorising

Quadratic equations can often be solved by factorising.

Remember

There are three types of quadratic equations with $a = 1$.

- If $b = 0$
$$x^2 - c = 0$$
$$x^2 = c$$
$$x = \pm\sqrt{c}$$
- If $c = 0$
$$x^2 + bx = 0$$
$$x(x + b) = 0$$
$$x = 0 \text{ or } x = -b$$
- If $b \neq 0$ and $c \neq 0$
$$x^2 + bx + c = 0$$
$$(x + p)(x + q) = 0$$
$$x = -p \text{ or } x = -q$$

where $p \times q = c$ and $p + q = b$.

If c is positive then p and q have the same sign as b.
If c is negative then p and q have opposite signs to each other.

Example 1

Solve these quadratic equations.

a $x^2 - 81 = 0$
$$x^2 = 81$$
$$x = -9 \text{ or } x = 9$$

b $x^2 - 7x = 0$
$$x(x - 7) = 0$$
$$x = 0 \text{ or } x = 7$$

c $x^2 - 10x + 21 = 0$
$$(x - 7)(x - 3) = 0$$
$$x = 7 \text{ or } x = 3$$

(*Note*: there are *two* solutions)

Exercise 23

Solve these equations by factorising.

1 $x^2 + 3x + 2 = 0$ **2** $x^2 + 5x + 6 = 0$ **3** $x^2 + x - 6 = 0$

4 $x^2 + x - 2 = 0$ **5** $x^2 + 7x + 10 = 0$ **6** $x^2 + 7x + 12 = 0$

7 $x^2 - 2x - 15 = 0$ **8** $x^2 - 2x - 8 = 0$ **9** $x^2 - 6x + 9 = 0$

10 $x^2 - 4x + 4 = 0$ **11** $x^2 + 4x - 12 = 0$ **12** $x^2 + 5x - 24 = 0$

13 $x^2 + x = 0$ **14** $x^2 + 2x = 0$ **15** $x^2 - 4x = 0$

16 $x^2 - 3x = 0$ **17** $x^2 - 4 = 0$ **18** $x^2 - 9 = 0$

19 $x^2 - 36 = 0$ **20** $x^2 - 49 = 0$

Exercise 23★

Solve these equations by factorising.

1 $x^2 + 6x + 5 = 0$ **2** $x^2 + 6x + 8 = 0$ **3** $x^2 - 3x - 4 = 0$

4 $x^2 + 3x - 18 = 0$ **5** $x^2 + 15x + 56 = 0$ **6** $x^2 + 15x + 54 = 0$

7 $x^2 - 4x - 45 = 0$ **8** $x^2 + 2x - 63 = 0$ **9** $x^2 - 14x + 49 = 0$

10 $x^2 - 10x + 25 = 0$ **11** $x^2 - 3x - 40 = 0$ **12** $x^2 + 3x - 180 = 0$

13 $x^2 - 13x = 0$ **14** $x^2 - 11x = 0$ **15** $x^2 + 17x = 0$

16 $x^2 + 19x = 0$ **17** $x^2 - 81 = 0$ **18** $x^2 - 144 = 0$

19 $x^2 - 121 = 0$ **20** $x^2 - 169 = 0$

More difficult quadratic equations

When $a \neq 1$, factorisation may be harder. *Always* take out any common factors first.

Example 2

Solve these quadratic equations.

a $9x^2 - 25 = 0$

$$9x^2 = 25$$

$$x^2 = \frac{25}{9}$$

$$x = \pm\frac{5}{3}$$

b $3x^2 - 12x = 0$

$$3x(x - 4) = 0$$

$$x = 0 \text{ or } x = 4$$

c $12x^2 - 24x - 96 = 0$

$$12(x^2 - 2x - 8) = 0$$

$$12(x + 2)(x - 4) = 0$$

$$x = -2 \text{ or } x = 4$$

If there is no simple number factor, then the factorisation is harder.

Example 3

Solve $3x^2 - 13x - 10 = 0$.

$3x^2 - 13x - 10 = 0$

$(3x + 2)(x - 5) = 0$

$x = -\frac{2}{3}$ or $x = 5$

Exercise 24

Solve these equations by factorising.

1 $4x^2 - 49 = 0$
2 $25x^2 - 9 = 0$
3 $16x^2 - 81 = 0$
4 $9x^2 - 16 = 0$
5 $3x^2 + 6x = 0$
6 $2x^2 - 10x = 0$
7 $5x^2 - 5x = 0$
8 $4x^2 + 12x = 0$
9 $2x^2 - 10x + 12 = 0$
10 $2x^2 + 14x + 20 = 0$
11 $2x^2 - 5x + 2 = 0$
12 $2x^2 - 7x + 6 = 0$
13 $2x^2 + 5x + 3 = 0$
14 $2x^2 + 7x + 3 = 0$
15 $3x^2 + 9x + 6 = 0$
16 $3x^2 + 12x - 15 = 0$
17 $2x^2 - 18 = 0$
18 $2x^2 - 50 = 0$
19 $3x^2 - 6x = 0$
20 $3x^2 + 9x = 0$
21 $3x^2 + 7x + 2 = 0$
22 $3x^2 + 14x + 8 = 0$
23 $3x^2 - 5x - 2 = 0$
24 $3x^2 - 11x + 6 = 0$
25 $4x^2 - 4x = 24$
26 $4x^2 + 16x = 20$
27 $3x^2 + 8x + 4 = 0$
28 $4x^2 + 13x + 3 = 0$
29 $3x^2 + 10x = 8$
30 $4x^2 - 3x = 10$

Exercise 24★

Solve these equations by factorising.

1 $49x^2 - 25 = 0$
2 $9x^2 - 64 = 0$
3 $128 - 18x^2 = 0$
4 $75 - 12x^2 = 0$
5 $10x + 5x^2 = 0$
6 $28x - 7x^2 = 0$
7 $6x^2 - 9x = 0$
8 $14x^2 + 4x = 0$
9 $2x^2 - 6x + 4 = 0$
10 $2x^2 + 16x + 30 = 0$
11 $2x^2 - 7x + 6 = 0$
12 $2x^2 + 7x - 15 = 0$
13 $3x^2 + 31x + 36 = 0$
14 $3x^2 + 30x + 63 = 0$
15 $6x^2 - 7x - 3 = 0$
16 $6x^2 - 5x + 1 = 0$
17 $8x^2 + 6x + 1 = 0$
18 $4x^2 + 3x - 1 = 0$
19 $5x^2 - 27x + 10 = 0$
20 $4x^2 + 8x - 21 = 0$
21 $10x^2 - 23x + 12 = 0$
22 $10x^2 + 11x - 35 = 0$
23 $3x^2 = 17x + 28$
24 $2x^2 = x + 15$
25 $3x^2 - 48 = 0$
26 $4x^2 - 36 = 0$
27 $7x^2 - 21x = 0$

28 $8x^2 - 24x = 0$

29 $4x^2 + 40x + 100 = 0$

30 $4x^2 - 24x + 32 = 0$

31 $4x^2 = 29x - 7$

32 $4x^2 = 23x - 15$

33 $x(6x - 13) = -6$

34 $3x(2x - 9) = -30$

35 $9x^2 + 25 = 30x$

36 $6x^2 + 2 = 7x$

The quadratic formula

The quadratic formula is used to solve quadratic equations that may be awkward to solve by other means.

Key Point

If $ax^2 + bx + c = 0$ then $x = \dfrac{-b \pm \sqrt{b^2 - 4ac}}{2a}$

Example 4

Solve $3x^2 - 8x + 2 = 0$ giving your solution correct to 3 significant figures.

Here $a = 3$, $b = -8$ and $c = 2$. *Note*: b is a negative number.

Substituting into the formula gives

$$x = \frac{-(-8) \pm \sqrt{(-8)^2 - 4 \times 3 \times 2}}{2 \times 3} = \frac{8 \pm \sqrt{64 - 24}}{6}$$

So $x = \dfrac{8 + \sqrt{40}}{6} = 2.39$ or $x = \dfrac{8 - \sqrt{40}}{6} = 0.279$

Example 5

Solve $2.3x^2 + 3.5x - 4.8 = 0$ giving your solution correct to 3 significant figures.

Here $a = 2.3$, $b = 3.5$ and $c = -4.8$. Substituting into the formula gives

$$x = \frac{-3.5 \pm \sqrt{12.25 - 4 \times 2.3 \times (-4.8)}}{2 \times 2.3} = \frac{-3.5 \pm \sqrt{56.41}}{4.6}$$

So $x = \dfrac{-3.5 + \sqrt{56.41}}{4.6} = 0.872$ or $x = \dfrac{-3.5 - \sqrt{56.41}}{4.6} = -2.39$

The solutions are $x = 0.872$ or $x = -2.39$.

Exercise 25

Solve these equations using the quadratic formula.

Give your solutions correct to 3 significant figures.

1 $x^2 + 2x - 5 = 0$

2 $x^2 + 6x - 8 = 0$

3 $x^2 - 2x - 6 = 0$

4 $x^2 - 6x - 15 = 0$

5 $x^2 + 4x = 8$

6 $x^2 + 2x = 7$

7 $x^2 - 10x + 15 = 0$

8 $x^2 + 12x + 34 = 0$

9 $x^2 + 14x - 3 = 0$

10 $x^2 + 16x - 7 = 0$

11 $x^2 - 20x - 33 = 0$

12 $x^2 - 14x + 47 = 0$

13 $x^2 - 4x - 20 = 0$

14 $x^2 - 8x - 10 = 0$

15 $x^2 - 10x = 120$

16 $x^2 - 14x = 41$

17 $x^2 + 3x - 2 = 0$

18 $x^2 + x - 8 = 0$

19 $x^2 - 5x - 3 = 0$

20 $x^2 - 7x + 9 = 0$

21 $x^2 + x - 8 = 0$

22 $x^2 + 2x - 4 = 0$

23 $x^2 - 2x - 7 = 0$

24 $x^2 - x - 3 = 0$

25 $3x^2 + 6x + 2 = 0$

26 $3x^2 + 7x + 3 = 0$

27 $4x^2 + x - 4 = 0$

28 $3x^2 + 5x + 1 = 0$

29 $x^2 + x = 5$

30 $x^2 + 2x = 7$

31 $6x - 1 = x^2$

32 $16x - 3 = x^2$

33 $8 + 3x - 7x^2 = 0$

34 $1 + 5x - 3x^2 = 0$

35 $8x = 2 + 5x^2$

36 $2x = 7x^2 - 3$

Exercise 25★

Solve these equations using the quadratic formula.

Give your solutions correct to 3 significant figures.

1 $x^2 - 6x + 1 = 0$

2 $x^2 - 4x + 1 = 0$

3 $x^2 - 16x + 3 = 0$

4 $x^2 - 12x - 25 = 0$

5 $x^2 + 6x - 12 = 0$

6 $x^2 + 4x + 2 = 0$

7 $x^2 + 13x + 4 = 0$

8 $x^2 + 12x + 6 = 0$

9 $x^2 - 6x + 7 = 0$

10 $x^2 - 7x + 5 = 0$

11 $3x^2 - 5x = 2$

12 $3x^2 - 2x = 1$

13 $x^2 - 13 = 6x$

14 $x^2 + 14 = 8x$

15 $x^2 - 2x = 1$

16 $x^2 - 6x = 2$

17 $2x^2 - 16x + 4 = 0$

18 $2x^2 + 8x - 6 = 0$

19 $2x^2 - 5x = 7$

20 $2x^2 + 7x = 3$

21 $x(5x + 12) = -5$

22 $x(3x + 8) = -2$

23 $3 - 10x - 4x^2 = 0$

24 $3 - 4x - 6x^2 = 0$

25 $7x^2 = 4 + 4x$

26 $4x^2 = 3 + 6x$

27 $x(5x - 8) = -1$

28 $x(5x + 2) = 1$

29 $3x^2 = 7x + 2$

30 $4x^2 = 8x + 3$

31 $10 + 3x - 2x^2 = 0$

32 $5 - 2x - 4x^2 = 0$

33 $2.3x^2 - 12.6x + 1.3 = 0$

34 $3.7x^2 - 9.4x + 2.8 = 0$

35 $x(x + 1) + (x - 1)(x + 2) = 3$

36 $x(x + 1) + (x + 2)(x + 3) = 4$

Activity 11

- Use the graphs to find how many solutions there are to each of these equations.

  $x^2 + 8x + 15 = 0$ $x^2 + 8x + 16 = 0$ $x^2 + 8x + 17 = 0$

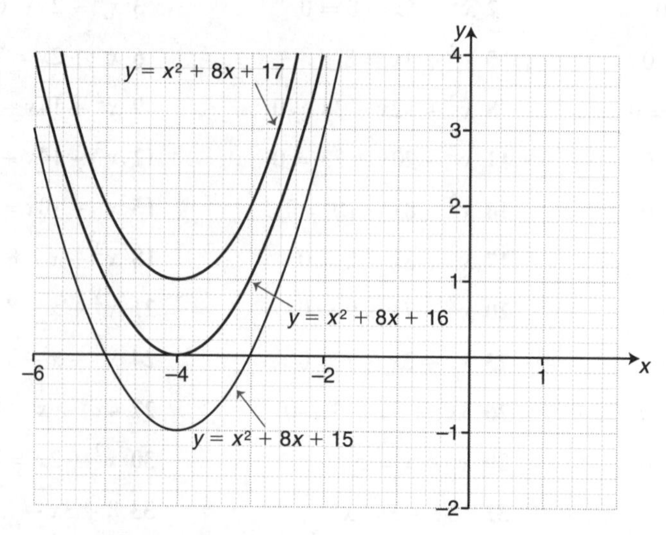

- For each of the quadratic equations, work out the values of $b^2 - 4ac$.
- Try to find a rule involving $b^2 - 4ac$ that tells you how many solutions a quadratic equation has.

Investigate

For what values of k does the equation $x^2 + 8x + k = 0$ have real solutions?

Exercise 26

State how many solutions there are to these equations. Do *not* solve them.

1 $x^2 - 2x + 1 = 0$

2 $x^2 - 9 = 0$

3 $x^2 + 4 = 0$

4 $x^2 + 2x + 1 = 0$

5 $x^2 - 2x + 5 = 0$

6 $x^2 - 4x + 4 = 0$

7 $x^2 + 6x + 1 = 0$

8 $x^2 - 2x - 3 = 0$

9 $x^2 - x + 1 = 0$

10 $x^2 + 8x + 12 = 0$

Exercise 26★

State how many solutions there are to these equations. Do *not* solve them.

1 $x^2 - 3 = 0$

2 $x^2 + 3x + 3 = 0$

3 $x^2 - x - 1 = 0$

4 $4x^2 - 4x + 5 = 0$

5 $4x^2 - 4x + 1 = 0$

6 $2x^2 + 3x + 2 = 0$

7 $4x^2 - 7x + 2 = 0$

8 $2x^2 - 4x + 9 = 0$

9 $3x^2 + 8x + 3 = 0$

10 $9x^2 + 6x + 1 = 0$

Problems leading to quadratic equations

Key Points

- Where relevant, draw a clear diagram and put all the information on it.
- Let x stand for what you are trying to find.
- Form a quadratic equation in x and simplify it.
- Solve the equation by either factorising or using the formula.
- Check that the answers make sense.

Example 6

The width of a rectangular photograph is 4 cm more than the height. The area is 77 cm^2. Find the height of the photograph.

Let x be the height in cm.
Then the width is $x + 4$ cm.
The diagram is shown on the right.

As the area is 77 cm^2,

$$x(x + 4) = 77$$
$$x^2 + 4x = 77$$
$$x^2 + 4x - 77 = 0$$
$$(x - 7)(x + 11) = 0$$

So $x = 7$ or -11 cm

The height cannot be negative, so the height is 7 cm.

Example 7

The chords of a circle intersect as shown.
Find the value of x.

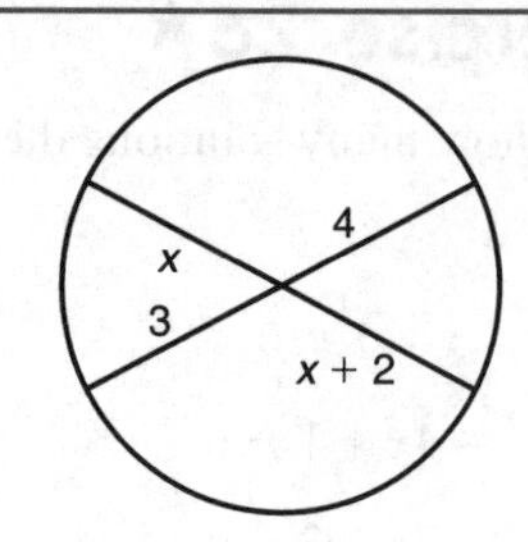

Using the intersecting chords theorem:

$$x(x+2) = 3 \times 4$$
$$x^2 + 2x = 12$$
$$x^2 + 2x - 12 = 0$$

This equation does not factorise.
Using the quadratic formula with $a = 1$, $b = 2$ and $c = -12$ gives

$$x = -2 \pm \sqrt{\frac{2^2 - 4 \times (-12)}{2}}$$
$$x = 2.61 \text{ or } -4.61 \text{ (3 s.f.)}$$

As x cannot be negative, $x = 2.61$ to 3 s.f.

Example 8

A rectangular fish pond is 6 m by 9 m. The pond is surrounded by a concrete path of constant width. The area of the pond is the same as the area of the path. Find the width of the path.

Let x be the width of the path.

The diagram is shown on the right.

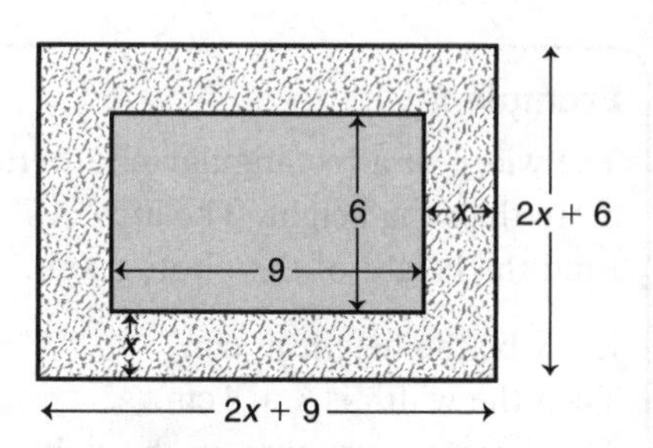

The area of the path is $(2x+9)(2x+6) - 9 \times 6$

$$= 4x^2 + 30x + 54 - 54$$
$$= 4x^2 + 30x$$

The area of the pond is $9 \times 6 = 54\,\text{m}^2$.
As the area of the path equals the area of the pond:

$$4x^2 + 30x = 54$$
$$4x^2 + 30x - 54 = 0$$
$$2x^2 + 15x - 27 = 0$$
$$(2x-3)(x+9) = 0$$

So $\quad x = 1.5 \text{ or } -9\,\text{m}$

As x cannot be negative, the width of the path is 1.5 m.

Exercise 27

1 The height, h m, of a rocket above the ground after t seconds is given by $h = 35t - 5t^2$. When is the rocket 50 m above the ground?

2 The distance, d m, that a scooter has rolled down a hill after t seconds is given by $d = 2t + t^2$. Find how long it takes the scooter to travel 48 m.

3 One number is four more than another number. The product of the numbers is 96. Find the numbers.

4 One number is two less than another number. The product of the numbers is 63. Find the numbers.

5 The width of a rectangle is 2 cm more than the height. The area is 12 cm^2. Find the height of the rectangle.

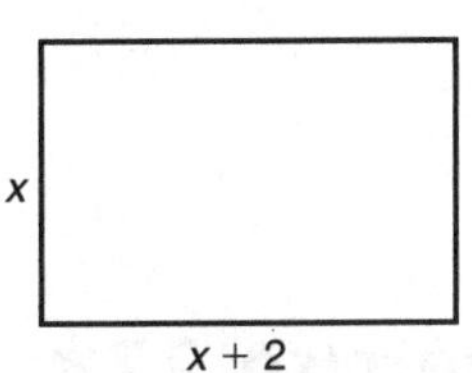

6 The height of a rectangle is 3 cm more than the width. The area is 30 cm^2. Find the width of the rectangle.

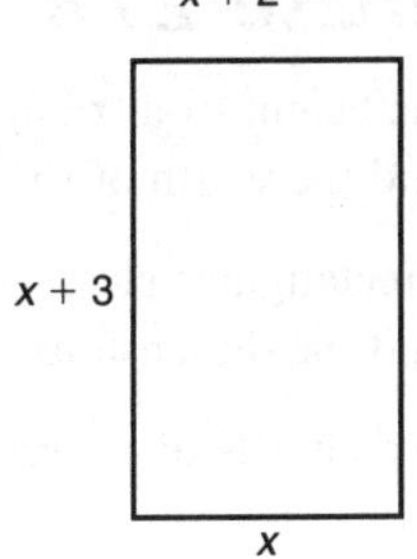

7 The height of a right-angled triangle is 3 cm more than the width. The area is 10 cm^2. Find the width of the triangle.

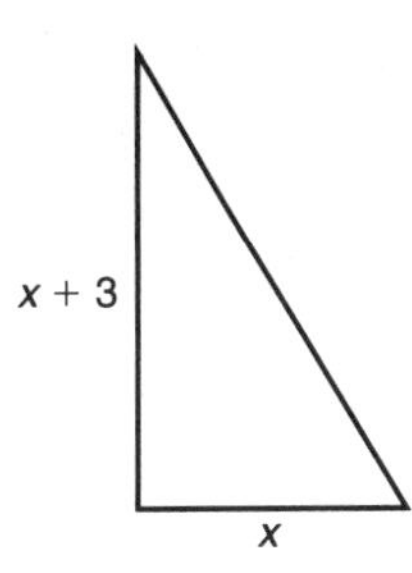

8 The height of a right-angled triangle is 1 cm more than the width. The area is 12 cm^2. Find the width of the triangle.

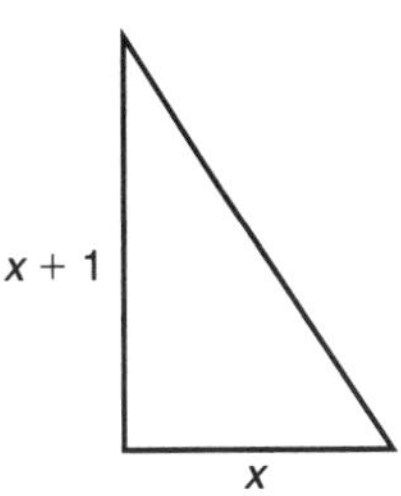

9 The chords of a circle intersect as shown. Find x.

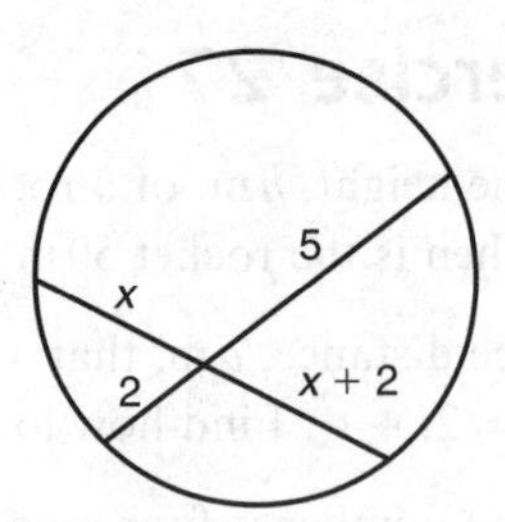

10 The chords of a circle intersect as shown. Find x.

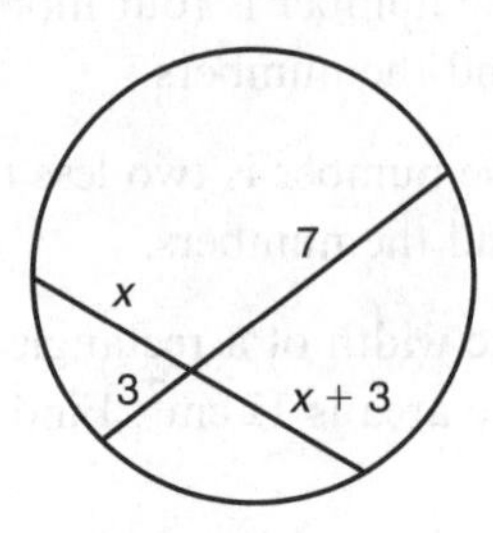

Exercise 27★

1 The height of a triangle is 3 cm more than the width. The area is $14\,\text{cm}^2$. Find the width of the triangle.

2 A rectangular classroom has a perimeter of 28 m and an area of $48\,\text{m}^2$. Find the dimensions of the classroom.

3 Two chords of a circle intersect as shown. Find x.

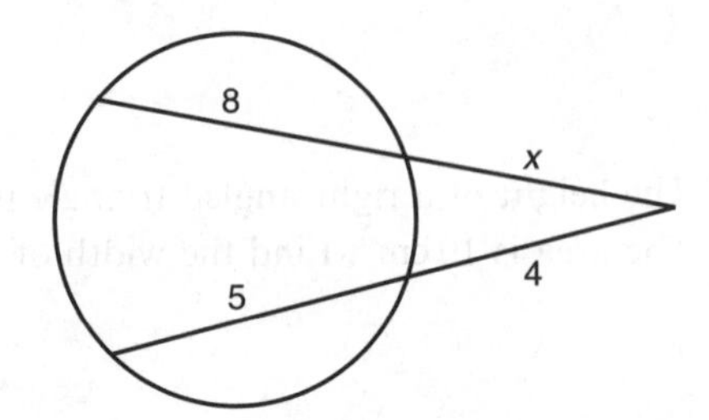

4 A chord and tangent of a circle intersect as shown. Find x.

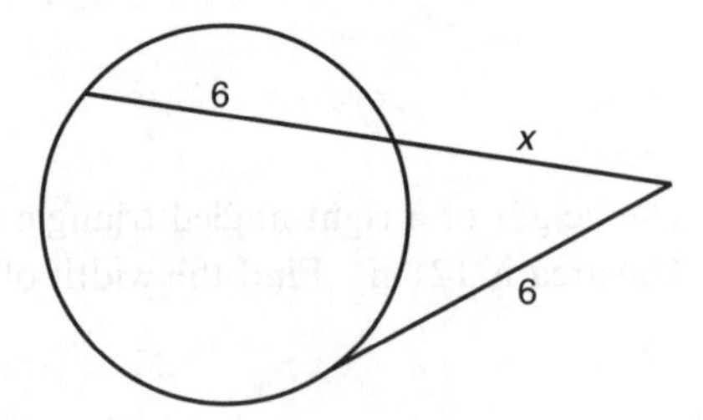

5 The sum of the squares of two consecutive integers is 145. Find the integers.

6 The sum of the squares of two consecutive odd integers is 130. Find the integers.

7 The perimeter of a rectangular room is 13.5 m and the length of a diagonal is 5 m. Find the dimensions of the room.

8 The perimeter of a rectangular room is 32 m. The length of a diagonal is 8 m more than the width. Find the dimensions of the room.

9 The sum of the first n integers $1 + 2 + 3 + \ldots + n = \frac{n(n+1)}{2}$.

a How many numbers must be taken to have a sum greater than one million?

b Why can't the sum ever equal 100 000?

10 An n-sided polygon has $\frac{n(n-3)}{2}$ diagonals.

a How many sides has a polygon with 665 diagonals?

b Why can't a polygon have 406 diagonals?

11 Lee spent \$1200 on holiday. If he had spent \$50 less per day, he would have been able to stay an extra two days. How long was his holiday?

12 One week a syndicate of x people won \$1000 in a lottery. If there had been two fewer people in the syndicate, each person would have received \$25 more. How many people are in the syndicate?

Solving quadratic inequalities

Squares and square roots in inequalities need care.
If you think the answer to $x^2 < 4$ is $x < 2$, then you are mistaken.
For example, try $x = -3$: $-3 < 2$ but $(-3)^2$ is not less than 4.

Remember

To solve a quadratic inequality, sketch the graph of the quadratic function.

Example 9

Solve $x^2 - 4 < 0$.

First sketch $y = x^2 - 4$.
To do this, find where the graph intersects the x-axis by solving $x^2 - 4 = 0$.
$x^2 - 4 = 0 \Rightarrow x^2 = 4 \Rightarrow x = -2$ or $x = 2$.

So the graph intersects the x-axis at $x = -2$ and $x = 2$.
(These are known as the critical values.)

Also, when $x = 0$, $y = -4$, so the graph cuts the y-axis at -4.

The graph is a parabola, which is ∪-shaped.

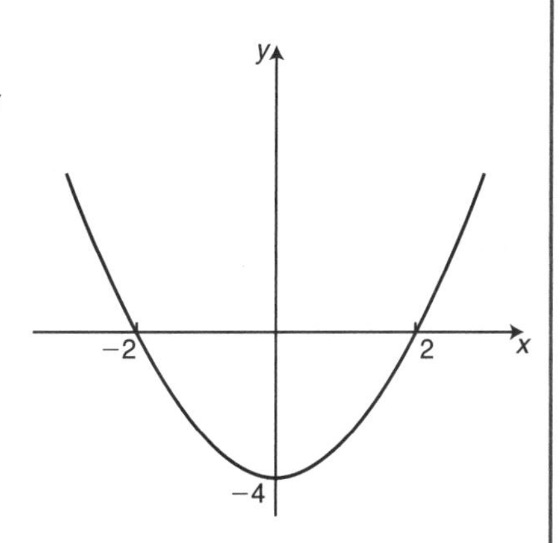

We want the region *below* the x-axis. As this is *one* region, the answer is *one* inequality.

The solution is $-2 < x < 2$.

UNIT 2 ◆ Algebra

Example 10

Solve $x^2 - x - 2 \leqslant 0$.

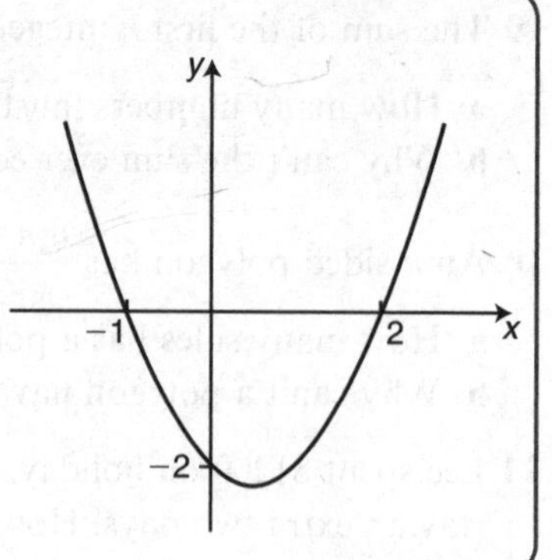

First sketch $y = x^2 - x - 2$.
As $y = (x - 2)(x + 1)$, the critical values are $x = 2$ and $x = -1$.
When $x = 0$, $y = -2$.

We want the region *below* the x-axis. As this is *one* region, the answer is *one* inequality.

The solution is $-1 \leqslant x \leqslant 2$.

Example 11

Solve $x^2 - 5x + 6 \geqslant 0$.

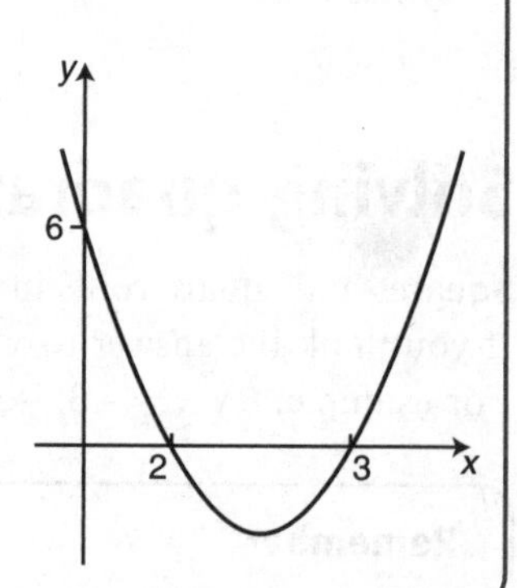

First sketch $y = x^2 - 5x + 6$.
As $y = (x - 2)(x - 3)$, the critical values are $x = 2$ and $x = 3$.

When $x = 0$, $y = 6$.

We want the region *above* the x-axis. As this has *two* components, the answer is *two* inequalities.

The solution is $x \leqslant 2$ or $x \geqslant 3$.

Exercise 28

Solve the following inequalities:

1 $x^2 < 16$

2 $x^2 < 9$

3 $x^2 \geqslant 25$

4 $x^2 \geqslant 36$

5 $x^2 + 3 \leqslant 84$

6 $x^2 - 7 \leqslant 29$

7 $3x^2 < 75$

8 $2x^2 < 72$

9 $4x^2 + 3 > 67$

10 $5x^2 - 6 > 14$

11 $(x - 1)(x + 3) \leqslant 0$

12 $(x - 3)(x + 1) \leqslant 0$

13 $(x + 3)(x + 4) > 0$

14 $(x + 2)(x + 7) > 0$

15 $(2x - 1)(x + 1) < 0$

16 $(3x + 1)(x - 1) < 0$

17 $x^2 + 7x + 10 \geqslant 0$

18 $x^2 + 8x + 12 \geqslant 0$

19 $x^2 + 2x - 15 \leqslant 0$

20 $x^2 + x - 6 \leqslant 0$

Exercise 28★

Solve the following inequalities:

1 $2x^2 \leqslant 50$

2 $3x^2 \leqslant 48$

3 $2x^2 - 1 \geqslant 31$

4 $5x^2 + 3 \geqslant 23$

5 $28 - x^2 > 3$

6 $79 - 2x^2 > 7$

7 $(x - 5)^2 > 4$

8 $(x + 3)^2 > 9$

9 $3(x + 2)^2 < 48$

10 $2(x - 1)^2 < 50$

11 $x^2 + 10x + 21 \leqslant 0$

12 $x^2 + 5x + 6 \leqslant 0$

13 $x^2 + x - 12 > 0$

14 $x^2 + 3x - 4 > 0$

15 $8 - 2x - x^2 < 0$

16 $20 - x - x^2 < 0$

17 $2x^2 - 3x - 2 \leqslant 0$

18 $3x^2 + 10x - 8 \leqslant 0$

19 $6x^2 + 17x - 3 \geqslant 0$

20 $3x^2 - 7x - 20 \geqslant 0$

21 Solve $(x + 1)^2 \leqslant 5x^2 + x + 1$.

22 Solve $(3x - 1)(x + 1) \geqslant x(2x - 3) - 5$.

23 Solve $(x - 5)^2 + 5(2x - 3) > 2x(x + 3) - 6$.

24 Two numbers differ by seven. The product of the two numbers is less than 78. Find the possible range of values for the smaller number.

25 The area of rectangle A is less than the area of rectangle B. Find the range of possible values of x.

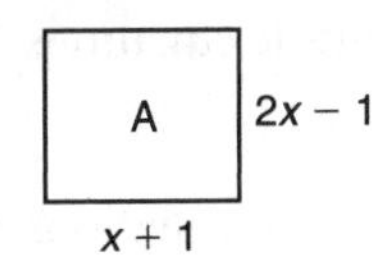

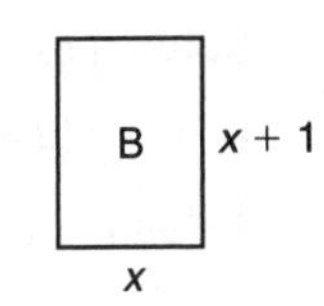

26 The area of the right-angled triangle A is greater than the area of rectangle B. Find the range of possible values of x.

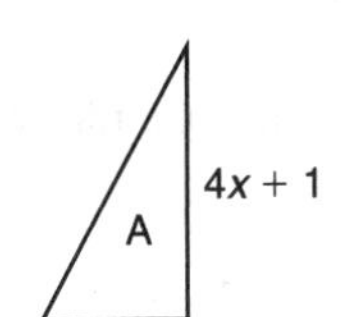

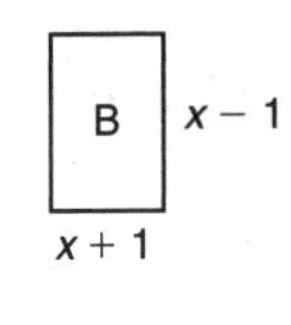

27 The perimeter of a rectangle is 28 cm. Find the range of possible values of the width of the rectangle if the diagonal is less than 10 cm.

28 The area of a rectangle is 12 cm^2. Find the range of possible values of the width of the rectangle if the diagonal is more than 5 cm.

Exercise 29 (Revision)

1 Solve these quadratic equations:

a $x^2 - 25 = 0$

b $x^2 + 4x = 0$

2 Solve these quadratic equations by factorisation:

a $x^2 + x - 12 = 0$

b $5x^2 - 5x - 30 = 0$

c $3x^2 + x - 2 = 0$

3 Use the quadratic formula to solve these equations:

a $x^2 - 2x - 4 = 0$ **b** $3x^2 - 5x + 1 = 0$

4 The height of a rectangle is 1.5 cm more than the width. The area is $10\,\text{cm}^2$. Find the width of the rectangle.

5 The chords of a circle intersect as shown in the diagram. Find x.

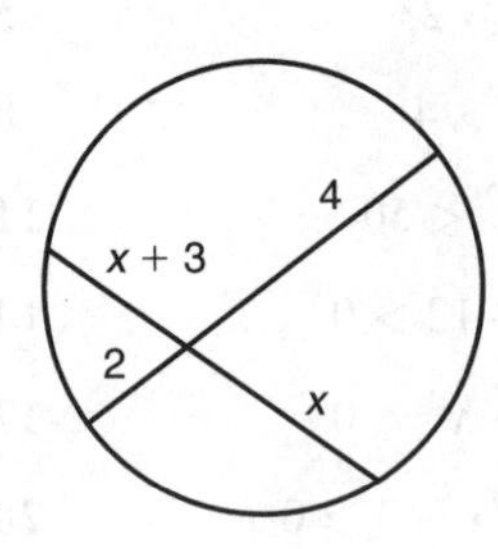

6 Solve the quadratic inequalities:

a $x^2 \geqslant 4$ **b** $x^2 + 2x - 15 < 0$

Exercise 29★ (Revision)

1 Solve these quadratic equations:

a $x^2 - 20 = 0$ **b** $x^2 - 9x = 0$

2 Solve these quadratic equations by factorisation:

a $x^2 + x - 72 = 0$ **b** $7x^2 = 14x + 168$ **c** $2x(4x + 7) = 15$

3 Use the quadratic formula to solve these equations:

a $3x^2 = 7x + 5$ **b** $2.1x^2 + 8.4x - 4.3 = 0$

4 A cereal packet is a cuboid with height 12 cm. The depth of the box is 4 cm more than the width, and the volume is $480\,\text{cm}^3$. Find the width of the box.

5 The area of a rectangular lawn is $30\,\text{m}^2$. During landscaping the length was decreased by 1 m and the width increased by 1 m, but the area did not change. Find the original dimensions of the lawn.

6 Two chords of a circle intersect as shown in the diagram. Find x.

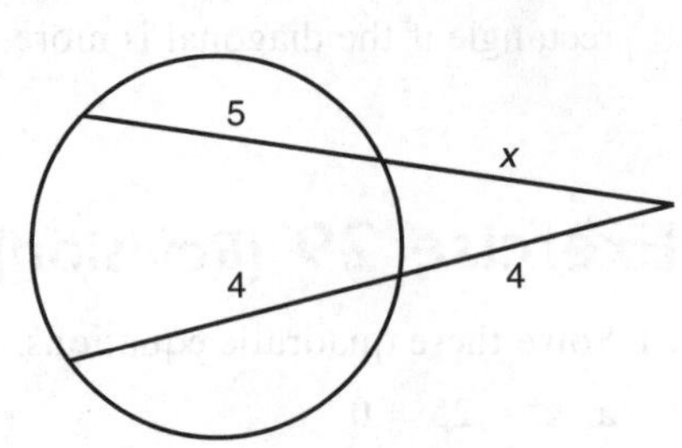

7 Solve the quadratic inequalities:

a $x^2 - 12x + 32 < 0$ **b** $x^2 - 2x \geqslant 10$

Using graphs to solve equations

Using the graph of $y = x^2$ to solve quadratic equations

An accurately drawn graph can be used to solve equations that may prove difficult to solve exactly by other methods.

The graph of $y = x^2$ is easy to draw and can be used to solve quadratic equations.

Example 1

Draw the graph of $y = x^2$. Use this graph to solve the equation $x^2 - x - 3 = 0$.

Rearrange $x^2 - x - 3 = 0$ as $x^2 = x + 3$.
This can be solved by finding the x-co-ordinates of the intersection points of the graphs $y = x^2$ and $y = x + 3$.
The graph on the right shows the solutions are approximately $x = -1.3$ or $x = 2.3$.

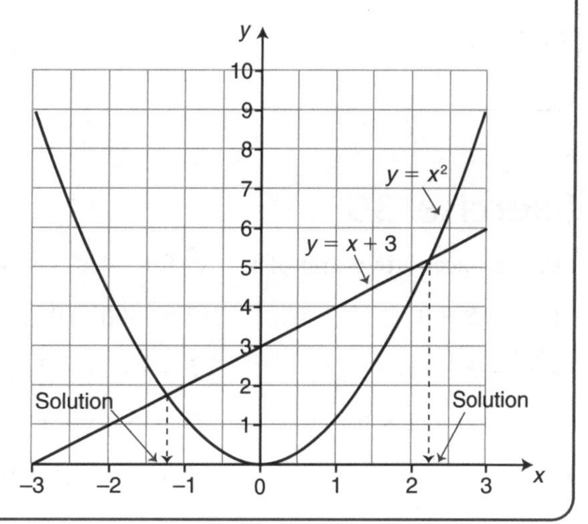

Example 2

Draw the graph of $y = x^2$. Use this graph to solve the equation $x^2 + 2x = 2$.

Rearrange $x^2 + 2x = 2$ as $x^2 = 2 - 2x$.
This can be solved by finding the x-co-ordinates of the intersection points of the graphs $y = x^2$ and $y = 2 - 2x$.
The graph on the right shows the solutions are approximately $x = -2.7$ or $x = 0.7$.

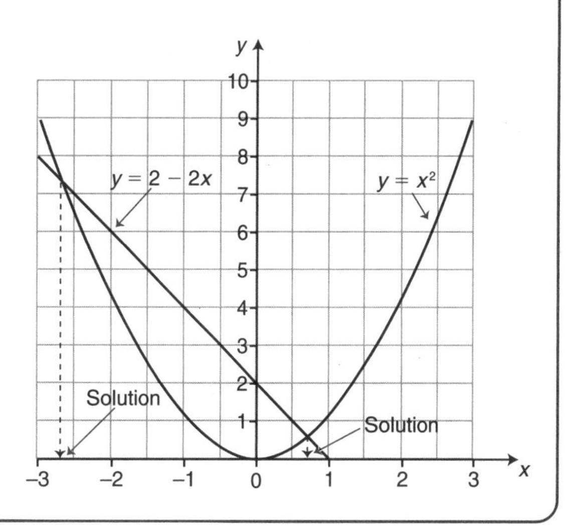

Example 3

Draw the graph of $y = x^2$. Use this graph to solve the equation $2x^2 + x - 8 = 0$.

Rearrange $2x^2 + x - 8 = 0$ as $x^2 = 4 - \frac{1}{2}x$.
This can be solved by finding the x-co-ordinates of the intersection points of the graphs $y = x^2$ and $y = 4 - \frac{1}{2}x$.
The graph on the right shows the solutions are approximately $x = -2.3$ or $x = 1.8$.

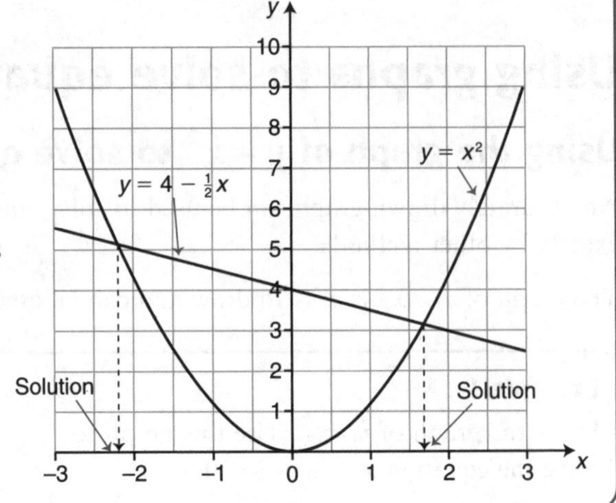

Exercise 30

Draw an accurate graph of $y = x^2$ for $-4 \leqslant x \leqslant 4$. Use this graph to solve these equations.

1 $x^2 - 5 = 0$ **2** $x^2 - 3 = 0$ **3** $x^2 - x - 2 = 0$

4 $x^2 + x - 3 = 0$ **5** $x^2 + 2x - 7 = 0$ **6** $x^2 - 2x - 6 = 0$

7 $x^2 - 4x + 2 = 0$ **8** $x^2 + 4x + 1 = 0$ **9** $2x^2 - x - 20 = 0$

10 $2x^2 + x - 16 = 0$ **11** $x^2 - x + 1 = 0$ **12** $x^2 + x + 2 = 0$

Exercise 30★

Draw an accurate graph of $y = x^2$ for $-4 \leqslant x \leqslant 4$. Use this graph to solve these equations.

1 $x^2 - x - 3 = 0$ **2** $x^2 + x - 4 = 0$ **3** $x^2 + 3x + 1 = 0$

4 $x^2 - 2x - 2 = 0$ **5** $x^2 - 4x + 4 = 0$ **6** $x^2 + 2x + 1 = 0$

7 $2x^2 + x - 12 = 0$ **8** $2x^2 - x - 10 = 0$ **9** $3x^2 - x - 27 = 0$

10 $3x^2 + x - 21 = 0$ **11** $3x^2 - 3x + 6 = 0$ **12** $4x^2 + 6x + 3 = 0$

Using other graphs to solve quadratic equations

Example 4

Draw the graph of $y = x^2 - 5x + 5$ for $0 \leqslant x \leqslant 5$. Use this graph to solve these three equations:

$$0 = x^2 - 5x + 5 \qquad 0 = x^2 - 5x + 3 \qquad 0 = x^2 - 4x + 4$$

To solve: $0 = x^2 - 5x + 5$

Find where the graph of $y = x^2 - 5x + 5$ cuts the line $y = 0$ (the x-axis).

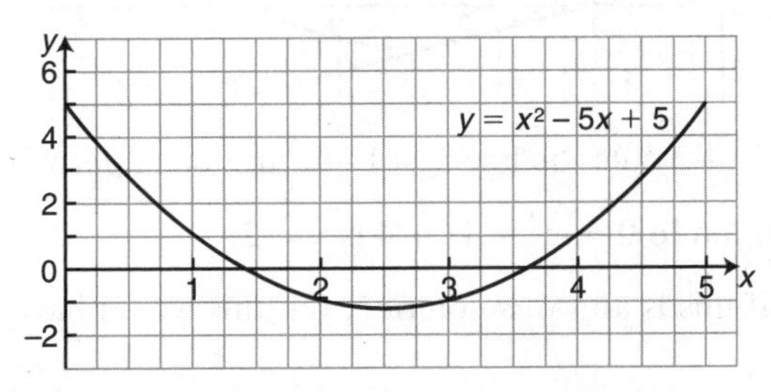

The graph cuts the x-axis at approximately $x = 1.4$ and $x = 3.6$.

So the approximate solutions to $0 = x^2 - 5x + 5$ are $x = 1.4$ or $x = 3.6$.

To solve: $0 = x^2 - 5x + 3$

$0 = x^2 - 5x + 3$ (Add 2 to both sides)

$2 = x^2 - 5x + 5$

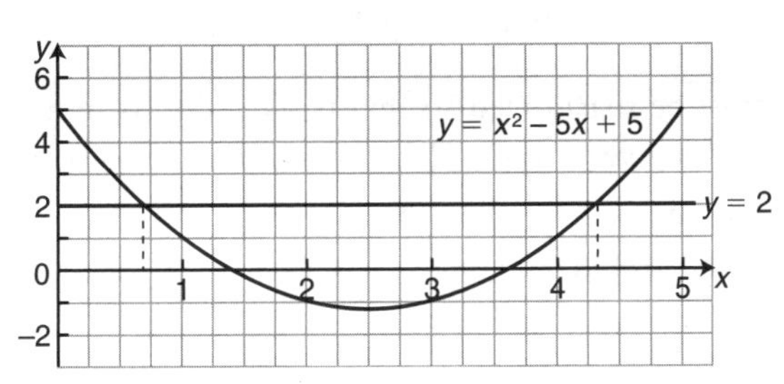

The graph of $y - x^2 - 5x + 5$ cuts the line $y = 2$ at $x = 0.7$ and $x = 4.3$ approximately.

So the approximate solutions to $0 = x^2 - 5x + 3$ are $x = 0.7$ or $x = 4.3$.

To solve: $0 = x^2 - 4x + 4$

$$0 = x^2 - 4x + 4 \quad \text{(Add } 1 - x \text{ to both sides)}$$

$$1 - x = x^2 - 5x + 5$$

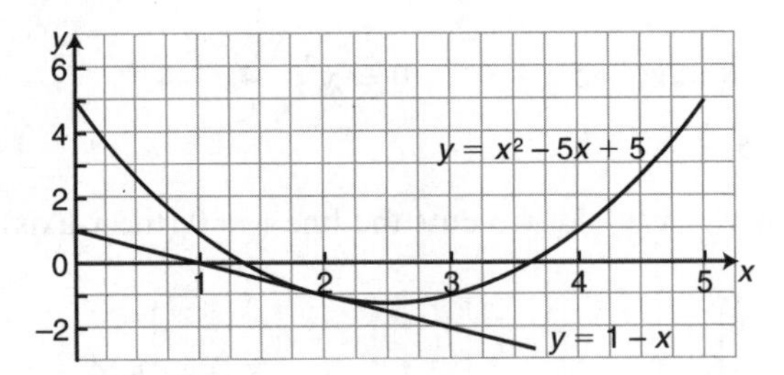

The graph of $y = x^2 - 5x + 5$ cuts the line $y = 1 - x$ at $x = 2$ approximately.

So the approximate solution to $0 = x^2 - 4x + 4$ is $x = 2$.

In this case, it looks as if this is an *exact* solution, but this would have to be checked by substitution.

Note: If the line had not cut the graph, there would be *no* real solutions.

Example 5

If the graph of $y = 6 + 2x - x^2$ has been drawn, find the equation of the line that should be drawn to solve

a $0 = 2 + 2x - x^2$ **b** $0 = 7 + x - x^2$

a $0 = 2 + 2x - x^2$ must be rearranged so that $6 + 2x - x^2$ is on the right-hand side. Adding 4 to both sides gives $4 = 6 + 2x - x^2$, so the line to be drawn is $y = 4$.

b $0 = 7 + x - x^2$ must be rearranged so that $6 + 2x - x^2$ is on the right-hand side. Adding $x - 1$ to both sides gives $x - 1 = 6 + 2x - x^2$, so the line to be drawn is $y = x - 1$.

The graphs are shown.

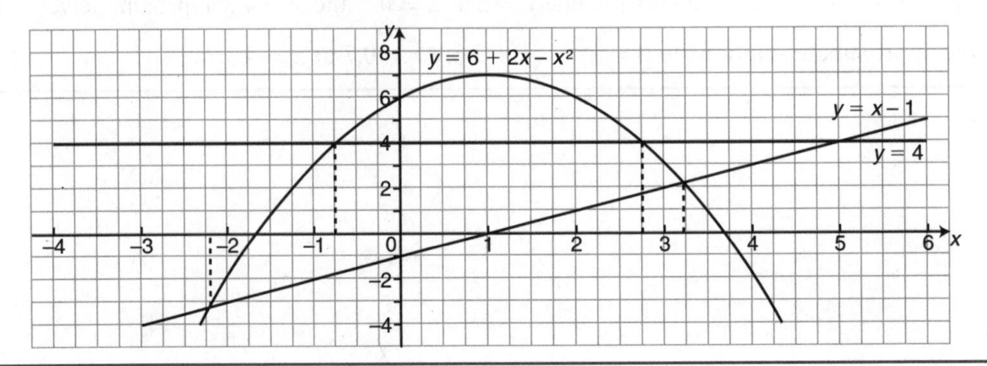

Exercise 31

1 Draw the graph of $y = x^2 - 3x$ for $-1 \leqslant x \leqslant 5$. Use your graph to solve these equations.

a $x^2 - 3x = 0$ **b** $x^2 - 3x = 2$ **c** $x^2 - 3x = -1$

d $x^2 - 3x = x + 1$ **e** $x^2 - 3x - 3 = 0$ **f** $x^2 - 5x + 1 = 0$

2 Draw the graph of $y = x^2 - 2x$ for $-2 \leqslant x \leqslant 4$. Use your graph to solve these equations.

a $x^2 - 2x = 0$ **b** $x^2 - 2x = 5$ **c** $x^2 - 2x = -\frac{1}{2}$

d $x^2 - 2x = 1 - x$ **e** $x^2 - 2x - 2 = 0$ **f** $x^2 - 4x + 2 = 0$

3 Draw the graph of $y = x^2 - 4x + 3$ for $-1 \leqslant x \leqslant 5$. Use your graph to solve these equations.

a $x^2 - 4x + 3 = 0$ **b** $x^2 - 4x - 3 = 0$ **c** $x^2 - 5x + 3 = 0$ **d** $x^2 - 3x - 2 = 0$

4 Draw the graph of $y = x^2 - 3x - 4$ for $-2 \leqslant x \leqslant 5$. Use your graph to solve these equations.

a $x^2 - 3x - 4 = 0$ **b** $x^2 - 3x + 1 = 0$ **c** $x^2 - 2x - 4 = 0$ **d** $x^2 - 4x + 2 = 0$

5 Find the equations solved by the intersection of these pairs of graphs.

a $y = 2x^2 - x + 2$, $y = 3 - 3x$ **b** $y = 4 - 3x - x^2$, $y = 2x - 1$

6 Find the equations solved by the intersection of these pairs of graphs.

a $y = 3x^2 + x - 5$, $y = 2x + 1$ **b** $y = 2x + 3 - x^2$, $y = 1 - 2x$

7 If the graph of $y = 3x^2 + 4x - 2$ has been drawn, find the equations of the lines that should be drawn to solve these equations.

a $3x^2 + 2x - 4 = 0$ **b** $3x^2 + 3x - 2 = 0$ **c** $3x^2 + 7x + 1 = 0$

8 If the graph of $y = 3x^2 - 3x + 5$ has been drawn, find the equations of the lines that should be drawn to solve these equations.

a $3x^2 - 4x - 1 = 0$ **b** $3x^2 - 2x - 2 = 0$ **c** $3x^2 + x - 3 = 0$

9 Romeo is throwing a rose up to Juliet's balcony. The balcony is 2 m away from him and 3.5 m above him. The equation of the path of the rose is $y = 4x - x^2$, where the origin is at Romeo's feet. Find by a graphical method where the rose lands.
The balcony has a 1 m high railing. Does the rose pass over the railing?

10 A cat is sitting on a 2 m high fence when it spots a mouse 1.5 m away from the foot of the fence. The cat leaps along the path $y = -0.6x - x^2$, where the origin is where the cat was sitting and x is measured in metres. Find, by a graphical method, whether the cat lands on the mouse.

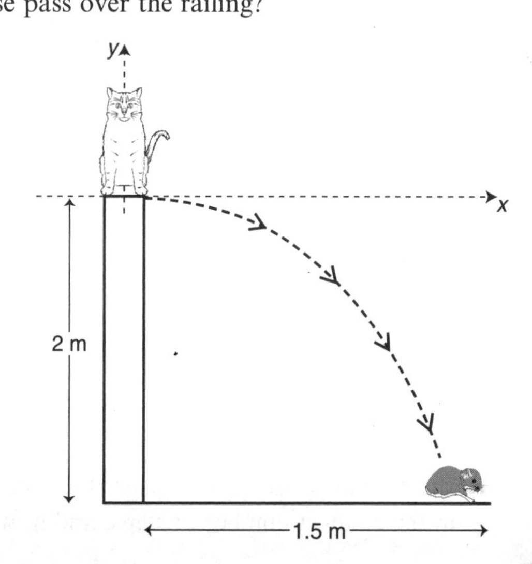

Exercise 31★

1 Draw the graph of $y = 5x - x^2$ for $-1 \leqslant x \leqslant 6$. Use your graph to solve these equations.

a $5x - x^2 = 0$ **b** $5x - x^2 = 3$ **c** $5x - x^2 = x + 1$ **d** $x^2 - 6x + 4 = 0$

2 Draw the graph of $y = x - 2x^2$ for $-2 \leqslant x \leqslant 3$. Use your graph to solve these equations.

a $x - 2x^2 = 0$ **b** $x - 2x^2 = -4$ **c** $x - 2x^2 = -x - 3$ **d** $2x^2 - 2x - 2 = 0$

3 Draw the graph of $y = 2x^2 + 3x - 1$ for $-3 \leqslant x \leqslant 2$. Use your graph to solve these equations.

a $2x^2 + 3x - 1 = 0$ **b** $2x^2 + 3x - 4 = 0$ **c** $2x^2 + 5x + 1 = 0$

4 Draw the graph of $y = 3x^2 - x - 2$ for $-2 \leqslant x \leqslant 3$. Use your graph to solve these equations.

a $3x^2 - x - 2 = 0$ **b** $3x^2 - x - 4 = 0$ **c** $3x^2 - 3x - 1 = 0$

5 Find the equations solved by the intersection of these pairs of graphs.

a $y = 6x^2 - 4x + 3$, $y = 3x + 5$ **b** $y = 7 + 2x - 5x^2$, $y = 3 - 5x$

6 Find the equations solved by the intersection of these pairs of graphs.

a $y = 4x^2 - 5x + 2$, $y = 2x + 7$ **b** $y = 3x + 1 - 3x^2$, $y = 3 - 4x$

7 If the graph of $y = 5x^2 - 9x - 6$ has been drawn, find the equations of the lines that should be drawn to solve these equations.

a $5x^2 - 10x - 8 = 0$ **b** $5x^2 - 7x - 5 = 0$

8 If the graph of $y = 4x^2 + 7x - 8$ has been drawn, find the equations of the lines that should be drawn to solve these equations.

a $4x^2 + 8x - 5 = 0$ **b** $4x^2 + 4x - 3 = 0$

9 Jason is serving at tennis. He hits the ball from a height of 2.5 m and the path of the ball is given by $y = -0.05x - 0.005x^2$, where the origin is the point where he hits the ball.

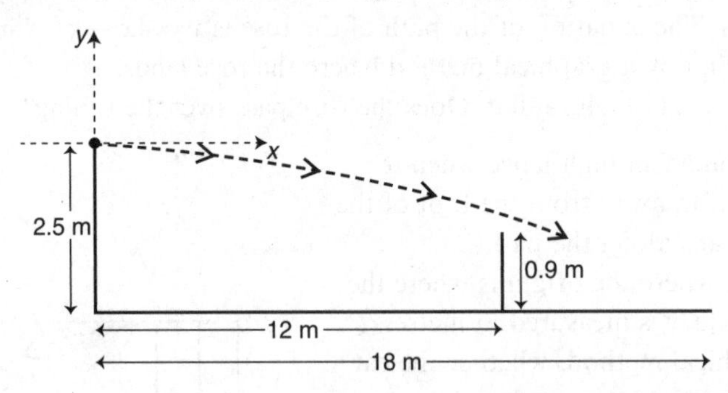

a The net is 0.9 m high and is 12 m away. Does the ball pass over the net?

b For the serve to be legal it must land between the net and the service line, which is 18 m away. Is the serve legal?

10 A young girl is playing a game, which consists of throwing marbles up a flight of stairs. Each step is 20 cm high and 25 cm wide. The path of the marble is given by $y = \frac{5}{2}x - \frac{2}{3}x^2$, where x and y are both measured in metres. Where should the girl stand to throw the marble up the greatest number of steps, and how many steps is this?

Using graphs to solve cubic and other equations

Example 6

Draw the graph of $y = x^3$.

Use this graph to solve these two equations:

a $x^3 + 2x - 4 = 0$ **b** $x^3 - 3x + 1 = 0$

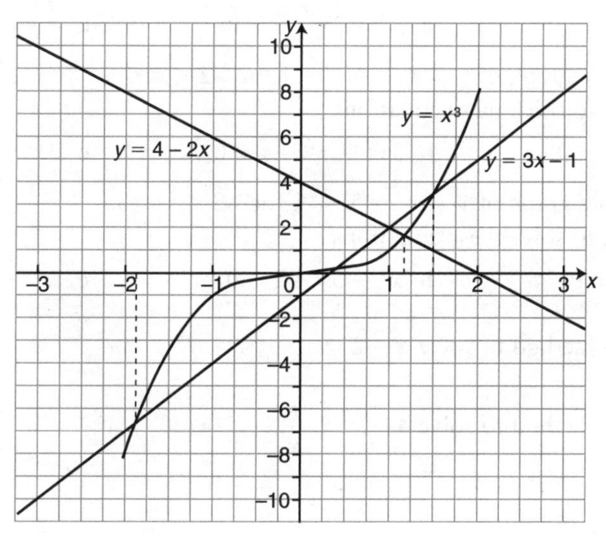

To solve **a**: $x^3 + 2x - 4 = 0$

$$x^3 + 2x - 4 = 0 \quad \text{(Rearrange)}$$

$$x^3 = 4 - 2x$$

This can be solved by finding the x-co-ordinates of the intersection points of the graphs $y = x^3$ and $y = 4 - 2x$.

From the graph the solution is approximately $x = 1.2$.
The graph shows there is only one solution.

To solve **b**: $x^3 - 3x + 1 = 0$

$$x^3 - 3x + 1 = 0 \quad \text{(Rearrange)}$$

$$x^3 = 3x - 1$$

This can be solved by finding the x-co-ordinates of the intersection points of the graphs $y = x^3$ and $y = 3x - 1$.

From the graph, the solutions are approximately $x = -1.9$, $x = 0.4$ or $x = 1.5$.

Exercise 32

1 Draw an accurate graph of $y = x^3$ for $-3 \leqslant x \leqslant 3$.
Use your graph to solve these equations.

a $x^3 - 3x = 0$ **b** $x^3 - 3x - 1 = 0$ **c** $x^3 - 2x + 1 = 0$

2 Draw an accurate graph of $y = x^4$ for $-2 \leqslant x \leqslant 2$.
Use your graph to solve these equations.

a $x^4 - 4x = 0$ **b** $x^4 - 2x - 3 = 0$ **c** $x^4 + x - 3 = 0$

3 Draw an accurate graph of $y = 3x^2 - x^3 - 1$ for $-2 \leqslant x \leqslant 3$.
Use your graph to solve these equations.

a $3x^2 - x^3 - 1 = 0$ **b** $3x^2 - x^3 - 4 = 0$ **c** $3x^2 - x^3 - 4 + x = 0$

4 Draw an accurate graph of $y = x^3 - 5x + 1$ for $-3 \leqslant x \leqslant 3$.
Use your graph to solve these equations.

a $x^3 - 5x + 1 = 0$ **b** $x^3 - 5x - 2 = 0$ **c** $x^3 - 7x - 1 = 0$

5 Use a graphical method to solve $x - 2 = \frac{3}{x}$.

6 Use a graphical method to solve $4 - x^2 = \frac{1}{x}$.

Exercise 32★

1 Draw an accurate graph of $y = \frac{12}{x^2}$ for $-4 \leqslant x \leqslant 4$. Use your graph to solve these equations.

a $\frac{12}{x^2} - x - 2 = 0$ **b** $\frac{12}{x^2} = 12 - x^2$ **c** $3x^3 + 10x^2 - 12 = 0$

2 Draw an accurate graph of $y = x^4 - 4x^2 + 2$ for $-3 \leqslant x \leqslant 3$. Use your graph to solve these equations.

a $x^4 - 4x^2 + 2 = 0$ **b** $x^4 - 4x^2 - 2x + 3 = 0$ **c** $2x^4 - 8x^2 + x + 2 = 0$

3 If the graph of $y = x^2 + \frac{16}{x}$ has been drawn, what graph must be drawn to solve $x^3 - 3x^2 - 8x + 16 = 0$?

4 If the graph of $y = x - 4 + \frac{3}{x}$ has been drawn, what graph must be drawn to solve $x^3 - x^2 + 5x - 3 = 0$?

5 Use a graphical method to solve $x^3 = x^2 + 2x - 1$.

6 Use a graphical method to solve $\frac{1}{x^2} = x^3 + 3$.

Investigate

For what values of k does the equation $x^3 - 12x + k = 0$ have

a one solution? **b** two solutions? **c** three solutions?

UNIT 2 ◆ Graphs

Using graphs to solve nonlinear simultaneous equations

Activity 12

Mary is watering her garden with a hose. Her little brother, Peter, is annoying her so she tries to squirt him with water.

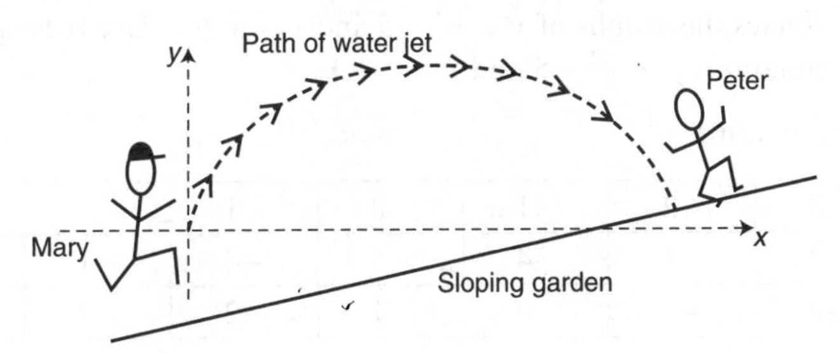

The path of the water jet is given by $y = 2x - \frac{1}{4}x^2$.

The slope of the garden is given by $y = \frac{1}{4}x - 1$.

Peter is standing at (8, 1).

The origin is the point where the water leaves the hose, and units are in metres.

- Copy and complete these tables.

x	0	2	4	6	8	10
$2x$			8			
$-\frac{1}{4}x^2$				−9		
$y = 2x - \frac{1}{4}x^2$		3				

x	0	2	4	6	8	10
$\frac{1}{4}x$					2	
$y = \frac{1}{4}x - 1$				0.5		

- On one set of axes, draw the two graphs representing the path of the water and the slope of the garden.
- Is Mary successful in making Peter wet?
- Mary alters the angle of the hose so that the path of the water is given by $y = x - 0.1x^2$. Draw in the new path. Is Peter made wet this time?

In Activity 12, the simultaneous equations $y = 2x - \frac{1}{4}x^2$ and $y = \frac{1}{4}x - 1$ were solved graphically by drawing both graphs on the same axes and finding the x-co-ordinates of the points of intersection.

Example 7

Draw on one set of axes the graphs of $y = x^2 - 5$ and $y = x + 1$. Use these graphs to solve the simultaneous equations $y = x^2 - 5$ and $y = x + 1$.

First make a table of values.

x	-3	-2	-1	0	1	2	3
$x^2 - 5$	4	-1	-4	-5	-4	-1	4
$x + 1$	-2	-1	0	1	2	3	4

Then draw the graphs.

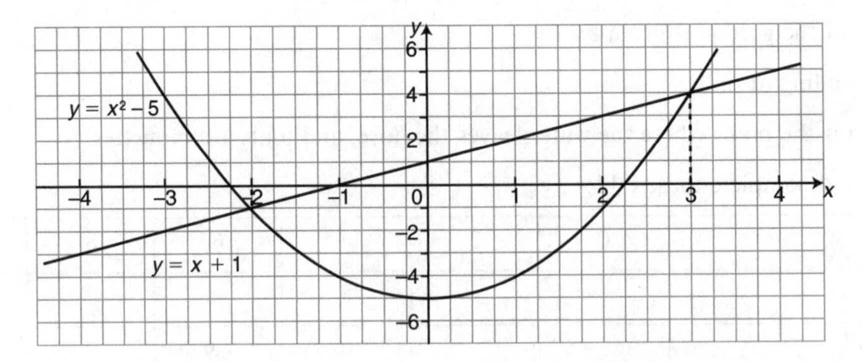

The co-ordinates of the intersection points are $(-2, -1)$ and $(3, 4)$, so the solutions are $x = -2$, $y = -1$ or $x = 3$, $y = 4$.

Key Point

To solve simultaneous equations graphically, draw *both* graphs on *one* set of axes.
The co-ordinates of the intersection points are the solutions of the simultaneous equations.

Exercise 33

For Questions 1–14, solve the simultaneous equations graphically, drawing graphs for $-4 \leqslant x \leqslant 4$.

1 $y = x^2 + 2$, $y = 5$

2 $y = x^2 - 1$, $y = 4$

3 $y = 4 - x^2$, $y = 1 + 2x$

4 $y = 1 - x^2$, $y = x - 1$

5 $y = x^2 + 2x - 1$, $y = 1 + 3x$

6 $y = x^2 + 4x - 3$, $y = x + 1$

7 $y = x^2 - 4x + 6$, $y = 2x - 2$

8 $y = x^2 - 2x + 4$, $y = 1 + 2x$

9 $x^2 + y^2 = 4$, $y = 1 - \dfrac{x}{4}$

10 $x^2 + y^2 = 9$, $y = 2x + 1$

11 $y = \dfrac{4}{x}$, $y = x - 1$

12 $y = 1 - \dfrac{6}{x}$, $y = 2 - x$

13 $y = x^3 + 2x^2$, $y = \frac{1}{2}x + 1$

14 $y = x - x^3$, $y = \frac{1}{2} - x$

15 During a match, Matthew kicks a football onto the roof of the stand. The path of the football is given by

$$y = 2.5x - \frac{x^2}{15}$$

The equation of the roof of the stand is given by

$$y = \frac{x}{2} + 10 \text{ for } 20 \leqslant x \leqslant 35$$

All units are in metres. Find by a graphical method where the football lands on the roof.

16 A volcano shaped like a cone ejects a rock from the summit. The path of the rock is given by

$$y = 2x - x^2$$

and the side of the volcano is given by

$$y = -\frac{2x}{3}$$

where the units are in kilometres and the origin is at the summit. Find by a graphical method where the rock lands.

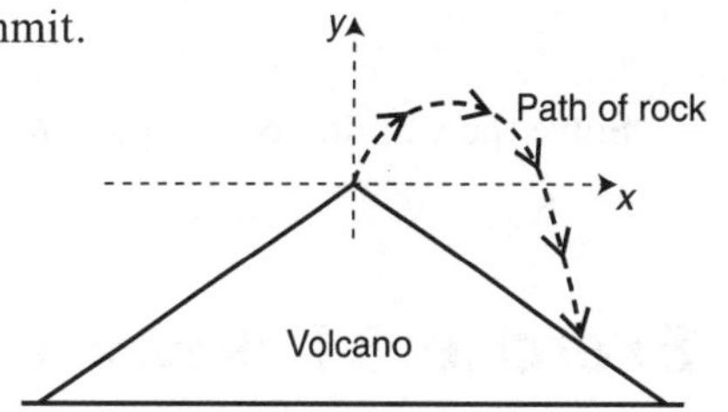

Exercise 33★

For Questions 1–14, solve the simultaneous equations graphically. (For Questions 1–8, draw graphs for $-4 \leqslant x \leqslant 4$.)

1 $y = x^2 + x + 1$, $y = 3$

2 $y = x^2 - 2x - 1$, $y = 2$

3 $y = x^2 - x - 5$, $y = 1 - 2x$

4 $y = x^2 + 4x - 5$, $y = 2x + 3$

5 $y = 2x^2 - 2x - 4$, $y = 6 - x$

6 $y = 2x^2 - 5x - 6$, $y = 3 - 2x$

7 $y = 10x^2 + 3x - 4$, $y = 2x - 2$

8 $y = 8x^2 + 3x - 4$, $y = 5 - 3x$

9 $(x + 1)^2 + (y - 2)^2 = 9$, $y = 2x + 3$

10 $(x - 3)^2 + (y + 4)^2 = 4$, $y = 1 - 2x$

11 $y = x^3 - 4x^2 + 5$, $y = 3 - 2x$, $-1 \leqslant x \leqslant 4$

12 $y = x^3 + 3x^2 - 7$, $y = 2x - 1$, $-4 \leqslant x \leqslant 2$

13 $y = \dfrac{10}{x} + 4$, $y = 5x + 2$, $-2 \leqslant x \leqslant 3$

14 $y = \dfrac{12}{x} - x$, $y = 2x - 3$, $-3 \leqslant x \leqslant 4$

15 In a ski jump, the path of the jumper is given by

$$y = 0.6x - \frac{x^2}{40}$$

while the landing slope is given by

$$y = -\frac{x}{2}$$

The origin is at the point of take-off of the jumper, and all units are in metres.

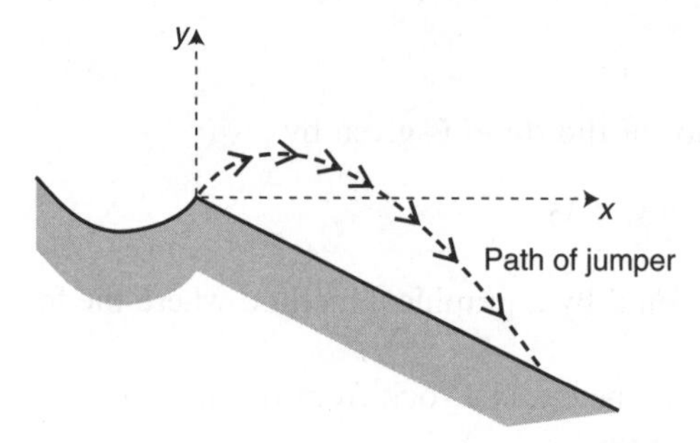

Solve the simultaneous equations by a graphical method and find how long the jump is.

Exercise 34 (Revision)

1 A distress rocket is fired out to sea from the top of a 50 m high cliff.
Taking the origin at the top of the cliff, the path of the rocket is given by $y = x - 0.01x^2$.
Use a graphical method to find where the rocket lands in the sea.

2 Draw the graph of $y = x^2 - 2x - 1$ for $-2 \leqslant x \leqslant 4$.
Use your graph to solve these equations.

a $x^2 - 2x - 1 = 0$ **b** $x^2 - 2x - 4 = 0$ **c** $x^2 - x - 3 = 0$

3 The graph of $y = x^2$ has been drawn. What lines should be drawn to solve the following equations?

a $x^2 = 4$
b $x^2 = x + 1$
c $x^2 - 2x + 1 = 0$

4 Solve the simultaneous equations $y = x^2$ and $y = x + 3$ graphically.

5 The area of a rectangle is $30\,\text{cm}^2$; the perimeter is 24 cm. If x is the length of the rectangle and y is the width, form two equations for x and y and solve them graphically.

Exercise 34★ (Revision)

1 Lauren is marketing a handbag. The shape of the handbag is a circle 24 cm in diameter and the handle is a parabola with equation

$$y = 22 - \frac{x^2}{4}$$

where the origin is at the centre of the circle. Lauren needs to tell a subcontractor where to attach the fasteners for the handle to the bag. Find by a graphical method the position of the fasteners.

2 Draw the graph of $y = 3x + 5 - 2x^2$ for $-2 \leqslant x \leqslant 4$.
Use your graph to solve these equations.

a $3x + 2 - 2x^2 = 0$ **b** $x + 7 - 2x^2 = 0$ **c** $2x + 2 - x^2 = 0$

3 The graph of $y = x^2 + 2x - 4$ has been drawn. What lines should be drawn to solve the following equations?

a $x^2 + 2x = 5$
b $x^2 + x - 4 = 0$
c $x^2 - x - 3 = 0$

4 Solve the simultaneous equations $y = x^3$ and $y = 4 - 4x^2$ graphically.

5 The hypotenuse of a right-angled triangle is 4 cm long; the sum of the lengths of the other two sides is 5 cm. If the lengths of the other two sides are x and y, form two equations for x and y and solve them graphically.

Shape and space 2

Converting measurements

Converting lengths

Remember

$$10\,\text{mm} = 1\,\text{cm} \qquad 1000\,\text{mm} = 1\,\text{m}$$
$$100\,\text{cm} = 1\,\text{m} \qquad 1000\,\text{m} = 1\,\text{km}$$

Example 1

Change 3 km to cm.

$$\begin{aligned} 3\,\text{km} &= 3 \times 1000\,\text{m} && (\text{as } 1\,\text{km} = 1000\,\text{m}) \\ &= 3 \times 1000 \times 100\,\text{cm} && (\text{as } 1\,\text{m} = 100\,\text{cm}) \\ &= 3 \times 10^5\,\text{cm} \end{aligned}$$

Example 2

Change 5×10^6 mm to km.

$$\begin{aligned} 5 \times 10^6 &= \frac{5 \times 10^6}{1000}\,\text{m} && (\text{as } 1000\,\text{mm} = 1\,\text{m}) \\ &= \frac{5 \times 10^6}{1000 \times 1000}\,\text{km} && (\text{as } 1000\,\text{m} = 1\,\text{km}) \\ &= 5\,\text{km} \end{aligned}$$

Exercise 35

Fill in all the gaps in the following table.

Question	km	m	cm	mm
1	5			
2	8			
3	3			
4	4			
5		2000		
6		7000		
7			5000	
8			6000	
9				10^6
10				10^7
11		1.5×10^4		
12		3.3×10^3		

Exercise 35★

Fill in all the gaps in the following table.

Question	km	m	cm	mm
1	500			
2	700			
3	2.5×10^4			
4	4.3×10^3			
5		5×10^6		
6		7×10^5		
7			50	
8			60	
9				9×10^9
10				8×10^{11}
11				4
12				7

13 A nanometre is 10^{-9} metres.

a How many nanometres are there in 200 km?

b How many km are there in 10^{18} nanometres?

14 A micrometre is 10^{-6} metres.

a How many micrometres are there in 5000 km?

b How many km are there in 10^{15} micrometres?

Converting areas

A diagram is useful, as shown in the following examples.

Example 3

A rectangle measures 1 m by 2 m. Find the area in mm^2.

1 m is 1000 m.
2 m is 2000 mm.
So the diagram is as shown on the right.

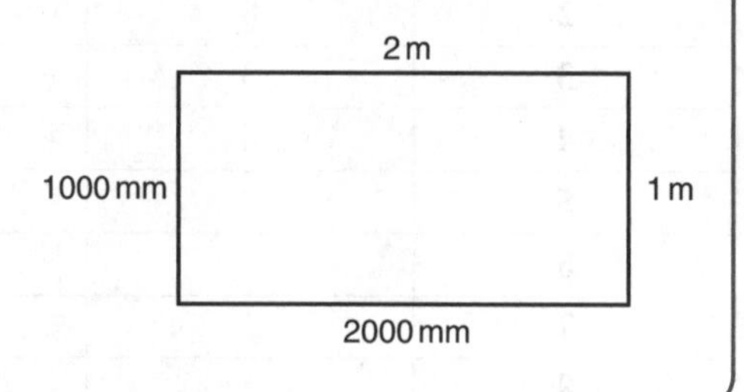

So the area is $1000 \times 2000\ mm^2$

$= 2\,000\,000\ mm^2$

$= 2 \times 10^6\ mm^2$

Example 4

Change $30\,000\ cm^2$ to m^2.

$1\ m^2 = 1\ m \times 1\ m$

$= 100\ cm \times 100\ cm$

$= 10\,000\ cm^2$

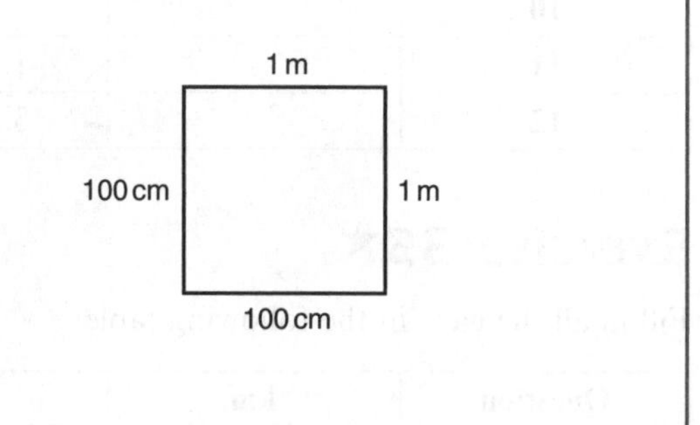

So $30\,000\ cm^2 = \frac{30\,000}{10\,000}\ m^2$

$= 3\ m^2$

Exercise 36

Fill in all the gaps in the following table.

Question	km^2	m^2	cm^2	mm^2
1	2			
2	4			
3	1			
4	3			
5		50		
6		80		
7			6×10^6	
8			3×10^7	
9				10^{13}
10				10^{14}
11		4×10^4		
12		5×10^5		

Exercise 36★

Fill in all the gaps in the following table.

Question	km^2	m^2	cm^2	mm^2
1	80			
2	90			
3		6000		
4		4000		
5			6×10^{10}	
6			3×10^{12}	
7				2×10^{21}
8				5×10^{19}
9	7×10^{-2}			
10	9×10^{-4}			
11			4	
12				5

Converting volumes

Remember

$$1 \text{ litre} = 1000\,\text{cm}^3$$

Again diagrams are very helpful.

Example 5

A cuboid measures 1 m by 2 m by 3 m. Find the volume in mm^3.

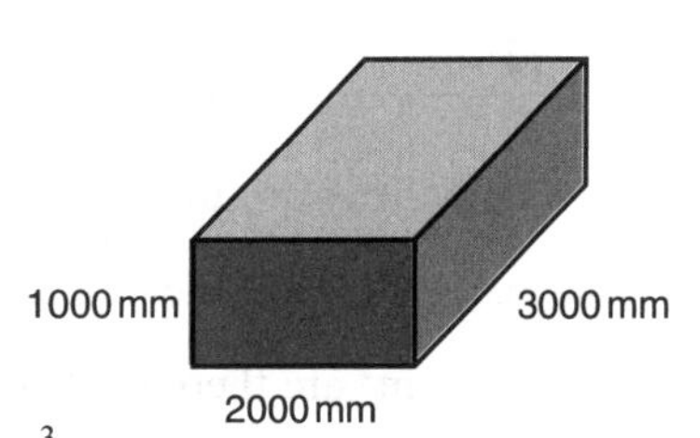

1 m is 1000 mm.
2 m is 2000 mm.
3 m is 3000 mm.

So the volume is $1000 \times 2000 \times 3000\,\text{mm}^3 = 6 \times 10^9\,\text{mm}^3$.

Example 6

Change 10^7 cm^3 to m^3.

$$1\ \text{m}^3 = 1\ \text{m} \times 1\ \text{m} \times 1\text{m}$$
$$= 100\ \text{cm} \times 100\ \text{cm} \times 100\ \text{cm}$$
$$= 10^6\ \text{cm}^3$$

100 cm

100 cm

100 cm

$$\text{So } 10^7\ \text{cm}^3 = \frac{10^7}{10^6}\ \text{m}^3$$
$$= 10\ \text{m}^3$$

Exercise 37

Fill in all the gaps in the following table.

Question	km^3	m^3	cm^3	mm^3
1	1			
2	2			
3	4			
4	3			
5		8		
6		6		
7			4×10^3	
8			5×10^5	
9				10^{15}
10				10^{14}
11		7×10^2		
12		3×10^3		

13 How many litres are there in 1 m^3?

14 How many litres are there in 1 km^3?

15 How many m^3 are there in 10 000 litres?

16 How many mm^3 are there in 10 litres?

Exercise 37★

Fill in all the gaps in the following table.

Question	km^3	m^3	cm^3	mm^3
1	70			
2	80			
3		600		
4		500		
5			3×10^8	
6			6×10^{10}	
7				5×10^{25}
8				2×10^{23}
9	4×10^{-6}			
10	8×10^{-8}			
11				3
12				8

13 How many litres are there in $512\,m^3$?

14 How many litres are there in $12\,km^3$?

15 How many m^3 are there in 10^6 litres?

16 How many mm^3 are there in 100 litres?

17 A picometre is 10^{-12} m. How many cubic picometres are there in $1\,km^3$?

18 Light travels at about 300 000 km/s. A light year is the distance light travels in one year. How many mm^3 are there in 1 cubic light year?

Circles, semicircles and quadrants

Remember

The perimeter of a shape is the distance all the way round the shape.

Circle

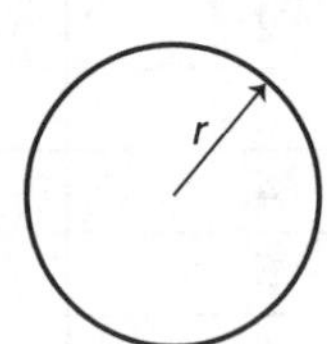

The perimeter of a circle is called the circumference.
If C is the circumference, A the area and r the radius, then

$$C = 2\pi r$$
$$A = \pi r^2$$

Semicircle

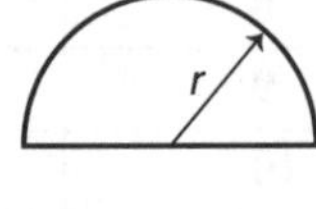

A semicircle is half a circle cut along a diameter.
The perimeter is half the circumference of the circle plus the diameter, so $P = \pi r + 2r$

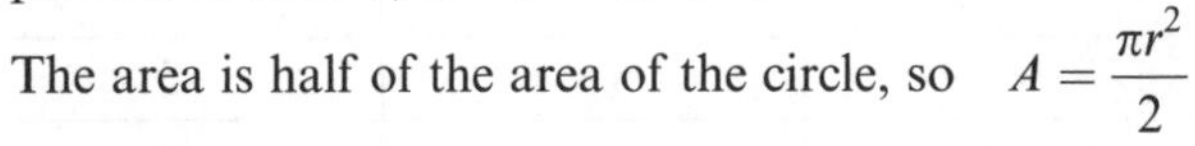
The area is half of the area of the circle, so $A = \dfrac{\pi r^2}{2}$

Quadrant

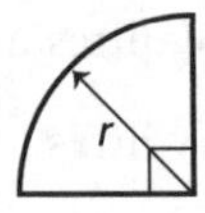

A quadrant is quarter of a circle.
The perimeter is a quarter of the circumference of the circle plus twice the radius, so $P = \dfrac{\pi r}{2} + 2r$

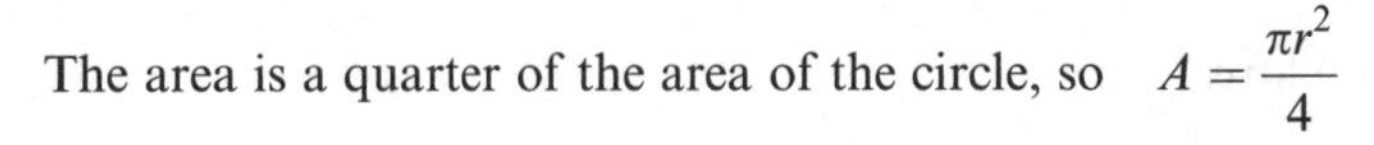
The area is a quarter of the area of the circle, so $A = \dfrac{\pi r^2}{4}$

Example 7

The circumference of a circle is 10 cm. Find the radius.

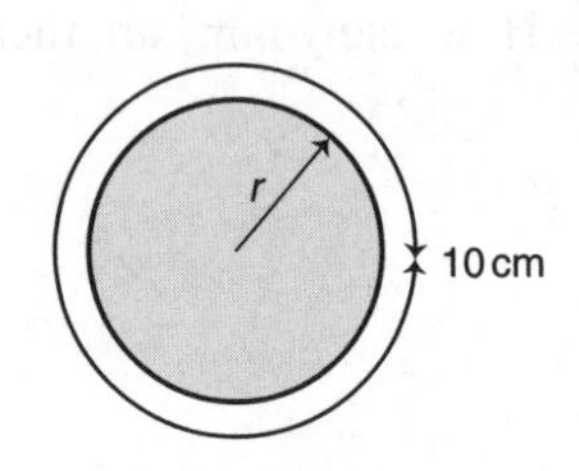

Using $C = 2\pi r$

$10 = 2\pi r$ (Make r the subject of the equation)

$r = \dfrac{10}{2\pi}$

$= 1.59$ cm to 3 s.f.

Example 8

The area of a circle is $24\,\text{cm}^2$. Find the radius.

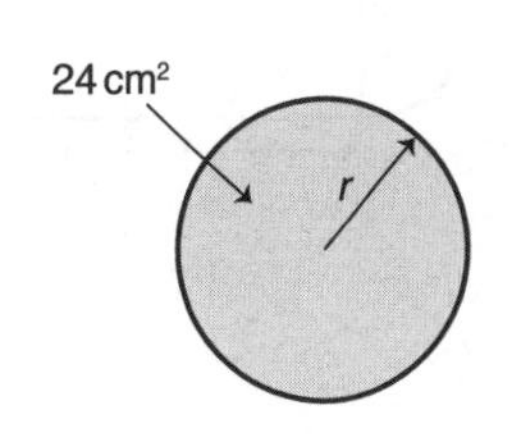

Using $A = \pi r^2$

$$24 = \pi r^2 \qquad \text{(Make } r \text{ the subject of the equation)}$$

$$r^2 = \frac{24}{\pi}$$

$$r = \sqrt{\frac{24}{\pi}}$$

$$= 2.76\,\text{cm to 3 s.f.}$$

Example 9

Find the perimeter and area of the shape shown.

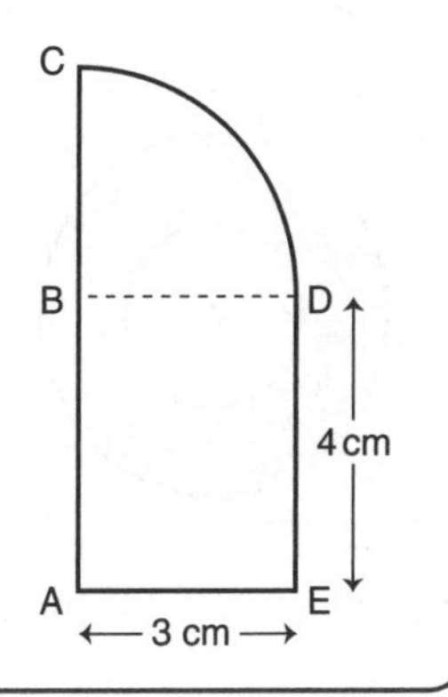

The radius of the quadrant BCD is 3 cm, so BC = 3 cm.

The perimeter = AB + BC + arc CD + DE + EA

$$= 4 + 3 + \tfrac{3\pi}{2} + 4 + 3$$

$$= 18.7\,\text{cm (to 3 s.f.)}$$

The area = area of quadrant BCD + area of rectangle ABDE

$$= \tfrac{9\pi}{4} + 12$$

$$= 19.1\,\text{cm}^2 \text{ (to 3 s.f.)}$$

Exercise 38

Find the perimeter and area of each of the following shapes, giving answers to 3 s.f. All dimensions are in cm. All arcs are parts of circles.

1

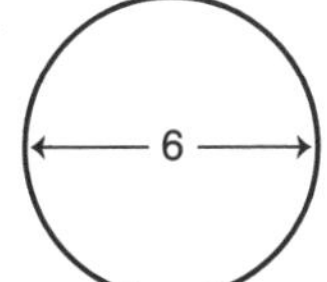

2

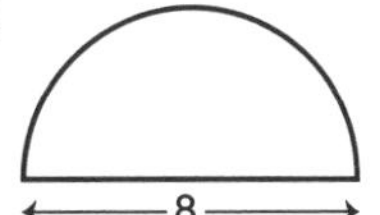

3

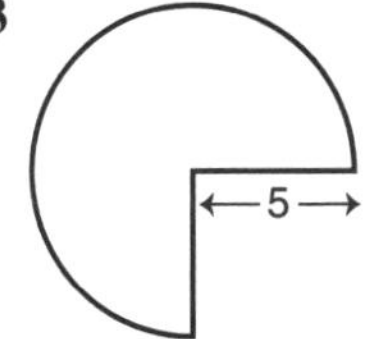

4

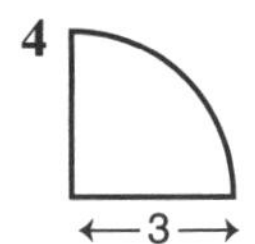

5

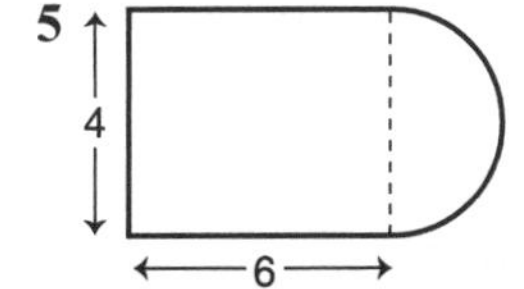

6

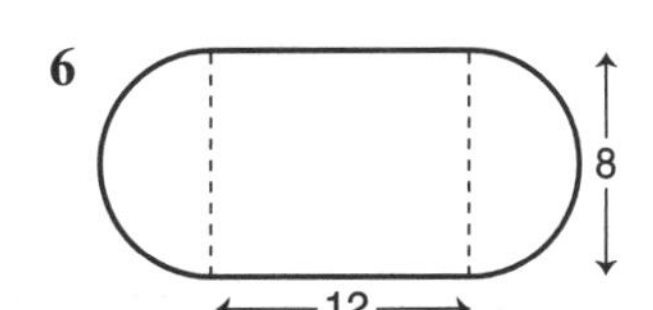

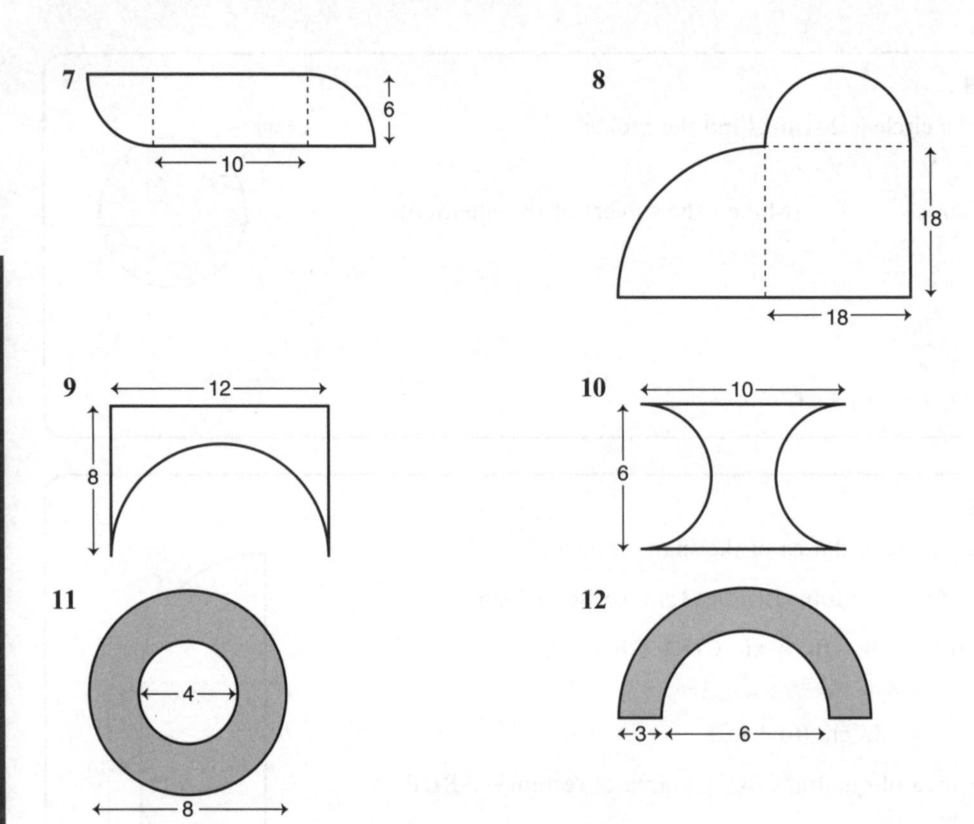

For Questions 13–20, fill in all the gaps in the following table:

Question	Radius in cm	Circumference in cm	Area in cm^2
13		6	
14		8	
15			14
16			9
17		52	
18		76	
19			84
20			68

21 A bicycle wheel has a diameter of 66 cm. How many km does the bicycle travel if the wheel rotates 1000 times?

22 A car wheel has a diameter of 48 cm. On a journey the wheel rotates 5000 times. How long is the journey in km?

23 The minute hand of a clock is 80 mm long. How many metres does the tip of the hand travel in 12 hours?

24 A CD is 120 mm in diameter. A speck of dust is on the edge of the CD. How many kilometres does the speck of dust travel when the CD rotates 10 000 times?

Exercise 38★

Find the perimeter and area of each of the following shapes, giving answers to 3 s.f.
All dimensions are in cm.

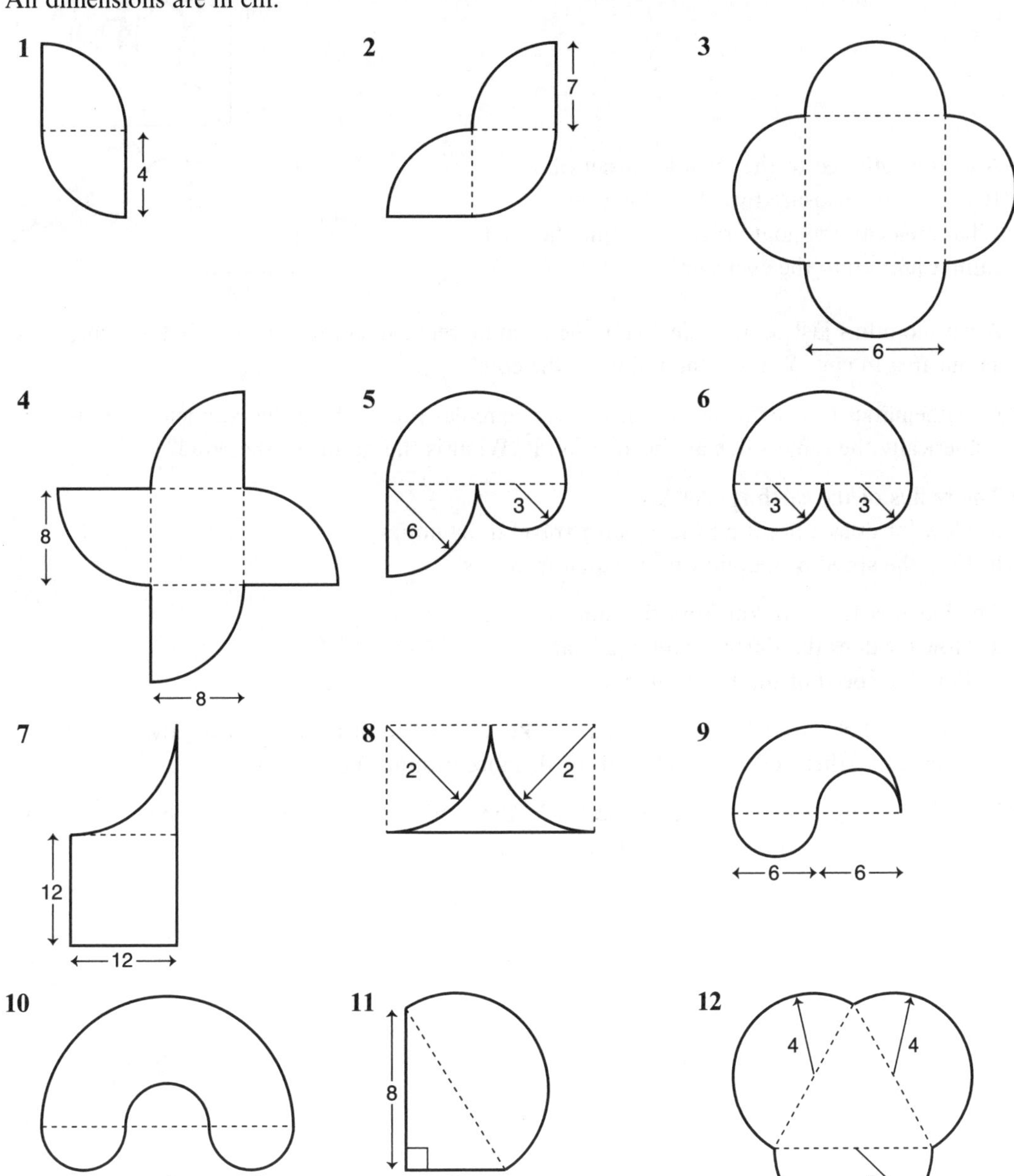

13 The area of a quadrant of a circle is $8\,\text{cm}^2$. Find the radius and perimeter.

14 The area of a semicircle is $22\,\text{cm}^2$. Find the radius and perimeter.

15 A cow is tethered by a rope to one corner of a 20 m square field. The cow can graze half the area of the field. How long is the rope?

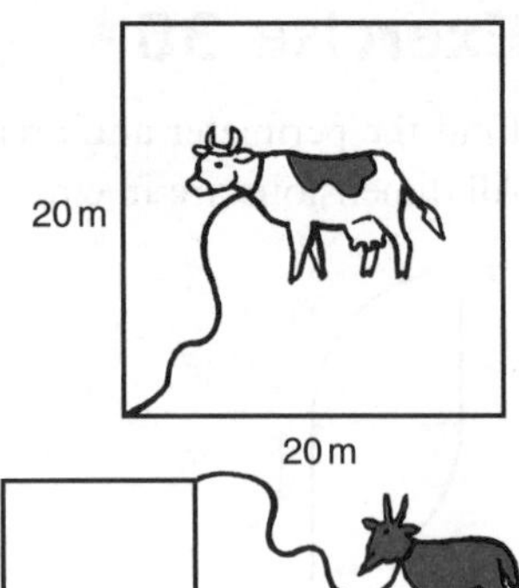

16 A goat is tethered to the outside corner of a 10 m square enclosure by a 15 m long rope. What area can the goat graze? (Assume the goat cannot jump into the enclosure.)

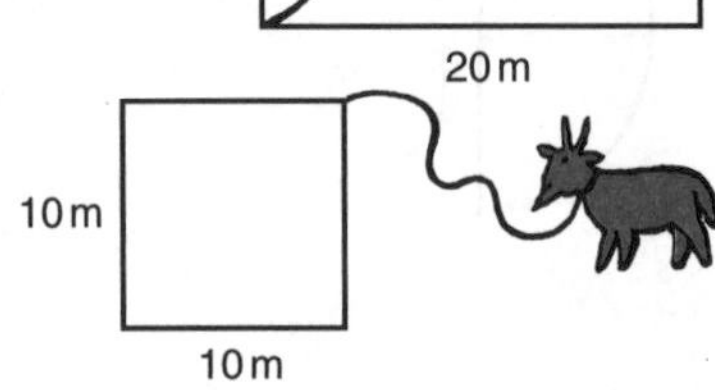

17 A new coin has just been made where the circumference in cm is numerically the same value as the area in cm^2. What is the radius of the coin?

18 A mathematics teacher wishes to make a semicircular pond where the perimeter in m is numerically the same value as the area in m^2. What is the radius of the pond?

19 The radius of the earth is 6380 km.

a How far does a point on the equator travel in 24 hours?

b Find the speed of a point on the equator in m/s.

20 The Earth is 1.5×10^8 km from the Sun.

a How far does the Earth travel in a year?

b Find the speed of the Earth in m/s.

21 A hot-air balloon travels round the Earth 1 km above the surface, following the equator. How much further does it travel than the distance around the equator?

22 Mala is running around a circular racetrack, 2 m further out than her friend Noz. Every time they run round the track, how much further does Mala run than Noz?

23 The perimeter of this shape is 12 cm. Find r and the area.

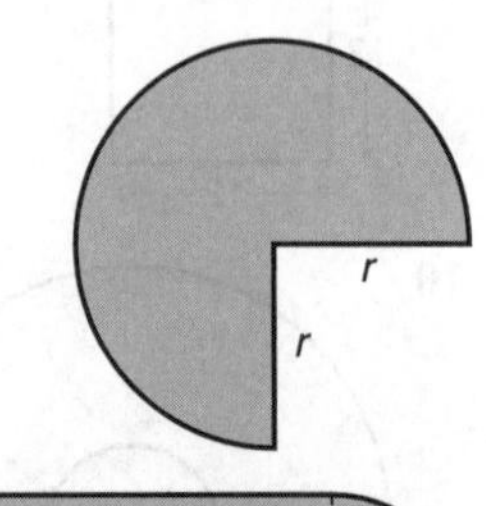

24 The shape shown consists of a square and a semicircle. The perimeter is 22 cm. Find the radius of the semicircle and the area of the shape.

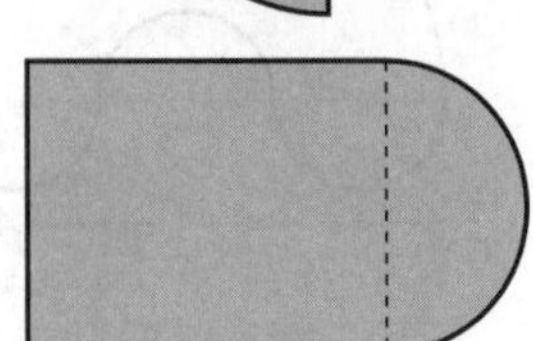

Arc of a circle

An **arc** is part of the circumference of a circle.

The arc shown is the fraction $\frac{x}{360}$ of the whole circumference.

So the arc length is

$$\frac{x}{360} \times 2\pi r$$

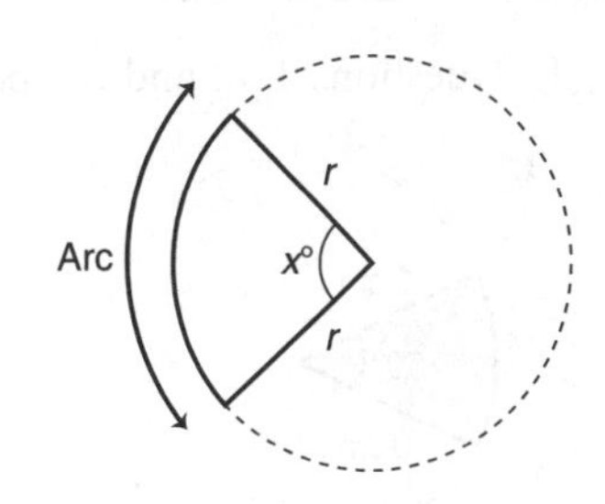

Remember

$$\text{Arc length} = \frac{x}{360} \times 2\pi r$$

Example 10

Find the perimeter of the shape shown.

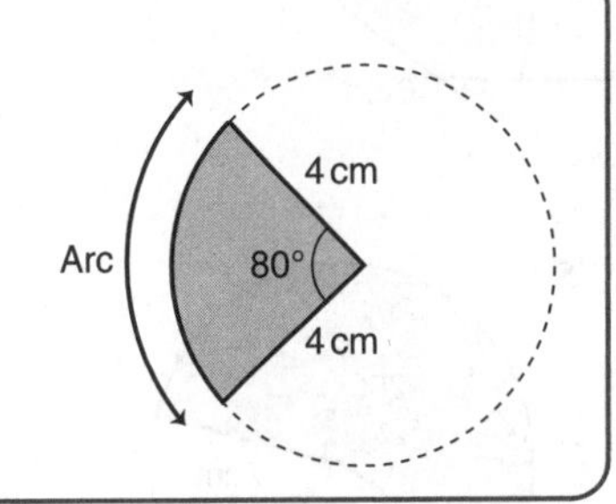

Using $\text{Arc length} = \frac{x}{360} \times 2\pi r$

$\text{Arc length} = \frac{80}{360} \times 2\pi \times 4 = 5.585\text{ cm}$

The perimeter $= 5.585 + 4 + 4 = 13.6\text{ cm}$ to 3 s.f.

Example 11

Find the angle marked x.

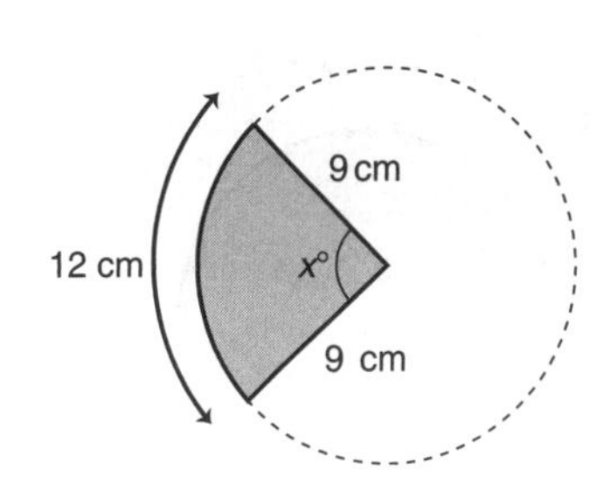

Using $\text{Arc length} = \frac{x}{360} \times 2\pi r$

$12 = \frac{x}{360} \times 2\pi \times 9$ (Make x the subject of the equation)

$x = \frac{12 \times 360}{2\pi \times 9}$

$= 76.4°$ to 3 s.f.

Example 12

Find the radius r.

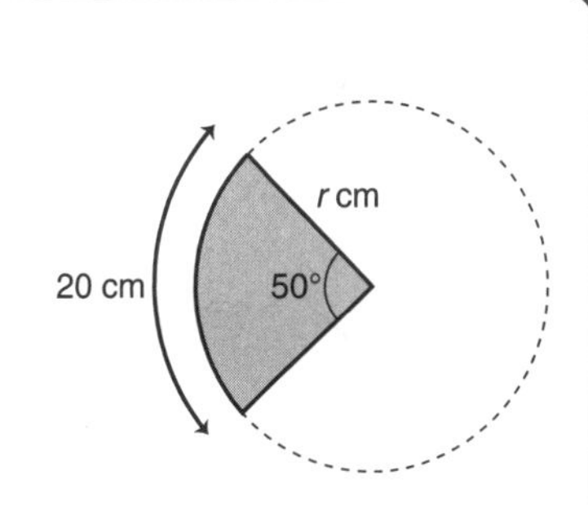

Using $\text{Arc length} = \frac{x}{360} \times 2\pi r$

$20 = \frac{50}{360} \times 2\pi r$ (Make r the subject of the equation)

$r = \frac{20 \times 360}{50 \times 2\pi}$

$= 22.9\text{ cm}$ to 3 s.f.

Exercise 39

In Questions 1–8, find the perimeter of the shape. Give answers to 3 s.f.

1

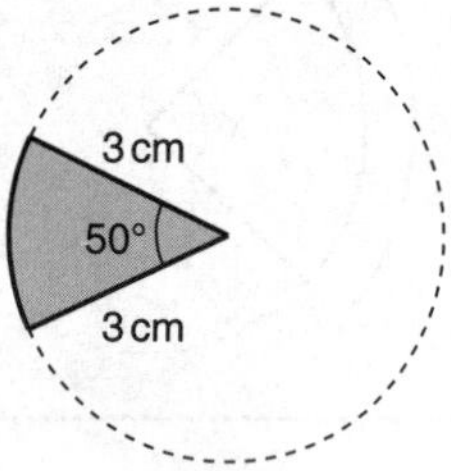

2

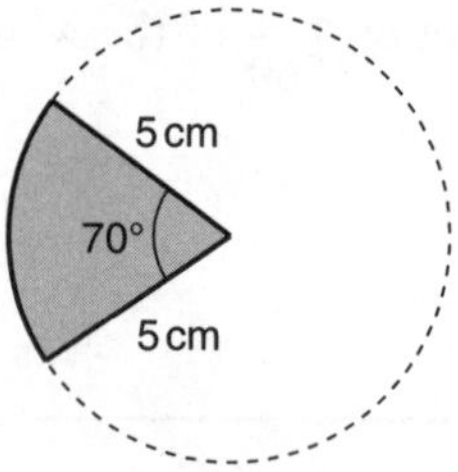

3

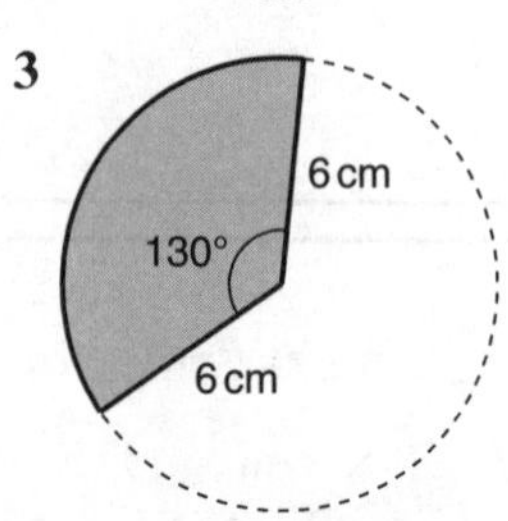

4

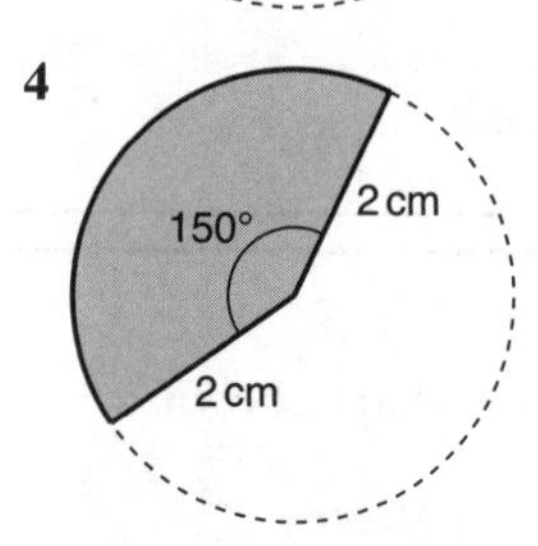

5

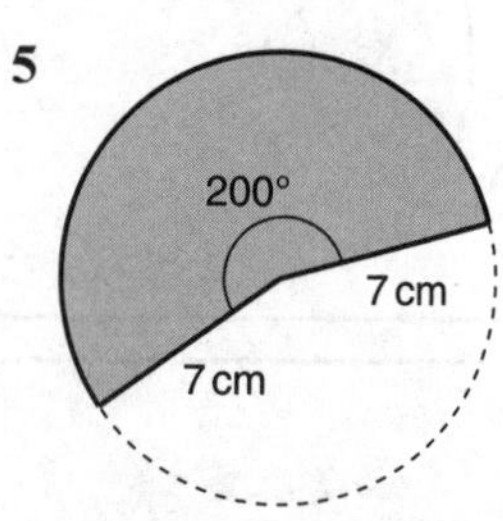

6

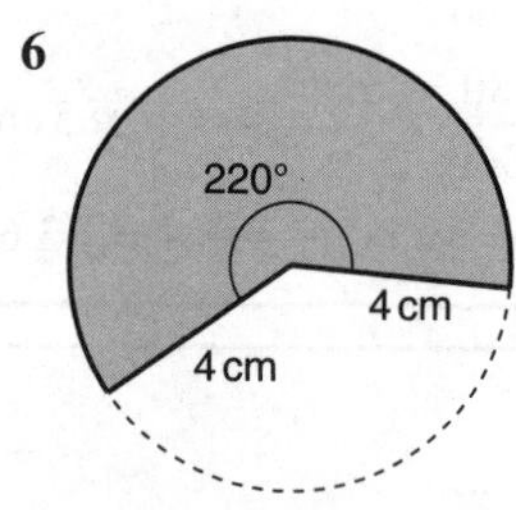

7

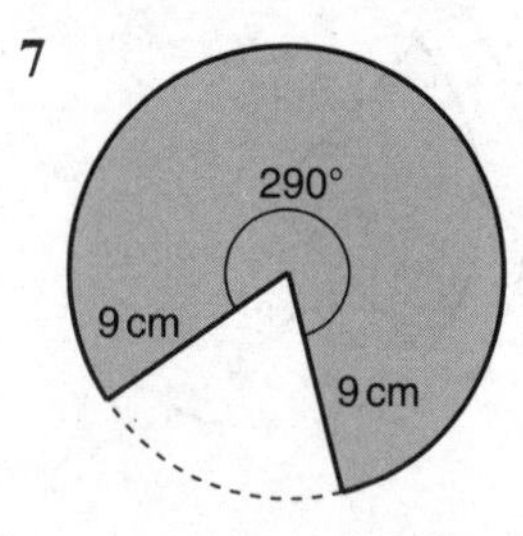

8

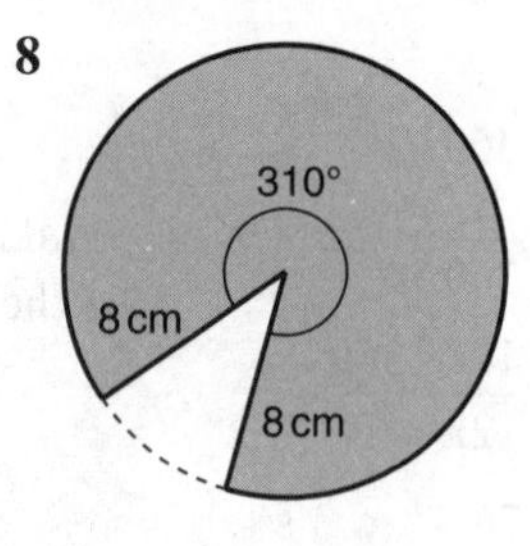

In Questions 9–12, find the angle marked x.

9

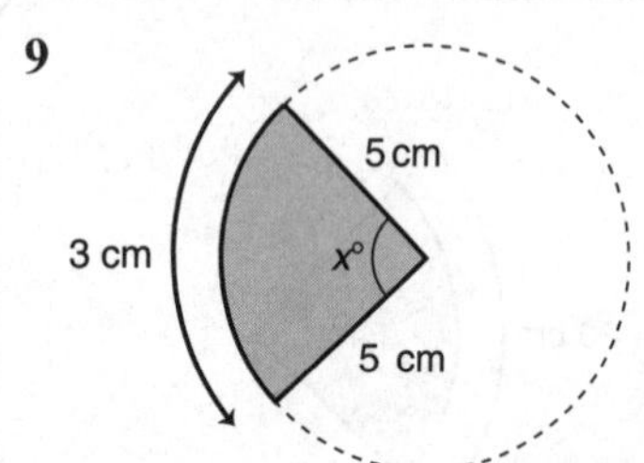

10

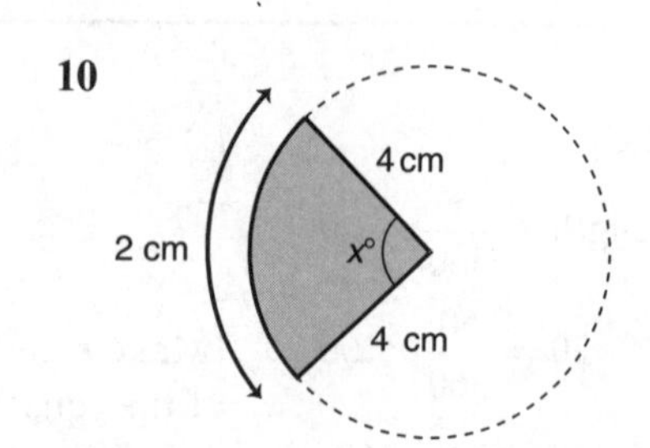

11

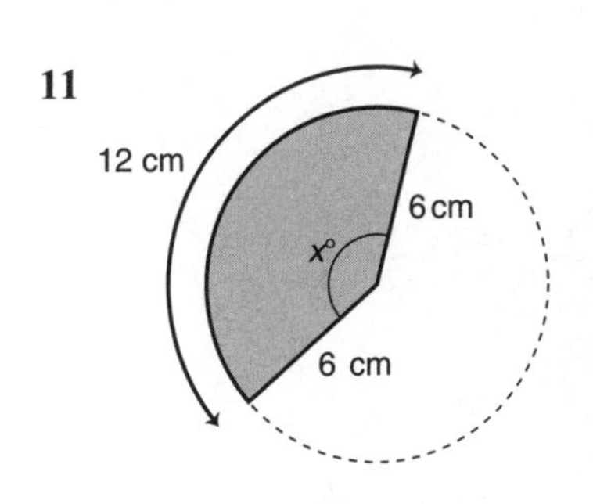

12

18 cm

8 cm

x°

8 cm

In Questions 13–16, find the radius r.

13

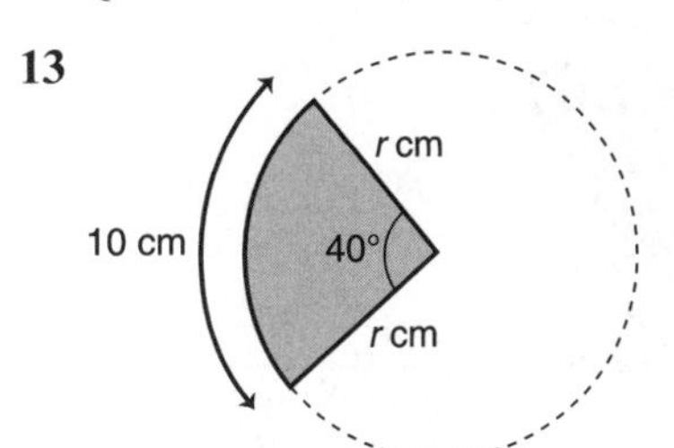

14

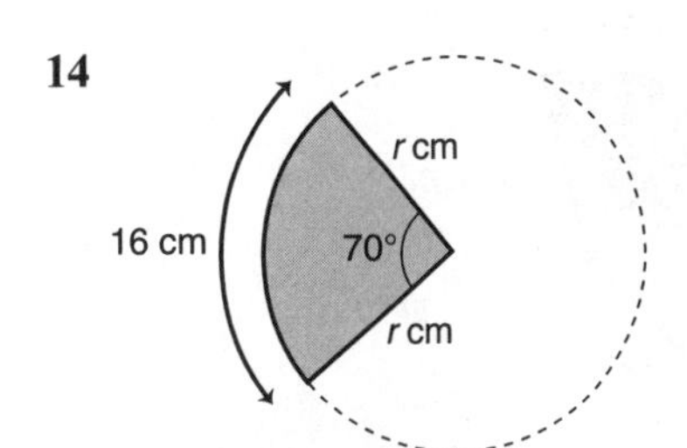

15

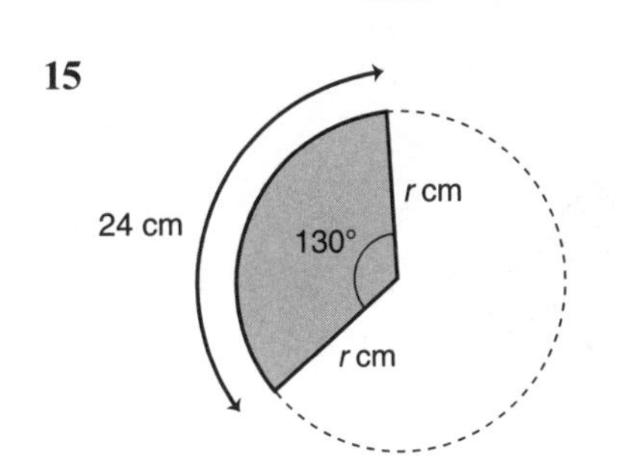

16

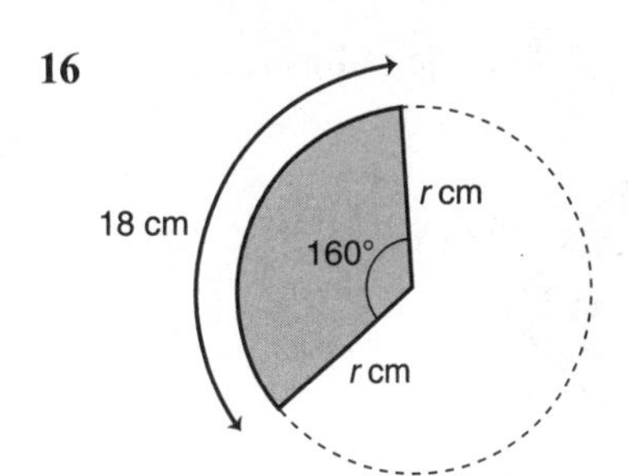

Exercise 39★

In Questions 1–4, find the perimeter of the shape. Give answers to 3 s.f.

1

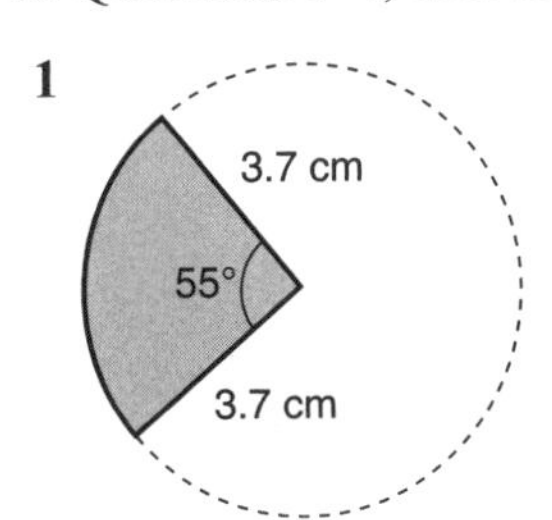

2

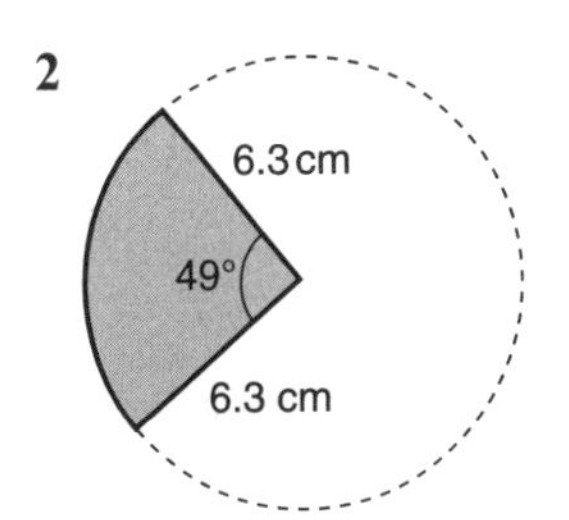

3

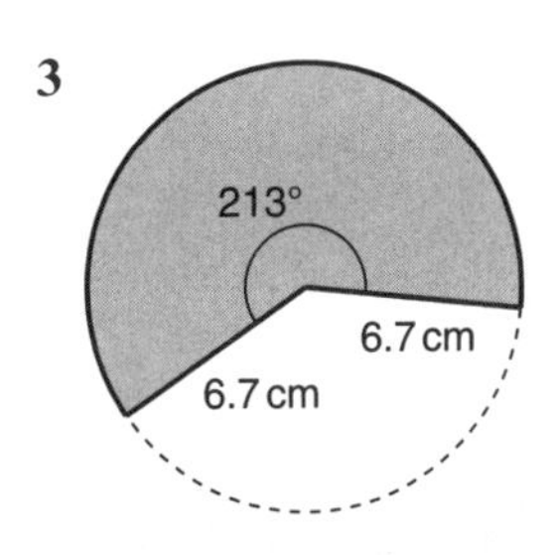

4

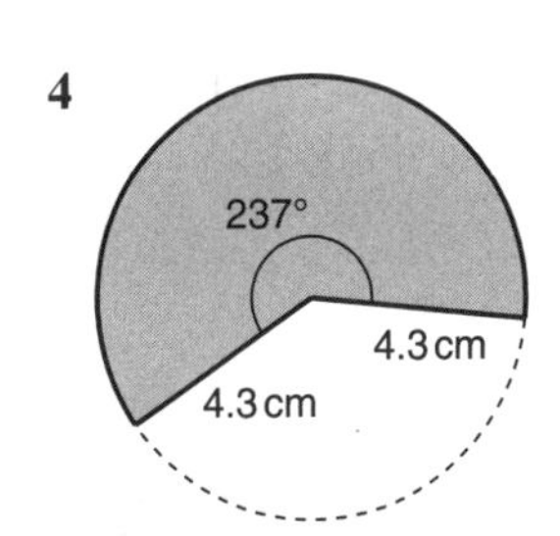

In Questions 5–8, find the angle marked x.

5

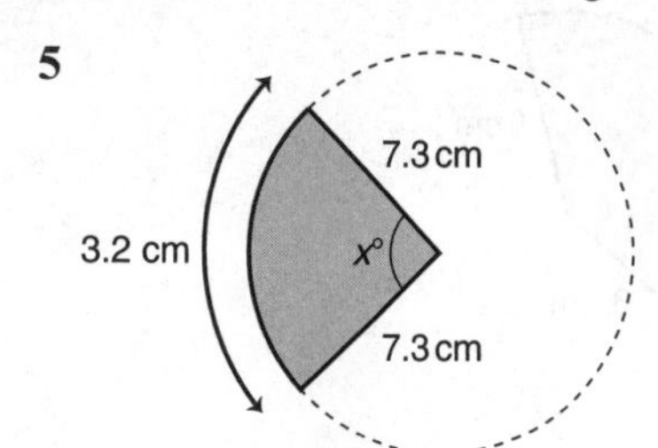

6

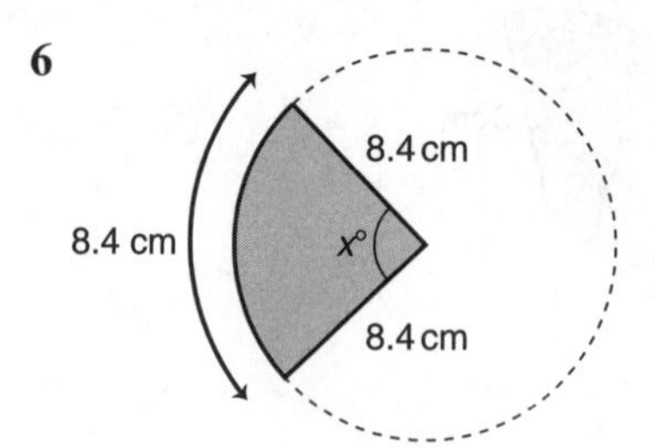

7

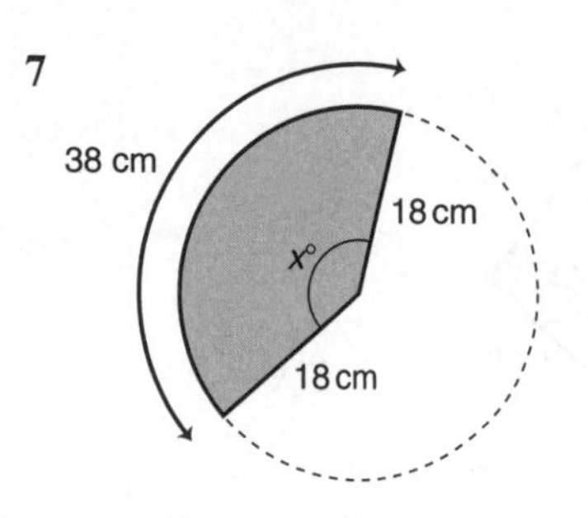

8

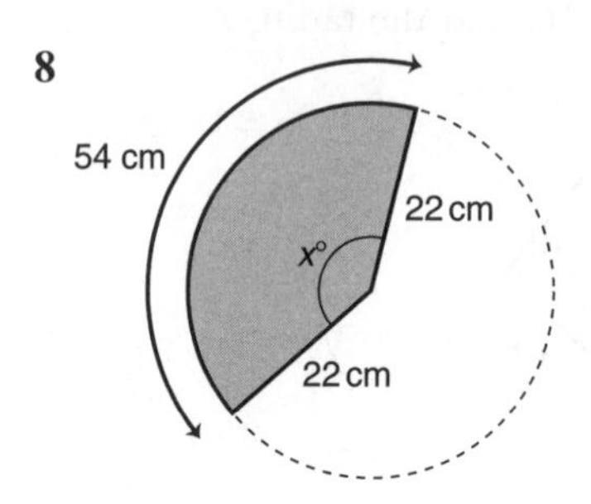

In Questions 9–12, find the radius r.

9

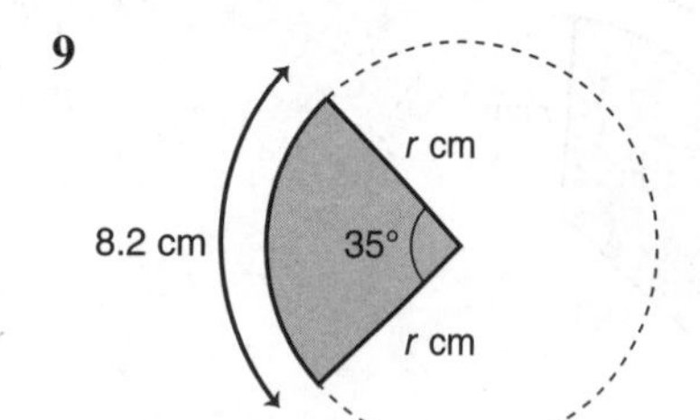

10

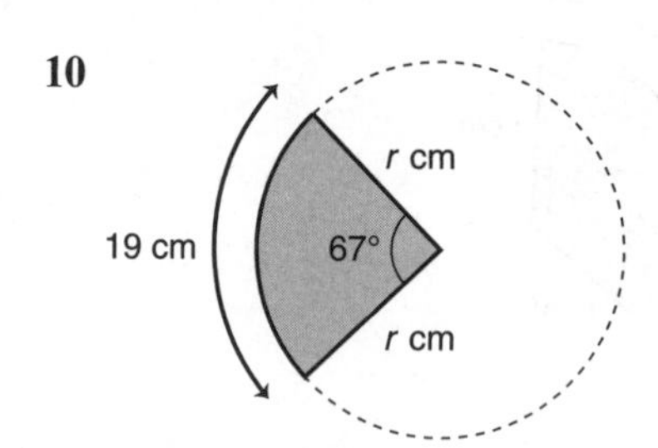

11

235 cm

r cm

115°

r cm

12

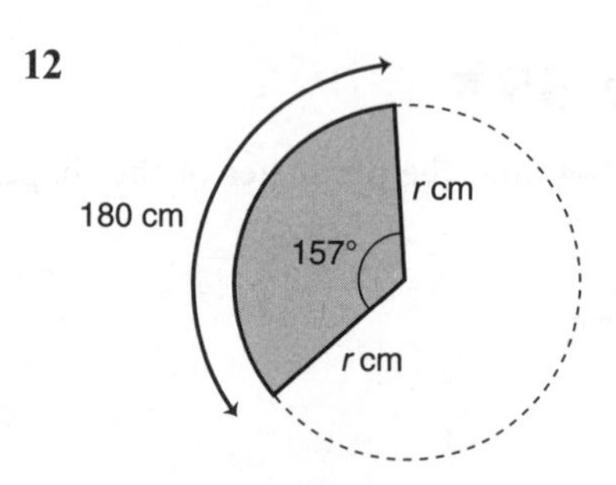

13 The minute hand of a watch is 9 mm long. How far does the tip travel in 35 minutes?

14 A pendulum of length 85 cm swings through an angle of 16°. How far does the pendulum bob travel?

15 Find the perimeter of the shape to 3 s.f.

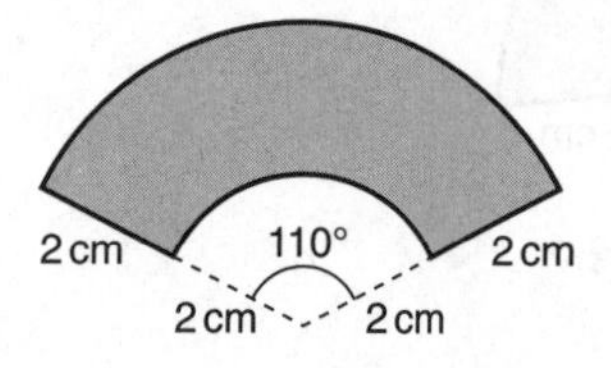

16 Find the perimeter of the shape to 3 s.f.

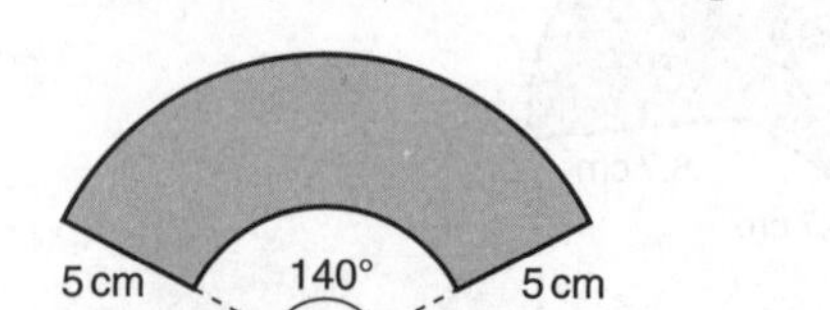

17 The perimeter of the shape is 28 cm.
Find the value of r.

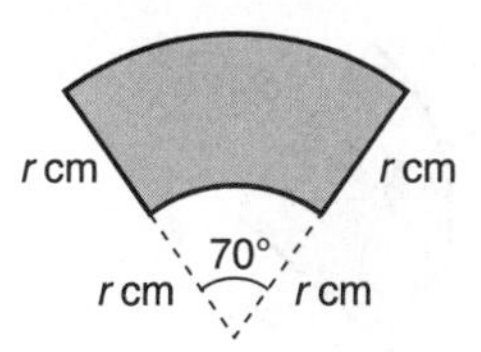

18 The perimeter of the shape is 20 cm.
Find r.

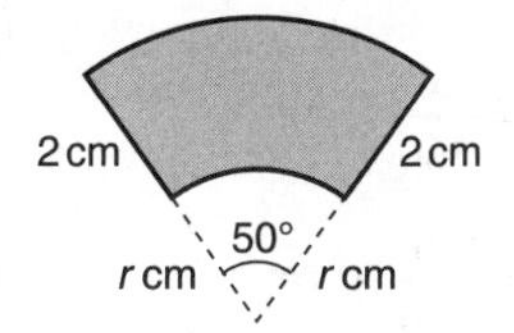

Sectors of circles

A **sector** of a circle is a region whose boundary is an arc and two radii.

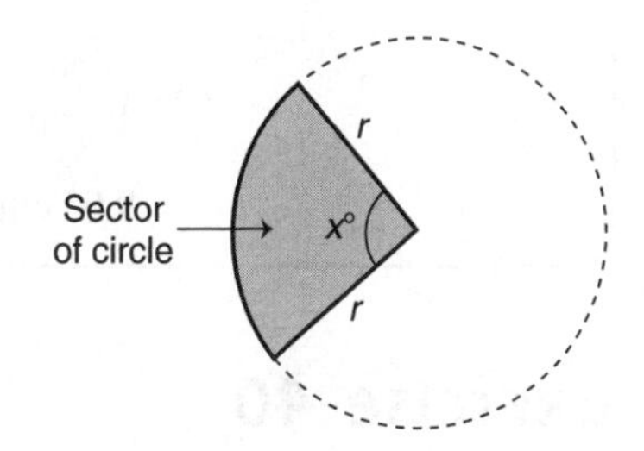

The sector shown is the fraction $\frac{x}{360}$ of the whole circle.

So the sector area is

$$\frac{x}{360} \times \pi r^2$$

Remember

$$\text{Sector area} = \frac{x}{360} \times \pi r^2$$

Example 13

Find the area of the sector shown.

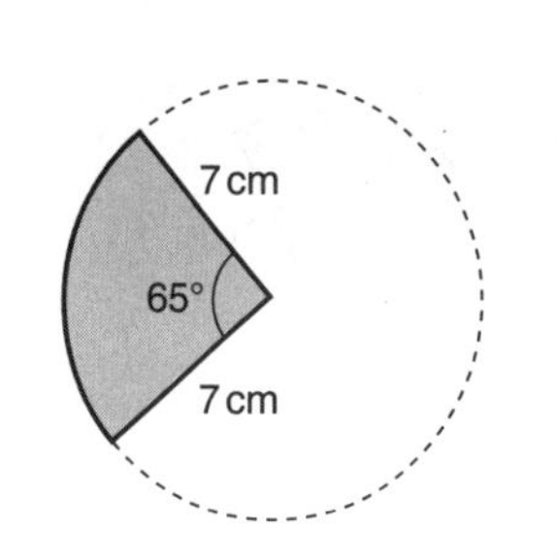

Using $\quad \text{Sector area} = \frac{x}{360} \times \pi r^2$

$$A = \frac{65}{360} \times \pi 7^2$$

$$= 27.8\ \text{cm}^2 \text{ to 3 s.f.}$$

Example 14

Find the angle marked x.

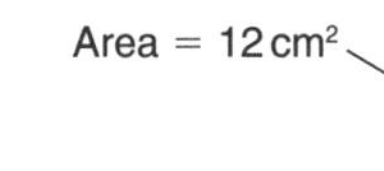

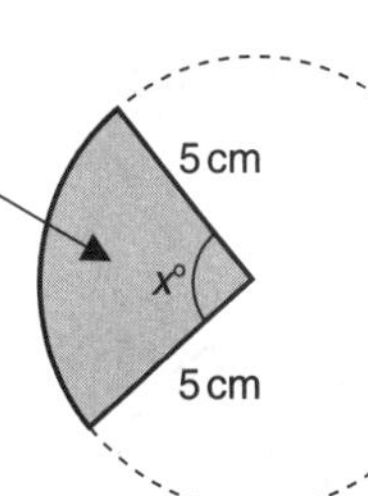

Using $\quad \text{Sector area} = \frac{x}{360} \times \pi r^2$

$$12 = \frac{x}{360} \times \pi \times 5^2$$ (Make x the subject of the equation)

$$x = \frac{12 \times 360}{\pi \times 5^2}$$

$$= 55.0° \text{ to 3 s.f.}$$

Example 15

Find the radius of the sector shown.

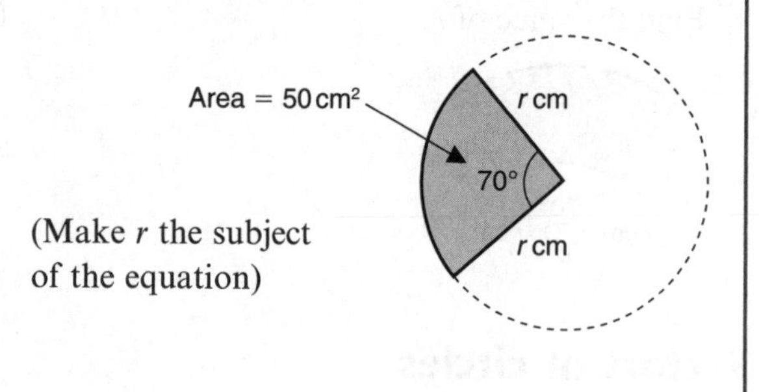

Using $\quad$ Sector area $= \dfrac{x}{360} \times \pi r^2$

$$50 = \frac{70}{360} \times \pi r^2 \qquad \text{(Make } r \text{ the subject of the equation)}$$

$$r^2 = \frac{50 \times 360}{70 \times \pi}$$

$$r = \sqrt{\frac{50 \times 360}{70 \times \pi}}$$

$$= 9.05\,\text{cm (to 3 s.f.)}$$

Exercise 40

In Questions 1–8, find the area of the shape. Give answers to 3 s.f.

1

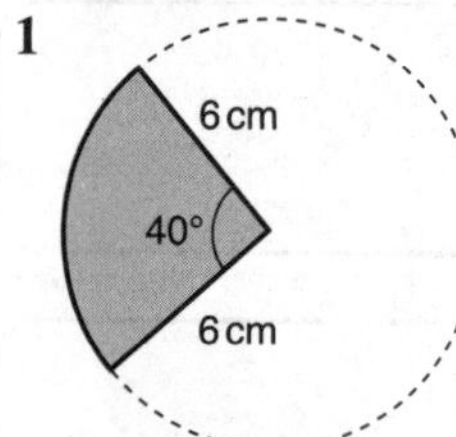

3

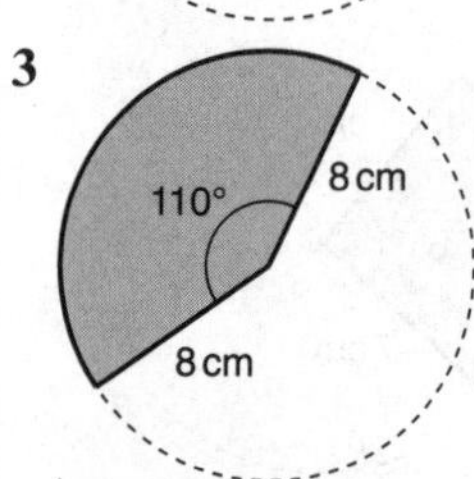

5

7

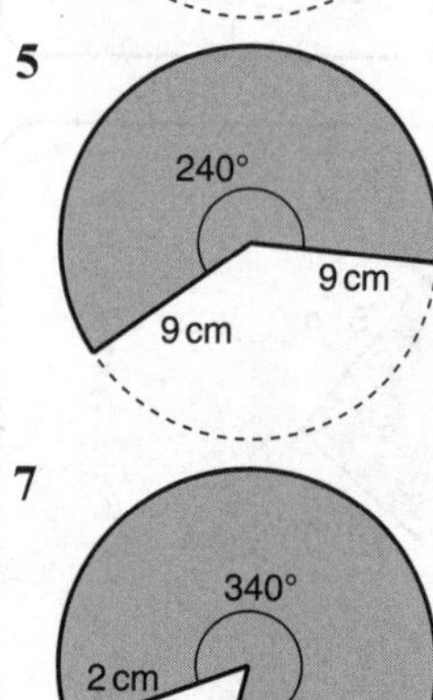

2

4

6

8

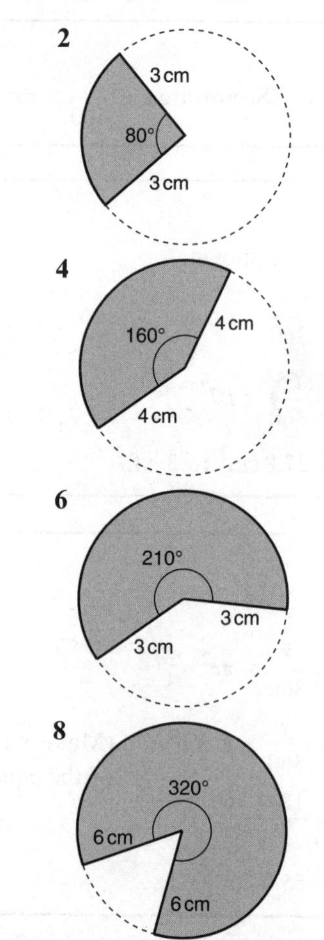

In Questions 9–12, find the angle marked x.

9

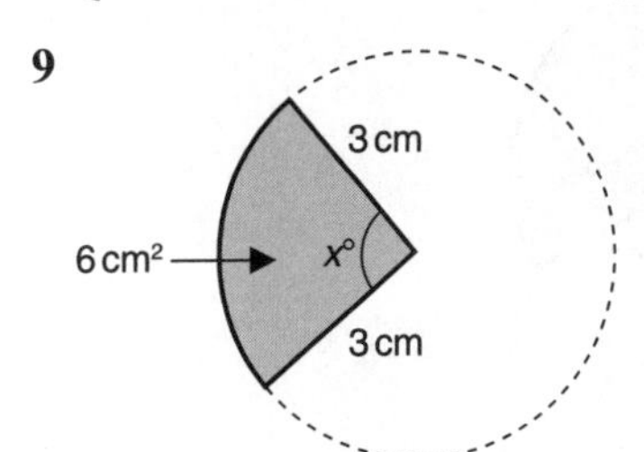

10

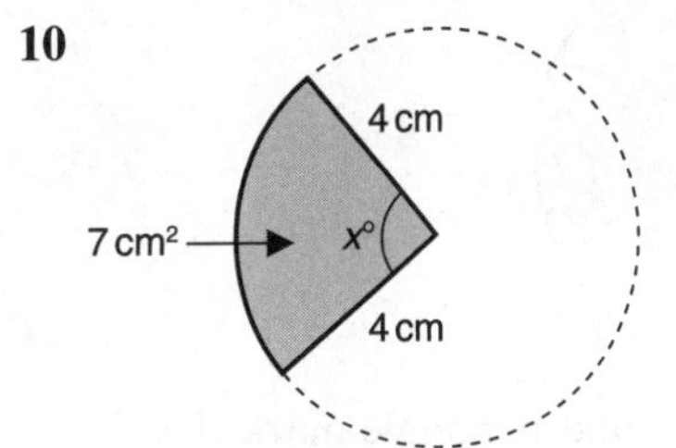

11

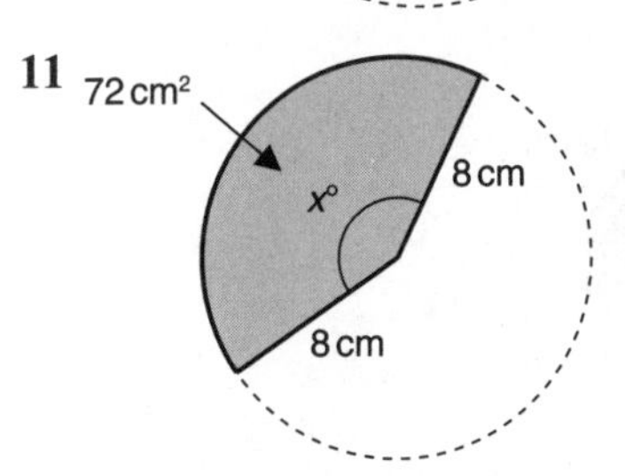

12

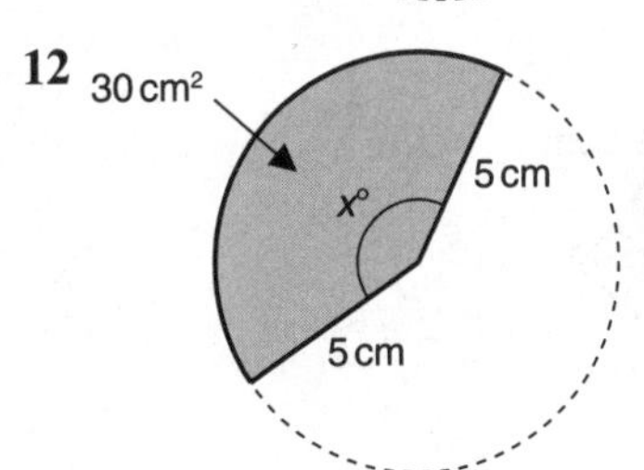

In Questions 13–16, find the radius r.

13

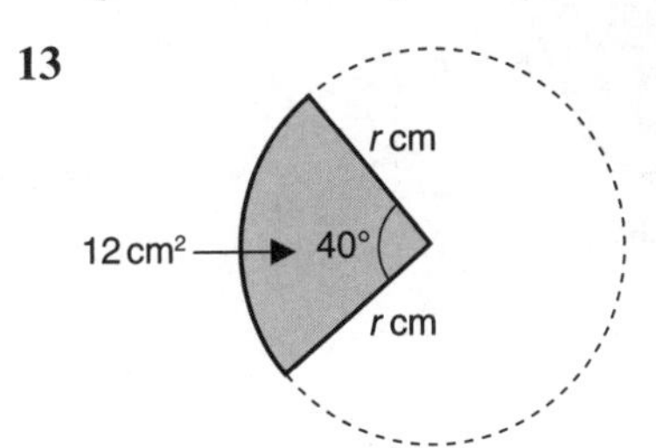

14

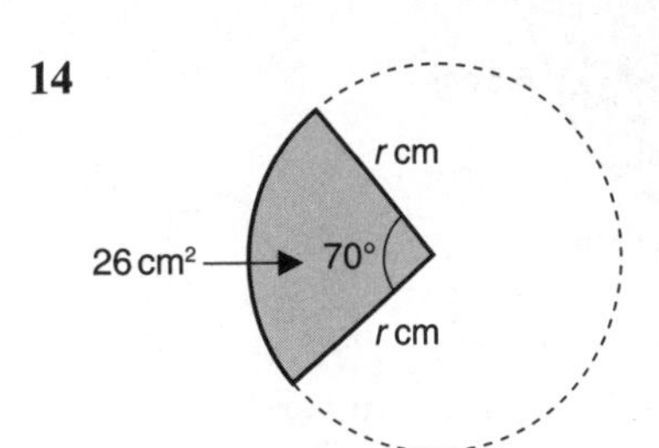

15

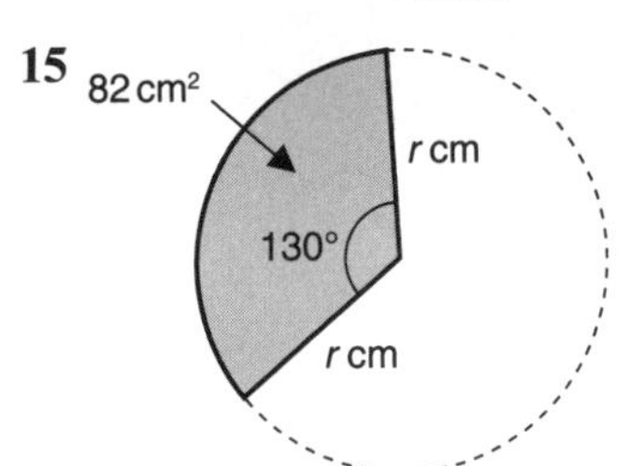

16

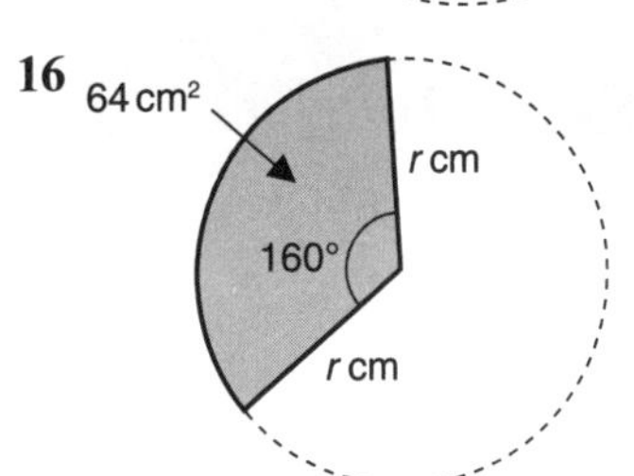

Exercise 40★

In Questions 1–4, find the area of the shape. Give answers to 3 s.f.

1

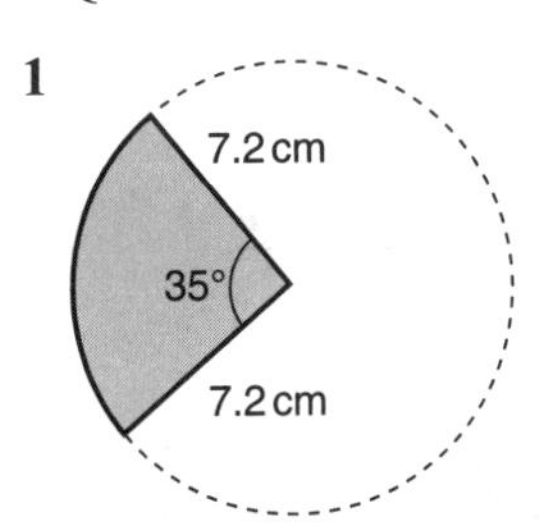

2

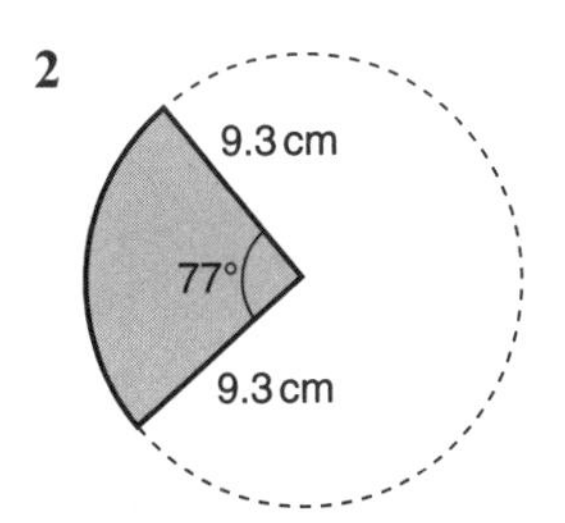

3

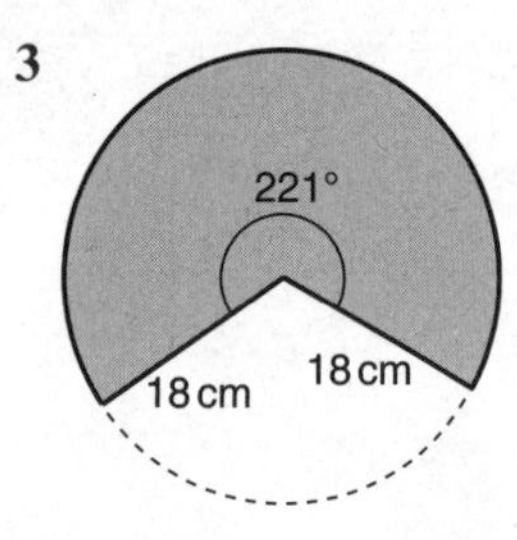

4

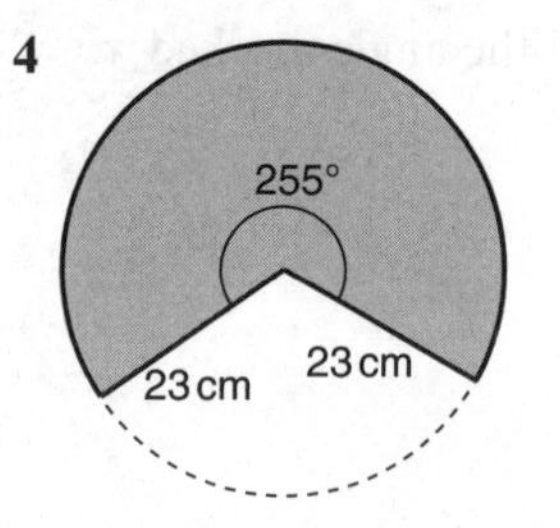

In Questions 5–8, find the angle marked x.

5

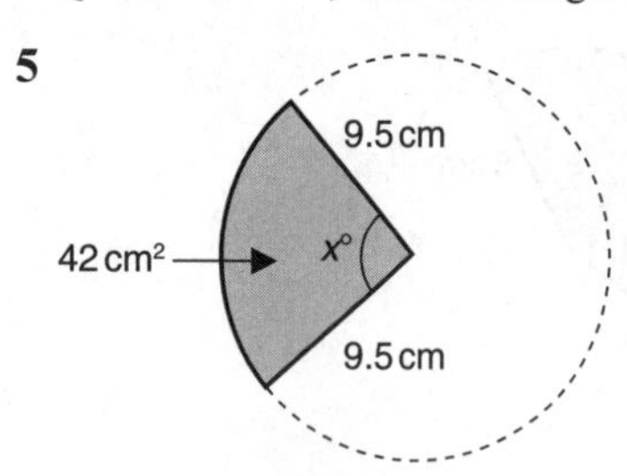

6

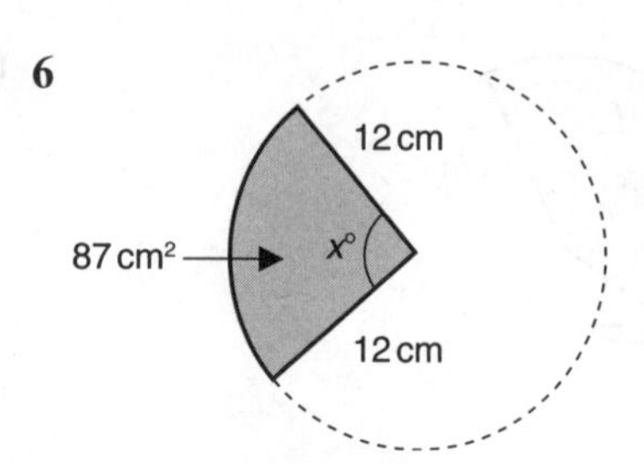

7

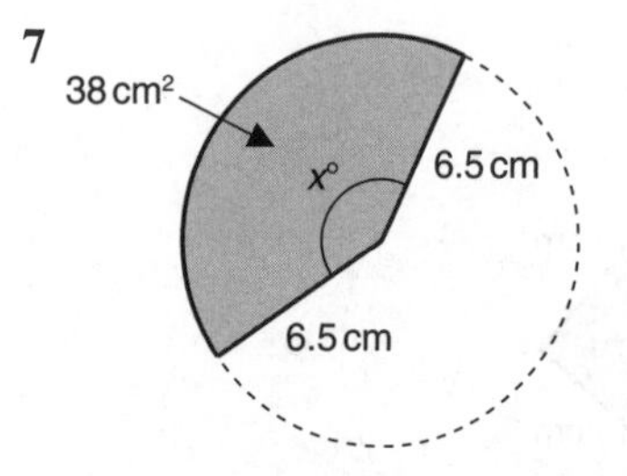

8

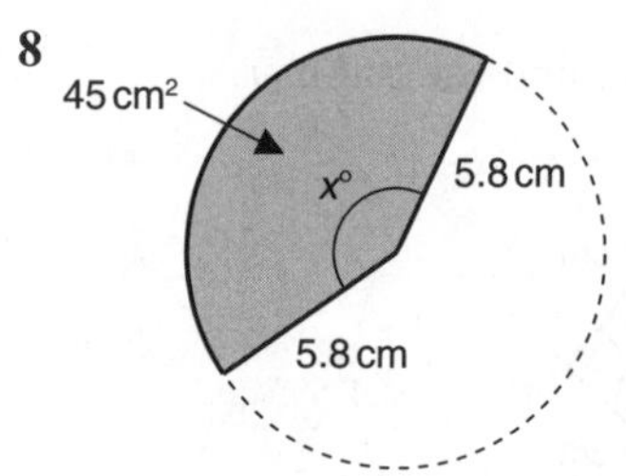

In Questions 9–12, find the radius r.

9

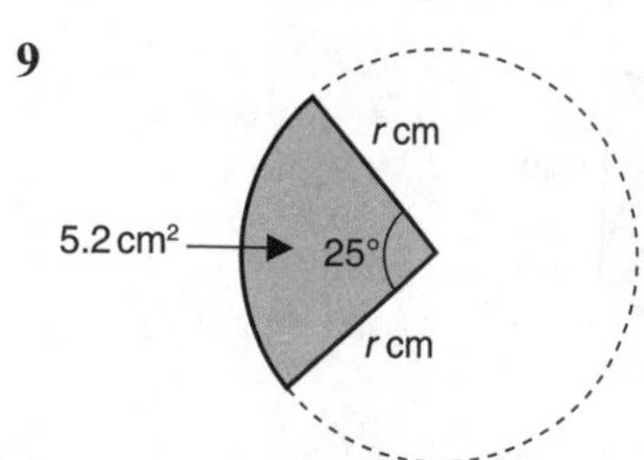

10

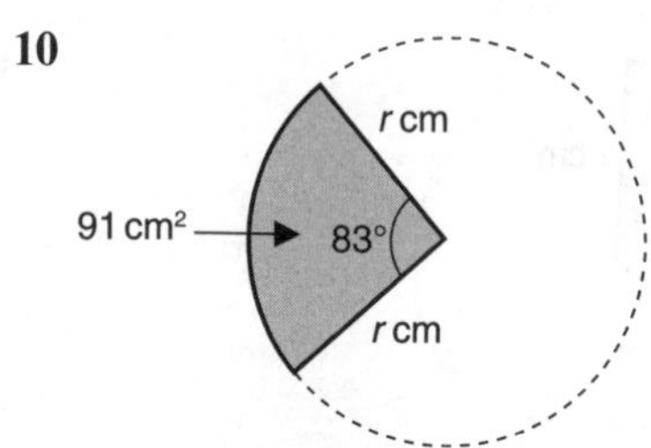

11

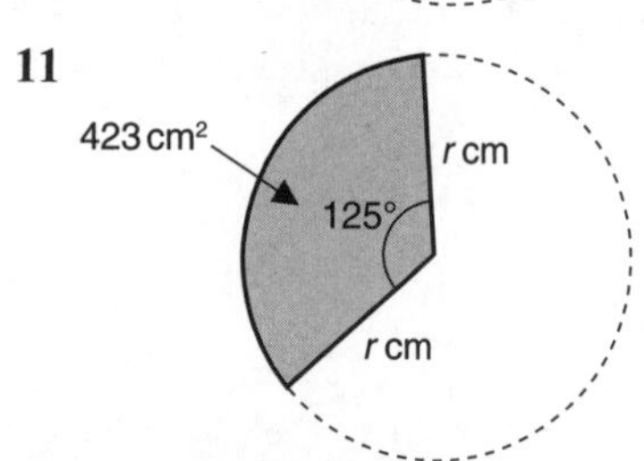

12

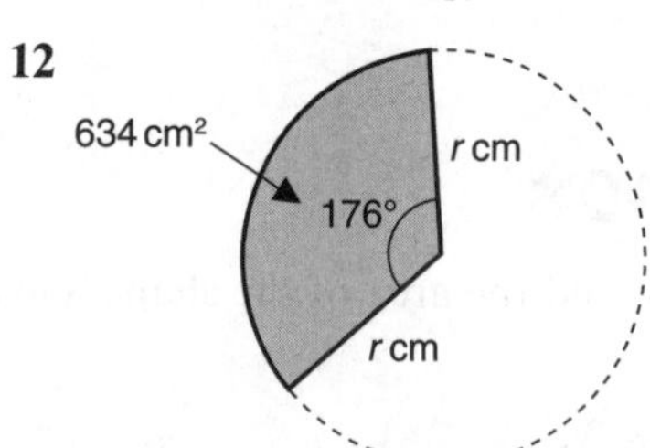

13 Find the area of the shape to 3 s.f.

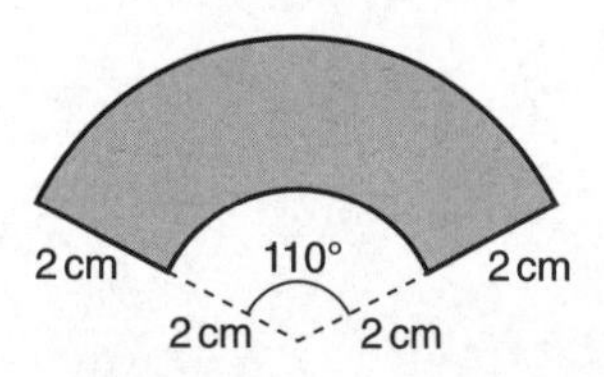

14 Find the area of the shape to 3 s.f.

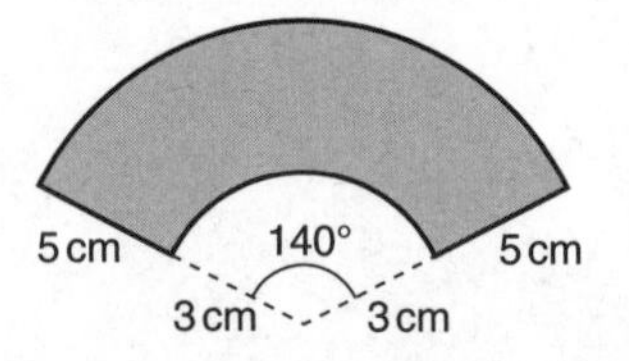

15 The area of the shape is $54\,\text{cm}^2$.
Find the value of r.

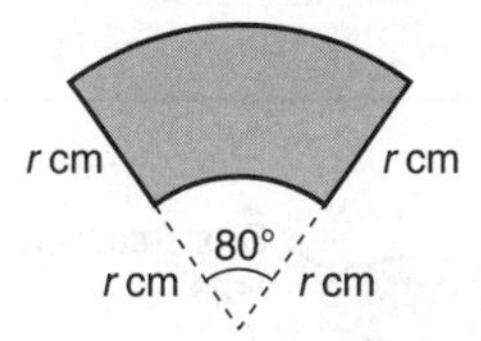

16 The area of the shape is $40\,\text{cm}^2$.
Find the value of r.

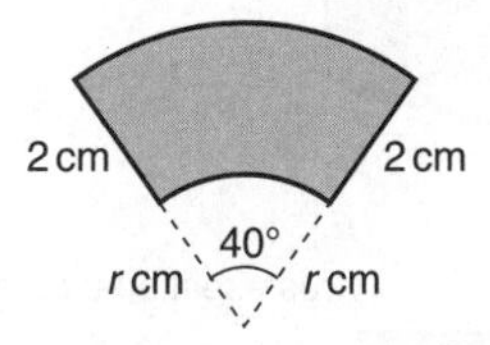

17 Find the shaded area.

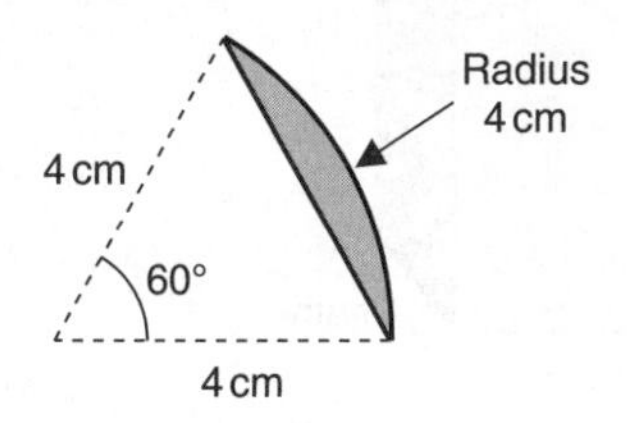

18 Find the shaded area.

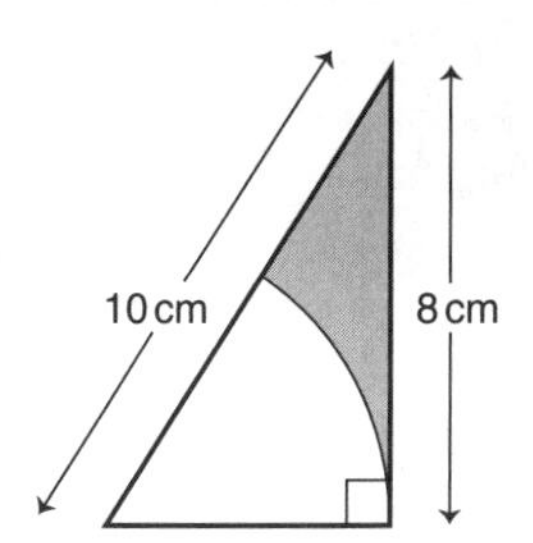

19 Three circular pencils, each with a diameter of 1 cm, are held together by an elastic band.
What is the (stretched) length of the band?

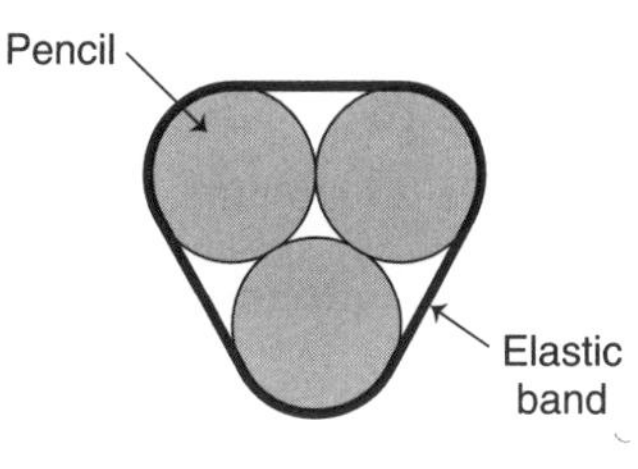

20 Three beer mats, each with a diameter of 8 cm, are placed on a table as shown. Find the shaded area.

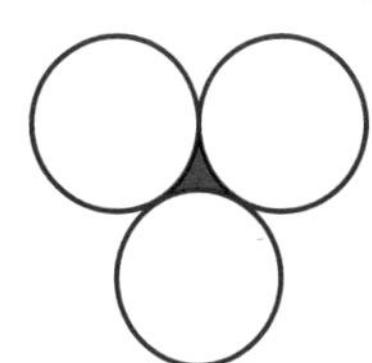

Surface areas and volumes of solids

Prisms

Remember

Any solid with parallel sides that has a constant cross-section is called a prism.

Volume of a prism = cross-sectional area $\times$ height

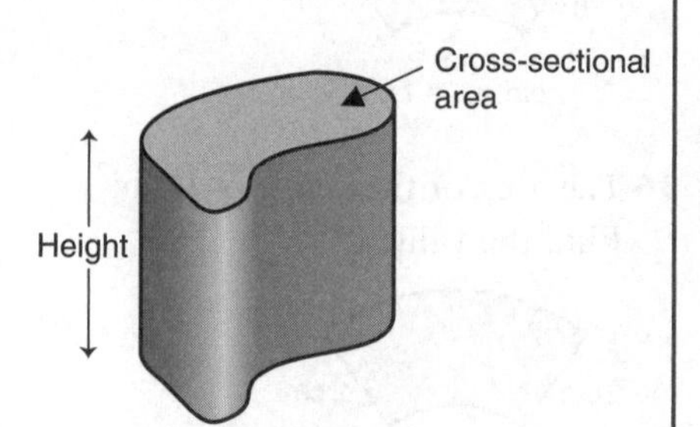

A **cuboid** is a prism with a rectangular cross-section.

Volume of a cuboid = width $\times$ depth $\times$ height

Surface area = sum of the area of the six rectangles making up the faces

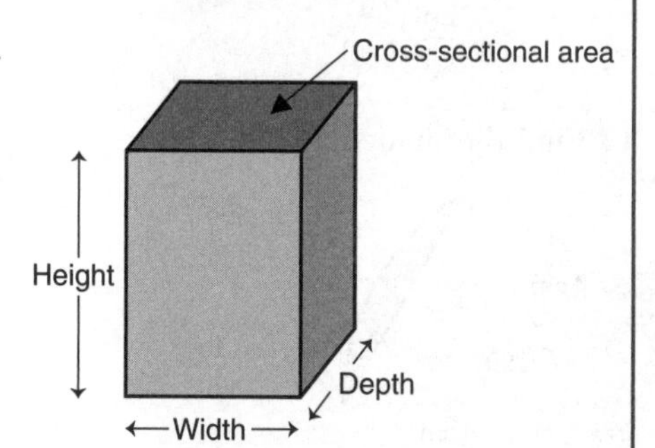

A **cylinder** is a prism with a circular cross-section.

If the height is h and the radius r, then

Volume of a cylinder = $\pi r^2 h$

Curved surface area of a cylinder = $2\pi rh$

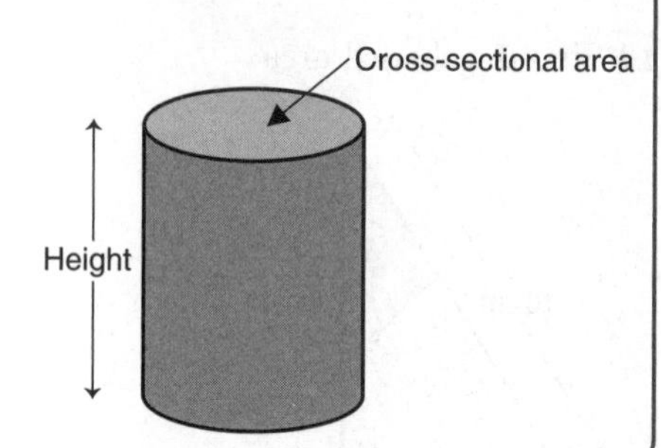

Example 16

Calculate the volume and surface area of the prism shown.

The cross-section is a right-angled triangle.

$$\text{Cross-sectional area} = \frac{1}{2} \times 3 \times 4 = 6\,\text{cm}^2$$

So Volume = $6 \times 8 = 48\,\text{cm}^3$

$$\begin{aligned}\text{Surface area} &= \text{two end triangles plus three rectangles}\\ &= 2 \times 6 + 5 \times 8 + 3 \times 8 + 4 \times 8\\ &= 108\,\text{cm}^2\end{aligned}$$

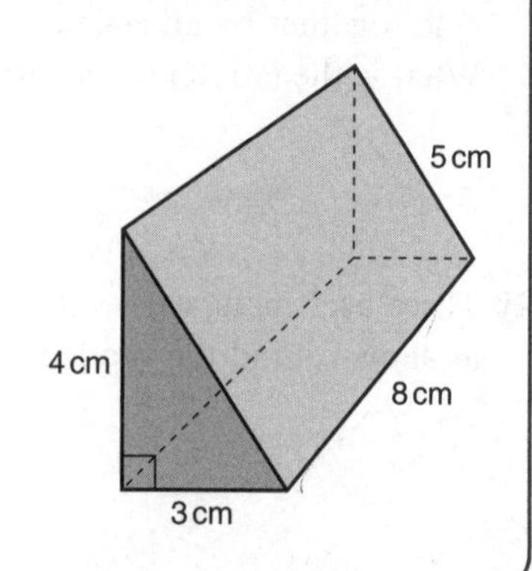

Example 17

Calculate the volume and surface area of a cola can that is a cylinder with diameter 6 cm and height 11 cm.

Using $V = \pi r^2 h$ with $r = 3$ and $h = 11$

$V = \pi \times 3^2 \times 11$

$= 311\text{ cm}^3$ to 3 s.f.

Surface area = two ends plus curved surface area

$A = 2 \times \pi r^2 + 2\pi rh$

$= 2 \times \pi \times 3^2 + 2 \times \pi \times 3 \times 11$

$= 264\text{ cm}^2$ to 3 s.f.

Exercise 41

1 Find the volume of the prism shown.

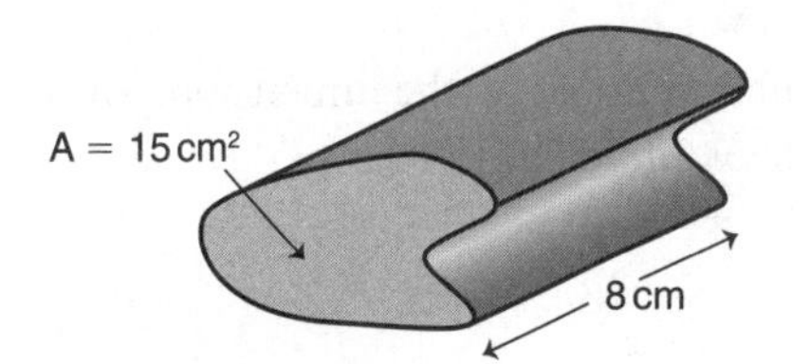

2 Find the volume of the prism shown.

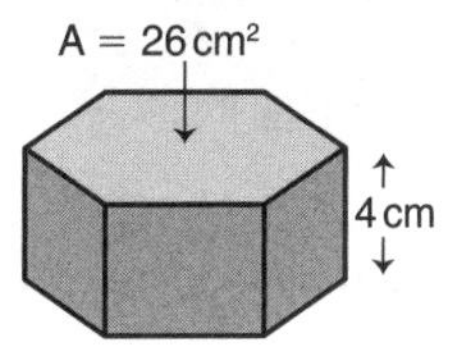

3 Find the volume and surface area of this wedge of cheese.

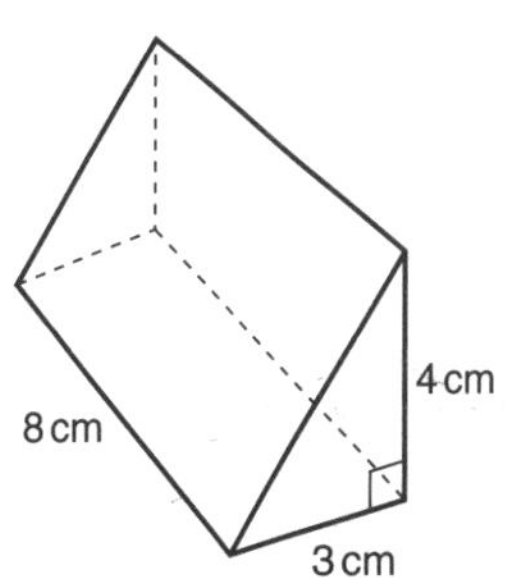

4 Find the volume and surface area of this pack of butter.

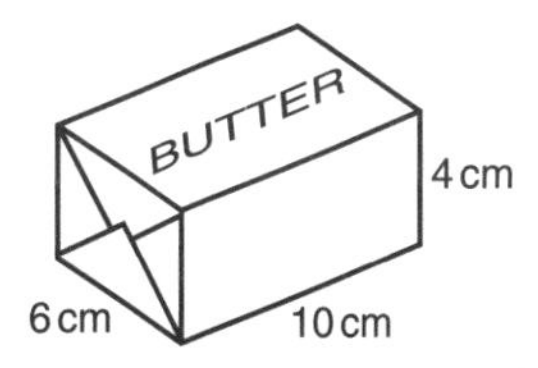

5 Find the volume and surface area of this fuel tank.

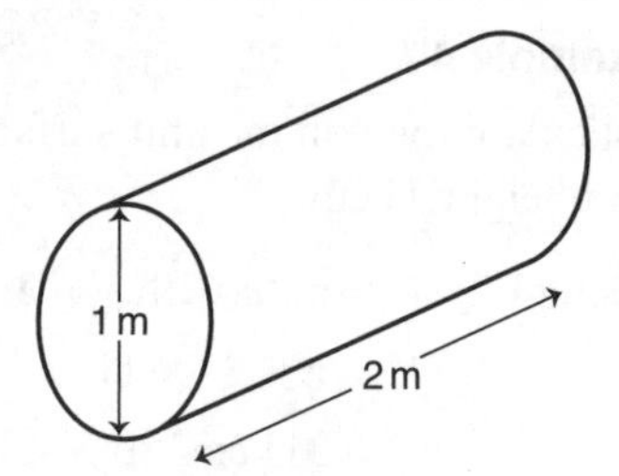

6 Find the volume and surface area of this can of drink.

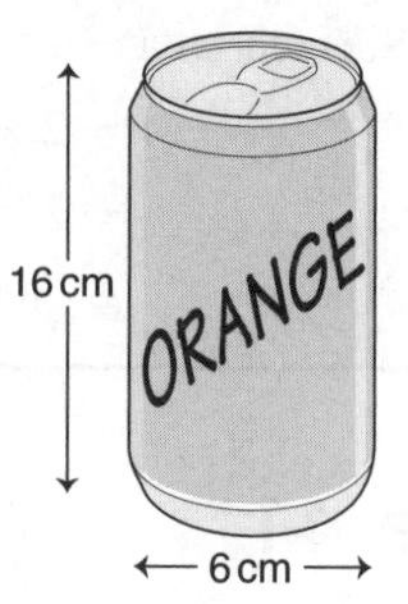

7 A swimming pool has the dimensions shown.
Find the volume in m^3.

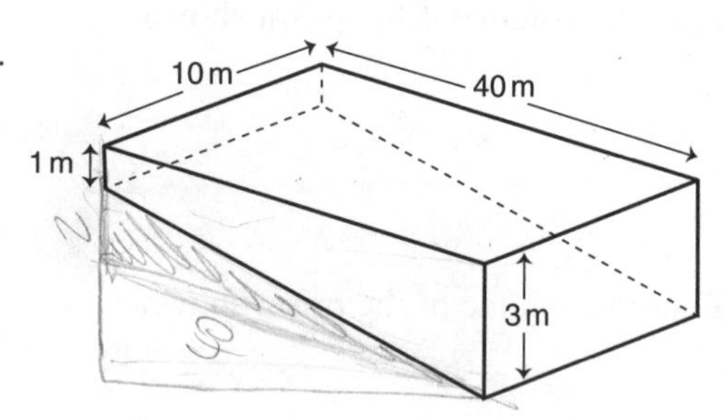

8 A penthouse shed has the dimensions shown.
Find the volume in m^3.

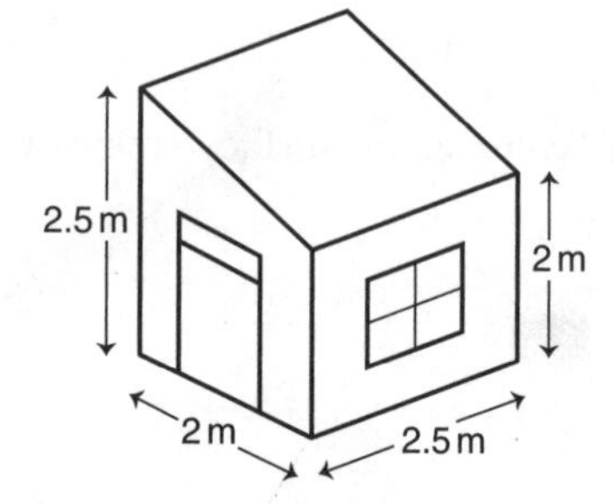

9 A water trough has the dimensions shown.
Find the volume in m^3.

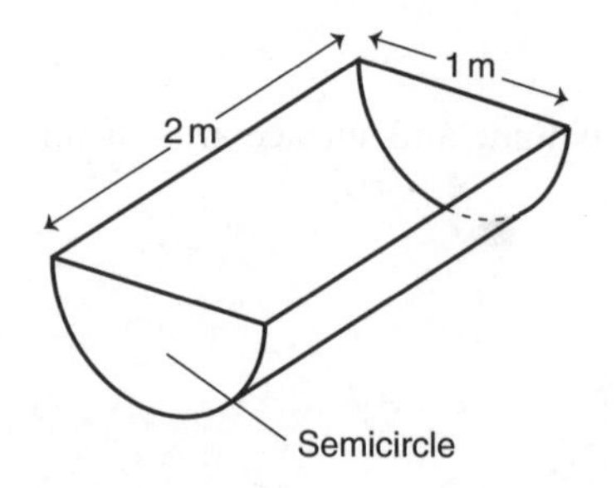

10 A barn has the dimensions shown.
Find the volume in m^3.

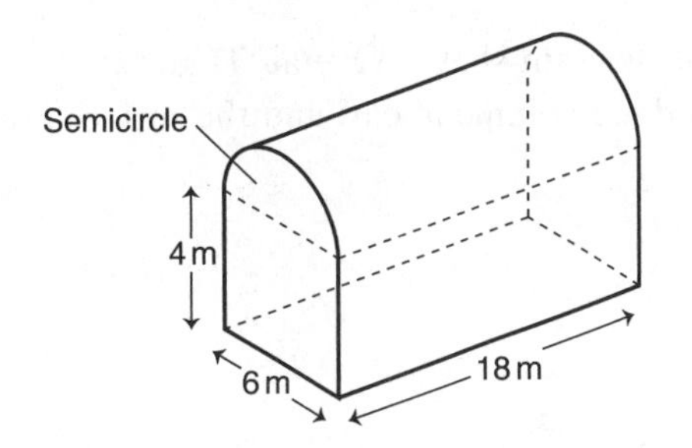

11 A carton contains 1 litre of orange juice.
The carton is 10 cm wide and 6 cm deep.
How tall is it?

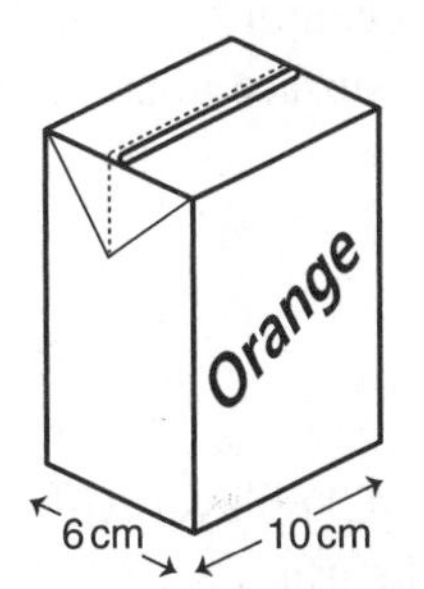

12 A half-litre jar of olive oil is 8 cm in diameter.
How tall is it?

Exercise 41★

1 Find the volume of the metal bar in cm^3.

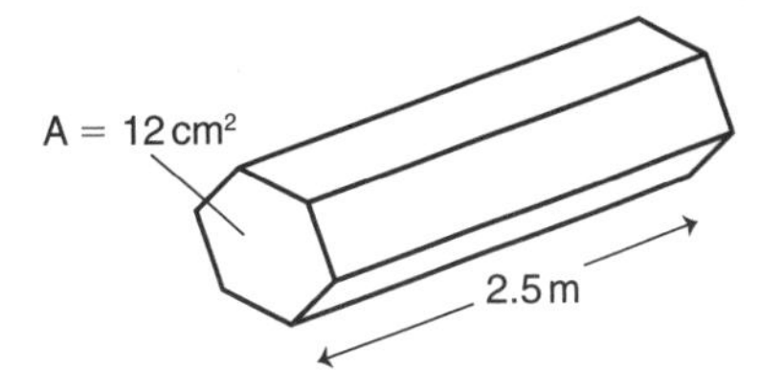

2 The diagram shows an extrusion for a conservatory roof.
Find the volume of the extrusion in cm^3.

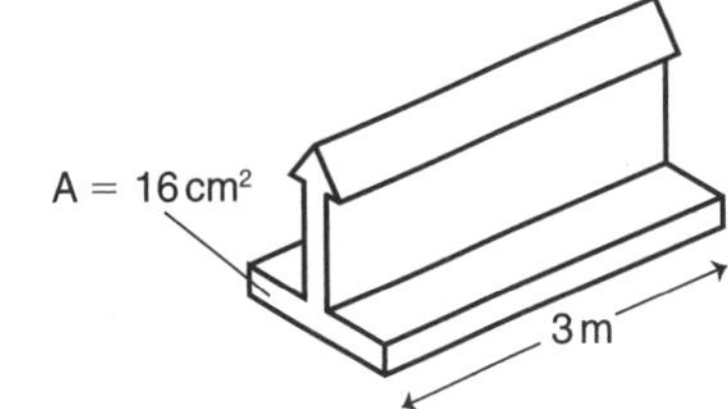

3 The diagram shows some stage steps.
Find the volume in cm^3 and the surface area in cm^2.

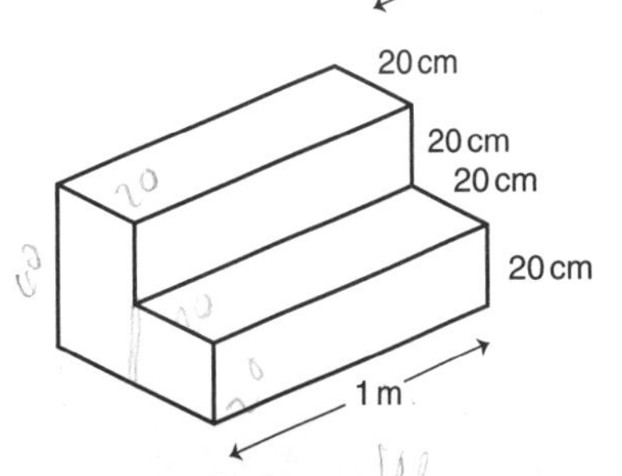

4 The diagram shows a metal 'T' girder.
Find the volume in cm^3 and the surface area in cm^2.

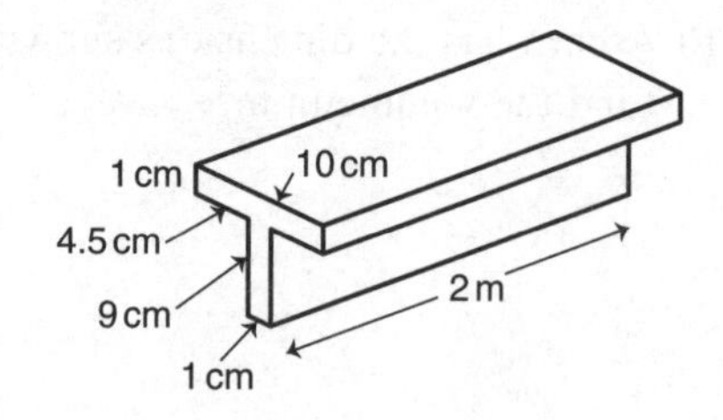

5 The diagram shows a can of food with semicircular ends.
Find the volume and surface area.

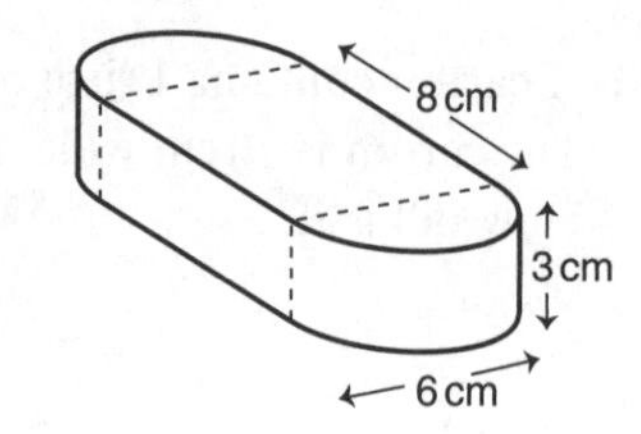

6 The diagram shows a CD player with semicircular ends.
Find the volume and surface area.

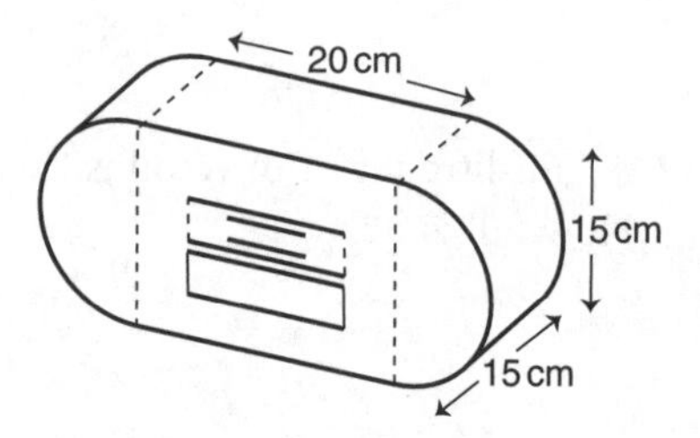

7 The diagram shows a sweet.
Find the volume in cm^3 and the surface area in cm^2.

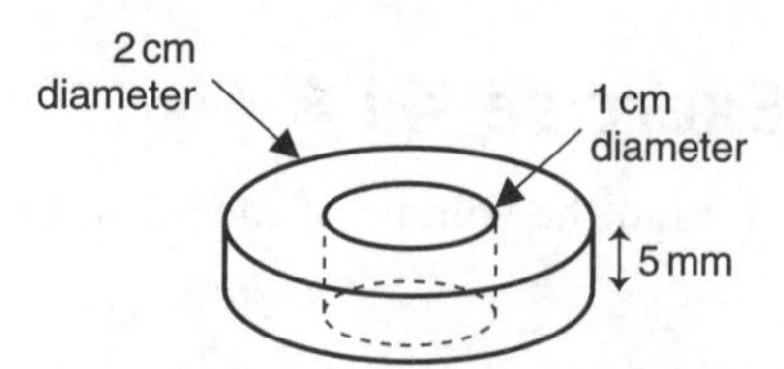

8 The diagram shows a concrete pipe.
Find the volume in m^3.

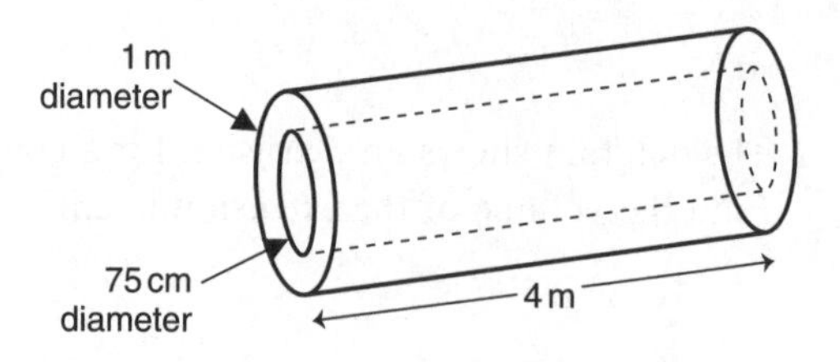

9 Find the volume and surface area of the object shown in the diagram.

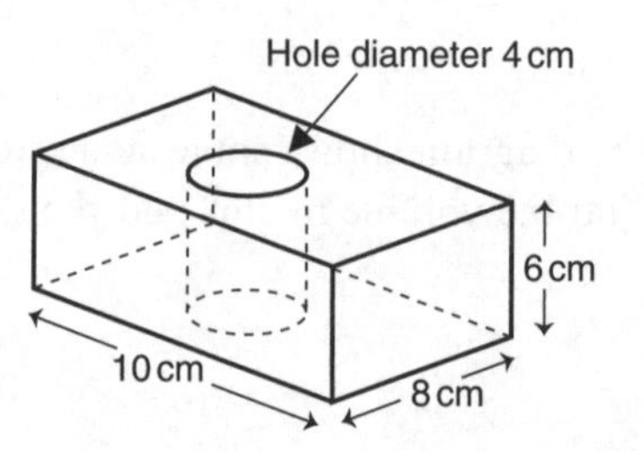

10 Find the volume and surface area of the object shown in the diagram.

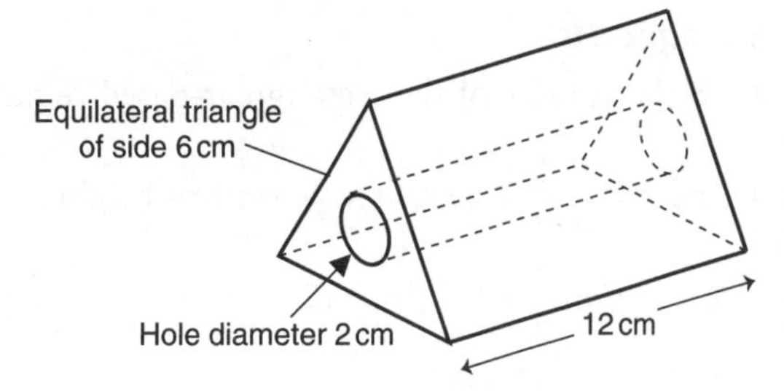

11 A toilet roll has the dimensions shown. If the thickness of the paper is $\frac{1}{5}$ mm, find the length of paper on the roll.

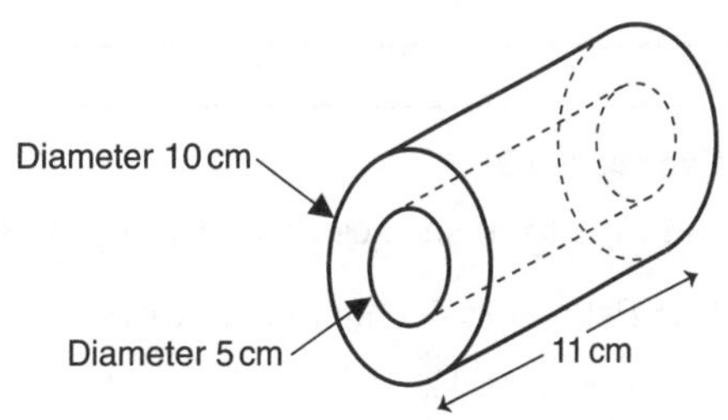

12 A reel of sticky tape has the dimensions shown. If the tape is 25 m long, how thick is the tape?

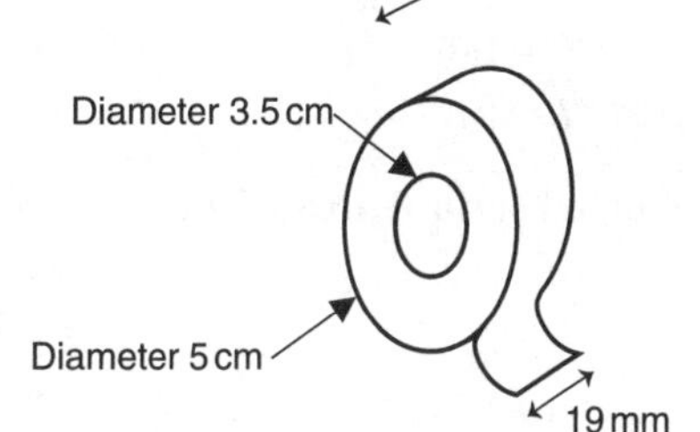

Pyramids, cones and spheres

Remember

Volume of a pyramid $= \frac{1}{3} \times$ base area $\times$ vertical height

Surface area = area of the base plus the triangular faces.

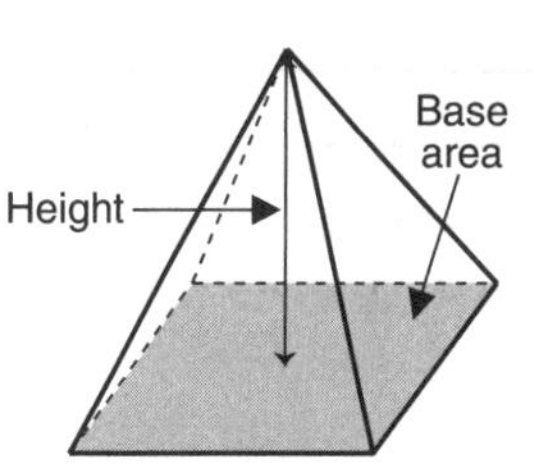

A **cone** is a pyramid with a circular base.

Volume of a cone $= \frac{1}{3} \times \pi r^2 \times h$

Curved surface area of a cone $= \pi r l$, where l is the slant height

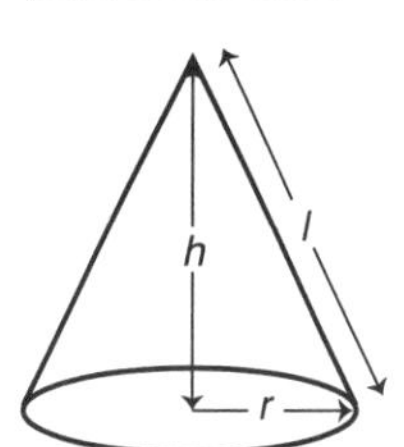

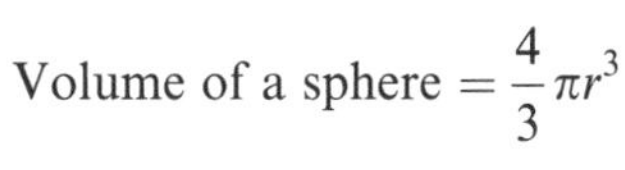

Volume of a sphere $= \frac{4}{3}\pi r^3$

Surface area of a sphere $= 4\pi r^2$

Example 18

Find the volume of the rectangular-based pyramid shown.

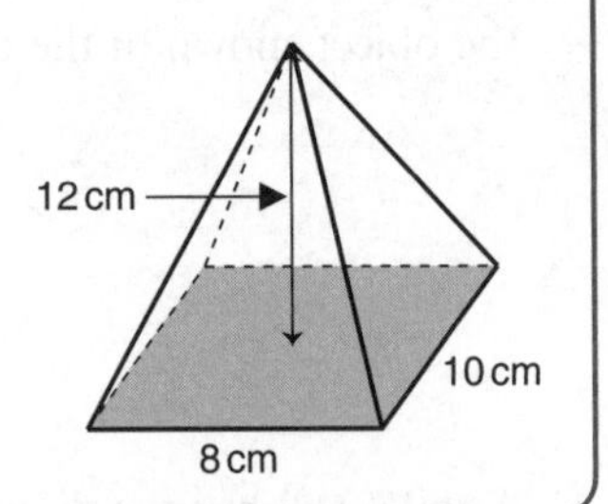

Using $V = \frac{1}{3} \times$ base area $\times$ vertical height

$$V = \frac{1}{3} \times 8 \times 10 \times 12$$
$$= 320\,\text{cm}^3$$

Example 19

Find the total surface area of the cone shown.

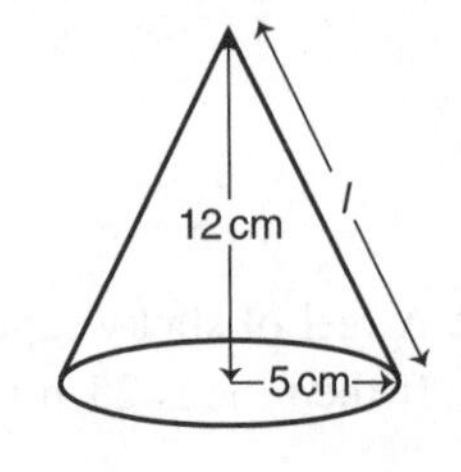

Use Pythagoras' Theorem to work out l.

$$l^2 = 5^2 + 12^2$$
$$l^2 = 169$$
$$l = 13$$

Curved surface area of a cone $= \pi r l$

$$= \pi \times 5 \times 13$$
$$= 204\,\text{cm}^2$$

The base is a circle with area $\pi r^2 = \pi \times 5 \times 5$

$$= 78.5\,\text{cm}^2$$

Total surface area = curved surface area plus base

$$= 204 + 78.5$$
$$= 283\,\text{cm}^2 \text{ to 3 s.f.}$$

Example 20

A squash ball has a volume of $33\,\text{cm}^3$. Find the radius and surface area.

Using volume of a sphere $= \frac{4}{3}\pi r^3$

$$33 = \frac{4}{3}\pi r^3 \qquad \text{(Make } r \text{ the subject of the equation)}$$

$$r^3 = \frac{33 \times 3}{4 \times \pi}$$

$$r = \sqrt[3]{\frac{33 \times 3}{4 \times \pi}}$$

$$= 1.99\,\text{cm}$$

Using surface area of a sphere $= 4\pi r^2$

$$A = 4 \times \pi \times 1.99^2$$
$$= 49.8\,\text{cm}^2 \text{ to 3 s.f.}$$

Exercise 42

1 The Great Pyramid of Giza is a square-based pyramid with dimensions as shown.
Find the volume in m^3.

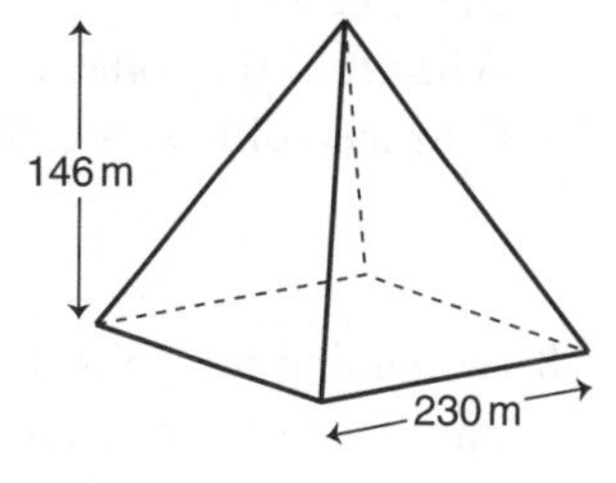

2 The glass pyramid at the Louvre is a square-based pyramid with dimensions as shown. Find the volume in m^3.

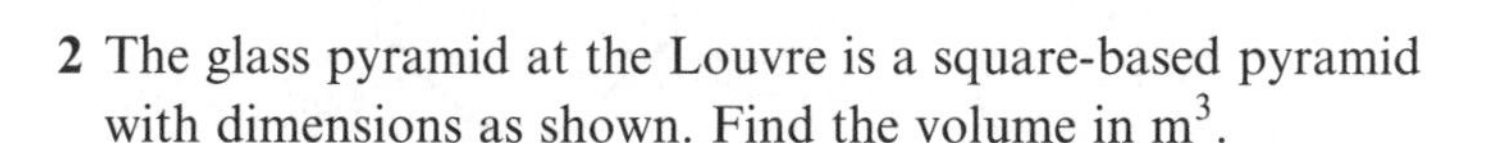

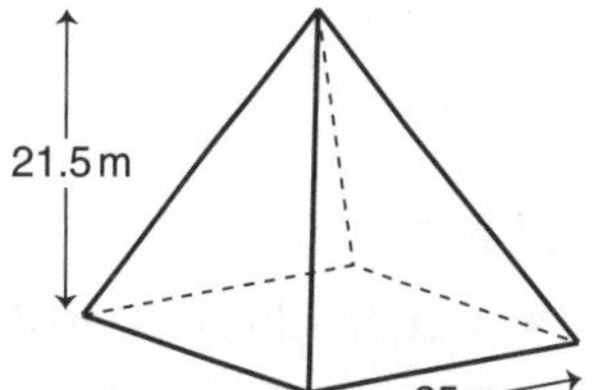

3 A traffic cone has the dimensions shown.
Find the volume in cm^3 and the curved surface area in cm^2.

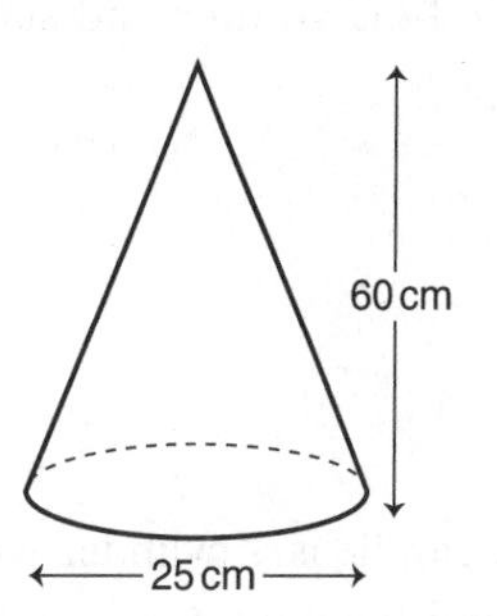

4 A funnel is an inverted cone with the dimensions shown.
Find the volume in cm^3 and the surface area in cm^2.

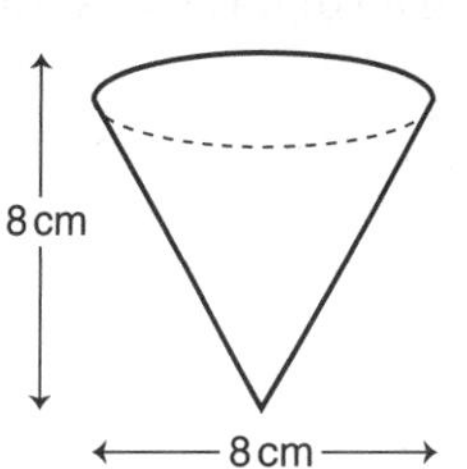

5 A food colander is a hemisphere with diameter 20 cm.
Find the volume and internal surface area.

6 A hanging flower basket is a hemisphere with diameter 30 cm.
Find the volume and surface area.

7 A scoop for ground coffee is a hollow hemisphere with diameter 4 cm.
When full, the coffee forms a cone on top of the scoop.
Find the volume of coffee.

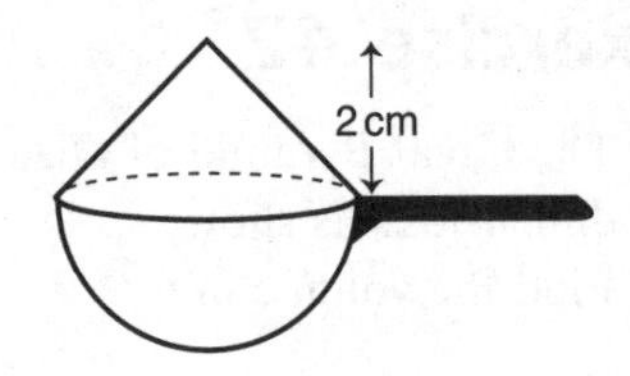

8 A spinning top consists of a hemisphere on top of a cone.
Find the volume of the toy.

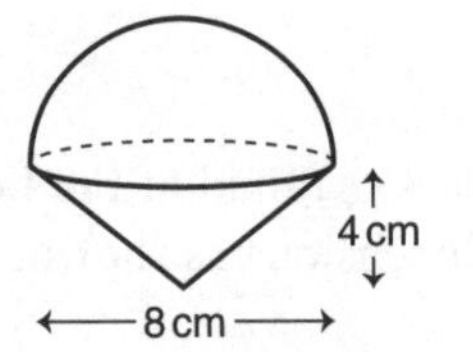

9 A grain silo is a cylinder with a hemisphere on top, with the dimensions shown.
Find the volume and surface area including the base.

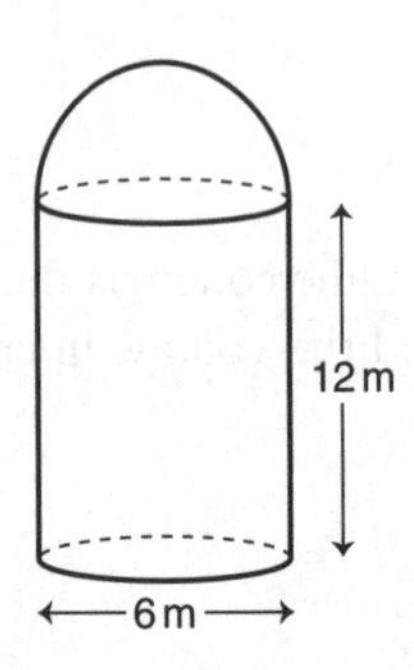

10 A candle is a cylinder with a cone on top.
Find the volume and surface area.

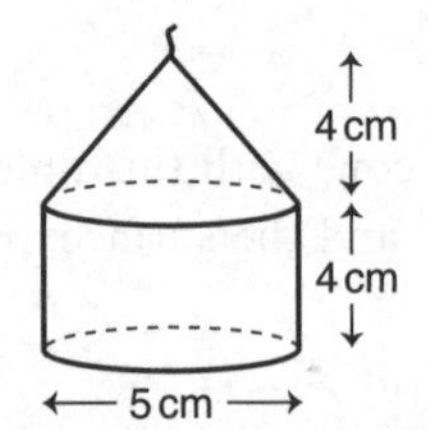

11 The volume of a football is 7240 cm^3.
Find the radius and surface area.

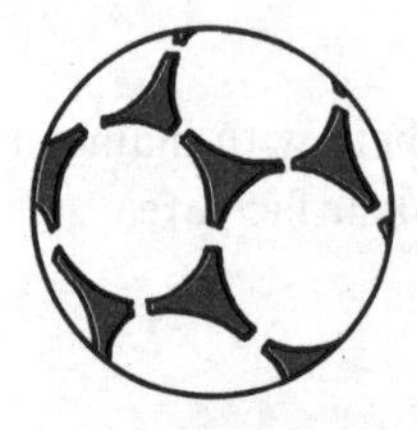

12 The volume of a cricket ball is 180 cm^3.
Find the radius and surface area.

13 A flat roof is a rectangle 6 m by 8 m.
The rain drains into a cylindrical water butt with radius 50 cm.
By how much does the water level in the butt rise if 1 cm of rain falls?
(Assume the butt does not overflow.)

14 A fuel tanker is pumping fuel into an aircraft's fuel tank. The tanker is a cylinder 2 m in diameter and 3 m long. The aircraft's tank is a cuboid $5\text{ m} \times 4\text{ m} \times 1\text{ m}$ high. Before pumping, the fuel tanker is full and the aircraft's fuel tank is empty. How deep is the fuel in the aircraft's tank after pumping is complete?

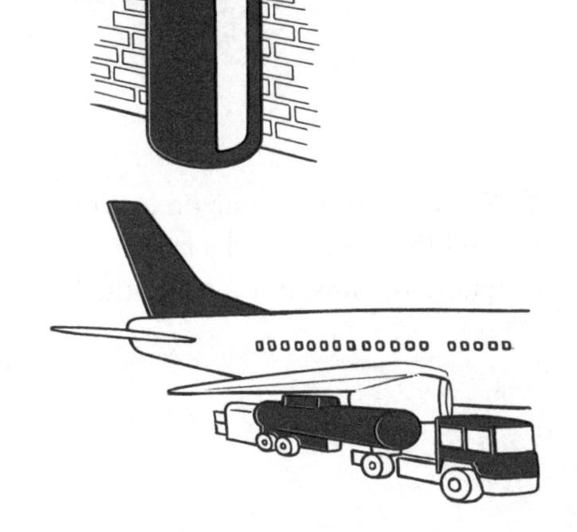

Exercise 42★

1 A television tube is in the shape of a pyramid with dimensions as shown.
What is the volume?

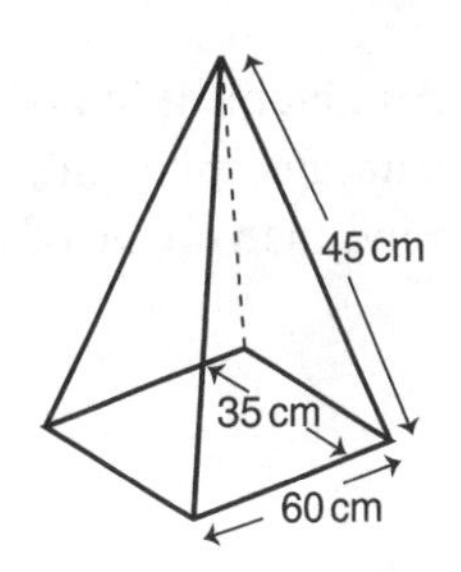

2 A crystal consists of two square-based pyramids as shown.
Calculate the volume of the crystal.

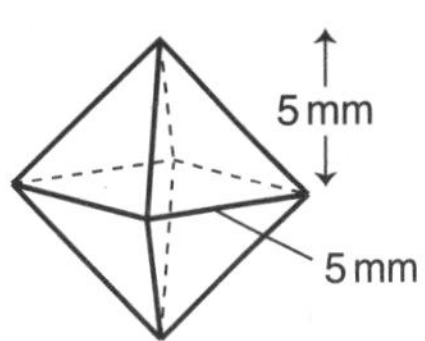

3 An ice-cream cone is full of ice cream as shown.
What is the volume of ice cream?

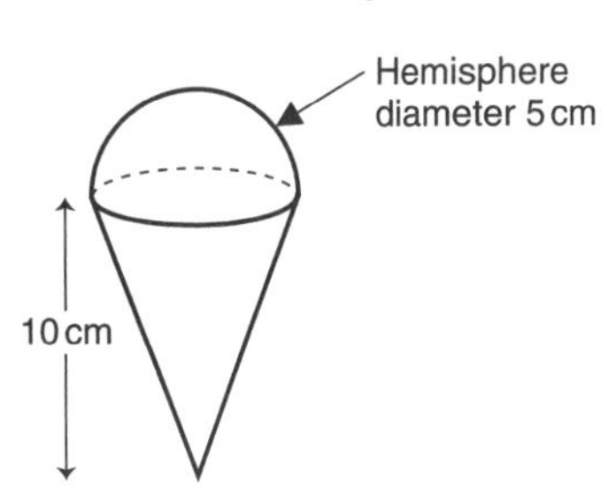

4 A toy consists of a solid hemisphere with a cone on top.
What is the volume of the toy?

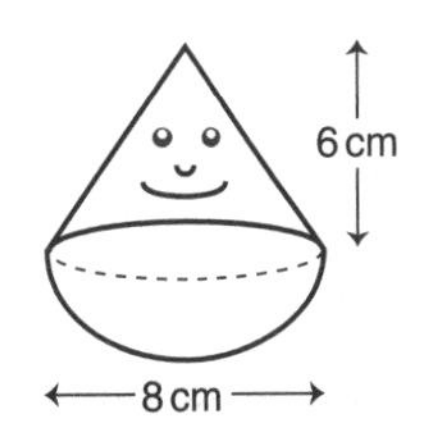

5 A water bottle is a cylinder with a cone at one end and a hemisphere at the other.
Find the volume and surface area.

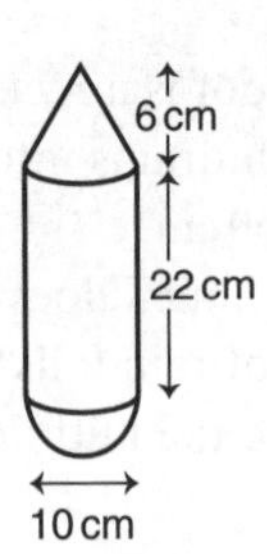

6 A toy rocket is a cone on a cylinder.
Find the volume and surface area.
Ignore the fins, but include the base.

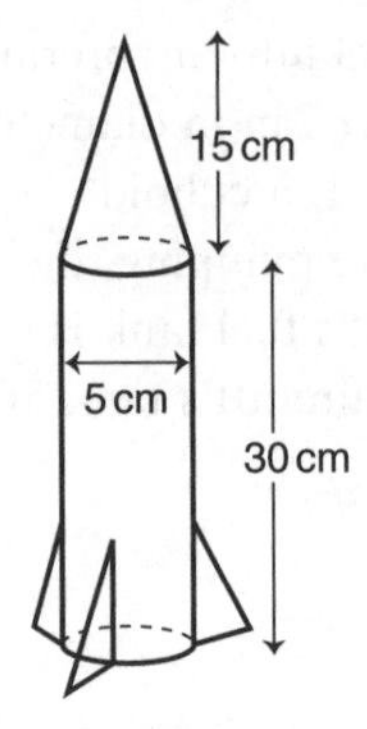

7 A monument in South America is in the shape of a truncated pyramid.
Find the volume of the monument.

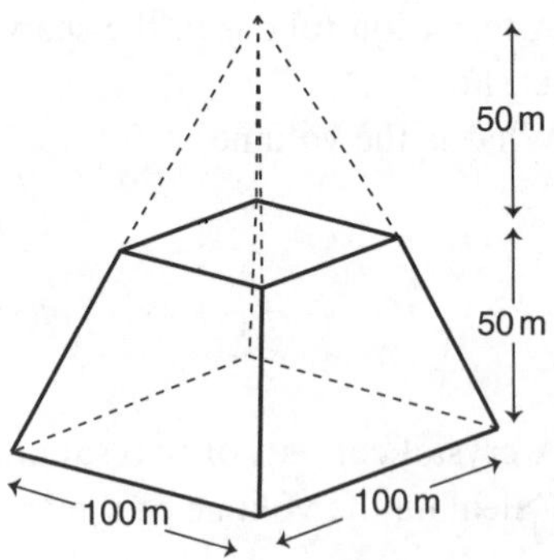

8 A vase is a truncated cone.
Find the volume of the vase.

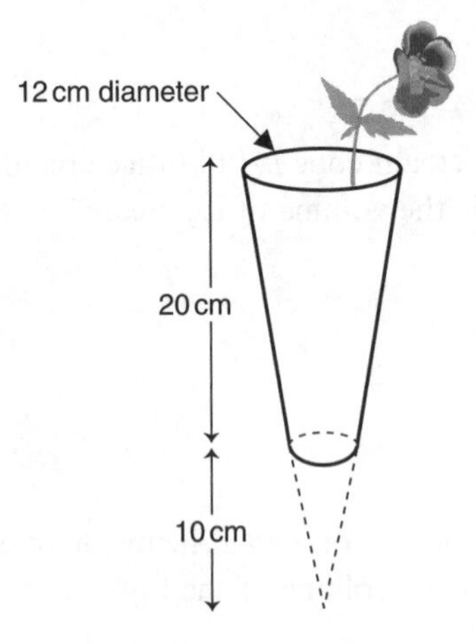

9 The volume of the Earth is 1.09×10^{12} km^3.
Find the surface area of the Earth.

10 The volume of the Sun is 1.41×10^{18} km^3.
Find the surface area of the Sun.

11 A stone ball is dropped into a barrel of water and sinks to the bottom. The ball is completely covered by water. By how much does the water rise in the barrel?

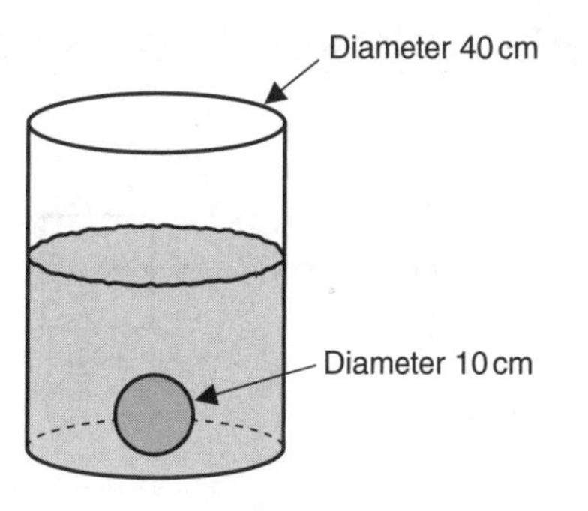

12 A cuboid of chocolate measures 12 cm by 8 cm by 6 cm. It is melted down and cast into chocolates that are spheres with diameter 2 cm. How many spheres can be made?

13 The sphere and the cone shown have the same volume.
Calculate the height of the cone.

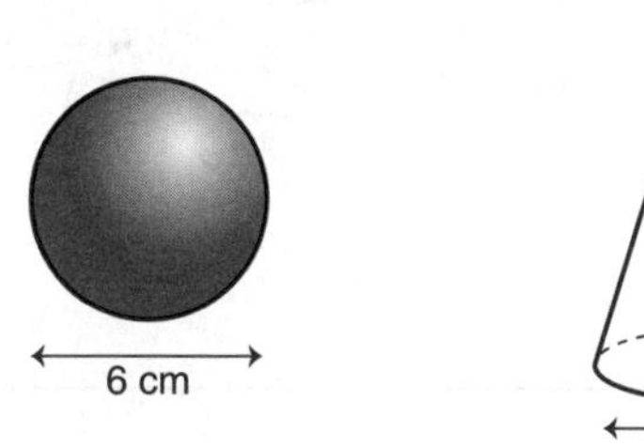

14 A spherical drop of oil with diameter 3 mm falls onto a water surface and produces a circular oil film of radius 10 cm.
Calculate the thickness of the oil film.

Areas of similar shapes

When a shape doubles in size, then the area does **NOT** double, but increases by a factor of four.

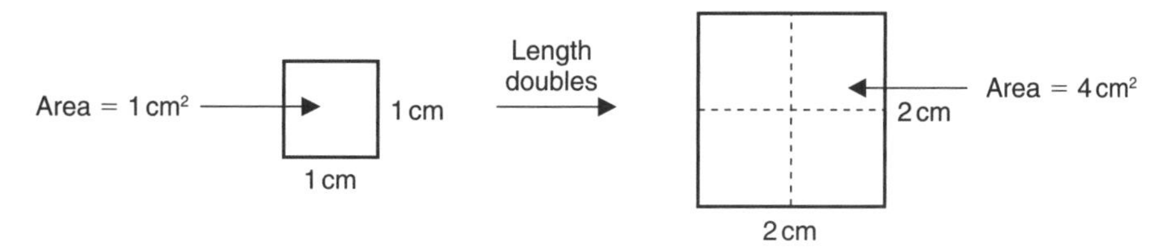

The Length Scale Factor is 2, and the Area Scale Factor is 4.

If the shape triples in size, then the area increases by a factor of nine.

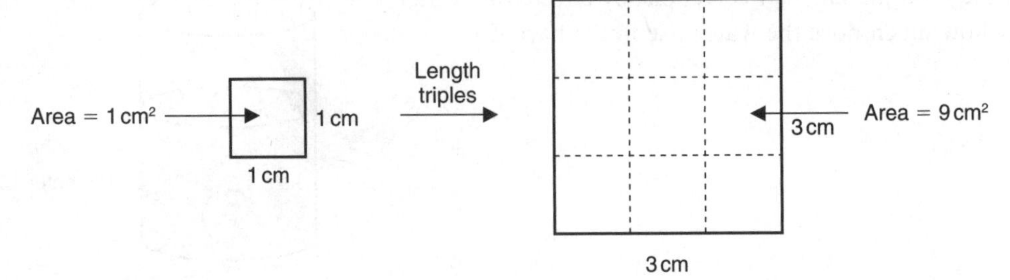

If a shape increases by a Length Scale Factor of k, then the Area Scale Factor is k^2.

This applies even if the shape is irregular.

Area = 2 cm²

Length doubles

Area Scale Factor = 4

Area = 2 × 4 = 8 cm²

Note: The two shapes must be similar.

Remember

If the Length Scale Factor is k, then the Area Scale Factor is k^2.

Example 21

The two shapes shown are similar. The area of the smaller shape is $10\,\text{cm}^2$. Find the area of the larger shape.

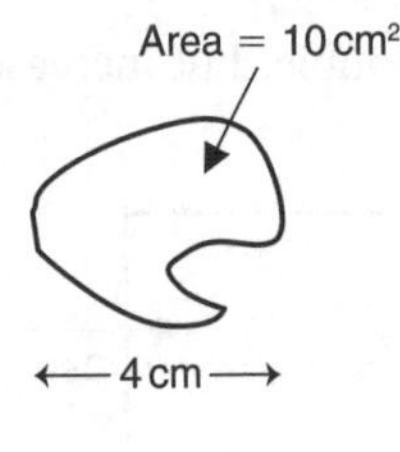

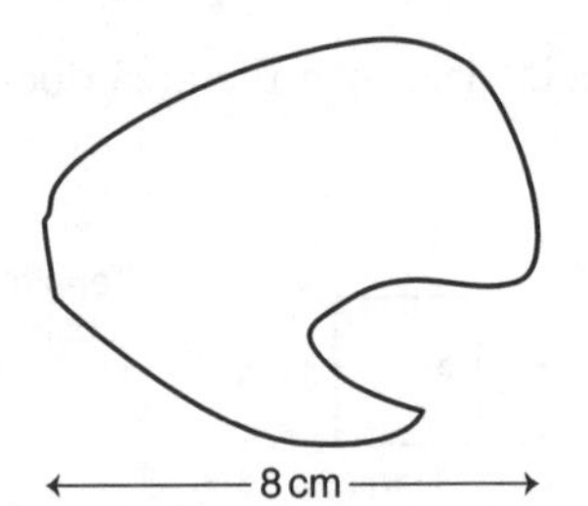

The Length Scale Factor $k = \frac{8}{4} = 2$ (Note: Divide the length of the second shape by the length of the first shape.)

The Area Scale Factor $k^2 = 2^2 = 4$

So the area of the larger shape is $10 \times 4 = 40\,\text{cm}^2$.

Example 22

The two shapes shown are similar. The area of the larger shape is 18 cm^2. Find the area of the smaller shape.

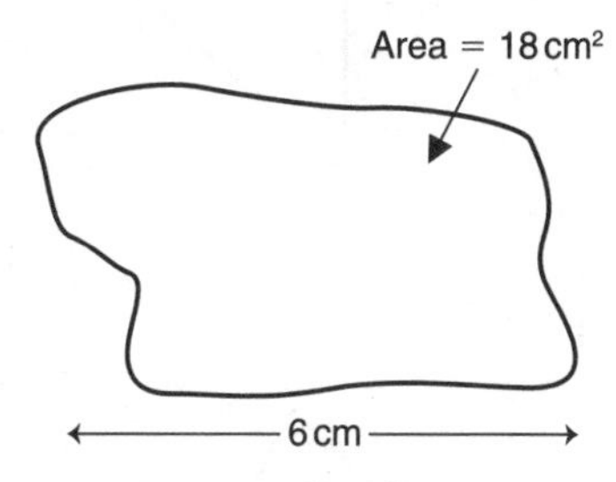

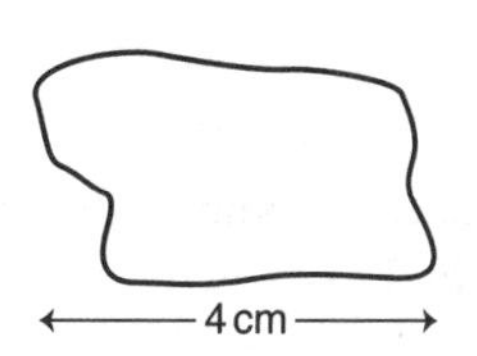

The Length Scale Factor $k = \frac{4}{6} = \frac{2}{3}$ (Note: Divide the length of the second shape by the length of the first shape.)

The Area Scale Factor $k^2 = \left(\frac{2}{3}\right)^2 = \frac{4}{9}$

So the area of the smaller shape is $18 \times \frac{4}{9} = 8\text{ cm}^2$.

Example 23

The two triangles are similar, with dimensions and areas as shown. What is the value of x?

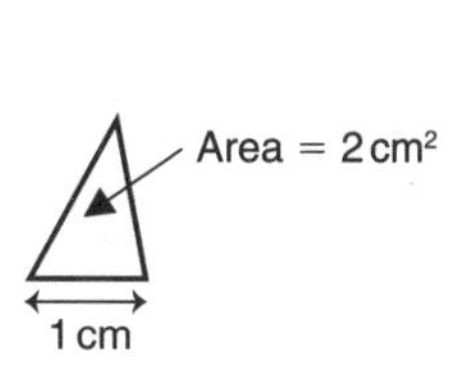

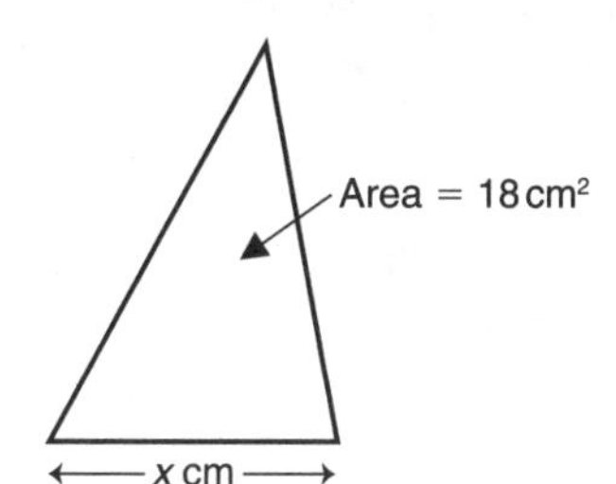

The Area Scale Factor $k^2 = \frac{18}{2} = 9$ (Note: Divide the area of the second shape by the area of the first shape.)

The Length Scale Factor $k = \sqrt{9} = 3$

So $x = 1 \times 3 = 3$ cm.

Exercise 43

1 A and B are similar shapes.
The area of A is 4 cm^2.
Find the area of B.

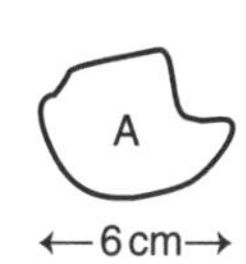

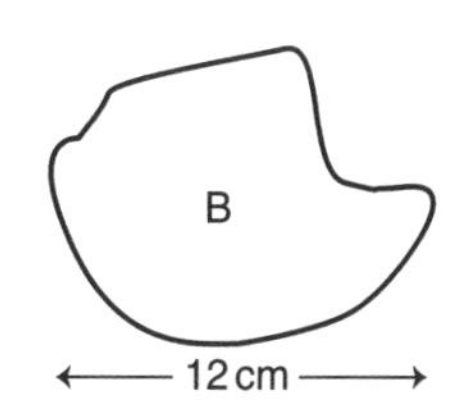

2 C and D are similar shapes.
The area of C is $6\,cm^2$.
Find the area of D.

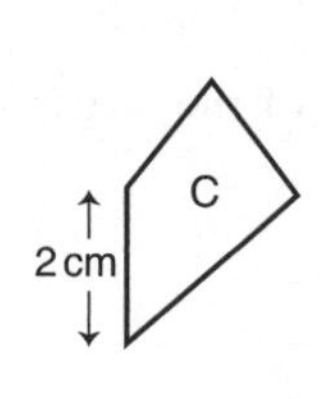

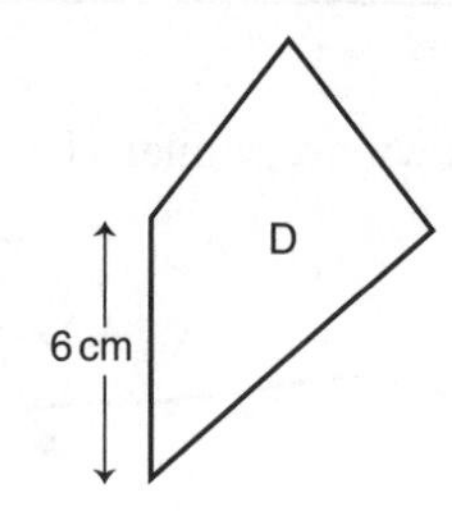

3 **a** Why are the two triangles shown similar?
b If the area of T_1 is $3.8\,cm^2$, find the area of T_2.

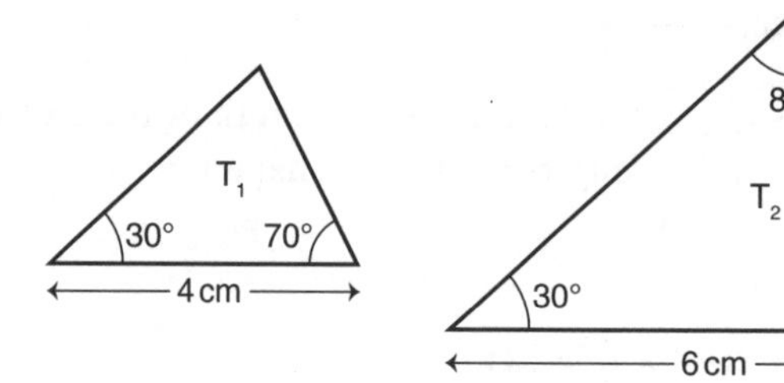

4 **a** Why are the two triangles shown similar?
b If the area of T_3 is $15.8\,cm^2$, find the area of T_4.

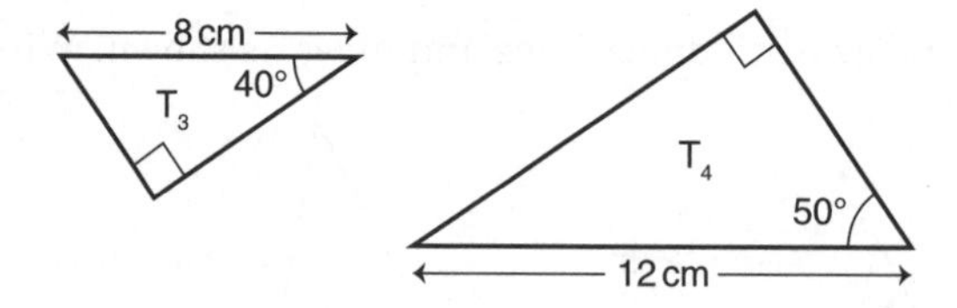

5 E and F are similar shapes.
The area of E is $480\,cm^2$.
Find the area of F.

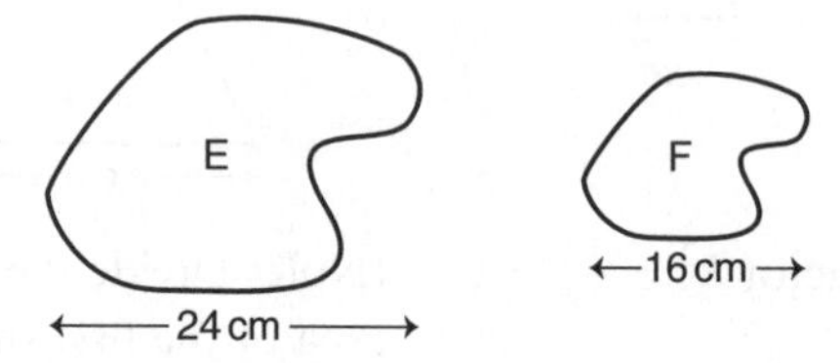

6 G and H are similar shapes.
The area of H is $36\,cm^2$.
Find the area of G.

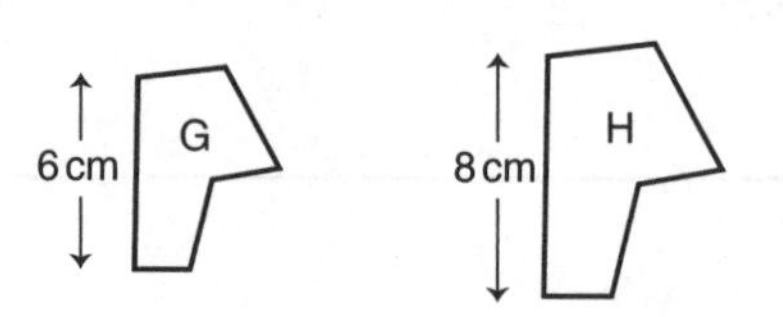

7 I and J are similar shapes.
The area of I is $150\,cm^2$.
Find the area of J.

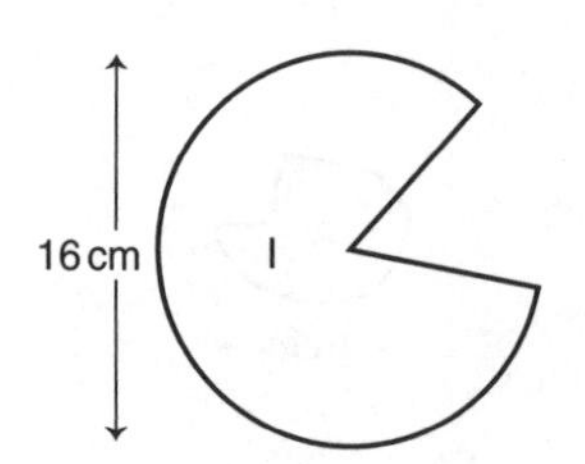

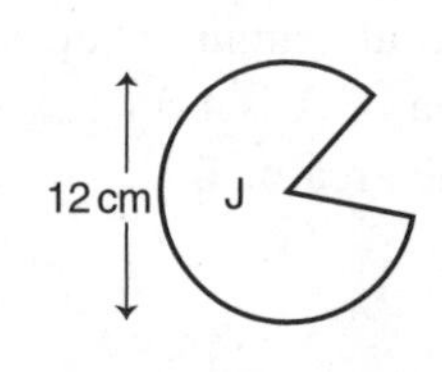

8 K and L are similar triangles.
The area of K is $24\,\text{cm}^2$.
Find the area of L.

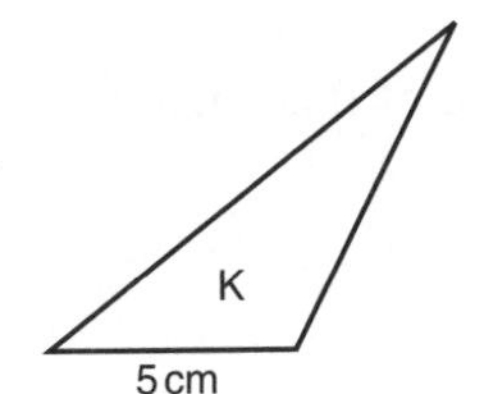

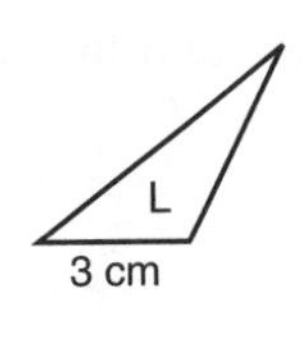

9 The shapes M and N are similar.
The area of M is $8\,\text{cm}^2$ and the area of N is $32\,\text{cm}^2$.
Find x.

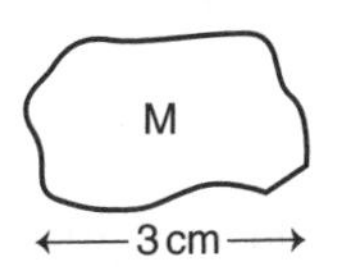

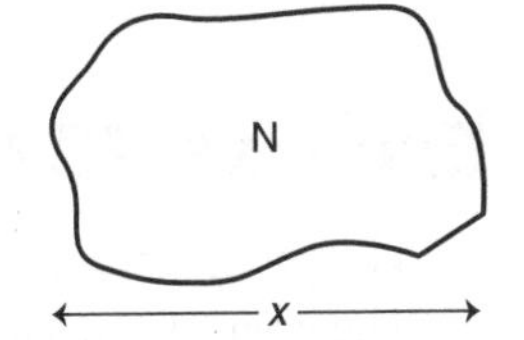

10 The shapes O and P are similar.
The area of O is $3\,\text{cm}^2$ and the area of P is $27\,\text{cm}^2$.
Find x.

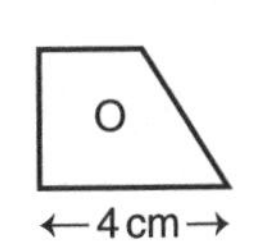

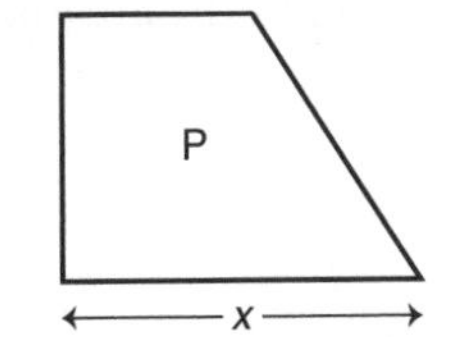

11 Q and R are similar shapes.
The area of Q is $5\,\text{cm}^2$ and the area of R is $11.25\,\text{cm}^2$.
Find x.

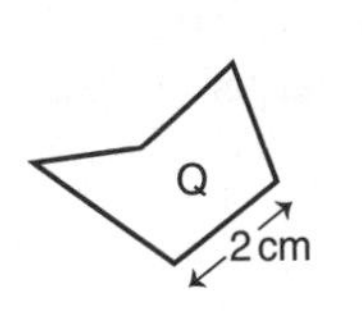

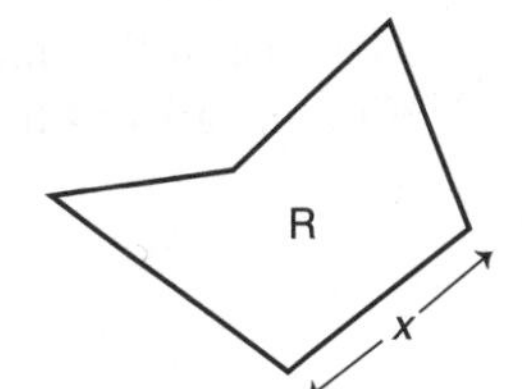

12 S and T are similar shapes.
The area of S is $9\,\text{cm}^2$ and the area of T is $16\,\text{cm}^2$.
Find x.

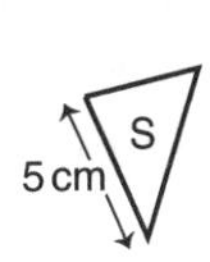

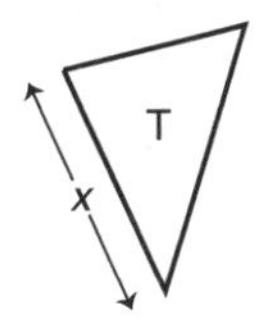

13 U and V are similar shapes.
The area of U is $48\,\text{cm}^2$ and the area of V is $12\,\text{cm}^2$.
Find x.

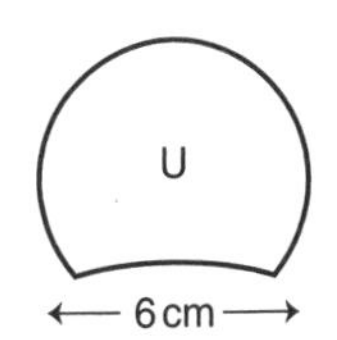

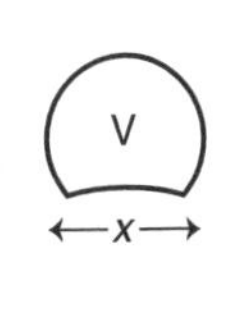

14 W and X are similar shapes.
The area of W is $36\,\text{cm}^2$ and the area of X is $4\,\text{cm}^2$.
Find x.

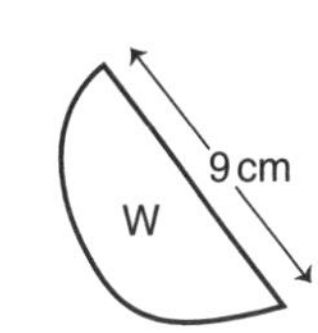

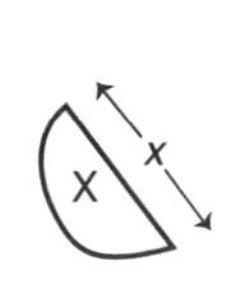

15 Y and Z are similar ellipses.
The area of Y is $225\,\text{cm}^2$ and the area of Z is $100\,\text{cm}^2$.
Find x.

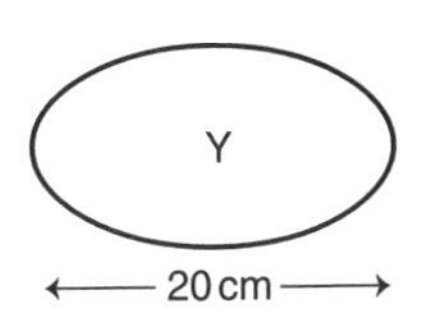

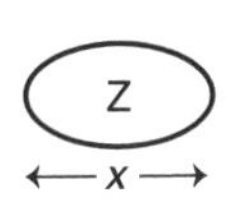

16 A and B are similar pentagons.
The area of A is $160\,cm^2$ and the area of B is $90\,cm^2$.
Find x.

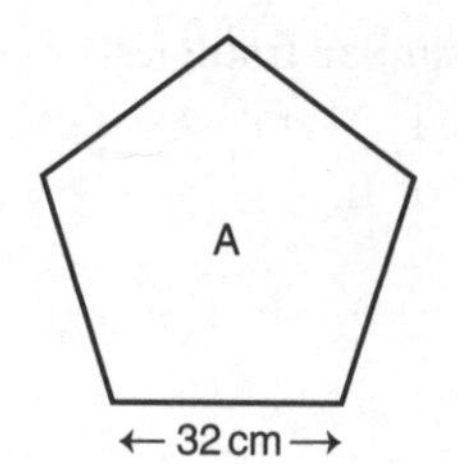

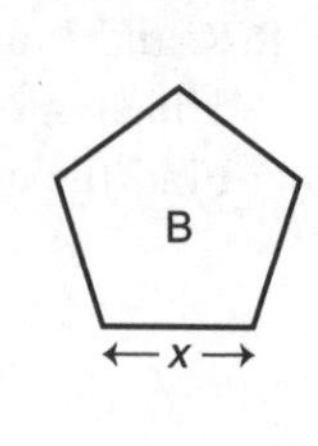

Exercise 43★

1 The two stars are similar in shape.
The area of the smaller star is $300\,cm^2$.
Find the area of the larger star.

2 The two shapes shown are similar.
The area of the smaller shape is $216\,cm^2$.
Find the area of the larger shape.

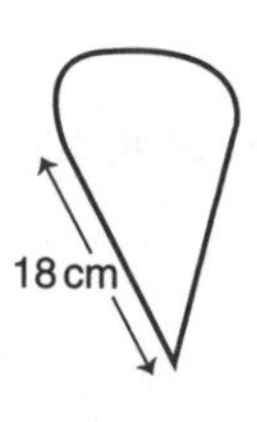

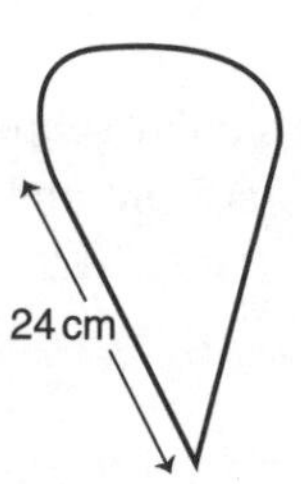

3 The two shapes shown are similar.
The area of the larger shape is $125\,cm^2$.
Find the area of the smaller shape.

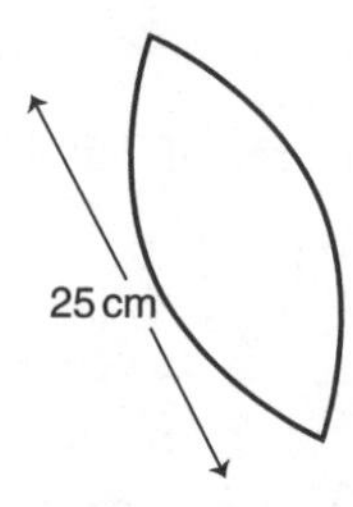

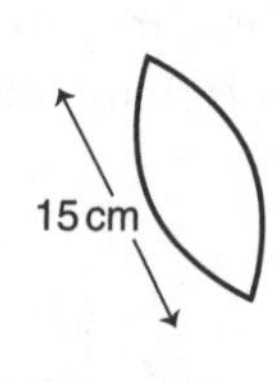

4 The two triangles are similar.
The area of the larger triangle is $75\,cm^2$.
Find the area of the smaller triangle.

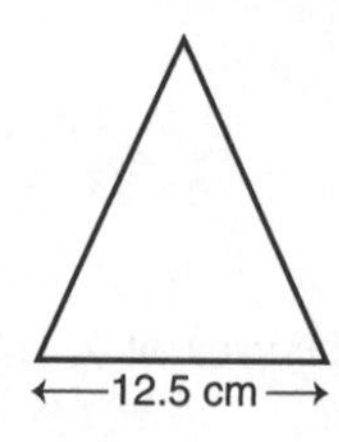

5 The two shapes are similar.
Find x.

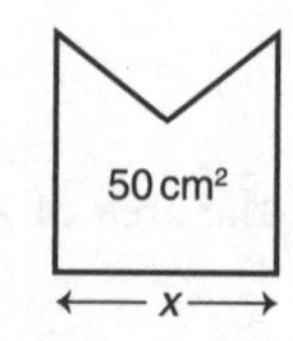

6 The two shapes are similar.
Find x.

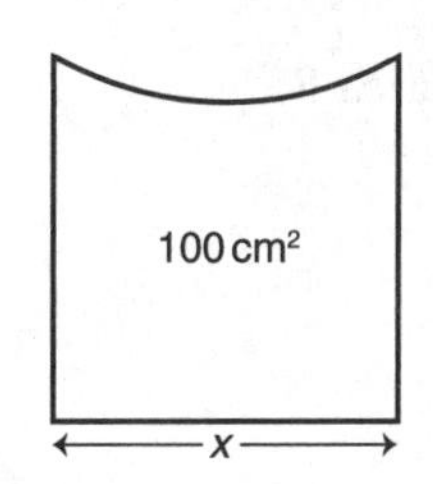

7 The two leaves shown are similar in shape.
Find x.

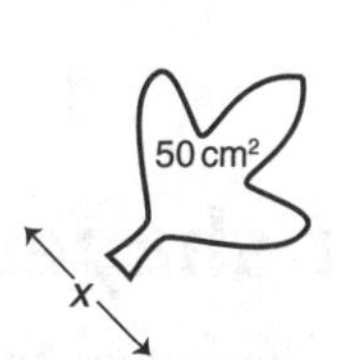

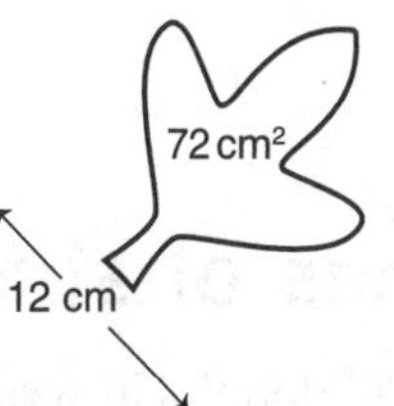

8 The two flower patterns are similar in shape.
Find x.

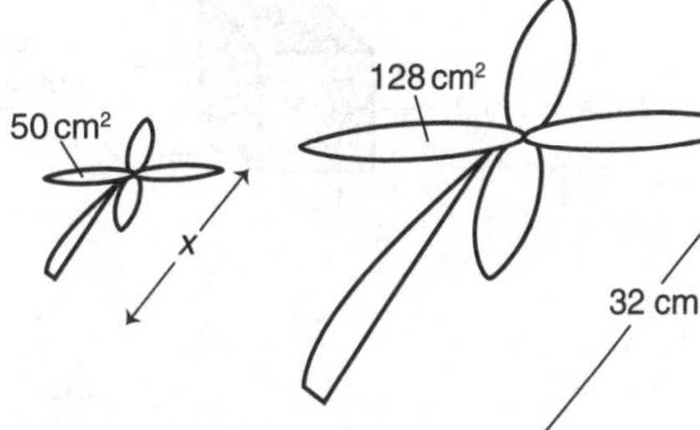

9 A model aeroplane is made to a scale of $\frac{1}{20}$ of the size of the real plane. The area of the wings of the real plane are $40\,\text{m}^2$. Find the area of the wings of the model in cm^2.

10 A map of a wood is drawn to a scale of $1:1000$. The area of the wood is $10^4\,\text{m}^2$. Find the area of the wood on the map in cm^2.

11 An oil slick increases in length by 20%. Assuming the shape is similar to the original shape, what is the percentage increase in area?

12 Abdul enlarges a digital photograph by 30%, keeping the photograph similar to the original. What is the percentage increase in area?

13 Meera washes some napkins in hot water and they shrink by 10%. What is the percentage reduction in area?

14 Jean reduces a document by 15% on a photocopier. What is the percentage reduction in area?

15 Calculate the shaded area A.

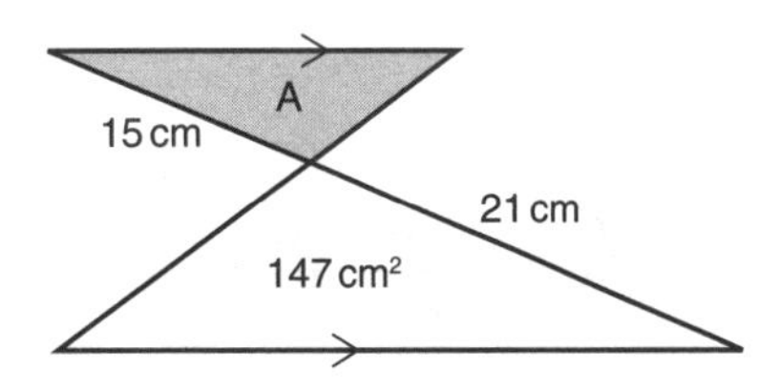

UNIT 2 ◆ Shape and space

16 Calculate the shaded area B.

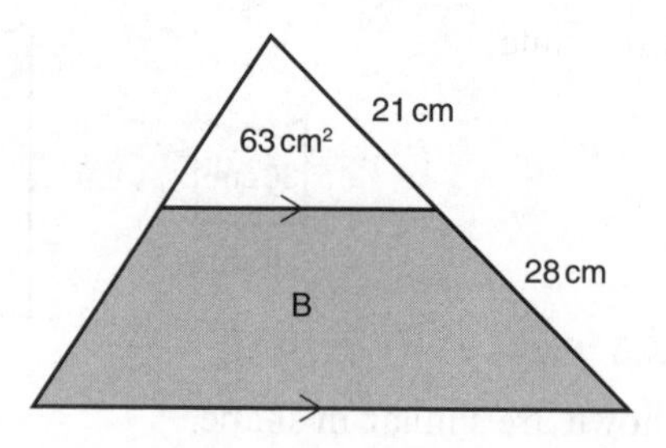

Volumes of similar shapes

When a solid doubles in size, the volume does **NOT** double, but increases by a factor of eight.

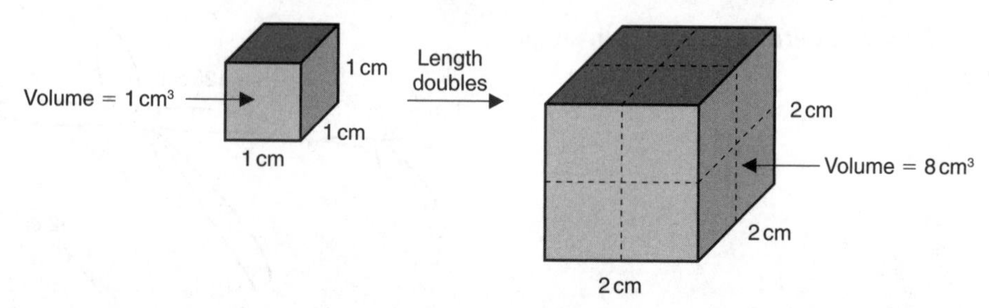

The Length Scale Factor is 2 and the Volume Scale Factor is 8.

If the solid triples in size, then the volume increases by a factor of 27.

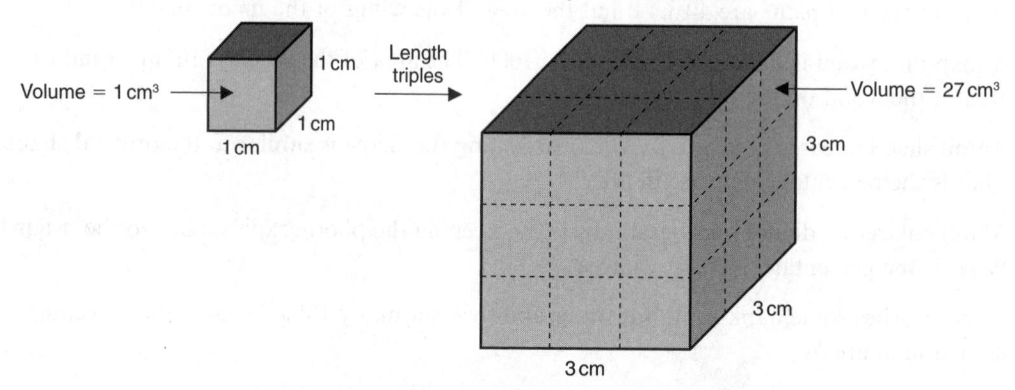

If a solid increases by a Length Scale Factor of k, then the Volume Scale Factor is k^3.

This applies even if the solid is irregular.

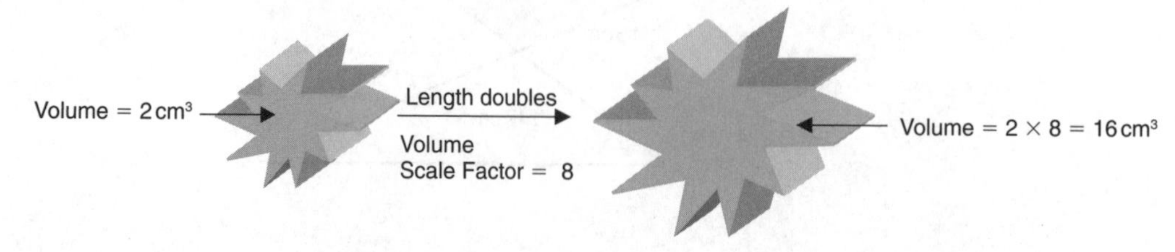

Note: The two solids must be similar.

Remember

If the Length Scale Factor is k, then the Volume Scale Factor is k^3.

Example 24

The two solids shown are similar. The volume of the smaller solid is $20\,\text{cm}^3$. Find the volume of the larger solid.

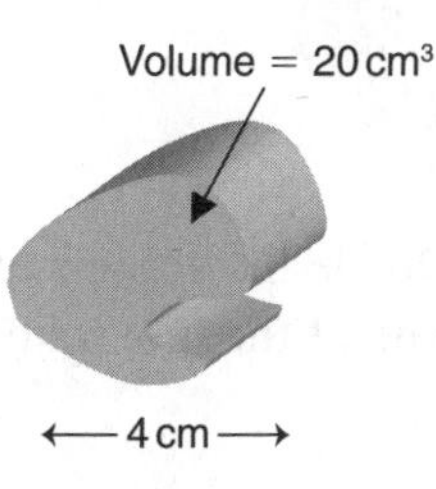

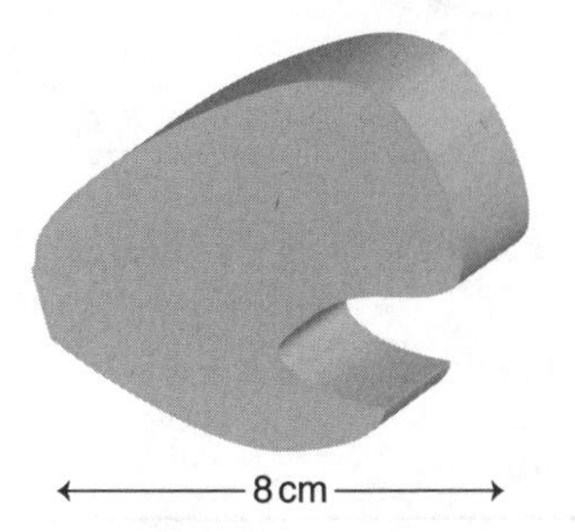

The Length Scale Factor $k = \frac{8}{4} = 2$ (Note: Divide the length of the second solid by the length of the first solid.)

The Volume Scale Factor $k^3 = 2^3 = 8$

So the volume of the larger solid is $20 \times 8 = 160\,\text{cm}^3$.

Example 25

The two cylinders shown are similar. The volume of the larger cylinder is $54\,\text{cm}^3$. Find the volume of the smaller cylinder.

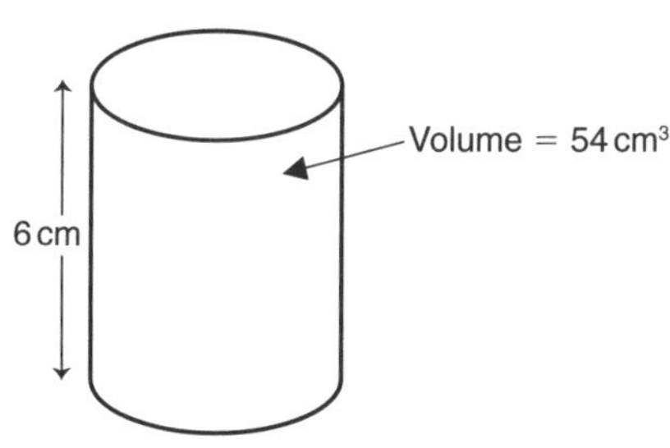

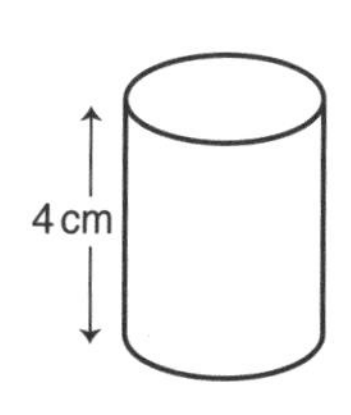

The Length Scale Factor $k = \frac{4}{6} = \frac{2}{3}$ (Note: Divide the height of the second cylinder by the height of the first cylinder)

The Volume Scale Factor $k^3 = \left(\frac{2}{3}\right)^3 = \frac{8}{27}$

So the volume of the smaller cylinder is $54 \times \frac{8}{27} = 16\,\text{cm}^3$.

ms are similar, with dimensions and volumes as shown. What is the value of x?

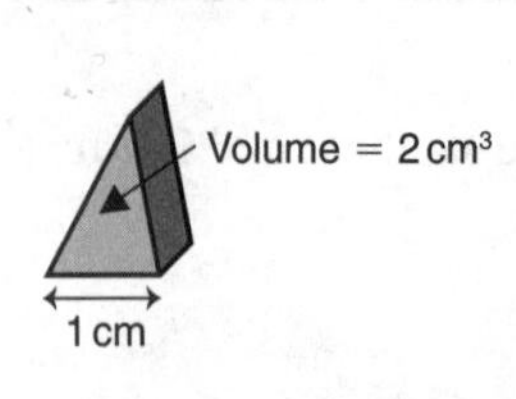

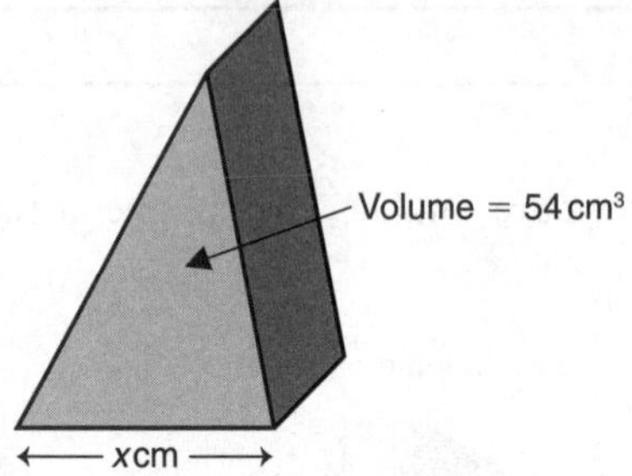

The Volume Scale Factor $k^3 = \frac{54}{2} = 27$ (Note: Divide the volume of the second solid by the volume of the first solid.)

The Length Scale Factor $k = \sqrt[3]{27} = 3$

So $x = 1 \times 3 = 3$ cm.

Exercise 44

1 The cylinders shown are similar.
Find the volume of the larger cylinder.

Volume = 100 cm³
8 cm
16 cm

2 The cones shown are similar.
Find the volume of the larger cone.

Volume = 2 cm³
2 cm

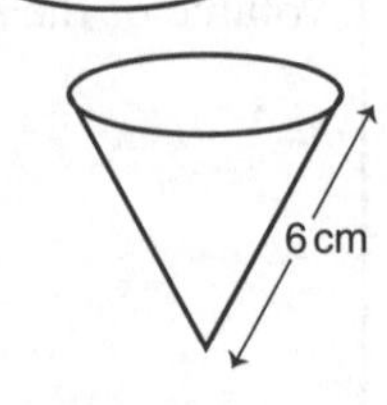

3 The solids shown are similar.
Find the volume of the larger solid.

6 cm
9 cm
Volume = 200 cm³

4 The statues shown are similar.
Find the volume of the larger statue.

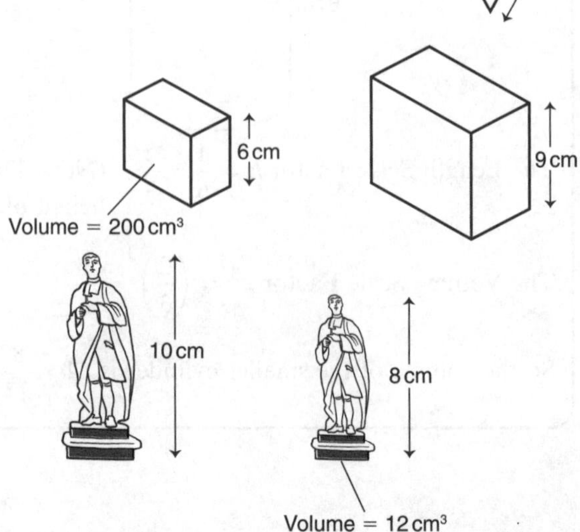

5 Find the volume of the smaller sphere.
(Note: All spheres are similar.)

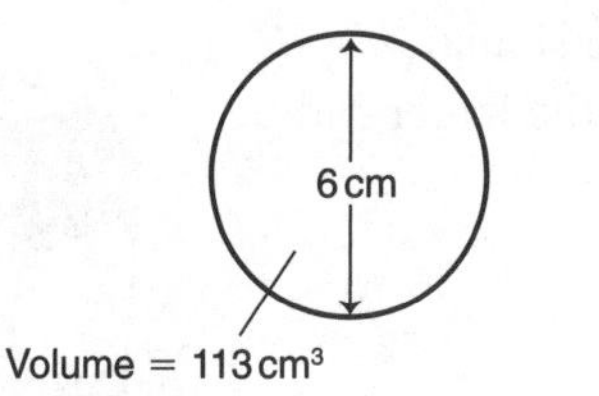

6 The two bottles are similar.
Find the volume of the smaller bottle.

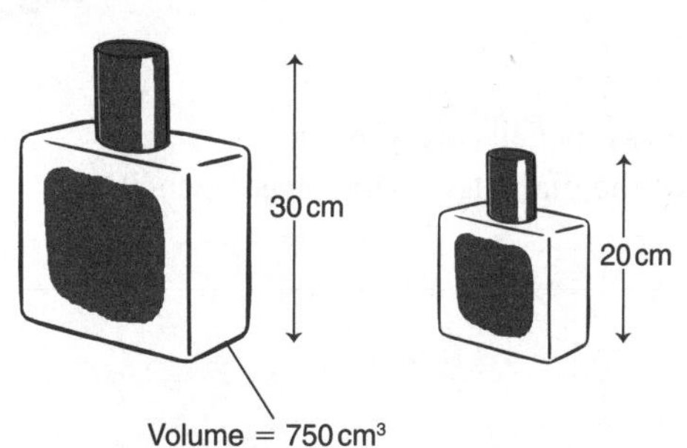

7 The two mobile phones are similar.
Find the volume of the smaller phone.

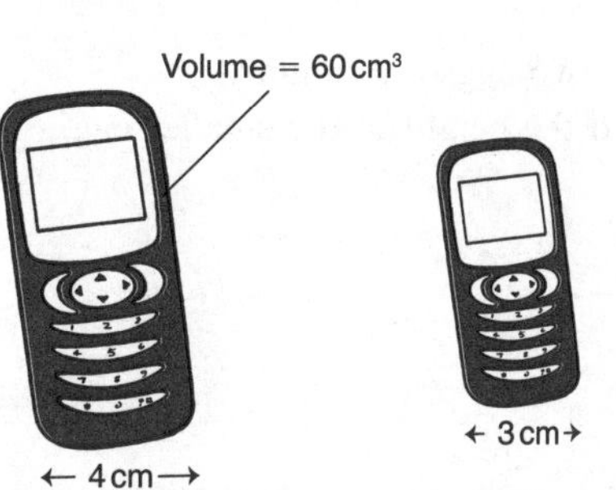

8 The two paperweights are similar.
Find the volume of the smaller weight.

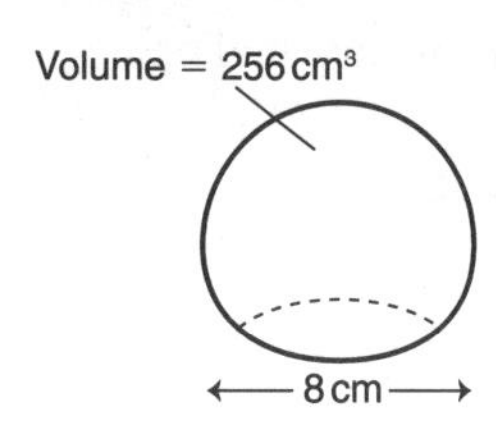

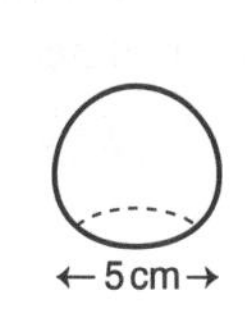

9 The two glasses are similar.
Find the height of the larger glass.

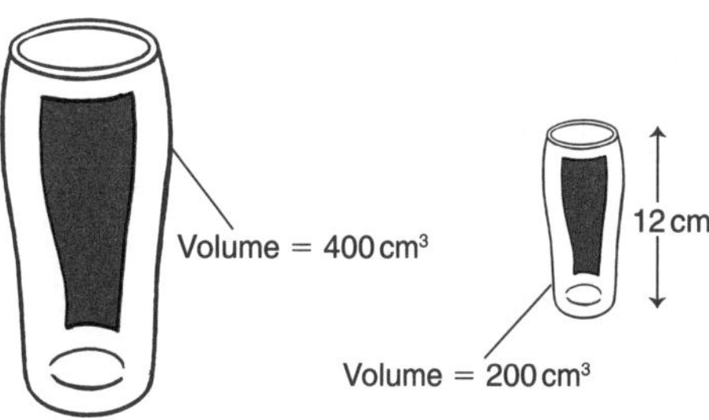

10 The two cats are similar.
Find the height of the larger cat.

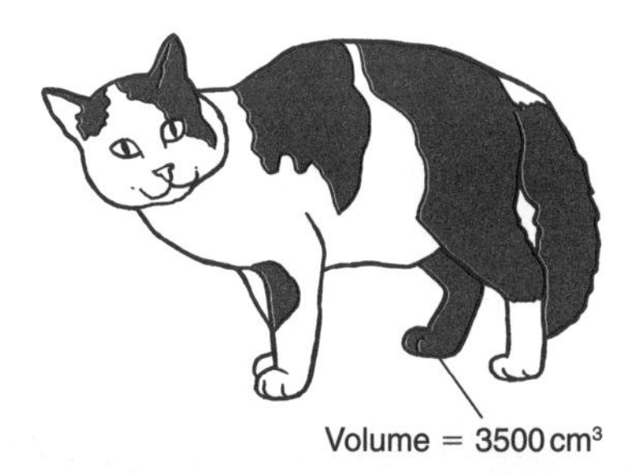

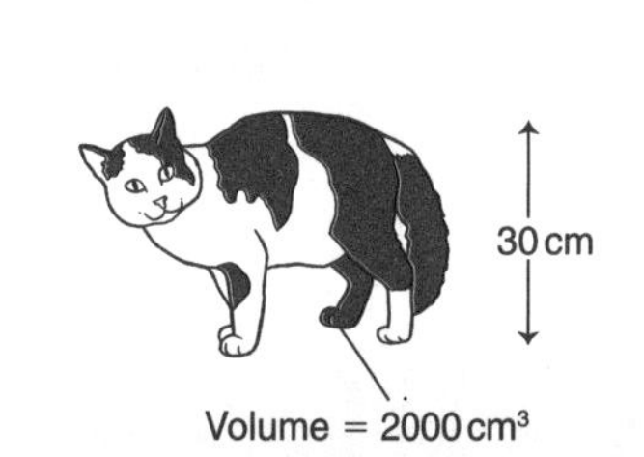

11 The two radios are similar.
Find the width of the larger radio.

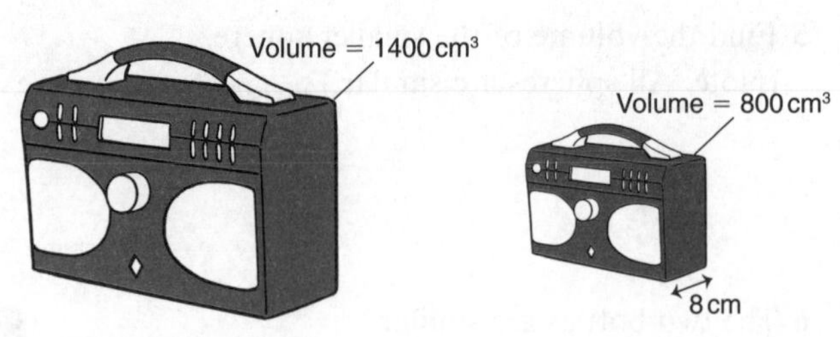

12 The two pencils are similar.
Find the diameter of the larger pencil.

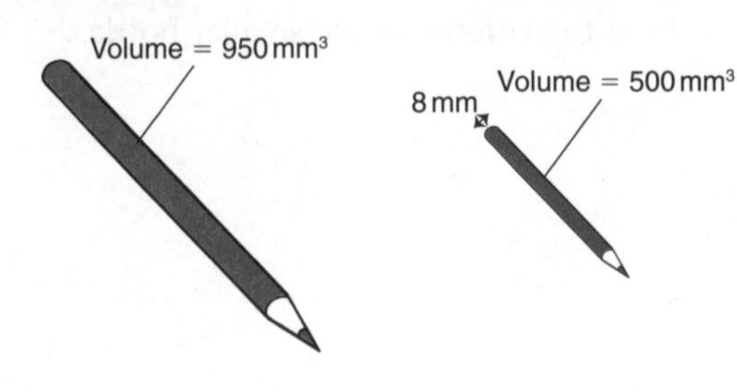

13 The two eggs are similar.
Find the height of the smaller egg.

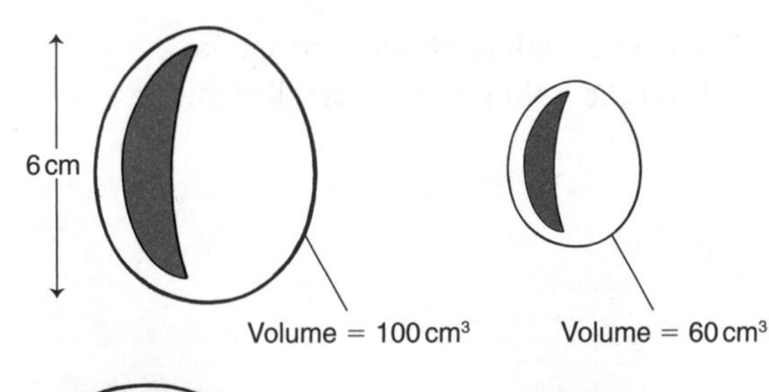

14 The two chocolates are similar.
Find the height of the smaller chocolate.

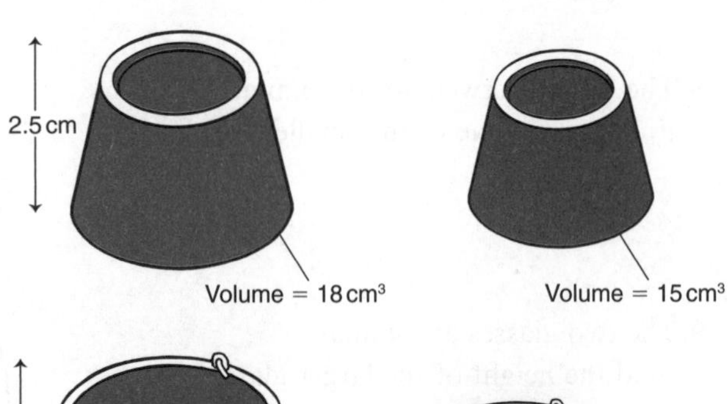

15 The two buckets are similar.
Find the height of the smaller bucket.

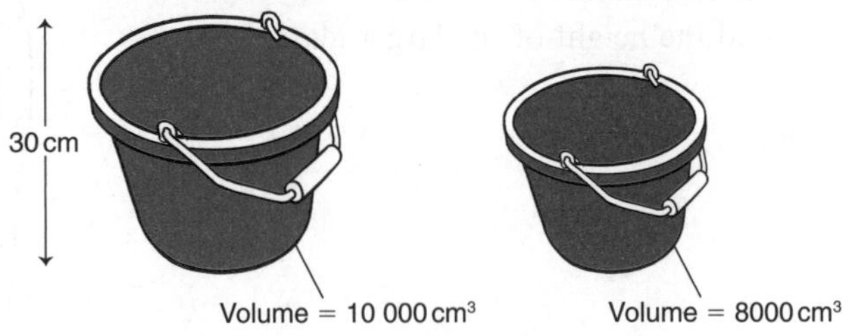

16 The two candles are similar.
Find the height of the smaller candle.

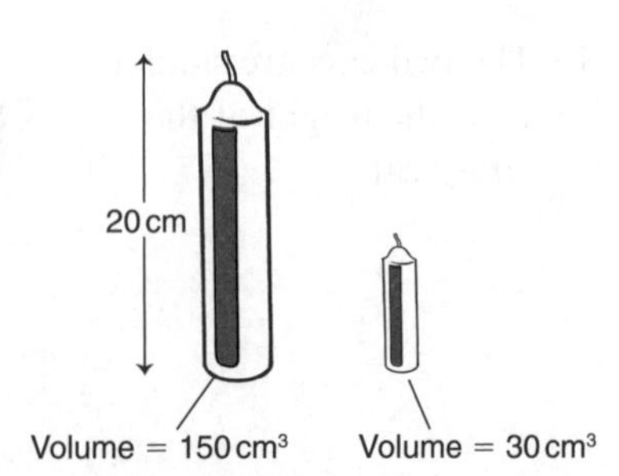

UNIT 2 ◆ Shape and space

Exercise 44★

1 X and Y are similar shapes. The volume of X is $40\,cm^3$. Find the volume of Y.

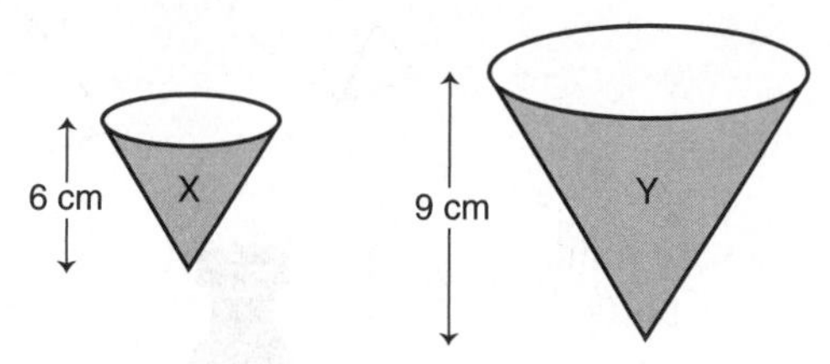

2 X and Y are similar shapes. The volume of Y is $50\,cm^3$. Find the volume of X.

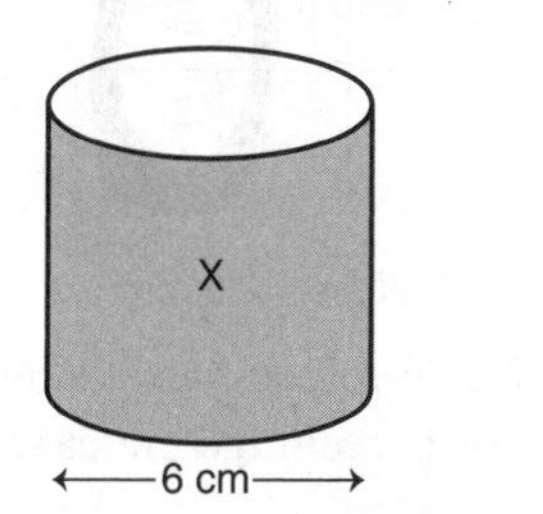

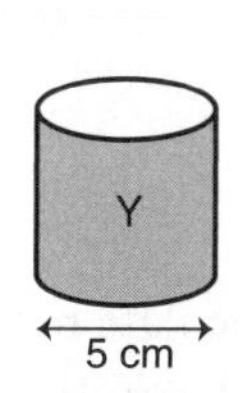

3 X and Y are similar shapes.
The volume of Y is $128\,cm^3$.
Find the volume of X.

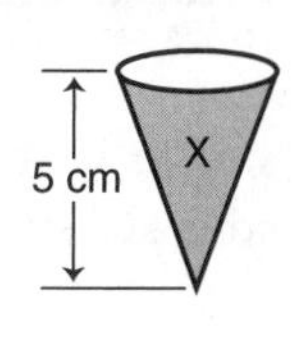

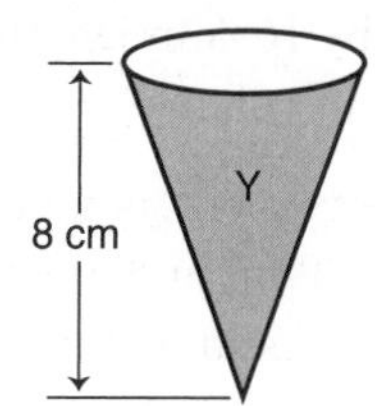

4 X and Y are similar shapes.
The volume of X is $108\,cm^3$.
Find the volume of Y.

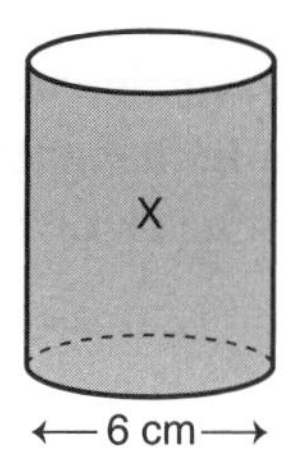

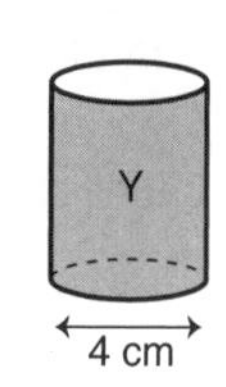

5 The two packets of cereal are similar.
Find the height of the larger packet.

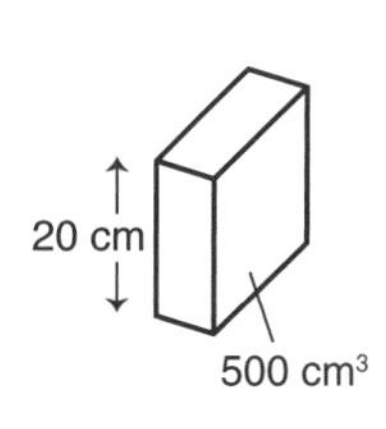

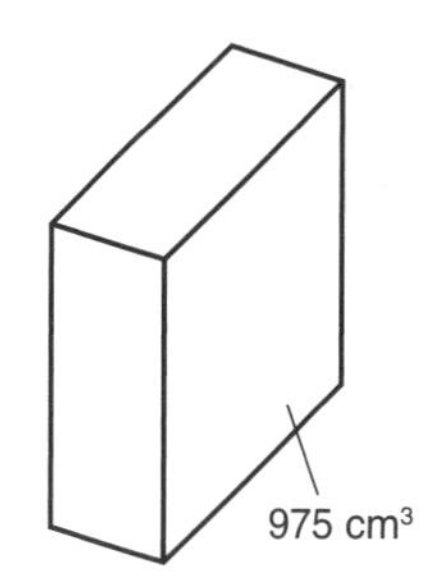

6 The two candlesticks are similar.
Find the height of the larger candlestick.

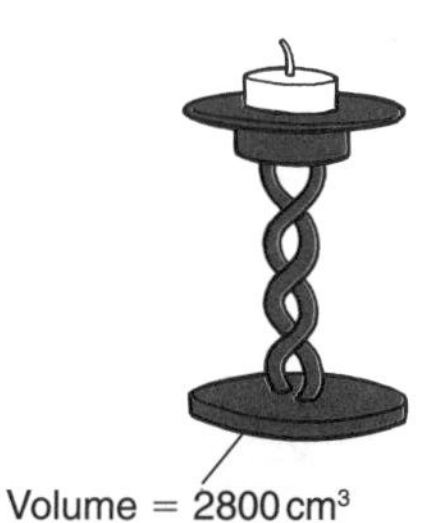

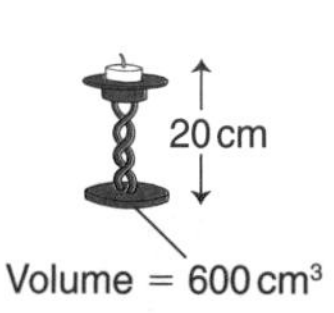

7 The two packets of sweets are similar.
Find the length of the smaller packet.

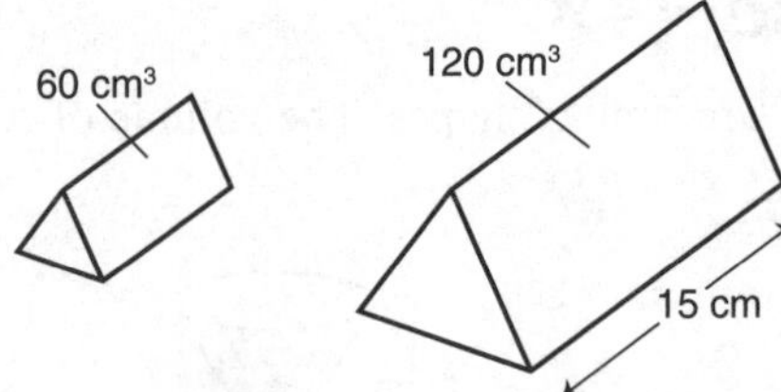

8 The two bottles of shampoo are similar.
Find the height of the smaller bottle.

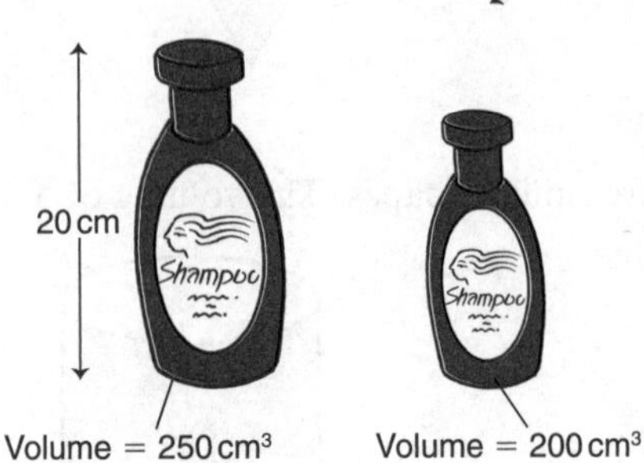

9 The manufacturers of a chocolate bar decide to produce a similar bar by increasing all dimensions by 20%. What would be a fair percentage increase in price?

10 A certain fish grew by 16% in length in a year and remained similar. What was the percentage increase in volume?

11 'ET Pizza' produces two pizzas that are similar in shape.
The smaller pizza is 20 cm in diameter and costs \$10.
The larger pizza is 30 cm in diameter.
What is a fair cost for the larger pizza?

12 Two wedding cakes are made from the same mixture and have similar shapes.
The larger cake costs \$135 and is 30 cm in diameter.
Find the cost of the smaller cake, which has a diameter of 20 cm.

13 A model aeroplane is made to a scale of $\frac{1}{20}$ of the real plane.

a The wing area of the model is $1000\,\text{cm}^2$. Find the wing area of the real plane in m^2.

b The volume of the real plane is $12\,\text{m}^3$. Find the volume of the model in cm^3.

14 A supermarket stocks similar small and large cans of beans. The areas of their labels are $63\,\text{cm}^2$ and $112\,\text{cm}^2$ respectively.

a The weight of the large can is 640 g. What is the weight of the small can?

b The height of the small can is 12 cm. What is the height of the large can?

15 Two solid statues are similar in shape and made of the same material.
One is 1 m high and weighs 64 kg. The other weighs 1 kg.

a What is the height of the smaller statue?

b If 3 g of gold is required to cover the smaller statue, how much is needed for the larger one?

16 A solid sphere weighs 10 g.

a What will be the weight of another sphere made from the same material but having three times the diameter?

b The surface area of the 10 g sphere is $20\,\text{cm}^2$. What is the surface area of the larger sphere?

17 Suppose that a grown-up hedgehog is an exact enlargement of a baby hedgehog on a scale of 3:2, and that the baby hedgehog has 2000 quills with a total length of 15 m and a skin area of $360\,cm^2$.

a How many quills would the grown-up hedgehog have?

b What would be their total length?

c What would be the grown-up hedgehog's skin area?

d If the grown-up hedgehog weighed 810 g, what would the baby hedgehog weigh?

Exercise 45 (Revision)

1 Change 3 km to mm.

2 Change $4 \times 10^4\,cm^2$ to m^2.

3 Change $1\,m^3$ to cm^3.

4 Find the area and perimeter of the shape shown.

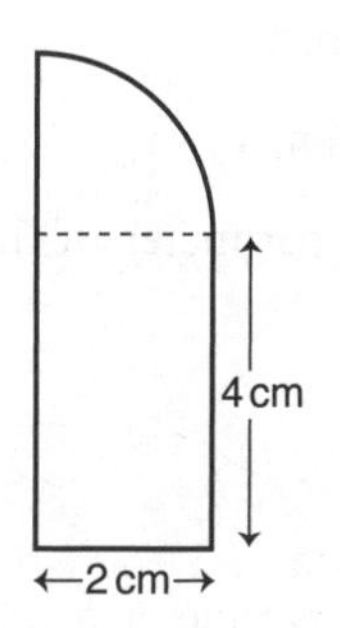

5 Find the area and perimeter of the shape shown.

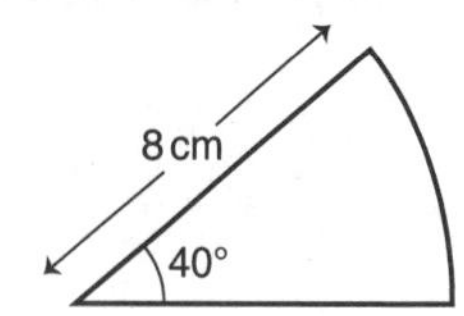

6 Find the volume and surface area of this prism.

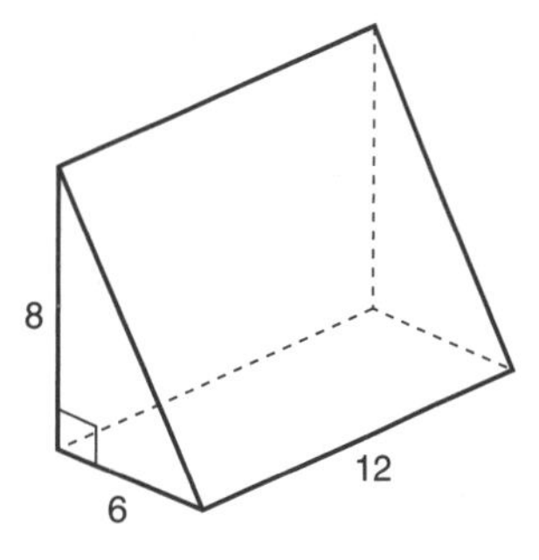

7 Find the volume of the garage shown.
The height of the top of the pyramid roof is 5 m from ground level.

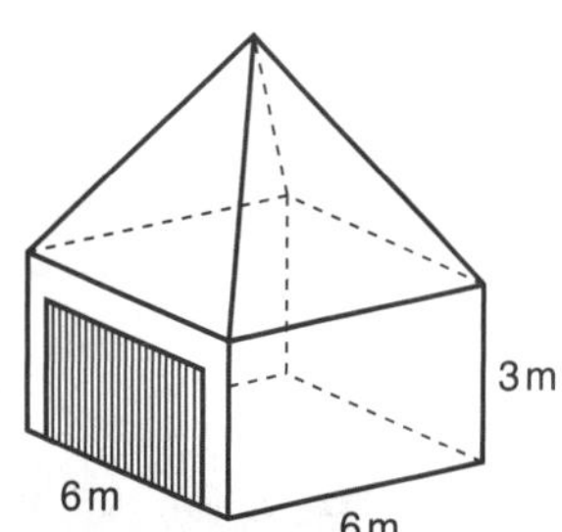

8 X and Y are similar shapes. The area of Y is $50\,\text{cm}^2$. Find the area of X.

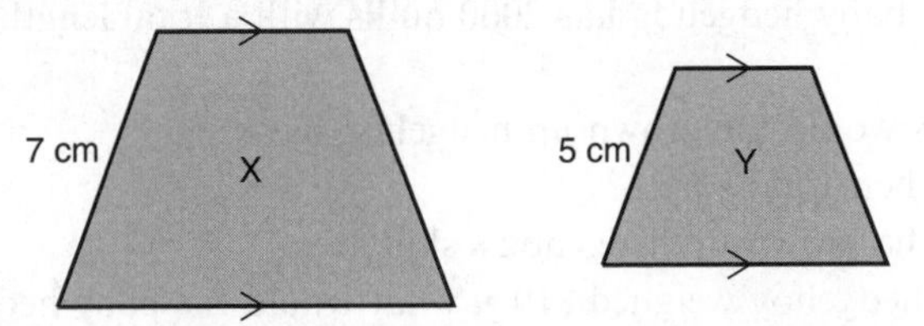

9 Two similar buckets have depths of 30 cm and 20 cm.
The smaller bucket holds 8 litres of water. Find the capacity of the larger bucket.

Exercise 45★ (Revision)

1 Change 4×10^6 mm to km.

2 Change $5\,\text{m}^2$ to mm^2.

3 Change $8 \times 10^{12}\,\text{mm}^3$ to m^3.

4 Find the area and perimeter of the shape shown shaded.

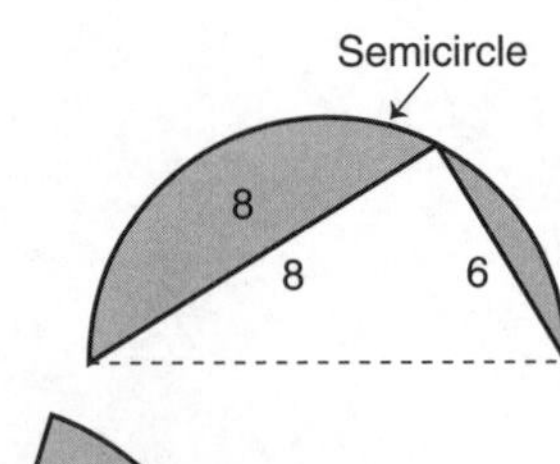

5 Find the area and perimeter of the shape shown.

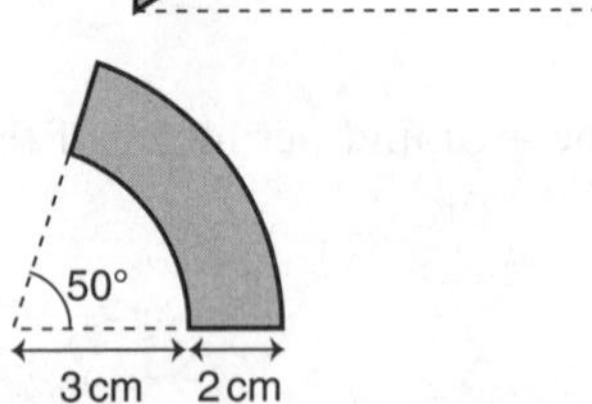

6 Find the volume and surface area of this prism.

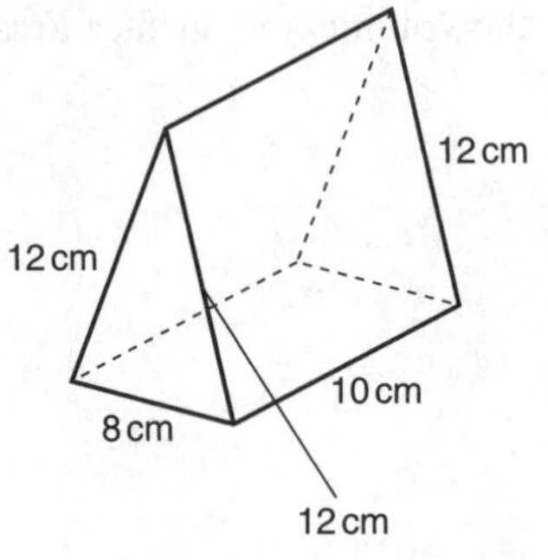

7 Find the volume and surface area of this partially burnt candle.

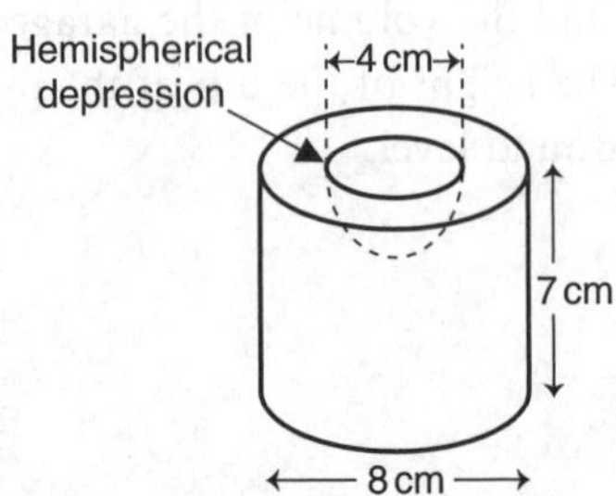

8 X and Y are similar shapes. The area of X is 81 cm^2; the area of Y is 49 cm^2. Find x.

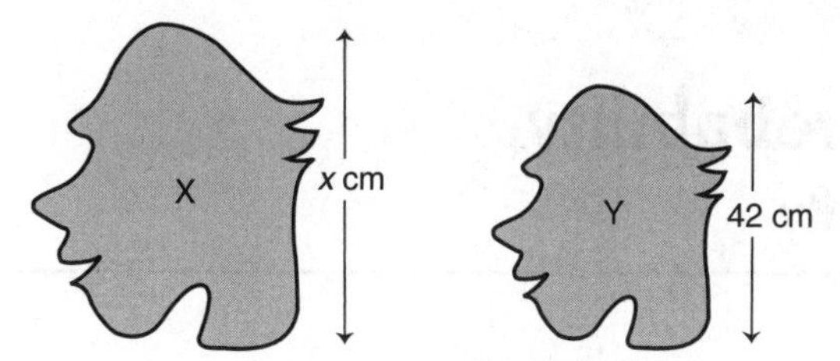

9 'McEaters' sells drinks in three similar cups, small, medium and large.

a The height of the small cup is 10 cm and the height of the large cup is 15 cm. The small cup costs \$2. What is a fair price for the large cup?

b The medium cup costs \$3.47. What is the height of a medium cup?

Handling data 2

Compound probability

Laws of probability

Remember

- $p(A)$ means the probability of event A occurring.
- $p(\overline{A})$ means the probability of event A *not* occurring.
- $0 \leqslant p(A) \leqslant 1$
- $p(A) + p(\overline{A}) = 1$, so $p(\overline{A}) = 1 - p(A)$.

Example 1

Calculate the probability that a prime number is not obtained when a fair 10-sided spinner that is numbered from 1 to 10 is spun.

Let A be the event that a prime number is obtained.

$p(\overline{A}) = 1 - p(A)$

$= 1 - \frac{4}{10}$ (There are 4 prime numbers: 2, 3, 5, 7)

$= \frac{6}{10} = \frac{3}{5}$

Independent events

If two events have no effect on each other, they are independent.

If it snowed in Moscow, it would be unlikely that this event would have any effect on your teacher winning the lottery on the same day. These events are said to be *independent*.

Mutually exclusive events

Some events exclude the outcome of another. A number rolled on a dice cannot be both odd and even. These events are said to be *mutually exclusive*.

Combined events

Multiplication ('and') rule

Two dice are thrown together.
One is a fair dice numbered 1 to 6.
On the other, each face is of a different colour: red, yellow, blue, green, white and purple.

All possible outcomes are shown in this possibility-space diagram.

	R	Y	B	G	W	P
1	•	•	•	•	•	⊙
2	•	•	•	•	•	•
3	•	•	•	•	•	⊙
4	•	•	•	•	•	•
5	•	•	•	•	•	⊙
6	•	•	•	•	•	•

What is the probability that the dice will show an odd number and a purple face?

Let this event be A. By inspection of the possibility space, $p(A) = \frac{3}{36} = \frac{1}{12}$.

Event O is that an odd number is thrown. $p(O) = \frac{1}{2}$.

Event P is that a purple colour is thrown. $p(P) = \frac{1}{6}$.

$p(A) = p(O \text{ and } P) = p(O) \times p(P) = \frac{1}{2} \times \frac{1}{6} = \frac{1}{12}$

Remember

For A and B, two independent events, the probability of both events occurring is:

$\mathbf{p(A \text{ and } B) = p(A) \times p(B)}$

Addition ('or') rule

A card is randomly selected from a pack of 52 playing cards.
What is the probability that it is an ace or a king?

Let this event be E. There are 8 cards which are either aces or kings, so $p(E) = \frac{8}{52}$.

Event A is that an ace is selected. $p(A) = \frac{4}{52}$. Event K is that a king is selected. $p(K) = \frac{4}{52}$.

$p(E) = p(A \text{ or } K) = p(A) + p(K) = \frac{4}{52} + \frac{4}{52} = \frac{8}{52}$

Remember

For A and B, two mutually exclusive events, the probability of event A or event B occurring is:

$\mathbf{p(A \text{ or } B) = p(A) + p(B)}$

Tree diagrams

Tree diagrams show all the possible outcomes. Together with the 'and' and 'or' rules, they can make problems easier to solve.

Example 2

A litter of Border Collie puppies contains four females and two males. A vet randomly removes one from its basket, and it is *not* replaced before another is chosen.
What is the probability that the vet removes two males?

Let event F be that a female is picked.
Let event M be that a male is picked.

Notice that when the second puppy is taken, there are only five left in the basket.

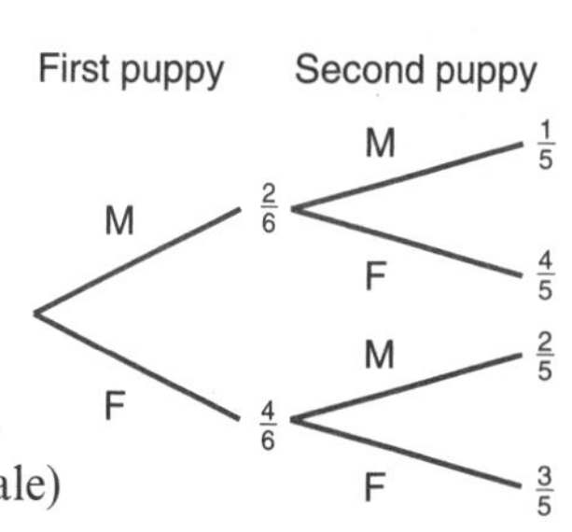

Let event A be that two males (dogs) are chosen.

$p(A) = p(M_1 \text{ and } M_2)$ (M_1 means the first puppy is a male. M_2 means the second puppy is a male)

$= p(M_1) \times p(M_2)$ (given by tree diagram route MM)

$= \frac{2}{6} \times \frac{1}{5} = \frac{2}{30} = \frac{1}{15}$

What is the probability that the vet removes one male and one female?

Let event B be that a male and a female are chosen.

$p(B) = p(M_1 \text{ and } F_2)$ or $p(F_1 \text{ and } M_2)$
(given by tree diagram routes MF and FM)

$= \frac{2}{6} \times \frac{4}{5} + \frac{4}{6} \times \frac{2}{5}$

$= \frac{8}{30} + \frac{8}{30} = \frac{16}{30} = \frac{8}{15}$

Exercise 46

Use tree diagrams to solve these problems.

1 A fair six-sided dice is thrown twice.
Calculate the probability of obtaining these scores.

a Two sixes
b No sixes
c A six and not a six, in that order
d A six and not a six, in any order

2 A box contains two red and three green beads.
One is randomly chosen, and replaced before another is chosen.
Calculate the probability of obtaining these combinations of beads.

a Two red beads
b Two green beads
c A red bead and a green bead, in that order
d A red bead and a green bead, in any order

3 A chest of drawers contains four yellow ties and six blue ties.
One is randomly selected and replaced before another is chosen.
Calculate the probability of obtaining these ties.

a Two yellow ties
b Two blue ties
c A yellow tie and a blue tie, in that order
d A yellow tie and a blue tie, in any order

4 A spice rack contains three jars of chilli and four jars of mint.
One is randomly selected, and replaced before another is chosen.

a Calculate the probability of selecting two jars of chilli.
b What is the probability of selecting a jar of chilli and a jar of mint?

5 In a game of basketball the probability of scoring from a free shot is $\frac{2}{3}$.
A player has two consecutive free shots.

a Calculate the probability that he scores two baskets.
b What is the probability that he scores one basket?
c What is the probability that he scores no baskets?

6 Each evening Dina either reads a book or watches television.
The probability that she watches television is $\frac{3}{4}$, and if she does this, the probability that she will fall asleep is $\frac{4}{7}$. If she reads a book, the probability that she will fall asleep is $\frac{3}{7}$.

a Calculate the probability that she does not fall asleep.
b What is the probability that she *does* fall asleep?

7 Two cards are picked at random from a pack of 52 playing cards with replacement. Calculate the probability that these cards are picked.

a Two kings
b A red card and a black card
c A picture card and an odd number card
d A heart and a diamond

8 Helga oversleeps one day in 5, and when this happens she breaks her shoelace 2 out of 3 times. When she does *not* oversleep, she breaks her shoelace only 1 out of 6 times. If she breaks her shoelace, she is late for school.

a Calculate the probability that Helga is late for school.
b What is the probability that she is not late for school?
c In the space of 30 school days, how many times would you expect Helga to be late?

The 'at least' situation

Example 3

A fruit basket contains two oranges (O) and three apples (A).
A fruit is selected at random and *not* replaced before another is randomly selected.
Calculate the probability of choosing *at least* one apple.

Let the number of apples be a.
'At least one apple' means that $a \geqslant 1$.

One method of calculating $p(a \geqslant 1)$ is to find all the possibilities.

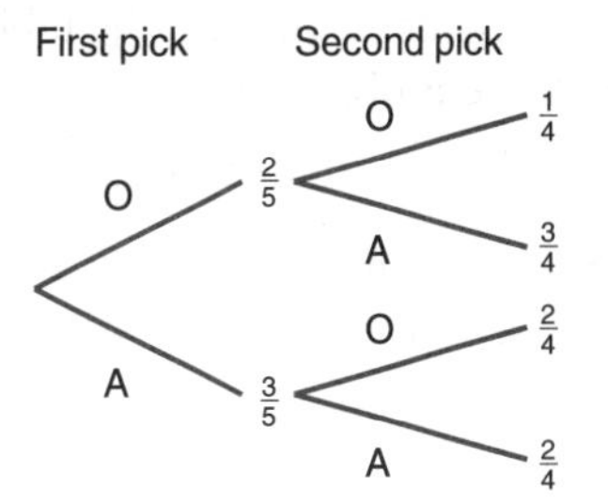

$$p(a \geqslant 1) = p(AO) + p(OA) + p(AA)$$
$$= \tfrac{3}{5} \times \tfrac{2}{4} + \tfrac{2}{5} \times \tfrac{3}{4} + \tfrac{3}{5} \times \tfrac{2}{4}$$
$$= \tfrac{3}{10} + \tfrac{3}{10} + \tfrac{3}{10} = \tfrac{9}{10}$$

A quicker method uses the rule $p(E) + p(\overline{E}) = 1$.

$$p(a \geqslant 1) = 1 - p(a = 0)$$
$$= 1 - p(OO)$$
$$= 1 - \tfrac{2}{5} \times \tfrac{1}{4} = 1 - \tfrac{1}{10} = \tfrac{9}{10}$$

Exercise 46★

1 A box contains two black stones and four white stones.
One is randomly selected and *not* replaced before another is randomly taken out.
Use a tree diagram to help you calculate the probability of selecting these stones:

a A black stone and a white stone, in that order
b A black stone and a white stone, in any order
c At least one black stone.

2 A bag contains three orange counters and four purple counters. One is randomly selected and replaced *together with a counter of the colour not picked out* (orange or purple). Another counter is then randomly selected. Use a tree diagram to help you to calculate these probabilities:

a p(Two orange counters)
b p(One counter of each colour)
c p(At least one purple counter)

3 An archer fires his arrows at a target.
The probability that he scores a bullseye in any one attempt is $\frac{1}{3}$. If he fires twice, calculate these probabilities:

a p(Two bullseyes)
b p(No bullseyes)
c p(At least one bullseye)

4 A netball shooter has two free shots.
The probability that she scores (or misses) with each shot can be found from this table.

	First shot	Second shot
Scores	$\frac{2}{3}$	
Misses		$\frac{3}{7}$

Copy and complete the table, and use it to construct a tree diagram to help you to calculate these probabilities:

a p(Both shots missed) **b** p(Scores once) **c** p(Scores at least once)

5 The spinner is spun twice.
Use tree diagrams to help you to calculate these probabilities:

a p(Two even numbers)
b p(An even number and an odd number)
c p(A black number and a white number)
d p(A white even number and a black odd number)

6 A marble is randomly taken from bag A and is then placed in bag B. A marble is then randomly selected from bag B.

a Use a tree diagram to help you to find the probability that this marble is black.
b What is the probability that the marble is white?

7 A box contains two red sweets and three green sweets. A sweet is selected at random and *not* replaced. If a red sweet is picked on the first attempt, then two extra reds are placed in the box. If a green sweet is picked on the first attempt, then three extra greens are placed in the box. Calculate these probabilities of selecting from two picks:

a p(Two red sweets)
b p(A red sweet and a green sweet)
c p(At least one green sweet)

8 Suzi has just taken up golf, and she buys a golf bag containing five different clubs. Unfortunately, she does not know when to use each club, and so chooses them randomly for each shot. The probabilities for each shot that Suzi makes are shown in this table.

	Good shot	Bad shot
Right club	$\frac{2}{3}$	
Wrong club		$\frac{3}{4}$

a Copy and complete the table, and use it to construct a tree diagram to calculate the probability that Suzi makes a good shot.

b At one short hole, she can reach the green in one shot if it is 'good'. If her first shot is 'bad', it takes one more 'good' shot to reach the green. Find the probability that she reaches the green in at most two shots.

Activity 13

Network XAB: XAB is a simple one-way road system. When there is a choice at a junction, a driver is *equally likely* to turn down any accessible road.

p(A) is the probability of reaching A starting at X.
p(B) is the probability of reaching B starting at X.

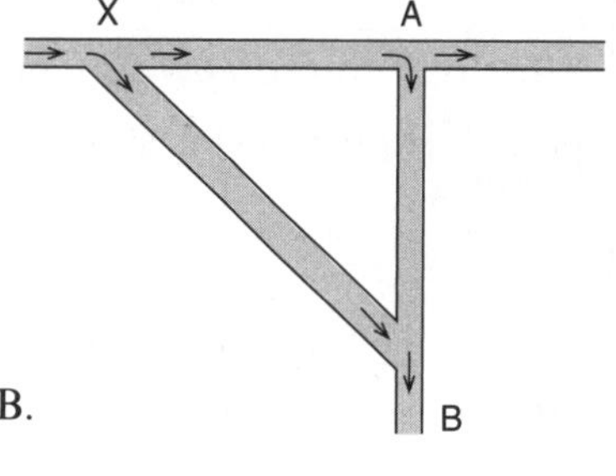

- Copy the network XAB.
  Write down the probability of going from X to B and A to B.
  Calculate p(A) and p(B).
- 60 vehicles pass through X.
  How many vehicles would you expect to pass through junction A?
  How many would you expect to pass through junction B?
  How many would you expect to go through B not via A?

Network XABCD: The road system is now extended as shown.

- Copy the network XABCD, and write down the probability of turning into each of the roads.
  Calculate p(A), p(B), p(C) and p(D).
- 60 vehicles pass through X.
  How many vehicles would you expect to pass through each junction: A, B, C and D?
- Compare these theoretical answers by devising a method to simulate 60 vehicles passing through the road network XABCD with each vehicle starting at X (hint: dice, counters or computers).

'Harder' problems

Example 4

A vet has a 90% probability of detecting a particular virus in a horse.
If he detects the virus, the operation has an 80% success rate the first time it is attempted.
If this operation is unsuccessful, it can be repeated, but with a success rate of only 40%.
Any subsequent operations have such a low chance of success that a vet will not attempt further operations.
Let event D be that the virus is detected.
Let event S be that the operation succeeds.
Let event F be that the operation fails.

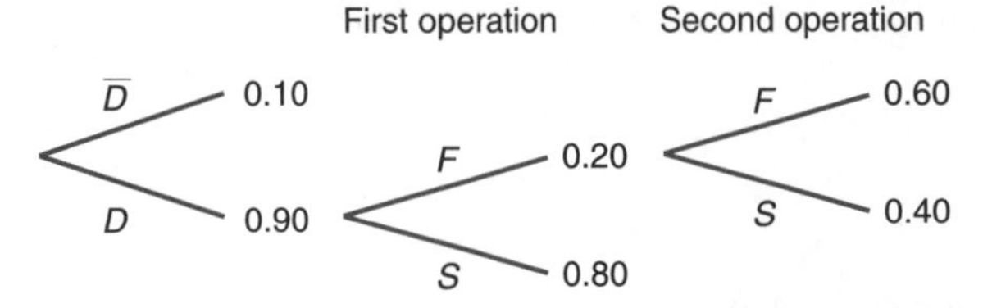

What is the probability that an infected horse will be operated on successfully once?

$$\begin{aligned}\text{p(the first operation is successful)} &= \text{p}(D \text{ and } S)\\ &= \text{p}(D) \times \text{p}(S)\\ &= 0.90 \times 0.80 = 0.72\end{aligned}$$

What is the probability that an infected horse will be cured at the second attempt?

$$\begin{aligned}\text{p(the second operation is successful)} &= \text{p}(D \text{ and } F \text{ and } S)\\ &= \text{p}(D) \times \text{p}(F) \times \text{p}(S)\\ &= 0.90 \times 0.20 \times 0.40 = 0.072\end{aligned}$$

What is the probability that an infected horse will be cured?

$$\begin{aligned}\text{p(the horse is cured)} &= \text{p(the first operation is successful or the second operation is successful)}\\ &= \text{p(the first operation is successful)} + \text{p(the second operation is successful)}\\ &= 0.72 + 0.072 = 0.792\end{aligned}$$

Exercise 47★

1 Two normal six-sided dice have their spots covered and replaced by the letters A, B, C, D, E and F, with one letter on each face.

a If two such dice are thrown, calculate the probability that they show two vowels. (Vowels are a, e, i, o, u.)

b What is the probability of throwing a vowel and a consonant? (Consonants are non-vowels.)

2 A pack of 20 cards is formed using the ace, ten, jack, queen and king of each of the four suits from an ordinary full pack of playing cards. This reduced pack is shuffled and then dealt one at a time *without replacement*. Calculate these probabilities:

a p(The first card dealt is a king)
b p(The second card dealt is a king)
c p(At least one king is dealt in the first three cards)

3 A box contains two white beads and five red beads.
A bead is randomly selected and its colour noted.
It is then returned to the box together with two more beads of the same colour.

a If a second bead is now randomly selected from the box, calculate the probability that it is the same colour as the first bead.
b What is the probability that the second bead is a different colour from the first bead?
c Find the probability that the second bead drawn is white.
d Nick says, 'However many beads of the same colour as the first bead withdrawn are added, the probability that the second bead selected is white stays the same!'
True or false? Justify your answer.

4 The probability that a washing machine will break down in the first 5 years of use is 0.27.
The probability that a television will break down in the first 5 years of use is 0.17.
Mr Khan buys a washing machine and a television on the same day.
By using a tree diagram or otherwise, calculate the probability that, in the five years after that day

a both the washing machine and the television will break down;
b at least one of them will break down. *EDEXCEL*

5 A virus is present in 1 in 250 of a flock of sheep. To make testing for the virus possible, a quick test is used on each sheep. However, the test is not completely reliable. An infected sheep tests positive in 85% of cases and a healthy sheep tests positive in 5% of cases.

a Use a tree diagram to help you to calculate the probability that a sheep will be infected and test positive.
b What is the probability that a sheep will be infected but test negative?
c What is the probability that a sheep will test positive?

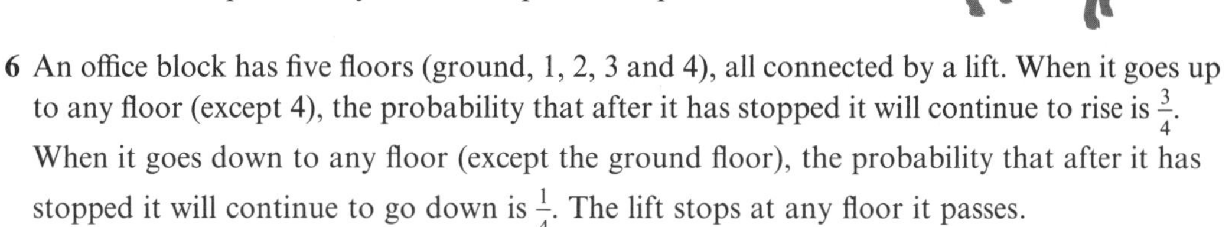

6 An office block has five floors (ground, 1, 2, 3 and 4), all connected by a lift. When it goes up to any floor (except 4), the probability that after it has stopped it will continue to rise is $\frac{3}{4}$.
When it goes down to any floor (except the ground floor), the probability that after it has stopped it will continue to go down is $\frac{1}{4}$. The lift stops at any floor it passes.
The lift is currently at the first floor having just descended.
Calculate the probability of these events.

a Its second stop is the third floor.
b Its third stop is the fourth floor.
c Its fourth stop is the first floor.

7 Two opera singers, Mario and Clarissa, both perform on the same night. The independent probabilities that two newspapers X and Y publish reviews of their recitals are given in this table.

	Mario's recital	Clarissa's recital
Probability of review in newspaper X	$\frac{1}{2}$	$\frac{2}{3}$
Probability of review in newspaper Y	$\frac{1}{4}$	$\frac{2}{5}$

a If Mario buys both newspapers, find the probability that both papers review his recital.
b If Clarissa buys both newspapers, find the probability that only one paper reviews her recital.
c Mario buys one of the newspapers at random. What is the probability that it has reviewed *both* recitals?

8 A school has an unreliable clock in its tower. The probabilities of gain or loss in the clock in any 24-hour period are given in this table.

	Gain of 1 minute	No change	Loss of 1 minute
Probability	$\frac{1}{2}$	$\frac{1}{3}$	$\frac{1}{6}$

If the clock is set to the correct time at noon on Sunday, find the probability of these events:
a The clock is correct at noon on Tuesday.
b The clock is *not* slow at noon on Wednesday.

9 A card is randomly taken from an ordinary pack of cards and not replaced. This process is repeated again and again. Explain, with calculations, why these probabilities are found:
a p(First card is a heart) $= \frac{1}{4}$
b p(Second card is a heart) $= \frac{1}{4}$
c p(Third card is a heart) $= \frac{1}{4}$
d p(Fourth card is a heart) $= \frac{1}{4}$
and so on.

10 Show that in a room of only 23 people, the probability of two of them sharing the same birth date is just over $\frac{1}{2}$.

Investigate

A television audience of 1024 people is mesmerised by a psychic who convinces them that there is someone with special telepathic powers among them. He sets about demonstrating this claim by simply tossing a fair coin.

The audience is split into two halves of 512. One half is asked to focus on heads, while the other half concentrates on tails. The coin is tossed, and the group who are wrong sit down. The audience is now split into two halves of 256, and this process is repeated time and again until just one person is left standing. This person is declared to be the gifted owner of the telepathic powers!

1 Find the probability of this person being picked out as the 'gifted' one.
2 Is this process misleading? Explain.

Activity 14

The data in the table below shows the risks of death from various causes in the USA in 2000.

Cause	Risk of death per person per year ($\times 10^{-7}$)
Rock climbing	400
Football	400
Drinking (1 bottle of wine per day)	750
Smoking (20 cigarettes per day)	50 000
Motor racing	200 000
Being run over by a vehicle	500
Floods	22
Earthquakes	17
Tornadoes	22
Lightning	3
Falling aircraft	1
Bites of venomous creatures	2
Influenza	2000
Being struck by a meteorite	0.0006

- Rank these activities in order of safety. Comment.
- How do you think these figures were obtained? Comment on their accuracy.

Exercise 48 (Revision)

Use tree diagrams to answer these questions.

1 A box contains two maths books and three French books. A book is removed and replaced before another is taken. Find the probability of these events:

a Two French books are selected.
b A maths and a French book are selected, in any order.

2 A hockey penalty taker has a $\frac{3}{4}$ probability of scoring a goal.

a If she takes two penalties, find the probability that she scores no goals.
b What is the probability that she scores one goal from two penalties?

3 The probability that a biased coin shows tails is $\frac{2}{5}$.

a Find the probability that, when the coin is thrown twice, it shows two heads.
b What is the probability of exactly one tail in two throws?

4 Assume that it is either sunny or rainy in Italy. If the weather is sunny one day, the probability that it is sunny the day after is $\frac{1}{5}$. If it rains one day, the probability that it rains the next is $\frac{3}{4}$.

a If it is sunny on Sunday, calculate the probability that it rains on Monday.
b What is the probability that it is sunny on Sunday and Tuesday?

5 The letters of the word HYPOTHETICAL appear on plastic squares that are placed in a bag and jumbled up. A square is randomly selected and replaced before another is taken. Calculate these probabilities.

a p(Two H) **b** p(Exactly one T) **c** p(A vowel)

6 All female chaffinches have the same patterns of laying eggs. The probability that any female chaffinch will lay a certain number of eggs is given in the table.

Number of eggs	0	1	2	3	4 or more
Probability	0.1	0.3	0.3	0.2	x

a Calculate the value of x.
b Calculate the probability that a female chaffinch will lay less than 3 eggs.
c Calculate the probability that two female chaffinches will lay a total of 2 eggs.

LONDON

Exercise 48★ (Revision)

Use tree diagrams where appropriate.

1 A chocolate box contains four milk chocolates and five Turkish delights. Gina loves milk chocolates, and hates Turkish delight. She takes one chocolate randomly, and *does not replace it* before picking out another one at random.

a Calculate the probability that Gina is very happy.
b What is the probability that Gina is very unhappy?
c What is the probability that she has at least one milk chocolate?

2 The probability that Mr Glum remembers his wife's birthday and buys her a present is $\frac{1}{3}$. The probability that he does not lose the present on the way home is $\frac{2}{3}$. The probability that Mrs Glum likes the present is $\frac{1}{5}$.

a Calculate the probability that Mrs Glum receives a birthday present.
b What is the probability that Mrs Glum receives a birthday present but dislikes it?
c What is the probability that she is happy on her birthday?
d What is the probability that she is not happy on her birthday?

3 A sleepwalker gets out of bed and is five steps away from his bedroom door. He is equally likely to take a step forward as he is to take one backwards.

a Calculate the probability that after five steps he is at his bedroom door.

b What is the probability that by then he is only one step closer to his bedroom door?

c What is the probability that, having taken five steps, he is closer to his bedroom door than to his bed?

4 There are 25 beads in a bag.
Some of the beads are red.
All the other beads are blue.
Kate picks two beads at random without replacement.
The probability that she will pick two red beads is 0.07.
Calculate the probability that the two beads she picks will be of different colours.

LONDON

5 The diagram shows a shooting target that is divided into three regions by concentric circles with radii that are in the ratio 1:2:3.

a Find the ratio of the areas of the three regions in the form a:b:c, where a, b and c are integers.

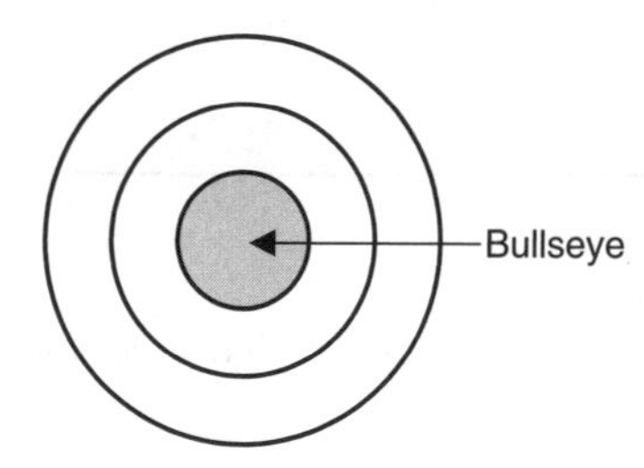

The probability that a shot will hit the target is $\frac{4}{5}$. If it does hit the target, the probability of it hitting any region is proportional to the area of that region.

b Calculate the probability that a shot will hit the bullseye.

c Two shots are fired. Calculate the probability of these events:

(i) They both hit the bullseye.

(ii) The first hits the bullseye and the second does not.

(iii) At least one hits the bullseye.

6 In the UK National Lottery, six numbers are chosen from 49, numbered consecutively from 1 to 49.
If these numbers match six chosen by any entrant, a huge sum of money is won.

a Show that the probability of winning the Lottery on a selection of six numbers is approximately 1 in 14 million.

b If the lottery could be played once a minute, after how many years could you reasonably expect to win?

Summary 2

Number

Negative and fractional indices

$a^m \times a^{-n} = a^{m-n}$ (Add indices)

$a^m \div a^{-n} = a^{m--n} = a^{m+n}$ (Subtract indices)

$(a^m)^{-n} = a^{-mn}$ (Multiply indices)

$a^{\frac{1}{2}} = \sqrt{a}$ $a^{\frac{1}{3}} = \sqrt[3]{a}$

$a^{-n} = \dfrac{1}{a^n}$ $a^{-\frac{1}{2}} = \dfrac{1}{a^{\frac{1}{2}}} = \dfrac{1}{\sqrt{a}}$

$a^{\frac{3}{2}} = \sqrt{a^3}$ $a^{-\frac{3}{2}} = \dfrac{1}{a^{\frac{3}{2}}} = \dfrac{1}{\sqrt{a^3}}$

Algebra

Solving quadratic equations

The first step is to rearrange the equation so that it equals zero.

Factorising

The different types are illustrated below.

- No x term: $x^2 - 4 = 0 \Rightarrow x^2 = 4 \Rightarrow x = \pm 2$
- No number term: $x^2 + 4x = 0 \Rightarrow x(x+4) = 0 \Rightarrow x = 0$ or -4
- Simple factorising: $x^2 - x - 2 = 0 \Rightarrow (x+1)(x-2) = 0 \Rightarrow x = -1$ or 2
- Number factor: $3x^2 - 3x - 6 = 0 \Rightarrow 3(x^2 - x - 2) = 0 \Rightarrow 3(x+1)(x-2) = 0 \Rightarrow$ $x = -1$ or 2
- Harder factorising: $2x^2 + 5x + 2 = 0 \Rightarrow (2x+1)(x+2) = 0 \Rightarrow x = -\frac{1}{2}$ or -2

Using the quadratic formula

If $ax^2 + bx + c = 0$ then $x = \dfrac{-b \pm \sqrt{b^2 - 4ac}}{2a}$

It is easy to make a mistake with the signs. Write down the values of a, b and c. Remember that if $b = -3$ then $-b = +3$, and that b^2 must be positive (it is easy to get this wrong with a calculator). If one of a or c is negative then $-4ac$ will be positive.

Problems leading to quadratic equations

- Where relevant, draw a clear diagram and put all the information on it.
- Let x stand for what you are trying to find.
- Form a quadratic equation in x and simplify it.
- Solve the equation by either factorising or using the formula.
- Check that the answers make sense.

Solving quadratic inequalities

To solve a quadratic inequality, sketch the graph of the quadratic function to find the critical values.
If *one* part of the number line is required, the answer is *one* inequality.
If *two* parts of the number line are required, the answer is *two* inequalities.

Solve

a $x^2 - x - 2 < 0$ **b** $x^2 - x - 2 \geqslant 0$

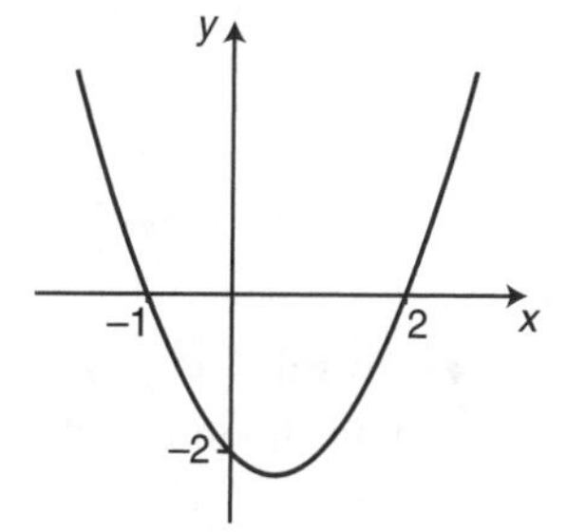

First sketch $y = x^2 - x - 2$ by finding where the graph intersects the x-axis.
$x^2 - x - 2 = 0 \Rightarrow (x + 1)(x - 2) = 0 \Rightarrow x = -1$ or $x = 2$

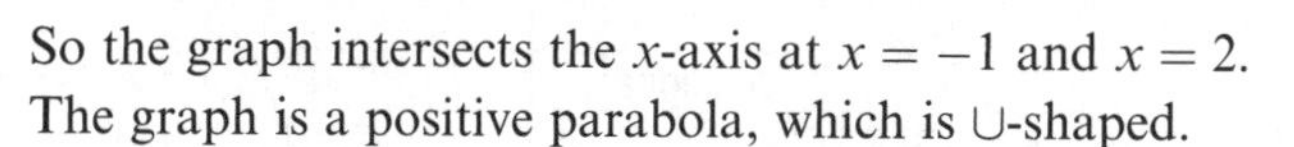

So the graph intersects the x-axis at $x = -1$ and $x = 2$.
The graph is a positive parabola, which is ∪-shaped.

a We want the region *below* the x-axis. As this is *one* part of the number line, the answer is *one* inequality, so $-1 < x < 2$.

b We want the region *above* the x-axis. As this is *two* parts of the number line, the answer is *two* inequalities, so $x \leqslant -1$ or $x \geqslant 2$.

Solving equations graphically

To solve an equation graphically, draw on the same set of axes the graphs of:

y_1 = Left-hand side of the equation, y_2 = Right-hand side of the equation.

The x-co-ordinates of any points of intersection will be solutions of the equation.

- To solve $x^2 = x + 3$, find the intersections of $y = x^2$ with $y = x + 3$.

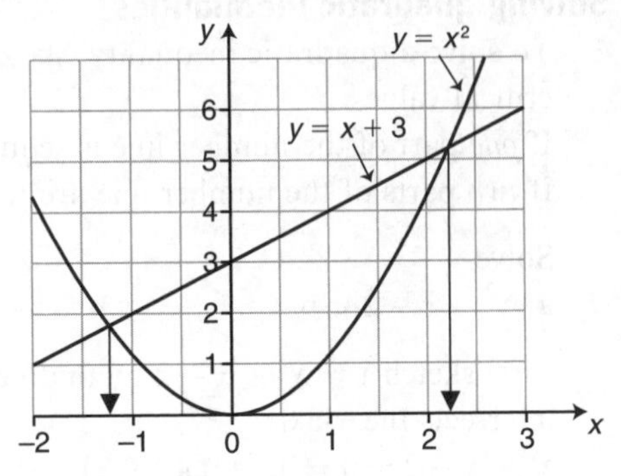

- To solve $x^2 + x - 3 = 0$, find the intersections of $y = x^2 + x - 3$ with $y = 0$.

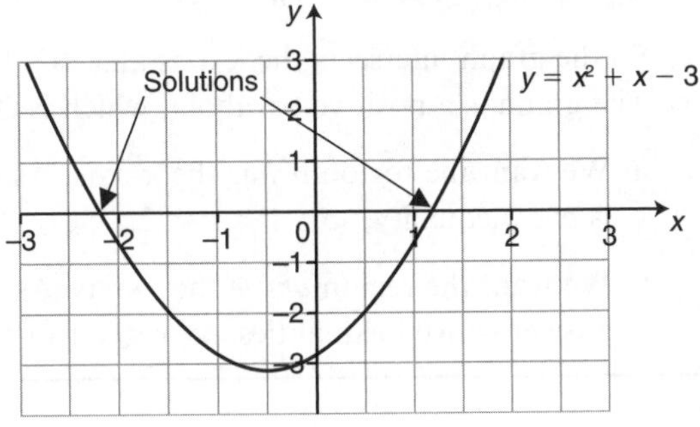

The equation can be rearranged so that the graphs are simpler to plot.

- To solve $x^2 + x - 3 = 0$, rearrange as $x^2 = 3 - x$, then find the intersections of $y = x^2$ with $y = 3 - x$, giving the same solutions as above.

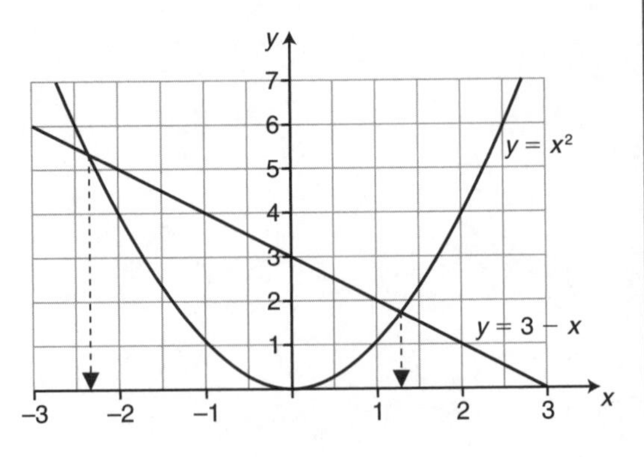

If the graph of $y = x^2 - 5x + 2$ has been plotted, then the graph can be used to solve other equations.

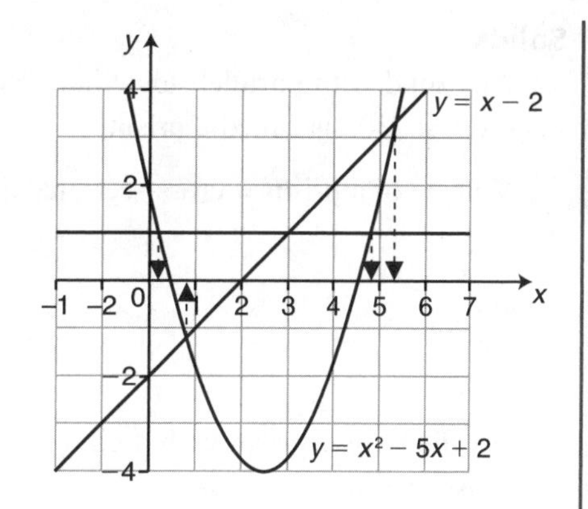

- To solve $x^2 - 5x + 1 = 0$, rearrange as $x^2 - 5x + 2 = 1$.
  Find the intersections of $y = x^2 - 5x + 2$ with $y = 1$.
- To solve $x^2 - 6x + 4 = 0$, rearrange as $x^2 - 5x + 2 = x - 2$.
  Find the intersections of $y = x^2 - 5x + 2$ with $y = x - 2$.

To solve **simultaneous** equations graphically, plot both graphs on one set of axes.
The co-ordinates of any points of intersection will be solutions of the equation.

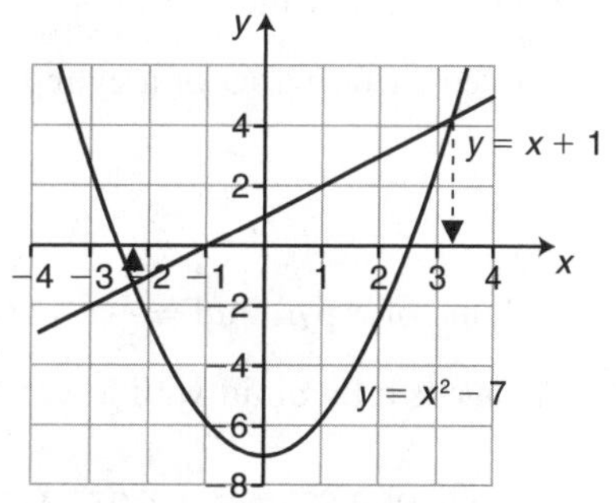

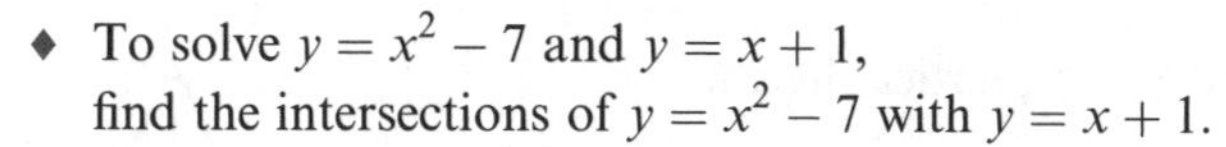

- To solve $y = x^2 - 7$ and $y = x + 1$, find the intersections of $y = x^2 - 7$ with $y = x + 1$.

Shape and space

Converting measurements

10 mm = 1 cm   1000 mm = 1 m   100 cm = 1 m   1000 m = 1 km   1000 cm^3 = 1 litre

The **perimeter** of a shape is the distance all the way round the shape.

Circles

The perimeter of a circle is called the **circumference**.

$$C = 2\pi r \qquad A = \pi r^2$$

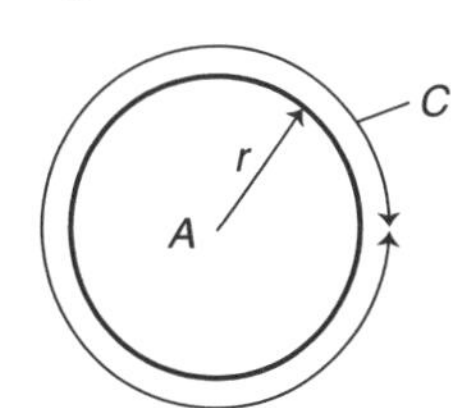

An **arc** is part of the circumference of a circle.
A **sector** is a region bounded by an arc and two radii.

$$\text{Arc length} = \frac{x}{360} \times 2\pi r \qquad \text{Sector area} = \frac{x}{360} \times \pi r^2$$

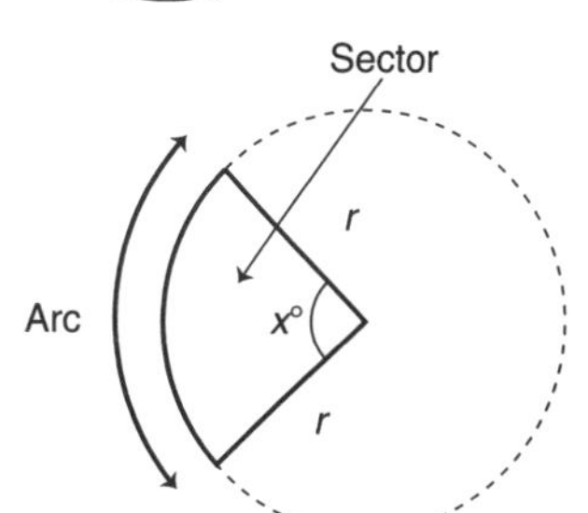

Solids

Any solid with parallel sides that has a constant cross-section is called a **prism**.

Volume of a prism = cross-sectional area × height

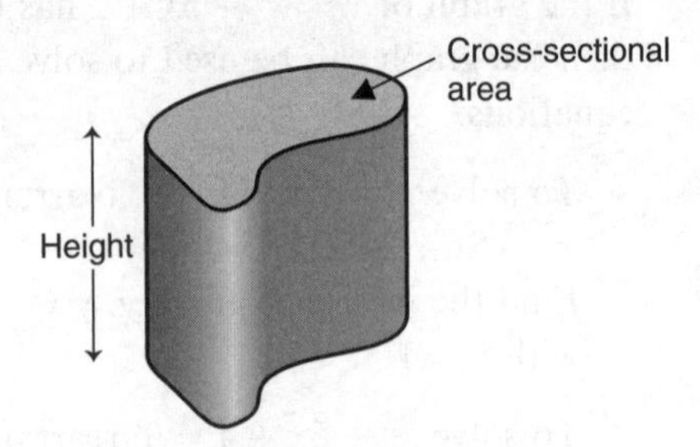

A **cylinder** is a prism with a circular cross-section.

Volume of a cylinder $= \pi r^2 h$

Curved surface area of a cylinder $= 2\pi rh$

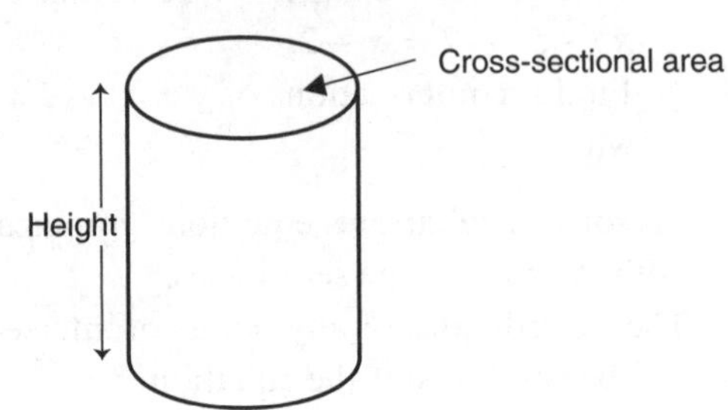

Volume of a pyramid $= \frac{1}{3} \times$ base area × vertical height

A **cone** is a pyramid with a circular base.

Volume of a cone $= \frac{1}{3} \times \pi r^2 \times h$

Curved surface area of a cone $= \pi rl$, where l is the slant height

Similar shapes and solids

If the Length Scale Factor is k, then the Area Scale Factor is k^2.
If the Length Scale Factor is k, then the Volume Scale Factor is k^3.

These statements are only true if the figures are **similar**.

Handling data

Independent events

If two events have no effect on each other, they are said to be independent.
For example, the fact that it rains on Tuesday is not going to have an effect on whether the QE2 liner sails from Portsmouth on Wednesday.

Mutually exclusive events

A light cannot be both on and off. Such events are said to be mutually exclusive.

Single event

From an ordinary pack of 52 cards, one card is selected at random.

p(Selecting a queen) $= \frac{4}{52} = \frac{1}{13}$ p(Not selecting a queen) $= 1 - \frac{4}{52} = \frac{12}{13}$

Event A has a probability of p(A), so that $0 \leqslant p(A) \leqslant 1$. Also, $p(A) + p(\overline{A}) = 1$.

Combined events

An ordinary dice is thrown twice.
Let event A be that a 6 is thrown. Then $\overline{A}$ is the event that a 6 is *not* thrown.
p(Throwing only one 6) $= [p(A) \times p(\overline{A})] + [p(\overline{A}) \times p(A)]$

Tree diagram

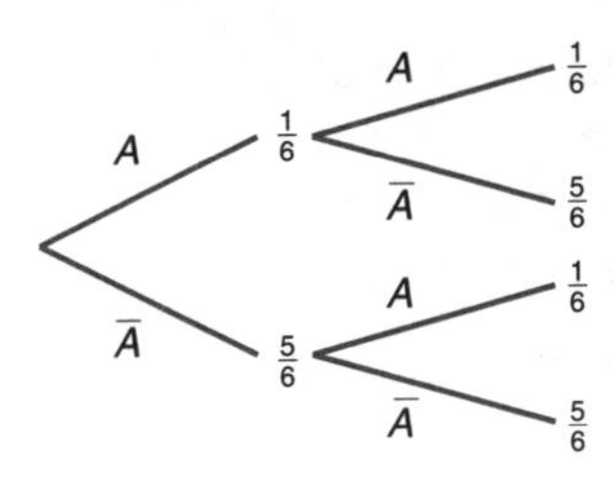

p(Throwing **only one** 6) $= \left(\frac{1}{6} \times \frac{5}{6}\right) + \left(\frac{5}{6} \times \frac{1}{6}\right)$

$= \frac{10}{36} = \frac{5}{18}$

p(Throwing **at least one** 6) $= p(AA) + p(A\overline{A}) + p(\overline{A}A)$

$= 1 - p(\overline{A}\,\overline{A})$

$= 1 - \frac{5}{6} \times \frac{5}{6}$

$= 1 - \frac{25}{36} = \frac{11}{36}$

Examination practice 2

1 Express in their simplest form:

a $x^2 \times x^5$ **b** $x^5 \div x^2$
c $(x^2)^5$ **d** $(x^{-1})^3$

2 Work these out and, where appropriate, leave your answer as a fraction.

a 3^{-2} **b** $6^{\frac{1}{3}}$
c $2^{-3} \times 2^2$ **d** $5^{-4} \div 5^{-2}$
e $(4^{-1})^2$

3 Simplify these.

a $b^2 \times b^{-1}$ **b** $c^{-1} \div c^2$
c $(b^2)^{-1}$ **d** $c^{\frac{1}{2}} \times c^{\frac{1}{2}}$
e $a^{\frac{1}{4}} \div a^{\frac{1}{4}}$

4 Solve for x:

a $2^x = 128$ **b** $4^x = \frac{1}{4}$
c $81^x = 9$ **d** $125^x = 5$

5 The width of a rectangular photograph is 5 cm more than the height. The area is 80 cm^2. Find the height of the photograph.

6 a Use the intersecting chords theorem to form an equation in x from the diagram.
b Solve your equation to find x, giving your answer correct to 3 s.f.

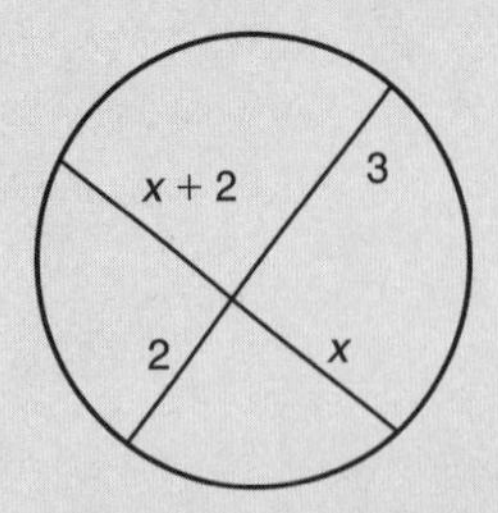

7 Show that the x-co-ordinates of the points of intersection of the graphs $y = 4x - 1$ and $y = 2x + \dfrac{1}{x}$ are the solutions of the equation $4x^2 - x - 1 = 0$.

8 The graph of $y = x^3 + 2x^2 - 4x + 1$ has been drawn.
Find the equations of the lines that must be drawn to solve the following equations:

a $x^3 + 2x^2 - 4x + 1 = 0$
b $x^3 + 2x^2 - 4x = 0$
c $x^3 + 2x^2 - 5x + 1 = 0$
d $x^3 + 2x^2 - 3x - 1 = 0$

9 Quadrilateral A is mathematically similar to quadrilateral B.

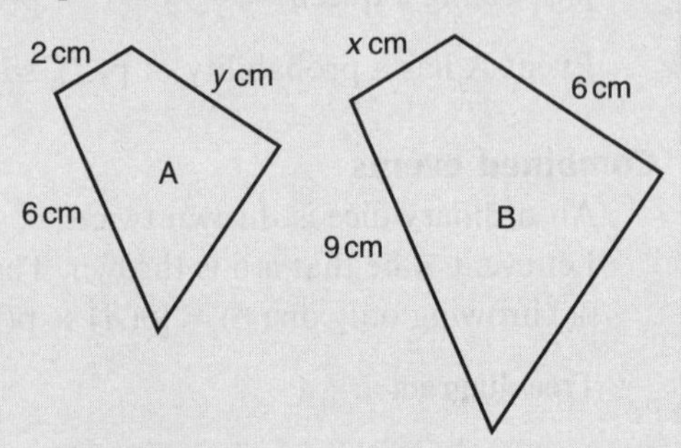

a Calculate the value of x.
b Calculate the value of y.
c The area of B is 36 cm^2.
Find the area of A.

10 a Calculate the surface area of the cone shown, giving your answer to 3 s.f.
b A smaller mathematically similar cone has a volume $\frac{1}{125}$ of the volume of the cone shown.
Calculate the surface area of the smaller cone, giving your answer to 3 s.f.

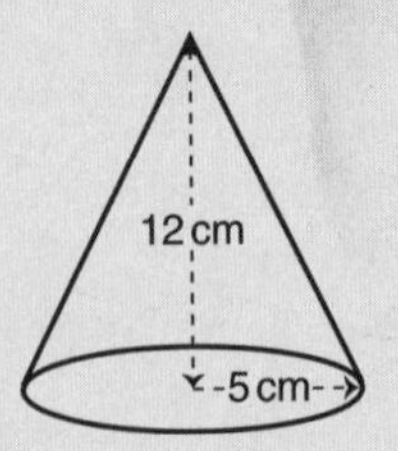

11 A cylindrical cake with radius 12 cm and height 10 cm has a slice cut out of it as shown. The shape of the top of the slice is a sector of the circle that forms the top of the cake.

a Calculate the area of the top of the slice that has been cut out.

b Calculate the volume of the cake that remains after the slice has been removed.

c Calculate the surface area of the cake that remains after the slice has been removed.

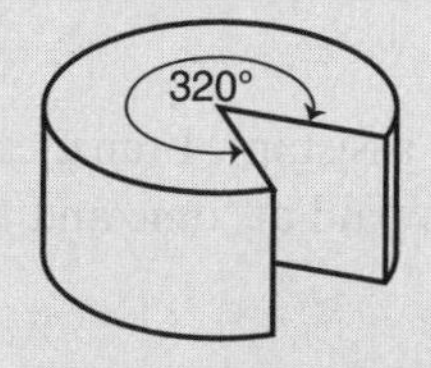

12 The probability that Richard beats John at badminton is 0.7. The probability that Richard beats John at squash is 0.6. These events are independent.
Calculate the probability that, in a week when they play one game of badminton and one of squash:

a Richard wins both games,

b Richard wins one game and loses the other.

MEI

13 A certain absent-minded teacher has a newspaper delivered to his home each morning, and he tries to remember to carry it from home to school, and back again, each day. However, in practice he often forgets, and the probability that he remembers to take his newspaper on either journey is $\frac{2}{3}$.
Also, the probability that he loses it on any journey (if he sets off with it) is $\frac{1}{5}$.
Work out the probabilities, on a particular day, that:

a he takes his newspaper with him to school and loses it on the way;

b he arrives at school with his newspaper;

c he sets off for home that evening with his newspaper;

d he arrives at home with his newspaper;

e he has that day's newspaper at home in the evening.

MEI

14 Alison, Brenda, Claire and Donna are the runners in a race.
The probabilities of Alison, Brenda, Claire and Donna winning the race are shown below.

Alison	Brenda	Claire	Donna
0.31	0.28	0.24	0.17

a Calculate the probability that either Alison or Claire will win the race.

Hannah and Tracy play each other in a tennis match.
The probability of Hannah winning the tennis match is 0.47.

b Copy and complete the probability tree diagram.

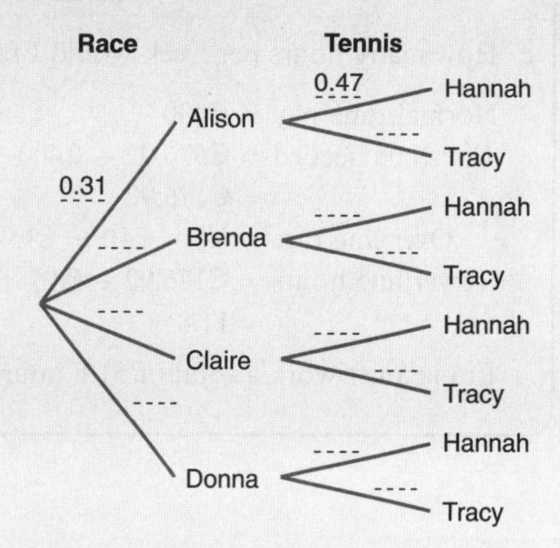

c Calculate the probability that Brenda will win the race and Tracy will win the tennis match.

LONDON

Number 3

Financial arithmetic

Wages and salaries

A **wage** is calculated at a certain rate for the number of hours per week and is usually paid weekly. Overtime – hours worked over and above the normal hours each week – is usually paid at a higher rate.

A **salary** is a fixed annual sum of money, usually paid each month.

(*Note*: 'per annum' means 'per year', and is often shortened to 'p.a.'.)

Example 1

The manager of a shoe shop earns a salary of €30 000 p.a.; an assistant, Franz, earns an hourly rate of €10 per hour for a 40-hour week, with overtime paid at 'time and a half'.

a Calculate Franz's pay for a 48-hour week.

Franz's wage = pay for normal time + pay for overtime
= 40 × €10 + 8 × (1.5 × €10)
= €400 + €120
= €520

b Calculate the manager's pay per week.

Manager's weekly pay = €30 000 ÷ 52
= €576.92

c How many hours per week would Franz have to work to earn as much as the manager?

Normal-time pay = €400
Overtime needed = €576.92 – €400
= €176.92
Overtime rate = 1.5 × €10 = €15 per hour
Overtime hours = €176.92 ÷ €15
= 11.8

Franz must work a total of 51.8 hours to earn as much as his manager.

Income tax and purchase tax

All governments collect money from their citizens and businesses through **taxation** to pay for public services: health, transport, policing, defence, etc.

Income tax is paid on the money earnt (*direct* taxation), and **purchase tax** is paid on the money spent (*indirect* taxation). In the European Union, there is a standardised form of purchase tax, called **Value Added Tax (VAT)**. There are many other taxes too, such as capital gains tax, corporation tax and inheritance tax. *Everyone* pays tax!

The philosophy of some governments is that wealthy people, who earn and spend more, should pay higher taxes than poorer people. Each year the Treasury predicts and calculates next year's expenditure and then adjusts the taxation rates so that enough money is collected to pay for everything.

Calculating income tax

A percentage of income is taken by the government as income tax. The higher the income, the higher the percentage rate!

Key Points

- **Gross income** = total income before tax and other deductions
- **Personal allowance** = the earnings that are 'tax-free'
- **Taxable income** = the earnings that are taxed
  = total income – personal allowance
- **Net income** = income that remains after tax and other deductions have been taken out

The Republic of Lexica provides a model for the tax rates in many countries.

Tax rates		Band of taxable income
Personal allowance	0%	\$0–\$5000 per annum
Basic rate	20%	\$5001–\$70 000 per annum
Higher rate	45%	Over \$70 000 per annum

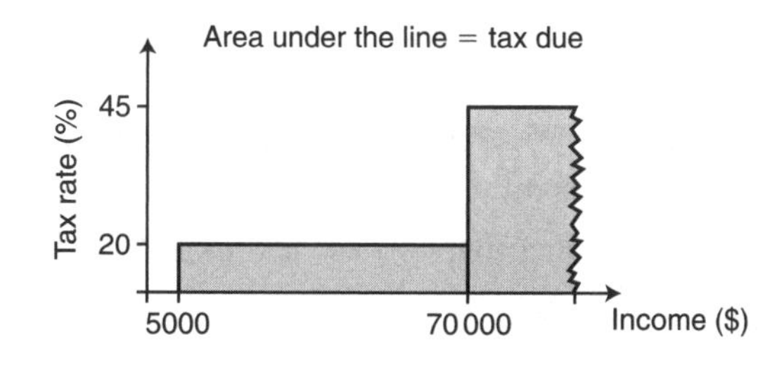

Example 2

Mrs Singh works in Lexica and earns a gross salary of \$100 000 p.a.

a Calculate her annual income tax bill.

b What percentage of her gross salary is deducted as tax?

Her income tax is deducted from her salary evenly each month.

c Calculate her net monthly pay, after deduction of income tax.

Note: Mrs Singh will also have other deductions from her gross pay (local tax, insurance, pension, ...). Only income and tax are considered in this example.

a Deduct personal allowance from gross salary to calculate the taxable income:

taxable income = \$100 000 – \$5000 = \$95 000

Calculate the tax in each income band as shown in this table.

Basic rate: 20%	**Higher rate: 45%**
\$70 000 – \$5000 = \$65 000	\$95 000 – \$70 000 = \$25 000
Tax = 20% of \$65 000	Tax = 45% of \$25 000
$= \frac{20}{100} \times \$65\,000 = \$13\,000$	$= \frac{45}{100} \times \$25\,000 = \$11\,250$

Total annual tax = \$13 000 + \$11 250 = \$24 250

b $\frac{24\,250}{100\,000} \times 100\% = 24.25\%$

24.25% of her gross salary is deducted in income tax.

c Net annual salary = \$100 000 – \$24 250 = \$75 750

$\therefore$ Net monthly salary = \$75 750 ÷ 12 = \$6312.50

Calculating purchase tax

Purchase tax is included in the price that is paid for an article. In the Republic of Lexica, the standard rate of purchase tax on most products is 15%, so the shopkeeper has to add 15% onto the original price to arrive at the selling price.

Some products, such as fruit, are zero-rated. Others are exempt and some, such as fuel, are taxed at a higher rate.

Key Points

- To *add* 15% purchase tax to the original price, *multiply* by 1.15.
- To *subtract* 15% purchase tax from the selling price, *divide* by 1.15.

Example 3

Belinda wants to buy a bicycle and finds a good deal on the Internet. At bike4sale.com, the price is \$98 + 15% purchase tax. Calculate the total price.

Total cost = $\$98 \times 1.15 = \$112.70$

Example 4

Postman Pat can claim back purchase tax at 15% on business expenses. At cycleshop.com, the price of a bicycle is \$126.50 (including purchase tax). Calculate the price *excluding* purchase tax.

$\$126.50 \div 1.15 = \$110$

Exercise 49

1 Calculate the weekly and annual gross pay of a shop assistant working a 40-hour week at \$8.65 per hour.

2 Calculate the weekly and annual gross pay of a secretary working a 35-hour week at \$9.05 per hour.

3 The manager of a fast-food outlet earns a gross salary of \$25 800 p.a. On average, she works 50 hours per week. Calculate her hourly rate of pay.

4 A tour guide receives a gross salary of \$24 650 p.a. Calculate his hourly rate of pay if he works an average of 48 hours each week.

5 A discount warehouse lists its prices 'excluding 15% purchase tax'. Calculate the full prices of these:

a An ink-jet printer at \$94.72
b Two pairs of designer jeans at \$46.00 each
c Six bottles of wine at \$8.95 each

6 A mail-order shopping catalogue lists its prices 'excluding 15% purchase tax'. Calculate the full prices of these:

a A fountain pen at \$15.00
b Three wallets at \$24.50 each
c Eight boxes of greetings cards at \$2.99 each

7 A building contractor can put the following purchases down as business expenses and thus reclaim the 15% purchase tax (which is included in the prices). Calculate his saving on each of these orders:

a One hammer drill at \$120
b Four boxes of nails at \$3.50 each
c Five UPC windows at \$345 each

8 How much purchase tax (at 15%) is included in the price of these?

a One tank of engine oil at $59.50
b 45 litres of petrol at $0.825 per litre
c Four new tyres at $95 each

For Questions 9–12, refer to the table of tax rates and tax bands for the Republic of Lexica on page 173, to calculate the annual income tax bill and net monthly pay on these gross salaries:

9 $50 000 **10** $60 000 **11** $95 000 **12** $110 000

Exercise 49★

For Questions 1–8, use the tax rates for the Republic of Lexica.

1 Companies in Lexica have to pay import duty of 17.5% on imported goods. Calculate the import duty on

a a television, valued at $850 **b** grain, valued at $45 per tonne.

2 A citizen of Lexica has to pay import duty of 17.5% on luxury goods bought abroad whilst on holiday. How much will he have to pay on

a a lap-top computer, valued at $950 **b** vintage wine, valued at $35 per bottle?

3 A tourist to Lexica can reclaim the purchase tax (at 15%) paid on goods when he leaves the country. How much can he reclaim on

a a suit, costing $150 **b** a set of skis, costing $400?

4 Charity organisations in Lexica can reclaim the purchase tax (at 15%) paid on business goods. How much can a cancer charity reclaim on

a office equipment, costing $2500 **b** a vehicle costing $9000?

5 A radio presenter has signed a $150 000 p.a. contract.

a Calculate the net salary per annum after tax has been deducted.
b Calculate the net pay per week.
c What percentage of the gross salary is paid in income tax?

6 It is known that a professional footballer is 'on $20 000 per week'.

a Calculate the gross pay per annum.
b Calculate the net salary per week after tax has been deducted.
c What percentage of the gross salary is paid in income tax?

7 A teacher earns a gross salary of $65 000 p.a.

a Calculate her annual income tax bill and her monthly 'take-home pay'.

She spends all her money and pays 15% purchase tax on everything she buys.

b Calculate the proportion of her gross earnings that is deducted as tax.

8 An office supervisor earns a monthly gross salary of $5500.

a Calculate her monthly income tax bill and her annual net pay.

She spends all her money and pays 15% purchase tax on everything she buys.

b Calculate the proportion of her gross earnings that is deducted as tax.

9 A person has 15% of his gross salary deducted in income tax. How much does he earn?

10 A person has 30% of his gross salary deducted in income tax. How much does he earn?

As part of the government's green policies, there is a special 'fuel tax' to discourage excessive use of fossil fuels and the production of greenhouse gases.

The 'fuel tax' on a litre of petrol is \$0.48. Purchase tax, at 15%, is then added to give the 'price at the pumps'.

11 a Show how, for a litre of petrol, an 'untaxed price' of \$0.25 gives a pump price of \$0.8395.

For this price, calculate the total tax as a percentage of the:

b pump price

c untaxed price.

12 When the 'pump price' is \$1 per litre, calculate:

a the 'untaxed price'

b a new figure for fuel tax that would result in a 10% reduction of this 'pump price'.

Activity 15

Investigate the tax rates for income tax and purchase tax in your country, and compare them with the model from the Republic of Lexica.

Calculate the income tax paid by a typical low-wage earner, average-wage earner and high-wage earner.

Savings, credit and loans

Banks, building societies and finance companies make their profit by buying and selling money. They 'buy it' from savers and pay them interest; they 'sell it' to borrowers and charge them a higher interest. The rates that banks advertise for savers and borrowers are, in fact, the prices that the bank or the customer pays for borrowing money. The banks will make their prices look as competitive and attractive as possible, so it is important to read the small print. Rates can be misleading!

Savings accounts

Banks pay compound interest, quoted at an annual rate (with or without tax deducted).

Example 5

Chi-Ho places \$500 in a savings account that pays a gross interest rate of 6%.

a How much will Chi-Ho have in his account after four years?

To add 6%, multiply by 1.06.

$\therefore$ savings = \$500 × $(1.06)^4$ = \$631.24

b Tax is deducted from the *interest* at 20% to give a net interest rate. Calculate the net interest rate.

To deduct 20% from the interest, calculate how much is kept.

$\therefore$ 80% of 6% = 0.8 × 6% = 4.8%

$\therefore$ net interest rate is 4.8%

c How much will Chi-Ho have in his account after four years at the net rate?

To calculate at a rate of 4.8%, multiply by 1.048 four times (once for each year).

$\therefore$ savings = \$500 × $(1.048)^4$ = \$603.14

Credit cards

Credit cards enable you to buy goods and then pay for them later. You receive a 'statement of your account' at the end of each month. You can pay the account in full without charge. If you do not pay the full amount, interest is added at a monthly rate to the amount that is not paid off.

Example 6

Falouka's credit card account is \$380 and the monthly interest charge is 1.5%. She only pays off \$50.

a Calculate the interest on the balance.

Balance = \$380 – \$50 = \$330

1.5% of \$330 = 0.015 × \$330 = \$4.95

b Calculate the interest charge as an annual rate.

To add 1.5%, multiply by 1.015.

For 12 months, multiply by $(1.015)^{12}$ = 1.1956

$\therefore$ interest rate is 19.56% (not 12 × 1.5% = 18%)

Credit plans and loans

A credit plan is a loan arranged by a shop or business to enable you to buy their goods. The interest on the loan is calculated using simple interest and then the monthly repayments are calculated to pay off the lump sum and interest over a fixed time period. However, the actual interest rate (APR) is much higher than the published rate because you do not borrow all the money for the whole time period. You will also be charged an 'arrangement fee' on top of the repayments.

Example 7

Nwanga is buying a car for \$10 000. He pays a 20% deposit and takes out a credit plan to borrow the remainder over three years at 8.5% p.a. The arrangement fee is \$80.

a Calculate the interest on the loan and the monthly repayments.

Deposit = 20% of \$10 000 = \$2000

$\therefore$ Loan = \$10 000 − \$2000 = \$8000

Interest = $3 \times 8.5\%$ of \$8000

= $3 \times 0.085 \times$ \$8000 = \$2040

Total to repay = \$8000 + \$2040 = \$10 040

Monthly repayments = \$10 040 ÷ 36 = \$278.89

b Calculate the total cost of the car.

Total cost = \$2000 + \$10 040 + \$80 = \$12 120

c Titi negotiates a 5% discount off the list price by offering to pay the full price 'up-front'. What is the real extra cost of the credit plan?

Interest = \$2040, fee = \$80, discount = 5% of \$10 000 = \$500

$\therefore$ total cost of credit = \$2040 + \$80 + \$500 = \$2620

Key Points

In these two formulae, P is the principal sum, R is the rate of interest, n is the number of years and A is the amount earned after n years.

Compound interest $A = P\left(1 + \frac{R}{100}\right)^n$ **Simple interest** $A = P\left(1 + \frac{Rn}{100}\right)$

Exercise 50

1 Calculate the interest in an account when interest is earned on these amounts:

a \$250 saved for one year at 8% p.a. **b** \$1240 saved for four years at 10% p.a.

2 Calculate the interest in an account when interest is earned on these amounts:

a \$600 saved for one year at 7.5% p.a. **b** \$945 saved for three years at 12% p.a.

For Questions 3–6, calculate the monthly rates as annual rates. Give your answers correct to 2 decimal places.

3 1% **4** 2% **5** 1.5% **6** 2.5%

For Questions 7–10, calculate the annual rates as monthly rates. Give your answers correct to 2 decimal places.

7 10% **8** 20% **9** 12% **10** 24%

11 Davina has \$600 owing on her credit account. The charge on uncleared balances is 1% per month. She pays off \$100. How much interest will be added to the account?

12 Fojir has \$250 owing on his credit account and forgets to make a payment at the end of the month. \$3.00 interest is added to his account. Calculate the interest charge as:

a a monthly rate **b** an annual rate.

13 Damon is taking out a credit plan to buy a new car, which costs \$12 000. He pays a deposit of 30% and decides to pay the rest over 36 months. The charge on the loan is calculated as simple interest at 8.0% p.a. and there is a fee of \$75.

a Calculate the monthly repayment.
b Calculate the total cost of the car.
c What is the true cost of the credit if the dealer offers a 10% discount for 'payment up-front'?

14 A television set is advertised at \$800. Trevor decides to pay a 20% deposit and the remainder in 24 monthly payments. There is a fee of \$50 for taking out a credit plan and the interest rate is 7.0%.

a Calculate the monthly repayment.
b Calculate the total cost of the television set.
c What is the true cost of Trevor's credit if Eve bought the same television for \$750?

Exercise 50★

1 Calculate the interest in an account when interest is earned on these amounts:

a \$280 saved for one year at 5.5% p.a. **b** \$1200 saved for three years at 6.4% p.a.

2 Calculate the interest in an account when interest is earned on these amounts:

a \$600 saved for two years at 7.5% p.a. **b** \$945 saved for five years at 6.25% p.a.

For Questions 3–6, calculate the monthly rates as annual rates. Give your answers correct to 2 decimal places.

3 1.2% **4** 1.8% **5** 1.75% **6** 2.25%

For Questions 7–10, calculate the annual rates as monthly rates. Give your answers correct to 2 decimal places.

7 18% **8** 36% **9** 22.5% **10** 27.5%

11 Bola has $638 owing on her credit account. The charge on uncleared balances is 1.7% per month. She pays off $65. How much interest will be added to the account?

12 Ntini has $258.95 owing on his credit account and forgets to make a payment at the end of the month. $3.80 interest is added to his account. Calculate the interest charge as:

a a monthly rate **b** an annual rate.

13 Read this quotation for a new car.

TraCar 1.25 LX 3-door	
On-the-road price	$10 050.00
Deposit	$3015.00
Balance	$7035.00
Total charge for credit*	$1811.08
Total amount payable	$11 861.09
36 monthly payments	$243.78
APR	16.9%

*Includes a finance facility fee of $70 payment with the first monthly payment.

a Calculate the rate of simple interest used to obtain the 'total charge for credit'.
b Calculate the monthly payments if the deposit had been $2000.

14 Read this quotation for a new computer.

Bell Q853z Computer	
Installed price	$3650.00
Deposit	$750.00
Balance	$2900.00
Total charge for credit*	$537.50
Total amount payable	$4187.50
24 monthly payments	$141.98
APR	17.2%

*Includes a finance facility fee of $30 payable with the first monthly payment.

a Calculate the rate of simple interest used to obtain the 'total charge for credit'.
b Calculate the monthly payments if the deposit had been $1100.

Activity 16

Investigate the three specimen tariffs for using a mobile phone on the 'Speedfone' network.

- For each tariff, form an expression for the monthly cost, C, of t minutes of air-time.
- Plot the graph of C against t for each tariff. (Horizontal axis: $0 \leqslant t \leqslant 300$; vertical axis: $0 \leqslant C \leqslant 60$)
- Use your graph to work out the range of times which make each tariff the cheapest.

Tariff	Monthly charge	Free air-time	Call charge
Pay-as-you-go	0	0	20 cents/min
Speakeasy	\$10	30 min	10 cents/min
Chatterbox	\$30	240 min	5 cents/min

There are other factors, which have not been included: the cost of the phone, text charges.

- Investigate the current tariffs available to you in your country.

Inflation and exchange rates

When prices rise, workers demand higher wages so that they can pay the higher prices. Of course, higher wages increase the cost of manufacturing goods and so that pushes prices up, which pushes wages up, ...

The **inflation rate** is the rate at which prices are rising. It is usually given as an annual rate, e.g. 4.3% p.a.

Low inflation generally means that a country has a strong economy. If a country has a strong economy, the international money dealers will buy that country's currency and that will push the exchange rate up.

The **exchange rate** is the price of buying and selling foreign currencies, e.g. 1 USD = €0.77, 1 CAD = €0.6, €1 = 1.7 AUD.

When the exchange rate for the pound goes up, the pound will buy more dollars, euros or yen. A higher exchange rate will make exports more expensive and imports cheaper. This weakens the economy because manufacturers have to cut their profit margins and drop the prices of their exported goods. So the exchange rate falls back.

Politicians and economists argue and debate the cause and effect of these issues endlessly!

Bureau de change

Large banks buy and sell currencies, in huge amounts, on international money markets. An individual visiting a foreign country for a holiday or business trip will buy its currency from a bureau de change.

The bureau de change will make money from the transaction by buying and selling at slightly different rates, and by charging a transaction fee.

The table below is an example of a bureau de change dealing in euros.

Airport Bureau de Change		
No transaction fee	*Exchange rates*	
EURO	**BUY**	**SELL**
American dollar	1.290	1.380
British pound	0.675	0.722
Chinese yuan	10.432	11.162
Hong Kong dollar	9.988	10.687
Malaysian ringgit	4.843	5.182
South African rand	8.121	8.689
Taiwan dollar	39.632	42.406
Thai baht	50.287	53.807
Venezuelan bolivar	2674.542	2861.760

Example 8

Before travelling to the Far East, I need to buy currency.

a How many Taiwan dollars will €50 buy?
b How much will it cost to buy 200 Malaysian ringgit?

a $50 \times 39.632 = \text{TD } 1981.60$
b $200 \div 4.843 = €41.297 \rightarrow €41.30$

Example 9

When I return, I convert my foreign currency back into euros.

a How many euros will I get for 2000 Taiwan dollars?
b How many euros will I get for 200 Malaysian ringgit?

a $2000 \div 42.406 = €47.163 \rightarrow €47.16$
b $200 \div 5.182 = €38.595 \rightarrow €38.60$

Notice that I lose money when I convert my money back into euros!

Exercise 51

Use the exchange rates from the airport bureau de change table on page 183 in this exercise.

In Questions 1–4, how much of the currency will €120 buy?

1 American dollar

2 Venezuelan bolivar

3 Hong Kong dollar

4 Chinese yuan

In Questions 5–8, how many euros will it cost to buy the given amount?

5 8000 Thai bahts

6 250 British pounds

7 10 000 Venezuelan bolivars

8 1200 South African rands

In Questions 9–12, calculate how much it will cost to convert €50 into the currency, and then to convert it back into euros.

9 American dollar

10 Thai baht

11 Hong Kong dollar

12 Venezuelan bolivar

13 What would it cost a tourist, in euros, to buy 1000 Chinese yuan, change his mind, and then sell them back?

14 During a working day the bank money bureau de change sells 25 000 American dollars and buys back exactly the same amount from other customers. How many euros does it make on the deals?

Many money exchange bureaus will also charge a fee, or commission, on a transaction.

Bank Money Bureau		
€5 fee on every transaction	***Exchange rates***	
EURO	**BUY**	**SELL**
American dollar	1.290	1.380
British pound	0.675	0.722
South African rand	8.121	8.689
Taiwan dollar	39.632	42.406

Example 10

The bureau de change charges a flat fee of €5 on every transaction. Using the rates from the bank money bureau:

a How many South African rand can be bought with €50?

b How many euros can be bought back for SAR 400?

c Calculate, for these amounts, the equivalent exchange rate if no fee was charged.

a Subtract fee, leaving €45:

$45 \times 8.121 = \text{SAR}\,365.445 \rightarrow \text{SAR}\,365.45$

b $400 \div 8.689 = €46.035 \rightarrow €46.04$

Subtract fee: $€46.04 - €5 = €41.04$

c Buying rate $= \dfrac{365.45}{50} = 7.31$

Selling rate $= \dfrac{400}{41.04} = 9.75$

Notice that the transaction fee significantly changes the true exchange rate. The fee will be less significant for larger amounts.

Exercise 51 ★

In Questions 1 and 2, using the exchange rates with no transaction fee, how much of the currency will €75 buy?

1 Venezuelan bolivar

2 Malaysian ringgit

In Questions 3 and 4, using the exchange rates with no transaction fee, how many euros will it cost to buy the given amount?

3 950 Hong Kong dollars

4 125 Chinese yuan

5 Using the exchange rates with no transaction fee, calculate the loss as a percentage when you change 150 euros into Thai bahts and then immediately change it back into euros.

6 Using the exchange rates with no transaction fee, calculate the loss as a percentage when you change 200 euros into Venezuelan bolivars and then immediately change it back into euros.

7 Use the bank money bureau rates, with the transaction fee, to calculate the loss when you change:

a 100 euros into Taiwan dollars and then immediately change it back into euros

b 200 euros into Taiwan dollars and then immediately change it back into euros.

c Express each loss as a percentage of the original amount.

8 Use the bank money bureau rates, with the transaction fee, to calculate the loss when you change:

a 100 euros into British pounds and then immediately change it back into euros

b 200 euros into British pounds and then immediately change it back into euros.

c Express each loss as a percentage of the original amount.

9 How many American dollars per euro do you actually get from the bank money bureau for:

a €50 **b** €500?

10 How many South African rands per euro do you actually get from the bank money bureau for:

a €50 **b** €500?

11 When the exchange rate between Europe and America is €1 = $1.30, an exported Porsche car costs $80 000 in America.

a Calculate the equivalent price in Europe.

Calculate its cost in America if the exchange rate moves to:

b €1 = $1.20 **c** €1 = $1.50

12 When the exchange rate between Europe and South Africa is €1 = SAR 8.00, a new Mercedes costs €50 000.

a Calculate the equivalent price in South Africa.

What would you expect the same car to cost when:

b the exchange rate is €1 = SAR 7.50 **c** the exchange rate is €1 = SAR 10.0?

13 Another money exchange bureau offers a rate of €1 = 1.20 American dollars with no transaction fee.

a Calculate the amount of dollars that €50 buys from the bank money bureau and from this money exchange.

b Calculate the amount of dollars that €100 buys from the bank money bureau and from this money exchange.

c How many euros would buy the same number of dollars from both deals?

Exercise 52 (Revision)

1 Calculate the annual gross salary of a part-time computer technician working an 18-hour week at 15.50 dollars per hour.

2 Using 15% as the rate of purchase tax, calculate the tax on these items:

a A new tyre advertised at $35.60 + purchase tax

b A new bicycle frame advertised at $65 including purchase tax

3 Use the tax rates and tax bands for the Republic of Lexica on page 173 to calculate the annual tax bill for:

a a mechanic earning $48 000 p.a. **b** a doctor earning $96 000 p.a.

4 Calculate the interest earned on \$500, invested for four years at 5.5% p.a.

5 Calculate 1.1% per month as an annual rate.

6 Use the exchange rate of €1 = 5.45 dollars to convert

a €360 into dollars

b 500 dollars into euros.

7 A computer is advertised at \$900. Lesley decides to pay a 20% deposit and the remainder in 24 monthly payments. There is no 'setting-up fee'. Simple interest is added to the loan at the rate of 9% p.a. before the monthly payments are calculated.

a Calculate the monthly payments.

b Calculate the total cost of the computer.

Exercise 52★ (Revision)

1 Calculate the interest gained on an investment of \$24 000 at 0.85% per month, over three years.

2 a Calculate 10% p.a. as a monthly rate, correct to 3 decimal places.

b How many months would it take an investment to double at this interest rate?

3 A builder in Lexica pays \$5000 in income tax for the year 2005–2006. Use the tax rates given on page 173 to calculate his gross earnings.

4 A tourist claims back the purchase tax on a luxury coat as he prepares to catch a flight out of Lexica. Calculate the amount if the coat cost 1000 dollars and the rate of tax is 35%.

5 A traveller at a European airport buys 3000 Mexican pesos at the bureau de change. The rate of exchange is €1 = 14.25 pesos, and there is a transaction fee of €8.

a Calculate the cost in euros.

b If there were no fee, calculate the exchange rate that would cost the same amount for 3000 pesos.

6 Read this finance quotation for an alarm system.

FINANCE QUOTATION	
Spectra Alarm System	
Installation price:	\$3995
Deposit:	\$1000
Balance:	\$2995
Total charge for credit (including a finance facility fee of \$50 payable with the first monthly payment):	\$611.58
Total amount payable:	\$4606.58
36 monthly payments:	\$98.79
APR:	12.1%

a Calculate the rate of simple interest used to obtain the 'total charge for credit'.

b Use the same formula to calculate the monthly payments if the deposit had been \$1500, rather than \$1000.

Solving two simultaneous equations, one linear and one nonlinear

Graphically this corresponds to the intersection of a straight line and a curve.

Activity 17

- Use the graph below to solve the simultaneous equations $x + 2y = 10$ and $x^2 + y^2 = 25$.

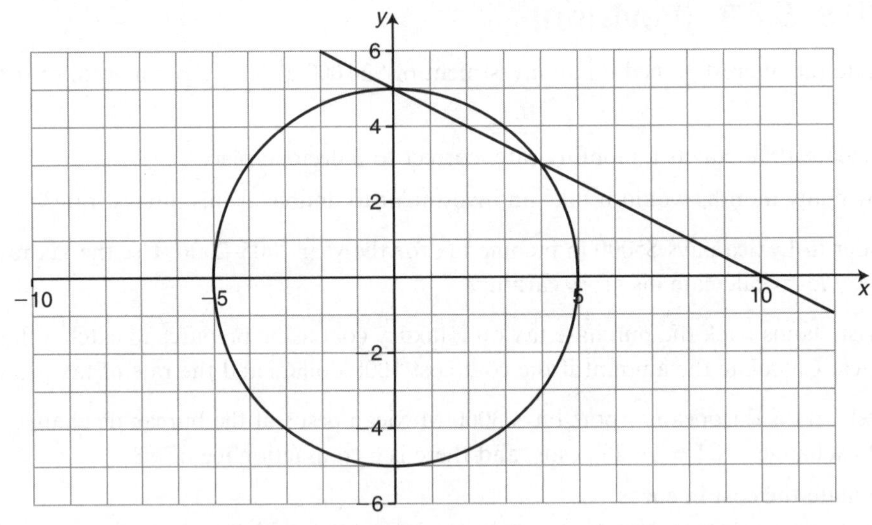

- Use the graph below to solve the simultaneous equations $3y = 4x - 25$ and $x^2 + y^2 = 25$. What is the connection between the line $3y = 4x - 25$ and the circle $x^2 + y^2 = 25$?

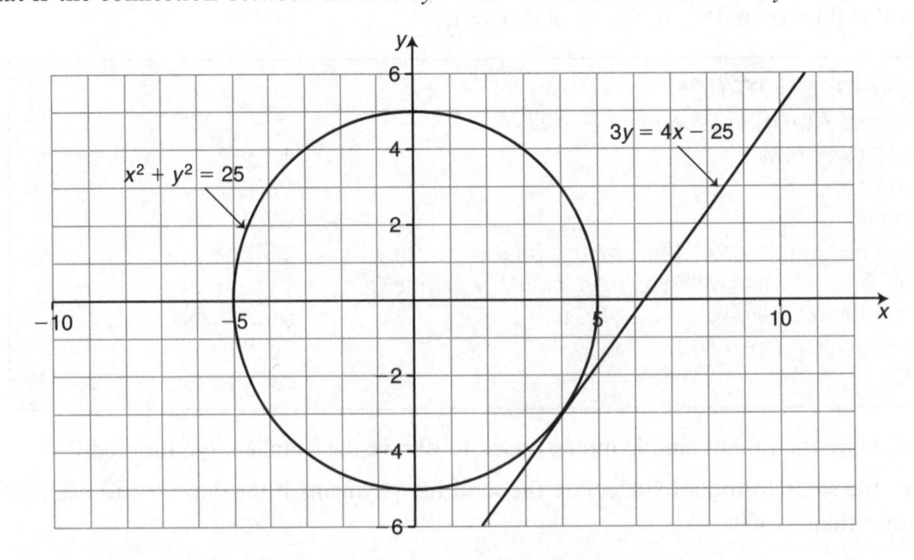

- Are there any real solutions to the simultaneous equations $3y = 18 - x$ and $x^2 + y^2 = 25$?

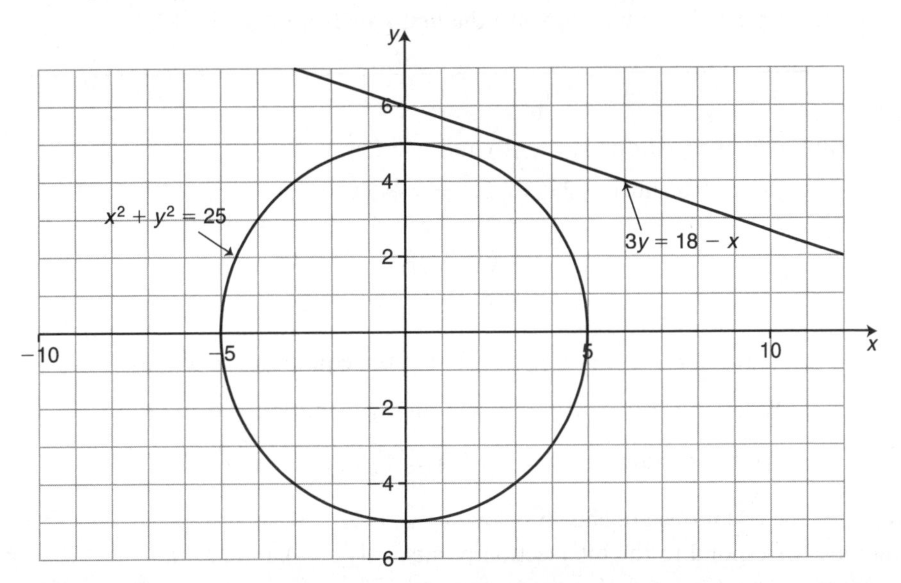

Key Point

When solving simultaneous equations where one equation is linear and the other is nonlinear:

- If there is one solution, the line is a tangent to the curve.
- If there is no solution, the line does not intersect the curve.

Drawing graphs is one way of solving simultaneous equations. Sometimes they can be solved algebraically. Example 1 shows how to solve the first pair from Activity 17.

Example 1

Solve the simultaneous equations $y = x + 6$ and $y = 2x^2$.

$$y = x + 6 \text{ and } y = 2x^2 \Rightarrow 2x^2 = x + 6$$
$$\Rightarrow 2x^2 - x - 6 = 0$$
$$\Rightarrow (2x + 3)(x - 2) = 0$$
$$\Rightarrow x = -1\tfrac{1}{2} \text{ or } x = 2$$

When $x = -1\frac{1}{2}$, $y = 4\frac{1}{2}$ (Using the equation $y = x + 6$)

When $x = 2$, $y = 8$ (Using the equation $y = x + 6$)

So the solutions are $x = -1\frac{1}{2}$, $y = 4\frac{1}{2}$ and $x = 2$, $y = 8$

The graphs of the equations are shown below.
The solutions correspond to the intersection points $(-1\frac{1}{2}, 4\frac{1}{2})$ and $(2, 8)$.

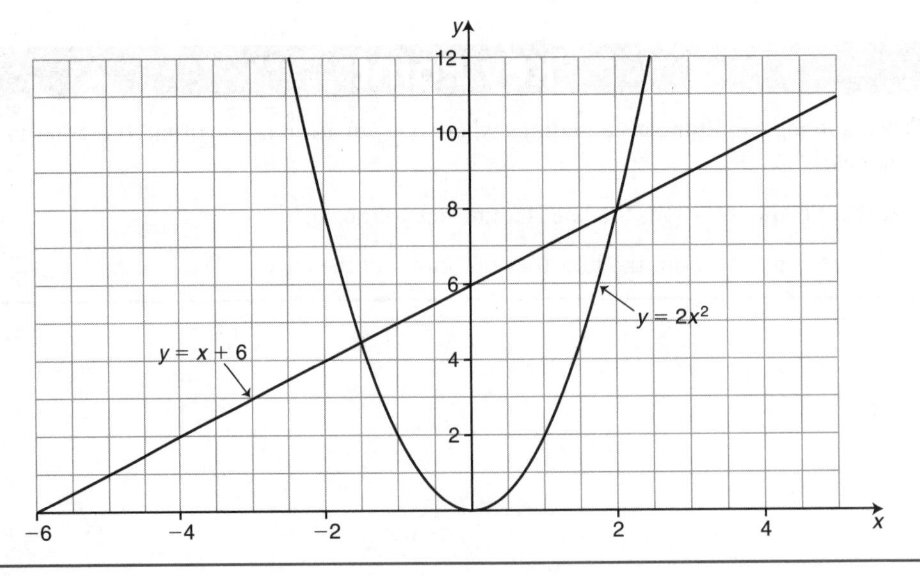

The algebraic method is more accurate than the graphical method of solving simultaneous equations but it is not always possible to solve the resulting equation.

Remember

- If the two equations are of the form

  $y =$ (some expression in x) and $y =$ (some other expression in x)

  then solve equation (some expression in x) = (some other expression in x) to find x.
- When x has been found, find y using the easier of the original equations.
- Write your solutions out in the correct pairs.

Example 2

Solve the simultaneous equations $x + 2y = 10$ and $x^2 + y^2 = 25$.

Make x the subject of the linear equation: $x = 10 - 2y$

Substitute this into the nonlinear equation: $(10 - 2y)^2 + y^2 = 25$

Simplify and solve this equation:

$$100 - 40y + 4y^2 + y^2 = 25$$
$$5y^2 - 40y + 75 = 0$$
$$y^2 - 8y + 15 = 0$$
$$(y - 3)(y - 5) = 0$$

So $y = 3$ or $y = 5$

When $y = 3$, $x = 4$ (using the equation $x = 10 - 2y$)
When $y = 5$, $x = 0$ (using the equation $x = 10 - 2y$)

So the solutions are $x = 0$, $y = 5$ and $x = 4$, $y = 3$.

The graphs of the equations are shown below.
The solutions correspond to the intersection points (0, 5) and (4, 3).

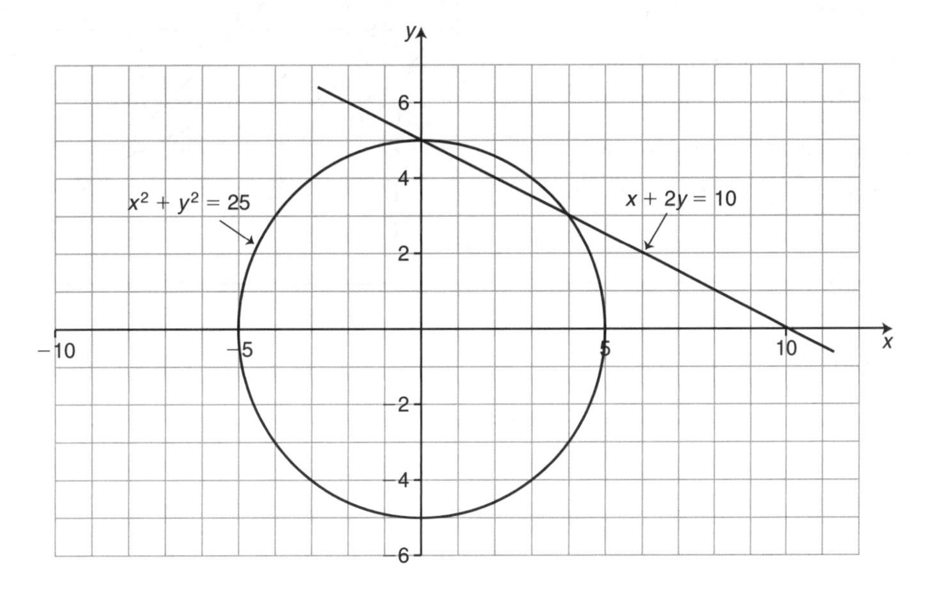

Remember

- *Always* substitute the linear equation into the nonlinear equation.
- Look at the equations carefully and see which is the easier unknown to eliminate.
- When one unknown has been found, find the other using the linear equation.
- Pair the solutions correctly.

Example 3

Solve the simultaneous equations $x + y = 4$ and $x^2 + 2xy = 2$.

Substituting for y in the second equation will make the working easier.

Make y the subject of the linear equation:

$$y = 4 - x$$

Substitute this into the nonlinear equation:

$$x^2 + 2x(4 - x) = 2$$

Simplify:

$$x^2 + 8x - 2x^2 = 2$$
$$x^2 - 8x + 2 = 0$$

Solve this equation using the quadratic formula:

$$x = 0.258 \text{ or } x = 7.74 \text{ (to 3 s.f.)}$$

Using $y = 4 - x$ gives $y = 3.74$ or $y = -3.74$ (to 3 s.f.)

So the solutions are $x = 0.258$ and $y = 3.74$ or $x = 7.74$ and $y = -3.74$, to 3 s.f.

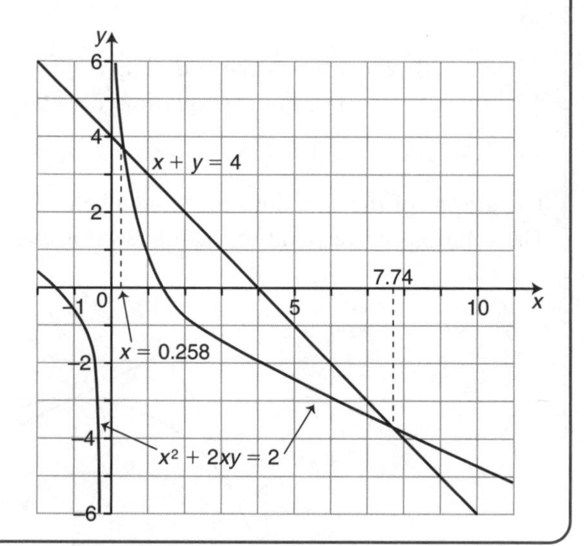

Exercise 53

For Questions 1–18, solve the simultaneous equations, giving your answers correct to 3 significant figures where appropriate.

1 $y = x + 6$, $y = x^2$

2 $y = 2x + 3$, $y = x^2$

3 $y = 3x + 4$, $y = x^2$

4 $y = 2x + 8$, $y = x^2$

5 $y = x + 1$, $y = x^2 - 2x + 3$

6 $y = x - 1$, $y = x^2 + 2x - 7$

7 $y = 2x - 1$, $y = x^2 + 4x - 6$

8 $y = 3x + 1$, $y = x^2 - x + 2$

9 $2x + y = 1$, $x^2 + y^2 = 2$

10 $x + 2y = 2$, $x^2 + y^2 = 1$

11 $x - 2y = 1$, $xy = 3$

12 $3x + y = 4$, $xy = -4$

13 $x + y = 2$, $3x^2 - y^2 = 1$

14 $x + y = 3$, $x^2 - 2y^2 = 4$

15 $x - 2y = 3$, $x^2 + 2y^2 = 3$

16 $2x + y = 2$, $4x^2 + y^2 = 2$

17 $x - y = 2$, $x^2 + xy - 3y^2 = 5$

18 $x + y = 4$, $2x^2 - 3xy + y^2 = 4$

19 The rim of my bicycle wheel has a radius of 30 cm, and the inner hub has a radius of 3 cm. The spokes are tangents to the inner hub. The diagram shows just one spoke. The x and y axes are positioned with the origin at the centre of the wheel as shown. The equation of the rim is $x^2 + y^2 = 900$.

a Write down the equation of the spoke.

b Solve the equations simultaneously.

c What is the length of the spoke?

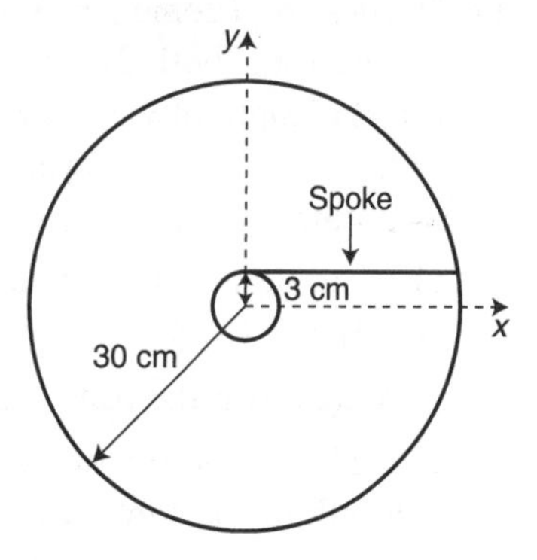

20 The shape of the cross-section of a vase is given by $y^2 - 24y - 32x + 208 = 0$, with units in cm. The vase is 20 cm high. Find the radius of the top of the vase.

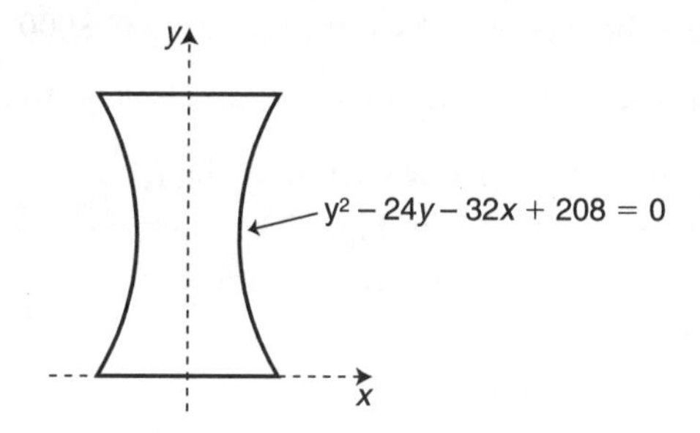

Exercise 53★

Solve these simultaneous equations, giving your answers correct to 3 significant figures where appropriate.

1 $y = 4x + 2$, $y = x^2 + x - 5$

2 $y = 1 - 3x$, $y = x^2 - 7x + 3$

3 $2x + y = 2$, $3x^2 - y^2 = 3$

4 $x + 2y = 1$, $x^2 + 2y^2 = 3$

5 $y - x = 4$, $2x^2 + xy + y^2 = 8$

6 $x + y = 3$, $x^2 + 3xy - 2y^2 = 8$

7 $x + y = 1$, $\frac{x}{y} + \frac{y}{x} = 2.5$

8 $x - y = 10$, $\frac{1}{y} - \frac{1}{x} = 5$

9 Find the points of intersection of the circle $x^2 - 6x + y^2 + 4y = 12$ and the line $4y = 3x - 42$. What is the connection between the line and the circle?

UNIT 3 ◆ Algebra

10 a Find the intersection points A and B of the line $4y + 3x = 22$ and the circle $(x - 2)^2 + (y - 4)^2 = 25$.

b Find the distance AB.

11 The design of some new spectacle frames is shown in the diagram. They consist of two circular rims, both 2 cm in radius, which are held 2 cm apart by a curved bridge piece and a straight length of wire AB.

Axes are set up as shown, and AB is 1.5 cm above the x-axis.

The equation of the left-hand circle is $(x + 3)^2 + y^2 = 4$.

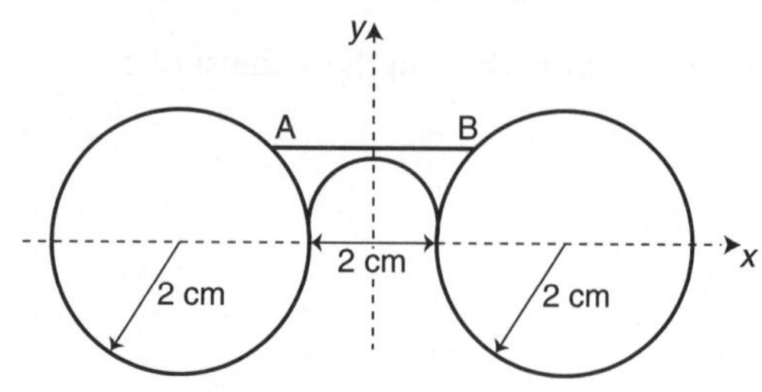

a Write down the equation of the line AB.

b Find the co-ordinates of A and B and the length of the wire AB.

12 A rocket is launched from the surface of the Earth. The surface of the Earth can be modelled by the equation $x^2 + y^2 = 6400^2$ where the units are in km.

The path of the rocket can be modelled by the equation $y = 8000 - \dfrac{x^2}{2500}$.

Find the co-ordinates of where the rocket takes off and where it lands.

13 Tracy is designing a paperweight which is part of a sphere 8 cm in diameter. The paperweight is 7 cm high. The diagram shows a cross-section of the paperweight. The equation of the circle is $x^2 + y^2 = 16$.

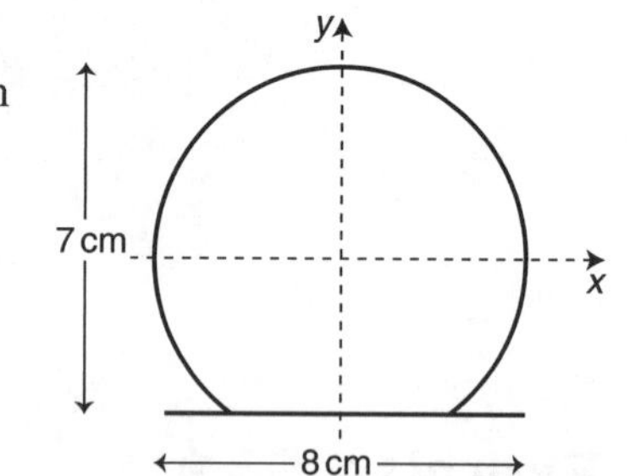

a Write down the equation of the base.

b Solve these simultaneous equations and find the diameter of the base.

14 The handle on a kitchen cabinet door is an arc of a circle of radius 14 cm. The handle sticks out by 2 cm (see diagram). The equation of the circle is $x^2 + y^2 = 196$.

a Write down the equation of the surface of the door.

b Solve these simultaneous equations to find the height of the handle (length AB).

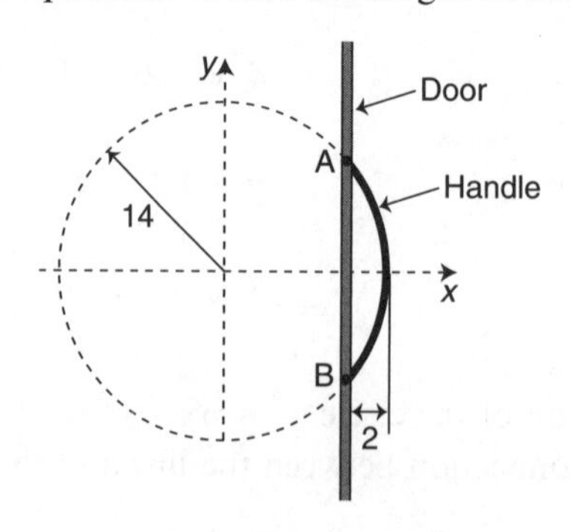

UNIT 3 ◆ Algebra

15 Maria is watering her garden with a hose. Her little brother, Peter, is annoying her and she tries to squirt him with water.

The path of the water jet is given by $y = 2x - \frac{1}{4}x^2$ and the slope of the garden is given by $y = \frac{1}{4}x - 1$. Peter is standing at (8, 1). The origin is the point where the water leaves the hose and units are in metres. Solve the simultaneous equations to find where the water hits the ground. Does Peter get wet?

16 During a match José kicks a football onto the roof of the stand. The path of the football is given by $y = 2.5x - \frac{x^2}{15}$. The equation of the roof of the stand is given by $y = \frac{x}{2} + 10$ for $20 \leqslant x \leqslant 35$. All units are in metres. Solve the simultaneous equations and find where the football lands on the roof.

17 A volcano shaped like a cone ejects a rock from the summit. The path of the rock is given by $y = 2x - x^2$ and the side of the volcano is given by $y = -\frac{2x}{3}$ where the units are in km and the origin is at the summit. Solve the simultaneous equations and find where the rock lands.

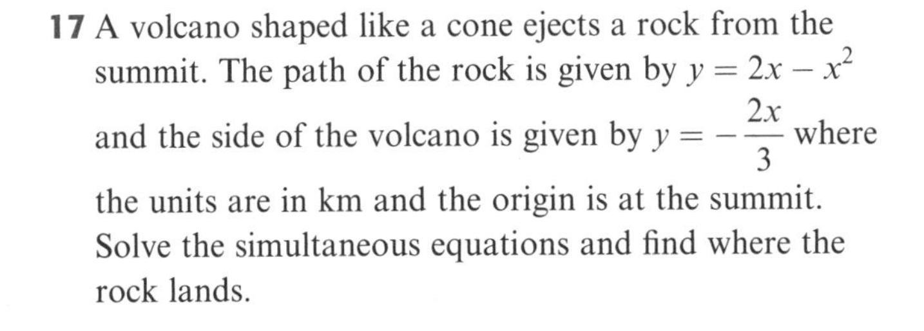

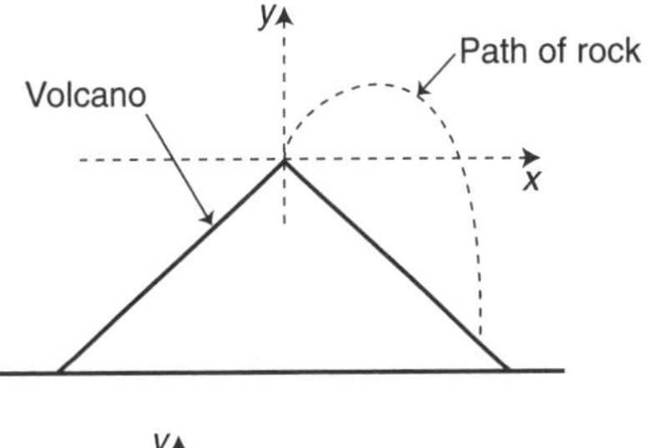

18 In a ski-jump the path of the jumper is given by $y = 0.6x - \frac{x^2}{40}$, while the landing slope is given by $y = -\frac{x}{2}$. The origin is at the point of take-off of the jumper, and all units are in metres. Solve the simultaneous equations and find how long the jump is.

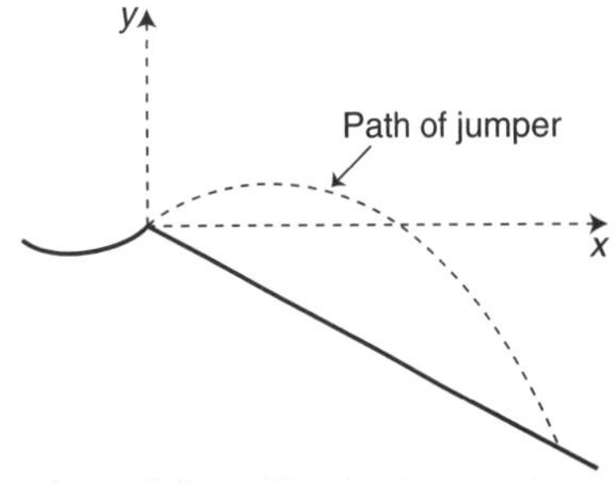

19 The lines $y = 2x$ and $2y + x = 20$ intersect the circle $(x - 4)^2 + (y - 3)^2 = 25$ at a common point, B.

a Find the co-ordinates of B and the other two points (A and C) where the lines intersect the circle.

b Find the angle between the two lines.

c What is the relationship between AC and the circle?

For Questions 20–22, solve the simultaneous equations.

20 $xy = 12$, $(x - 1)(y + 2) = 15$

21 $x^2 - y^2 = 16$, $x + y = 2$

22 $\frac{1}{x} + \frac{1}{y} = (x - 4)(y - 4) = 2$

Functions

A function is a set of rules for turning one number into another. Functions are very useful and, for example, they are much used in computer spreadsheets. In effect a function is a mathematical computer, an imaginary box that turns an input number into an output number.

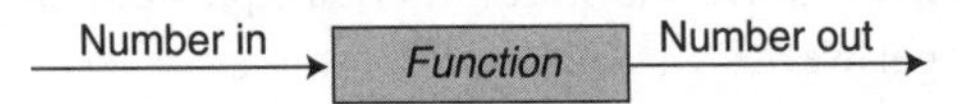

If the function doubles every number input, then

2 → Double input → 4

A letter can be used to stand for the rule. If we call the doubling function f, then

5 → f → 10

f has operated on 5 to give 10, so we write $f(5) = 10$.

If x is input then $2x$ is output, so we write $f(x) = 2x$ or $f : x \to 2x$.

Example 4

If $h : x \to 3x - 2$, find **a** $h(2)$ **b** $h\left(\frac{1}{3}\right)$ **c** $h(y)$

a $h(2) = 3 \times 2 - 2 = 4$ **b** $h\left(\frac{1}{3}\right) = 3 \times \frac{1}{3} - 2 = -1$ **c** $h(y) = 3y - 2$

Example 5

If $g(x) = \sqrt{5 - x}$, find **a** $g(1)$ **b** $g(5)$ **c** $g(-4)$

($\sqrt{x}$ means the *positive* square root of x.)

a $g(1) = \sqrt{5 - 1} = \sqrt{4} = 2$

b $g(5) = \sqrt{5 - 5} = \sqrt{0} = 0$

c $g(-4) = \sqrt{5 - (-4)} = \sqrt{9} = 3$

Example 6

a If $f(x) = 2 + 3x$ and $f(x) = 8$, find x.

b If $g(x) = \frac{1}{x-3}$ and $g(x) = \frac{1}{2}$, find x.

a $2 + 3x = 8 \Rightarrow 3x = 6 \Rightarrow x = 2$

b $\frac{1}{x-3} = \frac{1}{2} \Rightarrow x - 3 = 2 \Rightarrow x = 5$

Exercise 54

For Questions 1–8, calculate

a $f(2)$ **b** $f(-3)$ **c** $f(0.5)$ **d** $f(0)$

1 $f(x) = 2x + 1$

2 $f(x) = 4x - 1$

3 $f: x \to 3x - 2$

4 $f: x \to 5x + 3$

5 $f(x) = x^2 + 2x$

6 $f(x) = 2x^2 + x$

7 $f: x \to x^3 + 1$

8 $f: x \to x^3 - 1$

9 If $g(x) = \sqrt{x+1}$, calculate **a** $g(3)$ **b** $g(-1)$ **c** $g(99)$

10 If $g(x) = \sqrt{2x-1}$, calculate **a** $g(5)$ **b** $g\left(\frac{1}{2}\right)$ **c** $g(13)$

11 If $f: x \to \frac{1}{1+2x}$, calculate **a** $f(2)$ **b** $f(-1)$ **c** $f(a)$

12 If $f: x \to \frac{1}{2+x}$, calculate **a** $f(1)$ **b** $f(-3)$ **c** $f(z)$

13 If $h(x) = 2 - \frac{1}{x}$, calculate **a** $h(2)$ **b** $h(-2)$ **c** $h(y)$

14 If $h(x) = 1 + \frac{2}{x}$, calculate **a** $h(3)$ **b** $h(-3)$ **c** $h(p)$

15 If $f(x) = 2x + 2$ and $f(x) = 8$, find x.

16 If $h(x) = 3x - 1$ and $h(x) = 11$, find x.

17 If $p(x) = \frac{1}{x+1}$ and $p(x) = \frac{1}{4}$, find x.

18 If $q(x) = \frac{1}{3-x}$ and $q(x) = \frac{1}{2}$, find x.

19 If $g(x) = \sqrt{5x+1}$ and $g(x) = 4$, find x.

20 If $f(x) = \sqrt{13-x}$ and $f(x) = 3$, find x.

Exercise 54★

For Questions 1–8, calculate

a $f(-2)$ **b** $f(0.5)$ **c** $f(0)$ **d** $f(p)$

1 $f: x \to 8 - 2x$

2 $f: x \to 7 - 3x$

3 $f(x) = x^2 - 2x + 3$

4 $f(x) = 2x^2 + 3x - 1$

5 $f: x \to x(x+2)$

6 $f: x \to x(x-1)$

7 $f(x) = x^3 + 2$

8 $f(x) = x^3 - x$

9 If $g(x) = \dfrac{1}{3 - 2x}$, calculate a $g(3)$ b $g(-1)$ c $g(99)$

10 If $g(x) = \dfrac{1}{4 + 5x}$, calculate a $g(5)$ b $g\left(\frac{1}{2}\right)$ c $g(13)$

11 If $f: x \to \sqrt{x^2 + 2x}$, calculate a $f(2)$ b $f(-2)$ c $f(2a)$

12 If $f: x \to \sqrt{x^2 - 3}$, calculate a $f(2)$ b $f(-3)$ c $f(3z)$

13 If $h(x) = \dfrac{3x + 2}{x - 4}$, calculate a $h(2)$ b $h(-2)$ c $h(3y)$

14 If $h(x) = \dfrac{2x - 1}{3 - x}$, calculate a $h(4)$ b $h(-3)$ c $h(4p)$

15 If $f(x) = \dfrac{2}{3x + 1}$ and $f(x) = \frac{1}{2}$, find x.

16 If $h(x) = \dfrac{5}{2x - 1}$ and $h(x) = 2$, find x.

17 If $p(x) = x^2 - x - 4$ and $p(x) = 2$, find x.

18 If $q(x) = x^2 - 5x + 2$ and $q(x) = -2$, find x.

19 If $g(x) = \dfrac{2x + 17}{5 - x}$ and $g(x) = 4$, find x.

20 If $f(x) = \dfrac{12 - 4x}{3x + 16}$ and $f(x) = 2$, find x.

Example 7

Given $f(x) = 7x + 5$ and $h(x) = 6 + 2x$, find the value of x such that $f(x) = h(x)$.

$7x + 5 = 6 + 2x \Rightarrow 5x = 1 \Rightarrow x = \frac{1}{5}$

Example 8

If $f(x) = 3x + 1$, find:

a $f(2x)$ **b** $2f(x)$ **c** $f(x - 1)$ **d** $f(-x)$

a $f(2x) = 3(2x) + 1 = 6x + 1$
b $2f(x) = 2(3x + 1) = 6x + 2$
c $f(x - 1) = 3(x - 1) + 1 = 3x - 2$
d $f(-x) = 3(-x) + 1 = 1 - 3x$

Exercise 55

1 If $f(x) = 2x + 1$, find a $f(-x)$ b $f(x + 2)$ c $f(x) + 2$

2 If $f(x) = 3x - 2$, find a $f(-x)$ b $f(x + 1)$ c $f(x) + 1$

3 If $f(x) = 4x - 3$, find a $f(x + 1)$ b $f(2x)$ c $2f(x)$

4 If $f(x) = 2x - 4$, find a $f(x - 1)$ b $f(4x)$ c $4f(x)$

5 If $f(x) = 3 - x$, find a $f(-x)$ b $f(-3x)$ c $-3f(x)$

6 If $f(x) = 5 - x$, find a $f(-x)$ b $f(-5x)$ c $-5f(x)$

7 If $f(x) = x^2 + 2x$, find a $f(x - 1)$ b $f(x - 1) + 1$ c $1 - f(x - 1)$

8 If $f(x) = x^2 - 2x$, find a $f(x + 1)$ b $f(x + 1) + 1$ c $1 - f(x + 1)$

9 If $f(x) = x^2 - x$, find a $f(3x)$ b $3f(x)$ c $f(-x)$

10 If $f(x) = x^2 + x$, find a $f(2x)$ b $2f(x)$ c $f(-x)$

11 Given $f(x) = 4 + 3x$ and $g(x) = 8 - x$, find the value of x such that $f(x) = g(x)$.

12 Given $g(x) = 2x - 5$ and $h(x) = 4x + 3$, find the value of x such that $g(x) = h(x)$.

13 Given $f(x) = x^2 + 2x$ and $g(x) = x^2 + 5x - 3$, find the value of x such that $f(x) = g(x)$.

14 Given $g(x) = 2x^2 + 3x$ and $f(x) = x^2 + 3x + 4$, find the values of x such that $g(x) = f(x)$.

15 Given $p(x) = \dfrac{1}{3 - x}$ and $q(x) = \dfrac{1}{x + 4}$, find the value of x such that $p(x) = q(x)$.

16 Given $r(x) = \dfrac{2}{x + 1}$ and $s(x) = \dfrac{3}{x - 3}$, find the value of x such that $r(x) = s(x)$.

Exercise 55★

1 If $f(x) = 2 - x$, find	**a** $f(-x)$	**b** $f(-2x)$	**c** $-2f(x)$
2 If $f(x) = 4 - x$, find	**a** $f(-x)$	**b** $f(-4x)$	**c** $-4f(x)$
3 If $f(x) = x^2 + 1$, find	**a** $f(x + 2)$	**b** $f(x) + 2$	**c** $f(-x)$
4 If $f(x) = x^2 + 2$, find	**a** $f(x + 1)$	**b** $f(x) + 1$	**c** $f(-x)$
5 If $f(x) = 2x^2 - x$, find	**a** $f(-x)$	**b** $-f(x)$	**c** $f(2x)$
6 If $f(x) = 4x^2 - x$, find	**a** $f(-x)$	**b** $-f(x)$	**c** $f(4x)$
7 If $f(x) = 3 - x^2$, find	**a** $f(3x)$	**b** $3f(x)$	**c** $3f(-x)$
8 If $f(x) = 2 - x^2$, find	**a** $f(2x)$	**b** $2f(x)$	**c** $2f(-x)$
9 If $f(x) = \dfrac{1}{x^2}$, find	**a** $\dfrac{1}{f(x)}$	**b** $f\left(\frac{1}{x}\right)$	**c** $\dfrac{1}{f\left(\frac{1}{x}\right)}$
10 If $f(x) = \dfrac{1}{x}$, find	**a** $\dfrac{1}{f(x)}$	**b** $f\left(\frac{1}{x}\right)$	**c** $\dfrac{1}{f\left(\frac{1}{x}\right)}$

11 Given $f(x) = x^2 + x$ and $h(x) = 2x^2 + x - 18$, find the values of x such that $f(x) = h(x)$.

12 Given $g(x) = 4x^2 + x$ and $h(x) = x^2 + x + 12$, find the values of x such that $g(x) = h(x)$.

13 Given $p(x) = (x + 3)^2$ and $q(x) = 3 - x$, find the values of x such that $p(x) = q(x)$.

14 Given $r(x) = (x + 1)(x - 5)$ and $s(x) = x - 11$, find the values of x such that $r(x) = s(x)$.

15 Given $g(x) = \sqrt{2x + 1}$ and $h(x) = x - 1$, find the values of x such that $g(x) = h(x)$.

16 Given $f(x) = \dfrac{x}{x + 1}$ and $h(x) = \dfrac{6}{x + 5}$, find the values of x such that $f(x) = h(x)$.

Domain and range

Another way to picture a function is as a **mapping** from one set to another.

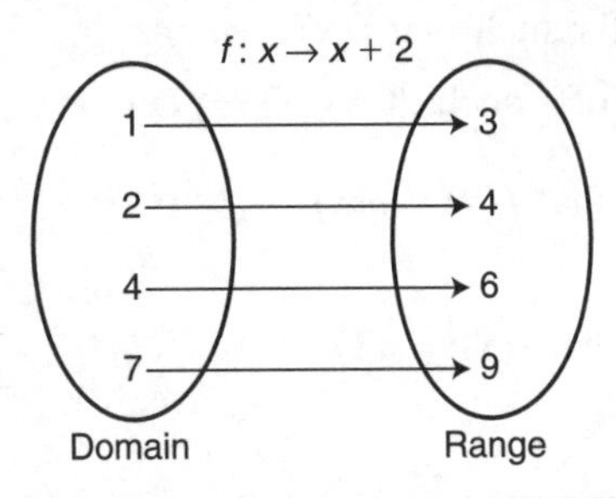

In this example the only numbers the function can use are from the set {1, 2, 4, 7}. This set is called the **domain** of the function.

The set {3, 4, 6, 9} produced by the function is called the **range** of the function.

Example 9

Find the range of the function $f: x \to 2x + 1$ if the domain is $\{-1, 0, 1, 2\}$.

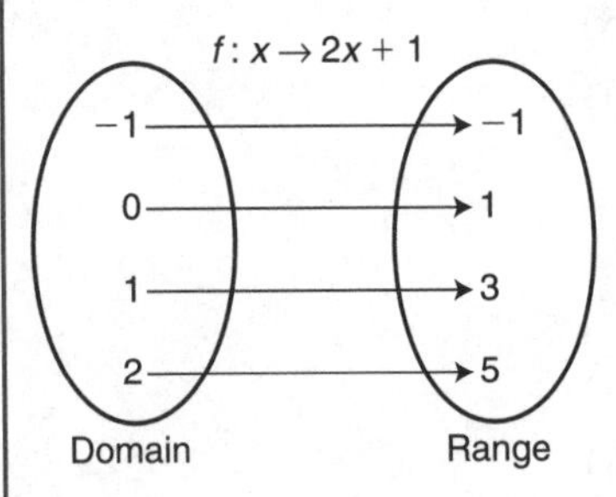

The diagram shows that the range is $\{-1, 1, 3, 5\}$.

Example 10

Find the range of the function $g(x) = x + 2$ if the domain is $\{x : x \geqslant -2, x \text{ is an integer}\}$.

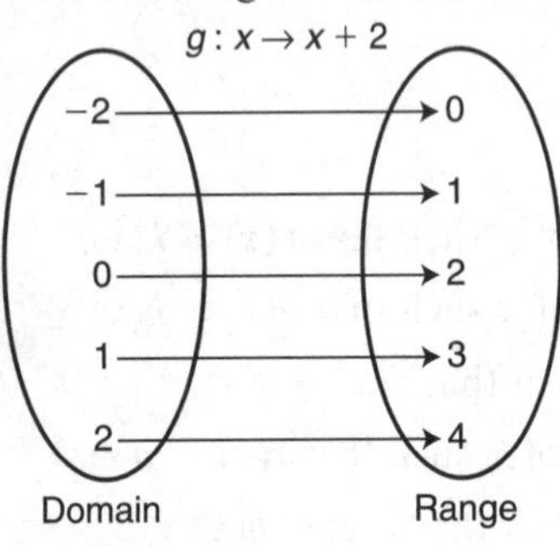

The diagram shows some of the domain. As the function changes an integer into another integer, the range is $\{y : y \geqslant 0, y \text{ is an integer}\}$.

Example 11

Find the range of:

a $h: x \to x^2$ **b** $f: x \to x^2 + 2$

if the domain of both functions is all real numbers.

a When a positive number is squared, the answer is positive.
When zero is squared, the answer is zero.
When a negative number is squared, the answer is positive.
This function cannot produce negative numbers.
So the range is $\{y : y \geqslant 0, y \text{ a real number}\}$.

b Since $x^2 \geqslant 0$, then $x^2 + 2 \geqslant 2$, so the range is $\{y : y \geqslant 2, y \text{ a real number}\}$.

The graph of the function gives a useful picture of the domain and range. The domain corresponds to the x-axis, and the range to the y-axis. The graphs of the functions used in Example 11 are shown below.

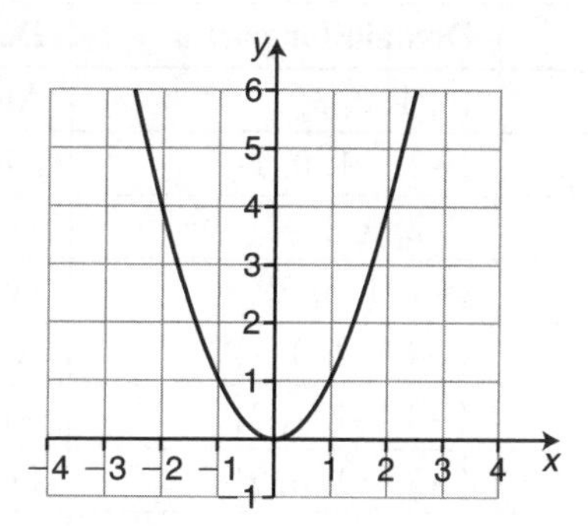

The first graph ($y = x^2$) shows that all the y values are greater than or equal to zero; that is, the range is $\{y : y \geqslant 0,\ y \text{ a real number}\}$.

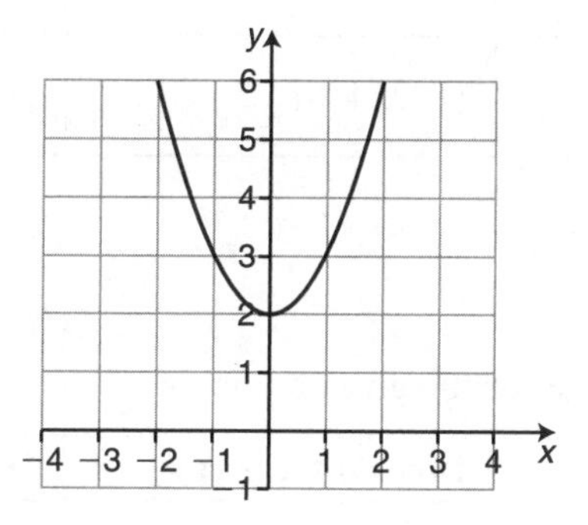

The second graph ($y = x^2 + 2$) shows that all the y values are greater than or equal to 2; that is, the range is $\{y : y \geqslant 2,\ y \text{ a real number}\}$.

Example 12

Find the range and domain for the function whose graph is shown below.

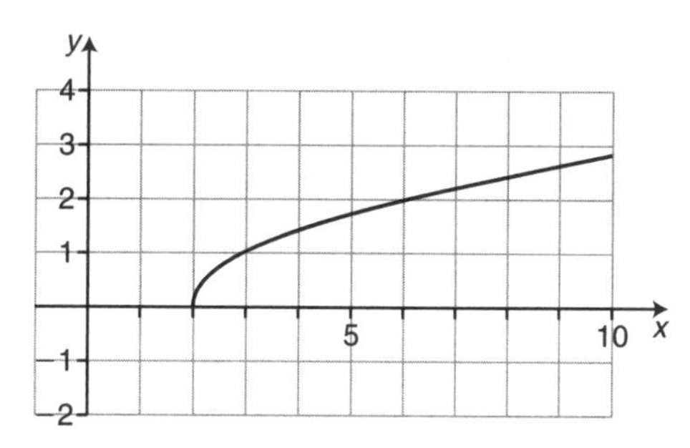

The x-axis shows that only numbers greater than or equal to 2 can be used in this function, so the domain is $\{x : x \geqslant 2,\ x \text{ a real number}\}$.

The y-axis shows that only numbers greater than or equal to zero are produced by this function, so the range is $\{y : y \geqslant 0,\ y \text{ a real number}\}$.

Exercise 56

For each function, find the range for the given domains.

	Function	Domain for part a	Domain for part b
1	$3x - 2$	$\{0, 1, 2, 3\}$	All real numbers
2	$4 - x$	$\{-2, -1, 0, 1\}$	All real numbers
3	$x^2 + x$	$\{-2, 0, 2, 4\}$	$\{x : x \geqslant 0, x \text{ a real number}\}$
4	x^2	$\{-1, 1, 3, 5\}$	$\{x : x \geqslant 0, x \text{ a real number}\}$
5	$x^2 + 1$	$\{-4, -2, 0, 2\}$	All real numbers
6	$x^2 - 1$	$\{-3, -1, 1, 3\}$	All real numbers
7	$x^3 - 3$	$\{-2, -1, 0, 2\}$	$\{x : x \geqslant 0, x \text{ a real number}\}$
8	$x^3 + x$	$\{-3, -1, 0, 2\}$	All real numbers
9	$\frac{12}{x}$	$\{2, 3, 4, 6\}$	$\{x : x \geqslant 1, x \text{ a real number}\}$
10	$2 + \frac{8}{x}$	$\{1, 2, 4, 8\}$	$\{x : x \geqslant 2, x \text{ a real number}\}$

Exercise 56★

For each function, find the range for the given domains.

	Function	Domain for part a	Domain for part b
1	$2 - 3x$	$\{-2, -1, 0, 1\}$	All real numbers
2	$5 + 4x$	$\{-1, 0, 1, 2\}$	All real numbers
3	$x^2 + 2x$	$\{-2, 0, 2, 4\}$	$\{x : x \geqslant 0, x \text{ a real number}\}$
4	$x^2 - x$	$\{-2, -1, 0, 1\}$	$\{x : x \geqslant 4, x \text{ a real number}\}$
5	$(x - 1)^2 + 2$	$\{-4, -2, 0, 2\}$	All real numbers
6	$(x + 1)^2 - 2$	$\{-4, -2, 0, 2\}$	All real numbers
7	$x^3 + x$	$\{-2, 0, 2, 4\}$	$\{x : x \geqslant 1, x \text{ a real number}\}$
8	$(x - 1)^3$	$\{-4, -2, 0, 2\}$	$\{x : x \geqslant 0, x \text{ a real number}\}$
9	$\frac{1}{x + 1}$	$\{0, 1, 2, 3\}$	$\{x : x \geqslant 0, x \text{ a real number}\}$
10	$x + \frac{12}{x}$	$\{2, 4, 6, 12\}$	$\{x : x \geqslant 6, x \text{ a real number}\}$

Sometimes there are numbers which cannot be used for the domain as they lead to impossible operations, usually division by zero or taking the square root of a negative number.

Example 13

Which numbers must be excluded from the domain of

a $f(x) = \frac{1}{x}$ **b** $g(x) = \frac{1}{x-2}$?

a Division by zero is not allowed, so zero must be excluded from the domain of f. The domain is $\{x : x \neq 0, x \text{ a real number}\}$.

b Division by zero is not allowed, so $x - 2 \neq 0$, which means $x = 2$ must be excluded from the domain of g. The domain is $\{x : x \neq 2, x \text{ a real number}\}$.

Example 14

State the domain and range of these functions:

a $f(x) = \sqrt{x}$ **b** $g(x) = 1 + \sqrt{x-2}$

a The square root of a negative number is not allowed, though it is possible to take the square root of zero. So the domain of f is $\{x : x \geqslant 0, x \text{ a real number}\}$. Since $\sqrt{}$ means the *positive* square root, the range of f is $\{y : y \geqslant 0, y \text{ a real number}\}$.

b The number under the square root sign must be greater than or equal to zero, so $x - 2 \geqslant 0$, which means $x \geqslant 2$. So the domain of g is $\{x : x \geqslant 2, x \text{ a real number}\}$. The range is $\{y : y \geqslant 1, y \text{ a real number}\}$.

Exercise 57

State which values (if any) cannot be included in the domain of the following functions.

1 $f : x \rightarrow \frac{1}{x+1}$

2 $g : x \rightarrow \frac{1}{x-1}$

3 $h : x \rightarrow \sqrt{x-2}$

4 $f : x \rightarrow \sqrt{2-x}$

5 $g(x) = x - \frac{1}{x}$

6 $h(x) = x + \frac{1}{x^2}$

7 $p(x) = x^2 + 3$

8 $q(x) = 5x - 1$

9 $r : x \rightarrow \sqrt{x^2 - 4}$

10 $s(x) = \sqrt{9 - x^2}$

Exercise 57★

State which values (if any) cannot be included in the domain of the following functions.

1 $h(x) = \frac{5}{2x-1}$

2 $g(x) = \frac{3}{4x-3}$

3 $f : x \rightarrow \sqrt{9-x}$

4 $h : x \rightarrow \sqrt{x+4}$

5 $p(x) = \frac{1}{(x+1)^2}$

6 $q(x) = \frac{1}{(1-x)^2}$

7 $r: x \to \dfrac{1}{x^2 - 1}$

8 $s: x \to \dfrac{1}{x^2 + 1}$

9 $f(x) = \dfrac{1}{\sqrt{x + 2}}$

10 $g(x) = \dfrac{1}{\sqrt{2 - x}}$

Composite functions

When one function is followed by another, the result is a composite function.

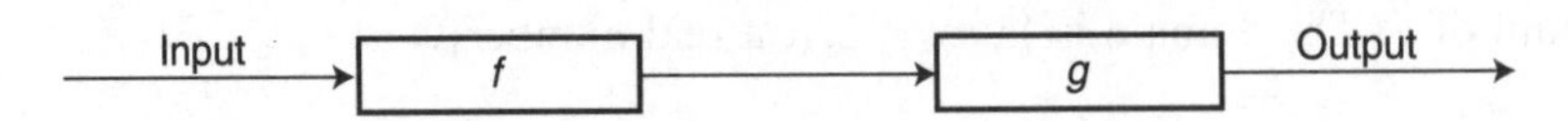

If $f: x \to 2x$ and $g: x \to x + 3$, then

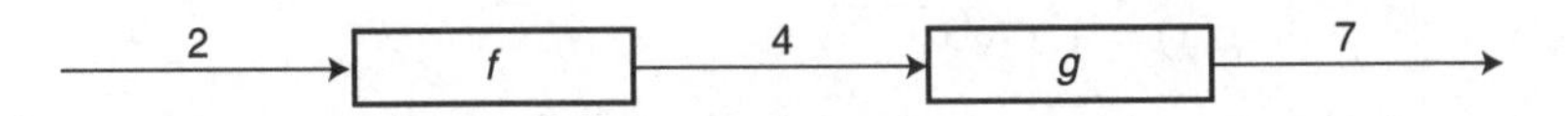

If the order of these functions is changed, then the output is different:

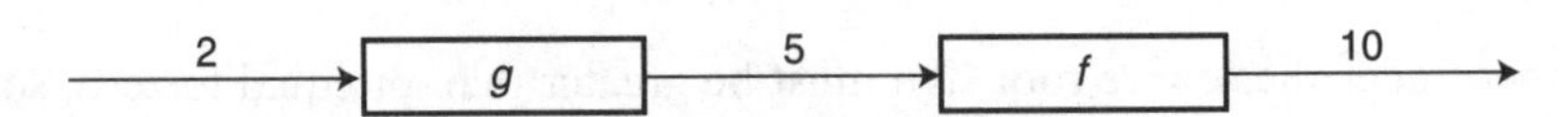

If x is input, then:

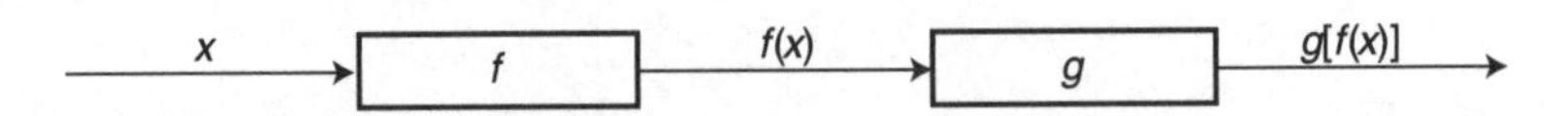

$g[f(x)]$ is usually written without the square brackets as $gf(x)$.
Remember $gf(x)$ means do f **first**, followed by g.
Note that the domain of g is the range of f.
In the same way, $fg(x)$ means do g **first**, followed by f.

Example 15

If $f(x) = x^2$ and $g(x) = x + 2$, find:

a $fg(3)$ **b** $gf(3)$ **c** $fg(x)$ **d** $gf(x)$

a $g(3) = 5$, so $fg(3) = f(5) = 25$

b $f(3) = 9$, so $gf(3) = g(9) = 11$

c $g(x) = x + 2$, so $fg(x) = f(x + 2) = (x + 2)^2$

d $f(x) = x^2$, so $gf(x) = g(x^2) = x^2 + 2$

Example 16

If $f(x) = x - 2$ and $g(x) = \sqrt{3x}$, what is the domain of:

a $gf(x)$ **b** $fg(x)$?

a The domain of $g(x)$ is $\{x : x \geqslant 0$, x a real number$\}$, so the range of $f(x)$ is $\{x : x \geqslant 0$, x a real number$\}$. This means the domain is of f is $\{x : x \geqslant 2$, x a real number$\}$.

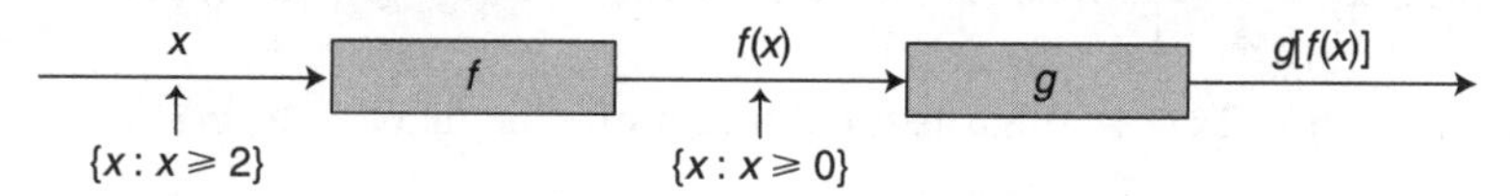

Alternative solution:

$gf(x) = g(x - 2) = \sqrt{3(x - 2)}$ which means the domain is $\{x : x \geqslant 2$, x a real number$\}$.

b The domain of $f(x)$ is any real number, so the range of $g(x)$ is any real number. This means the domain is of g is $\{x : x \geqslant 0$, x a real number$\}$.

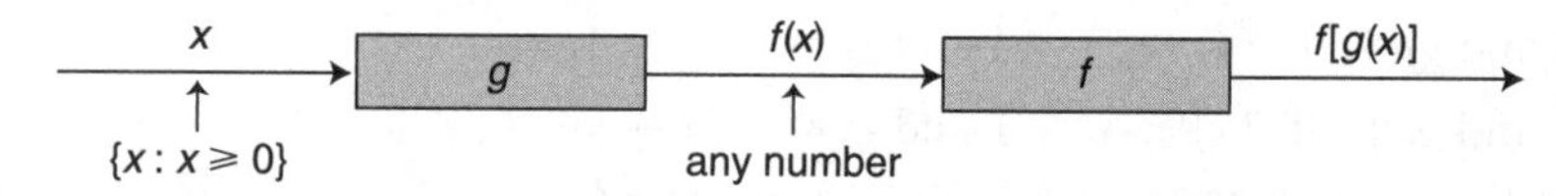

Alternative solution:

$fg(x) = f(\sqrt{3x}) = \sqrt{3x} - 2$, which means the domain is $\{x : x \geqslant 0$, x a real number$\}$.

Exercise 58

1 Find $fg(3)$ and $gf(3)$ if $f(x) = x + 5$ and $g(x) = x - 2$.

2 Find $fg(2)$ and $gf(2)$ if $f(x) = 2x - 1$ and $g(x) = x + 1$.

3 Find $fg(1)$ and $gf(1)$ if $f(x) = x^2$ and $g(x) = x + 2$.

4 Find $fg(-1)$ and $gf(-1)$ if $f(x) = x^3$ and $g(x) = x - 2$.

5 Find $fg(4)$ and $gf(4)$ if $f(x) = \dfrac{1}{x}$ and $g(x) = \dfrac{1}{x+1}$.

6 Find $fg(3)$ and $gf(3)$ if $f(x) = \dfrac{1}{x+2}$ and $g(x) = \dfrac{1}{x-2}$.

For Questions 7–14, find

a $fg(x)$ **b** $gf(x)$ **c** $ff(x)$ **d** $gg(x)$

7 $f(x) = x - 4$, $g(x) = x + 3$

8 $f(x) = x + 1$, $g(x) = x - 5$

9 $f(x) = 2x$, $g(x) = x + 2$

10 $f(x) = x + 4$, $g(x) = 4x$

11 $f(x) = x^2$, $g(x) = x + 2$

12 $f(x) = (x + 2)^2$, $g(x) = 2x$

13 $f(x) = x - 6$, $g(x) = x + 6$

14 $f(x) = 3x$, $g(x) = x \div 3$

15 $f(x) = \frac{x}{2}$ and $g(x) = x + 1$. Find x if **a** $fg(x) = 4$ **b** $gf(x) = 4$

16 $f(x) = x - 4$ and $g(x) = 2x$. Find x if **a** $fg(x) = 8$ **b** $gf(x) = 8$

17 $f(x) = x^2$ and $g(x) = \frac{1}{x+3}$. What is the domain of **a** $fg(x)$ **b** $gf(x)$?

18 $f(x) = x^3$ and $g(x) = \frac{1}{x-1}$. What is the domain of **a** $fg(x)$ **b** $gf(x)$?

19 $f(x) = \sqrt{x}$ and $g(x) = 2x - 1$. What is the domain of **a** $fg(x)$ **b** $gf(x)$?

20 $f(x) = \sqrt{x-4}$ and $g(x) = x^2$. What is the domain of **a** $fg(x)$ **b** $gf(x)$?

Exercise 58★

1 Find $fg(-3)$ and $gf(-3)$ if $f(x) = 2x + 3$ and $g(x) = 5 - x$.

2 Find $fg(5)$ and $gf(5)$ if $f(x) = 3x - 4$ and $g(x) = 5x + 1$.

3 Find $fg(2)$ and $gf(2)$ if $f(x) = x^2 + 1$ and $g(x) = (x + 1)^2$.

4 Find $fg(-1)$ and $gf(-1)$ if $f(x) = x^3 - 1$ and $g(x) = x^2 + 2$.

5 Find $fg(-3)$ and $gf(-3)$ if $f(x) = x + \frac{2}{x}$ and $g(x) = \frac{2}{x-1}$.

6 Find $fg(4)$ and $gf(4)$ if $f(x) = \frac{1}{x+2}$ and $g(x) = \frac{1}{x-2}$.

For Questions 7–14, find

a $fg(x)$ **b** $gf(x)$ **c** $ff(x)$ **d** $gg(x)$

7 $f(x) = \frac{x-4}{2}$, $g(x) = 2x$

8 $f(x) = 3x + 1$, $g(x) = \frac{x-5}{3}$

9 $f(x) = 2x^2$, $g(x) = x - 2$

10 $f(x) = x^2 + 4$, $g(x) = 4x$

11 $f(x) = \frac{1}{x-2}$, $g(x) = x + 2$

12 $f(x) = 6 - x^2$, $g(x) = x^2 - 6$

13 $f(x) = 4x$, $g(x) = \sqrt{\frac{x}{4} + 4}$

14 $f(x) = \sqrt{3x + 1}$, $g(x) = x \div 3$

15 $f(x) = 1 + \frac{x}{2}$ and $g(x) = 4x + 1$. Find x if **a** $fg(x) = 4$ **b** $gf(x) = 4$

16 $f(x) = \frac{1}{x-4}$ and $g(x) = x + 8$. Find x if **a** $fg(x) = 9$ **b** $gf(x) = 9$

17 $f(x) = 1 + x^2$ and $g(x) = \frac{1}{x-5}$. What is the domain of **a** $fg(x)$ **b** $gf(x)$?

18 $f(x) = x^3 - 1$ and $g(x) = \dfrac{1}{x+1}$. What is the domain of **a** $fg(x)$ **b** $gf(x)$?

19 $f(x) = \sqrt{2x+4}$ and $g(x) = 4x + 2$. What is the domain of **a** $fg(x)$ **b** $gf(x)$?

20 $f(x) = \sqrt{3x-9}$ and $g(x) = 2x^2$. What is the domain of **a** $fg(x)$ **b** $gf(x)$?

Inverse function

Consider the functions $f: x \rightarrow x + 1$ and $g: x - 1$.
If f is followed by g, then whatever number is input is also the output.

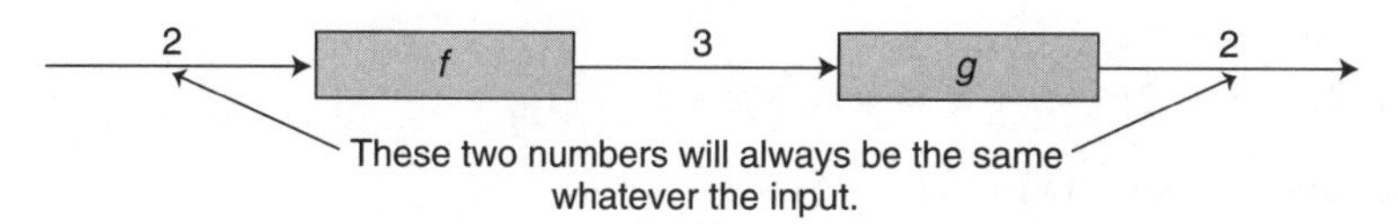

If x is the input, then x is also the output.

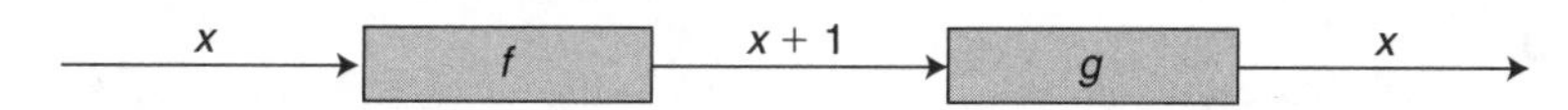

The function g is called the **inverse** of the function f. The inverse of f is the function that undoes whatever f has done. The notation f^{-1} is used for the inverse of f.

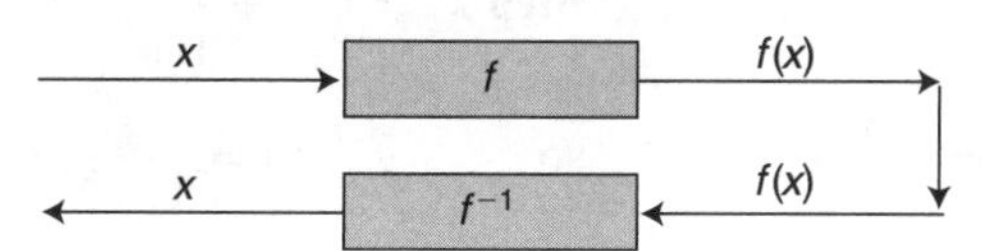

Activity 18

a Using the functions f and g defined above, show that f is the inverse of g.

b If $f(x) = 2x$, show that $g(x) = \dfrac{x}{2}$ is the inverse of f. Is f also the inverse of g?

c If $f(x) = 4 - x$, show that f is the inverse of f. (Functions like this are called 'self-inverse'.)

Example 17

$f(x) = \sqrt{3x-1}$

Find **a** $f^{-1}f(4)$ **b** $ff^{-1}(4)$

a As f^{-1} undoes what f has done, $f^{-1}f(4) = 4$.

b Similarly, f undoes what f^{-1} has done, so $ff^{-1} = 4$.

Finding the inverse function

If the inverse function is not obvious, the following steps will find it.

Step 1 Write the function as $y = \dots$
Step 2 Change any x to y and any y to x.
Step 3 Make y the subject, giving the inverse function.

Example 18

Find the inverse of $f(x) = 2x - 5$.

Step 1 $y = 2x - 5$
Step 2 $x = 2y - 5$
Step 3 $x = 2y - 5 \Rightarrow 2y = x + 5 \Rightarrow y = \frac{1}{2}(x + 5)$

The inverse function is $f^{-1}(x) = \frac{1}{2}(x + 5)$.

Example 19

Find the inverse of $g(x) = 2 + \dfrac{3}{x}$.

Step 1 $y = 2 + \dfrac{3}{x}$ **Step 2** $x = 2 + \dfrac{3}{y}$

Step 3 $x - 2 = \dfrac{3}{y} \Rightarrow y = \dfrac{3}{x - 2}$ so $g^{-1}(x) = \dfrac{3}{x - 2}$

Exercise 59

1 If $f(x) = 3x - 4$, find $f^{-1}f(7)$.
2 If $f(x) = 5 - 4x$, find $ff^{-1}(11)$.
3 If $f(x) = x^2 - 4x + 1$, find $ff^{-1}(4)$.
4 If $f(x) = \sqrt{2x^2 - 3}$, find $f^{-1}f(3)$.

For Questions 5–16, find the inverse of the function given.

5 $f(x) = 6x + 4$
6 $f(x) = \frac{1}{2}x - 3$
7 $f(x) = 9 - \frac{1}{3}x$
8 $f(x) = 12 - 5x$
9 $f(x) = 3(x - 6)$
10 $f(x) = 6(x + 5)$
11 $g(x) = \dfrac{1}{3x + 4}$
12 $h(x) = \dfrac{2}{5 - x}$
13 $p(x) = 4 - \dfrac{3}{x}$
14 $q(x) = \dfrac{5}{x} + 2$
15 $f(x) = x^2 + 7$
16 $f(x) = (x + 4)^2$

17 If $f(x) = 2x - 5$, find **a** $f^{-1}(3)$ **b** $f^{-1}(0)$ **c** $f^{-1}(-3)$
18 If $f(x) = 3(x - 2)$, find **a** $f^{-1}(2)$ **b** $f^{-1}(0)$ **c** $f^{-1}(-5)$
19 $f(x) = 2x + 5$. Solve the equation $f(x) = f^{-1}(x)$.
20 $f(x) = 3 - 4x$. Solve the equation $f(x) = f^{-1}(x)$.

Exercise 59★

1 If $f(x) = 7(2x + 3)$, find $f^{-1}f(17)$.

2 If $f(x) = 9(5 - 3x)$, find $ff^{-1}(13)$.

3 If $f(x) = (x - 7)^2 + 3x$, find $ff^{-1}(3.2)$

4 If $f(x) = \sqrt{16x + 2x^2}$, find $f^{-1}f(2.5)$.

For Questions 5–16, find the inverse of the function given.

5 $f(x) = 8(4 - 3x)$

6 $f(x) = 5(9 + 2x)$

7 $f(x) = \dfrac{3}{4 - 2x}$

8 $f(x) = \dfrac{5}{3x + 1}$

9 $f(x) = 4 - \dfrac{7}{x}$

10 $f(x) = \dfrac{6}{x} - 5$

11 $g(x) = \sqrt{x^2 + 7}$

12 $h(x) = \sqrt{(x + 1)^2 + 2}$

13 $p(x) = 2x^2 + 16$

14 $q(x) = (x + 3)^2 + 5$

15 $r(x) = \dfrac{2x + 3}{4 - x}$

16 $s(x) = \dfrac{5 - 7x}{x + 2}$

17 If $f(x) = x^2 - 5$, find **a** $f^{-1}(11)$ **b** $f^{-1}(44)$ **c** $f^{-1}(-5)$

18 If $f(x) = 2 - x^2$, find **a** $f^{-1}(-2)$ **b** $f^{-1}(2)$ **c** $f^{-1}(-23)$

19 $f(x) = 2(4x - 7)$. Solve the equation $f(x) = f^{-1}(x)$.

20 $f(x) = 3(5x + 14)$. Solve the equation $f(x) = f^{-1}(x)$.

21 $f(x) = 3 - \dfrac{2}{x}$. Solve the equation $f(x) = f^{-1}(x)$.

22 $f(x) = \dfrac{3}{x + 2}$. Solve the equation $f(x) = f^{-1}(x)$.

Exercise 60 (Revision)

1 If $g : x \to 3x + 7$, calculate **a** $g(2)$ **b** $g(-3)$ **c** $g(0)$

2 If $g(x) = 3 - 4x$, calculate x if **a** $g(x) = 5$ **b** $g(x) = -2$

3 If $f(x) = 5x - 2$, find **a** $f(x) + 1$ **b** $f(x + 1)$

4 Given $f(x) = 3x - 4$ and $g(x) = 2(x + 3)$, find the value of x such that $f(x) = g(x)$.

5 Given $f(x) = x^2$ and $g(x) = x + 6$, find the values of x such that $f(x) = g(x)$.

6 State which values of x cannot be included in the domain of:

a $f(x) = \dfrac{1}{x - 1}$ **b** $g : x \to \dfrac{3}{2x - 1}$ **c** $h(x) = \sqrt{x + 1}$ **d** $p : x \to \sqrt{3x - 6}$

7 Find the range of the following functions:

a $f(x) = 2x + 1$ **b** $g(x) = x^2 + 1$ **c** $h(x) = (x + 1)^2$ **d** $f(x) = x^3$

8 $f(x) = x^2 + 1$ and $g(x) = \dfrac{1}{x}$.

a Find (i) $fg(x)$ (ii) $gf(x)$

b What values should be excluded from the domain of: (i) $fg(x)$ (ii) $gf(x)$?

c Find and simplify $gg(x)$.

9 Find the inverse of:

a $p : x \to 4(2x + 3)$ **b** $q(x) = 7 - x$ **c** $r : x \to \dfrac{1}{x + 3}$ **d** $s(x) = x^2 + 4$

10 $f(x) = 4x - 3$ and $g(x) = \dfrac{x + 3}{4}$.

a Find the function $fg(x)$.

b Hence describe the relationship between the functions f and g.

c Write down the exact value of $fg(\sqrt{3})$.

In Questions 11–14 solve the simultaneous equations.

11 $y = x^2$, $y = x + 12$

12 $y + x^2 = 6x$, $y = 2x$

13 $y = x + 2$, $y = x^2 + 4x - 8$

14 $y = 1 - x$, $x^2 + y^2 = 4$

15 Robin Hood is designing a new bow in the shape of an arc of a circle of radius 2 m. Robin wants the distance between the string and the bow to be 30 cm. Use axes as shown in the diagram.

The equation of the circle is $x^2 + y^2 = 4$.

a Write down the equation of AB.

b Find the co-ordinates of A and B and the length of string Robin needs.

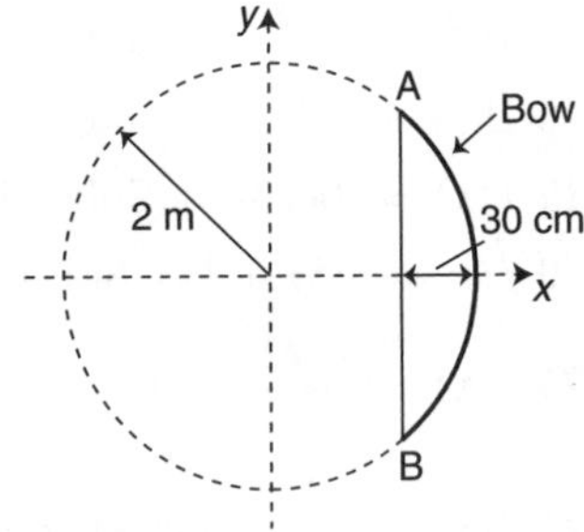

Exercise 60★ (Revision)

1 If $h : x \to \sqrt{x + 9}$, calculate **a** $h(7)$ **b** $h(0)$ **c** $h(-9)$

2 If $g(x) = x^2 - 2x$, calculate x if **a** $g(x) = 6$ **b** $g(x) = 56$

3 If $f(x) = 5 - 2x$, find **a** $f(x) - 1$ **b** $f(x - 1)$

4 Given $f(x) = \frac{1}{2}(x - 2)$ and $g(x) = 3(2 - x)$, find the value of x such that $f(x) = g(x)$.

5 Given $f(x) = (x + 3)^2$ and $g(x) = 2x^2 + 8x + 1$, find the values of x such that $f(x) = g(x)$.

6 State which values of x cannot be included in the domain of:

a $f(x) = \dfrac{5}{4 - 3x}$ **b** $g : x \to \dfrac{7}{(x + 1)^2}$ **c** $h(x) = \sqrt{2 + 5x}$ **d** $p : x \to \sqrt{x^2 - 9}$

7 Find the range of the following functions:

a $f(x) = 2x^2 + 3$ **b** $g(x) = (x - 2)^2$ **c** $h(x) = \sqrt{x + 2}$ **d** $f(x) = x^3 - 1$

8 $f(x) = x^3$ and $g(x) = \dfrac{1}{x - 8}$.

a Find: (i) $fg(x)$ (ii) $gf(x)$

b What values should be excluded from the domain of: (i) $fg(x)$ (ii) $gf(x)$?

c Find and simplify $gg(x)$.

9 Find the inverse of:

a $p: x \to 4(1 - 2x)$ **b** $q(x) = 2 - \dfrac{3}{4 - x}$ **c** $r: x \to \sqrt{2x - 3}$ **d** $s(x) = (x - 2)^2$

10 $p(x) = \dfrac{1}{x - 2}$ and $q(x) = \dfrac{1}{x} + 2$.

a Find the function $pq(x)$.

b Hence describe the relationship between the functions p and q.

c Write down the exact value of $pq(\sqrt{7})$.

In Questions 11–14 solve the simultaneous equations.

11 $y = 2x^2$, $y = 5x + 3$

12 $xy = 4$, $y = 2x + 2$

13 $y = 2x - 1$, $y = 2x^2 + 7x - 13$

14 $x + y = 2$, $2x^2 - x + y^2 = 6$

15 The path of the comet Fermat is an ellipse whose equation relative to the Earth is $x^2 + 36y^2 = 324$, where the units are in AU (1 AU, called an astronomical unit, is the distance from the Earth to the Sun).
The comet can be detected by eye at a distance of 3 AU from the Earth, and the equation of this circle is $(x - 17.5)^2 + y^2 = 9$.
Find the co-ordinates of the points where the comet can be seen from the Earth.

Graphs 3

Tangents to a curve

Remember

For a straight line the gradient $= \frac{\text{'rise'}}{\text{'run'}}$

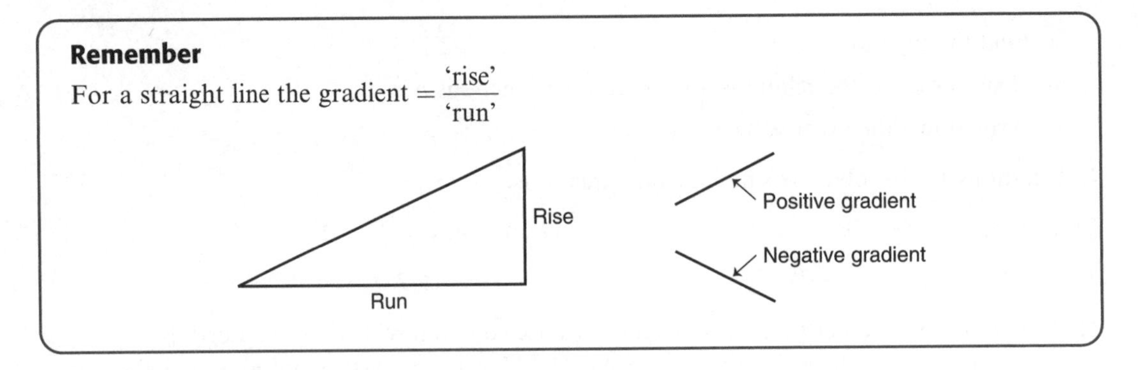

The gradient of a graph often provides useful information.

Graph	Gradient
Distance (m) – time (s)	Speed (m/s)
Speed (m/s) – time (s)	Acceleration (m/s^2)
Temperature (°C) – time (min)	Rate of change of temperature (°C/min)
Population (ants) – time (weeks)	Rate of change of population (ants/week)
Financial profit ($) – time (years)	Rate of change of profit ($/year)

Most graphs of real-life situations are curves rather than straight lines, but information on rates of change can still be found by drawing a tangent to the curve and using this to estimate the gradient of the curve at that point.

Remember

To find the gradient of a curve at a point:
Draw the tangent to the curve at the point. Do this by pivoting a ruler about the point until the angles between the ruler and the curve are as equal as possible.

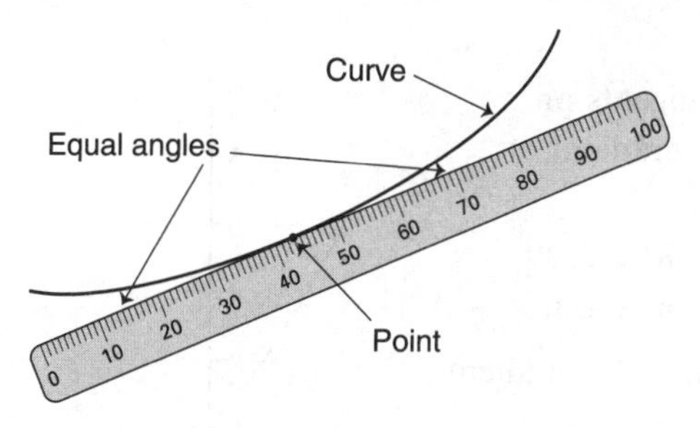

The gradient of the tangent is the gradient of the curve at the point.

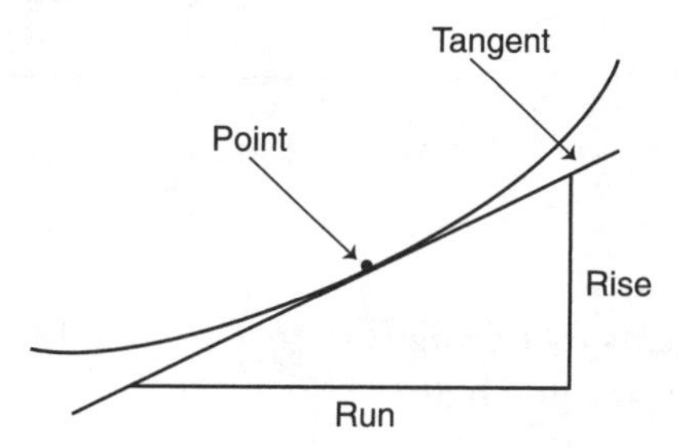

The gradient of the tangent is found by working out $\frac{\text{'rise'}}{\text{'run'}}$.

Example 1

Find the gradient of the curve $y = x^2 - x$ at the point where $x = 2$.
First draw the graph for x values around 2.

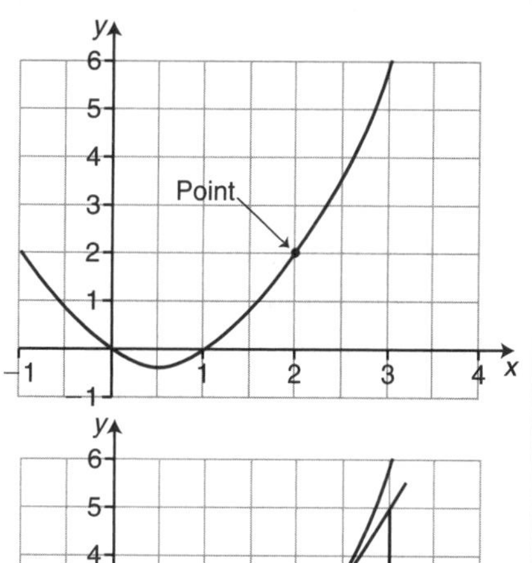

Next draw in the tangent.

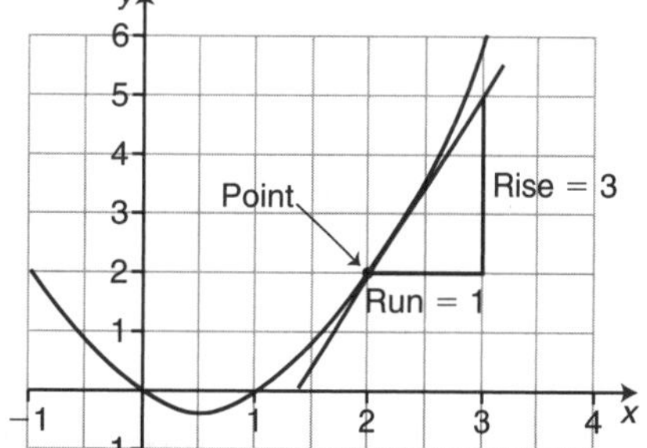

The gradient of the curve at $x = 2$ is 3.

UNIT 3 ♦ Graphs

Because the tangent is judged by eye, different people will get different answers for the gradient. The answers given are calculated using a different technique which you will learn in the next unit, so don't expect your answers to be exactly the same as those in the back of this book.

Exercise 61

1 By drawing suitable tangents on tracing paper, find the gradient of the graph at:

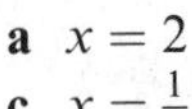

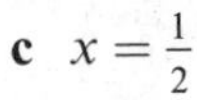

a $x = 2$ **b** $x = 2\frac{1}{2}$

c $x = \frac{1}{2}$ **d** $x = 0$

e Where on the graph is the gradient equal to 0?

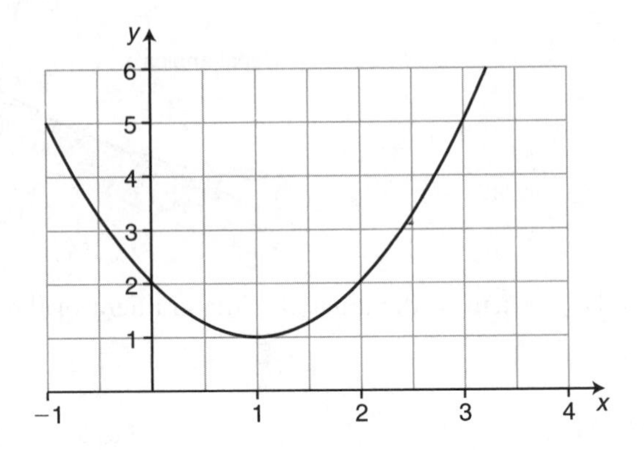

2 By drawing suitable tangents on tracing paper, find the gradient of the graph at:

a $x = 0$ **b** $x = 2$

c $x = 0.4$ **d** $x = 2.5$

e Where on the graph is the gradient equal to 1?

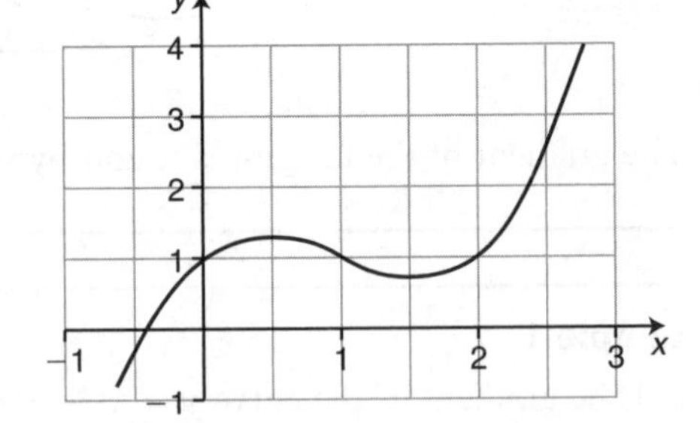

3 By drawing suitable tangents on tracing paper, find the gradient of the graph at:

a $x = -2$ **b** $x = -1$

c $x = 1\frac{1}{2}$ **d** $x = \frac{1}{2}$

e Where on the graph is the gradient equal to -2?

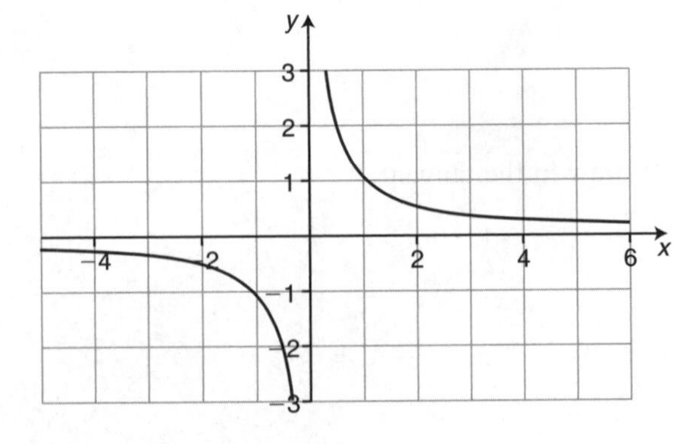

4 By drawing suitable tangents on tracing paper, find the gradient of the graph at:

a $x = -2$ **b** $x = -\frac{1}{2}$ **c** $x = \frac{1}{2}$ **d** $x = 1.5$

e Where on the graph is the gradient equal to 0?

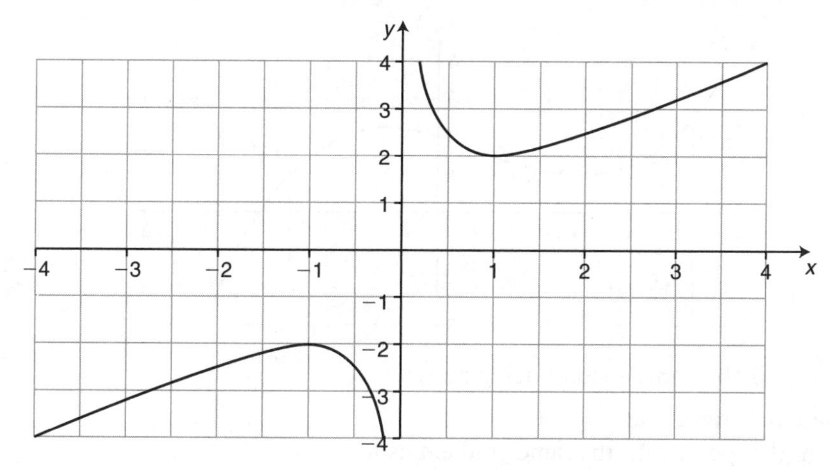

5 Plot the graph of $y = x(6 - x)$ for $0 \leqslant x \leqslant 6$. By drawing suitable tangents, find the gradient of the graph at:

a $x = 1$ **b** $x = 2$ **c** $x = 5$

d Where on the curve is the gradient equal to 0?

6 Plot the graph of $y = x(x - 4)$ for $-1 \leqslant x \leqslant 5$. By drawing suitable tangents, find the gradient of the graph at:

a $x = 0$ **b** $x = 2$ **c** $x = 3$

d Where on the curve is the gradient equal to 4?

Exercise 61★

1 By drawing suitable tangents on tracing paper, estimate the gradient of the graph at:

a $x = \frac{1}{2}$ **b** $x = -\frac{1}{2}$

c $x = 2$ **d** $x = 3$

e Where on the graph is the gradient equal to 0?

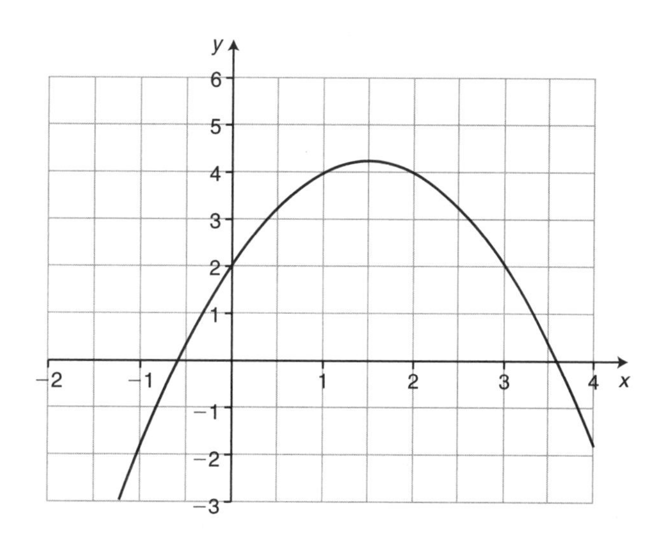

2 By drawing suitable tangents on tracing paper, estimate the gradient of the graph at:

a $x = -1\frac{1}{2}$ **b** $x = -\frac{1}{2}$ **c** $x = \frac{1}{2}$ **d** $x = 1$

e Where on the graph is the gradient equal to 0?

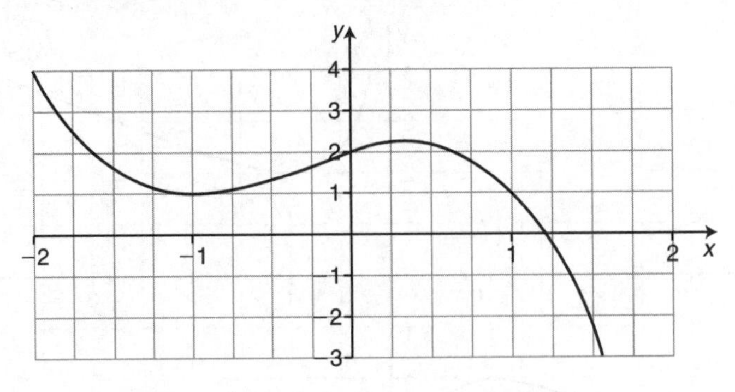

3 By drawing suitable tangents on tracing paper:

a Estimate the gradient at $x = -1$.

b Find another point with the same gradient as at $x = -1$.

c Find two points where the gradient is -2.

d Where on the graph is the gradient equal to 0?

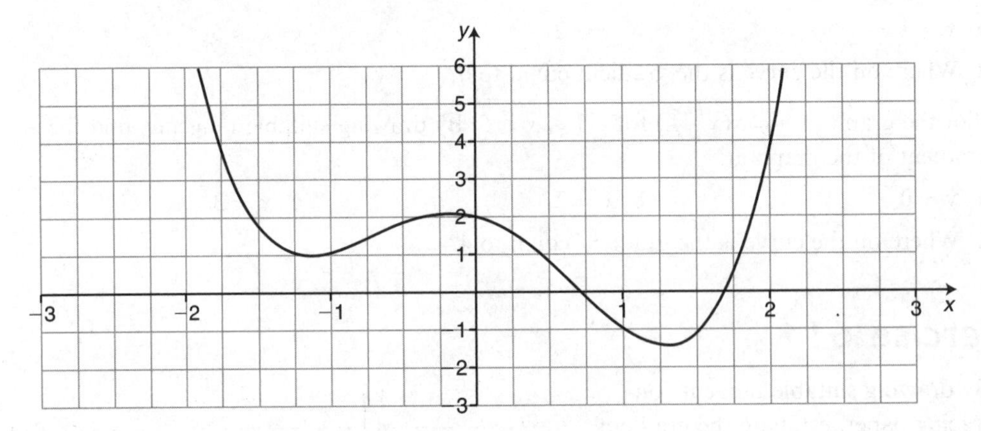

UNIT 3 ◆ Graphs

4 By drawing suitable tangents on tracing paper:

a Estimate the gradient at $x = 4$.

b Find another point with the same gradient as at $x = 4$.

c Find two points where the gradient is $\frac{1}{2}$.

d Where on the graph is the gradient equal to 0?

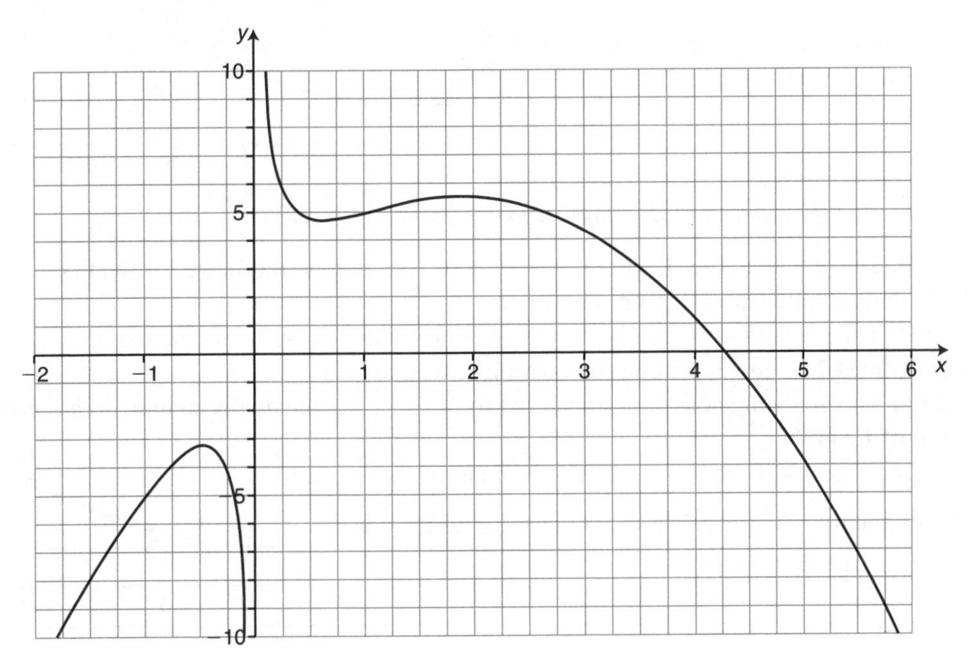

5 a Draw an accurate graph of $y = x^2$ for $-4 \leqslant x \leqslant 4$.

b Use your graph to complete the following table:

***x*-co-ordinate**	−4	−3	−2	−1	0	1	2	3	4
Gradient	−8								8

c Plot a graph showing your answer to part **b**. What do you notice?

6 a Plot the graph of $y = 2^x$ for $0 \leqslant x \leqslant 5$ by first copying and completing the table.

x	0	1	2	3	4	5
y						

b Use this graph together with suitable tangents to estimate the gradient of the curve at $x = 1$ and $x = 3$.

c Where on the curve is the gradient equal to 12?

Example 2

A dog is running in a straight line away from its owner. Part of the distance–time graph describing the motion is shown.

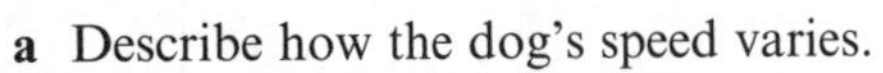

a Describe how the dog's speed varies.
b Estimate the dog's speed after 30 seconds.

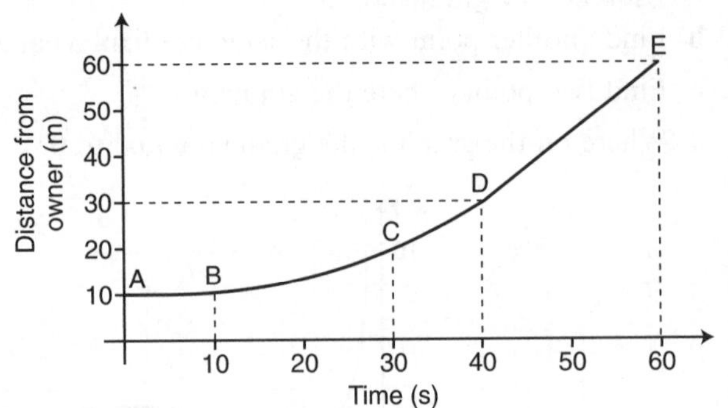

Remember that the gradient of a distance–time graph gives the speed.

a A to B: The gradient is zero, so the speed is zero. The dog is stationary for the first 10 seconds, 10 metres away from its owner.
B to D: The gradient is gradually increasing, so the speed is gradually increasing. For the next 30 seconds the dog runs with increasing speed.
D to E: The gradient is constant and equal to $\frac{30}{20}$ or 1.5, so the dog is running at a constant speed of 1.5 m/s.

b Draw a tangent at C and calculate the gradient of the tangent.
The gradient is $\frac{15\,\text{m}}{20\,\text{s}} = 0.75\,\text{m/s}$, so the speed of the dog is approximately 0.75 m/s.

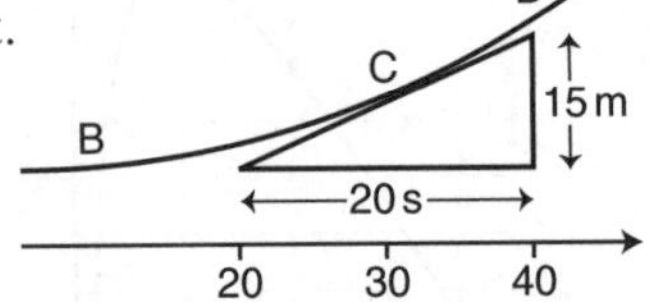

Exercise 62

1 The graph shows part of the distance–time graph for a car caught in a traffic jam.

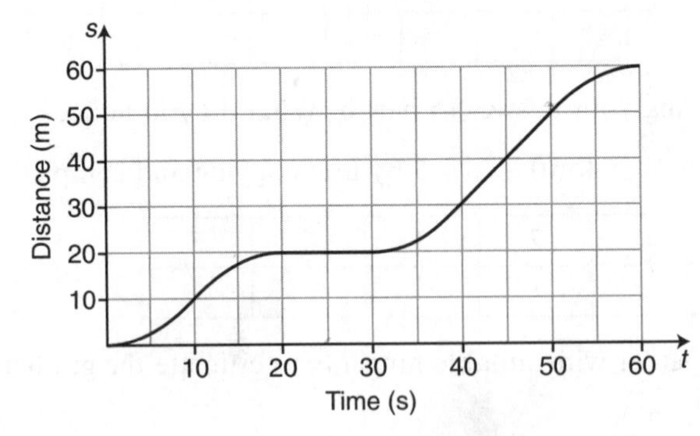

a By drawing suitable tangents on tracing paper, estimate the speed of the car when
(i) $t = 15$ s (ii) $t = 25$ s (iii) $t = 45$ s

b Describe the car's journey.

2 The graph shows the speed–time graph for a young girl in a race.

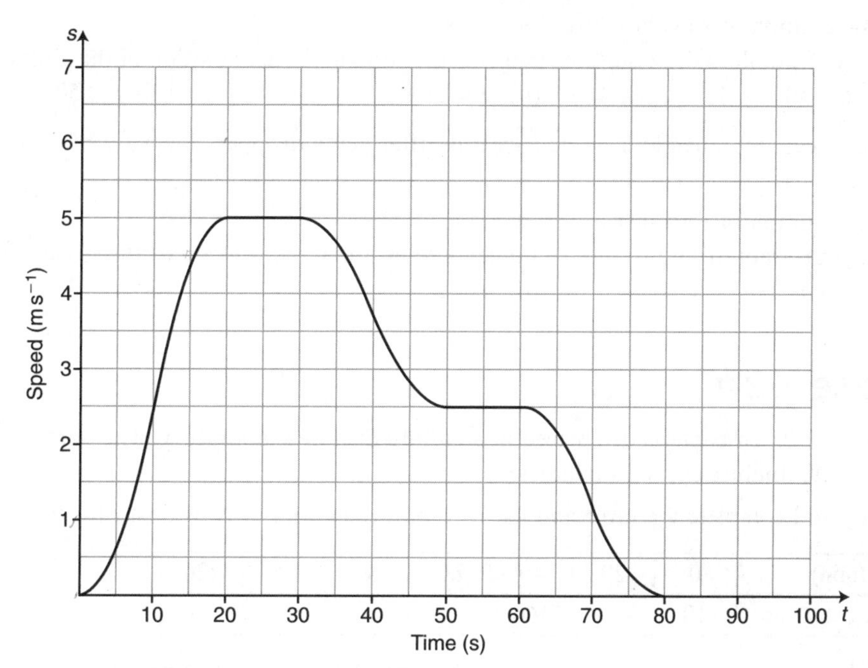

a By drawing suitable tangents on tracing paper, estimate the acceleration of the girl when
(i) $t = 15$ s (ii) $t = 25$ s (iii) $t = 70$ s

b Describe how the girl ran the race.

3 The temperature of a cup of coffee (T °C) after t minutes is given in this table:

***t* (min)**	0	1	2	3	4	5	6
***T*(°C)**	80	71	62	55	49	43	38

a Draw the temperature–time graph of this information.

b By drawing suitable tangents, estimate the rate of change of temperature in °C/min when:
(i) $t = 0$ (ii) $t = 3$ (iii) $t = 5$

4 The depth, d cm, of water in Ahmed's bath t minutes after he has pulled the plug out is given in this table:

***t* (min)**	0	1	2	3	4	5	6
***d* (cm)**	30	17	10	6	3.3	1.4	0

a Draw the graph showing depth against time.

b By drawing suitable tangents, estimate the rate of change of depth in cm/min when:
(i) $t = 0.5$ (ii) $t = 2.5$ (iii) $t = 5.5$

5 The velocity, v m/s, of a vintage aircraft t seconds after it starts to take off is given by $v = 0.025t^2$.

a Draw a graph of v against t for $0 \leqslant t \leqslant 60$.

b Use your graph to draw suitable tangents to estimate the acceleration of the aircraft when:

(i) $t = 10$ (ii) $t = 30$ (iii) $t = 50$

6 The distance, s m, fallen by a stone t seconds after being dropped down a well is given by $s = 5t^2$.

a Draw a graph of s against t for $0 \leqslant t \leqslant 4$.

b Use your graph to draw suitable tangents to estimate the velocity of the stone when:

(i) $t = 1$ (ii) $t = 2$ (iii) $t = 3$

Exercise 62★

1 At time $t = 0$, ten bacteria are placed in a culture dish in a laboratory. The number of bacteria, N, doubles every 10 minutes.

a Copy and complete the table and use it to plot the graph of N against t for $0 \leqslant t \leqslant 120$.

***t* (min)**	0	20	40	60	80	100	120
N	10		160				

b By drawing suitable tangents, estimate the rate of change in bacteria per minute when

(i) $t = 0$ (ii) $t = 60$ (iii) $t = 100$

2 The sales, N, in a mobile phone network are increasing at a rate of 5% every month. Present sales are two million.

a Copy and complete the following table showing sales forecasts for the next nine months and use it to plot the graph of N against t for $0 \leqslant t \leqslant 9$.

***m* (months)**	0	1	2	3	4	5	6	7	8	9
***N* (millions)**	2		2.21							

b By drawing suitable tangents, estimate the rate of increase of sales in numbers per month when:

(i) $t = 0$ (ii) $t = 4$ (iii) $t = 8$

3 A party balloon of volume 2000 cm^3 loses 15% of its air every 10 minutes.

a Copy and complete the table and use it to plot the graph of the volume, V, against t for $0 \leqslant t \leqslant 90$.

***t* (min)**	0	10	20	30	40	50	60	70	80	90
***V* (cm^3)**	2000		1445							

b By drawing suitable tangents, estimate the rate of change of the balloon's volume in cm^3/min when:

(i) $t = 10$ (ii) $t = 80$

c When was the rate of change of the balloon's volume at its maximum value, and what was this value?

4 A radioactive isotope of mass 120 g decreases its mass, M, by 20% every 10 seconds.

a Copy and complete the table and use it to plot the graph of M against t for $0 \leqslant t \leqslant 90$.

t (s)	0	10	20	30	40	50	60	70	80	90
M (g)	120				49.2					

b By drawing suitable tangents, estimate the rate of change of the isotope's mass in g/s when:
(i) $t = 20$ (ii) $t = 70$

c When was the rate of change of the isotope's mass at its maximum value, and what was this value?

5 The graph shows the depth, y metres, of water at Brigstock Harbour t hours after 1200 hrs.

a By drawing suitable tangents, estimate the rate of change of depth in m/h at:
(i) 1300 hrs
(ii) 1700 hrs
(iii) 2200 hrs

b At what times is the rate of change of depth a maximum, and at what times does this occur?

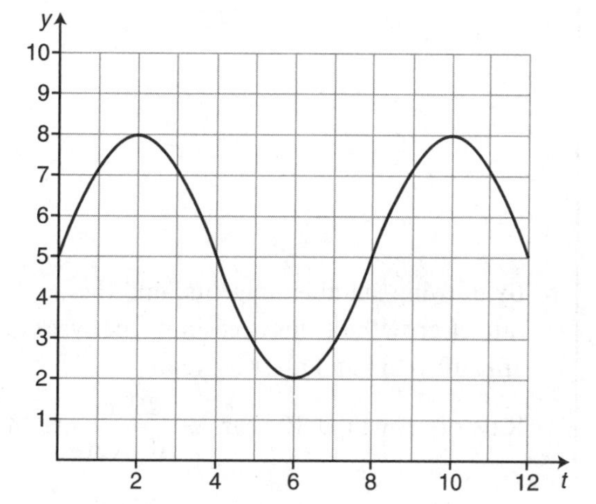

6 Steve performs a 'bungee-jump' from a platform above a river. His height, h metres, above the river t seconds after he jumps is shown on the graph.

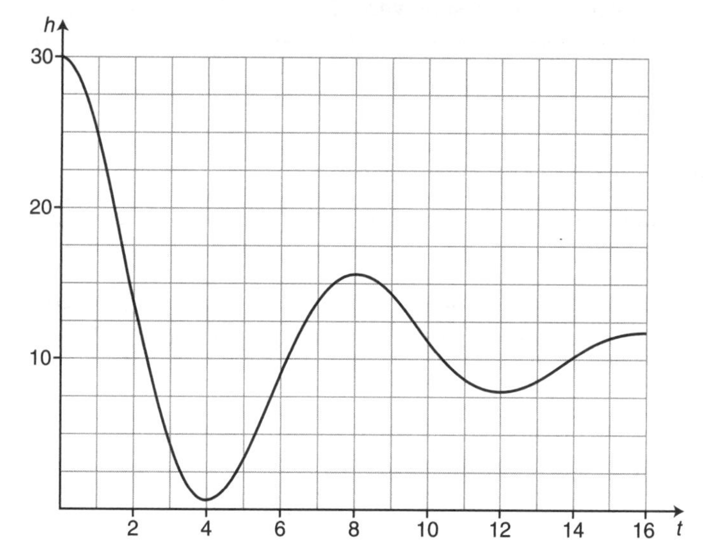

a By drawing suitable tangents, estimate Steve's velocity in m/s after:
(i) 1 s (ii) 8 s (iii) 14 s

b Estimate Steve's maximum velocity and the time it occurs.

Example 3

The area of weed covering part of a field doubles every 10 years. The area now covered is $100\,m^2$.

a Given that the area of weed, $A\,m^2$, after n years, is given by $A = 100 \times 2^{0.1n}$, draw the graph of A against n for $0 \leqslant n \leqslant 40$.

n (years)	0	10	20	30	40
A (m^2)	100	200	400	800	1600

This is the graph of area of weed against time.

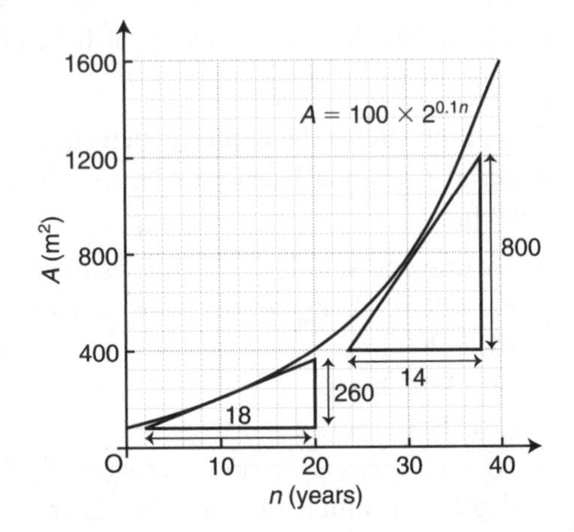

b By drawing suitable tangents, find the rate of growth of the weed in m^2 per year after 10 years and after 30 years.

Rate of growth at 10 years $\simeq \dfrac{260\,m^2}{18\text{ years}} \simeq 14\,m^2/\text{year}$.

Rate of growth at 30 years $\simeq \dfrac{800\,m^2}{14\text{ years}} \simeq 57\,m^2/\text{year}$.

The rate of growth is clearly increasing with time.

Exercise 63 (Revision)

1 a Draw the curve $y = x(x - 5)$ for $0 \leqslant x \leqslant 6$.

b By drawing suitable tangents, find the gradient of the curve at the points $x = 1$, $x = 2.5$ and $x = 5$.

2 a Plot the graph of $y = x^2(x - 4)$ for $-2 \leqslant x \leqslant 4$. Use your graph together with suitable tangents to estimate the gradient of the curve at $x = -1$ and $x = 3$.

b Where on the curve is the gradient equal to zero?

3 The depth, d mm, of fluid poured into a conical beaker at a constant rate, after t seconds, is shown in this table.

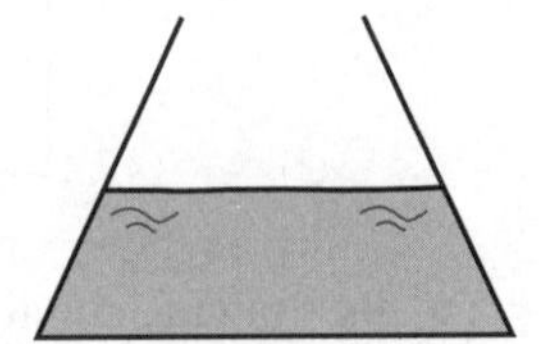

t (s)	0	4	8	12	16	20
d (mm)	1	2.5	6.3	15.8	39.8	100

a Draw the graph of depth against time from this table.

b Estimate the rate of change of the depth in mm/s when $t = 10$ s and when $t = 15$ s.

4 The height of a walnut tree, y cm after t months, is given in the table.

t (months)	0	6	12	18	24	30	36
y (cm)	1	2.5	6.3	15.8	39.8	100	251

a Draw the graph of y against t.

b By drawing suitable tangents to the curve, estimate the rate of growth of the walnut tree in cm *per year* when $t = 15$ and when $t = 30$.

Exercise 63★ (Revision)

1 a Plot the graph of $y = 3^x$ for $0 \leqslant x \leqslant 4$ by first copying and completing the table.

x	0	1	2	3	4
y	1		9		

b Use this graph together with suitable tangents to estimate the gradient of the curve at $x = 1$ and $x = 2$.

2 A catapult fires a stone vertically upwards. The height, h metres, of the stone, t seconds after firing, is given by the formula $h = 40t - 5t^2$.

a Draw the graph of h against t for $0 \leqslant t \leqslant 8$.

b Draw tangents to this curve and measure their gradients. Then copy and complete this table.

t (s)	0	1	2	3	4	5	6	7	8
Velocity (m/s)	40		20			–10			–40

c Draw the velocity–time graph for the stone for $0 \leqslant t \leqslant 8$.

d What can you say about the stone's acceleration?

3 The point A(1, 3) is on the graph of $y = x(4 - x)$ as shown.

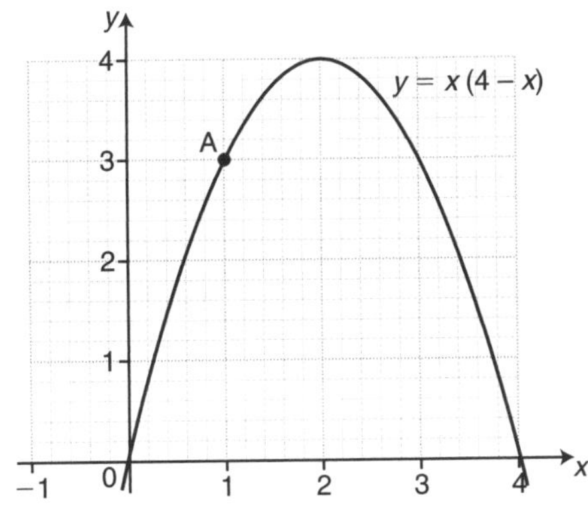

a By drawing a suitable tangent on tracing paper, find the gradient of the curve at point A.

b What is the equation of the tangent to the curve at A?

4 a By drawing the graph of $y = x^3 - 12x + 2$ for $-5 \leqslant x \leqslant 5$, find where on the curve the gradient is zero.

b Find the equation of the tangent to the curve at the point where $x = 1$.

Shape and space 3

Vectors

A **vector** has both size and direction. In contrast, a **scalar** has size but no direction. Vectors are very useful tools in mathematics and physics, helping to make calculations more direct. In 1881, American mathematician J. W. Gibbs published the book *Vector Analysis*, which established vectors as they are known today.

Activity 19

Identify these quantities as vector or scalar quantities:

- Volume
- Acceleration
- A pass in hockey
- Area
- Temperature
- Velocity
- Price
- Rotation of 180°
- Force
- Length
- Density
- 10 km on a bearing of 075°

Notation

In this book, vectors are written as bold letters (**a**, **p**, **x**, ...) or capitals covered by an arrow ($\overrightarrow{AB}$, $\overrightarrow{PQ}$, $\overrightarrow{XY}$, ...). In other books, you might find vectors written as bold italic letters (***a***, ***p***, ***x***, ...). When hand-writing vectors, they are written with a wavy or straight underline ($\underset{\sim}{a}$, $\underset{\sim}{p}$, $\underset{\sim}{x}$, or $\underline{a}$, $\underline{p}$, $\underline{x}$).

On co-ordinate axes, a vector can be described by a **column vector**, which can be used to find the **magnitude** and **angle** of the vector.

Example 1

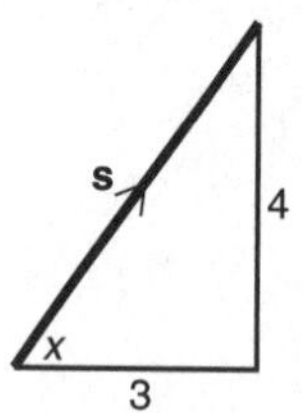

a Express vector **s** as a column vector.

$$\mathbf{s} = \begin{pmatrix} 3 \\ 4 \end{pmatrix}$$

b Find the magnitude of vector **s**.

Length of $\mathbf{s} = \sqrt{3^2 + 4^2} = \sqrt{25} = 5$

c Calculate the size of angle x.

$$\tan x = \frac{4}{3} \Rightarrow x = 53.1° \text{ (to 3 s.f.)}$$

Example 2

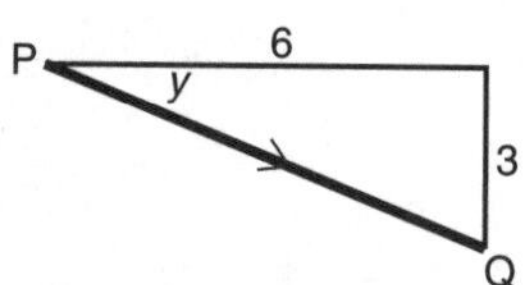

a Express vector $\overrightarrow{PQ}$ as a column vector:

$$\overrightarrow{PQ} = \begin{pmatrix} 6 \\ -3 \end{pmatrix}$$

b Find the magnitude of vector $\overrightarrow{PQ}$.

Length of $\overrightarrow{PQ} = \sqrt{6^2 + 3^2}$

$= \sqrt{45}$

$= 6.71$ (to 3 s.f.)

c Calculate the size of angle y.

$$\tan y = \frac{3}{6} \Rightarrow y = 26.6° \text{ (to 3 s.f.)}$$

Activity 20

Franz and Nina play golf. Their shots to the hole are shown as vectors.

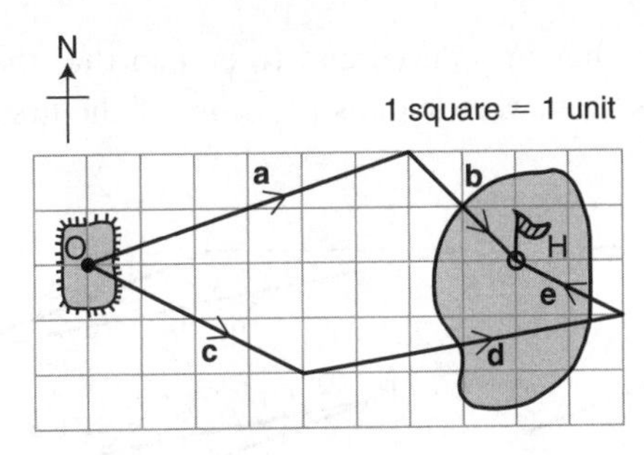

- Copy and complete this table by using the grid.

Vector	Column vector	Magnitude (to 3 s.f.)	Bearing
a	$\begin{pmatrix} 6 \\ 2 \end{pmatrix}$	6.32	072°
b			
c			
d			
e			

- Write down the vector $\overrightarrow{OH}$ and state if there is a connection between $\overrightarrow{OH}$ and the vectors **a**, **b** and the vectors **c**, **d**, **e**.

Vector geometry

Parallel and equivalent vectors

Two vectors are **parallel** if they have the same direction but not necessarily equal length.

For example, these vectors **a** and **b** are parallel.

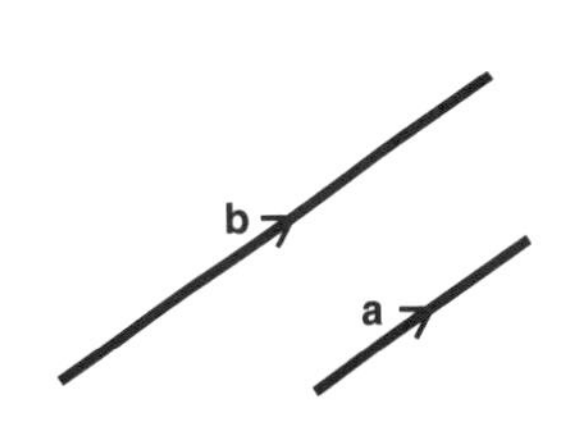

$\mathbf{a} = \begin{pmatrix} 3 \\ 2 \end{pmatrix} \quad \mathbf{b} = \begin{pmatrix} 6 \\ 4 \end{pmatrix}$

Two vectors are **equivalent** if they have the same direction and length.

Addition of vectors

The result of adding a set of vectors is the vector representing their total effect. This is the **resultant** of the vectors.

To add a number of vectors, they are placed end to end so that the next vector starts where the last one finished. The resultant vector joins the *start* of the first vector to the *end* of the last one.

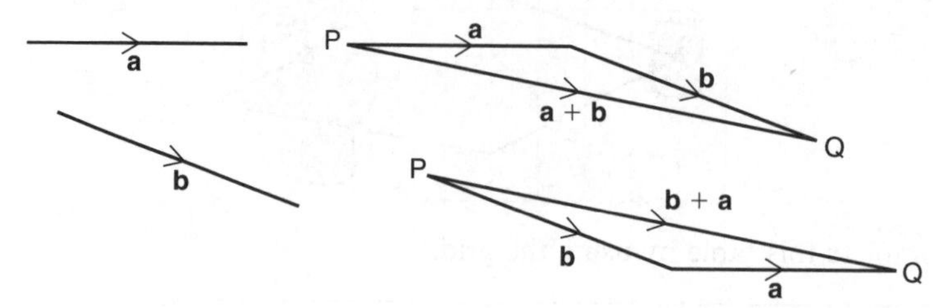

Vector $\overrightarrow{PQ} = \mathbf{a} + \mathbf{b} = \mathbf{b} + \mathbf{a}$

Multiplication of a vector by a scalar

When a vector is multiplied by a scalar, its length is multiplied by this number; but its direction is unchanged, unless the scalar is *negative*, in which case the direction is *reversed*.

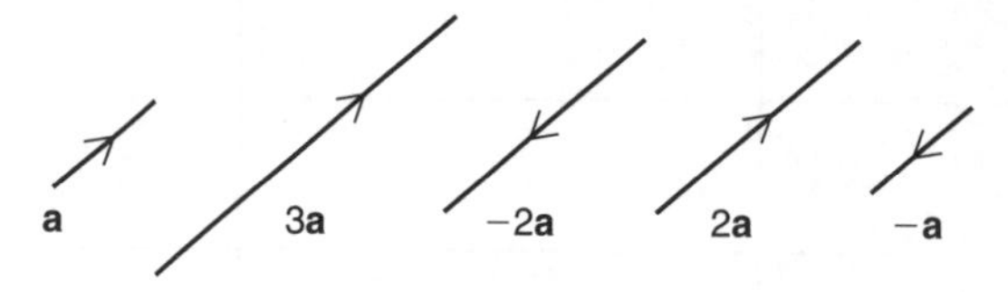

Example 3

Given vectors **a**, **b** and **c** as shown, draw the vector **r** where $\mathbf{r} = \mathbf{a} + \mathbf{b} - \mathbf{c}$.

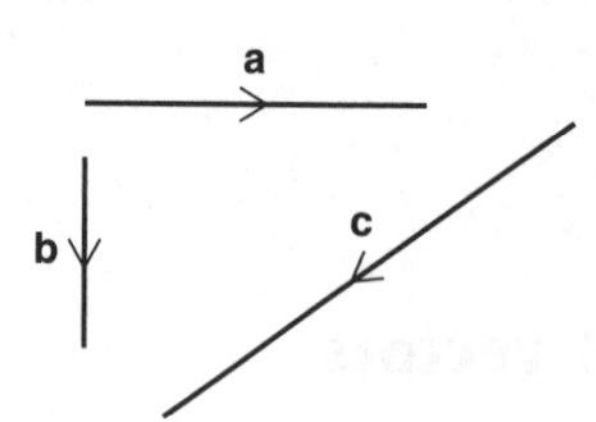

Here is the resultant of $\mathbf{a} + \mathbf{b} - \mathbf{c} = \mathbf{r}$.

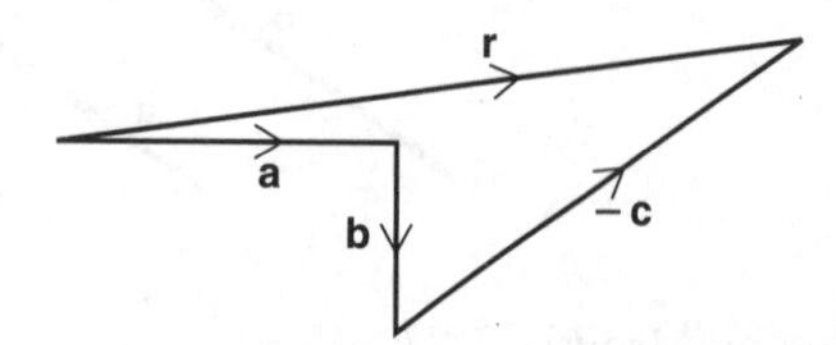

UNIT 3 ♦ Shape and space

Example 4

ABCDEF is a regular hexagon with centre O. $\overrightarrow{AB} = \mathbf{x}$ and $\overrightarrow{BC} = \mathbf{y}$. Express the vectors $\overrightarrow{ED}$, $\overrightarrow{DE}$, $\overrightarrow{FE}$, $\overrightarrow{AC}$, $\overrightarrow{FA}$ and $\overrightarrow{AE}$ in terms of $\mathbf{x}$ and $\mathbf{y}$.

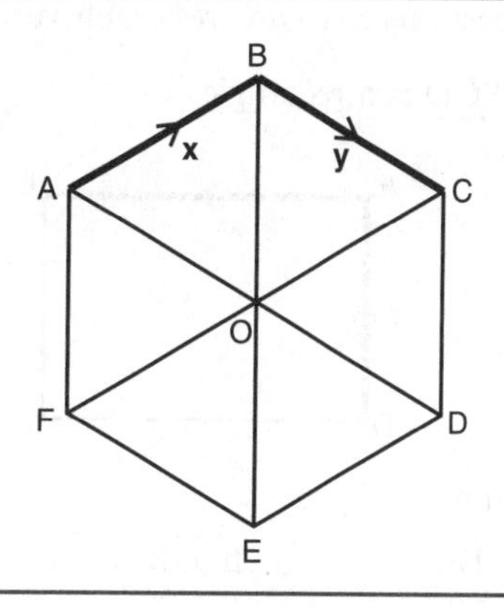

$\overrightarrow{ED} = \mathbf{x}$ $\qquad$ $\overrightarrow{DE} = -\mathbf{x}$

$\overrightarrow{FE} = \mathbf{y}$ $\qquad$ $\overrightarrow{AC} = \mathbf{x} + \mathbf{y}$

$\overrightarrow{FA} = \mathbf{x} - \mathbf{y}$ $\qquad$ $\overrightarrow{AE} = 2\mathbf{y} - \mathbf{x}$

Exercise 64

Use this diagram to answer Questions 1–4.

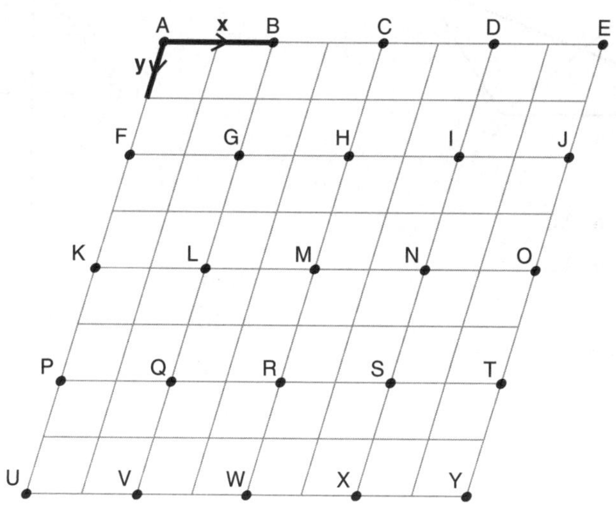

For Questions 1 and 2, express each vector in terms of $\mathbf{x}$ and $\mathbf{y}$.

1 a $\overrightarrow{XY}$ **b** $\overrightarrow{EO}$ **c** $\overrightarrow{WC}$ **d** $\overrightarrow{TP}$

2 a $\overrightarrow{KC}$ **b** $\overrightarrow{VC}$ **c** $\overrightarrow{CU}$ **d** $\overrightarrow{AS}$

For Questions 3 and 4, write the vector formed when the vectors given are added to point H as capital letters (e.g. $\overrightarrow{HO}$).

3 a $2\mathbf{x}$ **b** $\mathbf{x} + 2\mathbf{y}$ **c** $2\mathbf{y} - \mathbf{x}$ **d** $2\mathbf{x} + 2\mathbf{y}$

4 a $4\mathbf{y} + 2\mathbf{x}$ **b** $4\mathbf{y} - 2\mathbf{x}$ **c** $\mathbf{x} - 2\mathbf{y}$ **d** $2\mathbf{x} + 6\mathbf{y}$

In Questions 5–10, express each vector in terms of **x** and **y**.

5 ABCD is a rectangle.

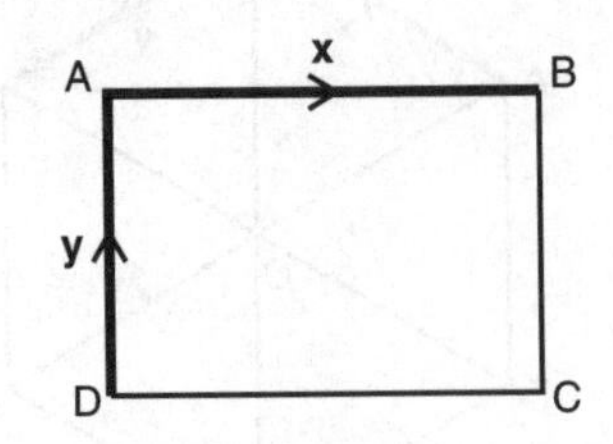

Find

a $\overrightarrow{DC}$ **b** $\overrightarrow{DB}$

c $\overrightarrow{BC}$ **d** $\overrightarrow{AC}$

6 ABCD is a trapezium.

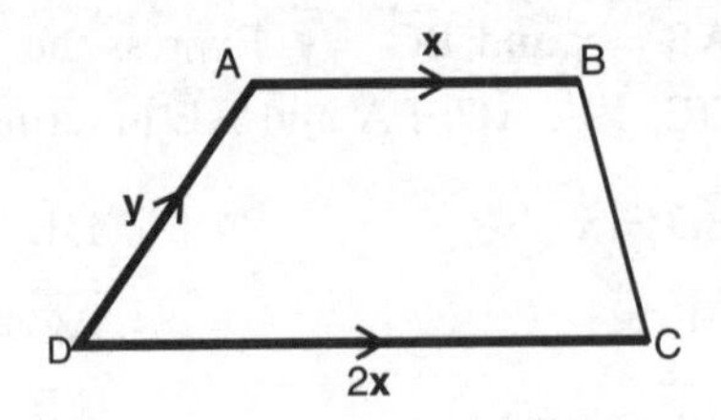

Find

a $\overrightarrow{AC}$ **b** $\overrightarrow{DB}$

c $\overrightarrow{BC}$ **d** $\overrightarrow{CB}$

7 ABCD is a parallelogram.

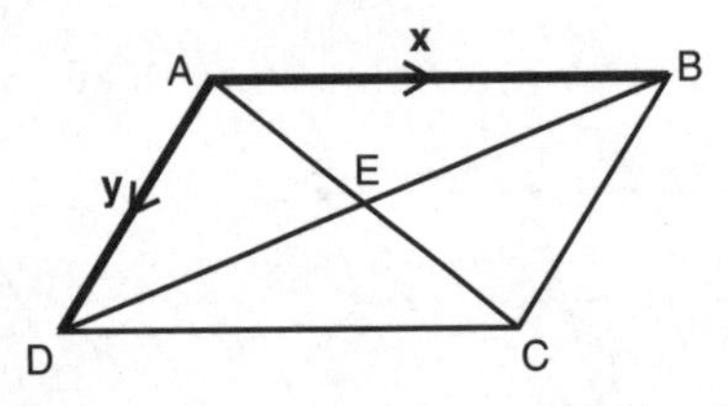

Find

a $\overrightarrow{DC}$ **b** $\overrightarrow{AC}$

c $\overrightarrow{BD}$ **d** $\overrightarrow{AE}$

8 ABCD is a rhombus.

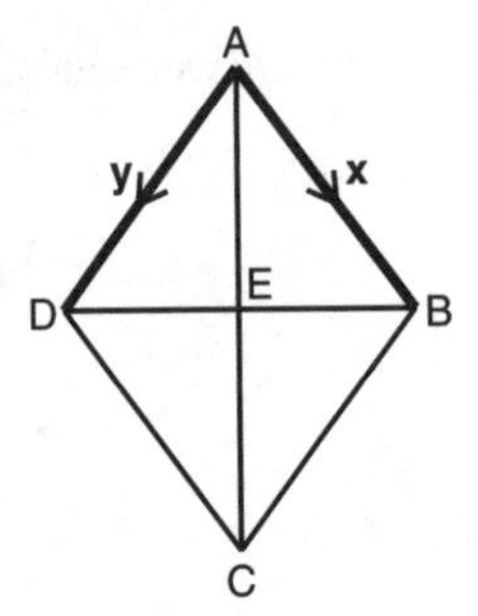

Find

a $\overrightarrow{BD}$ **b** $\overrightarrow{BE}$

c $\overrightarrow{AC}$ **d** $\overrightarrow{AE}$

9

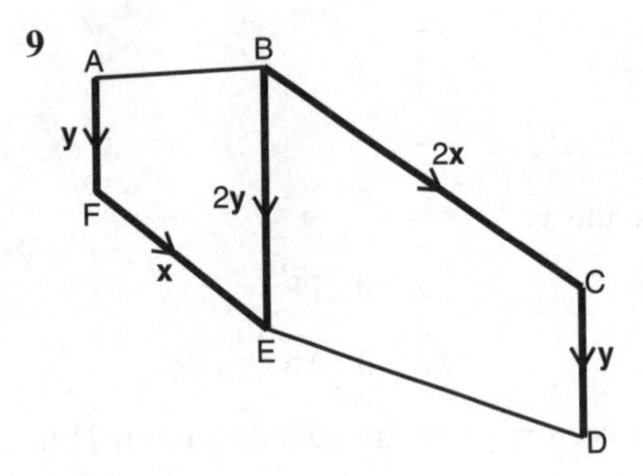

Find

a $\overrightarrow{AB}$ **b** $\overrightarrow{AD}$

c $\overrightarrow{CF}$ **d** $\overrightarrow{CA}$

10 ABC is an equilateral triangle, with $\overrightarrow{AC} = 2\mathbf{x}$, $\overrightarrow{AB} = 2\mathbf{y}$.

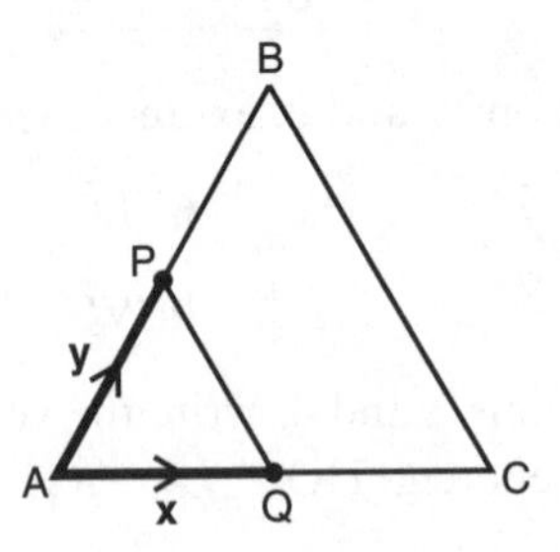

Find

a $\overrightarrow{PQ}$ **b** $\overrightarrow{PC}$

c $\overrightarrow{QB}$ **d** $\overrightarrow{BC}$

UNIT 3 ◆ Shape and space

11 The gear stick of a car is shown. The lever can only shift along the grooves shown to reach each gear. N is neutral and R is reverse.

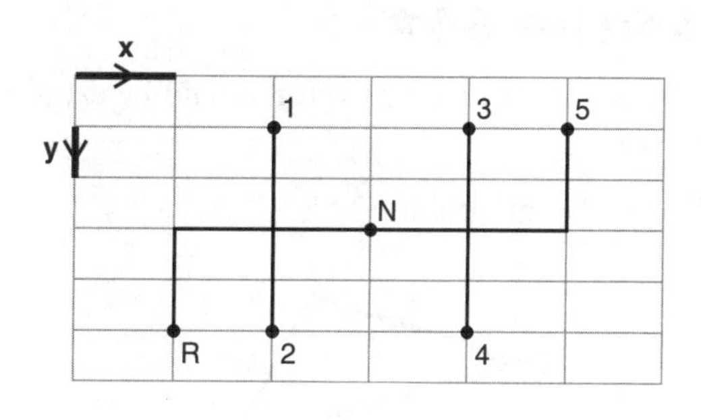

Express these gear changes in terms of **x** and **y**.

a 1st to 4th **b** 3rd to 2nd

c N to R **d** 2nd to 5th

12 A rectangular biscuit tin OABCDEFG is shown. $\overrightarrow{OA} = \mathbf{x}$, $\overrightarrow{AB} = \mathbf{y}$ and $\overrightarrow{OD} = \mathbf{z}$.

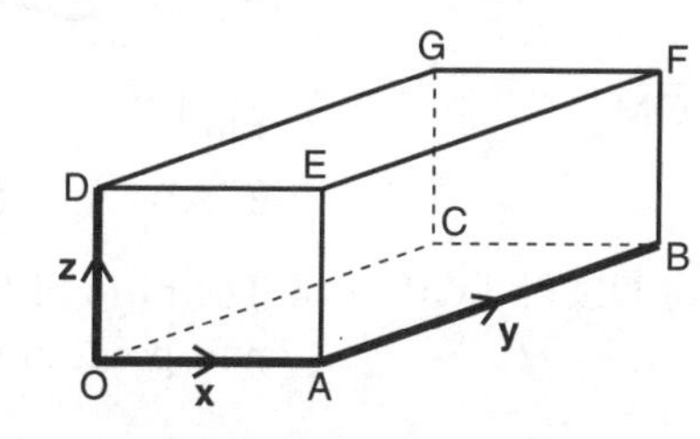

An ant crawls these direct journeys in search of crumbs. Express each journey as a vector in terms of **x**, **y** and **z**.

a $\overrightarrow{OE}$ **b** $\overrightarrow{OB}$ **c** $\overrightarrow{OF}$ **d** $\overrightarrow{EC}$

Example 5

In $\triangle$OAB, the mid-point on AB is M.
$\overrightarrow{OA} = \mathbf{x}$, $\overrightarrow{OB} = \mathbf{y}$, $\overrightarrow{OD} = 2\mathbf{x}$ and $\overrightarrow{OC} = 2\mathbf{y}$.

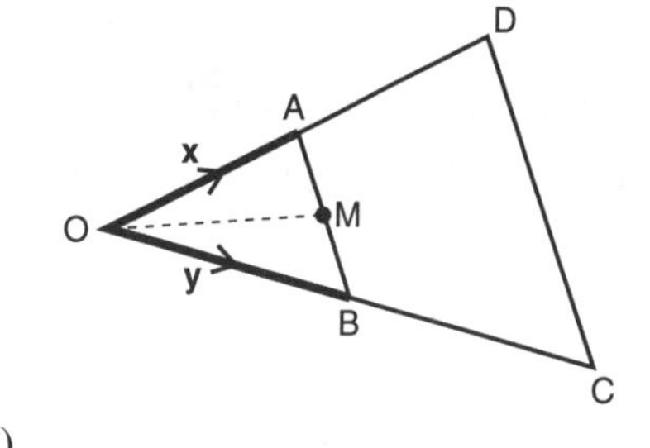

a Express $\overrightarrow{AB}$, $\overrightarrow{OM}$ and $\overrightarrow{DC}$ in terms of **x** and **y**.

$$\overrightarrow{AB} = \overrightarrow{AO} + \overrightarrow{OB} = -\mathbf{x} + \mathbf{y} = \mathbf{y} - \mathbf{x}$$

$$\overrightarrow{DC} = \overrightarrow{DO} + \overrightarrow{OC} = -2\mathbf{x} + 2\mathbf{y} = 2\mathbf{y} - 2\mathbf{x} = 2(\mathbf{y} - \mathbf{x})$$

$$\overrightarrow{OM} = \overrightarrow{OA} + \overrightarrow{AM} = \overrightarrow{OA} + \tfrac{1}{2}\overrightarrow{AB} = \mathbf{x} + \tfrac{1}{2}(\mathbf{y} - \mathbf{x}) = \mathbf{x} + \tfrac{1}{2}\mathbf{y} - \tfrac{1}{2}\mathbf{x} = \tfrac{1}{2}\mathbf{x} + \tfrac{1}{2}\mathbf{y} = \tfrac{1}{2}(\mathbf{x} + \mathbf{y})$$

b How are AB and DC related?

$\overrightarrow{DC} = 2\overrightarrow{AB}$, so AB is parallel to DC and half its magnitude.

Exercise 64★

In Questions 1–10, each vector should be expressed in terms of **x** and **y**, where $\overrightarrow{OA} = \mathbf{x}$ and $\overrightarrow{OB} = \mathbf{y}$.

1 M is the mid-point of AB.

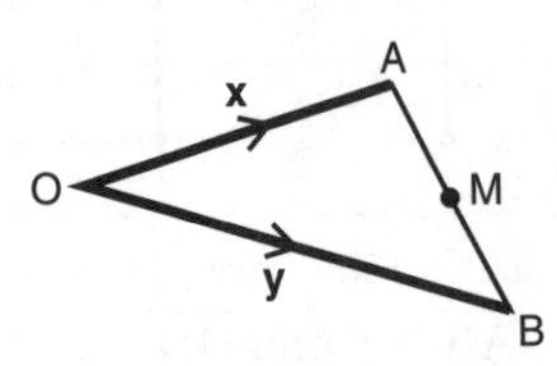

Find

a $\overrightarrow{AB}$ **b** $\overrightarrow{AM}$ **c** $\overrightarrow{OM}$

2 The ratio of AM:MB = 1:2.

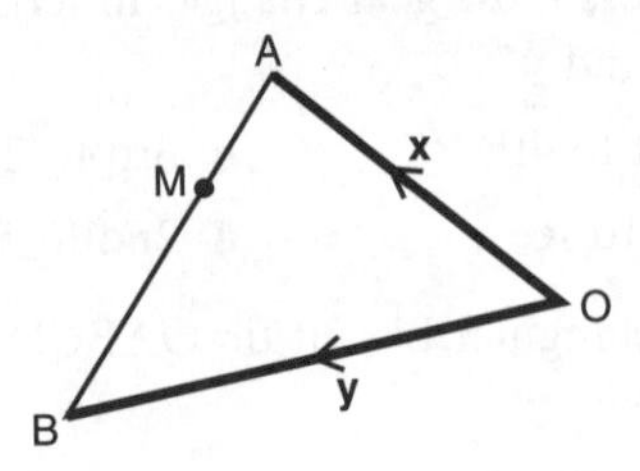

Find

a $\overrightarrow{AB}$ **b** $\overrightarrow{AM}$ **c** $\overrightarrow{OM}$

3 A and B are the mid-points of OD and OC respectively.

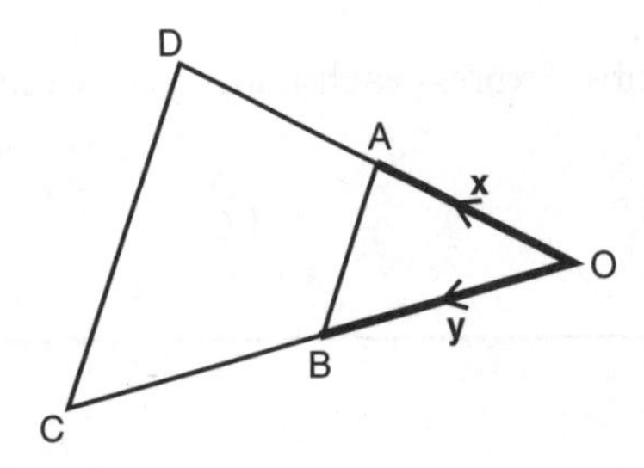

a Find $\overrightarrow{AB}$, $\overrightarrow{OD}$ and $\overrightarrow{DC}$.

b How are AB and DC related?

4 The ratio of OA:AD = 1:2 and B is the mid-point of OC.

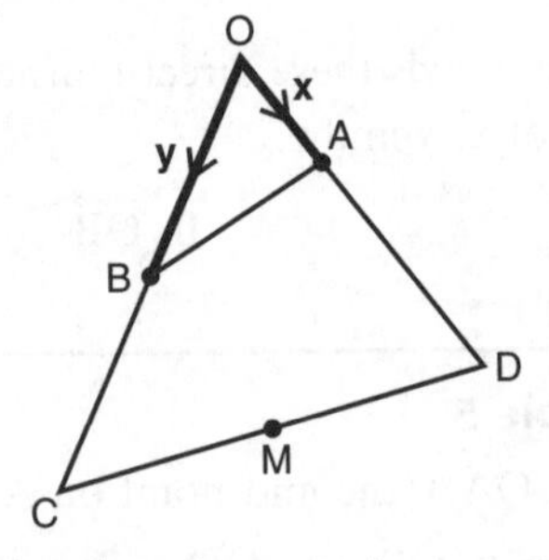

a Find $\overrightarrow{AB}$, $\overrightarrow{OD}$ and $\overrightarrow{DC}$.

b M is the mid-point of CD. Find $\overrightarrow{OM}$.

5 The ratio of OA:OC = 1:2 and OB:OD = 1:2.

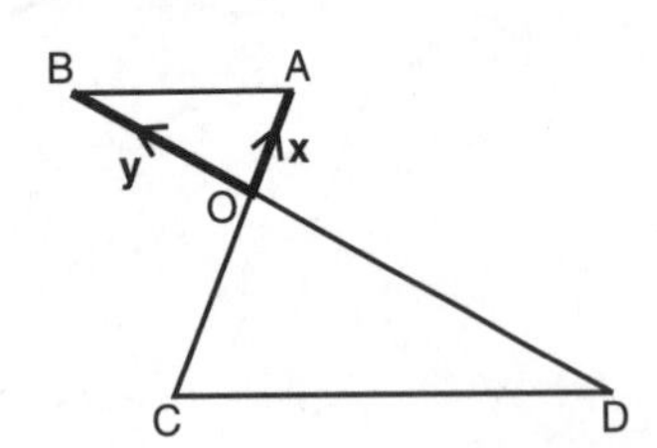

a Find $\overrightarrow{AB}$, $\overrightarrow{OC}$, $\overrightarrow{OD}$ and $\overrightarrow{DC}$.

b How are AB and DC related?

6 OABC is a parallelogram. The ratio of OD:DC = 1:2.

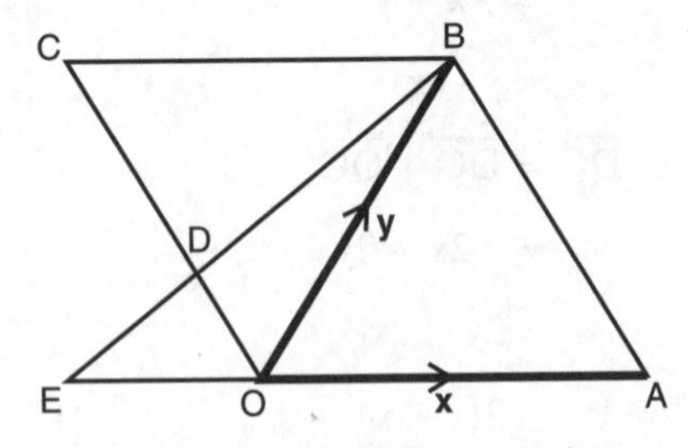

Find $\overrightarrow{AB}$, $\overrightarrow{OD}$, $\overrightarrow{BD}$, $\overrightarrow{OE}$ and $\overrightarrow{DE}$.

UNIT 3 ◆ Shape and space

7 OABCDE is a regular hexagon.

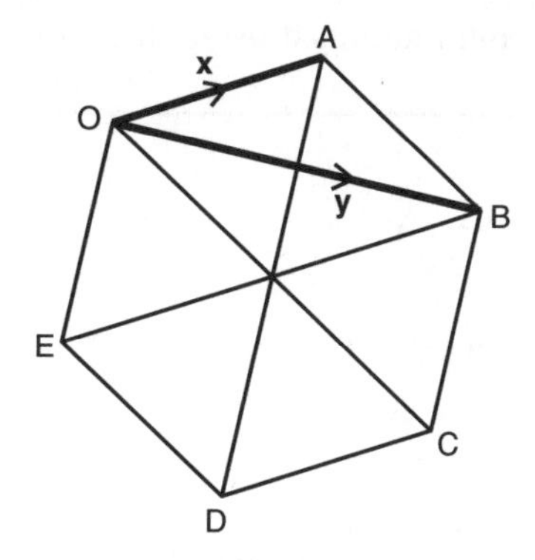

Find $\overrightarrow{AB}$, $\overrightarrow{BC}$, $\overrightarrow{AD}$ and $\overrightarrow{BD}$.

8 ABP is an equilateral triangle.
OAB is an isosceles triangle, where OA = AB.

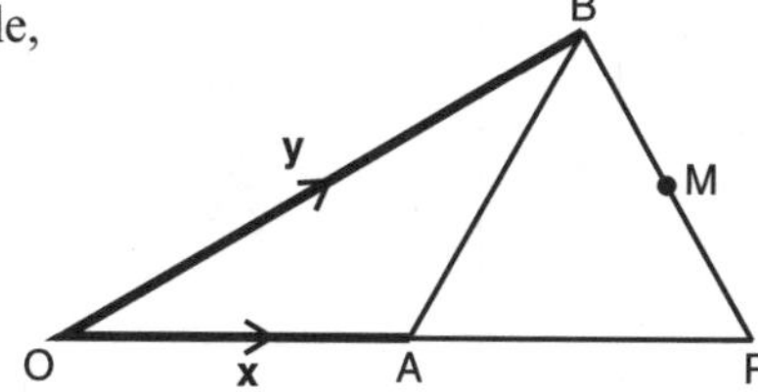

a Find $\overrightarrow{OP}$, $\overrightarrow{AB}$ and $\overrightarrow{BP}$.

b M is the mid-point of BP. Find $\overrightarrow{OM}$.

9 OM:MA = 2:3 and AN = $\frac{3}{5}$AB.

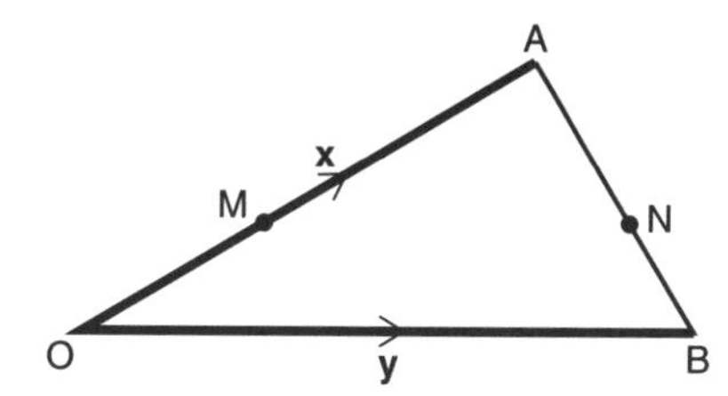

a Find $\overrightarrow{MA}$, $\overrightarrow{AB}$, $\overrightarrow{AN}$ and $\overrightarrow{MN}$.

b How are OB and MN related?

10 OM:MB = 1:2 and AN = $\frac{1}{3}$AB.

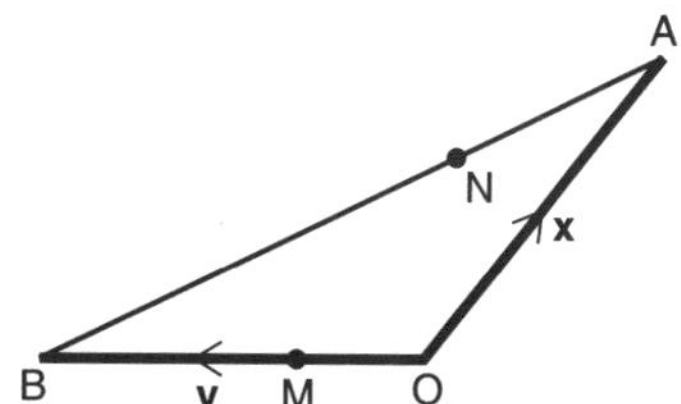

a Find $\overrightarrow{AB}$ and $\overrightarrow{MN}$.

b How are OA and MN related?

UNIT 3 ◆ Shape and space

More addition, subtraction and multiplication

Vectors can be added, subtracted and multiplied using their components.

Example 6

$\mathbf{s} = \begin{pmatrix} 1 \\ 2 \end{pmatrix}$, $\mathbf{t} = \begin{pmatrix} 3 \\ 0 \end{pmatrix}$ and $\mathbf{u} = \begin{pmatrix} -2 \\ 5 \end{pmatrix}$

a Express as column vectors: $\mathbf{p} = \mathbf{s} + \mathbf{t} + \mathbf{u}$, $\mathbf{q} = \mathbf{s} - 2\mathbf{t} - \mathbf{u}$ and $\mathbf{r} = 3\mathbf{s} + \mathbf{t} - 2\mathbf{u}$

b Sketch the resultants **p**, **q** and **r** accurately.

c Find their magnitudes.

a Calculation	b Sketch	c Magnitude
$\mathbf{p} = \mathbf{s} + \mathbf{t} + \mathbf{u}$ $= \begin{pmatrix} 1 \\ 2 \end{pmatrix} + \begin{pmatrix} 3 \\ 0 \end{pmatrix} + \begin{pmatrix} -2 \\ 5 \end{pmatrix}$ $= \begin{pmatrix} 2 \\ 7 \end{pmatrix}$	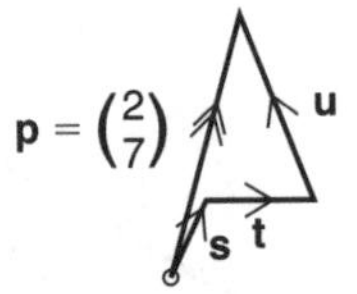 $\mathbf{p} = \begin{pmatrix} 2 \\ 7 \end{pmatrix}$ u s t	Length of **p** $= \sqrt{2^2 + 7^2}$ $= \sqrt{53}$ $= 7.3$ to 1 d.p.
$\mathbf{q} = \mathbf{s} - 2\mathbf{t} - \mathbf{u}$ $= \begin{pmatrix} 1 \\ 2 \end{pmatrix} - 2\begin{pmatrix} 3 \\ 0 \end{pmatrix} - \begin{pmatrix} -2 \\ 5 \end{pmatrix}$ $= \begin{pmatrix} 1 \\ 2 \end{pmatrix} + \begin{pmatrix} -6 \\ 0 \end{pmatrix} + \begin{pmatrix} 2 \\ -5 \end{pmatrix}$ $= \begin{pmatrix} -3 \\ -3 \end{pmatrix}$	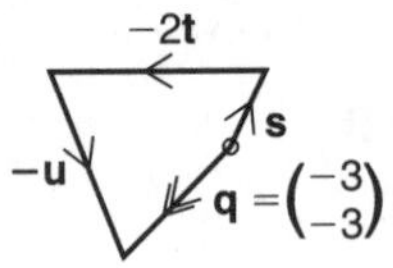 −2t s −u $\mathbf{q} = \begin{pmatrix} -3 \\ -3 \end{pmatrix}$	Length of **q** $= \sqrt{3^2 + 3^2}$ $= \sqrt{18}$ $= 4.2$ to 1 d.p.
$\mathbf{r} = 3\mathbf{s} + \mathbf{t} - 2\mathbf{u}$ $= 3\begin{pmatrix} 1 \\ 2 \end{pmatrix} + \begin{pmatrix} 3 \\ 0 \end{pmatrix} - 2\begin{pmatrix} -2 \\ 5 \end{pmatrix}$ $= \begin{pmatrix} 3 \\ 6 \end{pmatrix} + \begin{pmatrix} 3 \\ 0 \end{pmatrix} + \begin{pmatrix} 4 \\ -10 \end{pmatrix}$ $= \begin{pmatrix} 10 \\ -4 \end{pmatrix}$	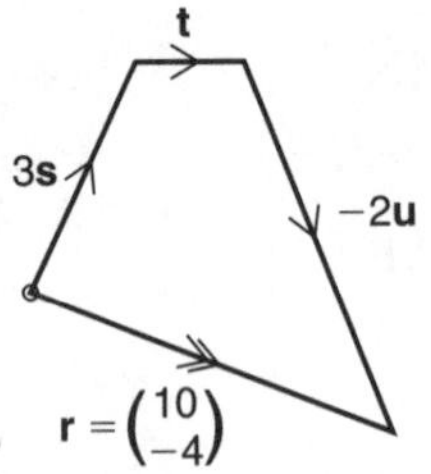 t 3s −2u $\mathbf{r} = \begin{pmatrix} 10 \\ -4 \end{pmatrix}$	Length of **r** $= \sqrt{10^2 + 4^2}$ $= \sqrt{116}$ $= 10.8$ to 1 d.p.

Exercise 65

1 Given that $\mathbf{p} = \begin{pmatrix} 2 \\ 3 \end{pmatrix}$ and $\mathbf{q} = \begin{pmatrix} 4 \\ 5 \end{pmatrix}$,

simplify and express $\mathbf{p} + \mathbf{q}$, $\mathbf{p} - \mathbf{q}$ and $2\mathbf{p} + 3\mathbf{q}$ as column vectors.

2 Given that $\mathbf{u} = \begin{pmatrix} 1 \\ 2 \end{pmatrix}$, $\mathbf{v} = \begin{pmatrix} -4 \\ 3 \end{pmatrix}$ and $\mathbf{w} = \begin{pmatrix} 2 \\ -5 \end{pmatrix}$,

simplify and express $\mathbf{u} + \mathbf{v} + \mathbf{w}$, $\mathbf{u} + 2\mathbf{v} - 3\mathbf{w}$ and $3\mathbf{u} - 2\mathbf{v} - \mathbf{w}$ as column vectors.

3 Given that $\mathbf{p} = \begin{pmatrix} 1 \\ 2 \end{pmatrix}$ and $\mathbf{q} = \begin{pmatrix} 3 \\ 4 \end{pmatrix}$,

simplify and express $\mathbf{p} + \mathbf{q}$, $\mathbf{p} - \mathbf{q}$ and $2\mathbf{p} + 5\mathbf{q}$ as column vectors.

4 Given that $\mathbf{s} = \begin{pmatrix} 1 \\ -3 \end{pmatrix}$, $\mathbf{t} = \begin{pmatrix} 2 \\ 3 \end{pmatrix}$ and $\mathbf{u} = \begin{pmatrix} 4 \\ -5 \end{pmatrix}$,

simplify and express $\mathbf{s} + \mathbf{t} + \mathbf{u}$, $2\mathbf{s} - \mathbf{t} + 2\mathbf{u}$ and $2\mathbf{u} - 3\mathbf{s}$ as column vectors.

5 Two vectors are defined as $\mathbf{v} = \begin{pmatrix} 3 \\ 1 \end{pmatrix}$ and $\mathbf{w} = \begin{pmatrix} 1 \\ 4 \end{pmatrix}$.

Express $\mathbf{v} + \mathbf{w}$, $2\mathbf{v} - \mathbf{w}$ and $\mathbf{v} - 2\mathbf{w}$ as column vectors, find the magnitude and draw the resultant vector triangle for each vector.

6 Two vectors are defined as $\mathbf{p} = \begin{pmatrix} 2 \\ -1 \end{pmatrix}$ and $\mathbf{q} = \begin{pmatrix} 3 \\ 5 \end{pmatrix}$.

Express $\mathbf{p} + \mathbf{q}$, $3\mathbf{p} + \mathbf{q}$ and $\mathbf{p} - 3\mathbf{q}$ as column vectors, find the magnitude and draw the resultant vector triangle for each vector.

7 Chloe, Leo and Max enter an orienteering competition. Each decides to take a different route, described using these column vectors, where the units are in km:

$\mathbf{s} = \begin{pmatrix} 1 \\ 1 \end{pmatrix}$ $\quad \mathbf{t} = \begin{pmatrix} 2 \\ 3 \end{pmatrix}$

They all start from the same point P, and take 3 hours to complete their routes:

Chloe	$\mathbf{s} + 2\mathbf{t}$
Leo	$2\mathbf{s} + \mathbf{t}$
Max	$5\mathbf{s} - \mathbf{t}$

a Express each journey as a column vector.

b Find the length of each journey, and hence calculate the average speed of each orienteer in km/hour.

8 Use the information in Question 7 to answer this question.
Chloe, Leo and Max were all aiming to be at their first marker position Q, which is at position vector $\begin{pmatrix} 1 \\ 5 \end{pmatrix}$ from point P.

a Find how far each of them was from Q after their journeys.

b Calculate the bearing of Q from each of the orienteers after their journeys.

Exercise 65★

1 Given that $\mathbf{p} = \begin{pmatrix} 2 \\ 1 \end{pmatrix}$ and $\mathbf{q} = \begin{pmatrix} 3 \\ -1 \end{pmatrix}$,
find the magnitude and bearing of the vectors $\mathbf{p} + \mathbf{q}$, $\mathbf{p} - \mathbf{q}$ and $2\mathbf{p} - 3\mathbf{q}$.

2 Given that $\mathbf{r} = \begin{pmatrix} 1 \\ -3 \end{pmatrix}$ and $\mathbf{s} = \begin{pmatrix} 4 \\ 1 \end{pmatrix}$,
find the magnitude and bearing of the vectors $2(\mathbf{r} + \mathbf{s})$, $3(\mathbf{r} - 2\mathbf{s})$ and $(4\mathbf{r} - 6\mathbf{s})\sin 30°$.

3 Given that $\mathbf{t} + \mathbf{u} = \begin{pmatrix} 1 \\ 1 \end{pmatrix}$, where $\mathbf{t} = \begin{pmatrix} m \\ 3 \end{pmatrix}$, $\mathbf{u} = \begin{pmatrix} 2 \\ n \end{pmatrix}$ and m and n are constants, find the values of m and n.

4 Given that $2\mathbf{v} - 3\mathbf{w} = \begin{pmatrix} 2 \\ -3 \end{pmatrix}$, where $\mathbf{v} = \begin{pmatrix} 5 \\ -m \end{pmatrix}$, $\mathbf{w} = \begin{pmatrix} n \\ 4 \end{pmatrix}$ and m and n are constants, find the values of m and n.

5 These vectors represent journeys undertaken by yachts in km.
Express each one in column vector form.

a

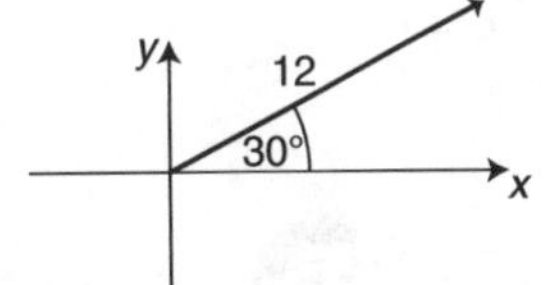

b

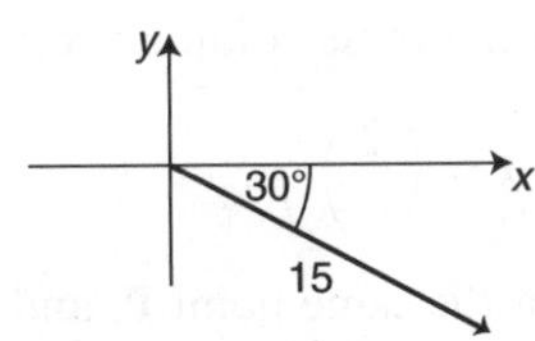

6 These vectors represent journeys undertaken by crows in km.
Express each one in column vector component form.

a

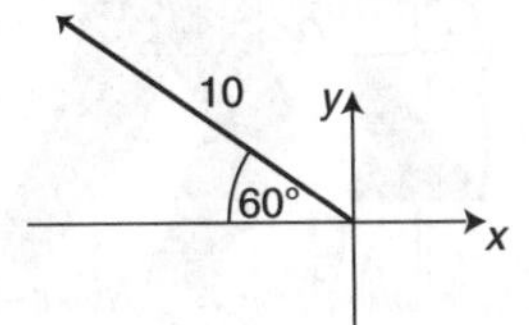

b

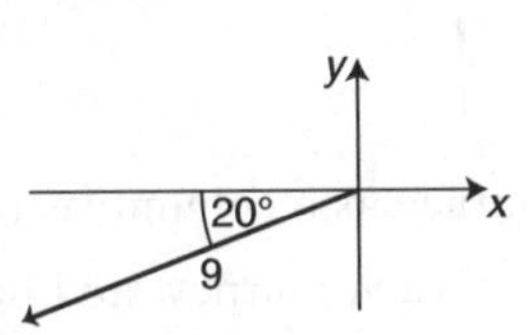

7 $\mathbf{s} = \begin{pmatrix} 2 \\ 3 \end{pmatrix}$ and $\mathbf{t} = \begin{pmatrix} -6 \\ 1 \end{pmatrix}$

a If $m\mathbf{s} + n\mathbf{t} = \begin{pmatrix} -16 \\ 6 \end{pmatrix}$, solve this vector equation to find the constants m and n.

b If $p\mathbf{s} + \mathbf{t} = \begin{pmatrix} 0 \\ q \end{pmatrix}$, solve this vector equation to find the constants p and q.

8 The centre spot O of a hockey pitch is the origin of co-ordinates, with the x-axis passing across the pitch, and the y-axis passing up the centre. All units are in metres.

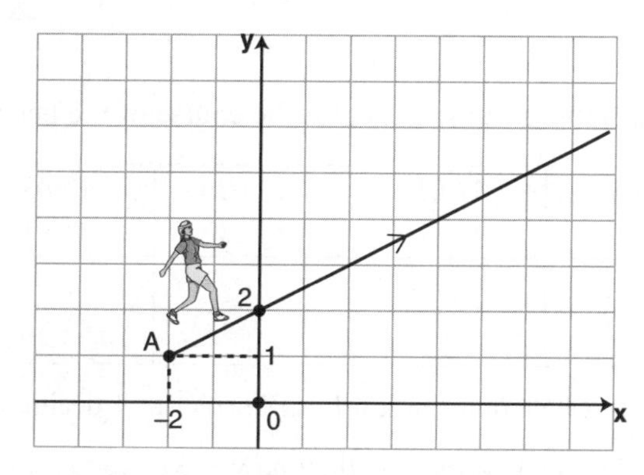

At $t = 0$, Anne is at A. When $t = 1$, Anne is at (0, 2). Anne's position after t seconds is shown by the vector starting at point A.

a Explain why after t seconds Anne's position vector $\mathbf{r}$ is given by

$$\mathbf{r} = \begin{pmatrix} -2 \\ 1 \end{pmatrix} + t\begin{pmatrix} 2 \\ 1 \end{pmatrix}$$

b Find Anne's speed in m/s.

At $t = 0$, Fleur, who is positioned on the centre spot, O, hits the ball towards Anne's path so that Anne receives it 5 s after setting off.

The position vector, **s**, of the ball hit by Fleur is given by

$$\mathbf{s} = (t - 3)\begin{pmatrix} a \\ b \end{pmatrix}$$

c Find the value of constants a and b and hence the speed of the ball in m/s.

Activity 21

A radar tracking station O is positioned at the origin of x and y axes, where the x-axis points due East and the y-axis points due North.
All the units are in km.

A helicopter is detected t minutes after midday with position vector **r**, where

$$\mathbf{r} = \begin{pmatrix} 12 \\ 5 \end{pmatrix} + t\begin{pmatrix} -3 \\ 4 \end{pmatrix}$$

- Copy and complete this table and use it to plot the course of the helicopter.

Time	12:00 $t = 0$	12:01 $t = 1$	12:02 $t = 2$	12:03 $t = 3$	12:04 $t = 4$
r	$\begin{pmatrix} 12 \\ 5 \end{pmatrix}$				

- Calculate the speed of this helicopter in km/hour correct to 1 decimal place, and its bearing.
- An international boundary is described by the line $y = 5x$.
  - Draw this boundary on your graph.
  - Estimate the time the helicopter crosses the boundary by careful inspection of your graph.
  - Considering the helicopter's position vector

  $$\mathbf{r} = \begin{pmatrix} x \\ y \end{pmatrix}$$

  express x and y in terms of t and use these equations with $y = 5x$ to find the time when the boundary is crossed, correct to the nearest second.

Exercise 66 (Revision)

1 Given that $\mathbf{p} = \begin{pmatrix} 3 \\ 4 \end{pmatrix}$ and $\mathbf{q} = \begin{pmatrix} -2 \\ 1 \end{pmatrix}$,

simplify $\mathbf{p} + \mathbf{q}$, $\mathbf{p} - \mathbf{q}$ and $3\mathbf{p} - 2\mathbf{q}$ as column vectors and find the magnitude of each vector.

2 Given that $\mathbf{r} = \begin{pmatrix} 2 \\ -5 \end{pmatrix}$ and $\mathbf{s} = \begin{pmatrix} 3 \\ 4 \end{pmatrix}$,

simplify $\mathbf{r} + \mathbf{s}$, $\mathbf{r} - \mathbf{s}$ and $3\mathbf{s} - 2\mathbf{r}$ as column vectors and find the magnitude of each vector.

3 In terms of vectors **x** and **y**, find these vectors.

a $\overrightarrow{AB}$ **b** $\overrightarrow{AC}$ **c** $\overrightarrow{CB}$

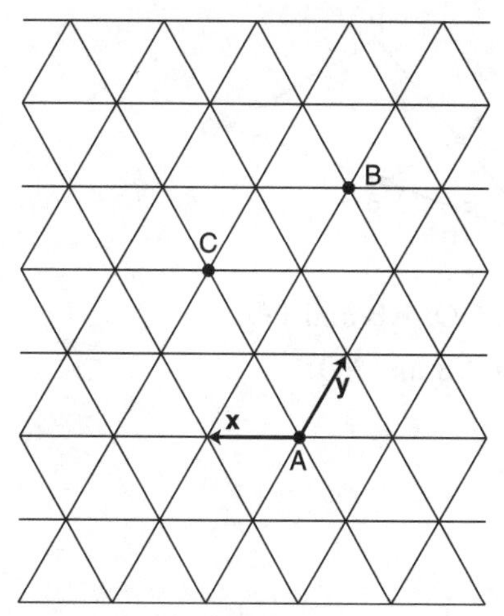

4 $\overrightarrow{OA} = \mathbf{v}$ and $\overrightarrow{OB} = \mathbf{w}$ and M is the mid-point of AB.
Find these vectors in terms of **v** and **w**:

a $\overrightarrow{AB}$ **b** $\overrightarrow{AM}$ **c** $\overrightarrow{OM}$

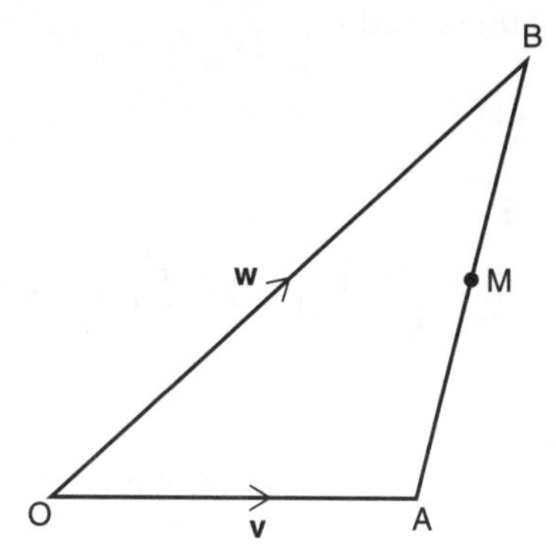

5 ABCDEF is a regular hexagon. Vectors $\overrightarrow{OA} = \mathbf{x}$ and $\overrightarrow{OB} = \mathbf{y}$.
Find these vectors in terms of **x** and **y**:

a $\overrightarrow{AB}$ **b** $\overrightarrow{FB}$ **c** $\overrightarrow{FD}$

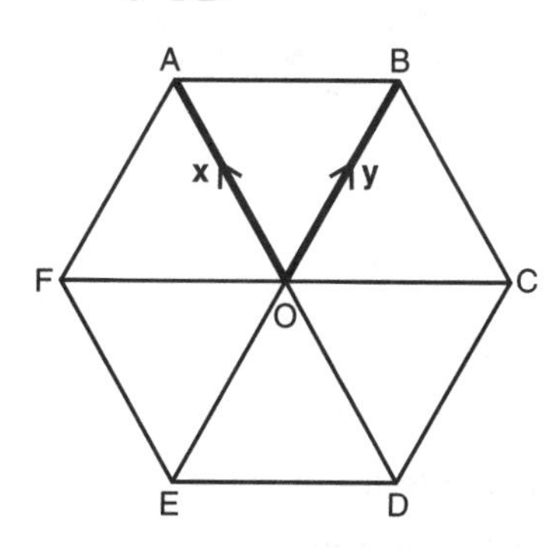

6 In the triangle OPQ, A and B are mid-points of the sides OP and OQ respectively, $\overrightarrow{OA} = \mathbf{a}$ and $\overrightarrow{OB} = \mathbf{b}$.

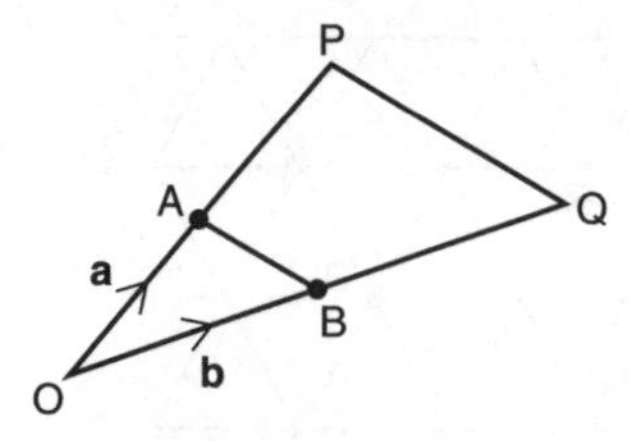

a Find in terms of **a** and **b**: $\overrightarrow{OP}$, $\overrightarrow{OQ}$, $\overrightarrow{AB}$ and $\overrightarrow{PQ}$.

b What can you conclude about AB and PQ?

7 $\mathbf{r} = \begin{pmatrix} 1 \\ 3 \end{pmatrix}$ and $\mathbf{s} = \begin{pmatrix} -2 \\ 5 \end{pmatrix}$.

a Calculate $2\mathbf{r} - \mathbf{s}$.

b Calculate $2(\mathbf{r} - \mathbf{s})$.

c Calculate the length of vector **s** in root form.

d $v\mathbf{r} + w\mathbf{s} = \begin{pmatrix} -3 \\ 13 \end{pmatrix}$

What are the values of the constants v and w?

Exercise 66★ (Revision)

1 Given that $2\mathbf{p} - 3\mathbf{q} = \begin{pmatrix} 5 \\ 15 \end{pmatrix}$, where $\mathbf{p} = \begin{pmatrix} 4 \\ m \end{pmatrix}$, $\mathbf{q} = \begin{pmatrix} n \\ -3 \end{pmatrix}$ and m and n are constants, find the values of m and n.

2 If $\mathbf{r} = \begin{pmatrix} 4 \\ -1 \end{pmatrix}$, $\mathbf{s} = \begin{pmatrix} 3 \\ 7 \end{pmatrix}$ and $m\mathbf{r} + n\mathbf{s} = \begin{pmatrix} 7 \\ 37 \end{pmatrix}$, find constants m and n.

3 In the diagram, OXYZ is a parallelogram. M is the mid-point of XY.

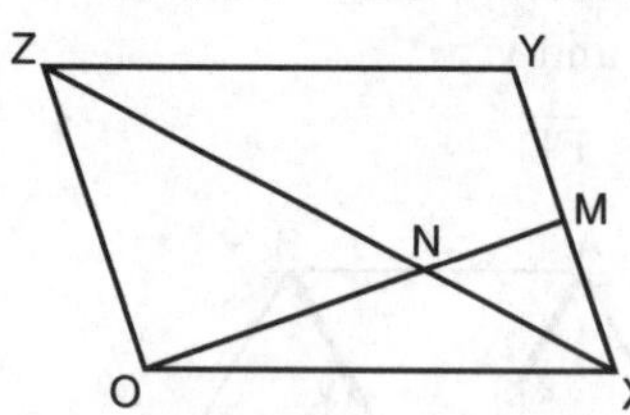

a Given that $\overrightarrow{OX} = \begin{pmatrix} 8 \\ 0 \end{pmatrix}$ and $\overrightarrow{OZ} = \begin{pmatrix} -2 \\ 6 \end{pmatrix}$,

write down the vectors $\overrightarrow{XM}$ and $\overrightarrow{XZ}$.

b Given that $\overrightarrow{ON} = v\overrightarrow{OM}$, write down in terms of v the vector $\overrightarrow{ON}$.

c Given that $\overrightarrow{ON} = \overrightarrow{OX} + w\overrightarrow{XZ}$, find in terms of w the vector $\overrightarrow{ON}$.

d Solve two simultaneous equations to find v and w.

4 ABCD is a parallelogram in which $\overrightarrow{AB} = \mathbf{x}$ and $\overrightarrow{BC} = \mathbf{y}$. AE:ED = 1:2.

a Express in terms of **x** and **y**, $\overrightarrow{AC}$ and $\overrightarrow{BE}$.

b AC and BE intersect at F, such that $\overrightarrow{BF} = v\,\overrightarrow{BE}$.

(i) Express $\overrightarrow{BF}$ in terms of **x**, **y** and v.

(ii) Show that $\overrightarrow{AF} = (1 - v)\mathbf{x} + \frac{1}{3}v\mathbf{y}$.

(iii) Use this expression for $\overrightarrow{AF}$ to find the value of v.

5 A shooting star's position vector **r**, t seconds after being detected on radar, is given by

$$\mathbf{r} = \begin{pmatrix} 12 \\ 7 \end{pmatrix} + t\begin{pmatrix} -11 \\ 3 \end{pmatrix}$$

where the units are in km.

a Find the position vector **r** of the star after $t = 0$, $t = 1$, $t = 2$ and $t = 3$ seconds.

b Plot the shooting star's route across the radar screen for the first 3 seconds.

c Calculate its speed in km/hour correct to 3 significant figures, and its bearing.

6 Amila and Winnie are playing basketball. During the game, their position vectors on the court are defined relative to the axes on the diagram. At time t seconds after the whistle, their position vectors are given by **r** and **s** respectively:

$$\mathbf{r} = \begin{pmatrix} 2 \\ -1 \end{pmatrix} + t\begin{pmatrix} 1 \\ 2 \end{pmatrix}$$

$$\mathbf{s} = \begin{pmatrix} -3 \\ 4 \end{pmatrix} + t\begin{pmatrix} 3 \\ 1 \end{pmatrix}$$

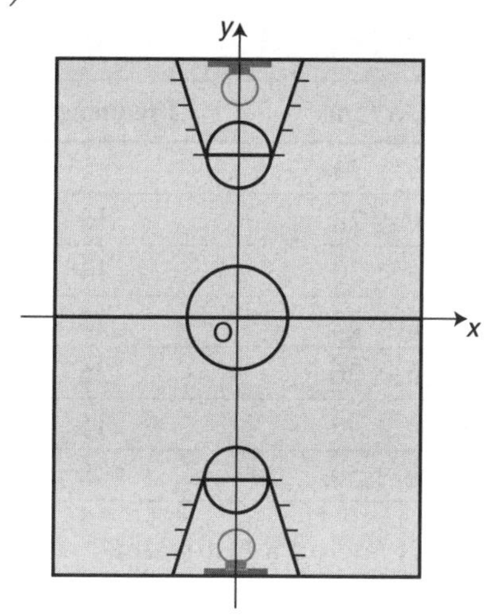

a Find the position vectors for Amila and Winnie after:

(i) 1 s (ii) 2 s

b Write down the vector from Amila to Winnie after 2 s and use it to find how far apart they are at this moment. Leave your answer in surd form.

c Calculate the speeds of the two girls in surd form.

Handling data 3

Histograms

Histograms look similar to bar charts, but . . .

- A bar chart measures *frequency* on the vertical axis.
- In a bar chart frequency is proportional to the *height* of the bar.
- A histogram measures *frequency density* on the vertical axis.
- In a histogram frequency is proportional to the *area* of the bar.

When data are divided into groups of *different* sizes, a histogram, rather than a bar chart, should be used to display the distribution.

Key Point

$$\text{Frequency density} = \frac{\text{frequency}}{\text{width of group}}$$

$$\therefore \quad (\text{frequency density}) \times (\text{width of group}) = \text{frequency}$$

Example 1

A horticulturist recorded the heights of a sample of 100 plants that he had grown from a new seed.
The results are shown in this table.

Height, h (cm)	Frequency, f
$16 \leqslant h < 24$	6
$24 \leqslant h < 28$	16
$28 \leqslant h < 30$	14
$30 \leqslant h < 32$	15
$32 \leqslant h < 36$	26
$36 \leqslant h < 40$	15
$40 \leqslant h < 50$	8

Represent the results on a bar chart and on a histogram.

The frequency densities are worked out in a calculation table. Four columns are needed.

Group	Frequency, f	Width	Frequency density
$16 \leqslant h < 24$	6	8	$6 \div 8 = 0.75$
$24 \leqslant h < 28$	16	4	$16 \div 4 = 4.00$
$28 \leqslant h < 30$	14	2	$14 \div 2 = 7.00$
$30 \leqslant h < 32$	15	2	$15 \div 2 = 7.50$
$32 \leqslant h < 36$	26	4	$26 \div 4 = 6.50$
$36 \leqslant h < 40$	15	4	$15 \div 4 = 3.75$
$40 \leqslant h < 50$	8	10	$8 \div 10 = 0.80$
	$\sum f = 100 = n$		

The bar chart displays the frequencies.

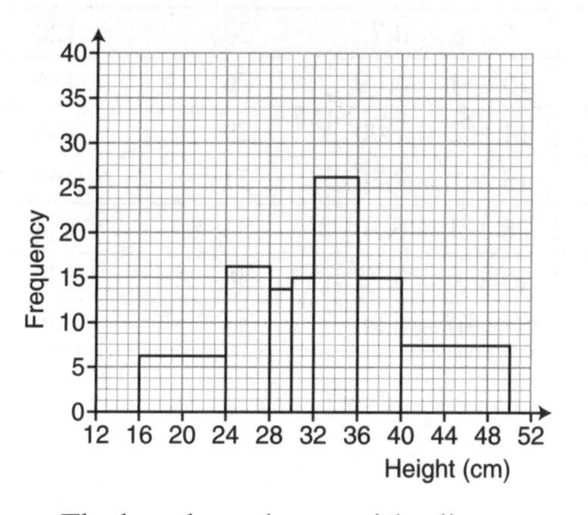

The histogram displays the frequency densities.

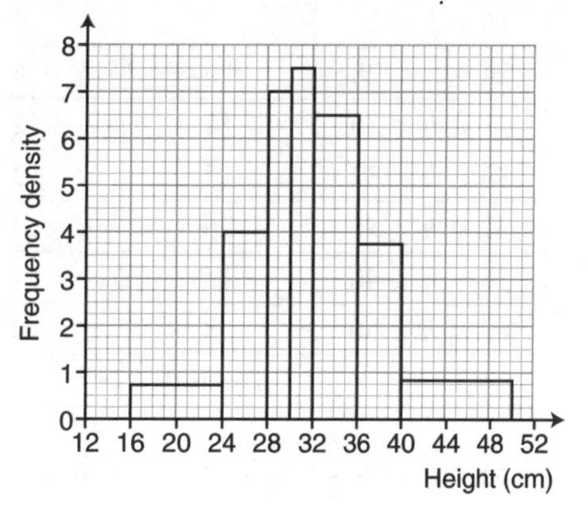

- The bar chart gives a misleading impression of the data because the groups are of different widths.
  A bar chart is *only* suitable when the data are divided into groups of the *same* size.

- In a histogram, the area of each 'bar' is:
  length × breadth = f.d. × (width of group)
  = frequency

  'Half as wide' ⟶ 'Twice as high'

UNIT 3 ♦ Handling data

Example 2

Use the horticulturist's data from Example 1.

a Calculate the mean height of the 100 plants.

b Estimate the percentage of plants which are longer than 38 cm.

a Extend the calculation table to include the mid-point (x cm) of each group and fx for each group.

Group	Freq., f	Width	Freq. density	Mid-point, x	fx
$16 \leqslant h < 24$	6	8	$6 \div 8 = 0.75$	20	120
$24 \leqslant h < 28$	16	4	$16 \div 4 = 4.00$	26	416
$28 \leqslant h < 30$	14	2	$14 \div 2 = 7.00$	29	406
$30 \leqslant h < 32$	15	2	$15 \div 2 = 7.50$	31	465
$32 \leqslant h < 36$	26	4	$26 \div 4 = 6.50$	34	884
$36 \leqslant h < 40$	15	4	$15 \div 4 = 3.75$	38	570
$40 \leqslant h < 50$	8	10	$8 \div 10 = 0.80$	45	360
	$n = 100$				$\Sigma fx = 3221$

$$\text{Mean} = \frac{\Sigma fx}{n} = \frac{3221}{100} = 32.21 \text{ cm}$$

So, the mean height is 32.2 cm (3 sig. fig.)

b From the histogram in Example 1:

$$\begin{aligned} \text{area of 'number over 38'} &= (2 \times 3.75) + (10 \times 0.8) \\ &= 7.5 + 8.0 \\ &= 15.5 \end{aligned}$$

Of the 100 plants in the sample, 15.5 are longer than 38 cm.
So 15.5% of the plants are longer than 38 cm.

Exercise 67

1 A survey revealed these results for the time spent on homework for a Friday night by a group of 60 school children.

Calculate the frequency density for each group and construct a histogram to display the results.

Time, t (min)	Number of children, f
$0 \leqslant t < 30$	6
$30 \leqslant t < 60$	12
$60 \leqslant t < 80$	18
$80 \leqslant t < 100$	12
$100 \leqslant t < 120$	9
$120 \leqslant t < 180$	3

Use these scales: horizontal axis, 1 cm = 10 min; vertical axis, 1 cm = 0.1.

2 The ages (in completed years) of the teaching staff at a school are given in this table.

Calculate the frequency density for each group and construct a histogram to display the results.

Age, x (years)	Number of teachers, f
$22 \leqslant x < 25$	12
$25 \leqslant x < 30$	10
$30 \leqslant x < 35$	14
$35 \leqslant x < 40$	15
$40 \leqslant x < 45$	13
$45 \leqslant x < 50$	9
$50 \leqslant x < 60$	17

Use these scales: horizontal axis, 2 cm = 5 years; vertical axis, 1 cm = 0.5.

3 A turkey farmer produced 89 turkeys for the Christmas market. Their weights are given in this table.

a Calculate the frequency density for each group and construct a histogram to display the results.

b What is the modal group for the weight of the turkeys?

c Estimate the percentage of turkeys weighing between 5 kg and 7.5 kg.

Weight (kg)	Frequency, f
2–4	7
4–5	7
5–6	10
6–6.5	12
6.5–7	19
7–8	16
8–10	18

Use these scales: horizontal axes, 1 cm = 1 kg; vertical axis, 1 cm = 0.5.

4 A fruit farmer checks the weights of 100 apples for quality control. The results are given in this table.

a Calculate the frequency density for each group and construct a histogram to display the results.

b What is the modal group for the weight of the apples?

c Estimate the percentage of apples weighing between 75 g and 100 g.

Weight (g)	Frequency, f
50–80	24
80–90	12
90–100	17
100–105	14
105–110	11
110–120	13
120–140	9

Use these scales: horizontal axis, 1 cm = 10 g; vertical axis, 1 cm = 0.2.

5 The ages (in completed years) of women giving birth in a local hospital during January 2000 are given in this table.

a Calculate the frequency density for each group in the table.

b Draw a histogram to illustrate the results.

c Calculate an estimate of the mean age of the mothers.

Age (years)	Frequency, f
14–16	7
16–20	38
20–25	60
25–30	68
30–35	52
35–45	25

Use these scales: horizonal axis, 1 cm = 2 years; vertical axis, 1 cm = 1.

6 The race times of cross-country runners are shown in this table.

a Calculate the frequency density for each group in the table.

b Draw a histogram to illustrate the results.

c Calculate an estimate of the mean time, correct to the nearest second.

Time, t (min)	Frequency, f
11–12	3
12–14	9
14–16	22
16–18	25
18–21	15
21–24	6

Use these scales: horizontal axis, 1 cm = 1 min; vertical axis, 1 cm = 1.

Exercise 67★

1 Fifty responses were received from a survey of French camp sites close to the Atlantic Coast. The 'size' of a camp site was defined by the number of mobile homes on it, and the results are shown in this table.

a Draw a histogram of the data.

b Estimate the percentage of sites with between 250 and 500 mobile homes.

c Calculate an estimate of the mean number of mobile homes per site.

Size of camp site	Number of sites
0–100	4
100–200	7
200–350	13
350–500	17
500–750	6
750–1000	3

2 This frequency table shows the distribution of ages for a cinema audience.

Age (years)	Frequency
18–20	24
20–25	42
25–30	24
30–40	16
40–50	30
50–60	30
60–80	34

a Draw a histogram of the data.

b Estimate the percentage of the audience aged between 35 and 55.

c Calculate an estimate of the mean age of the audience.

3 The ages of children attending a summer camp are given in this table.

a Calculate the frequency density for each group in the table.

b Calculate an estimate of the mean age of the children.

c Given that the height of the first bar of the histogram is 5 cm, calculate the heights of the other bars in the histogram.

Age (years)	Number
3–5	30
6–7	26
8–9	30
10	15
11	13
12–14	21
15–17	21

4 The ages (in completed years) of the members of a health and fitness club are shown in this table.

a Calculate the frequency density for each group in the table.

b Calculate an estimate of the mean age of the membership.

c Given that the height of the first bar is 5 cm, calculate the heights of the other bars.

Age (years)	Number
18–19	30
20–24	57
25–29	69
30–34	42
35–39	36
40–49	36
50–59	30

5 A snack bar manager has prepared a histogram that displays the trade on a typical Friday. This allows the manager to prepare the right quantity of snacks and to employ enough staff.

Read all parts of this question before preparing a calculation table.

a How many customers does the manager expect on a typical Friday?

b If each customer spends an average of $3, how much money does the manager expect to take?

c Calculate the mean number of customers per hour. Is this a useful statistic?

d Suggest suitable staffing rosters if one person can serve one customer every two minutes.

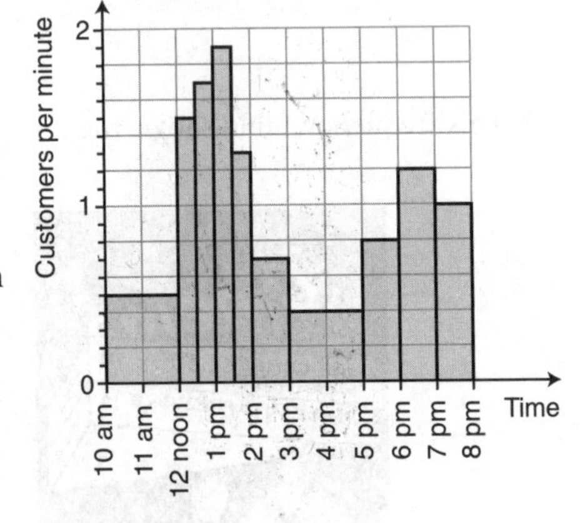

6 Traffic flow is recorded on a busy road as part of an investigation into a road improvement scheme. The results, for one weekday, are displayed on the histogram.

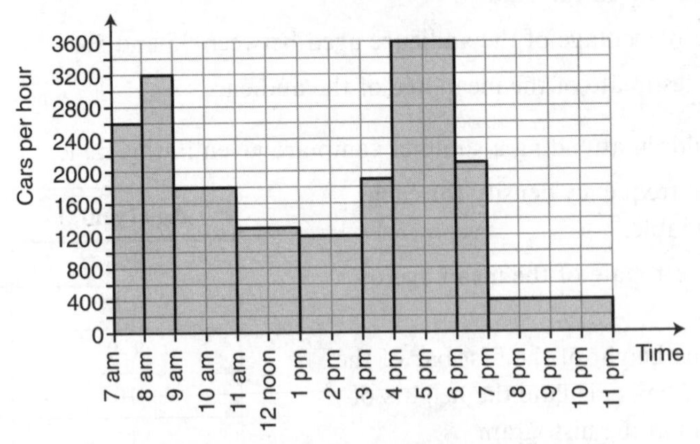

Read all parts of this question before preparing a calculation table.

a How many cars travelled on this road during the survey?

b Calculate the mean flow for the time of this survey.

c How many cars are passing through every five minutes at the busiest time?

d How might rate of traffic flow and total traffic flow affect the decisions of the planners?

7 This table and the unfinished histogram represent the playing times of a sample of video films.

Playing time (min)	Frequency, f
60–80	
80–90	
90–95	13
95–100	18
100–110	17
110–120	6
120–150	3

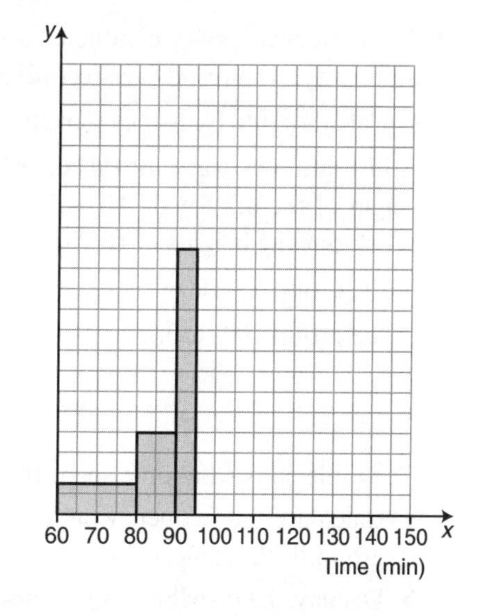

a Use the histogram to find the missing frequencies.

b Copy and complete the histogram, using the same scales and clearly labelling the vertical axis.

c Calculate the mean playing time of the videos.

8 A sample of batteries were tested by being continuously, used to power a toy train. This table and the unfinished histogram represent the times it took for the train to stop moving.

Time (hours)	Frequency, f
4–5	10
5–5.5	9
5.5–6	16
6–6.5	18
6.5–7	7
7–8	
8–10	

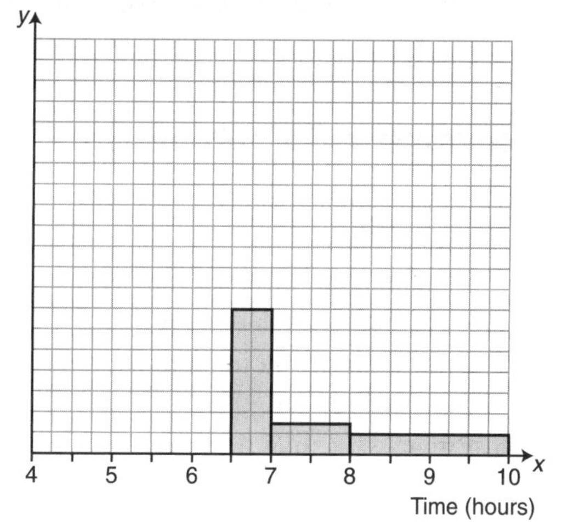

a Use the histogram to find the missing frequencies.

b Copy and complete the histogram, using the same scales and clearly labelling the vertical axis.

c Calculate the mean lifetime of the batteries.

Exercise 68 (Revision)

1 The motorway police conducted a survey on the distance between cars in the fast lane of a motorway. The results are recorded in this frequency table.

Distance (m)	Frequency, f
0–10	28
10–25	120
25–50	358
50–75	516
75–100	150
100–150	78

a Calculate the frequency densities of each group.

A histogram is drawn of this distribution, and the height of the bar for the '10–25' group is 12 cm. Work out the heights of the bars for:

b the group 75–100

c the group 100–150.

2 The table shows the time spent travelling to school by Year 10 pupils.

Time, t (min)	Number of pupils, f
0–10	12
10–20	23
20–40	32
40–60	18
60–90	9
90–120	6

a Calculate the frequency densities for each group in the table.

b Display the distribution in a histogram, and indicate on it the frequency represented by 1 cm^2.

c Calculate the total number of pupils in Year 10 and their mean travelling time.

d Estimate the percentage of pupils who take between 30 and 45 min to travel to school.

Use these scales: horizontal axis, 1 cm = 10 min; vertical axis, 1 cm = 2.

Exercise 68★ (Revision)

1 This table summarises Joseph's calls from the monthly itemised telephone bill for August. His father allows him $5 worth of calls but charges him for any extra.

Read all parts of this question before constructing a calculation table.

Duration of calls, t (min)	Number of calls, f
$0 \leqslant t < 5$	20
$5 \leqslant t < 15$	5
$15 \leqslant t < 30$	4
$30 \leqslant t < 40$	12
$40 \leqslant t < 50$	15
$50 \leqslant t < 60$	7
$60 \leqslant t < 75$	4
$75 \leqslant t < 90$	2
	$\Sigma f = 69$

a Draw a histogram to represent the information.

b Estimate the total time that Joseph spent on the phone.

c Calculate an estimate for the mean length, to the nearest second, of a call.

d Estimate the mean time, to the nearest second, that Joseph spent on the phone each day.

e If calls cost 0.8 cents per minute, estimate how much Joseph owes his father.

2 This frequency table and the histogram show the number of days that a sample of patients had to wait to see a specialist doctor after referral from their doctor in one particular region.

Waiting time, t (days)	f
$0 \leqslant t < 7$	14
$7 \leqslant t < 20$	65
$20 \leqslant t < 30$	
$30 \leqslant t < 60$	
$60 \leqslant t < 90$	342
$90 \leqslant t < 180$	207

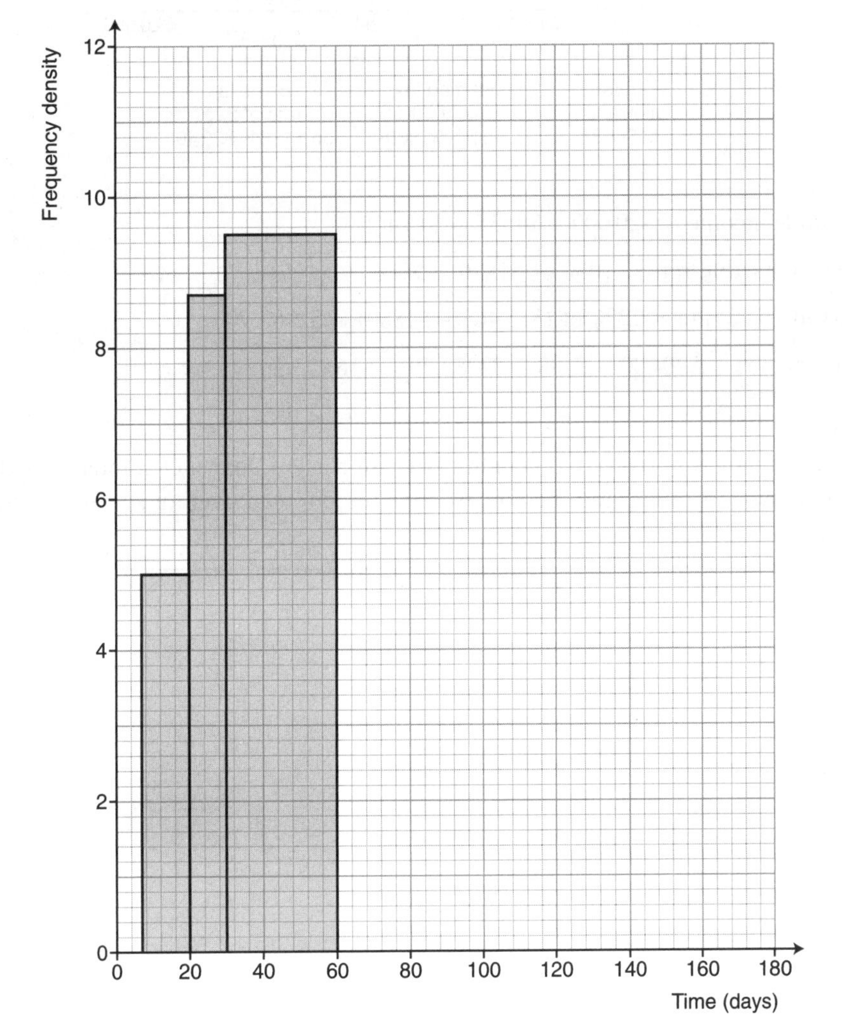

a Copy and complete the frequency table.

b Write down the heights of the missing bars in the histogram.

c Calculate estimates for the mean and median waiting times for referral to a specialist doctor in this region.

UNIT 3 ◆ Handling data

Summary 3

Number

Financial arithmetic

- Taxation (net income is that which remains after tax and other deductions)

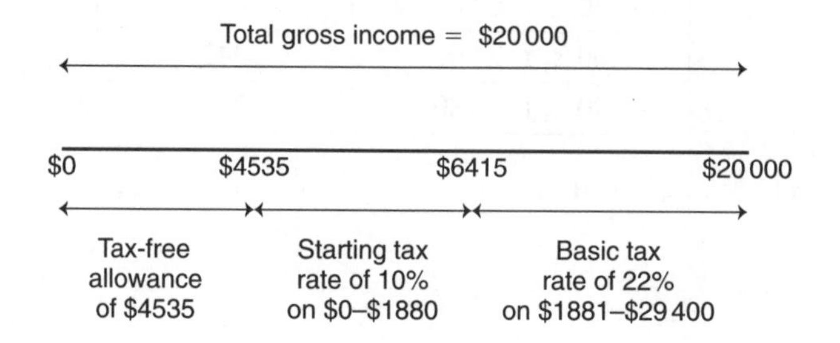

Taxable income = $20 000 – $4535 = $15 465

Tax at starting rate = 10% of $1880 = $188

Tax at basic rate = 22% of ($15 465 – $1880) = $2988.70

Net income = $20 000 – $188 – $2988.70 = $16 823.30

Loans

When money is borrowed, on credit cards, mortgages, loans, etc., interest is charged. The money borrowed and the interest have to be repaid over an agreed period of time.

Algebra

Solving simultaneous equations, one linear and one nonlinear

Graphically this corresponds to the intersection of a line and a curve.
Always substitute the linear equation into the nonlinear equation.

Solve the simultaneous equations $x^2 + y^2 = 13$ ①
$x - y + 1 = 0$ ②

The linear equation is equation ②.
Make y the subject of equation ② $y = x + 1 \Rightarrow y^2 = (x + 1)^2 \Rightarrow y^2 = x^2 + 2x + 1$ ③
Substitute ③ into ① $x^2 + x^2 + 2x + 1 = 13$

$2x^2 + 2x - 12 = 0$ (Divide by 2)

$x^2 + x - 6 = 0$

$(x + 3)(x - 2) = 0$

$x = -3$ or 2

Use ② to work out y, giving solutions as $(-3, 2)$ or $(2, 3)$.

Functions

- A function is a set of rules for turning one number into another.
- The **domain** of a function is the set of numbers the function can use.
  The domain of $f(x) = \frac{1}{x}$ is all real numbers except zero, because division by zero is not possible.
- The **range** of a function is the set of numbers produced by the function.
- The range of the function $f(x) = x^2$ is $\{y : y \geqslant 0, y \text{ a real number}\}$ because any number squared cannot be negative.
- The **inverse** function, $f^{-1}(x)$, undoes whatever the function has done.
  The inverse of $f(x) = x + 3$ is $f^{-1}(x) = x - 3$.
- $gf(x)$ is a **composite function**. $gf(x)$ means 'do f followed by g'.

Graphs

Tangents to a curve

To find the gradient of a curve at a point:

- Draw the tangent to the curve at the point using a ruler.

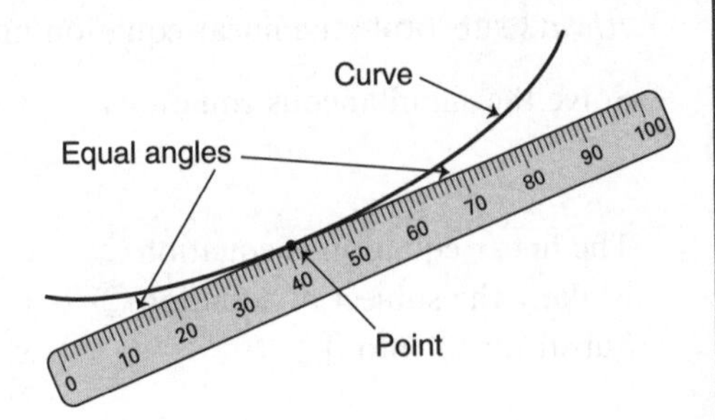

- The gradient of the tangent is the gradient of the curve at the point.

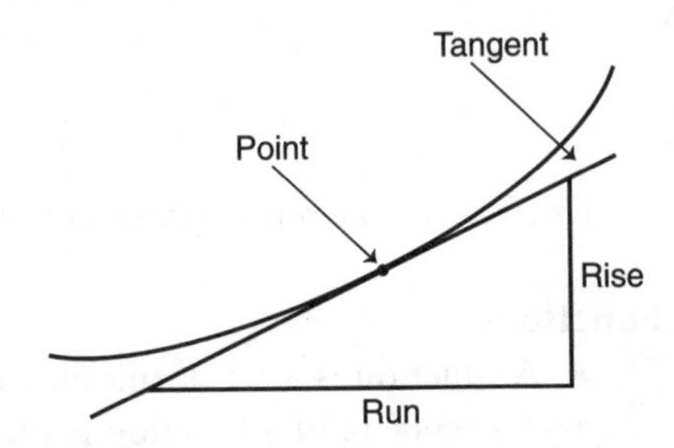

- The gradient of the tangent is found by working out $\frac{\text{'rise'}}{\text{'run'}}$.
- The gradient of a graph often provides useful information.

Graph	Gradient
Distance (m) – time (s)	Speed (m/s)
Speed (m/s) – time (s)	Acceleration (m/s^2)
Temperature (°C) – time (min)	Rate of change of temperature (°C/min)
Population (ants) – time (weeks)	Rate of change of population (ants/week)
Financial profit ($) – time (years)	Rate of change of profit ($/year)

Shape and space

Vector notation

Vectors have magnitude and direction and can be written as bold letters: **v**, **u**, …, with capitals covered by an arrow: $\overrightarrow{OP}$, $\overrightarrow{OQ}$, … or on co-ordinate axes as column vectors $\begin{pmatrix} 3 \\ -1 \end{pmatrix}$, $\begin{pmatrix} 0 \\ 6 \end{pmatrix}$, …

Multiplication of a vector by a scalar

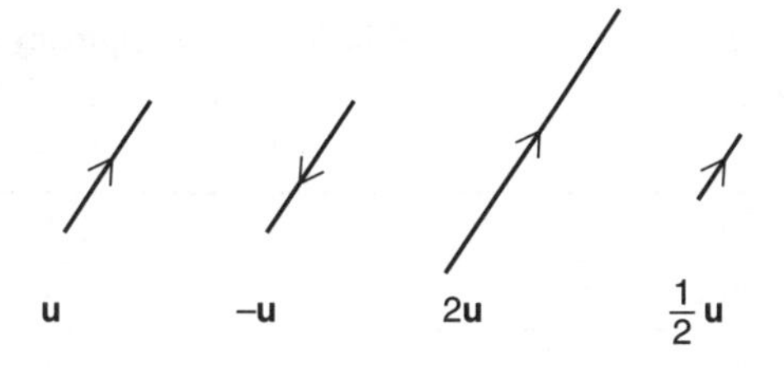

Vector geometry

- $\overrightarrow{AC} = \overrightarrow{AB} + \overrightarrow{BC} = 2\mathbf{a} + \mathbf{b}$

  $\therefore\ \overrightarrow{AD} = \overrightarrow{AC} + \overrightarrow{CD}$
  $= 2\mathbf{a} + \mathbf{b} + (-3\mathbf{a}) = \mathbf{b} - \mathbf{a}$

  AB is parallel to DC. ∴ ABCD is a trapezium

  Ratio of AB:DC = 2:3. ∴ 2DC = 3AB

- $\overrightarrow{OD} = \begin{pmatrix} 3 \\ 0 \end{pmatrix}$, $\overrightarrow{DC} = \begin{pmatrix} -2 \\ 4 \end{pmatrix}$, $\overrightarrow{CA} = \begin{pmatrix} -2 \\ -2 \end{pmatrix}$

  $\therefore\ \overrightarrow{OA} = \overrightarrow{OD} + \overrightarrow{DC} + \overrightarrow{CA}$

  $= \begin{pmatrix} 3 \\ 0 \end{pmatrix} + \begin{pmatrix} -2 \\ 4 \end{pmatrix} + \begin{pmatrix} -2 \\ -2 \end{pmatrix}$

  $= \begin{pmatrix} -1 \\ 2 \end{pmatrix}$

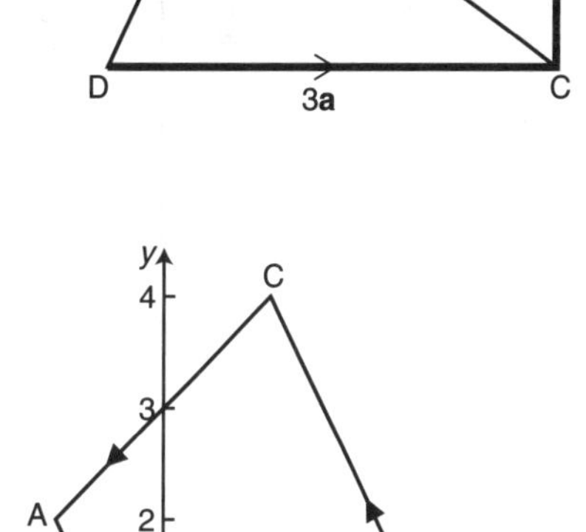

Handling data

Histograms

The horizontal axis is a continuous number line.
The **area** of each bar represents the frequency.
The **frequency density** is given by

$$\text{frequency density (height of bar)} = \frac{\text{frequency}}{\text{width of group}}$$

Group time (sec)	Frequency	Width	Frequency density (frequency ÷ bar width)
15–20	12	5	$12 \div 5 = 2.4$
20–28	16	8	$16 \div 8 = 2$
28–36	8	8	$8 \div 8 = 1$
36–40	5	4	$5 \div 4 = 1.25$

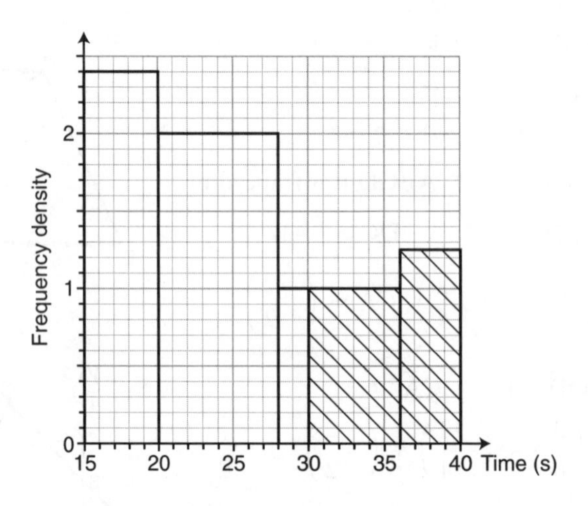

The percentage > 30 $= \dfrac{\left(\frac{3}{4} \times 8\right) + 5}{41} \times 100\% = 27\%$

Examination practice 3

1 Mr Jaipur earns a gross salary of \$25 000 p.a. He has a tax-free allowance of \$4500.

If the starting rate of tax is at 10% on the first \$1500 of taxable income and the basic rate of tax is at 25% on the remainder, find the amount of tax he has to pay on a year's salary.

2 The annual total profits of Butterworth's Bazaar are always shared out as follows: 36 people get a fixed sum of \$1000 each, and then the rest is split equally between 15 'shareholders'.

a In one year of trading, the shareholders got \$800 each. What were the profits?

b In the following year, the shareholders got a 50% increase each. What was the percentage increase in profits?

3 Solve these simultaneous equations, giving your answers to 3 s.f.

$$y = x^2 - 4x + 3 \quad \text{and} \quad y = 2x - 3$$

4 Solve these simultaneous equations:

$$2x - y = 4 \quad \text{and} \quad x^2 + y^2 = 16$$

5 $f : x \to \dfrac{1}{x - 1}$

a Find $f(-1)$.
b What value must be excluded from the domain of f?
c Express the inverse function f^{-1} in the form $f^{-1} : x \to \ldots$
d What value must be excluded from the domain of f^{-1}?

6 $f(x) = x + 3$ and $g(x) = x^2$

a State the domain and range of g.
b Find the composite function $gf(x)$.
c Find the inverse function $f^{-1}(x)$.
d Solve the equation $gf(x) = f^{-1}(x)$.

7 a For the graph shown, find, by drawing tangents, the gradients at:
(i) $x = -1$ (ii) $x = 1$

b Where is the gradient zero?

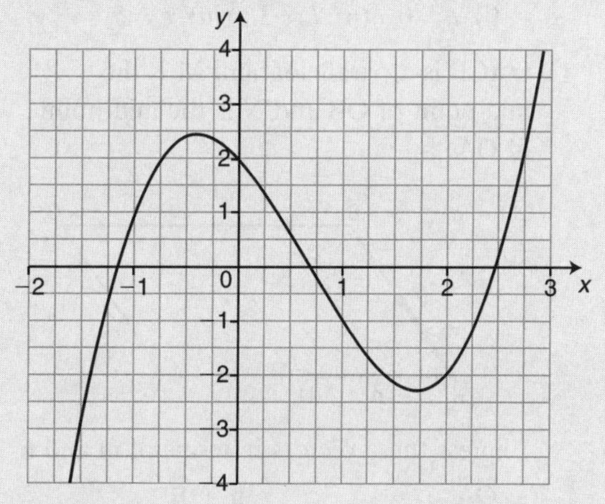

8 a Draw an accurate graph of $y = x^2 - 4x$ for $-1 \leqslant x \leqslant 5$.

b By drawing tangents, find the gradient when:
(i) $x = 0$ (ii) $x = 3$

c Where is the gradient zero?

9 The temperature of Sereena's bath, t °C, m minutes after the bath has been run, is given by the equation $t = \dfrac{1100}{m + 20}$, valid for $0 \leqslant m \leqslant 20$.

a Draw the graph of t against m for $0 \leqslant m \leqslant 20$.

b Sereena likes to be in the bath when the temperature is between 35 °C and 50 °C. Use your graph to find at what times Sereena should be in the bath.

c By drawing a suitable tangent, find the rate of change of temperature in °C/minute when $m = 7$.

10 After t seconds, the height, h metres, of an arrow fired by an archer is given by $h = 30t - 5t^2$ for $0 \leqslant t \leqslant 6$.

a Draw the graph of $h = 30t - 5t^2$ for $0 \leqslant t \leqslant 6$.

b By drawing tangents, find the velocity of the arrow in m/s when:
(i) $t = 0$ (ii) $t = 3$ (iii) $t = 5$

11 OACB is a parallelogram. M is the mid-point of OB and N is the mid-point of OA.

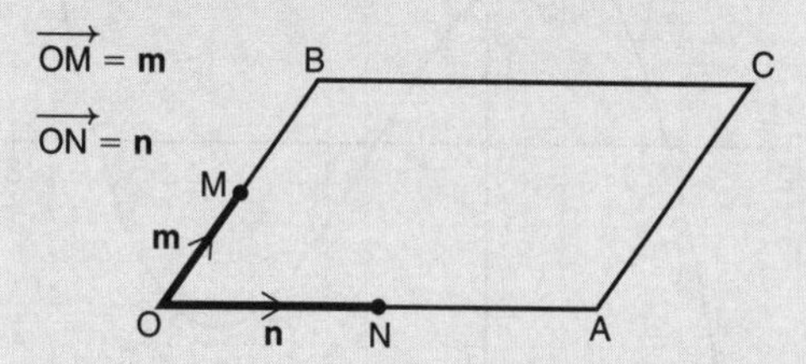

Express these vectors in terms of **m** and **n**.

a $\overrightarrow{OA}$ **b** $\overrightarrow{OB}$

c $\overrightarrow{AB}$ **d** $\overrightarrow{NM}$

e What can you deduce about $\overrightarrow{NM}$ and $\overrightarrow{AB}$?

12 AEBO is a rectangle and OBDC is a parallelogram.

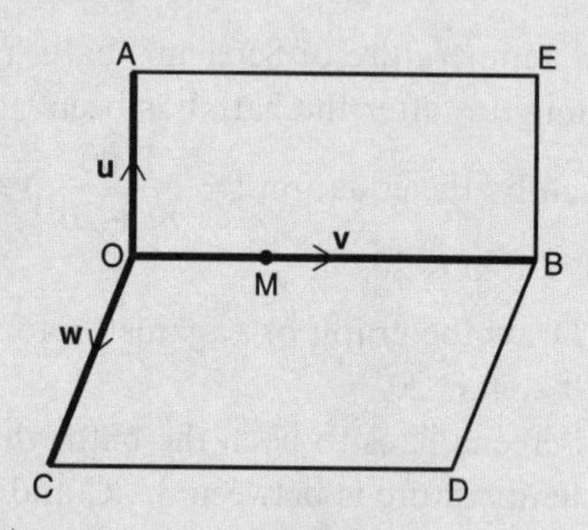

$\overrightarrow{OA} = \mathbf{u}$, $\overrightarrow{OB} = \mathbf{v}$, $\overrightarrow{OC} = \mathbf{w}$ and OM:MB = 1:2.
Express in terms of **u**, **v**, **w** in their simplest forms:

a $\overrightarrow{OM}$

b $\overrightarrow{AM}$

c $\overrightarrow{MD}$

13 The position of a point is determined by its position vector $\begin{pmatrix} x \\ y \end{pmatrix}$ relative to the origin O. A and B have position vectors $\overrightarrow{OA} = \begin{pmatrix} 20 \\ 15 \end{pmatrix}$ and $\overrightarrow{OB} = \begin{pmatrix} 30 \\ 40 \end{pmatrix}$.

a Write down $\overrightarrow{AB}$.

b Calculate the magnitude of the vector $\overrightarrow{AB}$.

c Calculate the angle that the vector $\overrightarrow{AB}$ makes with the x-axis.

d X is the mid-point of AB. Write $\overrightarrow{OX}$ as a column vector.

14 In the figure, OACB is a parallelogram.

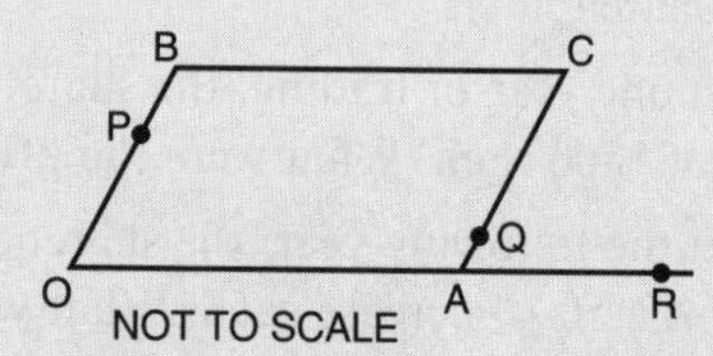

$\overrightarrow{OP} = \frac{2}{3}\overrightarrow{OB}$, $\overrightarrow{AQ} = \frac{1}{4}\overrightarrow{AC}$, $\overrightarrow{AR} = \frac{3}{5}\overrightarrow{OA}$

Denote the vector $\overrightarrow{OA}$ by **a**, the vector $\overrightarrow{OB}$ by **b**.

a Write down, in terms of **b**, the vector $\overrightarrow{OP}$.

b Write down, in terms of **a** and **b**, the vector $\overrightarrow{OQ}$.

c Deduce the vector $\overrightarrow{PQ}$.

d Write down the vector $\overrightarrow{OR}$.

e Deduce the vector $\overrightarrow{QR}$.

f Say, with reasons, whether PQR is a straight line.

MEI

15 Leon recorded the lengths, in minutes, of the films shown on television in one week. His results are shown in the histogram.

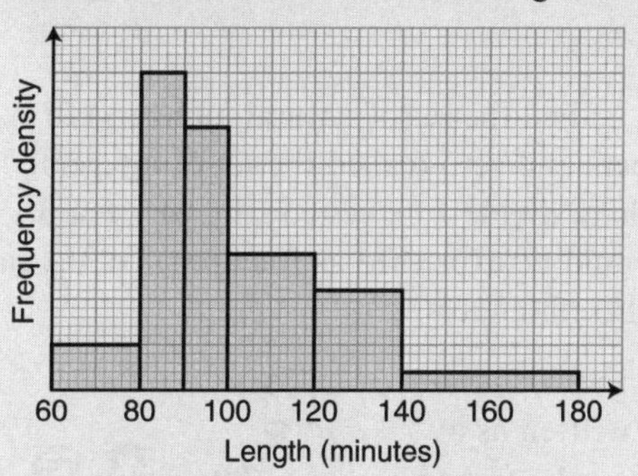

20 films had lengths from 60 minutes up to, but not including, 80 minutes.

a Use the information in the histogram to complete the table.

Length (minutes)	Frequency
60 up to but not including 80	20
80 up to but not including 90	
90 up to but not including 100	
100 up to but not including 120	
120 up to but not including 140	
140 up to but not including 180	

Leon also recorded the lengths, in minutes, of all the films shown on television in the following week. His results are given in the table below.

Length (minutes)	Frequency	Frequency density
60 up to but not including 90	72	48
90 up to but not including 140	x	
140 up to but not including 180	y	

b Copy and complete the table, giving your answers in terms of x and y.

LONDON

16 In a survey, 1000 people were asked to guess the weight of a turkey.

Weight (kg)	Frequency, f
5–10	80
10–20	360
20–30	380
30–40	120
40–45	60

a Calculate the frequency density for each group and construct a histogram to display the results.

b From your histogram, estimate the probability that a member of the sample chosen at random had guessed a weight of between 14 kg and 24 kg.

c Calculate the mean weight guess per person.

Number 4

Irrational numbers

Primitive people only needed the numbers 1, 2, 3, … (called the set of natural numbers) to count objects. Later it was found that negative numbers and zero (called the set of integers) were needed. Also, fractions were needed, for example to divide 4 loaves equally between 6 people. Early civilisations thought that all numbers could be expressed as fractions, i.e. the **ratio** of two integers, hence the word **rational** to describe these numbers.

In the sixth century BCE, the ancient Greeks proved that $\sqrt{2}$, the length of the diagonal of a unit square, could not be written as a fraction. They soon found that other numbers such as $\sqrt{3}$, $\sqrt{5}$ and $\sqrt[3]{7}$ could not be written as fractions. These numbers are called **irrational** numbers. Later it was shown that π is also an irrational number.

The decimal expansion of irrational numbers is infinite and shows no pattern. A decimal expansion that is infinite but where the digits recur is rational because it can be written as a fraction (see Number 1).

$\pi =$ 3.141 592 653 589 793 238 462 643 383 279 502 884 197 169 399 375 105 820 974 944 592 307 816 406 286 208 998 628 034 825 342 117 067 982 148 086 513 282 …

Together the rational numbers and the irrational numbers form the set of real numbers. All these sets can be shown on a Venn diagram where:

$\mathbb{N}$ is the set of natural numbers or positive integers {1, 2, 3, 4, …}.
$\mathbb{Z}$ is the set of integers {…, −2, −1, 0, 1, 2, …}.
$\mathbb{Q}$ is the set of rational numbers.
$\mathbb{R}$ is the set of real numbers.

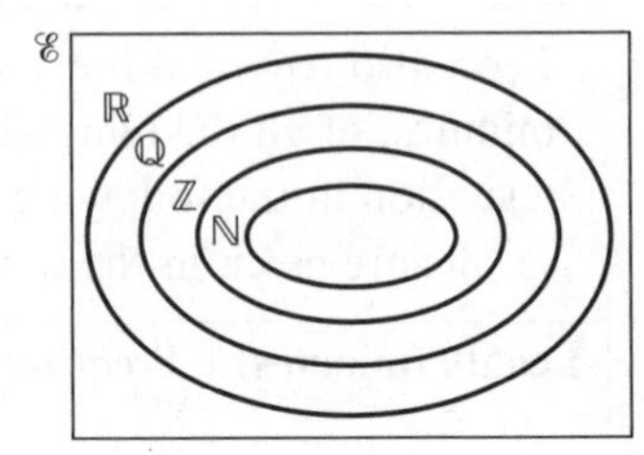

Example 1

Which of these numbers are rational and which are irrational?

$$2 \qquad 0.789 \qquad 0.\dot{6}\dot{7} \qquad \sqrt{2} \times \sqrt{2} \qquad \sqrt{2} + \sqrt{2}$$

$2 = \frac{2}{1}$ so it is rational.

$0.789 = \frac{789}{1000}$ so it is rational.

$0.\dot{6}\dot{7} = \frac{67}{99}$ so it is rational.

$\sqrt{2} \times \sqrt{2} = 2$ so it is rational.

$\sqrt{2} + \sqrt{2} = 2\sqrt{2}$ so it is irrational.

Example 2

a Find an irrational number between 4 and 5.

$4^2 = 16$ and $5^2 = 25$, so the square root of any integer between 17 and 24 inclusive will be irrational, for example $\sqrt{19}$.

b Find a rational number between $\sqrt{2}$ and $\sqrt{3}$.

$\sqrt{2} = 1.414213\ldots$ and $\sqrt{3} = 1.732050\ldots$ so any terminating or recurring decimal between these numbers will be rational, for example 1.5.

Remember

Good examples of irrational numbers are the square roots of prime numbers and π.

Exercise 69

For Questions 1–16, state which of the numbers are rational and which are irrational. Express the rational numbers in the form $\frac{a}{b}$ where a and b are integers.

1 5.7

2 7.3

3 $0.\dot{4}\dot{7}$

4 $0.\dot{2}\dot{3}$

5 $\sqrt{49}$

6 $\sqrt{25}$

7 2π

8 $3\sqrt{2}$

9 $\sqrt{3} \times \sqrt{3}$

10 $\sqrt{5} \times \sqrt{5}$

11 $\sqrt{3} \div \sqrt{3}$

12 $\sqrt{5} \div \sqrt{5}$

13 $\sqrt{5} + \sqrt{5}$

14 $\sqrt{3} + \sqrt{3}$

15 $\sqrt{3} - \sqrt{3}$

16 $\sqrt{5} - \sqrt{5}$

17 Find a rational number between $\sqrt{5}$ and $\sqrt{7}$.

18 Find a rational number between $\sqrt{3}$ and $\sqrt{5}$.

19 Find an irrational number between 7 and 8.

20 Find an irrational number between 3 and 4.

21 The circumference of a circle is 4 cm. Find the radius, giving your answer as an irrational number.

22 The area of a circle is $9\,\text{cm}^2$. Find the radius, giving your answer as an irrational number.

Exercise 69★

For Questions 1–14, state which of the numbers are rational and which are irrational. Express the rational numbers in the form $\frac{a}{b}$ where a and b are integers.

1 $\pi + 2$

2 $\sqrt{2} + 2$

3 $\sqrt{\frac{4}{25}}$

4 $\sqrt{\frac{9}{16}}$

5 $\sqrt{0.36}$

6 $\sqrt{0.81}$

7 $\sqrt{2\frac{1}{4}}$

8 $\sqrt{2\frac{7}{9}}$

9 $\sqrt{5} - \sqrt{3}$

10 $\sqrt{3} + \sqrt{7}$

11 $\sqrt{13} \div \sqrt{13}$

12 $\sqrt{13} \times \sqrt{13}$

13 $\sqrt{3} \times \sqrt{3} \times \sqrt{3}$

14 $\sqrt{2} \times \sqrt{9} \div \sqrt{2}$

15 Find a rational number between $\sqrt{11}$ and $\sqrt{13}$.

16 Find a rational number between $\sqrt{17}$ and $\sqrt{19}$.

17 Find an irrational number between 2 and 3.

18 Find an irrational number between 0 and 1.

19 Find two *different* irrational numbers whose product is rational.

20 Find an irrational number whose cube is rational.

21 The circumference of a circle is 6 cm. Find the area, giving your answer as an irrational number.

22 The area of a circle is $16\,\text{cm}^2$. Find the circumference, giving your answer as an irrational number.

23 Right-angled triangles can have sides with lengths that are rational or irrational numbers of units. Give an example of a right-angled triangle to fit each description below. Draw a separate triangle for each part.

a All sides are rational.

b The hypotenuse is rational and the other two sides are irrational.

c The hypotenuse is irrational and the other two sides are rational.

d The hypotenuse and one of the other sides is rational and the remaining side is irrational.

LONDON

Surds

A **surd** is a root that is irrational. The roots of all prime numbers are surds.

$\sqrt{2}$, $\sqrt{5}$ and $\sqrt[3]{7}$ are all surds.

$\sqrt{4} = 2$, so $\sqrt{4}$ is *not* a surd.

Note that $\sqrt{2}$ means the positive square root of 2.

When an answer is given using a surd, it is an *exact* answer.

For example, the diagonal of a square of side 1 cm is *exactly* $\sqrt{2}$ cm long, or sin 60° is *exactly* $\frac{\sqrt{3}}{2}$.

Simplifying surds

Example 3

Simplify the following, giving the exact answer:

a $2\sqrt{3}+3\sqrt{3}$ **b** $(2\sqrt{3})^2$ **c** $2\sqrt{3}\times 3\sqrt{3}$ **d** $(\sqrt{3})^3$

a $2\sqrt{3}+3\sqrt{3}=5\sqrt{3}$

b $(2\sqrt{3})^2=2\sqrt{3}\times 2\sqrt{3}=2\times 2\times\sqrt{3}\times\sqrt{3}=4\times 3=12$

c $2\sqrt{3}\times 3\sqrt{3}=2\times 3\times\sqrt{3}\times\sqrt{3}=6\times 3=18$

d $(\sqrt{3})^3=\sqrt{3}\times\sqrt{3}\times\sqrt{3}=3\sqrt{3}$

Exercise 70

Simplify the following, giving the exact answer:

1 $2\sqrt{5}+4\sqrt{5}$ **2** $3\sqrt{7}+2\sqrt{7}$ **3** $5\sqrt{3}-\sqrt{3}$

4 $6\sqrt{2}-3\sqrt{2}$ **5** $(4\sqrt{2})^2$ **6** $(5\sqrt{3})^2$

7 $(2\sqrt{5})^2$ **8** $(3\sqrt{7})^2$ **9** $4\sqrt{2}\times\sqrt{2}$

10 $3\sqrt{3}\times\sqrt{3}$ **11** $5\sqrt{7}\times 3\sqrt{7}$ **12** $8\sqrt{5}\times 2\sqrt{5}$

13 $(\sqrt{2})^3$ **14** $(\sqrt{5})^3$ **15** $4\sqrt{2}\div\sqrt{2}$

16 $3\sqrt{3}\div\sqrt{3}$ **17** $8\sqrt{5}\div\sqrt{5}$ **18** $2\sqrt{7}\div 4\sqrt{7}$

Exercise 70★

Simplify the following, giving the exact answer:

1 $4\sqrt{11}+\sqrt{11}$ **2** $3\sqrt{13}+2\sqrt{13}$ **3** $6\sqrt{7}-2\sqrt{7}$

4 $8\sqrt{17}-\sqrt{17}$ **5** $(3\sqrt{11})^2$ **6** $(2\sqrt{13})^2$

7 $8\sqrt{3}\times 4\sqrt{3}$ **8** $7\sqrt{5}\times 3\sqrt{5}$ **9** $(2\sqrt{7})^3$

10 $(3\sqrt{5})^3$ **11** $3\sqrt{2}\times 4\sqrt{2}\times 5\sqrt{2}$ **12** $2\sqrt{3}\times 3\sqrt{3}\times 4\sqrt{3}$

13 $(\sqrt{3})^4$ **14** $(\sqrt{7})^4$ **15** $4\sqrt{7}\div\sqrt{7}$

16 $5\sqrt{13}\div\sqrt{13}$ **17** $6\sqrt{11}\div\sqrt{11}$ **18** $4\sqrt{17}\div 12\sqrt{17}$

Example 3 part **b** showed that $(2\sqrt{3})^2=4\times 3=12$.

This means that $\sqrt{12}=\sqrt{4\times 3}=2\sqrt{3}$.

This is because $\sqrt{a\times b}=\sqrt{a}\times\sqrt{b}$.

Example 4

Simplify **a** $\sqrt{18}$ **b** $\sqrt{35}$

a Find two factors of 18, one of which is a perfect square, for example $18 = 9 \times 2$.
Then $\sqrt{18} = \sqrt{9} \times \sqrt{2} = 3\sqrt{2}$

b In this case a perfect square factor cannot be found.
As $35 = 5 \times 7$, then $\sqrt{35} = \sqrt{5} \times \sqrt{7} = \sqrt{5}\sqrt{7}$

Example 5

Simplify $\sqrt{72}$.

$72 = 36 \times 2$, so $\sqrt{72} = \sqrt{36} \times \sqrt{2} = 6\sqrt{2}$

In Example 5 you might have written $72 = 9 \times 8$ so $\sqrt{72} = 3\sqrt{8}$. This is correct, but it is not as simple as possible because $\sqrt{8}$ can be simplified further as $\sqrt{8} = \sqrt{4} \times \sqrt{2} = 2\sqrt{2}$. This gives $\sqrt{72} = 3 \times 2\sqrt{2} = 6\sqrt{2}$ as before.

So when finding factors, try to find the largest possible factor that is a perfect square.

Example 6

Simplify $\sqrt{18} + 2\sqrt{2}$.

$\sqrt{18} = 3\sqrt{2}$ (see Example 4)

So $\sqrt{18} + 2\sqrt{2} = 3\sqrt{2} + 2\sqrt{2} = 5\sqrt{2}$

Example 7

Express $5\sqrt{6}$ as the square root of a single number.

$5\sqrt{6} = \sqrt{25} \times \sqrt{6} = \sqrt{25 \times 6} = \sqrt{150}$

To simplify $\sqrt{\frac{16}{25}}$ note that $\left(\frac{4}{5}\right)^2 = \frac{4^2}{5^2} = \frac{16}{25}$ so this means that $\sqrt{\frac{16}{25}} = \frac{4}{5}$.

This is because $\sqrt{\frac{a}{b}} = \frac{\sqrt{a}}{\sqrt{b}}$.

Example 8

Simplify $\sqrt{\frac{81}{49}}$.

$$\sqrt{\frac{81}{49}} = \frac{\sqrt{81}}{\sqrt{49}} = \frac{9}{7}$$

Exercise 71

For Questions 1–12, simplify.

1 $\sqrt{12}$ **2** $\sqrt{20}$ **3** $\sqrt{18}$

4 $\sqrt{27}$ **5** $\sqrt{48}$ **6** $\sqrt{32}$

7 $\sqrt{45}$ **8** $\sqrt{75}$ **9** $\sqrt{12}+\sqrt{3}$

10 $\sqrt{20}+3\sqrt{5}$ **11** $\sqrt{32}+2\sqrt{2}$ **12** $\sqrt{45}+\sqrt{5}$

For Questions 13–18, express as the square root of a single number.

13 $5\sqrt{2}$ **14** $3\sqrt{2}$ **15** $3\sqrt{3}$

16 $2\sqrt{5}$ **17** $3\sqrt{6}$ **18** $2\sqrt{7}$

For Questions 19–24, simplify.

19 $\sqrt{\frac{1}{4}}$ **20** $\sqrt{\frac{1}{9}}$ **21** $\sqrt{\frac{4}{25}}$

22 $\sqrt{\frac{9}{16}}$ **23** $\sqrt{\frac{36}{81}}$ **24** $\sqrt{\frac{49}{64}}$

25 A rectangle has sides of length $\sqrt{18}$ and $\sqrt{8}$. Find exact values for the area, the perimeter and the length of a diagonal.

26 A rectangle has sides of length $\sqrt{12}$ and $\sqrt{27}$. Find exact values for the area, the perimeter and the length of a diagonal.

Exercise 71★

For Questions 1–12, simplify.

1 $\sqrt{28}$ **2** $\sqrt{63}$ **3** $\sqrt{99}$

4 $\sqrt{44}$ **5** $\sqrt{80}$ **6** $\sqrt{125}$

7 $\sqrt{117}$ **8** $\sqrt{52}$ **9** $\sqrt{48}+\sqrt{3}$

10 $\sqrt{75}+2\sqrt{3}$ **11** $\sqrt{27}+\sqrt{12}$ **12** $\sqrt{20}+\sqrt{80}$

For Questions 13–18, express as the square root of a single number.

13 $5\sqrt{3}$ **14** $2\sqrt{6}$ **15** $4\sqrt{5}$

16 $2\sqrt{8}$ **17** $3\sqrt{7}$ **18** $5\sqrt{11}$

For Questions 19–24, simplify.

19 $\sqrt{\frac{1}{36}}$ **20** $\sqrt{\frac{1}{49}}$ **21** $\sqrt{\frac{81}{100}}$

22 $\sqrt{\frac{121}{144}}$ **23** $\sqrt{\frac{49}{169}}$ **24** $\sqrt{\frac{196}{225}}$

25 The length of the diagonal of a square is 4 cm. Find exact values for the area and the perimeter.

26 A piece of wire is bent into the shape shown in the diagram. All dimensions are in cm. Find exact values for the length x, the perimeter and the area of the shape.

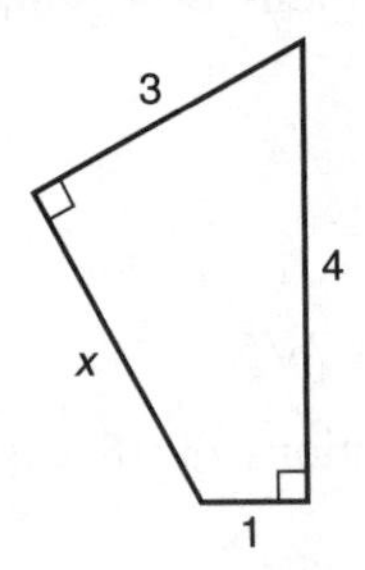

Activity 22

a The diagram shows an isosceles right-angled triangle. The two shorter sides are 1 cm in length.

(i) What is the exact value of x (the hypotenuse)?

(ii) What is the value of the angle marked a?

(iii) Use your answers to parts (i) and (ii) to fill in the table below, giving exact answers.

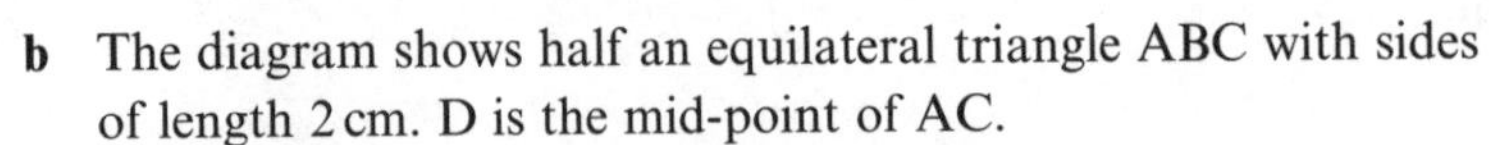

$\sin 45° =$	$\cos 45° =$	$\tan 45° =$

b The diagram shows half an equilateral triangle ABC with sides of length 2 cm. D is the mid-point of AC.

(i) Find the exact value of the length BD.

(ii) What are the values of the angles BAD and ABD?

(iii) Use your answers to parts (i) and (ii) to fill in the table below, giving exact answers.

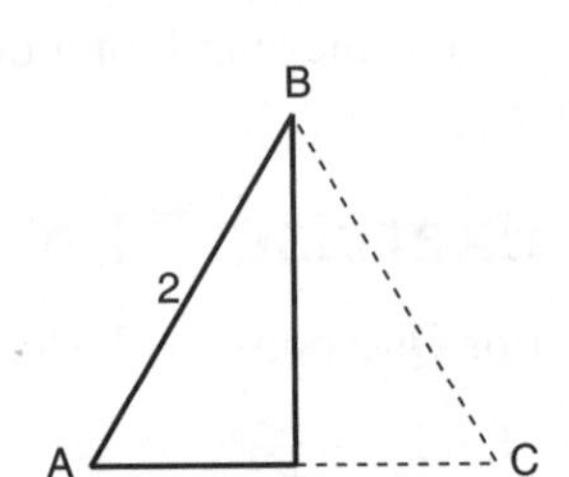

$\sin 30° =$	$\cos 30° =$	$\tan 30° =$
$\sin 60° =$	$\cos 60° =$	$\tan 60° =$

To simplify expressions like $(2 + 3\sqrt{2})^2$, multiply out the brackets using FOIL and then simplify.

Example 9

Simplify $(2 + 3\sqrt{2})^2$.

$(2 + 3\sqrt{2})^2 = (2 + 3\sqrt{2})(2 + 3\sqrt{2}) = 4 + 6\sqrt{2} + 6\sqrt{2} + 3\sqrt{2} \times 3\sqrt{2} =$
$4 + 12\sqrt{2} + 18 = 22 + 12\sqrt{2}$

Exercise 72

Simplify the following:

1 $(1+\sqrt{2})^2$

2 $(1+\sqrt{3})^2$

3 $(1-\sqrt{3})^2$

4 $(1-\sqrt{2})^2$

5 $(3+2\sqrt{3})^2$

6 $(2+3\sqrt{3})^2$

7 $(3-3\sqrt{2})^2$

8 $(2-2\sqrt{3})^2$

9 $(1+\sqrt{5})(1-\sqrt{5})$

10 $(1+\sqrt{3})(1-\sqrt{3})$

11 $(\sqrt{2}+\sqrt{3})^2$

12 $(\sqrt{5}-\sqrt{2})^2$

13 $(1+\sqrt{2})(1-\sqrt{5})$

14 $(1-\sqrt{3})(1+\sqrt{2})$

Exercise 72★

Simplify the following:

1 $(2+\sqrt{5})^2$

2 $(3+\sqrt{3})^2$

3 $(4-\sqrt{2})^2$

4 $(5-\sqrt{5})^2$

5 $(4-\sqrt{3})(4+\sqrt{3})$

6 $(5+\sqrt{5})(5-\sqrt{5})$

7 $(3+4\sqrt{2})^2$

8 $(4-5\sqrt{3})^2$

9 $(\sqrt{7}-\sqrt{3})^2$

10 $(\sqrt{2}+\sqrt{5})^2$

11 $(\sqrt{7}-\sqrt{5})(\sqrt{7}+\sqrt{5})$

12 $(\sqrt{11}-\sqrt{3})(\sqrt{11}+\sqrt{3})$

13 $(3+2\sqrt{2})(5-2\sqrt{7})$

14 $(2+3\sqrt{3})(4-3\sqrt{5})$

15 A rectangle has sides of length $\sqrt{3}+1$ and $\sqrt{3}-1$. Find exact values for the perimeter, the area and the length of a diagonal.

16 A right-angled triangle has a hypotenuse of length $2+\sqrt{2}$ and one other side of length $1+\sqrt{2}$. Find exact values for the length of the third side and the area.

Rationalising the denominator

When writing fractions it is not usual to write surds in the denominator (see Activity 23 for a reason). The surds can be cleared by multiplying the top and bottom of the fraction by the same amount. This is equivalent to multiplying the fraction by 1 and so does not change its value. The process is called 'rationalising the denominator'.

Example 10

Express $\frac{4}{\sqrt{8}}$ in the form $a\sqrt{2}$.

$$\frac{4}{\sqrt{8}} = \frac{4}{\sqrt{8}} \times \frac{\sqrt{8}}{\sqrt{8}} \quad \text{(Multiply top and bottom by } \sqrt{8}\text{)}$$

$$= \frac{4\sqrt{8}}{8}$$

$$= \frac{4 \times 2\sqrt{2}}{8} = \sqrt{2}$$

Exercise 73

Express these with rational denominators.

1 $\frac{1}{\sqrt{3}}$ **2** $\frac{1}{\sqrt{2}}$ **3** $\frac{1}{\sqrt{5}}$

4 $\frac{1}{\sqrt{7}}$ **5** $\frac{3}{\sqrt{3}}$ **6** $\frac{2}{\sqrt{2}}$

7 $\frac{4}{\sqrt{2}}$ **8** $\frac{6}{\sqrt{3}}$ **9** $\frac{4}{\sqrt{12}}$

10 $\frac{6}{\sqrt{8}}$ **11** $\frac{3}{2\sqrt{2}}$ **12** $\frac{5}{2\sqrt{3}}$

13 $\frac{1+\sqrt{2}}{\sqrt{2}}$ **14** $\frac{1+\sqrt{3}}{\sqrt{3}}$

Activity 23

a If you had no calculator but knew $\sqrt{2} = 1.414\,214$ to 6 d.p., how would you work out $\frac{1}{\sqrt{2}}$?

b Express $\frac{1}{\sqrt{2}}$ in the form $a\sqrt{2}$.

c Using your answer to part **b**, work out $\frac{1}{\sqrt{2}}$ to 6 d.p. without using your calculator.

d Rationalise the denominator in $\frac{1}{\sqrt{2}-1}$ by multiplying top and bottom by $\sqrt{2}+1$.

e Rationalise the denominator in $\frac{1}{\sqrt{2}+1}$.

Exercise 73★

Express these with rational denominators.

1 $\frac{1}{\sqrt{13}}$

2 $\frac{1}{\sqrt{11}}$

3 $\frac{a}{\sqrt{a}}$

4 $\frac{b^2}{\sqrt{b}}$

5 $\frac{6-\sqrt{3}}{\sqrt{3}}$

6 $\frac{4-\sqrt{2}}{\sqrt{2}}$

7 $\frac{\sqrt{2}}{\sqrt{6}}$

8 $\frac{\sqrt{8}}{\sqrt{12}}$

9 $\frac{5+2\sqrt{5}}{\sqrt{5}}$

10 $\frac{14+3\sqrt{7}}{\sqrt{7}}$

11 $\frac{1}{1-\sqrt{5}}$

12 $\frac{1}{1+\sqrt{7}}$

13 $\frac{1-\sqrt{2}}{3+\sqrt{2}}$

14 $\frac{2+\sqrt{3}}{2-\sqrt{3}}$

Exercise 74 (Revision)

1 Which of $0.\dot{3}$, π, $\sqrt{25}$ and $\sqrt{5}$ are rational?

2 Find a rational number between $\sqrt{3}$ and $\sqrt{5}$.

3 Find an irrational number between 3 and 4.

4 Write $3\sqrt{5}$ as the square root of a single number.

For Questions 5–15, simplify.

5 $3\sqrt{3}+2\sqrt{3}$

6 $3\sqrt{3}-2\sqrt{3}$

7 $3\sqrt{3}\times 2\sqrt{3}$

8 $3\sqrt{3}\div 2\sqrt{3}$

9 $\sqrt{8}$

10 $\sqrt{63}$

11 $\sqrt{3}+\sqrt{12}$

12 $\sqrt{\frac{4}{9}}$

13 $(1+\sqrt{5})^2$

14 $(2-\sqrt{2})^2$

15 $(1+\sqrt{2})(1-\sqrt{2})$

For Questions 16–19, rationalise the denominator.

16 $\frac{5}{\sqrt{5}}$

17 $\frac{6}{\sqrt{3}}$

18 $\frac{\sqrt{27}}{\sqrt{12}}$

19 $\frac{3+\sqrt{3}}{\sqrt{3}}$

20 A rectangle has sides of length $3\sqrt{2}$ and $5\sqrt{2}$. Find the exact values of the perimeter, the area and the length of a diagonal.

Exercise 74★ (Revision)

1 Which of $(\sqrt{3})^2$, $\sqrt{13}$, $\sqrt{5}+\sqrt{5}$ and $0.\dot{2}\dot{3}$ are rational?

2 Find a rational number between $\sqrt{7}$ and $\sqrt{11}$.

3 Find an irrational number between 6 and 7.

4 Write $4\sqrt{11}$ as the square root of a single number.

For Questions 5–15, simplify.

5 $5\sqrt{5}+3\sqrt{5}$

6 $5\sqrt{5}-3\sqrt{5}$

7 $5\sqrt{5}\times 3\sqrt{5}$

8 $5\sqrt{5}\div 3\sqrt{5}$

9 $\sqrt{48}$

10 $\sqrt{242}$

11 $7\sqrt{54}-\sqrt{24}$

12 $\sqrt{\frac{18}{50}}$

13 $(\sqrt{18}-\sqrt{2})^2$

14 $(2-\sqrt{2})^2$

15 $(\sqrt{7}-2\sqrt{2})(\sqrt{7}+3\sqrt{2})$

For Questions 16–20, rationalise the denominator.

16 $\frac{1}{2\sqrt{5}}$

17 $\frac{12}{\sqrt{6}}$

18 $\frac{6+2\sqrt{6}}{\sqrt{6}}$

19 $\frac{\sqrt{112}}{\sqrt{28}}$

20 $\sqrt{\frac{1}{2}}+\sqrt{\frac{1}{4}}+\sqrt{\frac{1}{8}}$

21 The two shorter sides of a right-angled triangle are $3\sqrt{3}$ and $4\sqrt{3}$. Find the exact values of the length of the hypotenuse and the area.

22 Find the exact height of an equilateral triangle with side length 2 units. Hence find the exact values of $\cos 60°$ and $\sin 60°$.

Algebra 4

Algebraic fractions

Simplifying algebraic fractions

To simplify algebraic fractions, factorise as much as possible and then cancel.

Example 1

Simplify $\dfrac{x^2+x}{x+1}$.

$$\frac{x^2+x}{x+1} = \frac{x(x+1)}{x+1}$$
$$= x$$

Example 2

Simplify $\dfrac{x^2+x-2}{x^2-4}$.

$$\frac{x^2+x-2}{x^2-4} = \frac{(x+2)(x-1)}{(x+2)(x-2)}$$
$$= \frac{x-1}{x-2}$$

Exercise 75

Simplify the following:

1 $\dfrac{3x+12}{2x+8}$

2 $\dfrac{8x+4y}{6x+3y}$

3 $\dfrac{x+x^2}{y+xy}$

4 $\dfrac{xy+y^2}{x^2+xy}$

5 $\dfrac{x^2-x-6}{x-3}$

6 $\dfrac{x^2-6x-7}{x+1}$

7 $\dfrac{x+1}{x^2+3x+2}$

8 $\dfrac{x+2}{x^2+4x+4}$

9 $\dfrac{x^2-y^2}{(x-y)^2}$

10 $\dfrac{(x+y)^2}{x^2-y^2}$

11 $\dfrac{x^2-x-2}{x^2-5x+6}$

12 $\dfrac{x^2+2x-3}{x^2+4x+3}$

13 $\dfrac{x^2-x-12}{x^2-2x-8}$

14 $\dfrac{x^2+9x+8}{x^2+7x-8}$

Exercise 75★

Simplify the following:

1 $\dfrac{6a+9b}{10a+15b}$

2 $\dfrac{14p+35q}{6p+15q}$

3 $\dfrac{x^2y+xy^2}{x^2y+x^3}$

4 $\dfrac{x^3y+xy}{xy^2+x^3y^2}$

5 $\dfrac{x^2-9x+20}{x^2-2x-15}$

6 $\dfrac{x^2+3x-18}{x^2+11x+30}$

7 $\dfrac{x^2-x-12}{x^2-16}$

8 $\dfrac{x^2+3x-10}{x^2-25}$

9 $\dfrac{3r^2+6r-45}{3r^2+18r+15}$

10 $\dfrac{2t^2+10t-28}{2t^2+18t+28}$

11 $\dfrac{a^3-ab^2}{a^3+2a^2b+ab^2}$

12 $\dfrac{2ab-a^2-b^2}{a^2b-b^3}$

13 $\dfrac{2x^2+12x-32}{xy-2y}$

14 $\dfrac{3x^3-3x^2-18x}{x^2y-3xy}$

Addition and subtraction

Algebraic fractions are added or subtracted in the same way as number fractions. As with number fractions, find the lowest common denominator, otherwise the working can become complicated.

Fractions with number denominators

Example 3

Express $\dfrac{x+1}{3}-\dfrac{x-3}{4}$ as a single fraction.

$$\frac{x+1}{3}-\frac{x-3}{4}=\frac{4(x+1)-3(x-3)}{12}$$

$$=\frac{4x+4-3x+9}{12}$$

$$=\frac{x+13}{12}$$

Example 4

Express $\frac{3(4x-1)}{2} - \frac{2(5x+3)}{3}$ as a single fraction.

$$\frac{3(4x-1)}{2} - \frac{2(5x+3)}{3} = \frac{9(4x-1) - 4(5x+3)}{6}$$
$$= \frac{36x - 9 - 20x - 12}{6}$$
$$= \frac{16x - 21}{6}$$

Note how in both examples the brackets in the numerator are not multiplied out until the second step. This will help avoid mistakes with signs, especially when the fractions are subtracted.

Remember

Find the lowest common denominator when adding or subtracting.

Exercise 76

Express as single fractions:

1 $\frac{x}{3} + \frac{x+1}{2}$

2 $\frac{x}{2} + \frac{x+2}{3}$

3 $\frac{x}{2} - \frac{x+2}{4}$

4 $\frac{x}{3} - \frac{x+1}{4}$

5 $\frac{x+4}{5} + \frac{x}{3}$

6 $\frac{x+3}{2} + \frac{x}{5}$

7 $x + \frac{x+3}{4}$

8 $x + \frac{x-2}{3}$

9 $\frac{x+1}{5} - x$

10 $\frac{x+3}{2} - x$

11 $\frac{x-1}{3} + \frac{x+2}{4}$

12 $\frac{x-2}{3} + \frac{x+1}{2}$

13 $\frac{x-3}{2} - \frac{x+1}{5}$

14 $\frac{x-4}{6} - \frac{x+2}{3}$

15 $\frac{2}{3} - \frac{2-3x}{6}$

16 $\frac{3}{4} - \frac{3-2x}{6}$

17 $\frac{2x+5}{4} - \frac{2(x-3)}{3}$

18 $\frac{x+3}{7} - \frac{3(x-1)}{2}$

19 $\frac{2(2x+1)}{5} + \frac{3(x-1)}{2}$

20 $\frac{3(x-2)}{4} + \frac{2(2x-1)}{5}$

Exercise 76★

Express as single fractions:

1 $\frac{2x-1}{5}+\frac{x+3}{2}$

2 $\frac{x-1}{4}+\frac{2x+2}{3}$

3 $\frac{x+1}{3}-\frac{2x+1}{4}$

4 $\frac{x-3}{5}-\frac{3x-2}{3}$

5 $\frac{2x+1}{7}-\frac{x-2}{2}$

6 $\frac{4x+3}{6}-\frac{x+2}{4}$

7 $\frac{3x-1}{5}+\frac{1-3x}{7}$

8 $\frac{2x+3}{4}+\frac{3-2x}{8}$

9 $\frac{2(x-5)}{3}-\frac{3(x+20)}{5}$

10 $\frac{3(x+4)}{8}-\frac{2(x-3)}{3}$

11 $\frac{x-3}{18}-\frac{x-2}{24}$

12 $\frac{3x+1}{20}-\frac{2x+1}{25}$

13 $\frac{3x}{4}+\frac{x-1}{5}-\frac{x+1}{6}$

14 $\frac{x-1}{2}+\frac{x-2}{3}+\frac{x-3}{4}$

15 $\frac{3x-2}{4}-\frac{x}{12}+\frac{x-5}{6}$

16 $\frac{2x-4}{15}+\frac{x}{5}-\frac{x-1}{3}$

17 $\frac{x-2}{3}-\frac{x-7}{6}+\frac{10x-1}{9}$

18 $\frac{7x-1}{20}-\frac{x-1}{30}-\frac{2(x+1)}{5}$

19 $\frac{2(3x-4)}{5}-\frac{5(4x-3)}{2}+x$

20 $\frac{2(4x-5)}{3}-\frac{4(2x-3)}{5}+x$

Fractions with x in the denominator

The same method is used as for fractions with numbers in the denominator. It is more important to find the lowest common denominator, otherwise the working can become very complicated.

Example 5

Express $\frac{3}{2x}-\frac{4}{3x}$ as a single fraction.

Notice that the lowest common denominator is $6x$, not $6x^2$.

$$\frac{3}{2x}-\frac{4}{3x}=\frac{3\times 3-2\times 4}{6x}$$

$$=\frac{1}{6x}$$

Example 6

Express $\frac{1}{x} + \frac{1}{x+1}$ as a single fraction.

$$\frac{1}{x} + \frac{1}{x+1} = \frac{(x+1)+x}{x(x+1)}$$

$$= \frac{2x+1}{x(x+1)} \longleftarrow \text{The denominator is best left factorised}$$

Example 7

Express $\frac{x+1}{x+2} - \frac{x-2}{x-1}$ as a single fraction.

$$\frac{x+1}{x+2} - \frac{x-2}{x-1} = \frac{(x-1)(x+1)-(x+2)(x-2)}{(x+2)(x-1)}$$

$$= \frac{(x^2-1)-(x^2-4)}{(x+2)(x-1)}$$

$$= \frac{x^2-1-x^2+4}{(x+2)(x-1)}$$

$$= \frac{3}{(x+2)(x-1)} \longleftarrow \text{The denominator is best left factorised}$$

Example 8

Express $\frac{2}{x-2} - \frac{6}{x^2-x-2}$ as a single fraction.

To find the lowest common denominator, factorise $x^2 - x - 2$ as $(x+1)(x-2)$, giving the lowest common denominator as $(x+1)(x-2)$. Because the denominator is left factorised it is easier to simplify the resulting expression.

$$\frac{2}{x-2} - \frac{6}{x^2-x-2} = \frac{2}{x-2} - \frac{6}{(x+1)(x-2)}$$

$$= \frac{2(x+1)-6}{(x+1)(x-2)}$$

$$= \frac{2x-4}{(x+1)(x-2)}$$

$$= \frac{2(x-2)}{(x+1)(x-2)}$$

$$= \frac{2}{x+1}$$

Remember

Always find the lowest common denominator.

Exercise 77

Express as single fractions:

1 $\frac{1}{2x} + \frac{1}{3x}$

2 $\frac{1}{4x} + \frac{1}{3x}$

3 $\frac{3}{4x} - \frac{1}{2x}$

4 $\frac{2}{3x} + \frac{1}{6x}$

5 $\frac{1}{2} - \frac{1}{x-2}$

6 $\frac{1}{x+2} - \frac{1}{3}$

7 $\frac{1}{x+1} + \frac{1}{x-1}$

8 $\frac{1}{x-1} - \frac{1}{x+2}$

9 $\frac{3}{x-1} - \frac{2}{x+2}$

10 $\frac{4}{x+1} - \frac{3}{x-3}$

11 $\frac{x}{x+2} + \frac{1}{x}$

12 $\frac{1}{x} + \frac{x}{x-3}$

13 $\frac{x+2}{x+1} - \frac{x+1}{x+2}$

14 $\frac{x+1}{x+3} - \frac{x-3}{x+1}$

Exercise 77★

Express as a single fraction:

1 $\frac{1}{x} + \frac{1}{3x} - \frac{1}{5x}$

2 $\frac{1}{2x} - \frac{1}{4x} + \frac{1}{6x}$

3 $\frac{4}{x+3} - \frac{3}{x+2}$

4 $\frac{5}{x+1} - \frac{2}{x+4}$

5 $\frac{1}{1+x} + x$

6 $1 - \frac{1}{x+1}$

7 $\frac{1}{x} - \frac{1}{x(x+1)}$

8 $\frac{1}{x(x+1)} + \frac{1}{x+1}$

9 $\frac{x+2}{x+1} - \frac{x+1}{x+2}$

10 $\frac{x+3}{x+2} - \frac{x-3}{x-2}$

11 $\frac{1}{x^2+3x+2} + \frac{1}{x+2}$

12 $\frac{4}{x^2+2x-3} + \frac{1}{x+3}$

13 $\frac{1}{x^2+2x+1} + \frac{3}{x+1}$

14 $\frac{1}{x^2+2x-3} + \frac{2}{x-1}$

15 $\frac{x+1}{x^2-4x+3} - \frac{x-3}{x^2-1}$

16 $\frac{x-2}{x^2-3x-4} - \frac{x-4}{x^2-6x+8}$

Equations with fractions

Equations with number denominators

If an equation involves fractions, clear the fractions by multiplying *both sides* of the equation by the lowest common denominator. Then simplify and solve in the usual way.

Example 9

Solve $\frac{3x}{7} = 2$.

$\frac{3x}{7} = 2$ (Multiply both sides by 7)

$7 \times \frac{3x}{7} = 7 \times 2$ (Simplify)

$3x = 14$ (Divide both sides by 3)

$x = \frac{14}{3}$

Example 10

Solve $\frac{x-1}{2} = \frac{x+2}{3}$.

$\frac{x-1}{2} = \frac{x+2}{3}$ (Multiply both sides by 6)

$6 \times \frac{x-1}{2} = 6 \times \frac{x+2}{3}$ (Simplify)

$3(x-1) = 2(x+2)$ (Multiply out brackets)

$3x - 3 = 2x + 4$ (Collect terms)

$3x - 2x = 4 + 3$ (Simplify)

$x = 7$

Example 11

Solve $\frac{x+1}{3} - \frac{x-3}{4} = 1$.

$\frac{x+1}{3} - \frac{x-3}{4} = 1$ (Multiply both sides by 12)

$12 \times \frac{x+1}{3} - 12 \times \frac{x-3}{4} = 12 \times 1$ (Notice *everything* is multiplied by 12)

$4(x+1) - 3(x-3) = 12$ (Multiply out, note sign change in 2nd bracket)

$4x + 4 - 3x + 9 = 12$ (Collect terms)

$4x - 3x = 12 - 4 - 9$ (Simplify)

$x = -1$

Remember

Clear the fractions by multiplying *both* sides of the equation by the lowest common denominator.

Exercise 78

Solve the following equations:

1 $\frac{x}{3} = 7$

2 $\frac{2x}{3} = 8$

3 $\frac{x}{2} = \frac{x+2}{4}$

4 $\frac{x}{3} = \frac{x+1}{4}$

5 $\frac{2-x}{3} = x$

6 $x = \frac{x+1}{5}$

7 $x = \frac{x+3}{2}$

8 $\frac{x+3}{4} = x$

9 $\frac{x-4}{6} = \frac{x+2}{3}$

10 $\frac{x-1}{3} = \frac{x+2}{4}$

11 $\frac{2-3x}{6} = \frac{2}{3}$

12 $\frac{3-2x}{6} = \frac{3}{4}$

13 $\frac{x+1}{3} = \frac{2x+1}{4}$

14 $\frac{x-3}{5} = \frac{3x-2}{3}$

15 $\frac{2(x-5)}{3} = \frac{3(x+20)}{5}$

16 $\frac{3(x+4)}{8} = \frac{2(x-3)}{3}$

17 $\frac{x+1}{7} - \frac{3(x-2)}{14} = 1$

18 $\frac{3(x-2)}{2} - \frac{x-5}{4} = 2$

19 $\frac{6-3x}{3} - \frac{5x+12}{4} = -1$

20 $\frac{16-5x}{8} - \frac{4(2x+5)}{5} = -2$

Exercise 78★

For Questions 1–20, solve the equation.

1 $\frac{3x}{7} = \frac{6}{35}$

2 $\frac{4x}{15} = \frac{8}{45}$

3 $\frac{x-3}{3} = 4$

4 $\frac{x+5}{2} = 3$

5 $\frac{x+1}{2} = 2x$

6 $\frac{x-1}{3} = x$

7 $\frac{3-x}{3} = \frac{2+x}{2}$

8 $\frac{2-3x}{4} = \frac{1+2x}{3}$

9 $\frac{x-3}{12} + \frac{x}{5} = 4$

10 $\frac{3x}{2} - \frac{x+1}{10} = \frac{11}{2}$

11 $\frac{2x+1}{3} = x-2$

12 $\frac{5x-1}{2} = 3x+1$

UNIT 4 ◆ Algebra

13 $\dfrac{7x-1}{6}+5x=6$

14 $\dfrac{x}{3}=\dfrac{3x+3}{2}-5$

15 $\dfrac{1+x}{2}=\dfrac{2-x}{3}+1$

16 $\dfrac{2(3-2x)}{3}-\dfrac{2+x}{6}=2$

17 $4-\dfrac{x-2}{2}=3+\dfrac{2-3x}{3}$

18 $\dfrac{x-2}{5}-\dfrac{2x+1}{3}+4=0$

19 $\dfrac{1-x}{2}-\dfrac{2+x}{3}+\dfrac{3-x}{4}=1$

20 $\dfrac{x+3}{2}+\dfrac{2x+3}{5}-\dfrac{3x-1}{7}+1=0$

21 Pedro does one-sixth of his journey to school by car and two-thirds by bus. He then walks the final kilometre. How long is his journey to school?

22 Meera is competing in a triathlon. She cycles half the course then swims one-thirtieth of the course. She then runs 14 km to the finish. How long is the course?

Equations with *x* in the denominator

These are solved in the same way as equations with numbers in the denominator. Clear the fractions by multiplying *both sides* of the equation by the lowest common denominator. Then simplify and solve in the usual way.

Example 12

Solve the equation $x+2=\dfrac{8}{x}$.

The lowest common denominator is x.

$$x+2=\frac{8}{x} \Rightarrow x\times x+x\times 2=x\times\frac{8}{x}$$
$$\Rightarrow \quad x^2+2x=8$$
$$\Rightarrow \quad x^2+2x-8=0$$
$$\Rightarrow (x-2)(x+4)=0$$
$$\Rightarrow \quad x=2 \text{ or } x=-4$$

UNIT 4 ◆ Algebra

Example 13

Solve the equation $\frac{1}{3} + \frac{1}{x+1} = \frac{x}{3}$.

The lowest common denominator is $3(x + 1)$.

$$\frac{1}{3} + \frac{1}{x+1} = \frac{x}{3} \Rightarrow 3(x+1) \times \frac{1}{3} + 3(x+1) \times \frac{1}{x+1} = 3(x+1) \times \frac{x}{3}$$

$$\Rightarrow \quad (x+1) + 3 = x(x+1)$$

$$\Rightarrow \quad x + 4 = x^2 + x$$

$$\Rightarrow \quad x^2 = 4$$

$$\Rightarrow \quad x = \pm 2$$

Example 14

Solve the equation $\frac{x}{x-2} - \frac{2}{x+1} = 3$.

The lowest common denominator is $(x - 2)(x + 1)$.

$$\frac{x}{x-2} - \frac{2}{x+1} = 3 \Rightarrow (x-2)(x+1) \times \frac{x}{x-2} - (x-2)(x+1) \times \frac{2}{x+1} = (x-2)(x+1) \times 3$$

$$\Rightarrow \quad x(x+1) - 2(x-2) = 3(x-2)(x+1)$$

$$\Rightarrow \quad x^2 + x - 2x + 4 = 3x^2 - 3x - 6$$

$$\Rightarrow \quad 2x^2 - 2x - 10 = 0$$

$$\Rightarrow \quad x^2 - x - 5 = 0$$

$$\Rightarrow \quad x = \frac{1 \pm \sqrt{1+20}}{2}$$

$$\Rightarrow \quad x = 2.79 \text{ or}$$

$$x = -1.79 \text{ to 3 s.f.}$$

Remember

Clear the fractions by multiplying *both* sides of the equation by the lowest common denominator.

Exercise 79

Solve the following equations:

1 $x + 5 = \frac{14}{x}$

2 $x + 2 = \frac{15}{x}$

3 $x + \frac{2}{x} = 3$

4 $x + \frac{6}{x} = 5$

5 $\frac{1}{2} - \frac{1}{x-2} = \frac{1}{4}$

6 $\frac{1}{x+4} - \frac{1}{3} = -\frac{1}{6}$

7 $\frac{x}{x+2} - \frac{1}{x} = 1$

8 $\frac{1}{x} + \frac{x}{x-3} = 1$

9 $\frac{6}{x-2} - \frac{6}{x+1} = 1$

10 $\frac{4}{x-1} - \frac{2}{x+2} = 1$

11 $\frac{x}{x-1} + \frac{8}{x+4} = 2$

12 $\frac{2x}{x-3} - \frac{4}{x+1} = 3$

13 $\frac{x-2}{x-1} = \frac{x+4}{2x+4}$

14 $\frac{2x-3}{x+1} = \frac{x+3}{x+5}$

15 $\frac{2x-1}{x+2} = \frac{4x+1}{5x+2}$

16 $\frac{2x+2}{x+8} = \frac{4x-1}{5x-2}$

Exercise 79★

For Questions 1–14, solve the equation.

1 $2x + 3 = \frac{7}{x}$

2 $4x + \frac{7}{x} = 29$

3 $1 + \frac{3}{x} = \frac{4}{x^2}$

4 $1 - \frac{3}{x} = \frac{10}{x^2}$

5 $\frac{7}{9} - \frac{x}{x+5} = \frac{1}{3}$

6 $\frac{1}{2} + \frac{x}{x+7} = \frac{4}{5}$

7 $\frac{7}{x-1} - \frac{4}{x+4} = 1$

8 $\frac{7}{x-3} - \frac{6}{x+1} = 1$

9 $\frac{6x}{x+1} - \frac{5}{x+3} = 3$

10 $\frac{4x}{x+1} - \frac{8}{x+5} = 2$

11 $\frac{2x+1}{x+1} = \frac{x+2}{2x+1}$

12 $\frac{x-3}{3x+1} = \frac{x+2}{x-5}$

13 $\frac{3x+2}{x+1} + \frac{x+2}{2x-5} = 4$

14 $\frac{4x-5}{x+2} + \frac{2x-4}{3x-8} = 3$

15 Mala drives the first 30 km of her journey at x km/h. She then increases her speed by 20 km/h for the final 40 km of her journey. Her journey takes 1 hour. Find x.

16 Lucas is running a race. He runs the first 800 m at x m/s.
He then slows down by 3 m/s for the final 400 m.
His total time is 130 seconds. Find x.

For Questions 17–19, solve the equation.

17 $\dfrac{3}{x+2}+\dfrac{4}{x+3}=\dfrac{7}{x+6}$

18 $\dfrac{4}{x-3}+\dfrac{3x-3}{x^2-x-6}=\dfrac{2-20x}{2x+4}$

19 $y-1=\dfrac{y^2+3}{y-1}+\dfrac{y-2}{y-6}$

Exercise 80 (Revision)

For Questions 1–4, simplify.

1 $\dfrac{3x+6}{x+2}$

2 $\dfrac{x^2+7x+10}{x+5}$

3 $\dfrac{x^2+6x+9}{x^2-9}$

4 $\dfrac{x^2-2x+1}{x^2+2x-3}$

For Questions 5–12, express as a single fraction.

5 $\dfrac{x-1}{2}-\dfrac{x}{3}$

6 $\dfrac{x-2}{3}-\dfrac{x}{4}$

7 $\dfrac{3x-1}{4}-\dfrac{3x+4}{5}$

8 $\dfrac{3x-2}{6}-\dfrac{3x+2}{9}$

9 $\dfrac{1}{x+1}-\dfrac{1}{x-2}$

10 $\dfrac{3}{x-1}-\dfrac{2}{x+1}$

11 $\dfrac{x}{x-2}+\dfrac{2}{x}$

12 $\dfrac{x+3}{x+2}-\dfrac{x+1}{x+4}$

For Questions 13–20, solve the equation.

13 $\dfrac{3x}{4}=2$

14 $\dfrac{x}{3}=\dfrac{x-2}{5}$

15 $1+\dfrac{x}{2}=\dfrac{x-3}{4}$

16 $\dfrac{x+2}{3}-\dfrac{x-2}{4}=1$

17 $x+1=\dfrac{2}{x}$

18 $\dfrac{x}{x-1}+\dfrac{1}{x}=1$

19 $\dfrac{3}{x-1}-\dfrac{8}{x+2}=1$

20 $\dfrac{x+3}{x+5}=\dfrac{x+1}{x+2}$

Exercise 80★ (Revision)

For Questions 1–4, simplify.

1 $\dfrac{2x+14}{3x+21}$

2 $\dfrac{x^2-12x+11}{x^2+4x-5}$

3 $\dfrac{x^2+11x+28}{x^2-49}$

4 $\dfrac{x^2-11x+10}{x^2-9x-10}$

For Questions 5–12, express as a single fraction.

5 $\dfrac{2x+5}{6}-\dfrac{x+7}{4}$

6 $\dfrac{3(x-1)}{6}-\dfrac{2(x+1)}{9}$

7 $\dfrac{x+1}{2}-\dfrac{x+2}{3}+\dfrac{x+3}{4}$

8 $\dfrac{2(3x-1)}{5}+1-\dfrac{x}{2}$

9 $\dfrac{x}{x+3}+\dfrac{8}{3x-1}$

10 $\dfrac{x+2}{x+1}-\dfrac{x+1}{x+2}$

11 $\dfrac{x}{x^2-3x-4}-\dfrac{1}{x-4}$

12 $\dfrac{3}{x-2}-\dfrac{6}{x^2-4}$

For Questions 13–20, solve the equation.

13 $\dfrac{x+3}{2}+\dfrac{x+2}{3}=\dfrac{4}{3}$

14 $\dfrac{4x+1}{3}-\dfrac{3x-1}{2}=0$

15 $\dfrac{2x+1}{7}-x=\dfrac{3x-1}{8}$

16 $\dfrac{3x-1}{7}=\dfrac{2x+1}{11}+1$

17 $\dfrac{x}{x-1}+\dfrac{1}{x}=\dfrac{5}{2}$

18 $\dfrac{2x-1}{x+3}=\dfrac{x}{2x+2}$

19 $\dfrac{x+1}{x-3}-\dfrac{1}{x}=2$

20 $\dfrac{2x+3}{x-5}=\dfrac{x-4}{x-3}$

Differentiation

The gradient of a curve

The gradient (or the slope, or direction) of a curve is constantly changing.

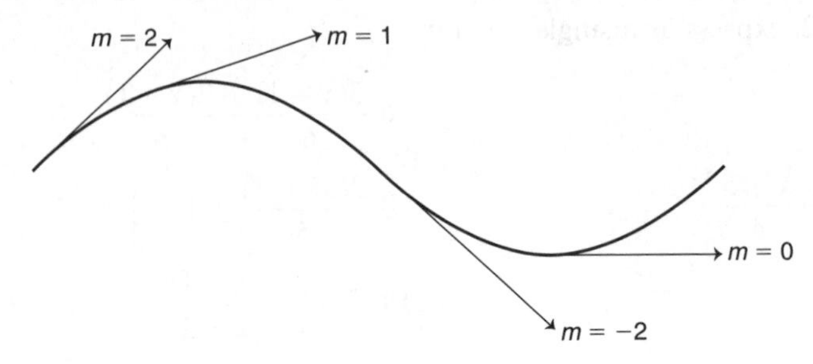

The gradient of the curve at any point is equal to the gradient of the tangent to the curve at that point.

Constructing the tangent

A tangent can be drawn 'by eye' by sliding a ruler up against the curve. However, this will only give an approximate result. The following process produces the exact tangent and thus the correct gradient.

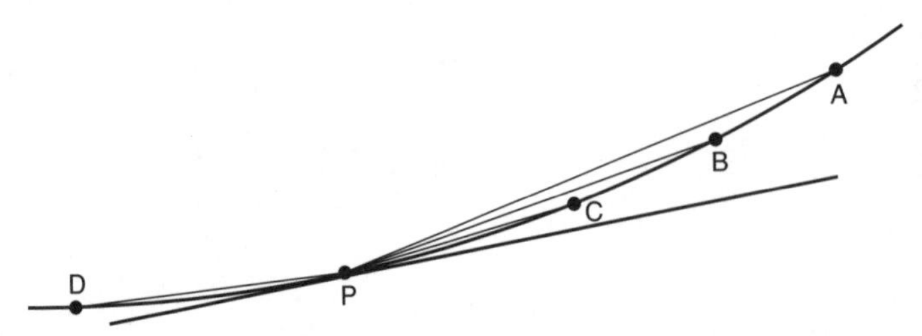

Place your ruler along line PA. Rotate the ruler around the point P, moving through positions PB, PC and PD. Your ruler will pass through the *tangent* at P, between the *chords* PC and PD.

gradient PA > gradient PB > gradient PC > tangent at P > gradient PD

This process can be developed into one that enables the gradient to be *precisely calculated*, without the need to draw the tangent.

The mathematicians Newton and Leibnitz first investigated this process in the seventeenth century. Newton called this branch of mathematics fluxions; nowadays, it is called calculus.

Calculus involves considering very small values (or increments) in x and y: δx and δy. δx is pronounced 'delta x' and means a very small distance (increment) along the x-axis. It does not mean δ multiplied by x. δx must be considered as one symbol.

This process of calculating a gradient (or a rate of change) is shown in Example 1.

Example 1

Consider the graph $y = x^2$.

The gradient of the curve at point A is equal to the tangent to the curve at this point.

A good 'first attempt' in finding this gradient is to consider a point B up the curve from A. The gradient of chord AB will be close to the required gradient if the steps along the x and y axes (δx and δy respectively) are small.

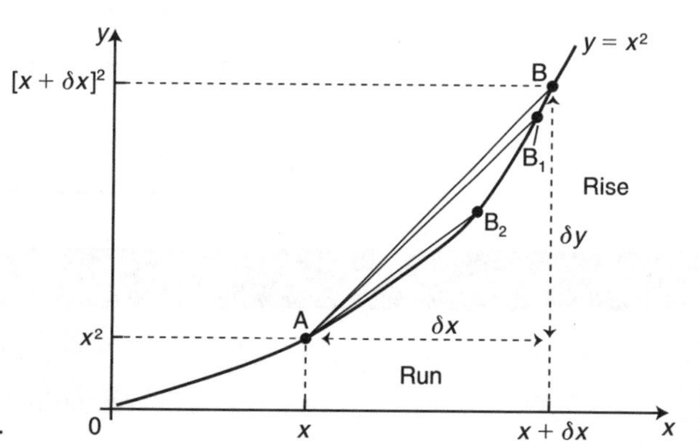

Let the *exact* gradient of $y = x^2$ at A be m. An estimate of m is found from:

$$\frac{\delta y}{\delta x} = \frac{\text{Rise}}{\text{Run}} = \frac{[x + \delta x]^2 - x^2}{\delta x} \qquad [\text{Expand } (x + \delta x)^2]$$

$$= \frac{[x^2 + 2x\delta x + (\delta x)^2] - x^2}{\delta x}$$

$$= \frac{2x\delta x + (\delta x)^2}{\delta x}$$

$$= \frac{\delta x(2x + \delta x)}{\delta x} = 2x + \delta x$$

So the estimate of m is $\frac{\delta y}{\delta x} = 2x + \delta x$

This estimate of m improves as point B slides down the curve to B_1, B_2, etc., closer and closer to A, resulting in δx and δy becoming smaller and smaller until, *at* point A, $\delta x = 0$.

What happens to $\frac{\delta y}{\delta x}$ as δx approaches zero?

Clearly, as δx gets smaller, $2x + \delta x$ approaches $2x$, and eventually, *at* A, $m = 2x$.

This is a beautifully simple result implying that the gradient of the curve $y = x^2$ at *any* point x is given by $2x$. So, at $x = 10$, the gradient of $y = x^2$ is 20, and so on.

Differentiating x^n with respect to x

This process of finding the gradient is called **differentiation** and can be applied to any power of x.
In all cases the gradient of x^n is nx^{n-1}.
The result is called the **derivative** or the gradient function.

The derivative is written $\frac{dy}{dx}$

curve	$\frac{dy}{dx}$	Gradient of curve at $x = 10$
$y = x^2$	$2x$	20
$y = x^3$	$3x^2$	300
$y = x^4$	$4x^3$	4 000
$y = x^5$	$5x^4$	50 000

Key Point

$$y = x^n$$

Differentiating:

$$\frac{dy}{dx} = nx^{n-1}$$

This result is true for all values of the index n.

Example 2

Find $\frac{dy}{dx}$ when $y = 5x$.

In index form:

$$y = 5x^1$$

Differentiating:

$$\frac{dy}{dx} = 5 \times 1x^0 = 5$$

Note that $y = 5x$ is a straight line with gradient $= 5$.

Example 3

Find $\frac{dy}{dx}$ when $y = 5x + 2$.

In index form:

$$y = 5x^1 + 2x^0$$

Differentiating:

$$\frac{dy}{dx} = 5 \times 1x^0 + 2 \times 0x^{-1}$$

$$\frac{dy}{dx} = 5 + 0 = 5$$

The '+ 2' lifts the line up 2 units. It does not change the gradient.

Example 4

Find $\frac{dy}{dx}$ when $y = 4x^6$.

Differentiating:

$$\frac{dy}{dx} = 4 \times 6x^5 = 24x^5$$

Example 5

Differentiate $y = 2x^2 + 3x^{-1}$.

Differentiating:

$$\frac{dy}{dx} = 2 \times 2x^1 + 3 \times -1x^{-2}$$

$$= 4x - 3x^{-2}$$

Exercise 81

In Questions 1–12, differentiate using the correct notation.

1 $y = x^3$ **2** $y = x^4$ **3** $y = x^5$

4 $y = x^6$ **5** $y = x^8$ **6** $y = x^7$

7 $y = x^{10}$ **8** $y = x^{11}$ **9** $y = x^{-1}$

10 $y = x^{-2}$ **11** $y = \frac{1}{x}$ **12** $y = \frac{1}{x^2}$

In Questions 13–20, find the gradient of the curve at the point where $x = 2$.

13 $y = x^2$ **14** $y = x^3$

15 $y = x^4$ **16** $y = x^5$

17 $y = x^7$ **18** $y = x^6$

Exercise 81★

In Questions 1–12, differentiate using the correct notation.

1 $y = x^{12}$ **2** $y = x^9$ **3** $y = x^1$

4 $y = x$ **5** $y = x^{-3}$ **6** $y = x^{-4}$

7 $y = x^{-6}$ **8** $y = x^{-5}$ **9** $y = x^{\frac{1}{2}}$

10 $y = x^{\frac{1}{3}}$ **11** $y = \sqrt{x}$ **12** $y = 1$

In Questions 13–20, find the gradient of the curve at the given value of x.

13 $y = x^{-3}$ at $x = 2$ **14** $y = x^{-4}$ at $x = 2$

15 $y = \frac{1}{x}$ at $x = 2$ **16** $y = \frac{1}{x^2}$ at $x = 2$

17 $y = x^{\frac{1}{2}}$ at $x = 4$ **18** $y = \sqrt[3]{x}$ at $x = 8$

19 $y = x^0$ at $x = 3$ **20** $y = x$ at $x = 2.5$

The rule for differentiation can also be extended when x^n terms are multiplied by a number and added together.

Key Points

$y = ax \equiv ax^1$

Differentiating:

$$\frac{dy}{dx} = ax^0 \equiv a$$

$y = c \equiv cx^0$

Differentiating:

$$\frac{dy}{dx} = 0$$

Exercise 82

Using the correct notation, differentiate the following.

1 $y = 4x^2$

2 $y = 3x^1$

3 $y = 2x^3$

4 $y = 5x^4$

5 $y = 3x^5$

6 $y = 4x^6$

7 $y = 3x^{-2}$

8 $y = 2x^{-3}$

9 $y = -2x^{-1}$

10 $y = -x^{-2}$

11 $y = x^5 + x^2$

12 $y = x^4 + x^3$

13 $y = 2x^3 + 4x$

14 $y = 5x^4 - 3x$

15 $y = 3x^3 + 2x^4$

16 $y = 5x^2 - x^3$

17 $y = 2x^2 + 3$

18 $y = 6x^3 - 5$

19 $y = 4 + x^{-1}$

20 $y = 2 + x^{-2}$

21 $y = 6 + 2x^{-3}$

22 $y = 1 + 5x^{-4}$

23 $y = x^2 + x^{-2}$

24 $y = x^3 + x^{-1}$

25 $y = x^3 - x^{-3}$

26 $y = x - x^{-1}$

Key Points

Before differentiating:

- multiply or divide composite terms to give individual terms
- express all algebraic fractions in index form.

Example 6

Find $\frac{dy}{dx}$ when $y = (2x + 3)^2$.

Multiply through:

$$y = 4x^2 + 12x + 9$$

Differentiating:

$$\frac{dy}{dx} = 8x + 12$$

Example 7

Find $\frac{dy}{dx}$ when $y = \frac{2}{x^3}$.

Express in index form:

$$y = 2x^{-3}$$

Differentiating:

$$\frac{dy}{dx} = -6x^{-4} \equiv -\frac{6}{x^4}$$

Exercise 82★

For Questions 1–20, differentiate using the correct notation.

1 $y = x^2(x + 2)$

2 $y = x(x^2 + 2)$

3 $y = (x + 1)(x + 3)$

4 $y = (x + 2)(x + 4)$

5 $y = (x - 3)(x + 2)$

6 $y = (x - 1)(x + 5)$

7 $y = (2x + 1)(x + 3)$

8 $y = (x + 2)(3x + 4)$

9 $y = (x - 6)(x + 6)$

10 $y = (x - 4)(x + 4)$

11 $y = (x + 3)^2$

12 $y = (x + 5)^2$

13 $y = (2x - 1)^2$

14 $y = (5 - 3x)^2$

15 $y = \left(2x - \frac{1}{x}\right)^2$

16 $y = \left(\frac{1}{x} + x\right)^2$

17 $y = \frac{2}{x^3}$

18 $y = \frac{1}{2x^3}$

19 $y = 2x^4 - \frac{1}{x^2}$

20 $y = 3x - \frac{1}{x}$

21 Find the gradient of the curve $y = 2x^2 - 3x$ at the point where $x = 2$.

22 Find the gradient of the curve $y = x^4 + 3x$ at the point where $x = 1$.

23 Find the co-ordinates of the point on the curve $y = x^2 - 6x$ where the gradient is zero.

24 Find the co-ordinates of the points on the curve $y = 4x + x^{-1}$ where the gradient is zero.

25 Find the co-ordinates of the point on the curve $y = 10 - 6x + x^2$ where $\frac{dy}{dx} = 0$. Sketch the curve and mark this point on your diagram.

26 The diagram shows a sketch of the curve $y = 4x + \frac{1}{x}$. Find the co-ordinates of points A and B.

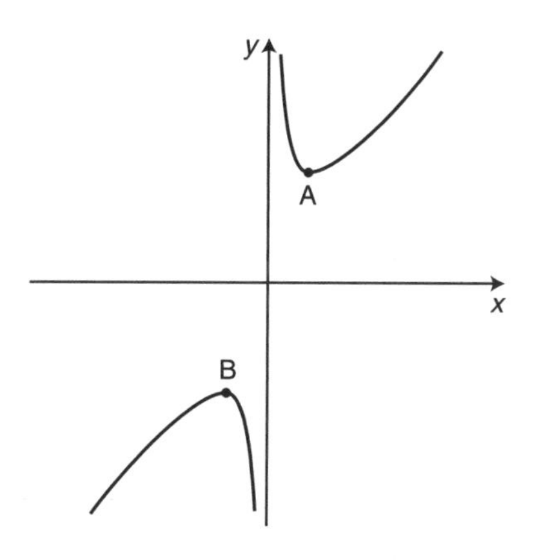

Tangents

For a given value of x:

- The equation gives the y value.
- $\frac{dy}{dx}$ gives the gradient of the tangent.
- Apply the technique for calculating the equation of a straight line.

Example 8

Calculate the equation of the tangent to the curve $y = x^3 - 2x^2 + 5$ at the point where $x = 2$.

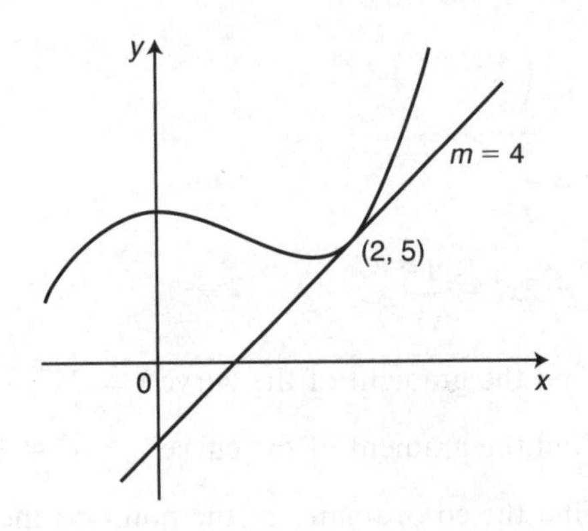

The y co-ordinate:

When $x = 2$, $y = 8 - 8 + 5 = 5$.

The gradient:

$$\frac{dy}{dx} = 3x^2 - 4x$$

And, when $x = 2$, $\frac{dy}{dx} = 12 - 8 = 4$.

$\therefore$ gradient of tangent $= 4$

The equation:

$$\frac{y-5}{x-2} = 4 \Rightarrow y = 4x - 3$$

Turning points

Turning points are points on the curve where the gradient is zero, i.e. where $\frac{dy}{dx} = 0$.

Maximum point

A turning point.

Gradient $= \frac{dy}{dx} = 0$

Gradient is **decreasing** from +ve, through zero, to −ve.

Gradient is *positive* just before and *negative* just after.

Minimum point

A turning point.

Gradient $= \frac{dy}{dx} = 0$

Gradient is **increasing** from −ve, through zero, to +ve.

Gradient is *negative* just before and *positive* just after.

Key Points

At a turning point, $\frac{dy}{dx} = 0$.

A turning point can be classified as a **maximum** or **minimum** by:

Knowing the shape of a familiar curve.

Or, by looking at the gradient immediately before and immediately after the stationary point.

Example 9

Find and classify the turning points on the curve $y = x^3 + 3x^2 - 9x - 7$.

$$\frac{dy}{dx} = 3x^2 + 6x - 9$$

$$\frac{dy}{dx} = 3(x^2 + 2x - 3)$$

$$\frac{dy}{dx} = 3(x - 1)(x + 3)$$

$$\therefore \frac{dy}{dx} = 0 \text{ when } x = 1 \text{ and } x = -3.$$

When $x = 1$, $y = 1 + 3 - 9 - 7 = -12$.
When $x = -3$, $y = -27 + 27 + 27 - 7 = 20$.
So there are turning points at $(1, -12)$ and $(-3, 20)$.

Consider the shape of the curve.

A 'positive cubic' is continuous and has this shape:

The maximum comes first and has the higher y co-ordinate.

Or, if you do not know the shape of the curve, look at the gradient 'before and after'.

At $(1, -12)$

x	0.9	1	1.1
$\frac{dy}{dx}$	−ve	0	+ve
	\	—	/

minimum

At $(-3, 20)$

−3.1	−3	−2.9
+ve	0	−ve
/	—	\

maximum

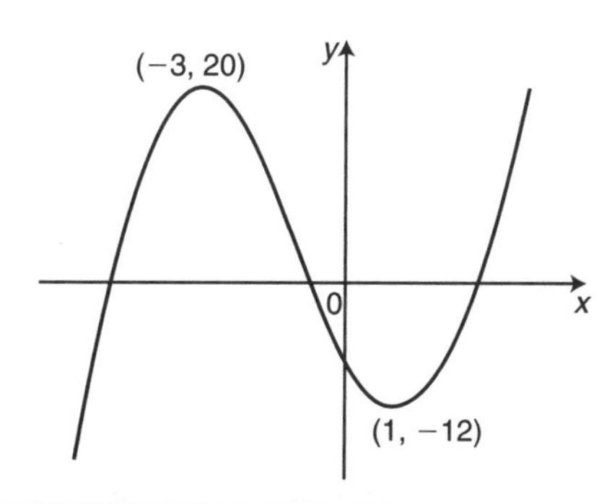

Exercise 83

1 For the curve $y = x^2 + 4x + 10$:

a Work out $\frac{dy}{dx}$.

b Solve the equation $\frac{dy}{dx} = 0$, and thus find the co-ordinates of the turning point.

2 For the curve $y = x^2 + 2x - 8$:

a Work out $\frac{dy}{dx}$.

b Solve the equation $\frac{dy}{dx} = 0$, and thus find the co-ordinates of the turning point.

3 For the curve $y = 15 + 6x - x^2$:

a Solve the equation $\frac{dy}{dx} = 0$.

b Thus find the maximum value of y.

4 For the curve $y = 6 - 10x - x^2$:

a Solve the equation $\frac{dy}{dx} = 0$.

b Thus find the maximum value of y.

5 For the curve $y = x^2 + 4x + 10$:

a Work out the value of y when $x = 1$.

b Work out the value of $\frac{dy}{dx}$ when $x = 1$.

c Thus find the equation of the tangent to the curve at the point where $x = 1$.

6 For the curve $y = x^2 + 2x - 8$:

a Work out the value of y when $x = 3$.

b Work out the value of $\frac{dy}{dx}$ when $x = 3$.

c Thus find the equation of the tangent to the curve at the point where $x = 3$.

7 For the curve $y = 15 + 6x - x^2$:

a Work out the value of y when $x = -2$.

b Work out the value of $\frac{dy}{dx}$ when $x = -2$.

c Thus find the equation of the tangent to the curve at the point where $x = -2$.

8 For the curve $y = 6 - 10x - x^2$:

a Work out the value of y when $x = -1$.

b Work out the value of $\frac{dy}{dx}$ when $x = -1$.

c Thus find the equation of the tangent to the curve at the point where $x = -1$.

9 An iron is being heated in a forge. The temperature, T, after t minutes is given by the formula $T = 4t^2 + 8t + 20$ for $0 \leqslant t \leqslant 10$.

a Work out $\dfrac{dT}{dt}$.

b Calculate the rate, in degrees per minute, at which the temperature is increasing:

(i) after one minute

(ii) after five minutes.

10 The number of registered voters, E, on the electoral roll of a regional district, t months after democracy was declared, is given by the formula $E = 100t^2 + 200t + 200$ for $0 \leqslant t \leqslant 6$.

a Work out $\dfrac{dE}{dt}$.

b Calculate the rate at which voters were registering:

(i) after one month

(ii) after four months.

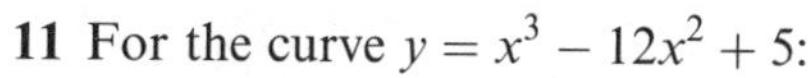

11 For the curve $y = x^3 - 12x^2 + 5$:

a Work out $\dfrac{dy}{dx}$.

b Solve the equation $\dfrac{dy}{dx} = 0$.

c Work out the co-ordinates of the two turning points.

d Determine whether each turning point is a maximum or minimum, and show your reasoning.

12 For the curve $y = x^3 - 6x^2 + 10$:

a Work out $\dfrac{dy}{dx}$.

b Solve the equation $\dfrac{dy}{dx} = 0$.

c Work out the co-ordinates of the two turning points.

d Determine whether each turning point is a maximum or minimum, and show your reasoning.

13 For the curve $y = 15 + 6x - x^2$:

a Find $\dfrac{dy}{dx}$.

b Find the co-ordinates of the point where $\dfrac{dy}{dx} = 4$.

c Thus find the equation of the tangent which is parallel to the line $y = 4x + 3$.

14 For the curve $y = 9 - 4x - x^2$:

a Find $\dfrac{dy}{dx}$.

b Find the co-ordinates of the point where $\dfrac{dy}{dx} = -6$.

c Thus find the equation of the tangent which is parallel to the line $6x + y = 0$.

15 For the curve $y = 11 + 6x - x^2$:

a Find $\frac{dy}{dx}$.

b Thus work out the maximum value of y.

16 For the curve $y = x^2 - 10x + 10$:

a Find $\frac{dy}{dx}$.

b Thus work out the minimum value of y.

17 For the curve $y = (4 + x)(2 - x)$:

a Find $\frac{dy}{dx}$.

b Thus work out the maximum value of y.

18 For the curve $y = (3 + x)(5 + x)$:

a Find $\frac{dy}{dx}$.

b Thus work out the minimum value of y.

19 The diagram shows a sketch of the graph of $y = 4x + \frac{1}{x}$.

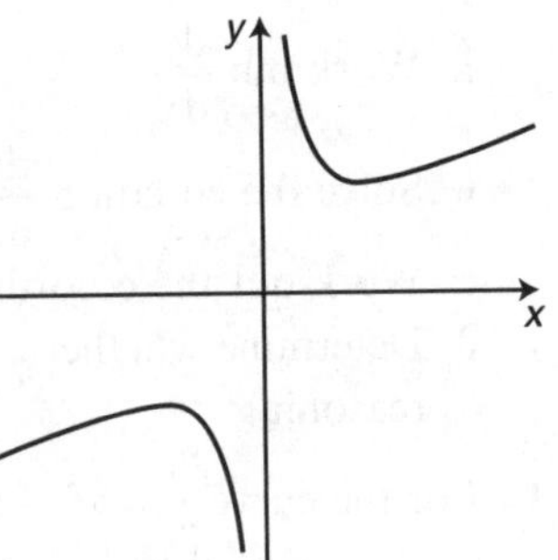

a Find $\frac{dy}{dx}$.

b Solve the equation $\frac{dy}{dx} = 0$.

c Calculate the co-ordinates of the turning points.

20 The diagram shows a sketch of the graph of $y = 1 - 9x - \frac{1}{x}$.

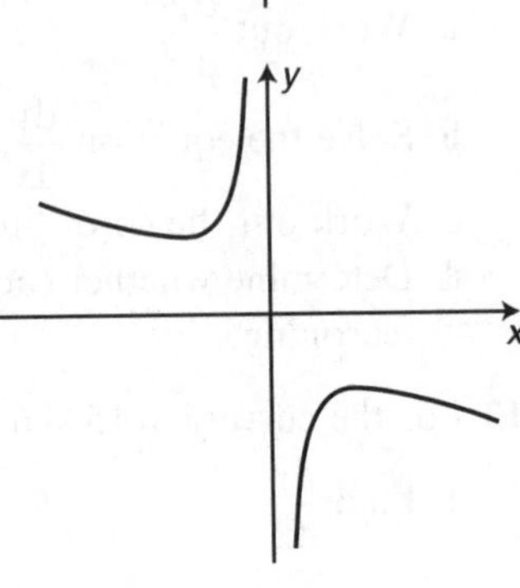

a Find $\frac{dy}{dx} = 0$.

b Solve the equation $\frac{dy}{dx} = 0$.

c Calculate the co-ordinates of the turning points.

UNIT 4 ♦ Graphs

Exercise 83★

1 For the curve $y = 3x^2 - 7x + 5$:

a Find $\frac{dy}{dx}$.

b Thus work out the equation of the tangent at the point (2, 3).

2 Find the equation of the tangent to the curve $y = x^2 - \frac{4}{x} + x$ at the point (−1, 4).

a Find $\frac{dy}{dx}$.

b Thus work out the equation of the tangent at the point (−1, 4).

3 For the curve $y = x^2 + 4x + 3$:

a Find the values of x for which $y = 0$.

b Find $\frac{dy}{dx}$.

c Find the gradients of the tangents at the points where the curve cuts the x-axis.

d Thus write down the equations of the tangents where the curve cuts the x-axis.

4 For the curve $y = 3 - 2x - x^2$:

a Find the values of x for which $y = 0$.

b Find $\frac{dy}{dx}$.

c Find the gradients of the tangents at the points where the curve cuts the x-axis.

d Thus write down the equations of the tangents where the curve cuts the x-axis.

5 Find the turning points of the function $y = 27x - x^3$ and determine their nature.

6 Find the turning points of the function $y = x^3 - 6x^2$ and determine their nature.

7 The population, P, of a new town is modelled by the formula $P = t^3 - t^2 + 25t + 10\,000$, where t is measured in years. t is set at 0 in the year 2000.
Use the formula to predict

a the population in the year 2010

b the rate of population growth in 2010.

8 The output of a company, \$$P$ in month t, is modelled by the formula $P = 800t^3 - 4000t^2 + 2000t + 480\,000$. January is set as $t = 1$.
Use the formula to predict

a the output in May

b the rate at which output is growing in May.

9 The temperature of a charcoal, T degrees, after t hours in the ashes of a fire is given by the formula $T = 270 + 80t - 20t^2$ for $0 \leqslant t \leqslant 6$.

a Work out $\frac{dT}{dt}$.

b Calculate the time at which the charcoal starts to cool down.

c Calculate the maximum temperature of the charcoal during the six hours.

d Calculate the rate, in degrees per hour, at which the charcoal is cooling after five hours.

10 The lead, L metres, of a runner in the last 75 minutes of a marathon is given by the formula $L = 1000 + 6t - \frac{t^2}{4}$, where t is the time in minutes.

a Work out $\frac{dL}{dt}$.

b Calculate the time at which the runner has the greatest lead.

c At what rate is the runner's lead being cut when $t = 60$ minutes?

11 For the curve $y = 2x^3 - x^2 - 4x + 10$:

a Work out $\frac{dy}{dx}$.

b Given that $\frac{dy}{dx} = 0$ when $x = 1$, find the other value of x for which $\frac{dy}{dx} = 0$.

c Work out the co-ordinates of the two turning points.

d Determine whether each turning point is a maximum or minimum, and show your reasoning.

12 For the curve $y = 2x^3 + 2x^2 - 16x - 12$:

a Work out $\frac{dy}{dx}$.

b Given that $\frac{dy}{dx} = 0$ when $x = -2$, find the other value of x for which $\frac{dy}{dx} = 0$.

c Work out the co-ordinates of the two turning points.

d Determine whether each turning point is a maximum or minimum, and show your reasoning.

13 The diagram shows a sketch of the graph of $y = x^2 + \frac{16}{x}$.

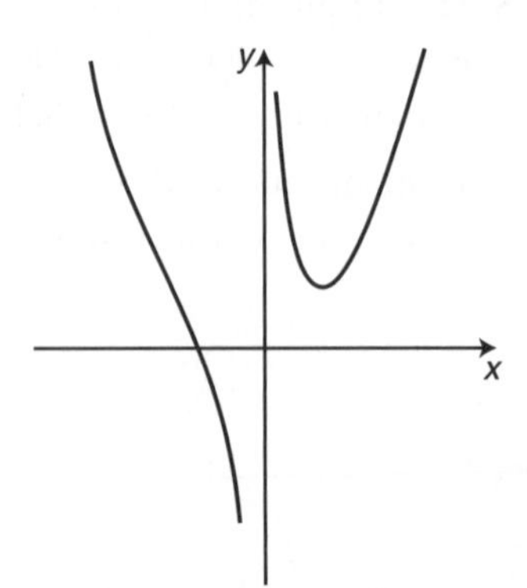

a Find $\frac{dy}{dx}$.

b Solve the equation $\frac{dy}{dx} = 0$.

c Calculate the co-ordinates of the turning point.

14 The diagram shows a sketch of the graph of

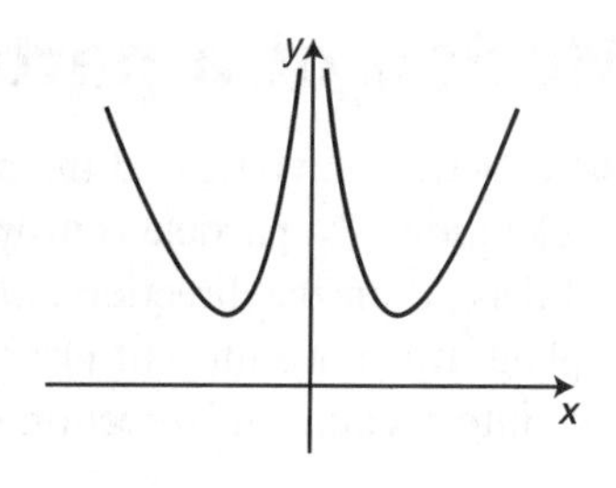

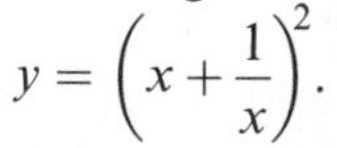

$y = \left(x + \frac{1}{x}\right)^2$.

a Expand y, and then find $\frac{dy}{dx}$.

b Solve the equation $\frac{dy}{dx} = 0$.

c Calculate the co-ordinates of the turning points.

15 Find the turning points on the curve $y = x^3 - x^2 - x + 10$. Thus show that the curve intersects with the x-axis only once.

16 Find the equations of the tangents to the curve $y = 4\left(x + \frac{1}{x}\right)$ which are parallel to the line $3x - y = 0$.

17 The temperature, $C°$, of the water in a lake, t months after it was constructed, is given by the formula $C = 3t + \frac{27}{t}$ for $1 \leqslant t \leqslant 6$.

a Work out $\frac{dC}{dt}$.

b Find the coldest temperature of the lake.

c Calculate the rate at which the temperature is warming after five months.

18 The population, P, of an endangered species was monitored over t years. P was modelled by the equation $P = 3t^2 + \frac{48}{t}$ for $1 \leqslant t \leqslant 5$.

a Work out $\frac{dP}{dt}$.

b Find the smallest population of the species.

c Calculate the rate at which the population is growing after four years.

19 The product of two positive numbers is 400. Find the least possible sum. Begin by letting one number $= x$.

20 An open-topped box is to be made out of a rectangular piece of card measuring 80 cm by 50 cm. Squares of side x cm are cut away from each corner of the card, which is then folded to form the box.

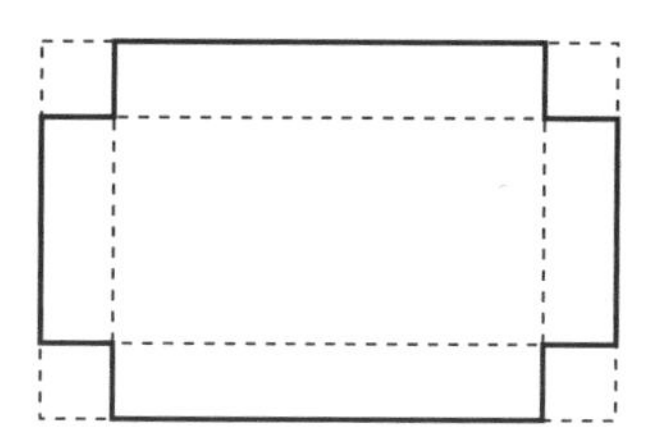

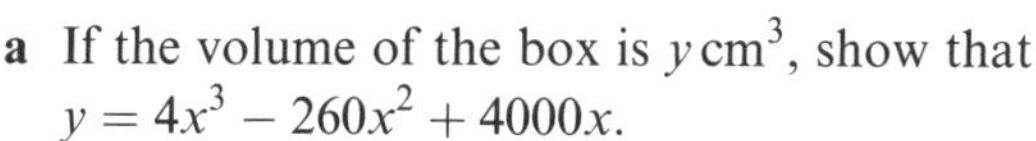

a If the volume of the box is y cm^3, show that $y = 4x^3 - 260x^2 + 4000x$.

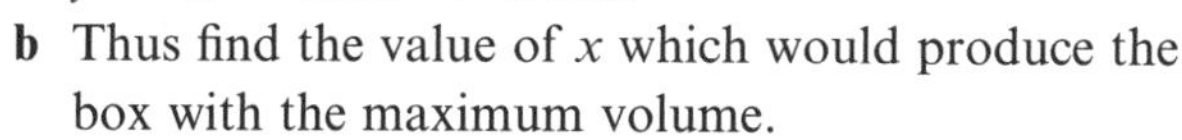

b Thus find the value of x which would produce the box with the maximum volume.

c State the dimensions and volume of the largest box.

Motion of a particle in a straight line

Isaac Newton investigated the motion of particles, and his second law states that the acceleration of a particle is proportional to the force applied to it. The behaviour of a particle will depend on the direction and magnitude of the forces and the time for which they are applied. It is not sufficient just to consider distance and speed, because these quantities do not take into account the direction of motion.

Key Points

s	**Displacement**	Distance; + or – indicates direction
v	**Velocity**	Speed; + or – indicates direction
a	**Acceleration**	+ or – indicates direction
t	**Time**	s, v and a are all expressed in terms of time, t

Velocity is the *rate* at which **displacement** changes over *time*. So, $v = \frac{ds}{dt}$.

Acceleration is the *rate* at which **velocity** changes over *time*. So, $a = \frac{dv}{dt}$.

$$\underset{s}{\text{displacement}} \xrightarrow{\textit{differentiate}} \underset{v = \frac{ds}{dt}}{\text{velocity}} \xrightarrow{\textit{differentiate}} \underset{a = \frac{dv}{dt}}{\text{acceleration}}$$

Example 10

The velocity of a ball, v m/s, after t seconds is given by $v = 8 + 10t - t^2$.

a Find the acceleration after t seconds.
b Work out when the acceleration is zero.
c Hence find the maximum velocity.

a $a = \frac{dv}{dt} = 10 - 2t$

b $0 = 10 - 2t$
$2t = 10$
$t = 5$ seconds

c The maximum velocity occurs when $a = 0$, when $t = 5$.
v (max) $= 8 + 50 - 25 = 33$ m/s

Example 11

A particle moves according to the formula $s = 3t^2 - t^3$, where s = distance from the starting point, O, at time t. Calculate:

a the time at which the particle momentarily comes to rest

b the time at which the particle returns to O

c the acceleration at time $t = 1$.

d Draw a velocity–time graph for $0 \leqslant t \leqslant 3$ and comment on the motion.

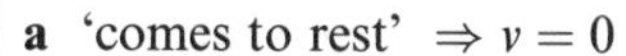

a 'comes to rest' $\Rightarrow v = 0$

$$v = \frac{\mathrm{d}s}{\mathrm{d}t} = 6t - 3t^2 = 3t(2 - t)$$

$v = 0$ when $t = 0, 2$

$\therefore\ t = 2$ or 0 seconds

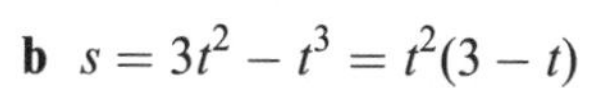

b $s = 3t^2 - t^3 = t^2(3 - t)$

$s = 0$ when $t = 0, 3$

$\therefore\ t = 3$ seconds

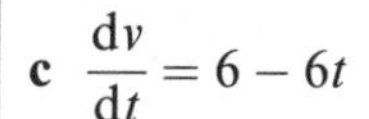

c $\frac{\mathrm{d}v}{\mathrm{d}t} = 6 - 6t$

$\therefore\ a = 0$ when $t = 1$

d

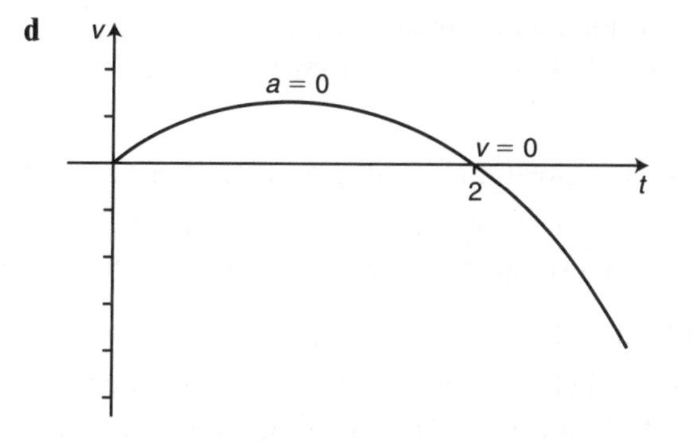

d The particle moves away for two seconds ($v \geqslant 0$) before changing direction and then arriving back at O ($s = 0$) after three seconds.

Exercise 84

1 The displacement, s metres, of a particle after t seconds is given by $s = 100 + 5t^2$.

Find an expression for the velocity, $v = \frac{\mathrm{d}s}{\mathrm{d}t}$.

2 The displacement, s metres, of a particle after t seconds is given by $s = 30 + 48t - 16t^2$.

Find an expression for the velocity, $v = \frac{\mathrm{d}s}{\mathrm{d}t}$.

3 The displacement, s metres, of a particle after t seconds is given by $s = 20 + 40t + 5t^2$.

a Find an expression for the velocity, v.

b Work out the velocity, in m/s, after 3 seconds.

4 The displacement, s metres, of a particle after t seconds is given by $s = 20 + 30t - 5t^2$.

a Find an expression for the velocity, v.

b Work out the velocity, in m/s, after 3 seconds.

5 The velocity, v m/s, of a particle after t seconds is given by $v = 32t + 100$.

Find an expression for the acceleration, $a = \frac{\mathrm{d}v}{\mathrm{d}t}$.

6 The velocity, v m/s, of a particle after t seconds is given by $v = 160 - 32t$.

Find an expression for the acceleration, $a = \frac{\mathrm{d}v}{\mathrm{d}t}$.

7 The displacement, s metres, of a particle after t seconds is given by $s = t^3 + 4t^2 - 5t + 2$.

a Find an expression for v.

b Find an expression for a.

c Work out, giving the correct units, the velocity and acceleration of the particle after one second.

8 The displacement, s metres, of a particle after t seconds is given by $s = t^3 - 2t^2 + 3t + 1$.

a Find an expression for v.

b Find an expression for a.

c Work out, giving the correct units, the velocity and acceleration of the particle after two seconds.

9 The velocity, v m/s, of a particle after t seconds is given by $v = t^2 + 10t + 5$.

a Find an expression for the acceleration, a.

b Work out the acceleration, in m/s, after 2 seconds.

10 The velocity, v m/s, of a particle after t seconds is given by $v = 24 + 6t - t^2$.

a Find an expression for the acceleration, a.

b Work out the acceleration, in m/s, after 2 seconds.

Exercise 84★

1 The displacement, s metres, of a particle after t seconds is given by $s = 4t^2 - \frac{2}{t} + 2$.

a Find an expression for $v = \frac{ds}{dt}$.

b Find an expression for $a = \frac{dv}{dt}$.

2 The displacement, s metres, of a particle after t seconds is given by $s = 6t + \frac{4}{t} + 10$.

a Find an expression for $v = \frac{ds}{dt}$.

b Find an expression for $a = \frac{dv}{dt}$.

3 The displacement, s metres, of a particle after t seconds is given by $s = 20t + 5t^2$.

a Find $\frac{ds}{dt}$.

b Calculate the velocity for $t = 1$, 2 and 3 seconds.

c Find the acceleration of the particle.

4 The displacement, s metres, of a particle after t seconds is given by $s = 4t^2 + 5t$.

a Find $\frac{ds}{dt}$.

b Calculate the velocity for $t = 1$, 2 and 3 seconds.

c Find the acceleration of the particle.

5 A ball is thrown vertically upwards with initial velocity 40 m/s. The height or displacement, s metres, after t seconds, is given by $s = 40t - 5t^2$.

a Find $\frac{ds}{dt}$.

b Find the velocity, v, when $t = 1$, 2, 3 and 4 seconds.

c Find the maximum height of the ball during the flight.

6 The displacement of a particle, s metres, from a point O at time t seconds is given by $s = 25t - t^2$, for $0 \leqslant t \leqslant 25$. Find

a $\frac{ds}{dt}$

b the value of t when $v = 0$

c the displacement of the particle when it is furthest from O.

7 A particle is moving in a straight line in a force field. Its distance, s metres, from point O after t seconds is given by $s = 40 - 2t - \frac{18}{t}$ for $1 \leqslant t \leqslant 20$.

a Find $\frac{ds}{dt}$ and $\frac{dv}{dt}$.

b Find the time at which the particle stops momentarily.

c Calculate the maximum distance from O during this period.

8 In a baseball game an outfielder moves in a straight line. His distance from the striker, s metres, t seconds after the strike is given by $s = 50 - \frac{t^2}{2} - \frac{8}{t}$ for $1 \leqslant t \leqslant 8$.

a Find $\frac{ds}{dt}$ and $\frac{dv}{dt}$.

b How fast is the outfielder running one second after the strike?

c Find the time at which the outfielder stops running.

d If the outfielder stops to field the ball, how far did the striker hit it?

9 A ball is hit vertically upwards from a cliff, which is 80 m above the sea. The displacement from the edge of the cliff, s, measured upwards in metres, after t seconds is given by $s = 30t - 5t^2$. Calculate:

a the time it takes for the ball to land in the sea

b the time it takes to reach the highest point

c the total distance travelled during the motion.

10 The displacement of a particle, s metres, from a fixed point O at time t seconds is given by $s = 4t^2 - 20t$ for $t \geqslant 0$. It starts at O.

a Calculate T, the time it takes to return to O.

b Calculate the distance it covers in the first T seconds.

Exercise 85 (Revision)

1 Find $\frac{dy}{dx}$ when $y = x^5 + 2x^3$.

2 Find $\frac{dy}{dx}$ when $y = x^3 + 2x^2 - x + 3$.

3 $y = x^2 - 3x + 4$

a Calculate the values of y and $\frac{dy}{dx}$ when $x = 3$.

b Thus calculate the equation of the tangent to the curve at the point where $x = 3$.

4 $y = x^{-2} - x^{-1}$. Calculate the value of $\frac{dy}{dx}$ when $x = 1$.

5 For the curve $y = x^3 + 3x^2 - 9x + 5$:

a Work out $\frac{dy}{dx}$.

b Solve the equation $\frac{dy}{dx} = 0$.

c Thus give the co-ordinates of the turning points and identify each as a maximum or a minimum.

6 The displacement, s metres, of a moving particle, P, at time t seconds is given by $s = 3t^3 - 4t + 5$ for $t \geqslant 0$.

a Work out $v = \frac{ds}{dt}$ and $a = \frac{dv}{dt}$.

b Find the time at which the particle stops.

Exercise 85★ (Revision)

1 Find $\frac{dy}{dx}$ when $y = 2x^4 - 3x^2 + 1$.

2 $y = (x^2 - 4)(x + 3)$. Calculate the value of $\frac{dy}{dx}$ when $x = 2$.

3 $y = \frac{1}{x^2} - \frac{1}{x}$. Calculate the value of the gradient when $x = 2$.

4 For the curve $y = 1 + 4x - x^2$:

a Work out $\frac{dy}{dx}$.

b Work out the equations of the tangents to the curve at the points (1, 4) and (4, 1).

c Find the point where the two tangents intersect.

5 $y = x(2x^2 + 3x - 36)$

a Work out $\frac{dy}{dx}$.

b Calculate the co-ordinates of the two turning points.

c Draw a sketch of the curve and label each turning point clearly.

6 The number of hairs, H, on the body of a mammal over a lifetime of 60 years is modelled by the equation $H = 80t^2 - \frac{4t^3}{3}$ for $0 \leqslant t \leqslant 60$, where t is the time, measured in years.

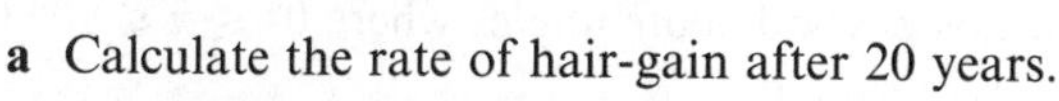

a Calculate the rate of hair-gain after 20 years.
b Calculate the rate of hair-loss after 50 years.
c Calculate the age at which the mammal has most hair.

7 The displacement, s metres, of a particle P at time t seconds is given by $s = (t^3 + 5t)\,\text{m}$.

a Find the expression for the velocity, $\frac{\mathrm{d}s}{\mathrm{d}t}$, and the acceleration, $\frac{\mathrm{d}v}{\mathrm{d}t}$, at time t.
b Find the distance travelled when $t = 2$ and when $t = 3$.
c Thus find the average velocity during the third second.

Trigonometric ratios for angles up to 180°

This section extends the $\sin x$, $\cos x$ and $\tan x$ ratios beyond acute angles where $0° \leqslant x \leqslant 90°$ to consider angles where $0° \leqslant x \leqslant 180°$.

Remember

- To work out the sine of 100°, press

$\sin 100° = 0.985$ (to 3 s.f.)

- To work out x in $\sin x = 0.766$, press

SHIFT sin 0.766 

$x = 50.0°$ (to 3 s.f.)

Activity 24

Sine ratio

This is the graph of $y = \sin x$, drawn for $0° \leqslant x \leqslant 180°$.

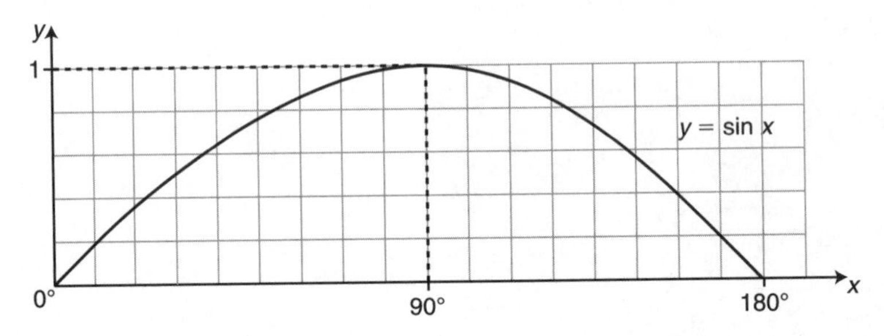

- Copy this table, and use the graph and your calculator to complete it.

x	36°	72°	108°	144°	180°
$\sin x$ (graph)					
$\sin x$ (calc. to 3 s.f.)					

- Copy and complete this table, for $0° \leqslant x \leqslant 90°$.

$\sin x$	0	0.2	0.5	0.8	1.0
x (graph)					
x (calc. to 3 s.f.)					

Cosine ratio

This is the graph of $y = \cos x$, drawn for $0° \leqslant x \leqslant 180°$.

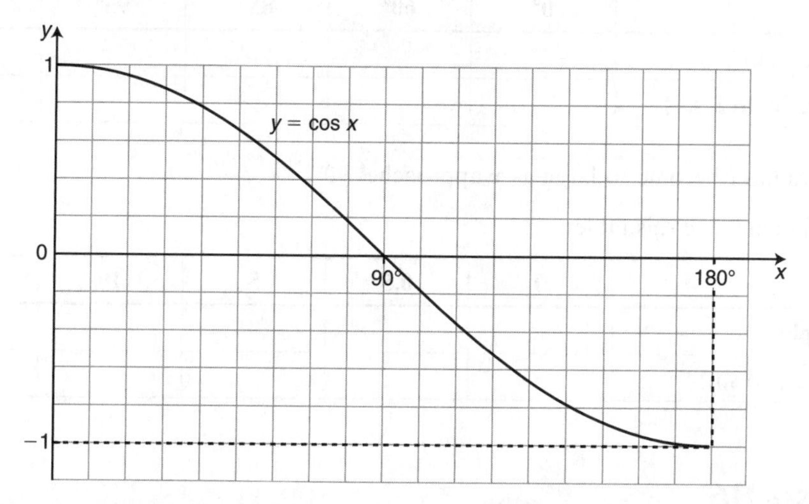

♦ Copy this table, and use the graph and your calculator to complete it.

x	36°	72°	108°	144°	180°
$\cos x$ (graph)					
$\cos x$ (calc. to 3 s.f.)					

♦ Copy and complete this table.

$\cos x$	0	0.2	0.5	0.8	–1.0
x (graph)					
x (calc. to 3 s.f.)					

Tangent ratio

This is the graph of $y = \tan x$, drawn for $0° \leqslant x \leqslant 180°$.

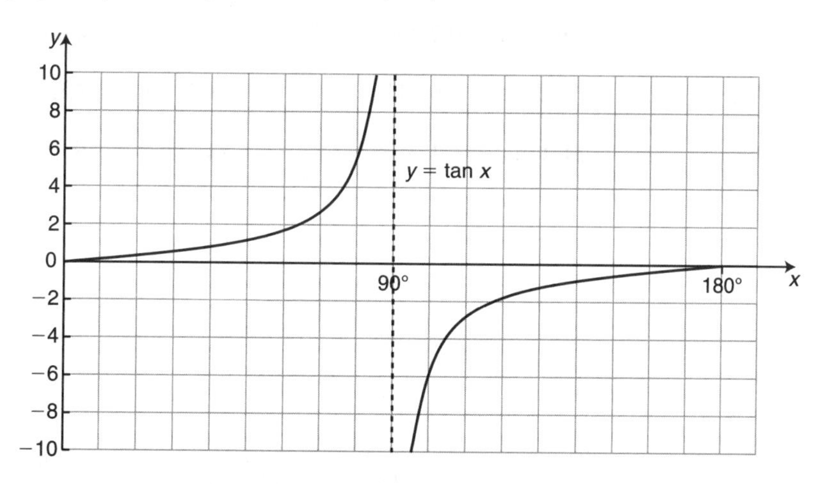

Unit 4 ♦ Shape and space

- Copy this table, and use the graph and your calculator to complete it.

x	0°	60°	85°	95°	180°
$\tan x$ (graph)					
$\tan x$ (calc. to 3 s.f.)					

- Why does $\tan x$ become so large as x approaches 90°?
- Copy and complete this table.

$\tan x$	0	0.5	5	10	−10
x (graph)					
x (calc. to 3 s.f.)					

Exercise 86

Use the accurately drawn graphs in Activity 24 to find *all* possible values of x, for $0° \leqslant x \leqslant 180°$. Use your calculator to check these values.

1 $\sin x = 0.3$
2 $\sin x = 0.7$
3 $\cos x = 0.3$
4 $\cos x = 0.6$
5 $\tan x = 3$
6 $\tan x = 2$
7 $\cos x = -0.3$
8 $\cos x = -0.6$
9 $\sin x = -0.3$
10 $\sin x = -0.7$
11 $\tan x = -3$
12 $\tan x = -2$

Exercise 86★

Use the accurately drawn graphs in Activity 24 to find all possible values of x, for $0° \leqslant x \leqslant 180°$. Use your calculator to check these values.

1 $\sin x = 0.4567$
2 $\sin x = 0.6543$
3 $\sin x = -0.4567$
4 $\sin x = -0.6543$
5 $\cos x = 0.2468$
6 $\cos x = 0.6428$
7 $\cos x = -0.3789$
8 $\cos x = -0.8416$
9 $\tan x = 2.458$
10 $\tan x = 0.6789$
11 $\tan x = -1.453$
12 $\tan x = -0.7342$

Sine rule

The **sine rule** is a method to calculate sides and angles for all triangles. It is useful for finding the length of one side when all angles and one other side are known, or finding an angle when two sides and the included angle are known.

Key Points

The sine rule

In triangle ABC, notice that side a is opposite angle A, side b is opposite angle B, etc.

- To find a side, use the sine rule written as

$$\frac{a}{\sin A} = \frac{b}{\sin B} = \frac{c}{\sin C}$$

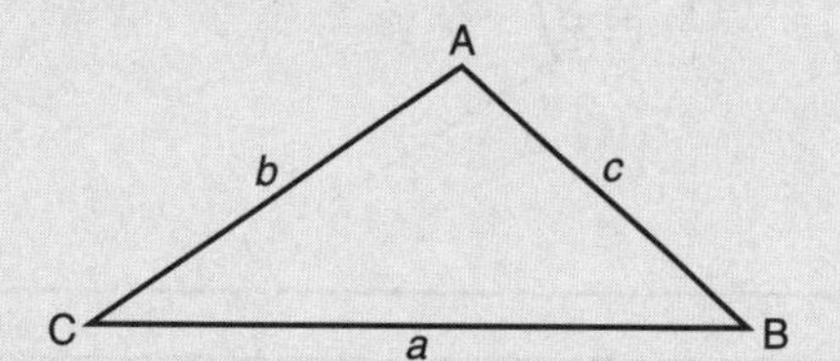

- To find an angle, use the sine rule written as

$$\frac{\sin A}{a} = \frac{\sin B}{b} = \frac{\sin C}{c}$$

Example 1

In triangle ABC, find the length of side a, correct to 3 significant figures.

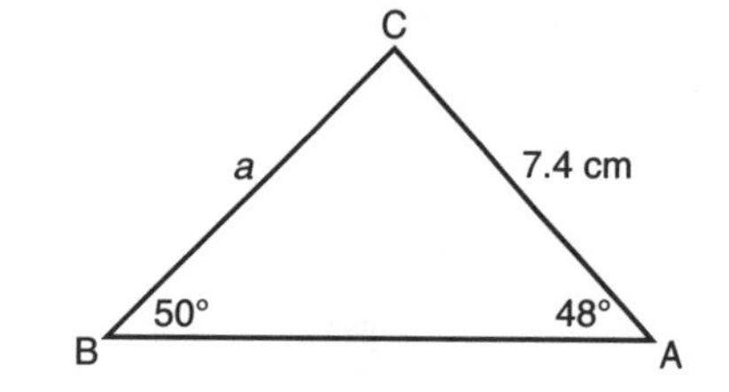

$$\frac{a}{\sin 48^\circ} = \frac{7.4\,\text{cm}}{\sin 50^\circ}$$

$$a = \frac{7.4\,\text{cm}}{\sin 50^\circ} \times \sin 48^\circ$$

$$a = 7.1787\ldots\ \text{cm}$$

$$a = 7.18\,\text{cm} \quad \text{(to 3 s.f.)}$$

Example 2

In triangle ABC, find angle B correct to 3 significant figures.

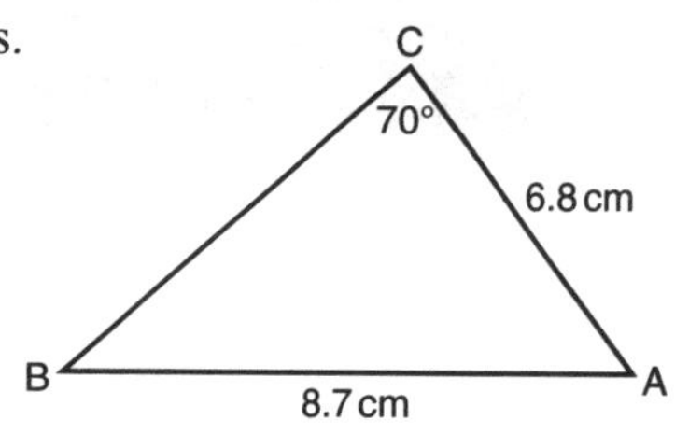

$$\frac{\sin B}{6.8} = \frac{\sin 70^\circ}{8.7}$$ (Multiply both sides by 6.8)

$$\sin B = \frac{\sin 70^\circ}{8.7} \times 6.8$$

$$B = 47.262\ldots^\circ$$

$$B = 47.3^\circ \quad \text{(to 3 s.f.)}$$

Remember

- Bearings are measured clockwise from North.
- When drawing bearings, start by drawing an arrow to indicate North.
- When calculating angles on a bearings diagram, look for 'alternate angles'.
- 

e is angle of elevation.
- 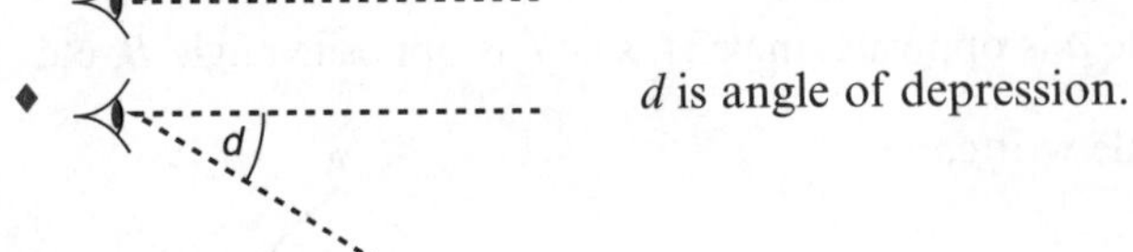

d is angle of depression.

N

Example 3

A yacht crosses the start line of a race at C, on a bearing of 026°. After 2.6 km, it rounds a buoy B and sails on a bearing of 335°. When it is due North of its start, at A, how far has it sailed altogether?

- Draw a diagram and include all the facts.

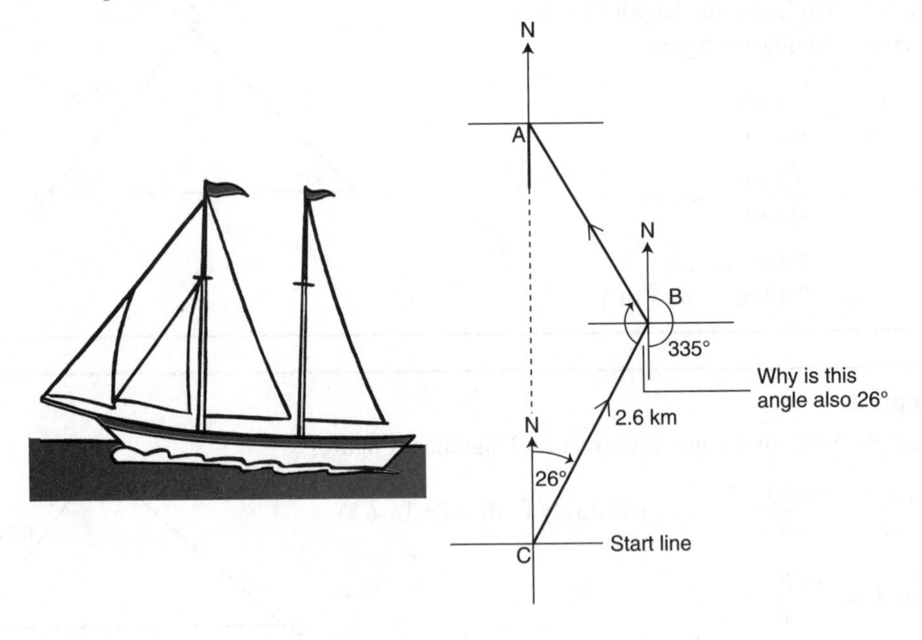

- Work out any necessary angles and redraw the triangle and include only the relevant facts.

  $\angle CBA = 335° - 26° - 180° = 129°$

  $\therefore \quad \angle BAC = 25°$

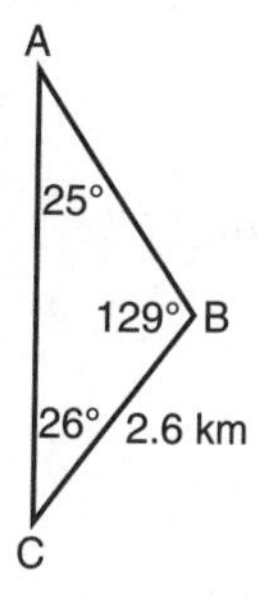

- Use the sine rule: in triangle ABC, the length AB has to be calculated.

  $$\frac{AB}{\sin 26°} = \frac{2.6\,\text{km}}{\sin 25°} \qquad \text{(Multiply both sides by } \sin 26°\text{)}$$

  $$AB = \frac{2.6\,\text{km}}{\sin 25°} \times \sin 26°$$

  $$AB = 2.6969\ldots\,\text{km}$$

$\therefore$ total distance travelled $= CB + BA$

$= 2.6\,\text{km} + 2.697\,\text{km} = 5.30\,\text{km}$ (to 3 s.f.)

Exercise 87

Write your answers correct to 3 significant figures.

1 Find x.

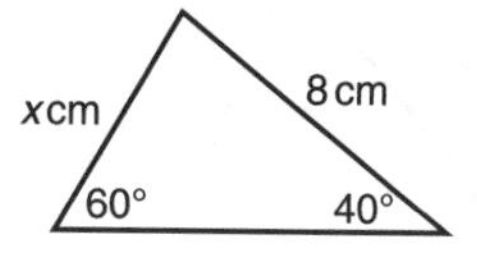

2 Find y.

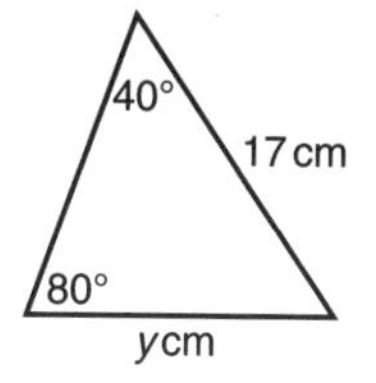

3 Find MN.

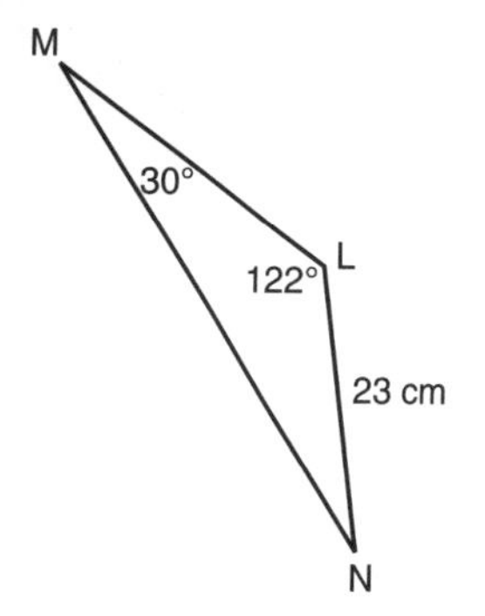

4 Find RT.

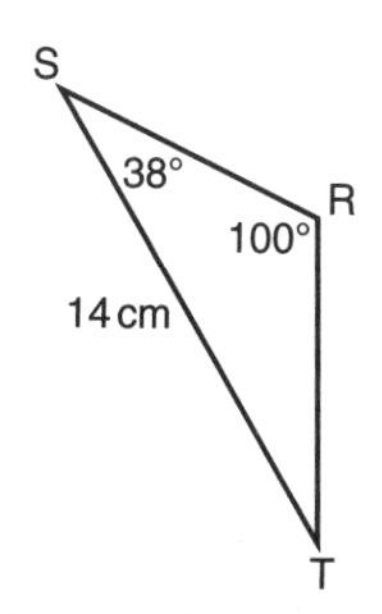

5 Find AC.

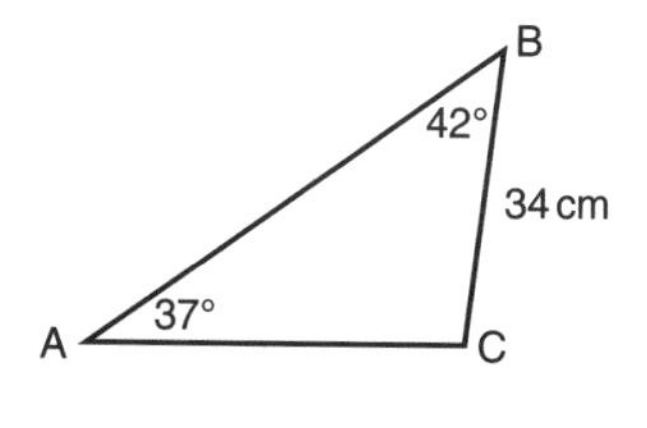

6 Find YZ.

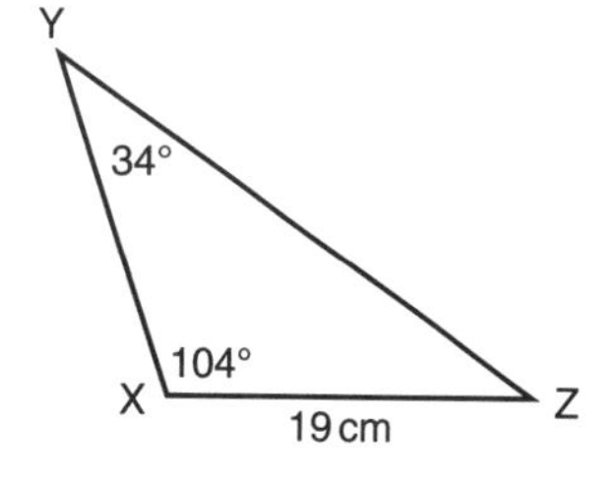

7 Find x.

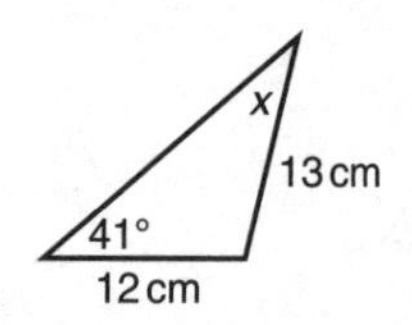

8 Find y.

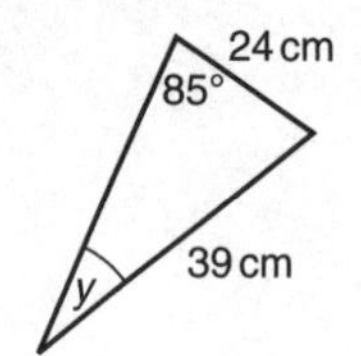

9 Find ∠ABC.

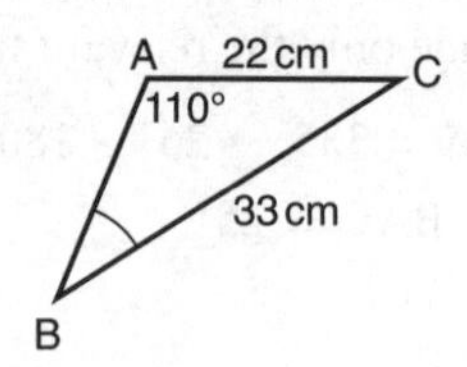

10 Find ∠XYZ.

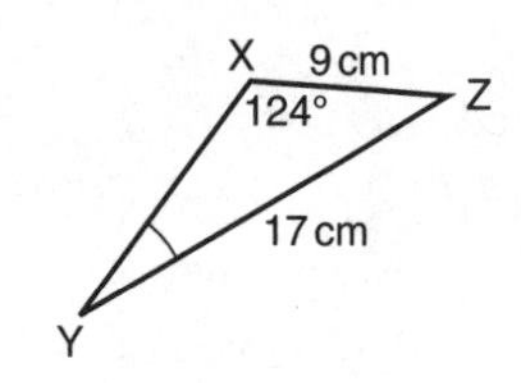

11 Find ∠ACB.

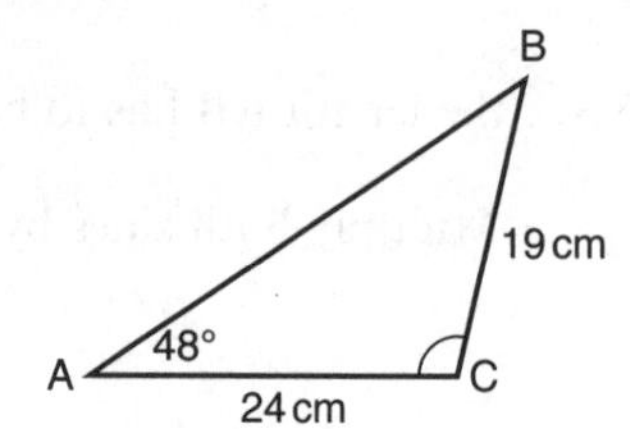

12 Find ∠DCE.

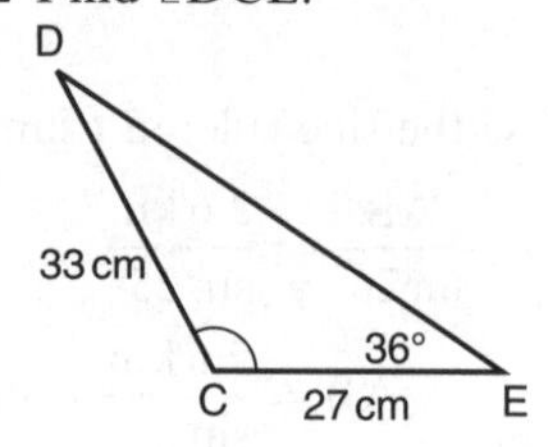

Activity 25

The ambiguous case

- Use a pair of compasses to construct triangle ABC, with AB = 7 cm, BC = 6 cm and ∠BAC = 50°.

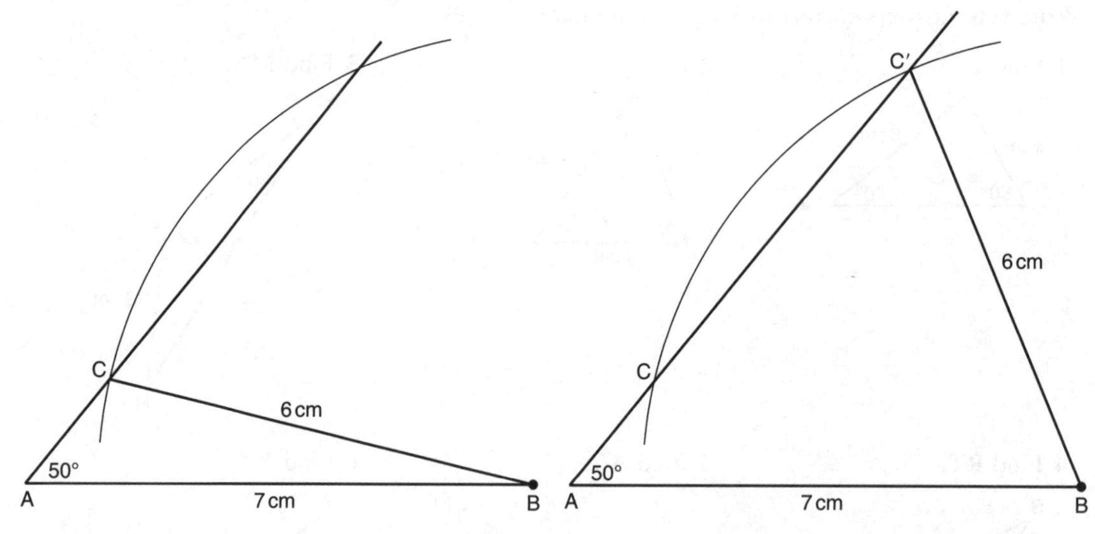

- Note that two triangles can be constructed: ABC and ABC′. The arc centred at B cuts the line from A at C *and* C′. This is an example of the 'ambiguous' case where two triangles can be constructed from the same facts.
- Show by calculation that AC ≈ 1.8 cm and AC′ ≈ 7.2 cm.

Exercise 87★

Write your answers correct to 3 significant figures.

1 Find x.

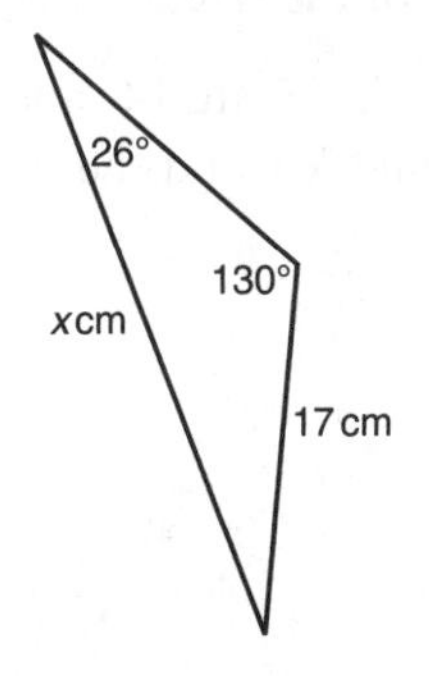

2 Find y.

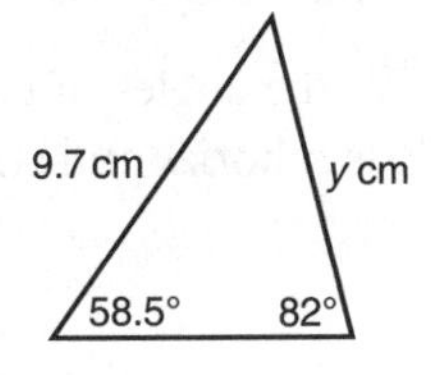

3 Find ∠LMN.

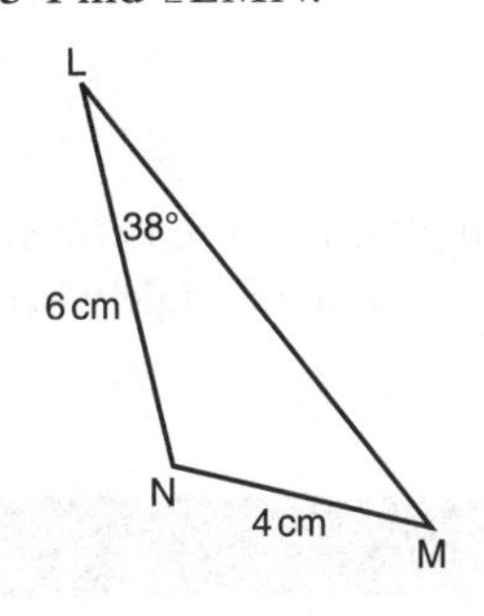

4 Find ∠RST.

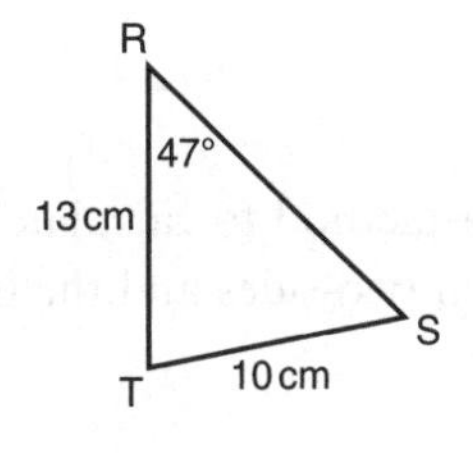

5 Find EF, ∠DEF and ∠FDE.

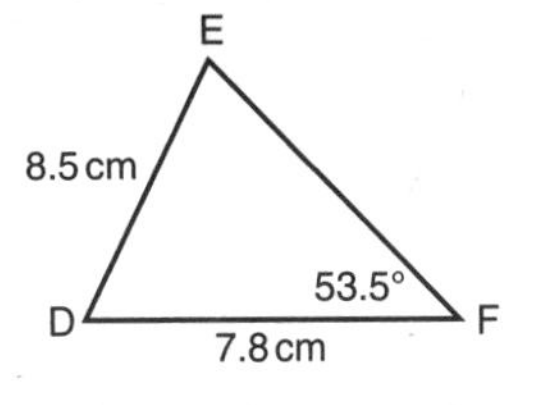

6 Find MN, ∠MLN and ∠LNM.

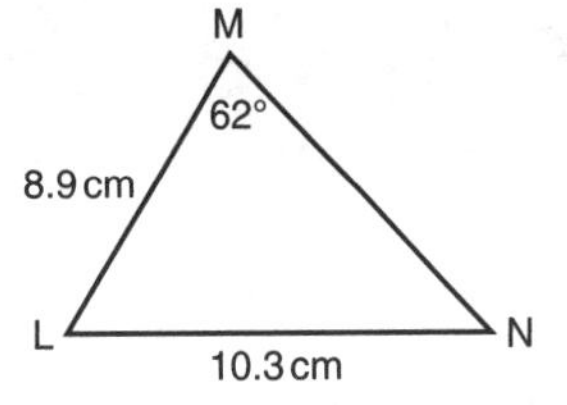

7 A yacht crosses the start line of a race on a bearing of 031°. After 4.3 km, it rounds a buoy and sails on a bearing of 346°. When it is due North of its start, how far has it sailed altogether?

8 A boat crosses the start line of a race on a bearing of 340°. After 700 m, it rounds a buoy and sails on a bearing of 038°. When it is due North of its start, how far has it sailed altogether?

9 The bearing of B from A is 065°. The bearing of C from B is 150°, and the bearing of A from C is 305°. If AC = 300 m, find BC.

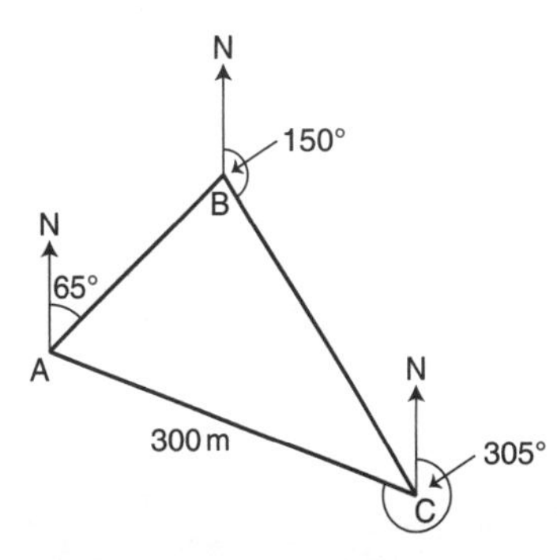

10 The bearing of Y from X is 205°. The bearing of Z from Y is 315°, and the bearing of X from Z is 085°. If XZ = 4 km, find the distance XY.

11 P and Q are points 80 m apart on the bank of a straight river. R is a point on the opposite bank where $\angle QPR = 76°$ and $\angle PQR = 65°$. Calculate PR and the width of the river.

12 From two points X and Y, the angles of elevation of the top T of a church TZ are 14° and 19°, respectively. If XYZ is a horizontal straight line and XY = 120 m, find YT and the height of the tower.

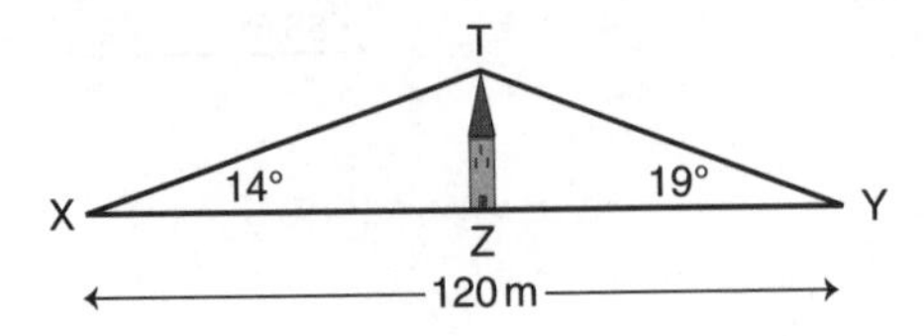

Cosine rule

The **cosine rule** is another method to calculate sides and angles of all triangles. It is used either to find the third side when two sides and the included angle are given, or to find an angle when all three sides are given.

Key Point

The cosine rule

In triangle ABC:

$$a^2 = b^2 + c^2 - (2bc \cos A)$$

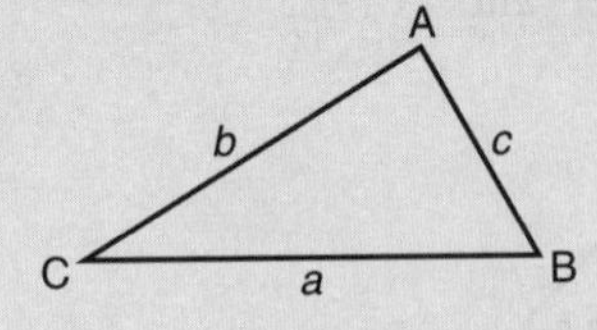

Key Points

In triangle ABC

- To find side a, use the cosine rule written as

$$a^2 = b^2 + c^2 - 2bc \cos A$$

- To find angle A, use the cosine rule written as

$$\cos A = \frac{b^2 + c^2 - a^2}{2bc}$$

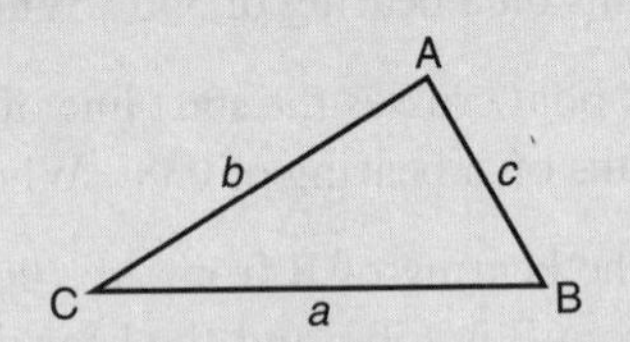

Example 4

In triangle ABC find a correct to 3 significant figures.

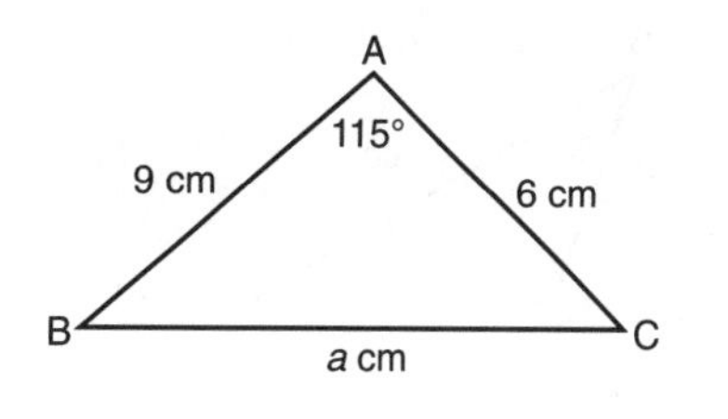

$$a^2 = b^2 + c^2 - 2bc\cos A$$
$$= 6^2 + 9^2 - 2 \times 6 \times 9 \times \cos 115°$$
$$= 36 + 81 - 108 \times (-0.4226)$$
$$= 36 + 81 + 45.64$$
$$a = \sqrt{162.6}$$
$$= 12.8 \quad \text{(to 3 s.f.)}$$

Example 5

In triangle ABC, find angle A correct to 3 significant figures.

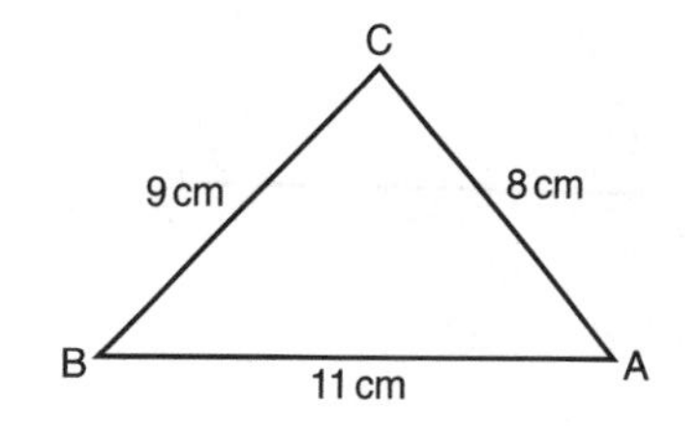

$$\cos A = \frac{b^2 + c^2 - a^2}{2bc}$$
$$= \frac{8^2 + 11^2 - 9^2}{2 \times 8 \times 11}$$
$$= \frac{104}{176}$$
$$A = 53.8° \quad \text{(to 3 s.f.)}$$

Exercise 88

Write your answers correct to 3 significant figures.

1 Find a.

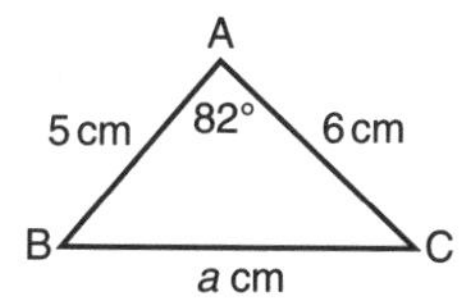

2 Find b.

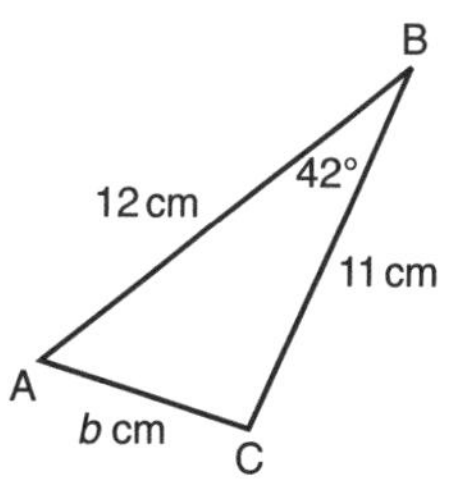

3 Find AB.

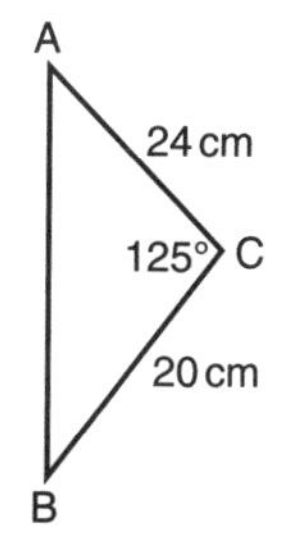

4 Find XY.

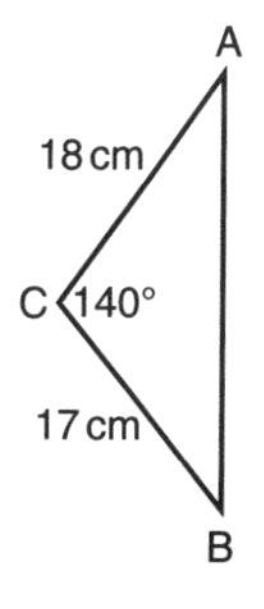

UNIT 4 ◆ Shape and space

5 Find RT.

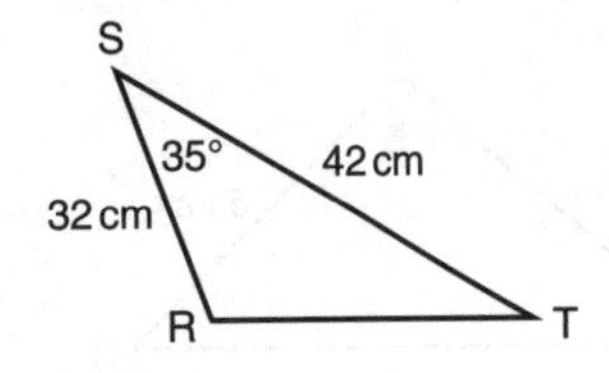

6 Find MN.

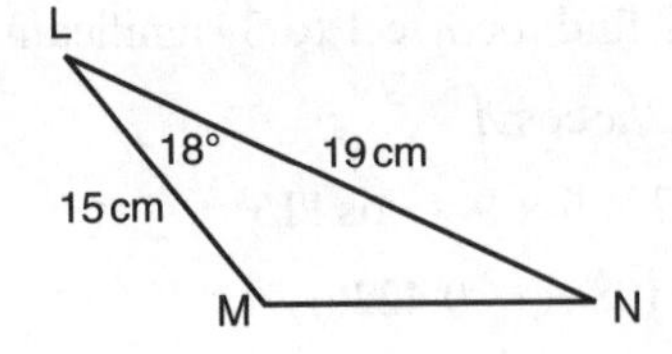

7 Find X.

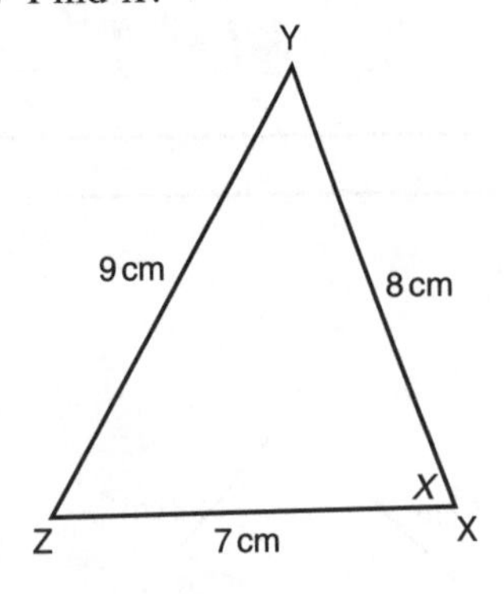

8 Find Y.

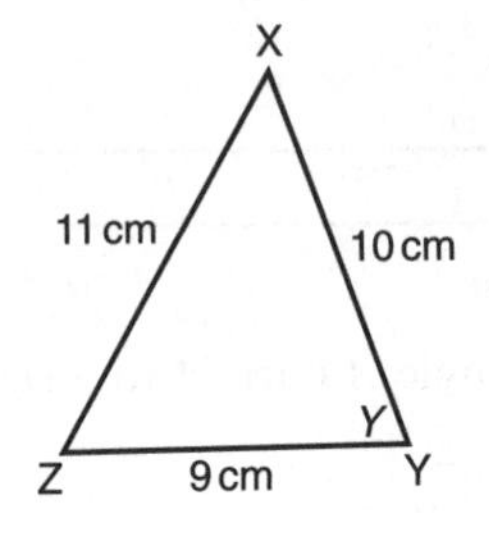

9 Find ∠ABC.

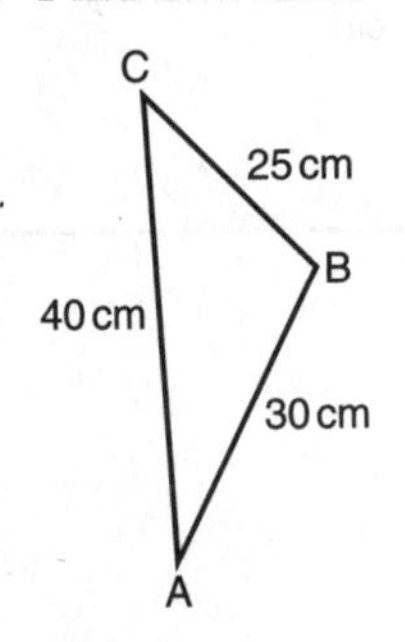

10 Find ∠XYZ.

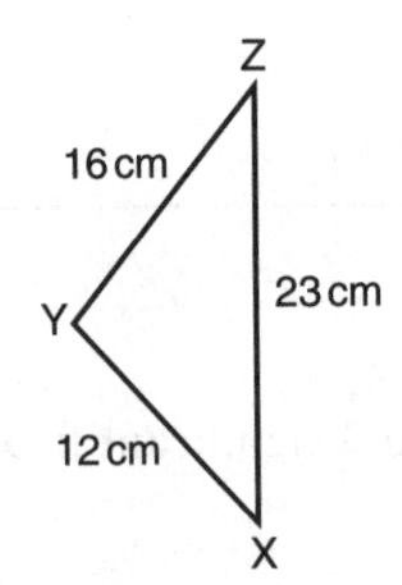

Exercise 88★

Write your answers correct to 3 significant figures.

1 Find x.

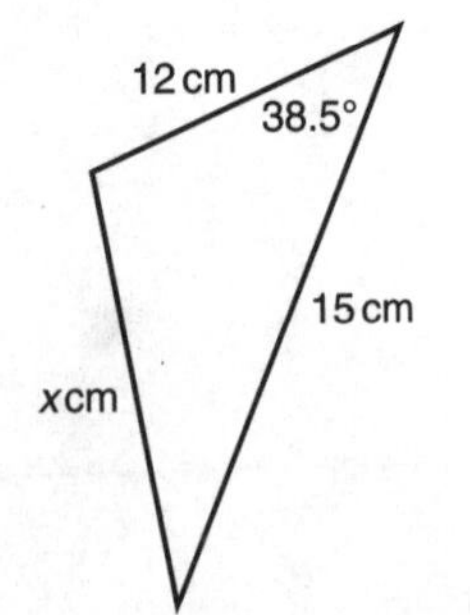

2 Find y.

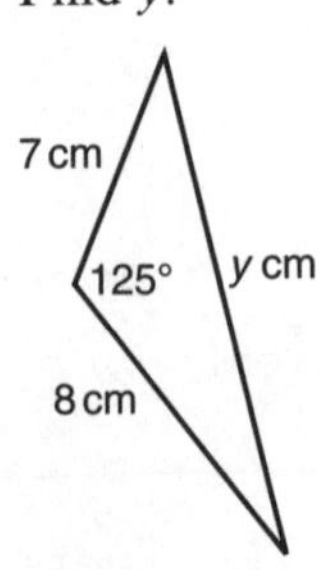

3 Find ∠XYZ.

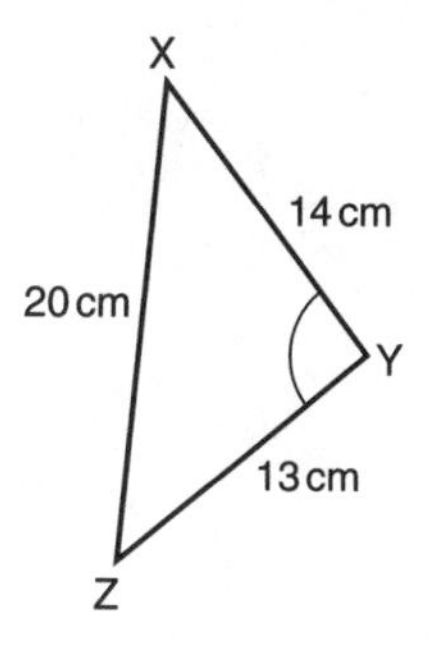

4 Find ∠ABC.

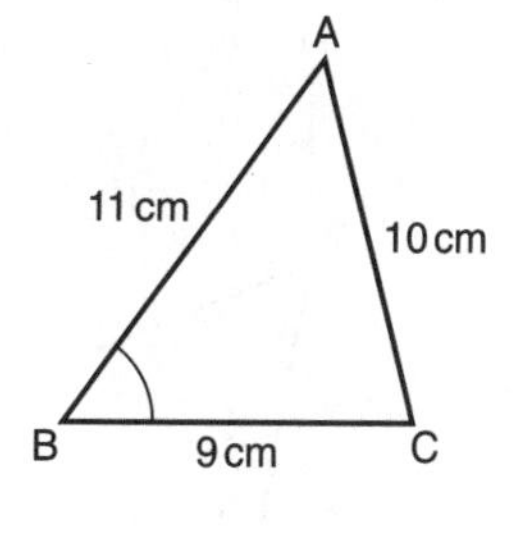

5 Find ∠BAC.

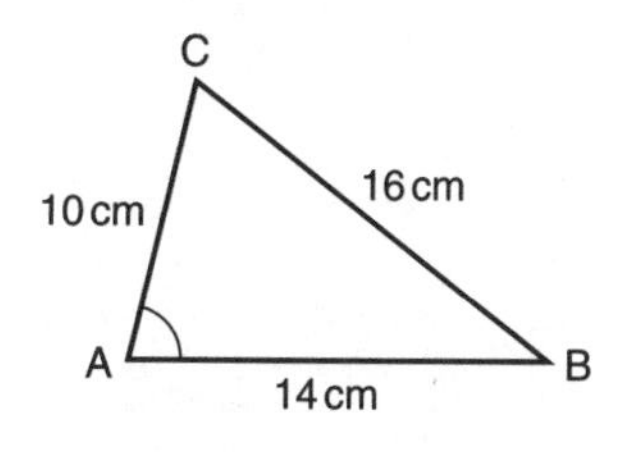

6 Find ∠RST.

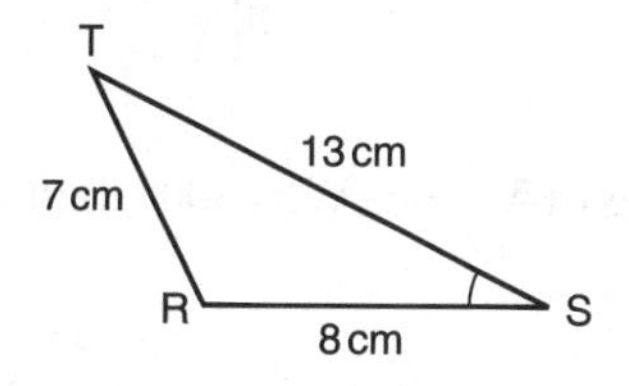

7 Find QR, ∠PQR and ∠QRP.

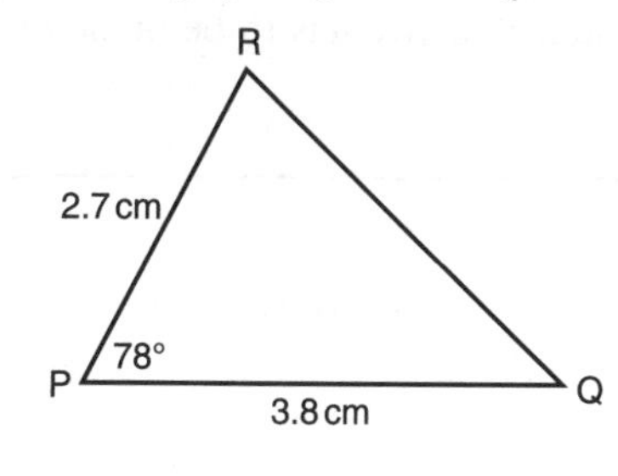

8 Find LM, ∠NLM and ∠LMN.

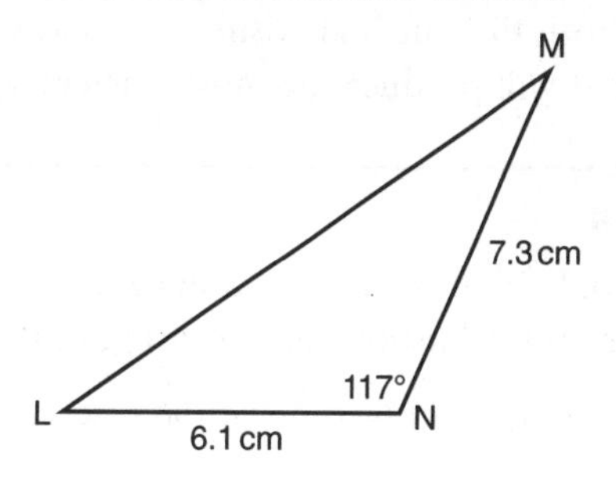

9 From S, a yacht sails on a bearing of 040°. After 3 km, at buoy A, it sails on a bearing of 140°. After another 4 km, at buoy B, it heads back to the start S. Calculate the total length of the course.

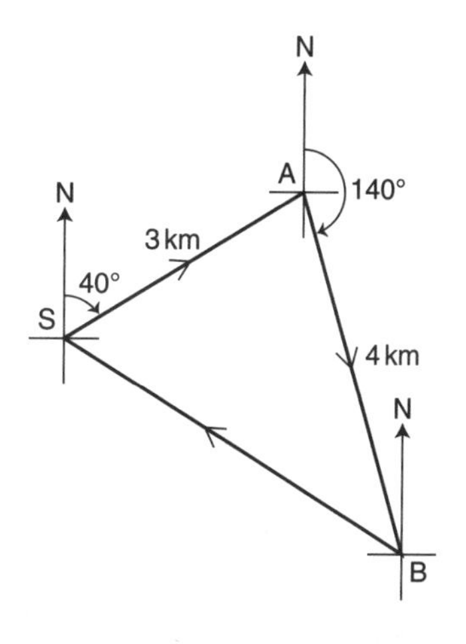

UNIT 4 ◆ Shape and space

10 Copy and label this diagram using the facts: VW = 50 km, WU = 45 km, VU = 30 km.

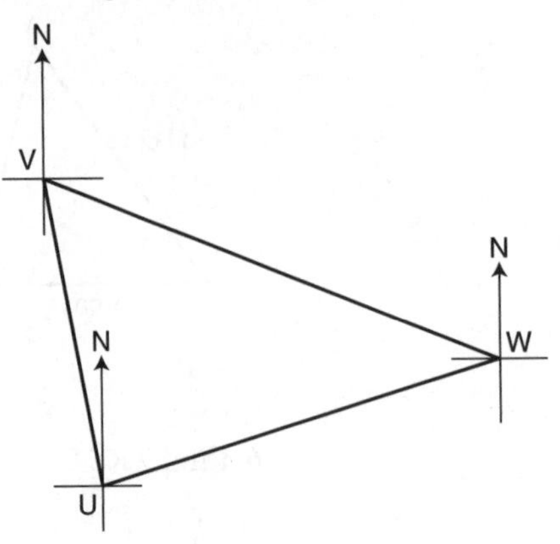

a Find ∠VWU.

b If the bearing of V from W is 300°, find the bearing of U from W.

Mixed questions

Sometimes both the sine and cosine rules have to be used or a careful choice has to be made of which method will produce the most efficient solution.

Example 6

A motorboat, M, is 8 km from a harbour, H, on a bearing of 125°, whilst a rowing boat, R, is 16 km from the harbour on a bearing of 074°.

Find the distance and bearing of the rowing boat from the motorboat.

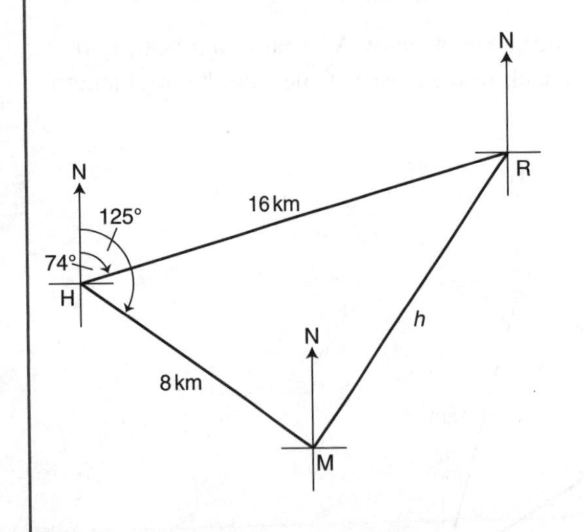

Cosine rule:

$\angle RHM = 125° - 74° = 51°$

$h^2 = 8^2 + 16^2 - 2 \times 8 \times 16 \times \cos 51°$

$= 158.89\ldots$

$\Rightarrow h = 12.6\,\text{km}$ (3 s.f.)

Sine rule:

$$\frac{\sin 51°}{12.6} = \frac{\sin M}{16}$$

$\sin M = 0.986\ldots$

$M = 80.7°$ (3 s.f.)

Bearing of R from M = 80.7° – 55°

= 025.7° (3 s.f.)

Exercise 89

Write your answers correct to 3 significant figures.

1 Find

a length a

b angle C.

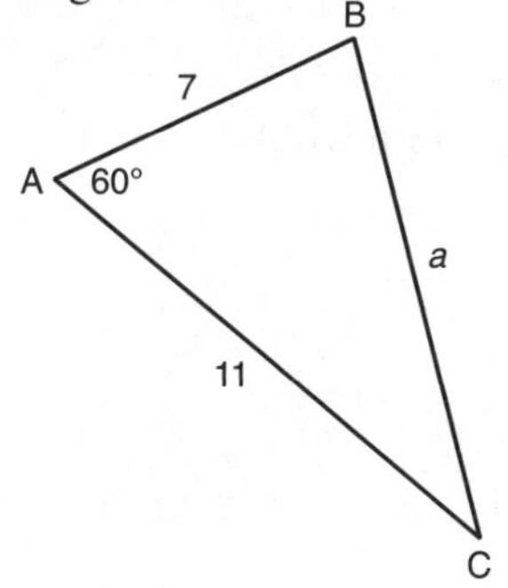

2 Find

a length r

b angle Q.

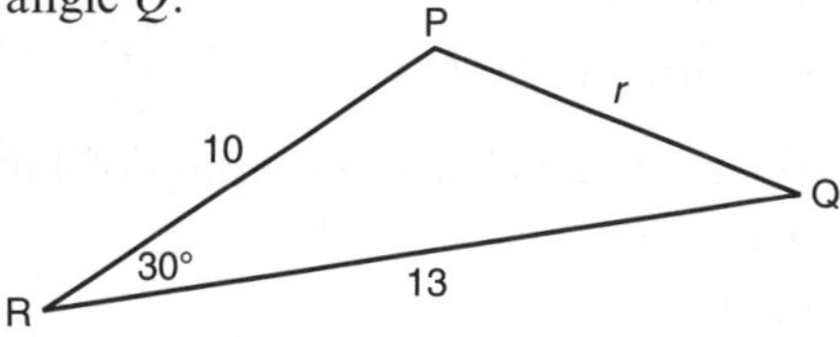

3 Find

a angle M

b angle N

c angle O.

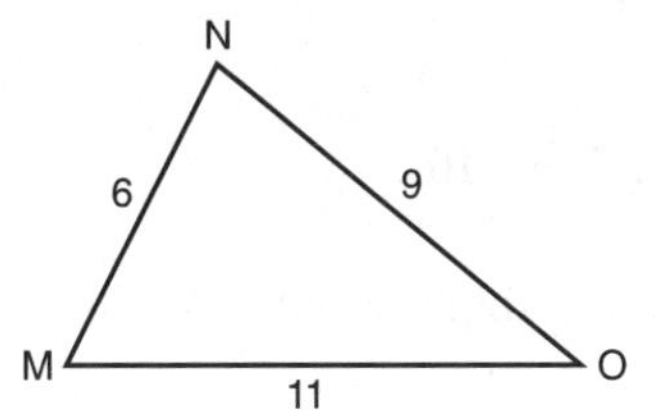

4 Find

a angle X

b angle Y

c angle Z.

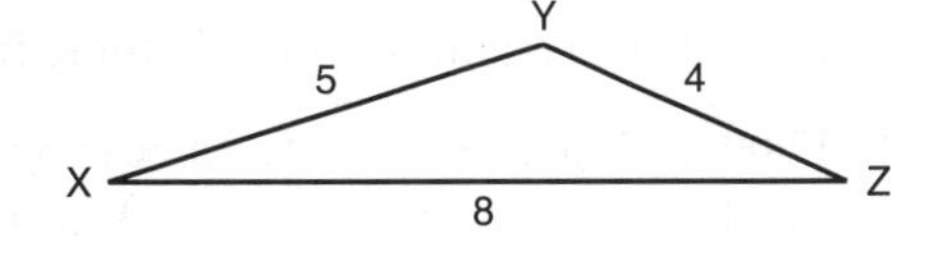

5 Find

a length a

b length b.

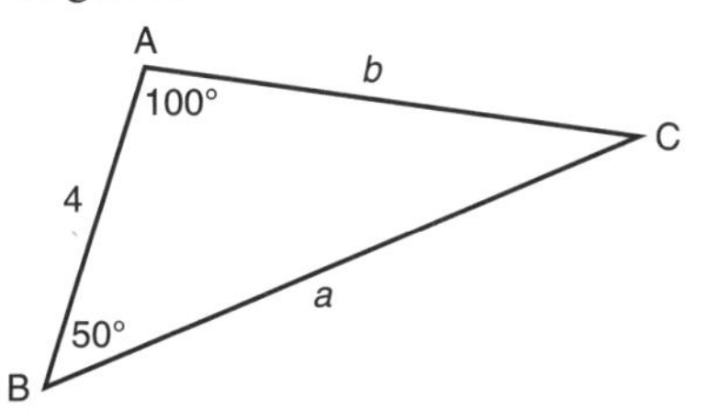

6 Find

a length p

b length r.

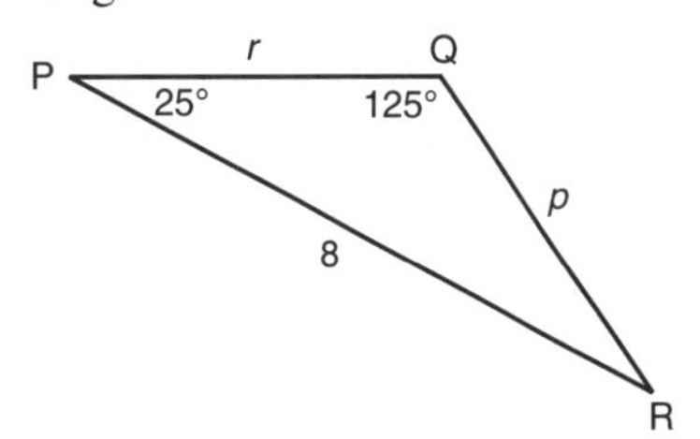

7 Find

a angle O

b length m.

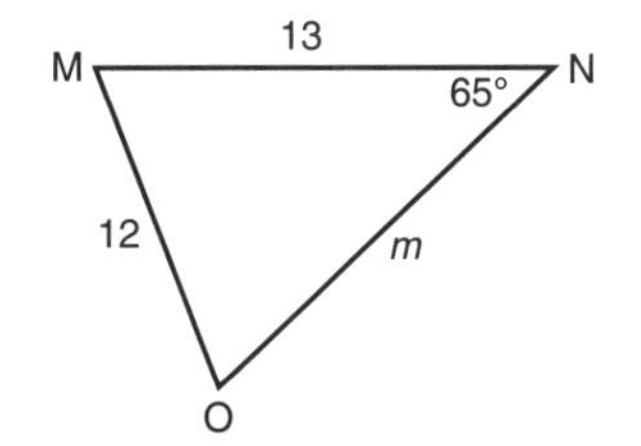

8 Find

a angle Z

b length x.

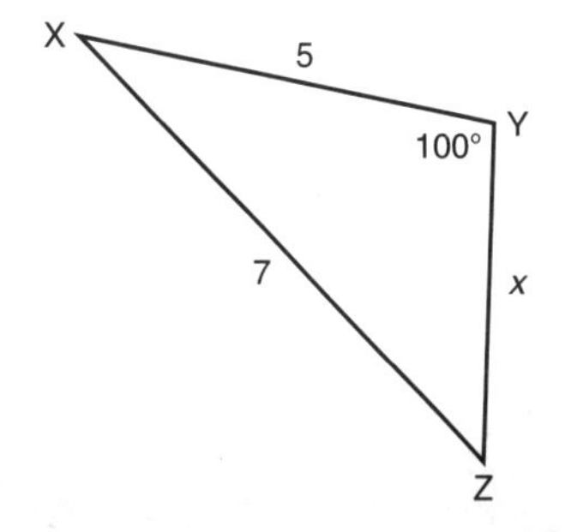

9 Point P is on a bearing of 060° from port O.
Point Q is on a bearing of 130° from port O.

OQ = 17 km, OP = 11 km.

Find the

a distance PQ

b bearing of Q from P.

10 Towns B and C are on bearings of 140° and 200° respectively from town A.
AB = 7 km and AC = 10 km.

Find the

a distance BC

b bearing of B from C.

Exercise 89★

Write answers to 3 significant figures.

1 Napoli is 170 km from Rome on a bearing of 130°.
Foggia is 130 km from Napoli on a bearing of 060°.
Find the distance and bearing of Rome from Foggia.

2 At 12:00, a ship is at X where its bearing from Trondheim, T, is 310°.
At 14:00, the ship is at Y where its bearing from T is 063°.
If XY is a straight line, TX = 14 km and TY = 21 km, find the ship's speed in km/h and the bearing of the ship's journey from X to Y.

3 Calculate

a angle BAE

b length CD

c angle ACD.

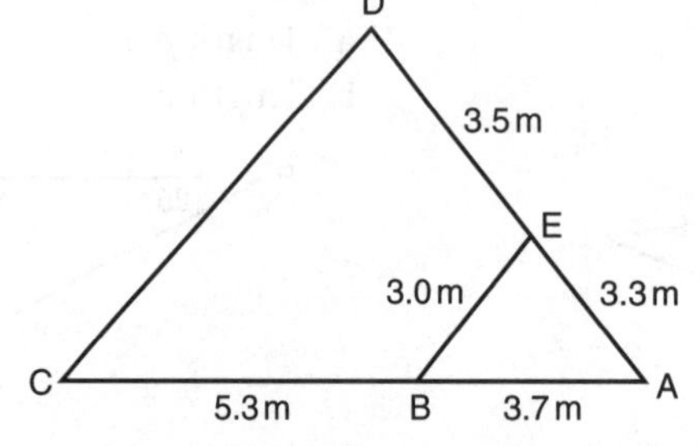

4 The diagonals of a parallelogram have lengths 12 cm and 8 cm, and the angle between them is 120°. Find the side lengths of the parallelogram.

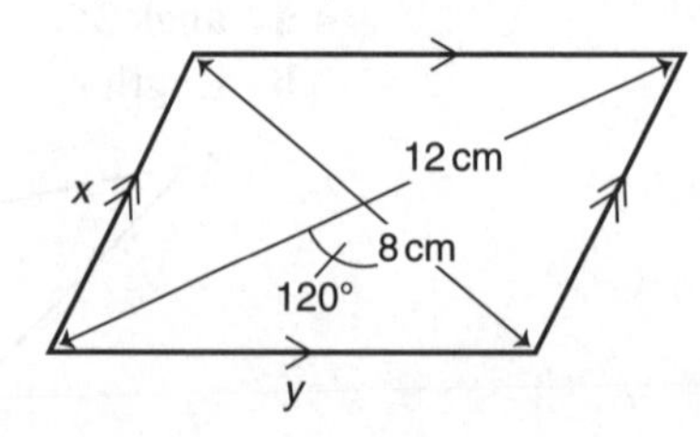

5 Two circles, centres X and Y, have radii 10 cm and 8 cm and intersect at A and B. XY = 13 cm. Find ∠BXA.

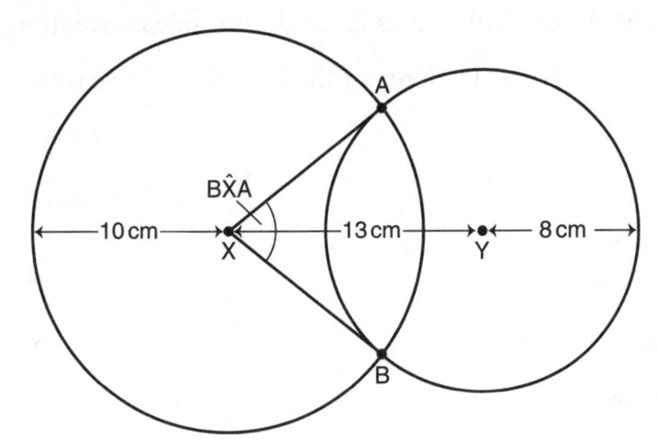

6 From a point A, the bearings of B and C are 040° and 160°, respectively. AB = 12 km and AC = 15 km. Find BC and the bearing of C from B.

7 SBC is a triangular orienteering course, with details as in this table.

	Bearing	Distance
Start S to B	195°	2.8 km
B to C	305°	1.2 km

Find the distance CS, and the bearing of S from C.

8 In the figure, find

a ∠XZY **b** WX

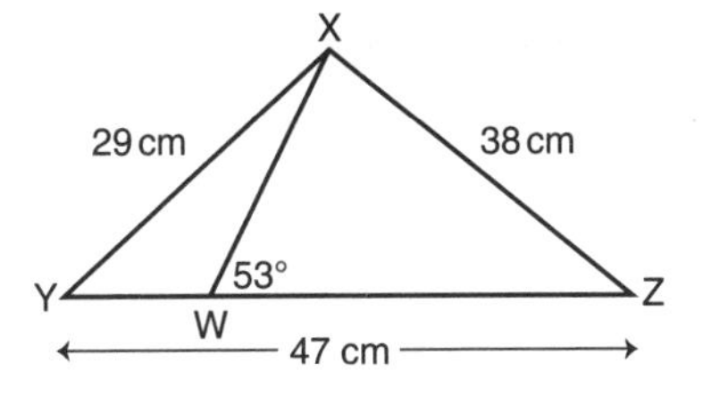

9 In the figure, find

a BD **b** ∠BCD

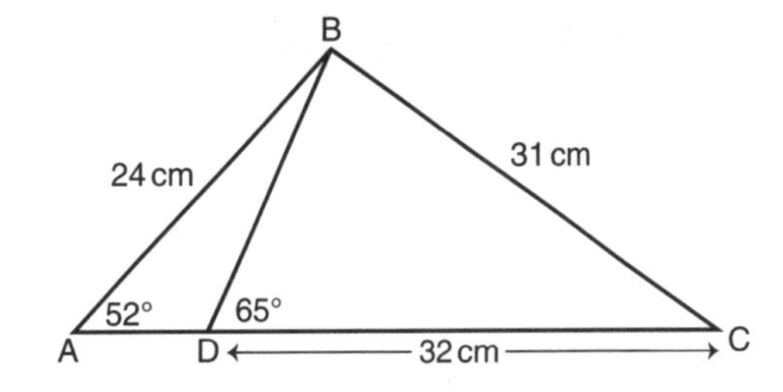

10 In triangle XYZ, YZ = 8 cm, XZ = 5 cm and angle $X = 60°$. If XY = x, use the cosine rule to show that $x^2 - 5x - 39 = 0$. Find x.

11 In triangle ABC, AB = 7.2 cm, BC = 9.4 cm, ∠ABC = 104.2°. Find ∠ACB.

Area of a triangle

Consider the triangle ABC below with the usual angle and side convention.

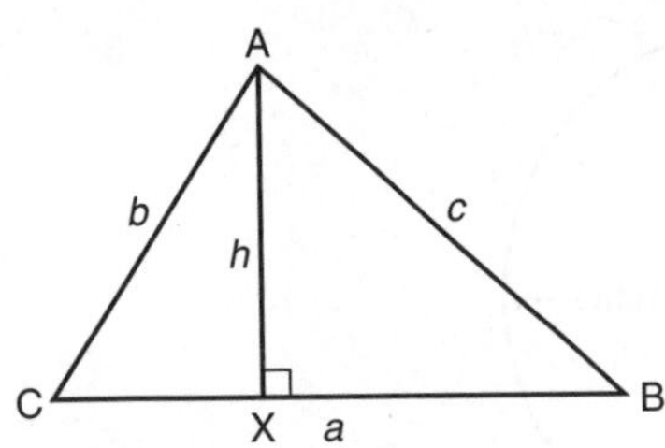

The area of the triangle $= \frac{1}{2} \times$ base $\times$ perpendicular height

$$= \frac{1}{2} \times a \times h$$

$$= \frac{1}{2} a(b \sin C)$$

So the area of any triangle $= \frac{1}{2} ab \sin C$

Simpler to recall is perhaps:

'Area = half the product of two sides × sine of the included angle'

Example 7

Find the area of the triangle ABC.

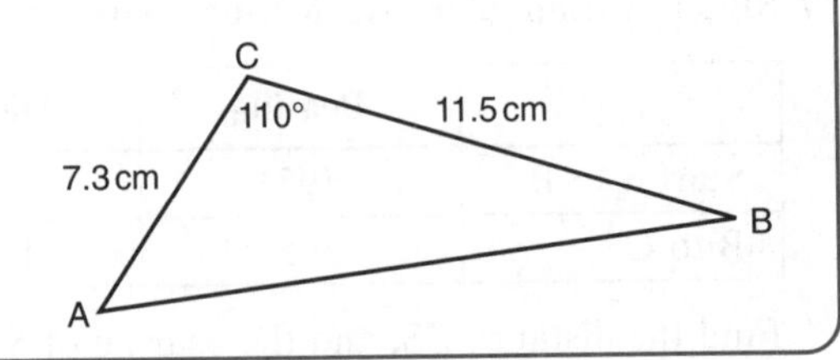

Area $= \frac{1}{2} ab \sin C$

$= \frac{1}{2} \times 11.5 \times 7.3 \times \sin 110°$

$= 39.4\,\text{cm}^2$ (3 s.f.)

Exercise 90

1 Find the area of triangle ABC.

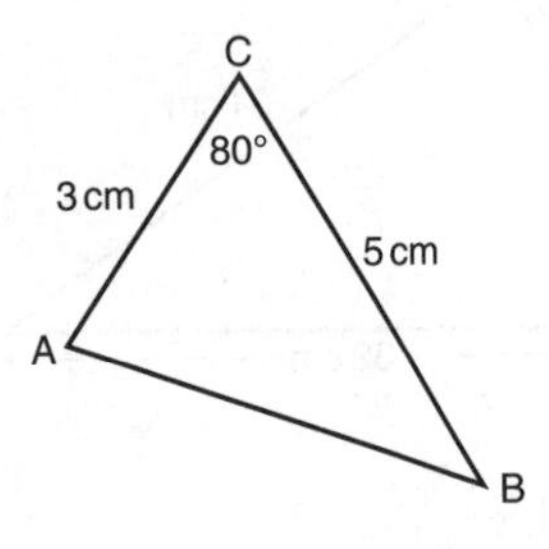

2 Find the area of triangle XYZ.

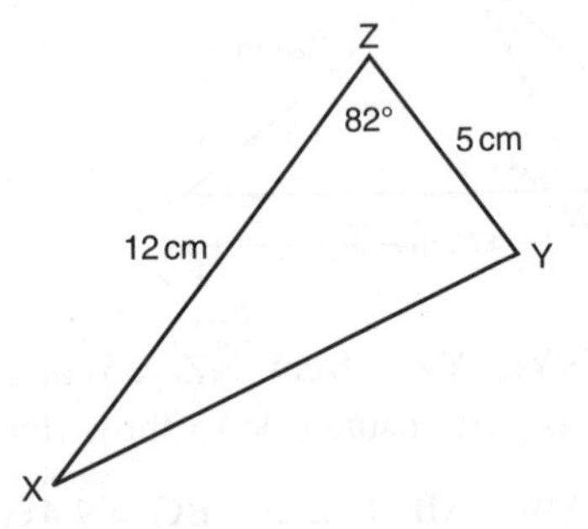

3 Find the area of triangle PQR.

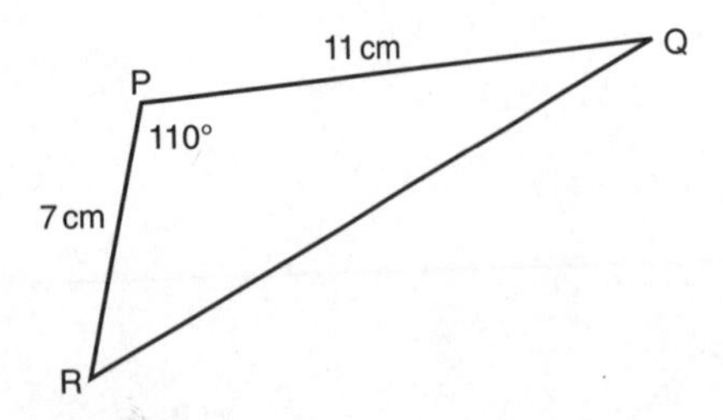

4 Find the area of triangle LMN.

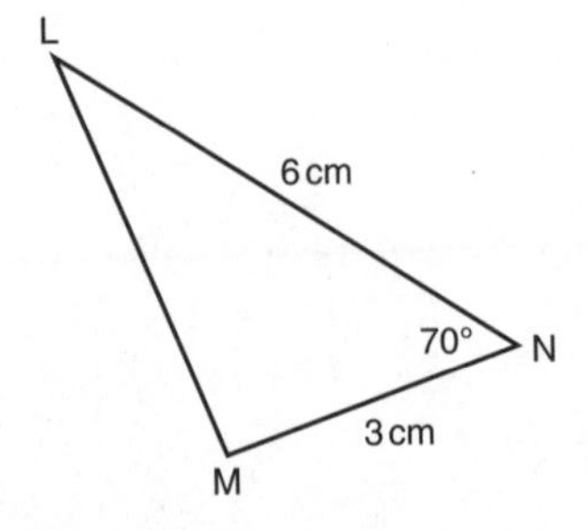

UNIT 4 ◆ Shape and space

5 Find the area of the triangle ABC if AB = 15 cm, BC = 21 cm and $B = 130°$.

6 Find the area of the triangle XYZ if XY = 10 cm, YZ = 17 cm and $Y = 65°$.

Exercise 90★

1 Find the area of an equilateral triangle of side 20 cm.

2 Find the side length of an equilateral triangle of area 1000 cm^2.

3 Find side length a if the area of triangle ABC is 75 cm^2.

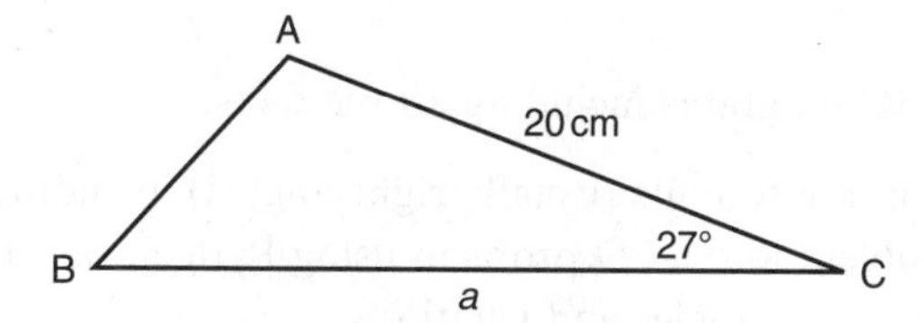

4 Find the angle ABC if the area of the isosceles triangle is 100 cm^2.

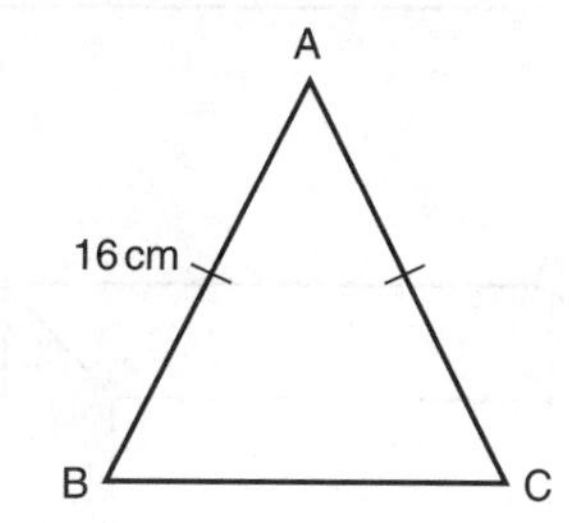

5 Find the area of triangle ABC.

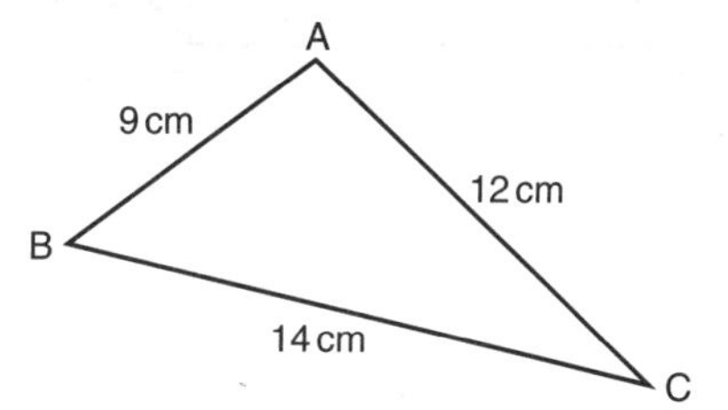

6 Find the perimeter of triangle ABC if its area is 200 cm^2.

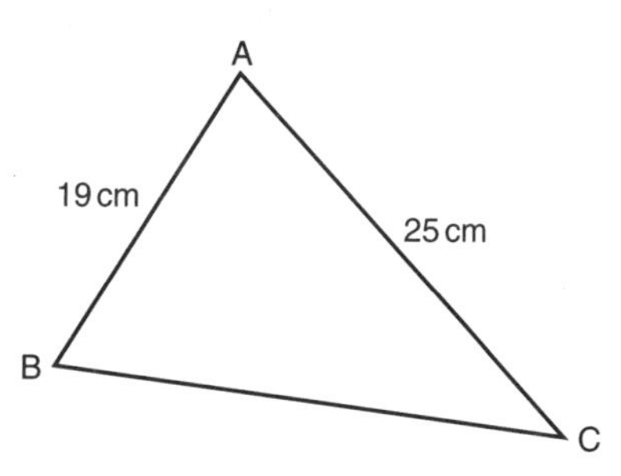

3D trigonometry

Place one end of a pencil B on a flat surface and put your ruler against the other end T as shown. The angle TBA is the angle between the straight line TB and the plane WXYZ.

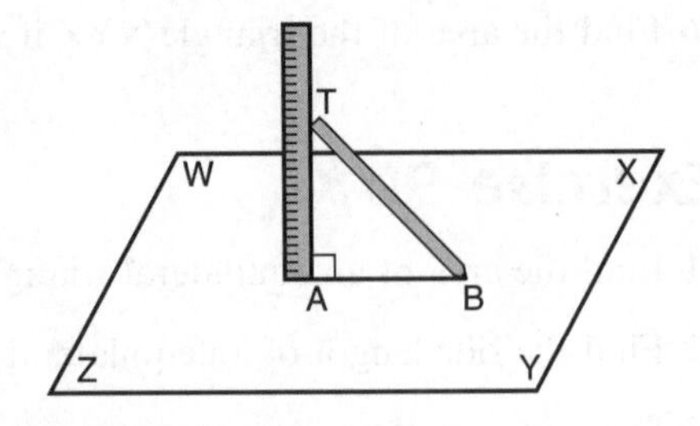

Remember

- Draw clear, large 3D diagrams including all the facts.
- Redraw the appropriate triangle (usually right-angled) including all the facts. This simplifies a 3D problem into a 2D problem using Pythagoras' Theorem and trigonometry to solve for angles and lengths.

Example 8

ABCDEFGH is a rectangular box.

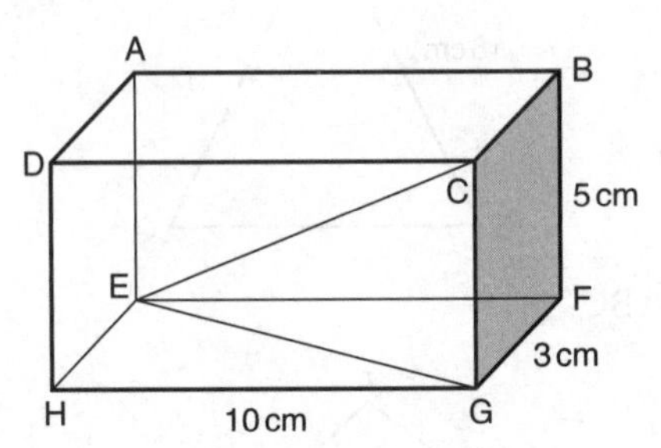

Find to 3 significant figures:

a length EG

b length CE

c the angle CE makes with plane EFGH (angle CEG).

a Draw triangle EGH in 2D.

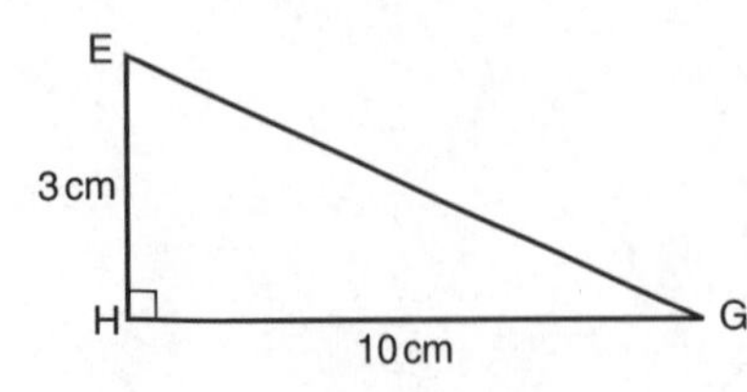

Pythagoras' Theorem

$$EG^2 = 3^2 + 10^2$$
$$= 109$$
$$EG = \sqrt{109}$$
$$EG = 10.4\,\text{cm (3 s.f.)}$$

b Draw triangle CEG in 2D.

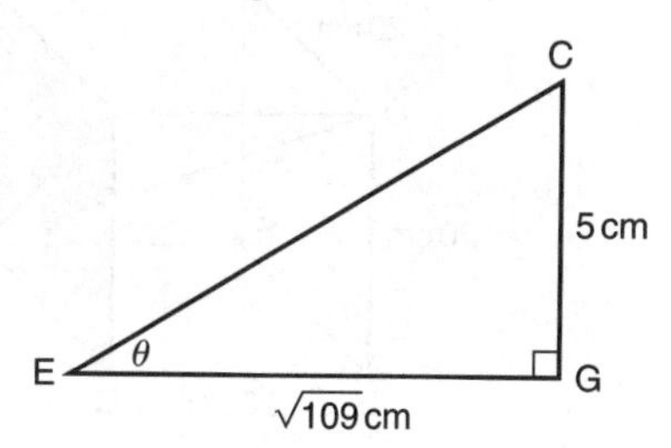

Pythagoras' Theorem

$CE^2 = 5^2 + 109$
$= 134$
$CE = \sqrt{134}$
$CE = 11.6\,\text{cm}$ (3 s.f.)

c Let angle CEG $= \theta$.

$\tan\theta = \dfrac{5}{\sqrt{109}} \Rightarrow \theta = \text{angle CEG} = 25.6°$ (3 s.f.)

Exercise 91

Give all answers to 3 significant figures.

1 ABCDEFGH is a rectangular box.

Find:

a EG

b AG

c the angle between AG and plane EFGH (angle AGE).

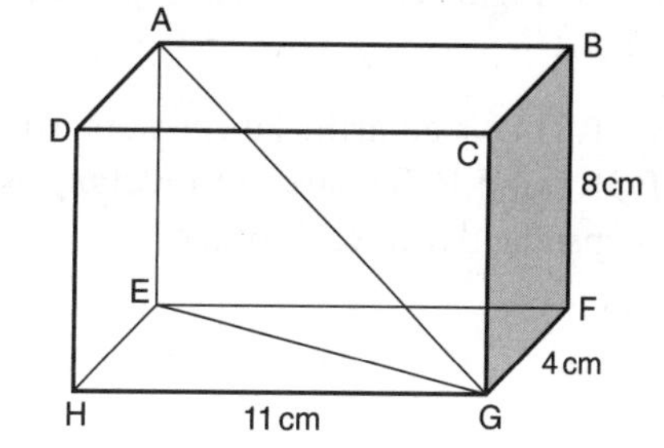

2 STUVWXYZ is a rectangular box.

Find:

a SU

b SY

c the angle between SY and plane STUV (angle YSU).

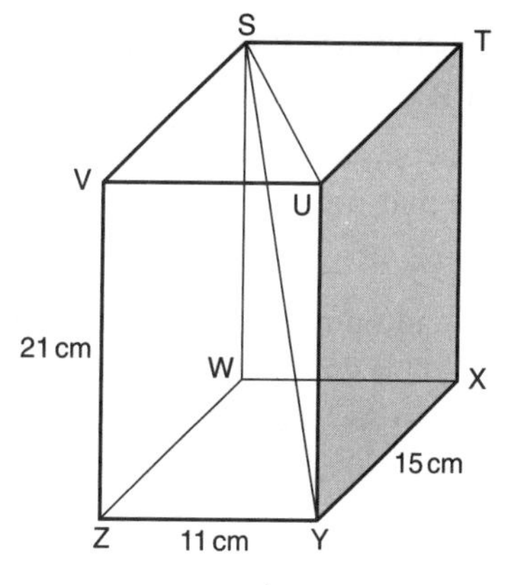

3 LMNOPQRS is a cube of side 10 cm.

Find:

a PR

b LR

c the angle between LR and plane PQRS (angle LRP).

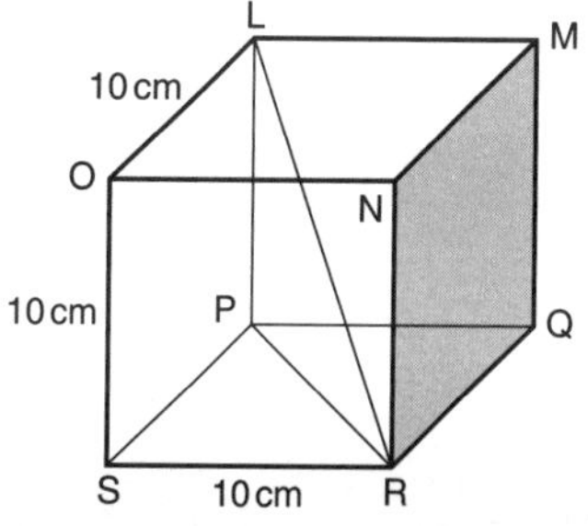

UNIT 4 ◆ Shape and space

4 ABCDEFGH is a cube of side 20 cm.

Find:

a CF

b DF

c the angle between DF and plane BCGF (angle DFC)

d the angle MHA, if M is the mid-point of AB.

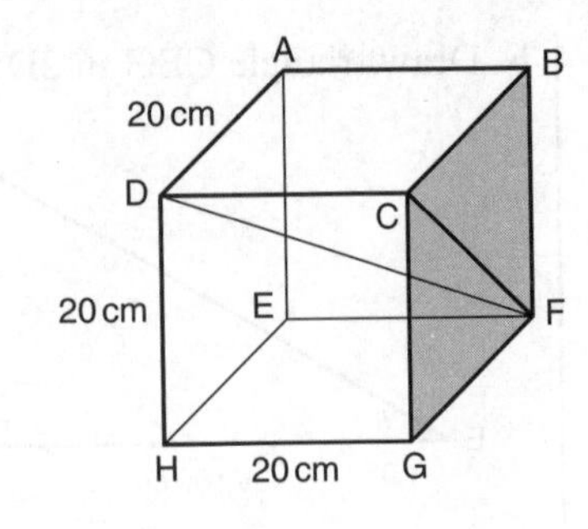

5 ABCDEF is a small ramp where ABCD and CDEF are both rectangles and perpendicular to each other.

Find:

a AC

b AF

c angle FAB

d the angle between AF and plane ABCD (angle FAC).

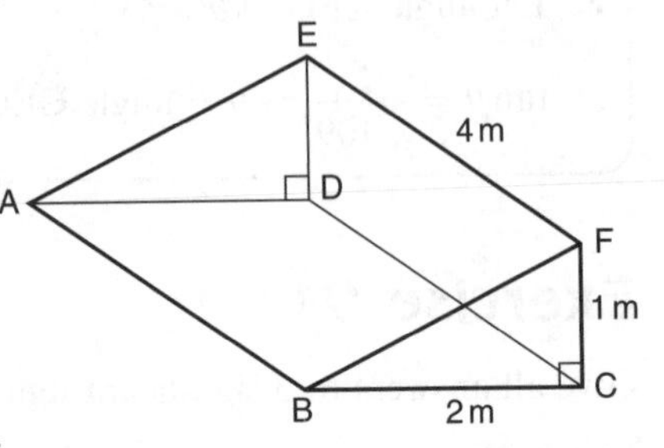

6 PQRSTU is an artificial ski-slope where PQRS and RSTU are both rectangles and perpendicular to each other.

Find:

a UP

b PR

c the angle between UP and plane PQRS (angle UPR)

d the angle between MP and plane PQRS, if M is the mid-point of TU.

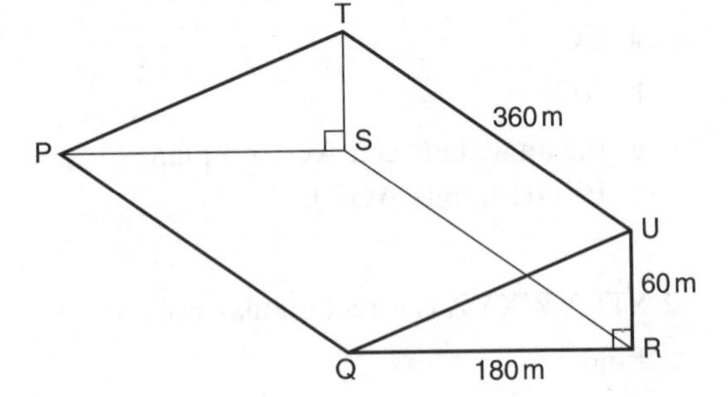

7 ABCD is a solid on a horizontal triangular base ABC. Edge AD is 25 cm and vertical. AB is perpendicular to AC. Angles ABD and ACD are equal to 30° and 20° respectively.

Find:

a AB

b AC

c BC.

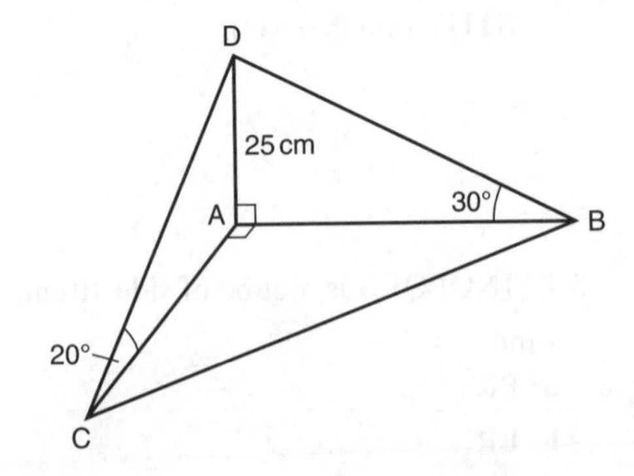

8 PQRS is a solid on a horizontal triangular base PQR. S is vertically above P. Edges PQ and PR are 50 cm and 70 cm respectively. PQ is perpendicular to PR. Angle SQP is 30°.

Find:

a SP

b RS

c angle PRS.

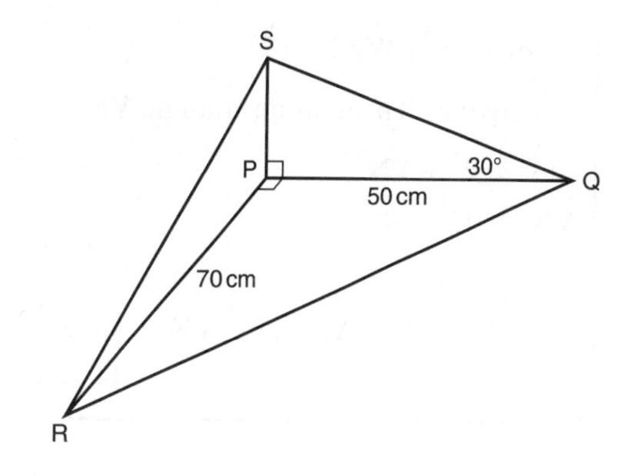

Example 9

VWXYZ is a solid regular pyramid on a rectangular base WXYZ where WX = 8 cm and XY = 6 cm. The vertex of the pyramid V is 12 cm directly above the centre of the base.

Find:

a VX

b the angle between VX and the base WXYZ (angle VXZ)

c the area of pyramid face VWX.

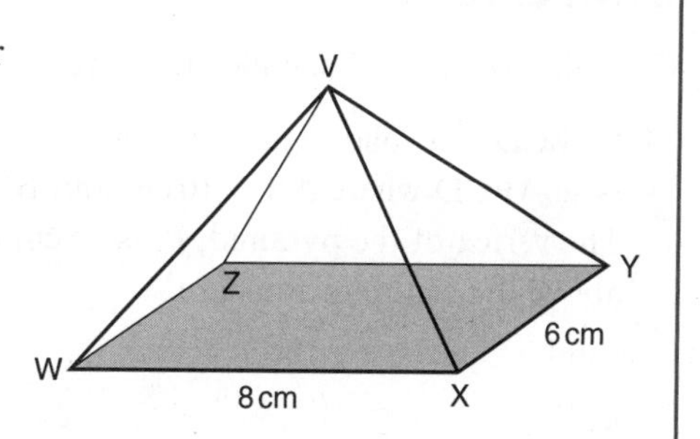

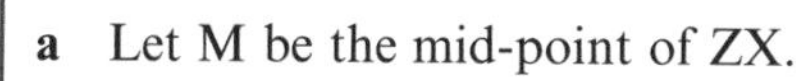

a Let M be the mid-point of ZX.

Draw WXYZ in 2D.

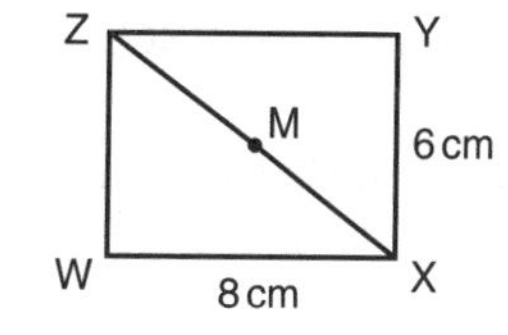

Pythagoras' Theorem on triangle ZWX:

$ZX^2 = 6^2 + 8^2$
$\quad\;\; = 100$
$ZX \;= 10\text{ cm} \Rightarrow MX = 5\text{ cm}$

Draw triangle VMX in 2D.
Pythagoras' Theorem on triangle VMX:

$VX^2 = 5^2 + 12^2$
$\quad\;\; = 169$
$VX \;= 13\text{ cm}$

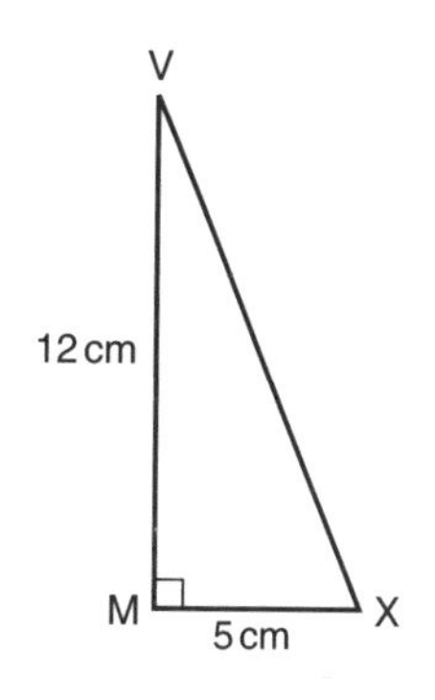

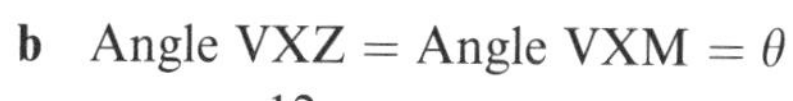

b Angle VXZ = Angle VXM = θ

$\tan\theta = \dfrac{12}{5} \Rightarrow \theta = 67.4°$ (3 s.f.)

$\Rightarrow$ Angle VXZ = 67.4° (3 s.f.)

c Let N be the mid-point of WX.
Area of triangle VWX = $\frac{1}{2} \times$ base $\times$ perpendicular height

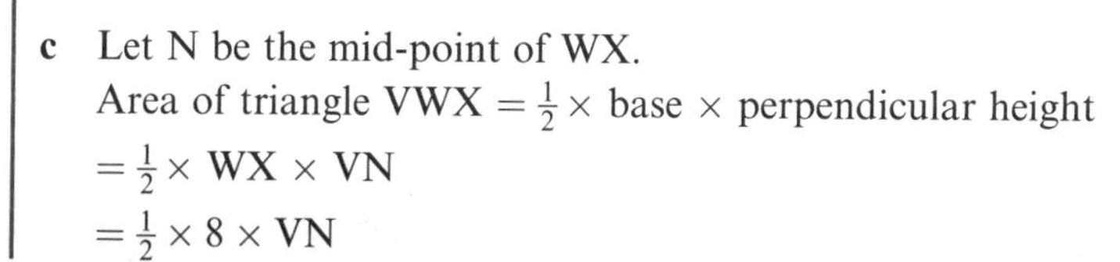

$= \frac{1}{2} \times \text{WX} \times \text{VN}$
$= \frac{1}{2} \times 8 \times \text{VN}$

Draw triangle VNX in 2D.

Pythagoras' Theorem on triangle VNX:

$13^2 = 4^2 + VN^2$

$VN^2 = 13^2 - 4^2$

$= 153$

$VN = \sqrt{153} \Rightarrow$ Area of VWX $= \frac{1}{2} \times 8 \times \sqrt{153}$

$= 49.5\,cm^2$ (3 s.f)

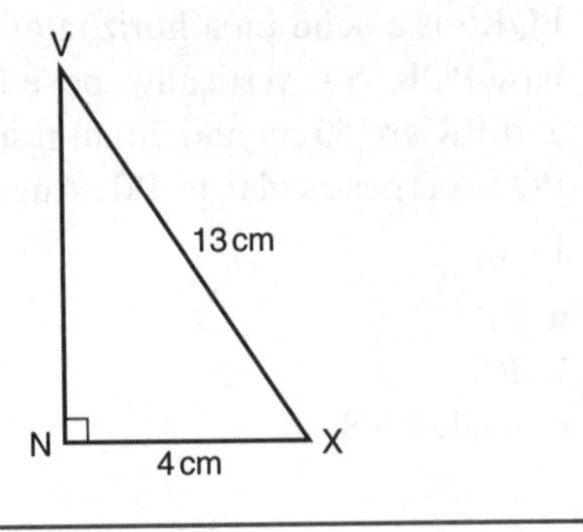

Exercise 91★

Give all answers to 3 significant figures.

1 PABCD is a solid regular pyramid on a rectangular base ABCD where AB = 10 cm and BC = 7 cm. The vertex of the pyramid, P, is 15 cm directly above the centre of the base.

Find:

a PA

b the angle between PA and base ABCD (angle PAC)

c the area of pyramid face PBC.

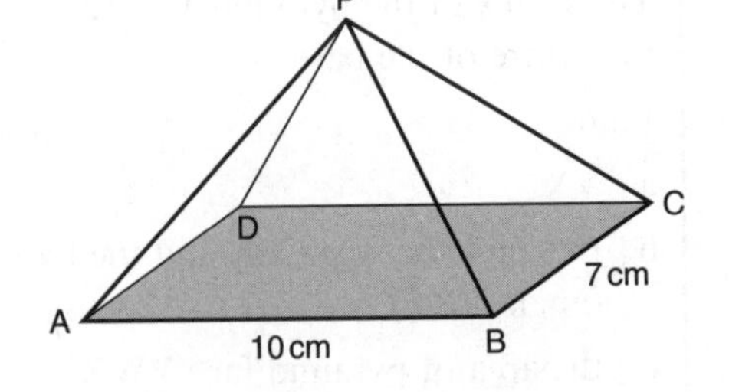

2 PQRST is a solid regular pyramid on a square base QRST where QR = 20 cm and edge PQ = 30 cm.

Find:

a the height of P above base QRST

b the angle PS makes with the base QRST

c the total external area of the pyramid including the base.

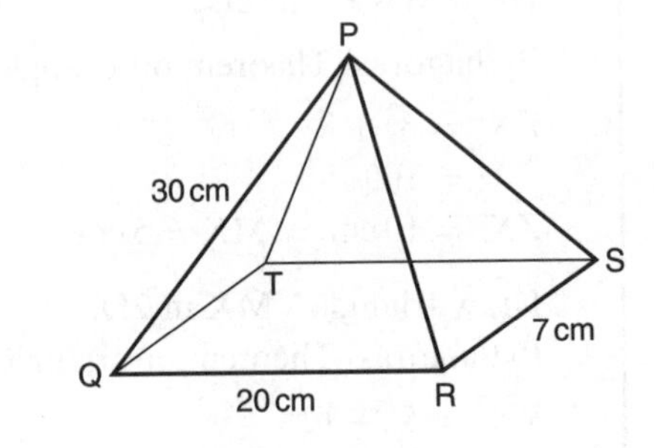

3 STUVWXYZ is a rectangular box. M and N are the mid-points of ST and WZ respectively.

Find angle:

a SYW

b TNX

c ZMY.

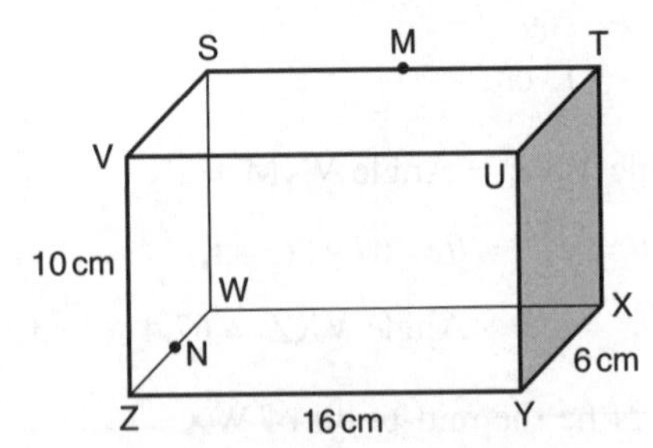

4 ABCDEFGH is a solid cube of volume $1728\,\text{cm}^3$.
P and Q are the mid-points of FG and GH respectively.
Find:

a angle QCP

b the total surface area of the solid remaining after pyramid PGQC is cut off.

5 A church is made from two solid rectangular blocks with a regular pyramidal roof above the tower, with V being 40 m above ground level.
Find:

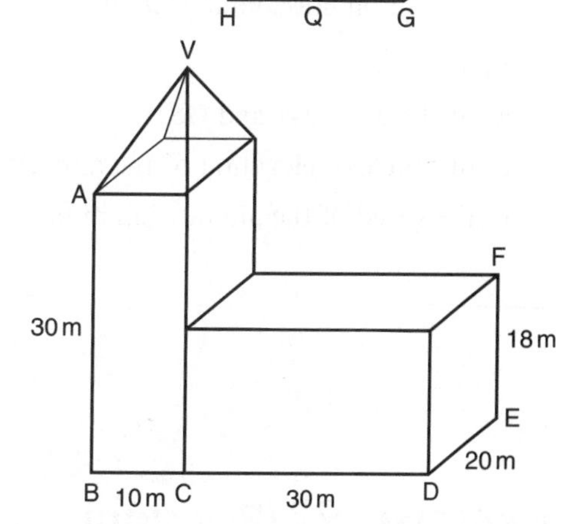

a VA

b the angle of elevation of V from E

c the cost of tiling the tower roof if the church is charged €250/m^2.

6 A hemispherical lampshade of diameter 40 cm is hung from a point by four chains, each of length 50 cm. If the chains are equally spaced on the rim of the hemisphere, find:

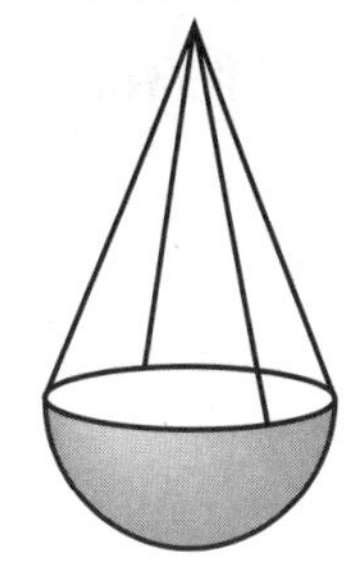

a the angle each chain makes with the horizontal

b the angle between two adjacent chains.

7 The angle of elevation to the top of a church tower is measured from A and from B.

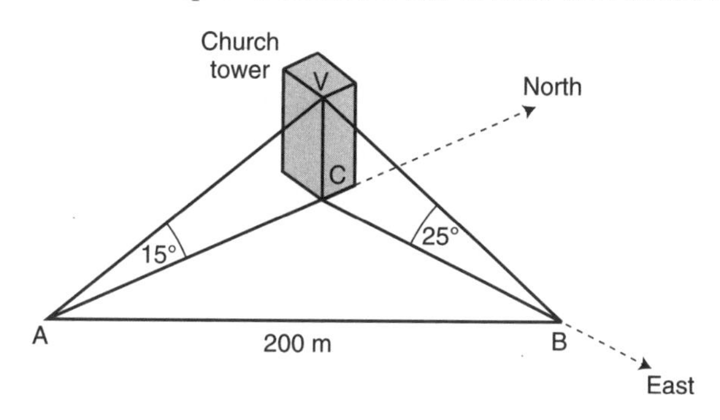

From A, due South of the church tower VC, the angle of elevation $\angle VAC = 15°$.
From B, due East of the church, the angle of elevation $\angle VBC = 25°$. AB = 200 m.
Find the height of the tower.

8 An aircraft is flying at a constant height of 2000 m. It is flying due East at a constant speed. At T, the plane's angle of elevation from O is 25°, and on a bearing from O of 310°. One minute later, it is at R and due North of O.

RSWT is a rectangle and the points O, W and S are on horizontal ground.

Find

a the lengths OW and OS

b the angle of elevation of the aircraft, ∠ROS

c the speed of the aircraft in km/h.

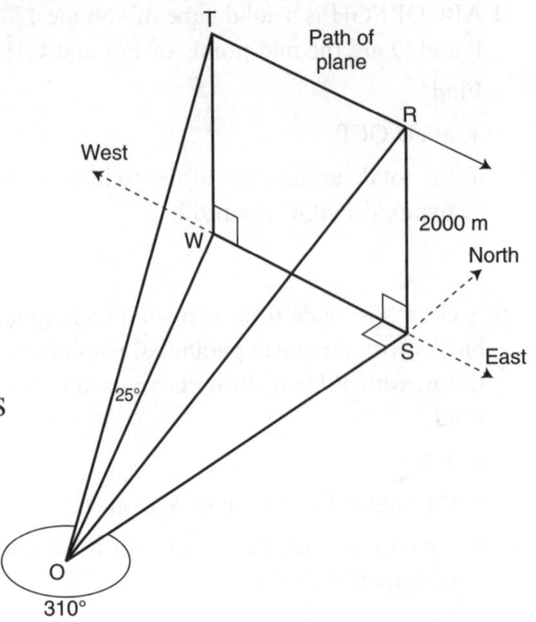

Exercise 92 (Revision)

Give all answers to 3 significant figures.

1 Find:

a angle C

b b.

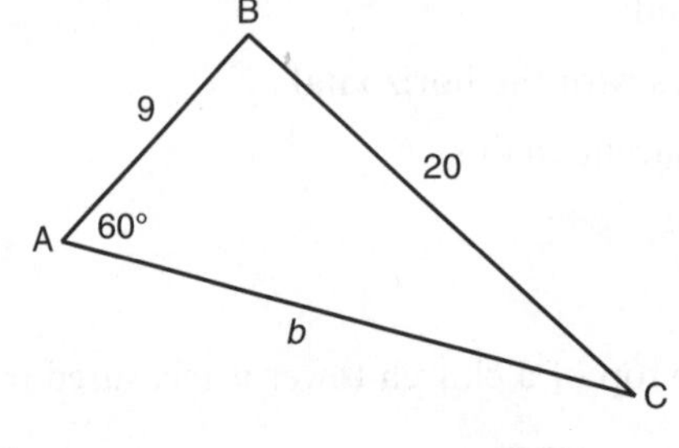

2 Find the area of △XYZ.

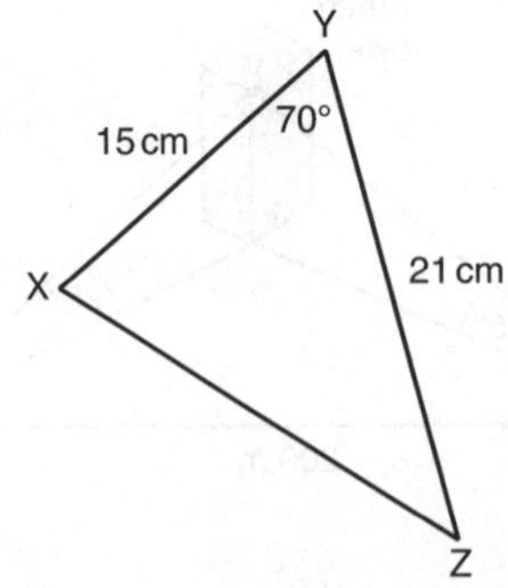

UNIT 4 ♦ Shape and space

3 The bearing of B from A is 150°, the bearing of C from B is 280°, the bearing of A from C is 030°.

a Find the angles of $\triangle ABC$.

b The distance AC is 4 km. Use the sine rule to find the distance AB.

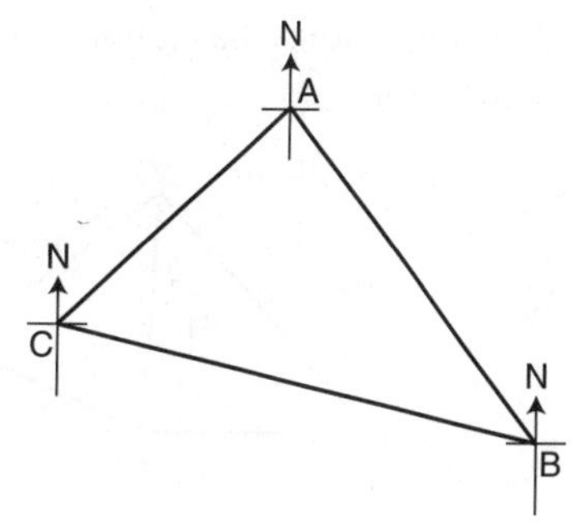

4 In $\triangle ABC$, AB = 6 cm, BC = 5 cm and AC = 10 cm. Use the cosine rule to find $\angle A$.

5 Copy this diagram. It shows a pyramid on a square base, where V is vertically above the centre of ABCD. Calculate

a the length AC

b the height of the pyramid

c the angle that VC makes with BC

d the angle that VC makes with ABCD.

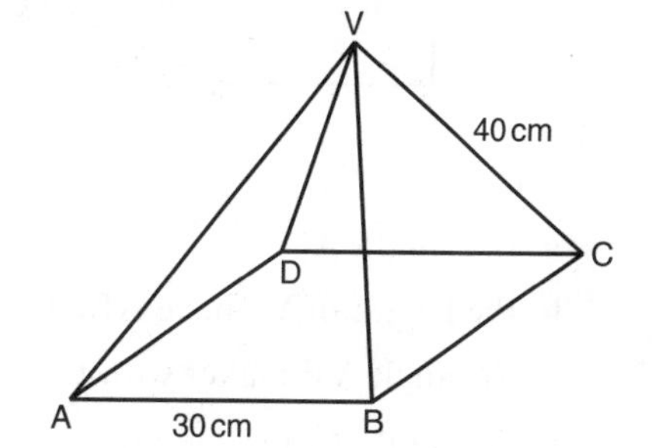

6 Copy this diagram of a ski slope and include these facts: DE = 300 m, AD = 400 m, CE = 100 m. Find

a the angle that CD makes with ADEF

b AE

c the angle that CA makes with ADEF.

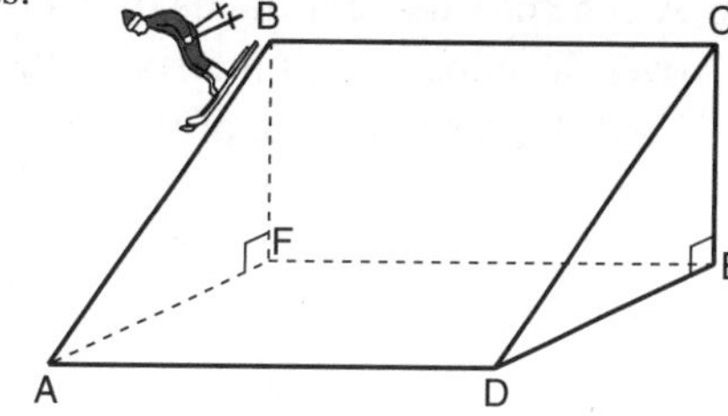

Exercise 92★ (Revision)

Where appropriate, give answers correct to 3 significant figures.

1 A yacht at point A is due West of a headland H. It sails on a bearing of 125°, for 800 m, to a point B. If the bearing of H from B is 335°, find the distance BH.

2 The sides of a parallelogram are 4.6 cm and 6.8 cm, with an included angle of 116°. Find the length of each diagonal.

3 A cuboid is a metres long, b metres wide and c metres high. Show that the length of the longest diagonal is given by $\sqrt{a^2 + b^2 + c^2}$.

4 Find the area of $\triangle XYZ$.

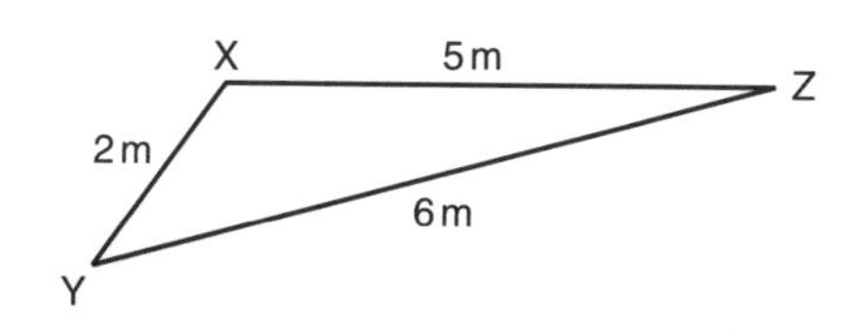

5 A doll's house has a horizontal square base ABCD and V is vertically above the centre of the base.

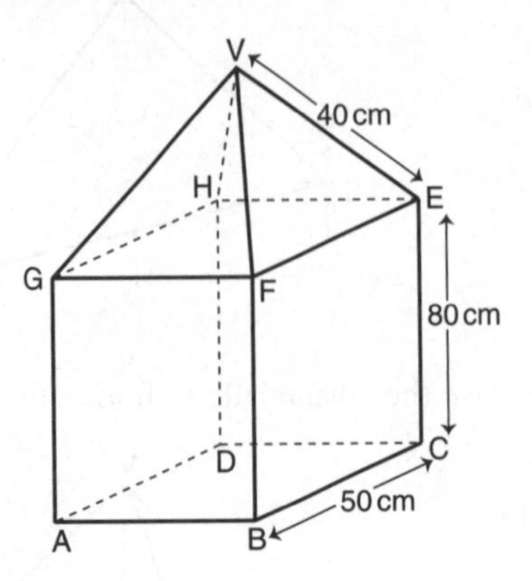

Calculate

a the length AC

b the height of V above ABCD

c the angle VE makes with the horizontal

d the total volume.

6 A cube of side 8 cm stands on a horizontal table. A hollow cone of height 20 cm is placed over the cube so that it rests on the table and touches the top four corners of the cube. Find the vertical angle of the cone.

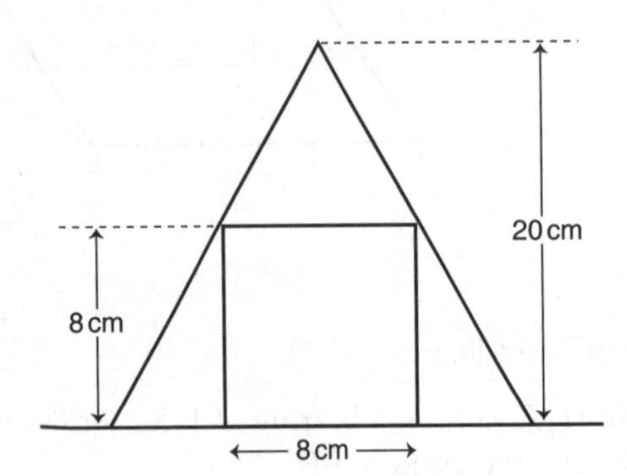

Handling data 4

More probability

The rules of probability are quite simple; more challenging questions require a deeper understanding and an efficient use of these laws.

Key Points

- $p(E)$ means the probability of event E occurring.
- $p(\bar{E})$ means the probability of event E *not* occurring.
- All probabilities have values between 0 and 1 inclusive, therefore

$$0 \leqslant p(E) \leqslant 1$$

- $p(\bar{E}) + p(E) = 1$, therefore $p(\bar{E}) = 1 - p(E)$

Multiplication ('and') rule

If two events A and B can occur without being affected by one another (for example, a dice is thrown and it starts to rain!), they are **independent**.

Key Point

If A and B are independent:

$$p(A \textit{ and } B) = p(A) \times p(B)$$

This makes sense, as multiplying fractions less than 1 produces a smaller fraction and the condition of two events happening together implies a smaller chance.

Example 1

A coin is flipped and a dice is thrown. Find the probability that the result is a tail and a multiple of three.

Event T is a tail thrown: $p(T) = \frac{1}{2}$

Event M is a multiple of three is rolled: $p(M) = \frac{2}{6}$

Probability of T and M: $p(T \text{ and } M) = p(T) \times p(M)$

$$= \frac{1}{2} \times \frac{2}{6}$$

$$= \frac{1}{6}$$

Addition ('or') rule

If two events A and B *cannot* occur at the same time (for example, a card drawn from a pack cannot be an Ace and a Queen) they are called **mutually exclusive**.

Key Point

If A and B are mutually exclusive:

$$p(A \text{ or } B) = p(A) + p(B)$$

This makes sense, as adding fractions produces a larger fraction and the condition of one or other event happening implies a greater chance.

Example 2

A card is randomly selected from a pack of 52 playing cards.

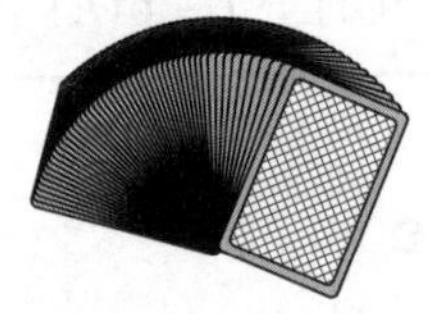

4 Aces

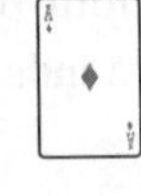

4 Queens

Find the probability that either an Ace or a Queen is selected.

Event A is an Ace is picked out: $p(A) = \frac{4}{52}$

Event Q is a Queen is picked out: $p(Q) = \frac{4}{52}$

Probability of A or Q:

$$\begin{aligned} p(A \text{ or } Q) &= p(A) + p(Q) \\ &= \frac{4}{52} + \frac{4}{52} \\ &= \frac{8}{52} \\ &= \frac{2}{13} \end{aligned}$$

Probability tree diagrams

Tree diagrams can be used to show all possible outcomes of a sequence of events. Together with the 'and' and 'or' rules, they can make problems easier to solve.

Example 3

Ghost Boy is an albino koala bear in San Diego Zoo. Albinos are born without melanin in their skin, which results in them having a very pale complexion and white fur in this case.

The probability of a koala having the albino gene is $\frac{1}{75}$. The albino gene must be inherited from *both* parents, but then only one out of four cubs produced by a pair of albino-gene carriers is an albino cub.

Find the probability of two randomly paired koalas producing an albino cub like Ghost Boy.

Let event C represent an albino-gene carrier and event $\overline{C}$ be a non-albino-gene carrier.

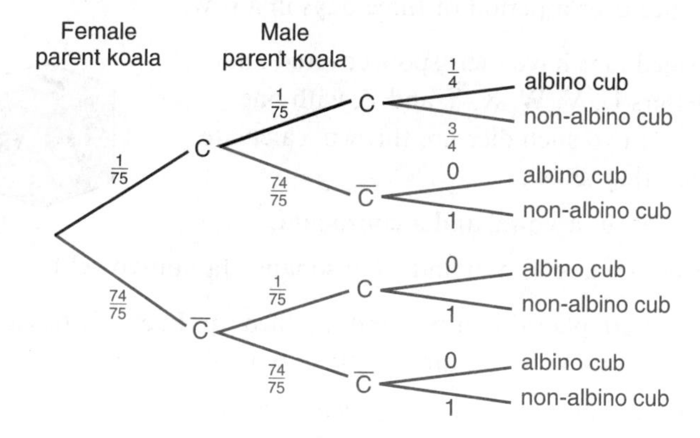

Let event A be that an albino koala cub is born. Then

$$\begin{aligned} p(A) &= p(C) \times p(C) \times \frac{1}{4} \\ &= \frac{1}{75} \times \frac{1}{75} \times \frac{1}{4} \\ &= \frac{1}{22\,500} \end{aligned}$$

Clearly, Ghost Boy is a rare creature!

Note: Only part of the tree diagram needed to be drawn to work out the probabilities. It is not always necessary to draw *all* the branches; this can overcomplicate the picture. A selection of branches may be sufficient to solve a problem, but you must ensure that all 'routes' are included.

Exercise 93

Use tree diagrams in these questions where appropriate.

1 Teresa loves playing tennis. If the day is sunny, the probability that she plays tennis is 0.7. If it is not a sunny day, the probability that she plays tennis is 0.4. Also, the probability that Saturday will not be sunny is 0.15. Use a tree diagram to find the probabilities of these outcomes:

a Teresa plays tennis on Saturday.

b Teresa does not play tennis on Saturday.

2 The probability of the Bullet Train to Tokyo Station arriving late is 0.3. The probability that it arrives on time is 0.6. What is the probability that the train

a is early? **b** is not late? **c** is on time for two days in a row?

d is late on a given Friday, but not late on the next two days?

e is late exactly once over a period of three days in a row?

3 Two normal six-sided dice have their spots covered and replaced by the letters U, V, W, X, Y and Z, with one letter on each face. If two such dice are thrown, calculate the probability that they show:

a two vowels **b** a vowel and a consonant.

(Remember: Vowels are a, e, i, o, u and a consonant is a non-vowel.)

4 Six small plastic tiles are placed in a bag and jumbled up. Each tile has a single letter on it, the letters being B, O and Y. There are two 'B's, two 'O's and two 'Y's.

a If one tile is removed at a time *without replacement* and the letters are placed in order from left to right, calculate the probability that the word formed is

(i) BOY (ii) YOB (iii) BOB.

b If two letter 'S's are placed in the bag from the start, what is the probability that after four selections, the word BOYS is *not* formed?

5 A box contains five distinctly different pairs of gloves. Kiril is in a hurry and randomly selects two gloves *without replacement*. What is the probability that

a the gloves are both right-handed?

b the gloves will be a right- and a left-handed glove?

c they will be a matching pair?

6 A drawer contains four pairs of black socks, three pairs of blue, two pairs of green, one pair of yellow and one odd red sock. Two socks are randomly selected *without replacement*. What is the probability that

a they are both black?

b they are both the same colour?

c one of the socks is a red one?

Example 4

Female elephants are called cows; male elephants are called bulls.

The probability that a particular elephant gives birth to a cow is $\frac{3}{5}$.
Use a tree diagram to find the probability that from three births she has produced exactly two bulls.

Let event C be the birth of a cow, and event B the birth of a bull. Consider a series of three births, being careful only to plot the paths required. There are three combinations that result in exactly two bulls from three births: BBC, BCB or CBB.

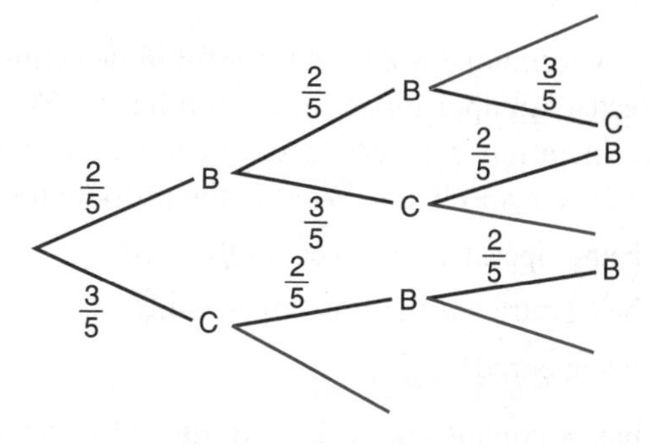

Note: Only part of the tree diagram is needed.

$$\begin{aligned}
\text{p(exactly 2 bulls from 3 births)} &= \text{p}(BBC) + \text{p}(BCB) + \text{p}(CBB) \\
&= \left(\tfrac{2}{5} \times \tfrac{2}{5} \times \tfrac{3}{5}\right) + \left(\tfrac{2}{5} \times \tfrac{3}{5} \times \tfrac{2}{5}\right) + \left(\tfrac{3}{5} \times \tfrac{2}{5} \times \tfrac{2}{5}\right) \\
&= \frac{12}{125} + \frac{12}{125} + \frac{12}{125} \\
&= \frac{36}{125}
\end{aligned}$$

Alternatively, because multiplication is commutative (independent of the order), one combination could have been considered (BBC say) and multiplied by 3, as the fractions do not change.

Exercise 93★

1 A bag contains three bananas, four pears and five kiwi-fruits. One piece of fruit is randomly withdrawn from the bag and *eaten before the next is taken*. Use a tree diagram to find these probabilities:

a That the first two fruits picked out are pears

b That the second fruit picked out is a pear

c That the first three fruits withdrawn are all different

d That neither of the first two fruits picked out is a banana, but the third is a banana

2 A box contains three \$1 coins, five 50-cent coins and four 20-cent coins. Two are selected randomly from the bag *without replacement*. What is the probability that:

a the two coins are of equal value?

b the two coins will total less than \$1?

c at least one of the coins will be a 50-cent coin?

3 Mr and Mrs Hilliam plan a family of four children. If babies of either sex are equally likely and assuming that only single babies are born, what is the probability of the Hilliam children being

a four girls? **b** three girls? **c** at least one girl?

4 A new technique is 80% successful in detecting cancer in infected patients. If cancer is detected, an appropriate operation has a 75% success rate, the first time it is attempted. If this operation is unsuccessful, it can be repeated only twice more, with success rates of 50% and 25% respectively. What is the probability of an infected patient

a being operated on successfully once?

b being cured at the third operation?

c being cured?

5 A bag X contains ten coloured discs of which four are white and six are red. A bag Y contains eight coloured discs of which five are white and three are red. A disc is drawn at random from X and placed in Y. A second disc is now drawn at random from X and placed in Y.

a Using a tree diagram, show that the probability that the two discs drawn are both red is $\frac{1}{3}$.

b Copy and complete this table.

Outcome		Probability
Bag X	Bag Y	
4R + 4W	5R + 5W	$\frac{1}{3}$
5R + 3W		
	3R + 7W	

c A disc is now drawn at random from the ten in Y and placed in X, so that there are now nine discs in each bag. Find the probability that there are:

(i) four red discs in bag X

(ii) seven red discs in bag X

(iii) six red discs in bag X

(iv) five red discs in bag X.

6 Two gamblers, Omar and Yosef, decide to play a simple game with a single fair die. The winner is the first player to roll a six! Omar goes first. What is the probability that:

a Omar wins the game? **b** Yosef wins the game?

Activity 26

The randomised response questionnaire technique was pioneered in America to find out the proportion of people who have participated in antisocial, embarrassing or illegal activities. Most people would not answer questions truthfully, but this method involves responding truthfully depending on the scores on two dice. There is no reason for anyone *not* to tell the truth, because each person can keep their dice score to themselves.

Two dice are used for each person: one black, one white. The black dice is rolled first. If it is even, the survey question must be answered truthfully. If it is odd, the white dice is rolled and the respondent must answer the question, 'Is the white dice number even?' truthfully.

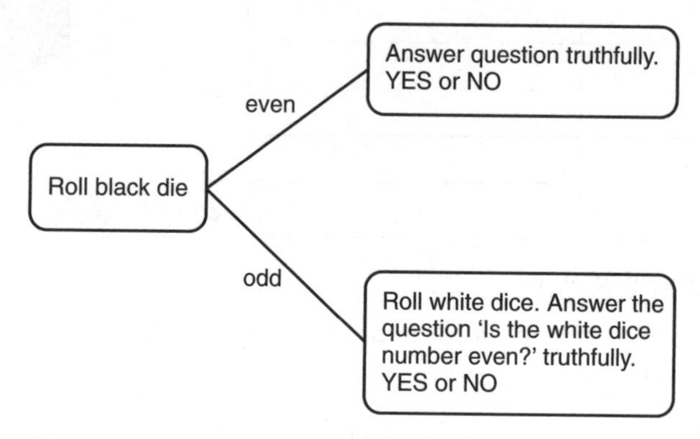

The number of respondents who answer YES is noted.

If 60 people responded to the question, 'Do you copy your homework?' and 24 answered 'yes', it is possible to estimate how many of this sample of 60 cheat with their homework.

- Show that the proportion of the group who copy homework is probably $\frac{3}{20}$
- Set a suitable question (not of an embarrassing nature!) which people may be reluctant to admit openly to answering truthfully and then carry out your own randomised response survey. Why would you wish to make your sample size as large as practicable?

UNIT 4 ♦ Handling data

Investigate

Double six!!!

Scoring two sixes on two dice is a highly desirable score to obtain in many games played throughout the world. Backgammon and Monopoly are two examples where scoring the magical double six score gives the player a great advantage.

How many rolls of a single dice would you expect to make before you obtain a six?

- Gather data to examine this problem practically by copying and completing the table.

Number of throws to obtain a six, x	Frequency, f
1	
2	
3	
4	
5	
6	
7	
⋮	

- Calculate the mean value of x.
- Use a simple computer application to simulate rolling the dice many more times and compare your answers.
- Try to investigate (by spreadsheet?) the result by producing a series that sums to give the mean value of x.
- Compare all your answers and comment.

How many rolls of two dice would you expect before you obtain a double six?

- Use a simple computer application to simulate this practical activity many times.
- Try to investigate (by spreadsheet?) the result by producing a series that sums to give the mean value of x.
- Compare your answers and comment.

Exercise 94 (Revision)

1 Fran has a box containing a *very large* number of stamps. Two-thirds are Cuban; the remainder are from Brazil. She randomly picks out two without replacement.

a Calculate an estimate of the probability that she has selected

(i) two Cuban stamps

(ii) a Cuban and a Brazilian stamp.

b A third stamp is randomly withdrawn from the box.
Calculate the probability that Fran has at least two Brazilian stamps from her three selections.

2 Gina and Iona are involved in a Ski Instructor Test. They are only allowed one re-test should they fail the first. Their probabilities of passing are as shown.

	Gina	Iona
p(pass on 1st test)	0.7	0.4
p(pass on 2nd test)	0.9	0.6

These probabilities are independent of each other. Calculate the probability that:

a Iona becomes a ski instructor at the second attempt

b Gina passes at her first attempt whilst Iona fails her first test

c only one of the girls becomes a ski instructor, assuming that a re-test is taken if the first test is failed.

3 Melissa is growing apple and pear trees for her orchard. She plants seeds in her greenhouse, but forgets to label the seedling pots. She knows that the types of apple and pear trees are in the numbers shown.

	Eating	Cooking
Apple	24	76
Pear	62	38

She picks out an apple seedling at random.

a Estimate the probability that it is an eater.

b Estimate the probability that it is a cooker.

She picks out an eater seedling at random.

c Estimate the probability that it is an apple.

d Estimate the probability that it is a pear.

e She picks out three seedlings at random without replacement. Find the probability that she ends up with at least one eating-apple seedling.

4 The probability that a seagull lays a certain number of eggs is as shown.

Number of eggs	Probability
0	0.1
1	0.2
2	0.3
3	0.2
4	0.1
5 or more	0.1

What is the probability that

a a seagull lays more than four eggs?

b a seagull lays no eggs?

c two seagulls lay a total of at least four eggs?

5 A football player works out from his previous season's results the probabilities of his scoring goals in a game.

Number of goals	Probability
0	0.4
1	0.3
2	0.2
3 or more	0.1

Assume that these probabilities apply to the current season.

a Find the probability that in a game he scores at most two goals.

b Find the probability that in a game he scores four goals.

c What is the probability that in two games he scores at least three goals?

Exercise 94★ (Revision)

1 Christmas lights are produced in a large batch such that one-fifth are defective. If lights are removed in turn from this large batch, find the probability from the first three lights of

a one defective **b** two defectives **c** at least one defective

2 Batteries are made on a factory production line such that 15% are faulty. If batteries are removed for quality control, find the probability from the first three batteries of

a no defectives **b** two defectives **c** at least two defectives

3 A veterinary surgeon has three independent tests A, B and C for the presence of a virus in a cow. Tests are performed in the order A, B, C.

Probabilities of a positive test depending on the presence of the virus			
	Test A	**Test B**	**Test C**
Virus present	$\frac{2}{3}$	$\frac{4}{5}$	$\frac{5}{6}$
Virus *not* present	$\frac{1}{5}$	$\frac{1}{6}$	$\frac{1}{7}$

a Find the probability that two tests will be positive and one negative on an infected cow.

b If two positive tests is the minimum criterion for indicating an infected cow, find the probability that:

(i) after three tests on an infected cow, the criterion is *not* met

(ii) after three tests on an uninfected cow, the criterion *is* met.

4 In a multiple choice test Mimi has not revised and has no idea of the correct answer. She guesses the answers and has a probability of p of obtaining the right option. From the first three questions the probability of her getting only one correct is 0.243. Find p.

5 Cage X contains four hamsters, one white and three brown. Meanwhile cage Y has three hamsters, two white and one brown. One hamster is taken from cage X and placed in cage Y. Hamsters are then removed one by one from cage Y.

a Copy and complete this tree diagram.

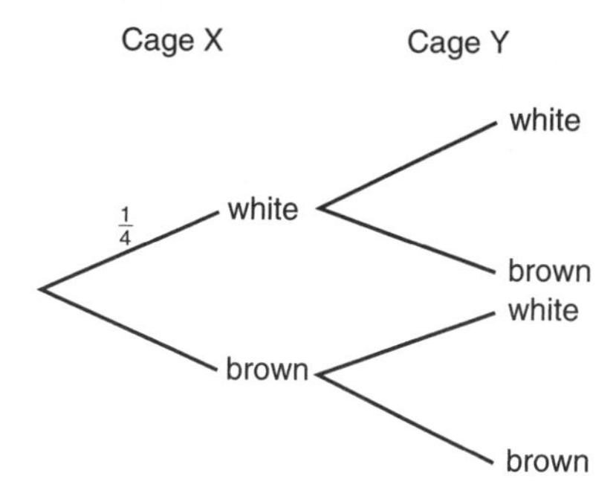

b Use your tree diagram to find the probability that the first hamster removed from cage Y is

(i) white (ii) brown

c What is the probability that, of the first two hamsters removed from Y, both are white?

UNIT 4 ◆ Handling data

Summary 4

Number

Irrational numbers

- Numbers such as $\sqrt{3}$ and π cannot be written as fractions and are called **irrational** numbers.
- The decimal expansion of irrational numbers is infinite and shows no pattern.
- Good examples of irrational numbers are the square roots of prime numbers and π.

Surds

- A surd is a root that is irrational. The roots of all prime numbers are surds.
- $\sqrt{2}$ means the *positive* square root of 2.
- When an answer is given using a surd, it is an *exact* answer.
- Surds can be cleared from the bottom of a fraction by multiplying the top and bottom of the fraction by the surd.

Examples of manipulating surds:

$\sqrt{2} \times \sqrt{2} = 2$

$3\sqrt{2} + 5\sqrt{2} = 8\sqrt{2}$

$(3 + \sqrt{2})^2 = (3 + \sqrt{2})(3 + \sqrt{2}) = 9 + 6\sqrt{2} + 2 = 11 + 6\sqrt{2}$

$\sqrt{18} = \sqrt{9 \times 2} = \sqrt{9} \times \sqrt{2} = 3\sqrt{2}$

$\frac{6}{\sqrt{3}} = \frac{6}{\sqrt{3}} \times \frac{\sqrt{3}}{\sqrt{3}} = \frac{6\sqrt{3}}{3} = 2\sqrt{3}$

Simplifying algebraic fractions

To simplify, factorise as much as possible and then cancel.

$$\frac{x^2+3x+2}{x^2-x-6}=\frac{(x+2)(x+1)}{(x+2)(x-3)}=\frac{x+1}{x-3}$$

Adding and subtracting fractions

Add or subtract in the same way as number fractions.

$$\frac{x}{3}+\frac{x-1}{6}=\frac{2x+x-1}{6}=\frac{3x-1}{6}$$

$$\frac{1}{x}-\frac{1}{x-1}=\frac{x-1-x}{x(x-1)}=\frac{-1}{x(x-1)}$$

Solving equations with fractions

Multiply *both sides* by the lowest common denominator to clear the fractions.

$$\frac{x+1}{3}=\frac{x-1}{2}\Rightarrow 2(x+1)=3(x-1)\Rightarrow 2x+2=3x-3\Rightarrow x=5$$

$$\frac{1}{x}-\frac{x}{x-2}=2\Rightarrow x-2-x^2=2x^2-4x\Rightarrow 3x^2-5x+2=0$$

$$\Rightarrow (3x-2)(x-1)=0\Rightarrow x=1 \text{ or } \tfrac{2}{3}$$

Graphs

Calculus

$y = x^n \Rightarrow \dfrac{dy}{dx} = nx^{n-1}$ Example: $y = x^4 - 5x^2 + 3x + 2 + x^{-2}$

$$\Rightarrow \frac{dy}{dx} = 4x^3 - 10x + 3 - 2x^{-3}$$

Tangents

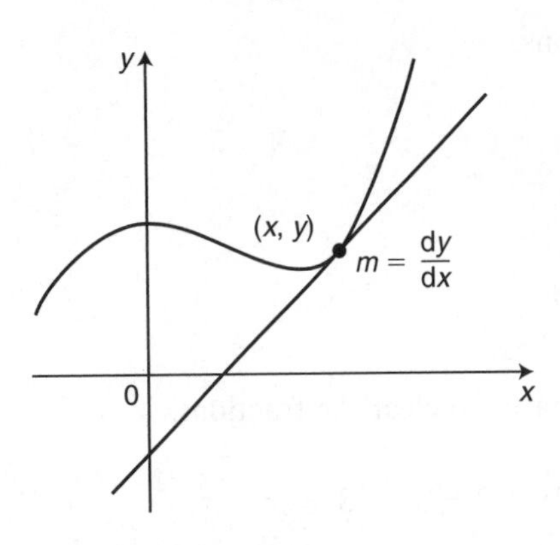

Turning points

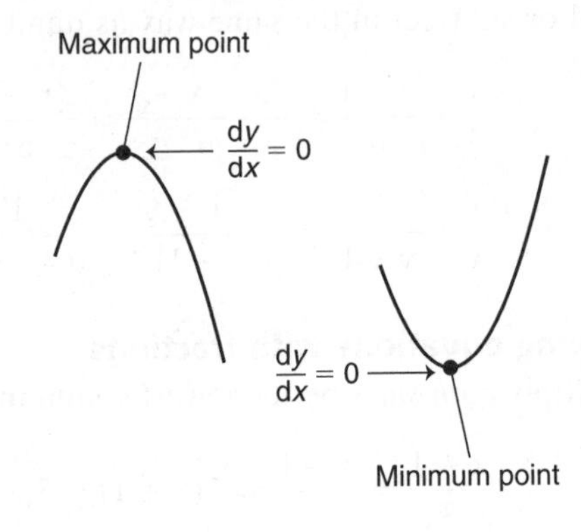

Linear kinematics

Velocity is the *rate* at which **displacement** changes over *time*. So, $v = \dfrac{ds}{dt}$.

Acceleration is the *rate* at which **velocity** changes over *time*. So, $a = \dfrac{dv}{dt}$.

displacement	⟶	velocity	⟶	acceleration
s	differentiate	$v = \dfrac{ds}{dt}$	differentiate	$a = \dfrac{dv}{dt}$

Shape and space

Trigonometric graphs ($0° \leqslant x \leqslant 180°$)

$y = \sin x$

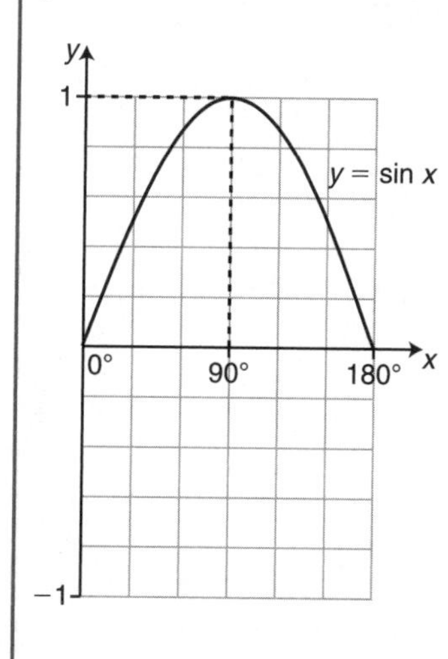

$y = \cos x$

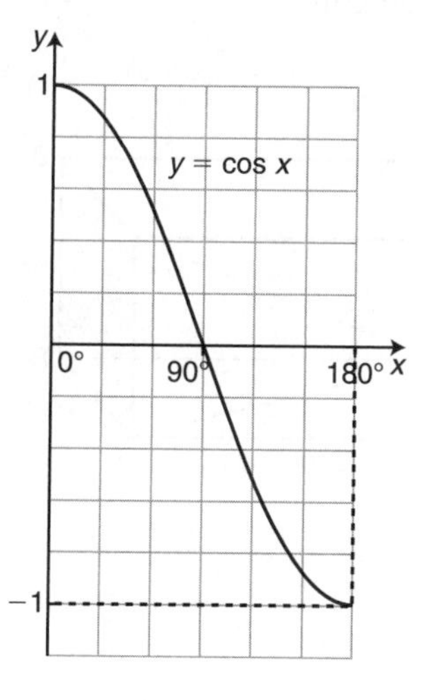

$y = \tan x$

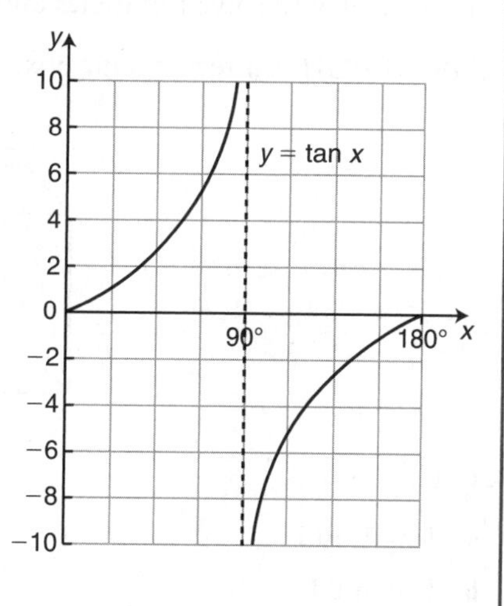

The area of a triangle

Area of any triangle $= \frac{1}{2}ab \sin C$

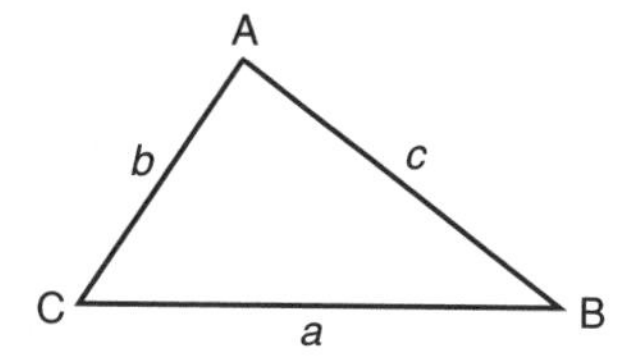

Simpler to recall is perhaps:

'Area = half the product of two sides × sine of the included angle'

Sine rule

$$\frac{a}{\sin A} = \frac{b}{\sin B} = \frac{c}{\sin C} \quad \text{or} \quad \frac{\sin A}{a} = \frac{\sin B}{b} = \frac{\sin C}{c}$$

Cosine rule

$a^2 = b^2 + c^2 - 2bc \cos A$

3D Trigonometry

- Draw clear, large 3D diagrams including all the facts.
- Redraw the appropriate triangle (usually right-angled) including all the facts.

This simplifies a 3D problem into a 2D problem using Pythagoras' Theorem and trigonometry to solve for angles and lengths.

ABCDEFGH is a rectangular box.

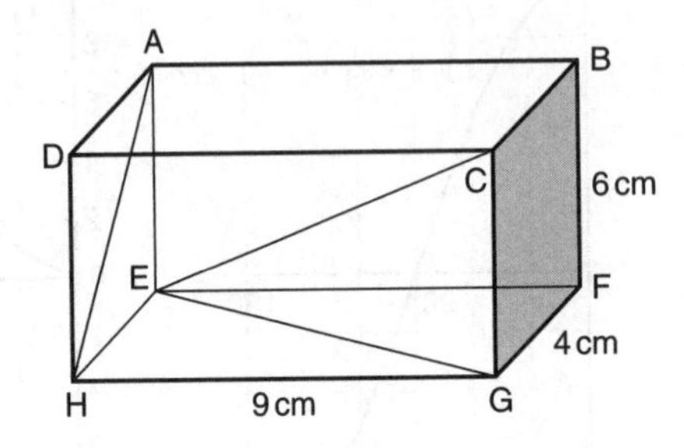

Find:

a length EG

b length CE

c the angle CE makes with plane EFGH (angle CEG).

a Draw triangle EGH in 2D.

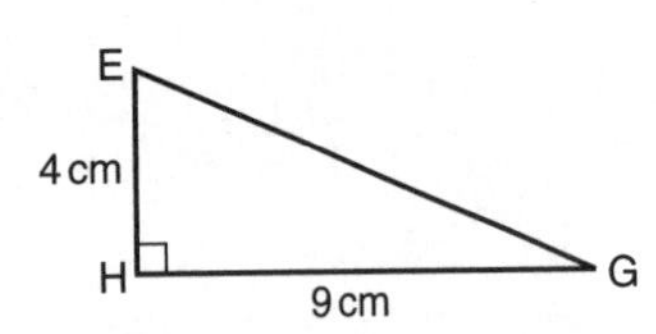

Pythagoras' Theorem:

$EG^2 = 4^2 + 9^2$

$= 97$

$EG = \sqrt{97}$

$= 9.85\,\text{cm}$ (3 s.f.)

b Draw triangle CEG in 2D.

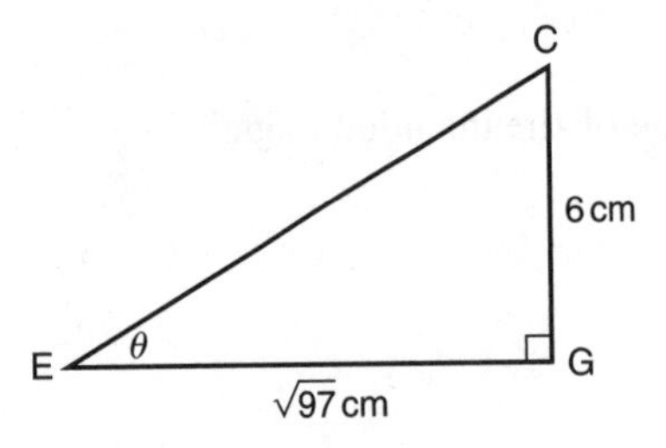

Pythagoras' Theorem:

$CE^2 = 6^2 + 97$

$= 133$

$CE = \sqrt{133}$

$CE = 11.5\,\text{cm}$ (3 s.f.)

c Let angle CEG $= \theta$.

$\tan\theta = \dfrac{6}{\sqrt{97}} \Rightarrow \theta = \text{angle CEG} = 31.4°$ (3 s.f.)

Handling data

Probability

- The **probability** of an event E happening, p(E)

$$p(E) = \frac{\text{number of desired outcomes}}{\text{total number of possible outcomes}}$$

(Impossible) $0 \leqslant p(E) \leqslant 1$ (Certain)

- The probability of an event not happening is $p(\bar{E})$.

$p(E) + p(\bar{E}) = 1$

Relative frequency

Experimental probability is measured by

$$\textbf{relative frequency} = \frac{\text{number of successes}}{\text{total number of trials}}$$

Independent events

If A and B are **independent** events

$$p(A \text{ and } B) = p(A) \times p(B)$$

$$p(A \text{ or } B) = p(A) + p(B)$$

To calculate the probability of an event occurring, it is necessary to consider *all* the ways in which that event can happen.

A bag contains three white and one black marble. Three marbles are removed from the bag individually, being replaced before each selection. Find the probability that from the three marbles there are at least two black marbles.

Let B and W be the number of black and white marbles respectively.

$p(B \geqslant 2) + p(B < 2) = 1$

so $\quad p(B \geqslant 2) = 1 - p(B < 2)$

$= 1 - p(B \leqslant 1)$

$= 1 - p(B = 0 \text{ or } B = 1)$

$= 1 - [p(B = 0) + p(B = 1)]$

$p(B = 0) = p(W = 3) = \left(\frac{3}{4}\right)^3 = \frac{27}{64}$

$p(B = 1) = \left(\frac{1}{4}\right)\left(\frac{3}{4}\right)^2 \times 3 = \frac{27}{64}$

$p(B \geqslant 2) = 1 - \left(\frac{27}{64} + \frac{27}{64}\right)$

so $\quad p(B \geqslant 2) = \frac{10}{64} = \frac{5}{32}$

Examination practice 4

1 Which of the following are rational and which are irrational?

a $0.\dot{7}$ **b** $\sqrt{3}+\sqrt{3}$ **c** $\sqrt{3}-\sqrt{3}$
d $\sqrt{3}\div\sqrt{3}$ **e** $\sqrt{3}\times\sqrt{3}$

2 a Find an irrational number between 5 and 6.
b Find a rational number between $\sqrt{3}$ and $\sqrt{5}$.

3 Simplify:

a $3\sqrt{2}+2\sqrt{2}$ **b** $(3\sqrt{2})^2$
c $3\sqrt{2}\times 2\sqrt{2}$ **d** $(\sqrt{2})^3$

4 Simplify:

a $\sqrt{48}$ **b** $\sqrt{48}+2\sqrt{3}$
c $\sqrt{50}$ **d** $\sqrt{\frac{9}{4}}$

5 Simplify:

a $(3+2\sqrt{3})^2$
b $(3+2\sqrt{3})(3-2\sqrt{3})$

6 Express $\frac{2}{\sqrt{8}}$ in the form $a\sqrt{2}$.

7 Find the exact area of the triangle shown, expressing your answer in the form $a\sqrt{3}$.

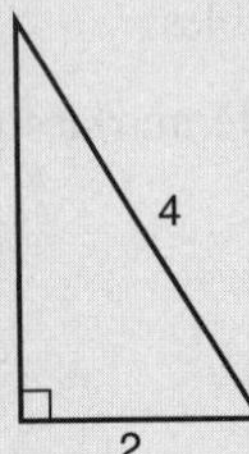

8 Simplify:

a $\frac{x^2+x}{x}$ **b** $\frac{x^2+x}{x+1}$ **c** $\frac{x^2-x-6}{2x-6}$

9 Simplify:

a $\frac{x^2+4x+3}{x+1}$ **b** $\frac{x^2+4x+4}{x^2-4}$
c $\frac{x^2+5x-6}{x^2+8x+12}$

10 Express as a single fraction:

a $\frac{x}{2}-\frac{x}{3}$
b $\frac{x+2}{3}-\frac{x+1}{4}$
c $\frac{3(x-1)}{4}-\frac{2(x+1)}{8}$

11 Express as a single fraction:

a $\frac{2}{x}-\frac{3}{2x}$
b $\frac{1}{x-2}-\frac{1}{x+3}$
c $\frac{x}{x^2-3x-4}-\frac{1}{x-4}$

12 Solve the equations:

a $\frac{x+1}{2}=\frac{x-2}{3}$
b $\frac{x+1}{3}-\frac{x}{6}=1$
c $\frac{3x-6}{5}+2x=4$

13 Solve the equations:

a $x+1=\frac{6}{x}$
b $\frac{2}{x}+\frac{2x}{x-1}=5$
c $\frac{x+1}{x-2}-\frac{x+1}{x+4}=1$

14 Work out $\frac{dy}{dx}$ when:

a $y=2x^3-6x^2+4$
b $y=(x^2+1)(2x-1)$
c $y=x^2-x+\frac{1}{x^2}$

15 The diagram shows a sketch of the curve $y = \frac{x^3}{3} + x^2 - 7x - 5$.

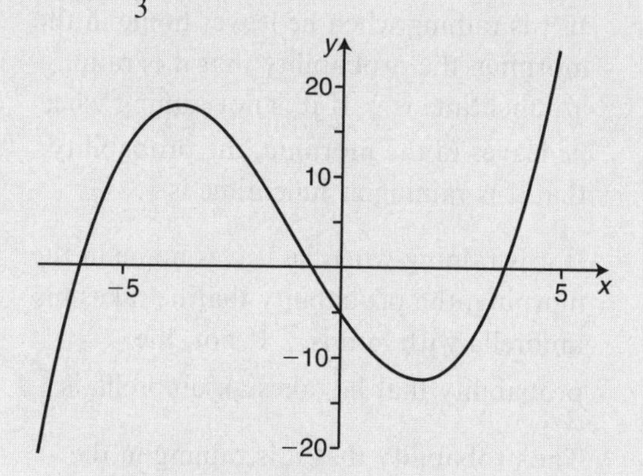

a Work out $\frac{dy}{dx}$.

b Hence find the points on the curve where the gradient is parallel to the line $y = x$.

16 a Show that the point (4, −4) on the curve $y = x^2 - 8x + 12$ is a turning point and determine its nature.

b Draw a sketch of the curve showing the turning point and the intersections with both axes.

17 A serving machine fires a tennis ball vertically into the air. The ball's height, h metres, after t seconds, is given by $h = 40t - 5t^2$ for $0 \leqslant t \leqslant 25$.

a Work out an expression for the velocity, v, after t seconds.

b Hence find the time when $v = 0$.

c Work out the maximum height reached by the ball.

18 Find length a and angle B.

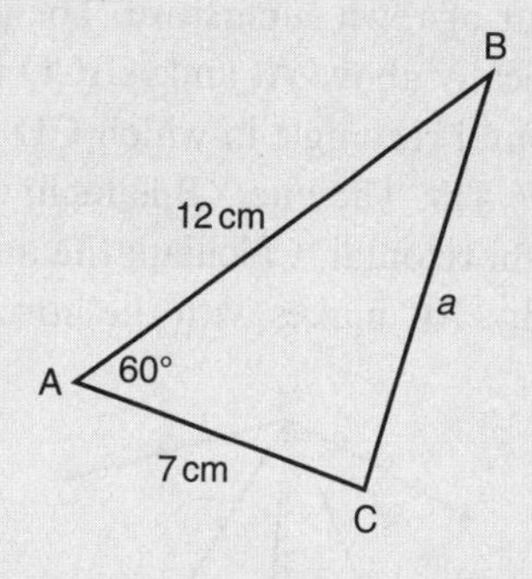

19 Find:

a angle C

b the area of triangle ABC.

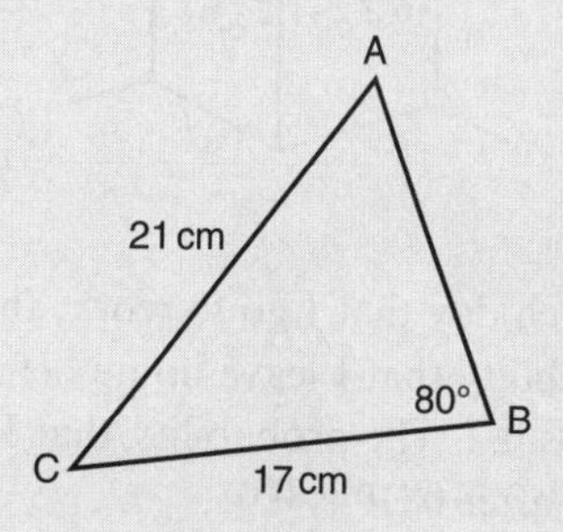

20 ABCDEFGH is a cube.

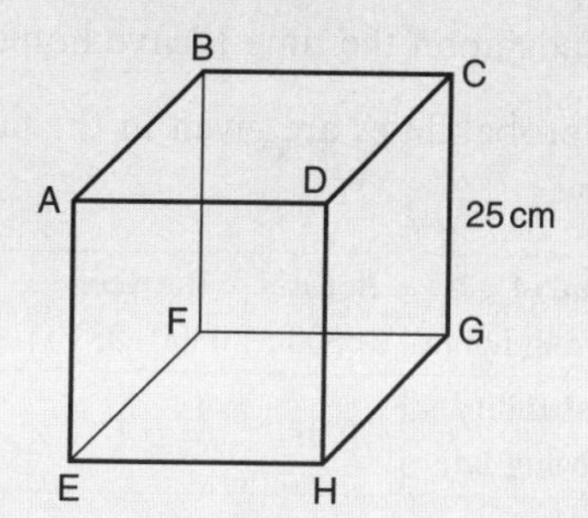

Find:

a length AG

b angle AGE.

UNIT 4 ♦ Examination practice

21 The diagram shows part of the roof of a new out-of-town superstore. The point X is vertically above A, and ABCD is a horizontal rectangle in which CD = 5.6 m, BC = 6.4 m. The line XB is inclined at 70° to the horizontal. Calculate the angle that the ridge XC makes with the horizontal.

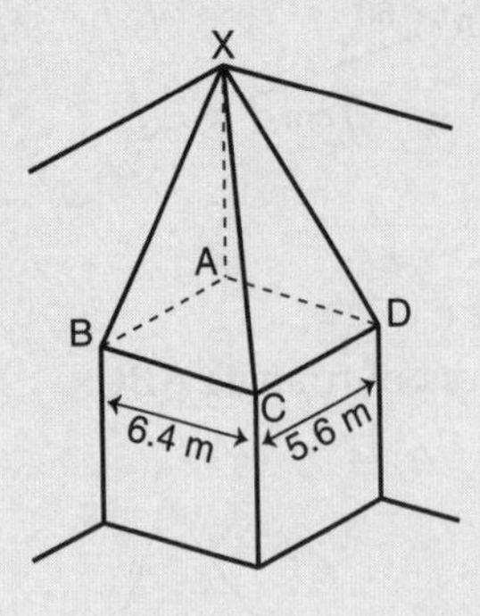

MEG

22 On each day that I go to work, the probability that I leave home before 0800 is 0.2. The probability that I leave home after 0810 is 0.05.

The probability that I am late for work depends upon the time I leave home.

The probabilities are given in the table below.

Time of leaving home	Before 0800	Between 0800–0810	After 0810
Probability of being late	0.01	0.1	0.2

I work 200 days each year. Estimate how many times I would expect to be late for work in a year.

NEAB

23 A man travels to work by car but walks from his place of work to a café for lunch.

If it is raining when he leaves home in the morning, the probability that it is raining at lunchtime is $\frac{5}{6}$. If it is not raining when he leaves in the morning, the probability that it is raining at lunchtime is $\frac{1}{8}$.

If it is raining when he leaves home in the morning, the probability that he takes his umbrella with him is $\frac{3}{4}$. If not, the probability that he takes an umbrella is $\frac{1}{5}$.

The probability that it is raining in the morning when he leaves home is $\frac{1}{3}$.

Calculate the probabilities of the following events, showing clearly (where appropriate) the fractions added or multiplied to arrive at your results.

a It is not raining when the man leaves home, and he does not take an umbrella.

b It is raining when he leaves home, but he does not take an umbrella.

c It is not raining when he leaves home, he does not take an umbrella, but it is raining at lunchtime.

d It is raining at lunchtime and he is without an umbrella.

MEI

Number 5 (Revision)

Directed numbers

$3 + (-4) = 3 - 4 = -1$
$3 - (-4) = 3 + 4 = 7$
$(-3) - (-4) = -3 + 4 = 1$
$(-3) + (-4) = -3 - 4 = -7$

$-6 \times (-2) = -12$
$(-6) \times (-2) = 12$
$6 \div (-2) = -3$
$(-6) \div (-2) = 3$

BIDMAS

Brackets, **I**ndices, **D**ivision/**M**ultiplication, **A**ddition/**S**ubtraction

Numbers

Multiples of 7 are 7, 14, 21, 28, 35, ...

Factors of 36 are 1, 2, 3, 4, 6, 9, 12, 18 and 36.

Prime factors of 36 are 2 and 3.

Prime numbers are 2, 3, 5, 7, 11, 13, 17, 19, ...

Square numbers are 1, 4, 9, 16, 25, 36, ...

Triangular numbers are 1, 3, 6, 10, 15, 21, ...

HCF and LCM

$90 = 2 \times 3^2 \times 5$
$105 = 3 \times 5 \times 7$
$\text{HCF} = 3 \times 5 = 15$
$\text{LCM} = 2 \times 3^2 \times 5 \times 7 = 630$

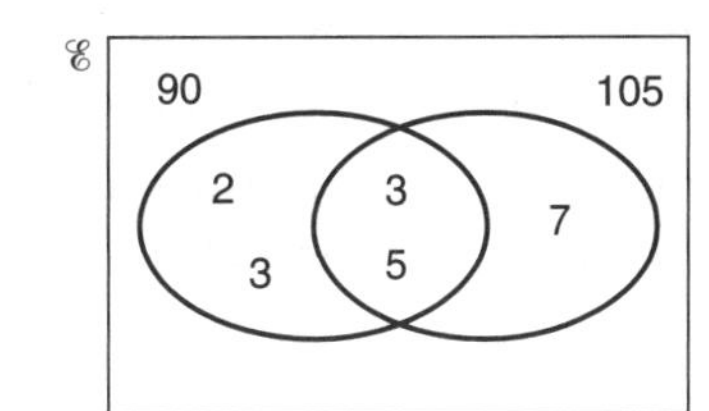

Fractions

To change $\frac{3}{8}$ into a decimal: divide 3 by 8 $\rightarrow$ 0.375

To change $\frac{7}{20}$ into a percentage: multiply by 100 $\rightarrow \frac{7}{20} \times 100 = 35\%$

To simplify $\frac{2.5}{40}$: multiply top and bottom by 2 $\rightarrow \frac{5}{80} = \frac{1}{16}$

Addition and subtraction: $\frac{1}{12} + \frac{3}{8} = \frac{2}{24} + \frac{9}{24} = \frac{11}{24}$

$\frac{5}{6} - \frac{3}{8} = \frac{20}{24} - \frac{9}{24} = \frac{11}{24}$

Mixed numbers and improper fractions: $4\frac{1}{5} = \frac{20}{5} + \frac{1}{5}$ or $\frac{(4\times5)+1}{5} = \frac{21}{5}$

Multiplication: $\frac{3}{7} \times \frac{2}{5} = \frac{6}{35}$

Division: $4\frac{1}{5} \div \frac{3}{4} = \frac{21}{5} \times \frac{4}{3} = \frac{7\times4}{5} = \frac{28}{5} = 5\frac{3}{5}$

Terminating fractions: $\frac{1}{2} = 0.5$, $\frac{3}{5} = 0.6$, $\frac{1}{16} = 0.0625$, $\frac{4}{25} = 0.16$, ...

The denominator contains only multiples of 2 and 5.

Recurring fractions: $\frac{1}{3} = 0.\dot{3}$, $\frac{3}{7} = 0.\dot{4}28\,57\dot{1}$, $\frac{5}{6} = 0.8\dot{3}$, ...

The denominator contains factors other than 2 or 5.

Recurring decimals

To convert $0.\dot{2}\dot{4}$ to a fraction:

Let $x = 0.242\,424\,24\ldots$

Then $100x = 24.242\,424\,24\ldots$

Subtracting: $99x = 24 \Rightarrow x = \frac{24}{99} = \frac{8}{33}$

Limits of accuracy

This carpet's dimensions are 2 m × 5 m (to the nearest 0.1 m).
What are the limits of accuracy for the carpet's perimeter and its area?

The largest possible dimensions are 2.05 m by 5.05 m.
The smallest possible dimensions are 1.95 m by 4.95 m.

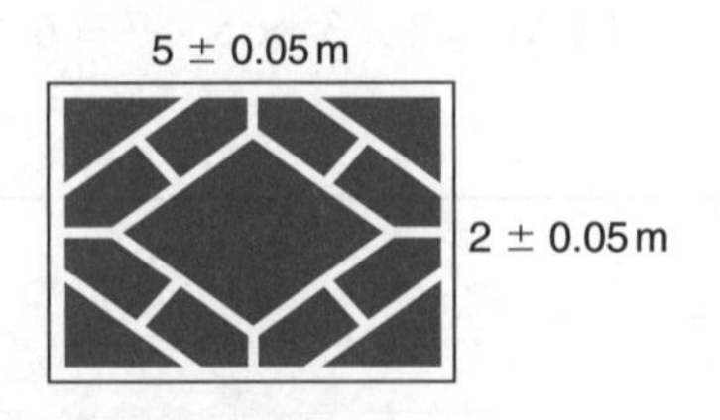

Largest perimeter = 2 × 2.05 m + 2 × 5.05 m = 14.2 m
Smallest perimeter = 2 × 1.95 m + 2 × 4.95 m = 13.8 m
So, the *exact* perimeter is between 13.8 m and 14.2 m.

Largest area = 2.05 m × 5.05 m = 10.35 m^2
Smallest area = 1.95 m × 4.95 m = 9.65 m^2
So, the *exact* area is between 9.65 cm^2 and 10.35 cm^2.

Significant figures

34.779 to 3 s.f. = 34.8
0.0659 to 2 s.f. = 0.066
582 175 to 3 s.f. = 582 000

Decimal places

2.0765 to 2 d.p. = 2.08
0.05296 to 3 d.p. = 0.053
0.71748 to 4 d.p. = 0.7175
0.71748 to 3 d.p. = 0.717

Estimation

$$\frac{5.56 \times 104.3 \times 0.51}{2.75} \approx \frac{6 \times 100 \times 0.5}{3} = 100$$

Rounding to 1 s.f.

$$\frac{25.7 \times 223.1}{20.3} \approx \frac{26 \times 220}{20} = 286 \approx 290$$

Rounding to 2 s.f.

Standard form

Numbers are written in standard form as $a \times 10^n$ where $1 \leqslant a < 10$ and n is a positive or negative integer.

Very large numbers (positive index): $670\,000 = 6.7 \times 10^5$
Very small numbers (negative index): $0.000\,325 = 3.25 \times 10^{-4}$

When multiplying and dividing, separate the parts:

$(7.93 \times 10^8) \times (4.2 \times 10^6) = (7.93 \times 4.2) \times (10^8 \times 10^6) = 33.3 \times 10^{14} = 3.33 \times 10^{15}$
$(7.93 \times 10^8) \div (4.2 \times 10^6) = (7.93 \div 4.2) \times (10^8 \div 10^6) = 1.89 \times 10^2$

When adding and subtracting, make sure that terms have matching indices:

$(7.93 \times 10^8) + (4.2 \times 10^6) = (793 \times 10^6) + (4.2 \times 10^6) = 797.2 \times 10^6 = 7.972 \times 10^8$
$(7.93 \times 10^8) - (4.2 \times 10^6) = (793 \times 10^6) - (4.2 \times 10^6) = 788.8 \times 10^6 = 7.888 \times 10^8$

Rational and irrational numbers

Rational numbers *can* be expressed in the form of a fraction $\frac{p}{q}$, where p and q are both integers.

Examples: 25, 37.21, $\frac{2}{3}$, $\frac{3}{7} = 0.\dot{4}28\,57\dot{1}$

Irrational numbers *cannot* be expressed exactly as a fraction.
Examples: $\sqrt{2}$, $\sqrt{3}$, $\sqrt{7}$, π, $(\sqrt{3}+1)$

Irrational number (+ or − or × or ÷) *irrational number* might be rational or irrational!
Examples: $\sqrt{2} \times \sqrt{8} = \sqrt{16} = 4$ $\quad (4-\sqrt{3}) + (4+\sqrt{3}) = 8$

Indices

Rules and properties:

$a^m \times a^n = a^{m+n}$ $\quad 2^3 \times 2^4 = 2^7$ $\quad [(2\times 2\times 2)\times(2\times 2\times 2\times 2)]$

$a^m \div a^n = a^{m-n}$ $\quad 2^6 \div 2^2 = 2^4$ $\quad \left[\frac{2\times 2\times 2\times 2\times 2\times 2}{2\times 2}\right]$

$(a^m)^n = a^{mn}$ $\quad (2^4)^2 = 2^8$ $\quad [(2\times 2\times 2\times 2)\times(2\times 2\times 2\times 2)]$

$a^{-n} = \frac{1}{a^n}$ $\quad 2^{-3} = \frac{1}{2^3} = \frac{1}{8}$

$a^{\frac{1}{n}} = \sqrt[n]{a}$ $\quad 9^{\frac{1}{2}} = \sqrt{9} = 3,\quad 64^{\frac{1}{3}} = \sqrt[3]{64} = 4,\quad 32^{\frac{1}{5}} = \sqrt[5]{32} = 2$

$a^{\frac{p}{q}} = (a^{\frac{1}{q}})^p$ or $(a^p)^{\frac{1}{q}}$ $\quad 27^{\frac{2}{3}} = \sqrt[3]{27^2} = \sqrt[3]{729} = 9$

or $= (\sqrt[3]{27})^2 = 3^2 = 9$

$a^1 = a$

$a^0 = 1$ $\quad$ Consider $a^m \div a^m = a^{m-m} = a^0 = 1$

Surds

$\sqrt{a \times b} = \sqrt{a} \times \sqrt{b}$ $\quad \sqrt{28} = \sqrt{4\times 7} = \sqrt{4}\times\sqrt{7} = 2\times\sqrt{7}$

$\sqrt{\frac{a}{b}} = \frac{\sqrt{a}}{\sqrt{b}}$ $\quad \sqrt{\frac{3}{4}} = \frac{\sqrt{3}}{\sqrt{4}} = \frac{\sqrt{3}}{2}$

$x\sqrt{a} + y\sqrt{a} = (x+y)\sqrt{a}$ $\quad 3\sqrt{2} + 4\sqrt{2} = 7\sqrt{2}$

Note: $\sqrt{18} + \sqrt{32} = \sqrt{98}$ and $18 + 32 \neq 98$

To **rationalise** the denominator means to clear the surd from the denominator.

$\frac{2}{\sqrt{3}} = \frac{2\times\sqrt{3}}{\sqrt{3}\sqrt{3}} = \frac{2\times\sqrt{3}}{3}$ $\quad \frac{2}{\sqrt{3}} + \sqrt{3} = \frac{2}{3}\sqrt{3} + \sqrt{3} = \frac{5}{3}\times\sqrt{3}$

Percentages

Percentage of

35% is 35 *per cent* is 35 *per hundred* is $\frac{35}{100}$.

35% of \$5000 $= \frac{35}{100} \times \$5000 = \$1750$.

Percentage change

Multiplying factor method:

to add R%, multiply by $1 + \dfrac{R}{100}$

to subtract R%, multiply by $1 - \dfrac{R}{100}$

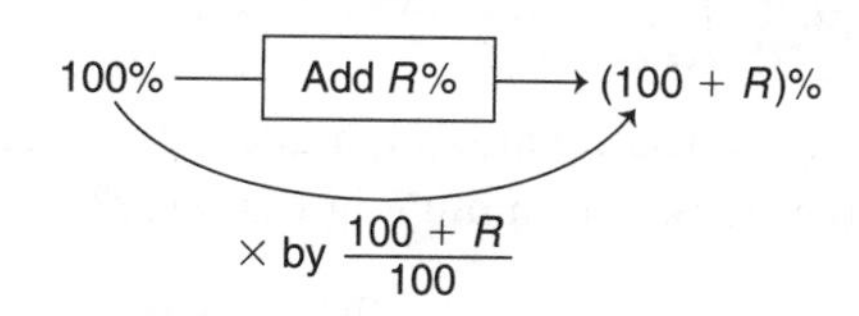

If the original price is \$60:

For a **profit** of 30% $\quad x = \$60 + \left(\frac{30}{100} \times \$60\right) = \$60 + \$18 = \$78 \qquad 1.30 \times \$60 = \$78$

For a **loss** of 30% $\quad x = \$60 - \left(\frac{30}{100} \times \$60\right) = \$60 - \$18 = \$42 \qquad 0.70 \times \$60 = \$42$

Inverse percentage

To find the original amount, divide by the multiplying factor.

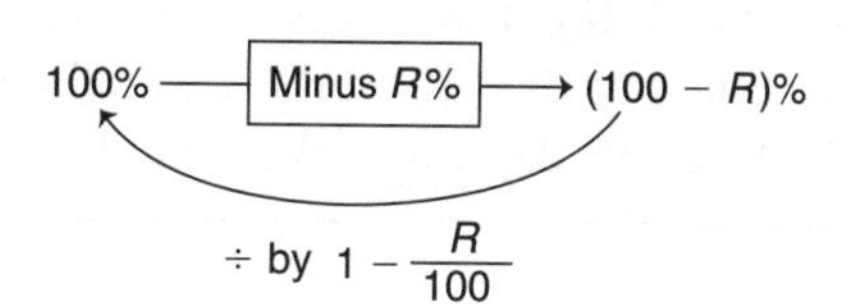

- A shirt is sold at a discount of 15% for \$20.40. Calculate the original price.

  Multiplying factor $= \frac{85}{100} = 0.85$

  $\therefore$ original price $= \$20.40 \div 0.85 = \$24$

- A trader sells apples at \$1.80 per kg. He makes a 20% profit. How much does he pay for them?

  Multiplying factor $= \frac{120}{100} = 1.2$

  $\therefore$ cost $= \$1.80 \div 1.20 = \$1.50$

Compound percentages

\$1000 is invested for four years at 5% compound interest.

Multiplying factor = 1.05 $\qquad$ Final amount $= \$1000 \times 1.05^4 = \$1215.51$

A new motorcycle costs \$10 000, but it loses 30% of its value in the first year, 20% in the second year and 10% in the third year. How much is it worth after three years?

Multiplying factors = 0.7, 0.8, 0.9 $\qquad$ Value $= \$10\,000 \times 0.7 \times 0.8 \times 0.9 = \$5040$

Ratio

Divide 84 kg in the ratio 3 : 4.

The sum of the ratios is $3 + 4 = 7$.

One part $= \frac{84}{7} = 12$ kg, three parts $= 3 \times 12 = 36$ kg and four parts $= 4 \times 12 = 48$ kg.

Direct proportion

An international bank made a profit of five billion dollars in 2004. How much profit did they make per minute in that year?

$$\text{Profit per minute} = \frac{\text{profit in year}}{\text{minutes in year}} = \frac{5 \times 10^9}{365 \times 24 \times 60} = \$9500 \text{ per minute (2 s.f.)}$$

Inverse proportion

It takes 4 days for 3 workers to build a fence.
To take one day, 12 workers would be needed. (days $\div 4$ and workers $\times 4$)
To take two days, 6 workers would be needed. (days $\times 2$ and workmen $\div 2$)

Financial arithmetic

A **wage** is earnings calculated at an hourly rate. Overtime is generally paid at a higher rate (e.g. 'time and a half') for working more than the normal time.

A **salary** is earnings calculated at a monthly or annual rate.

Gross pay has income tax deducted to leave **net** pay.

High earners have **income tax** deducted at a higher rate than low earners.

Governments levy **purchase tax** on the selling price of goods.

Banks and moneylenders **charge interest** on loans and **pay interest** on deposits.

Repayments of loans may be calculated using **simple interest** rates and/or **compound interest rates**.

Currencies can be bought and sold according to the **exchange rate**. Money exchange bureaus may charge a **commission**.

Exercise 95 (Revision)

1 Work out these.

a $2\frac{1}{3} - 1\frac{3}{4}$ **b** $2\frac{1}{3} \div 1\frac{3}{4}$ **c** $4 + 2 \times 10$ **d** $12^0 \div 12^{-1}$

2 Work out these.

a $4\frac{1}{5} \div \frac{7}{15}$ **b** $8\frac{3}{7} - 6\frac{4}{5}$

3 Place the following numbers in ascending order.

14.3, 14.25, 14.532, 14.235

4 Express 945 as a product of prime factors, using indices where necessary.

5 a Express 504 as a product of prime factors, using indices where necessary.
b Find the LCM and the HCF of 30 and 21.

6 a Change $0.\dot{3}\dot{4}$ to a fraction. **b** Change $\frac{3}{8}$ to a decimal.
c Find 2.8% of $40. **d** What is the reciprocal of 0.5?

7 Change $0.\dot{5}4\dot{3}$ to a fraction in its lowest terms.

8 Jupiter's diameter at its equator is $(89\,400 \pm 50)$ miles. The planet rotates once every 10 hours, correct to the nearest hour.

a Find the maximum speed of rotation at the equator in mph.
b Find the minimum speed of rotation at the equator in mph.

9 If $a = 6.7 \pm 0.03$, $b = 8.1 \pm 0.05$ and $c = 0.05 \pm 0.02$, find the maximum value of

$$\frac{b - a}{c}$$

10 Use standard form to work out an estimate of these, giving your answers in standard form correct to 1 significant figure.

a $(4.7 \times 10^6) \times (2.1 \times 10^{-4})$ **b** $(4.6 \times 10^6) \div (2.1 \times 10^{-4})$
c $(6.7 \times 10^{-5}) + (4.2 \times 10^{-6})$ **d** $\sqrt{(3.8 \times 10^{-5})}$

11 Express the following numbers to the number of decimal places indicated.

a 37.6248 (3 d.p.) **b** 37.6248 (2 d.p.) **c** 1.3979 (3 d.p.) **d** 1.3979 (2 d.p.)

12 Express the following numbers to the number of significant figures indicated.

a 52.38 (3 s.f.) **b** 0.5748 (2 s.f.) **c** 38 547 (3 s.f.) **d** 0.002 649 (2 s.f.)

13 Use standard form to work out an estimate of the following, and give your answer in standard form correct to 1 significant figure.

a $(6 \times 10^7) \times (8 \times 10^{-4})$ **b** $(3 \times 10^6) \div (7 \times 10^4)$

14 Work out these correct to 3 significant figures and show a rough check.

a $\dfrac{12.8}{4.5 + 8.1}$ **b** $\sqrt{7.8^2 + 3.7^2 - 2 \times 7.8 \times 3.7 \cos 47°}$ **c** $10^4 \times 9^{-4}$

15 The length of a bacterium is 0.003 05 mm. Write this number in standard form.

16 Calculate the value of $\dfrac{5.1^2 + 3.4^2}{5.80 - \sqrt{2}}$, and express your answer correct to 3 significant figures.

17 Work out these.

a $2^{-3} \times 2^5 \div 2^{-2}$ b $(3^2)^{-\frac{1}{2}}$ c $(8)^{-\frac{2}{3}}$ d $1 \div 3^{-2}$

18 Work out these.

a $2^{-4} \div 2^{-3}$ b $(5^3)^{\frac{1}{3}}$ c $(8)^{-\frac{1}{3}}$

19 Simplify these.

a $2\sqrt{3} + 2\sqrt{3}$ b $2\sqrt{3} \div 2\sqrt{3}$ c $2\sqrt{3} \times 2\sqrt{3}$

20 a Express $2\sqrt{6}$ in the form $\sqrt{n}$. b Simplify $\sqrt{20} + \sqrt{45}$.

c Simplify $4\sqrt{5} - 3\sqrt{5}$. d Rationalise the denominator of $\dfrac{2}{\sqrt{2}}$.

21 The annual total profits of Butterworth's Bazaar are always shared out as follows: 36 people get a fixed sum of \$1000 each, and then the rest is split equally between 15 'shareholders'.

a In one year of trading, the shareholders got \$800 each. What were the profits?

b In the following year, the shareholders got a 50% increase each. What was the percentage increase in profits?

22 A music CD rotates at different rates in order to maintain a constant linear velocity of 1.3 m/s. Its playing time is 75 minutes and etched into its surface is a continuous spiral track containing 1500 million pits.

a Find the total length of the spiral track in kilometres.

b Find the mean number of pits

(i) per millimetre of spiral track

(ii) detected by the CD player's laser pickup per second.

23 The world population in 2000 was 6 billion. By 2050, it is expected to have increased by an average of 0.9% per annum.

a Find the expected population in 2050.

b Find the average percentage increase per annum if, by 2050, the population were to have doubled.

24 Find 2.5% of $\$4 \times 10^4$.

25 Blood comprises 8% of the weight of a human. Amy weighs 40 kg.
Find the weight of her blood.

26 A CD is bought for \$13, which includes tax at 17.5%.
What was the price before tax was added?

27 In the UK, in the 1990s, the poorest half of the population paid 11% income tax compared with 20% in the 1970s. Find this decrease as a percentage.

28 A rare stamp, bought for \$500, increases in value at 5% per annum. Find its value after 6 years.

29 A garden seat is bought for \$80. When it is sold, a loss of \$12 is made.
Find this loss as a percentage.

30 An adult has 5 litres of blood and 5 million red blood cells, each of a diameter of 8 μm (1 μm $= 10^{-6}$ m). Each cell contains 6.4×10^8 haemoglobin molecules.

a Find the number of haemoglobin molecules in an adult.
b Find the number of red blood cells per mm^3.

31 The Great Pyramid of Giza took 20 years to build with 2.4 million blocks of stone, each of an average weight of 2.5 tonnes.

a Find the average number of blocks laid per day.
b Find the average weight of blocks laid per day, correct to 2 significant figures.
c How long would it have taken to build another pyramid, with 1.5 million blocks, with the same number of men?

32 In January 1999, by working out $2^{3021377} - 1$, a computer took 46 hours to find the largest prime number, and it contained 909 526 digits. Rebecca tries to write out this number at one digit per second, with each 10 digits taking up 5 cm of space.

a Find the time it would take her.
b Find the length of the number, when written out, in kilometres.

33 The largest sewage works in the world is near Washington DC in the USA. It treats 740 million gallons per day. Using 1 gallon $\approx$ 4.55 litres, find the rate of treatment in

a gallons per second
b m^3 per second

34 Four identical machines produce 1000 identical clamps in 1 hour.
How many clamps can one machine produce in 8 hours?

35 In the year 2000, fifty Chinese students set a new domino record when 2.7 million dominoes fell in 7 minutes. On average, how many dominoes fell per second?

36 Divide \$26.60 in the ratio 4 : 3.

37 Chi-Ho earns \$8.50 per hour for a 40-hour week. Overtime is paid at $1\frac{1}{3}$ of the normal rate. How much does he earn if he works for 46 hours in a week?

38 The exchange rate between dollars and euros is \$1 = €1.196.

a How many euros can be bought for \$75?
b How many dollars can be bought for €85?

39 Calculate the purchase tax for these purchases:

a Six bottles of Chateau Le Plonque at \$5.88 exclusive of 15% purchase tax
b 40 litres of petrol at 83.6 cents per litre inclusive of 15% purchase tax.

40 Jack needs to borrow \$7500 to help buy a new car. Calculate the interest charge for a loan over four years from these lenders:

a A finance company, which charges 8% simple interest
b A bank, which charges 7.2% compound interest.

Exercise 95★ (Revision)

1 Find the lowest common denominator of the following fractions, and thus list them in ascending order.

$\frac{3}{4}, \frac{13}{18}, \frac{7}{9}, \frac{2}{3}$

2 What is the highest prime factor of 442?

3 **a** Express the numbers 180 and 84 as the product of prime factors.

b Thus work out the LCM and HCF of 180 and 84.

4 Express the number $0.\dot{2}1\dot{6}$ as a fraction in its lowest form.

5 A grain of sand weighs 10^{-4} g. How many million grains make up 1 kg?

6 Work out $0.\dot{2} + 0.\dot{2}8571\dot{4}$ as a fraction.

7 Use standard form to work out an estimate of the following, and give your answer in standard form correct to 1 significant figure.

a $(8.23 \times 10^{-3}) + (2.32 \times 10^{-4})$

b $\sqrt{8.3 \times 10^{-5}}$

8 The Mayan people, who lived in South America between AD 300 and AD 800, calculated the average length of one year to be 365.2 days. The accurate figure has been calculated to be 365.2422 days.

a Find the percentage error in the Mayan figure.

b Write your answer to part **a** as 'Accurate to about one part in ..... thousand'.

9 Simplify the ratio 234 : 1092.

10 To detect whether a star has a planetary system, astronomers measure the 'wobble' of the star. They can now detect a 'wobble' equivalent to a movement across the width of a human hair from a distance of 100 km away.

Taking the width as 0.06 mm, copy and complete this statement.

This is equivalent to detecting a movement of ..... cm on the Moon, which is 400 000 km away.

11 If $a = 9.7$ to 2 significant figures and $b = 3.4$ to 2 significant figures, find

a the maximum value of $a - b$

b the minimum value of $a \div b$

12 An oil slick is trebling in area every day. When first discovered, it covered an area of $9\,\text{m}^2$.

a What is the area of the slick a week after it was discovered?

b When will the slick reach an area of $3^{12}\,\text{m}^2$?

c When the area is $3^{14}\,\text{m}^2$, the oil slick is sprayed with a chemical that dissolves two-thirds of the slick. What area is left?

13 Write each of these expressions in the form a^x and hence solve the equation for a.

a $\frac{a^2}{a^{-2}}$, $16 = \frac{a^2}{a^{-2}}$

b $\frac{1}{\sqrt[3]{a}}$, $a^{-\frac{1}{3}} = \frac{1}{2}$

c $\frac{1}{a^2} + a^{-2}$, $32 = \frac{1}{a^2} + a^{-2}$

14 Solve these for x.

a $x^{\frac{1}{3}} = 6$

b $(x^{-6})^{-\frac{1}{3}} = 169$

c $(x)^{-\frac{2}{3}} = \frac{1}{9}$

15 Express these with rational denominators.

a $\dfrac{5}{\sqrt{2}}$ **b** $\dfrac{3}{2+\sqrt{3}}$

16 Simplify $\dfrac{3\sqrt{10}}{\sqrt{40}}$

17 Write down an irrational number between 3 and 4.

18 For $x = 4.385$ and $y = \sqrt{3}$, calculate the value of $\dfrac{x^2 + y^2}{3x - y}$, and express your answer correct to 3 significant figures.

19 $\sqrt{50} \approx 7.1$

a Express the error in this approximation, correct to 3 decimal places.
b Express the error as a percentage of $\sqrt{50}$, correct to 3 significant figures.

20 A telescope on Earth can just locate Pluto, at a distance of 5.59×10^9 km.
Use the formula $V = 4R^3$ to work out the approximate volume, V, of space that can be viewed from Earth, where R is the distance to Pluto.

21 Given that $11\,111 \times 11\,111 = 123\,454\,321$ find the *exact* value of

$$\frac{999\,999\,999 \times 999\,999\,999}{1+2+3+4+5+6+7+8+9+8+7+6+5+4+3+2+1}$$

22 Work out these correct to 3 significant figures and show a rough check.

a $23 \div 4.5^{\frac{3}{2}}$ **b** $48.5 - \left(\dfrac{1}{0.67}\right)^{-3}$ **c** $(0.3)^{\frac{4}{3}} \times (0.3)^{-\frac{5}{2}}$

23 The cradle used to raise the wreck of the 15th-century ship the *Mary Rose*, and the wreck itself, weighed 15 times more out of the water than in. The ratio of the cradle's weight to the wreck's weight was 1 : 5. The cradle and wreck weighed 540 tonnes out of the water. What was the apparent weight of the wreck under the water, without the cradle?

24 A car travels from X to Y at an average speed of 60 kph and returns to X at a speed of 40 kph. What is its average speed for the whole journey?

25 **a** According to a recent survey, one-third of boys and a quarter of girls exercise regularly by the age of 15. Find, as a percentage, how many more boys exercise than girls.
b Hamial's normal temperature is 37 °C.
(i) He catches an infection and his temperature rises by 4%.
Find his high temperature.
(ii) After taking antibiotics, his temperature drops by 4%.
Find his final temperature, correct to 2 significant figures.

26 A pair of shoes is sold for €67. If this price includes purchase tax at 17.5%, how much is the purchase tax?

27 Which is the larger result, and by how much: to increase 45 kg by 10% or reduce 56 kg by 10%?

28 A house valued at €150 000 is increased in price by 15%, but a prospective buyer is offered a 15% reduction on that price if he pays cash. How much does he pay?

29 In 1990, 1 in 4 of the population were likely to develop cancer.
In 2000, the proportion rose to 3 in 10. Find the percentage increase.

30 In the UK, the number of air passengers has grown at 4.9% per annum since 1960.
In 1960, there were 22 million. How many were there in 2000?

31 Between 1970 and 1990, developing countries with 'open economies' grew by 4.5% per year.
The old communist countries, with 'closed economies', grew by just 0.7% per year.
When would each have doubled?

32 The largest volcanic eruption in the last 10 000 years was at Thera in 1520 BCE. It brought about the end of the Minoan civilisation in Greece.

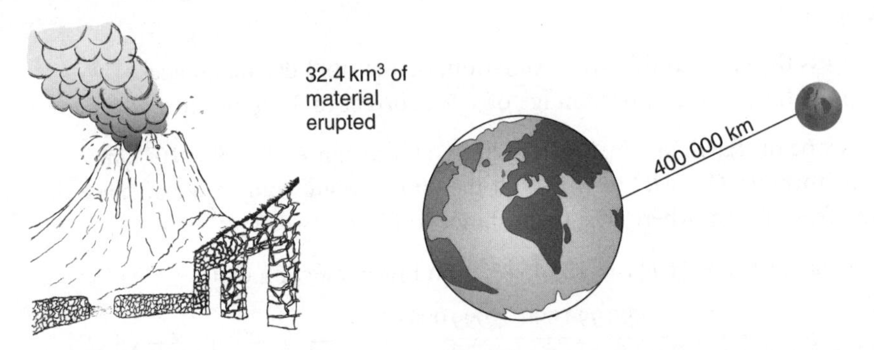

The volume erupted by Thera equals that of a 400 000 km column that reaches from the Earth to the Moon. If this column has a square cross-section, find its width.

33 At the end of the Ice Age, the water level in the Mediterranean rose by 120 metres in two years. 10 cubic miles of water cut through the Bosphorus into the Black Sea, flooding thousands of square miles.

a In the Mediterranean, what was the average increase in depth, in cm per day?

b Given that 1.609 km = 1 mile, change 10 cubic miles to cubic metres and write your answer in standard form.

c The area of the Isle of Wight in England is 38 000 hectares. (1 hectare = $10\,000\,\text{m}^2$)
Find h.

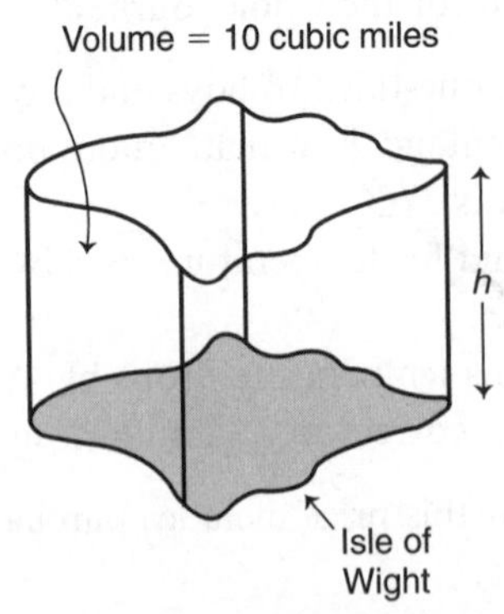

34 Three identical machines produce 40 identical tyres in 1 hour.
How many tyres can be produced by two machines in 6 hours?

35 The Earth has a mass of 5.98×10^{24} kg. Taking the Earth as a sphere of radius 6378 km, find its density in g/cm^3. (Vol. sphere $V = \frac{4}{3}\pi r^3$)

36 The cockroach, which has been around for 100 million years, is the fastest insect on six legs and can travel 1 metre in 1 second. How fast is this in km/hour?

37 The blades on a wind farm generator are 32 m long and rotate once every 4 seconds. Find the speed of the tip of a blade in mph. (1 mile $\approx$ 1.609 km)

38 The human gut has an area of about two tennis courts and on this live about 100 trillion bacteria. (One trillion is 10^{12}.)

a Taking the area of a tennis court as 260 m^2, calculate the number of bacteria per cm^2.

b Assume that each bacterium is in the shape of a cube and that a single layer covers the gut with no gaps. If these cubes were placed, touching each other, in a straight line, how long would the line be?

39 A spherical drop of mercury splashes and breaks into equal spheres with radii one-quarter that of the original sphere.

a What is the ratio of the volume of one of the new spheres to that of the original sphere?

b How many new spheres are there?

c What is the ratio of the surface area of one of the new spheres to that of the original sphere?

d What is the ratio of the total surface area of all the new spheres to the surface area of the original sphere?

40 How many mm^3 are there in a cube of external surface area 6 million km^2?

Algebra 5 (Revision)

Algebraic manipulation

Simplifying

$2xy$ means $2 \times x \times y$

x^3 means $x \times x \times x$

$2xy \times 3x^2y = 6x^3y^2$

Only **like terms** can be added or subtracted. $2a + 2a^2 - a = a + 2a^2$

$3a^2b - ab^2 + a^2b = 4a^2b - ab^2$

Multiply *every* term inside a bracket by the term outside. $2(3x - 4y) = 6x - 8y$

A negative sign outside the bracket means multiply by -1. $2 - (3x - 4y) = 2 - 3x + 4y$

Indices

$a^m \times a^n = a^{m+n}$ $\quad a^m \div a^n = a^{m-n}$ $\quad (a^m)^n = a^{mn}$ $\quad a^0 = 1\ (a \neq 0)$

Simplifying fractions

$\frac{2x}{9y} \times \frac{3y^2}{x} = \frac{2y}{3}$ $\quad \frac{2x}{y} \div \frac{x}{y} = \frac{2x}{y} \times \frac{y}{x} = 2$ $\quad \frac{3x}{4} + \frac{2y}{3} = \frac{9x + 8y}{12}$ $\quad \frac{3x}{4} - \frac{2y}{3} = \frac{9x - 8y}{12}$

Expanding brackets

Use **FOIL**

	First	Outside	Inside	Last	
$(x + 2)(x + 1) =$	x^2	$+x$	$+ 2x$	$+2$	$= x^2 + 3x + 2$
$(2x + 1)(x - 5) =$	$2x^2$	$-10x$	$+x$	-5	$= 2x^2 - 9x - 5$

	x	2
x	x^2	$2x$
1	x	2

Factorising

$x^2 + 3x = x(x + 3)$ $\qquad 3a^2b - 9ab^2 = 3ab(a - 3b)$

$x^2 + x - 2 = (x + 2)(x - 1)$ $\qquad 2x^2 - 9x - 5 = (2x + 1)(x - 5)$

An important factorisation is the **difference of two squares**.

$x^2 - 9 = (x - 3)(x + 3)$ $\qquad x^2 - a^2 = (x - a)(x + a)$

Further fractions

To simplify, factorise as much as possible, then cancel.

$$\frac{x^2 + 3x}{x} = \frac{x(x + 3)}{x} = x + 3$$

$$\frac{x^2 - 4x}{x^2 - x - 12} = \frac{x(x - 4)}{(x - 4)(x + 3)} = \frac{x}{x + 3}$$

Add or subtract in the same way as for number fractions.

$$\frac{x-1}{3}-\frac{x+3}{4}=\frac{4(x-1)-3(x+3)}{12}=\frac{4x-4-3x-9}{12}=\frac{x-13}{12}$$

$$\frac{3}{1+x}+\frac{2}{1-x}=\frac{3(1-x)+2(1+x)}{(1+x)(1-x)}=\frac{3-3x+2+2x}{(1+x)(1-x)}=\frac{5-x}{1-x^2}$$

$$\frac{x-1}{x+1}-\frac{x+3}{x-4}=\frac{(x-4)(x-1)-(x+1)(x+3)}{(x+1)(x-4)}$$

$$=\frac{(x^2-5x+4)-(x^2+4x+3)}{(x+1)(x-4)}=\frac{1-9x}{(x+1)(x-4)}$$

Formulae

Using formulae

Formulae describe how items are related to each other. Substitution and the use of the **BIDMAS** mnemonic will enable you to calculate their values.

A formula used in mechanics is $s = ut + \frac{1}{2}at^2$. Find s when $u = 4$, $a = 10$ and $t = 3$.

Substituting gives $s = 4 \times 3 + \frac{1}{2} \times 10 \times 3^2 = 12 + 45 = 57$.

Changing the subject

Use an identical process to solving equations: perform the *same* operation to *both* sides of the formula.

Make r the subject of $V = \frac{4}{3}\pi r^3$.

$$V = \frac{4}{3}\pi r^3$$

$$\frac{3V}{4\pi} = r^3$$

$$r = \sqrt[3]{\frac{3V}{4\pi}}$$

Make t the subject of $m = \frac{1+at}{1-at}$.

$$m = \frac{1+at}{1-at}$$

$$m(1-at) = 1+at$$

$$m - mat = 1 + at$$

$$m - 1 = at + mat$$

$$m - 1 = t(a + ma)$$

$$t = \frac{m-1}{a+ma}$$

Functions

- A function is a set of rules for turning one number into another.
- The **domain** of a function is the set of numbers the function can use.

  The domain of $f(x) = \frac{1}{x}$ is all real numbers except zero.

  (Division by zero is not allowed.)

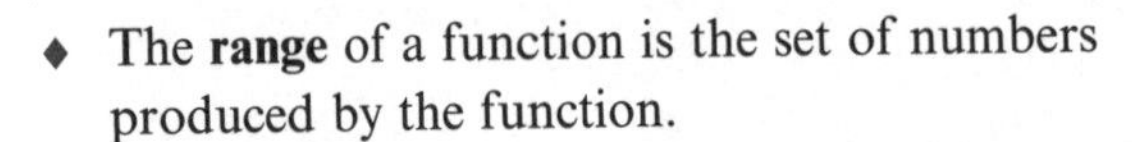

- The **range** of a function is the set of numbers produced by the function.

  The range of the function $f(x) = x^2$ is $\{y : y \geqslant 0$, y a real number$\}$.
  (Any number squared is always positive.)

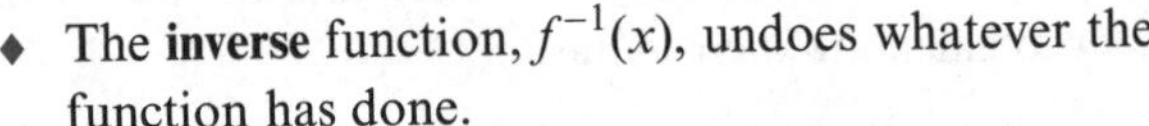

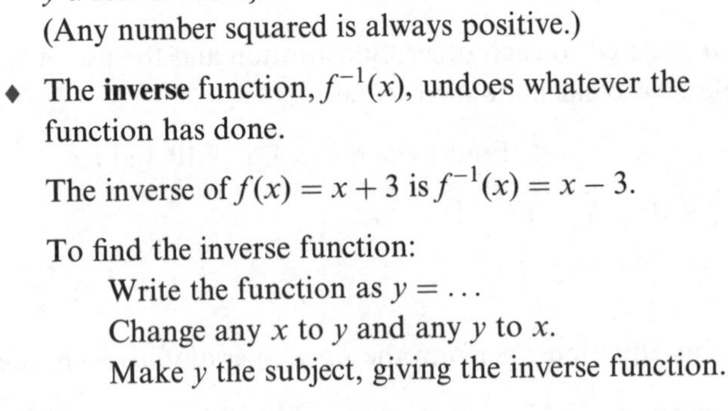

- The **inverse** function, $f^{-1}(x)$, undoes whatever the function has done.

  The inverse of $f(x) = x + 3$ is $f^{-1}(x) = x - 3$.

  To find the inverse function:
  Write the function as $y = \ldots$
  Change any x to y and any y to x.
  Make y the subject, giving the inverse function.

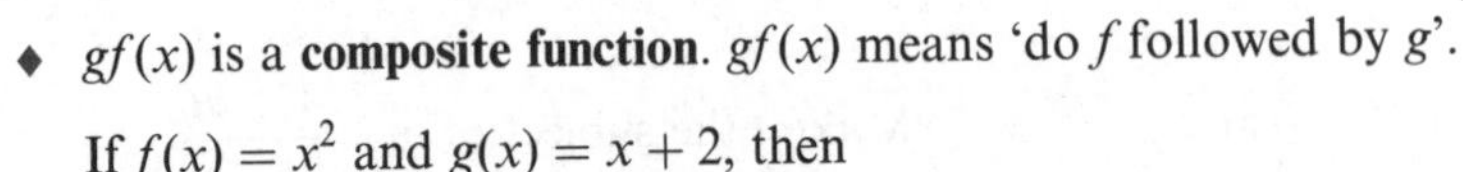

- $gf(x)$ is a **composite function**. $gf(x)$ means 'do f followed by g'.

  If $f(x) = x^2$ and $g(x) = x + 2$, then

  $gf(x) = x^2 + 2$ (do f first then g)
  $fg(x) = (x + 2)^2$ (do g first then f)

Linear equations

Solving linear equations

The way to solve linear equations is to isolate the unknown letter by systematically doing the same operation to both sides of the equation.

Solve $3(x + 1) - 2(x - 3) = 10$.

$$3(x + 1) - 2(x - 3) = 10$$
$$3x + 3 - 2x + 6 = 10$$
$$3x - 2x = 10 - 3 - 6$$
$$x = 1$$

Solve $3(x + 1) = 2(x - 3)$.

$$3(x + 1) = 2(x - 3)$$
$$3x + 3 = 2x - 6$$
$$3x - 2x = -6 - 3$$
$$x = -9$$

If the equation contains fractions, multiply everything by the LCM to clear the fractions.

Solve $\frac{5 - x}{3} = 1 - x$.

$\frac{5 - x}{3} = 1 - x$ (Multiply both sides by 3)

$5 - x = 3(1 - x)$ (Expand bracket)

$5 - x = 3 - 3x$ (Collect like terms)

$3x - x = 3 - 5$ (Simplify)

$2x = -2$ (Divide both sides by 2)

$x = -1$

Solve $\frac{x + 1}{6} + \frac{x - 2}{3} = \frac{5}{2}$.

The LCM of 6, 3 and 2 is 6. Multiply everything by 6.

$$\not{6} \times \frac{x + 1}{\not{6}} + {}^{2}\not{6} \times \frac{x - 2}{\not{3}} = {}^{3}\not{6} \times \frac{5}{\not{2}}$$

$x + 1 + 2(x - 2) = 3 \times 5$ (Expand bracket)

$x + 1 + 2x - 4 = 15$ (Collect like terms)

$3x = 18$ (Divide both sides by 3)

$x = 6$

Problems leading to linear equations

- Where relevant, draw a clear diagram and put all the information on it.
- Let x stand for what you are trying to find.
- Form an equation involving x.
- Solve the equation.
- Check that the answer makes sense.

The length of a rectangle is 1 cm more than the width.
The perimeter is 14 cm. Find the width.

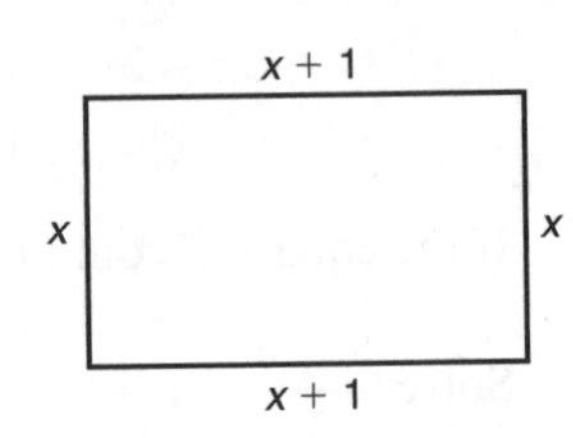

Let the width be x cm, then the length is $x + 1$ cm.

The diagram shows the lengths of the sides.

The equation is $x + (x + 1) + x + (x + 1) = 14$

Simplifying

$$4x + 2 = 14 \quad \text{(Subtract 2 from both sides)}$$
$$4x = 12 \quad \text{(Divide both sides by 4)}$$
$$x = 3$$

So the width is 3 cm.

Check: The width is 3 and the length is 4, so the perimeter is $3 + 4 + 3 = 4 = 14$.

Proportion

Direct proportion

All these statements mean the same thing.

- y is directly proportional to x.
- y varies directly with x.
- y varies as x.

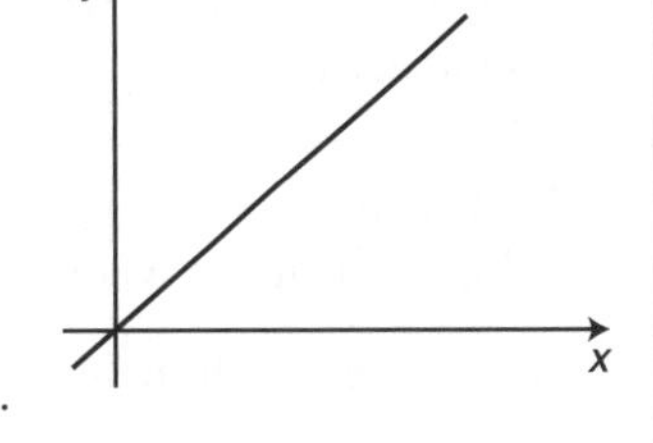

y is directly proportional to x means $y = kx$, for some fixed value of k.
The graph of y against x is a straight line through the origin.

If $y = 12$ when $x = 3$, then $12 = k \times 3 \Rightarrow k = 4$
So the equation is $y = 4x$.

- y is directly proportional to x^2 means $y = kx^2$.
- y is directly proportional to x^3 means $y = kx^3$.
- y is directly proportional to $\sqrt{x}$ means $y = k\sqrt{x}$.

Inverse proportion

y is inversely proportional to x means $y = \frac{k}{x}$ for some fixed value of k.

The graph of y plotted against x looks like this.
As x increases, y decreases.

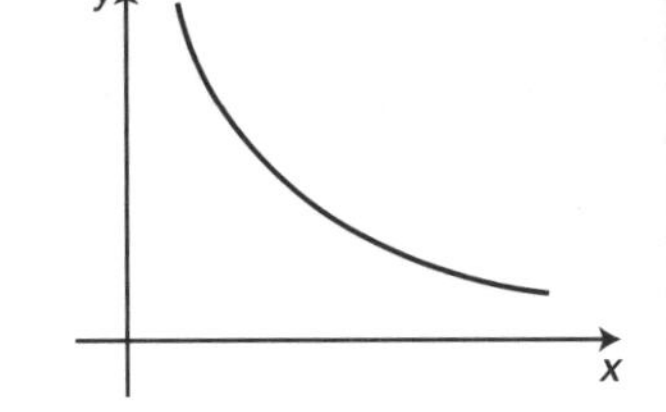

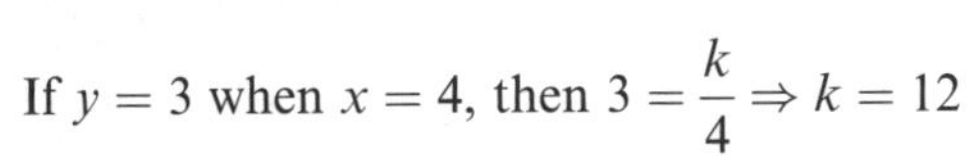

If $y = 3$ when $x = 4$, then $3 = \frac{k}{4} \Rightarrow k = 12$

So the equation is $y = \frac{12}{x}$.

- y is inversely proportional to x^2 means $y = \frac{k}{x^2}$.
- y is inversely proportional to x^3 means $y = \frac{k}{x^3}$.
- y is inversely proportional to $\sqrt{x}$ means $y = \frac{k}{\sqrt{x}}$.

Simultaneous linear equations

Solving simultaneous equations finds the point of intersection of the graphs of the equations.

There are two main ways to solve simultaneous linear equations: substitution and elimination.

Substitution

Use substitution when either x or y is the subject of at least one equation.

Solve the simultaneous equations $y = 2x - 1$ and $y = x + 2$.

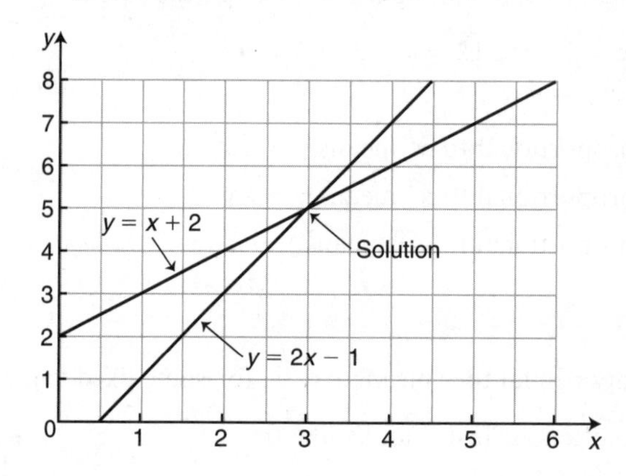

$y = 2x - 1$ and $y = x + 2$ means

$$2x - 1 = x + 2$$
$$2x - x = 2 + 1$$
$$x = 3$$

Substituting $x = 3$ into the first equation gives $y = 5$.
Check: $x = 3$, $y = 5$ satisfies the second equation, as $5 = 3 + 2$.
So the solution is $x = 3$, $y = 5$.

Solve the simultaneous equations $y = x + 1$ and $3x + 2y = 12$.

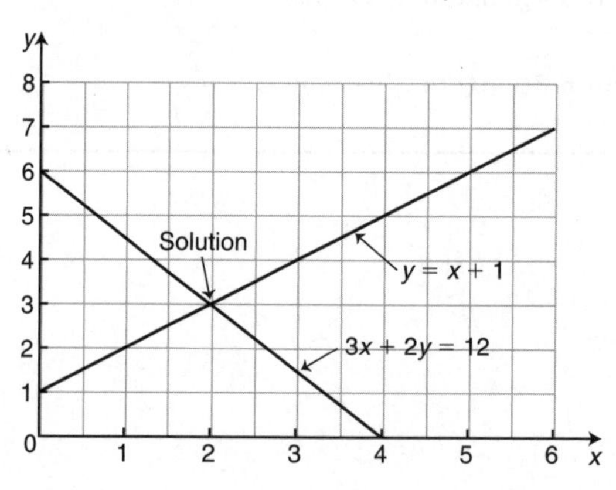

Substitute $y = x + 1$ into the second equation to give

$$3x + 2(x + 1) = 12$$
$$3x + 2x + 2 = 12$$
$$5x = 10$$
$$x = 2$$

Substituting $x = 2$ into the first equation gives $y = 3$.
Check: $x = 2$, $y = 3$ satisfies the second equation, as $3 \times 2 + 2 \times 3 = 12$.
So the solution is $x = 2$, $y = 3$.

Elimination

Use elimination when the numbers in front of either x or y are the same or the negative of each other. This means the equations can be added or subtracted to eliminate either x or y.

Solve the simultaneous equations $2x + y = 7$ and $x - y = 8$.

The numbers in front of y are $+1$ and -1, so elimination is suitable.

$$2x + y = 7$$
$$x - y = 8 \quad \text{(Add the equations together)}$$
$$3x \quad = 15$$
$$x \quad = 5$$

Substituting $x = 5$ into the second equation gives $5 - y = 8 \Rightarrow y = -3$
Check: $x = 5$, $y = -3$ satisfies the first equation, as $2 \times 5 - 3 = 7$.
So the solution is $x = 5$, $y = -3$.

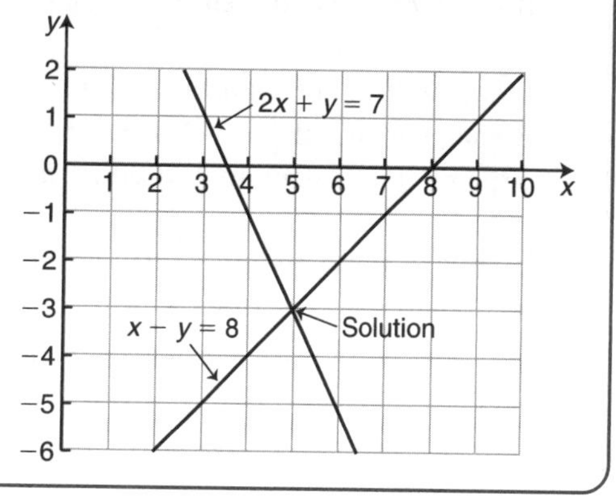

Quadratic equations

There are two main ways to solve quadratic equations: factorising and using the quadratic formula.

If the question asks for rounded solutions, for example to 3 s.f. or 3 d.p., then use the formula.

If you haven't managed to factorise the equation after about one minute, use the formula.

Always rearrange the equation so that it equals zero before you start.

Factorising

The different types are illustrated below:

- No x term — $x^2 - 16 = 0 \Rightarrow x^2 = 16 \Rightarrow x = \pm 4$
- No number term — $x - 5x = 0 \Rightarrow x(x - 5) = 0 \Rightarrow x = 0$ or 5
- Simple factorising — $x^2 - 2x - 8 = 0 \Rightarrow (x + 2)(x - 4) = 0 \Rightarrow x = -2$ or 4
- Number factor — $4x^2 - 8x - 32 = 0 \Rightarrow 4(x^2 - 2x - 8) = 0 \Rightarrow 4(x + 2)(x - 4) = 0$
  $\Rightarrow x = -2$ or 4
- Harder factorising — $3x^2 - 5x - 2 = 0 \Rightarrow (3x + 1)(x - 2) = 0 \Rightarrow x = -\frac{1}{3}$ or 2

Using the quadratic formula

If $ax^2 + bx + c = 0$ then $x = \dfrac{-b \pm \sqrt{b^2 - 4ac}}{2a}$

It is easy to make a mistake with the signs. Write down the values of a, b and c. Remember that if $b = -3$ then $-b = +3$ and b^2 must be positive (it is easy to get this wrong with a calculator). If either a or c is negative, then $-4ac$ will be positive.

Problems leading to quadratic equations

- Where relevant, draw a clear diagram and put all the information on it.
- Let x stand for what you are trying to find.
- Form a quadratic equation in x and simplify it.
- Solve the equation by either factorising or using the formula.
- Check that the answers make sense.

Solving simultaneous equations, one linear and one nonlinear

Graphically this corresponds to the intersection of a line and a curve.
Always substitute the linear equation into the nonlinear.

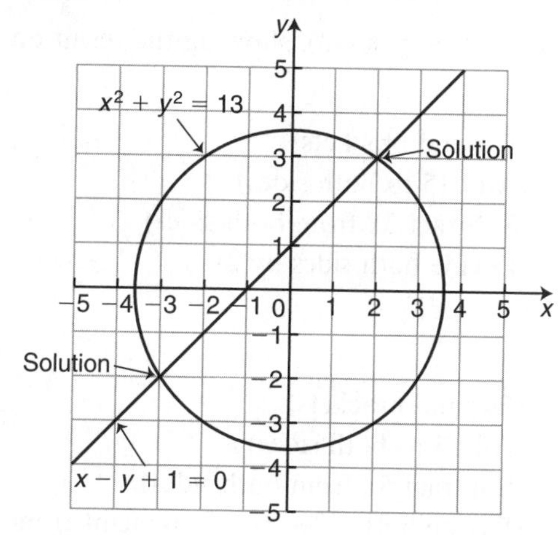

Solve the simultaneous equations $\quad x^2 + y^2 = 13 \quad ①$

$x - y + 1 = 0 \quad ②$

The linear equation is equation ②.
Make y the subject of equation ② $\quad y = x + 1$

$y^2 = (x + 1)^2$

$y^2 = x^2 + 2x + 1 \; ③$

Substitute ③ into ① $\quad x^2 + x^2 + 2x + 1 = 13$

$2x^2 + 2x - 12 = 0$ (Divide by 2)

$x^2 + x - 6 = 0$ (Factorise)

$(x + 3)(x - 2) = 0$

$x = -3$ or 2

Use ② to work out the y values, giving solutions as $(-3, -2)$ or $(2, 3)$.

Inequalities

Linear inequalities

These are solved in a similar way to equations, except that when both sides are multiplied or divided by a negative number the inequality is *reversed*. The following examples show the same inequality solved in two different ways.

Solve the inequality $3(x - 1) \leqslant 5(x - 3)$, showing the result on a number line.

Method 1

$3(x - 1) \leqslant 5(x - 3)$	(Expand brackets)
$3x - 3 \leqslant 5x - 15$	(Add 15 to both sides)
$3x + 12 \leqslant 5x$	(Subtract $3x$ from both sides)
$12 \leqslant 2x$	(Divide both sides by 2)
$6 \leqslant x$	

Method 2

$3(x - 1) \leqslant 5(x - 3)$	(Expand brackets)
$3x - 3 \leqslant 5x - 15$	(Add 3 to both sides)
$3x \leqslant 5x - 12$	(Subtract $5x$ from both sides)
$-2x \leqslant -12$	(Divide both sides by -2, remembering to *reverse* the sign)
$x \geqslant 6$	

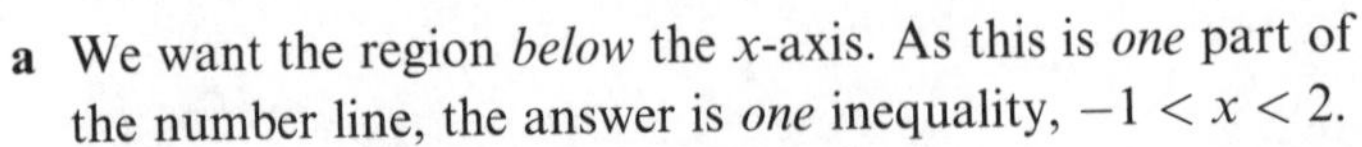

A solid circle means $\leqslant$ or $\geqslant$. An open circle means $<$ or $>$.

Quadratic inequalities

To solve a quadratic inequality, sketch the graph of the quadratic function to find the critical values.
If *one* part of the number line is required, the answer is *one* inequality.
If *two* parts of the number line are required, the answer is *two* inequalities.

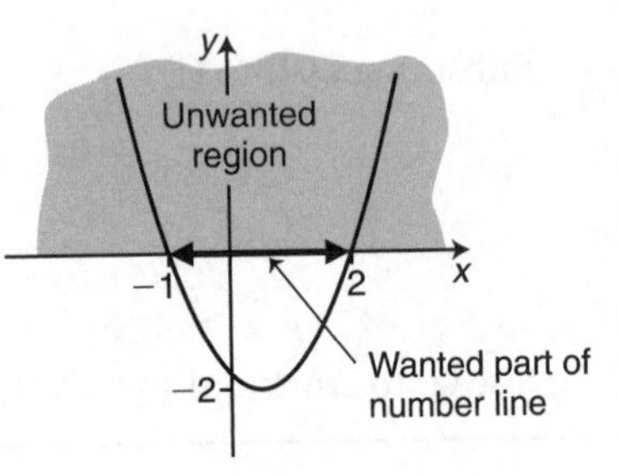

Solve **a** $x^2 - x - 2 < 0$ **b** $x^2 - x - 2 \geqslant 0$

First sketch $y = x^2 - x - 2$ by finding where the graph intersects the x-axis.
$x^2 - x - 2 = 0 \Rightarrow (x + 1)(x - 2) = 0 \Rightarrow x = -1$ or $x = 2$

So the graph intersects the x-axis at $x = -1$ and $x = 2$.
The graph is a positive parabola which is ∪-shaped.

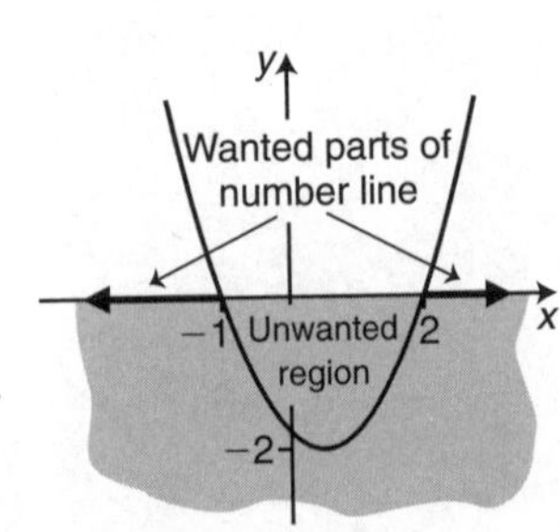

a We want the region *below* the x-axis. As this is *one* part of the number line, the answer is *one* inequality, $-1 < x < 2$.

b We want the region *above* the x-axis. As this is *two* parts of the number line, the answer is *two* inequalities, $x \leqslant -1$ or $x \geqslant 2$.

UNIT 5 ◆ Algebra

Exercise 96 (Revision)

1 Simplify:

a $x^6 \times x^4$ **b** $\dfrac{x^6}{x^4}$ **c** $\dfrac{y^4 \times y^5}{y^7}$ **d** $(2x^3y^2)^2$

2 Expand and simplify:

a $(x + 3)(x - 1)$ **b** $(2x - 1)(x + 4)$

3 Expand and simplify:

a $z(z + 3)$ **b** $3(x + 4) - 2(2x - 1)$

4 Factorise:

a $4p - 8$ **b** $x^2 + 3x$ **c** $3ab^2 + 6a^2b$

5 Factorise:

a $x^2 - 4$ **b** $x^2 + 3x + 2$

6 Simplify:

a $\dfrac{x}{4} + \dfrac{x}{3}$ **b** $\dfrac{x+1}{2} - \dfrac{x+1}{3}$

7 a Make u the subject of $m(v - u) = I$.

b Make r the subject of $\frac{1}{3}\pi r^2 h = 4$.

8 A formula used in mathematics is $v = u + at$.

a Work out v when $u = 6$, $a = -10$ and $t = 2$.

b Make a the subject of the formula.

9 A formula used in science is $T = 2\pi\sqrt{\dfrac{l}{g}}$.

a Work out T when $l = 2.45$ and $g = 9.81$.

b Make l the subject of the formula.

10 The cost, \$$C$, of using a mobile phone for one month is given by a fixed charge of \$30 plus \$0.15 for every minute spent on calls.

a One month Ashaf uses her phone for 560 minutes. What is her bill for that month?

b If Ashaf uses her phone for t minutes, find a formula giving the cost, \$$C$, in terms of t.

c Make t the subject of your formula.

d One month Ashaf's bill was \$127.50. How many minutes did she spend on calls that month?

11 Solve:

a $4x - 7 = 5$ **b** $5x - 4 = 2x + 2$ **c** $2(x + 3) - 3(x - 3) = 13$

12 Solve:

a $\dfrac{2+x}{5} = x - 2$ **b** $\dfrac{x+1}{2} + \dfrac{x-2}{3} = 4$

13 The length of a rectangle is 3 cm more than the width.
The perimeter is 46 cm. Find the width.

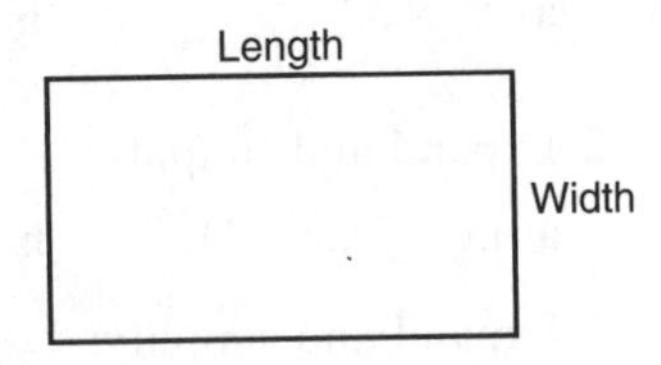

14 Tracy thought of a number. She then added 3, multiplied the result by 5 and then subtracted 8. Her answer was 42.

a Let x be the number Tracy first thought of. Form an equation in x.

b Solve the equation to find the number Tracy first thought of.

15 The angles of a triangle are as shown. Find the value of a.

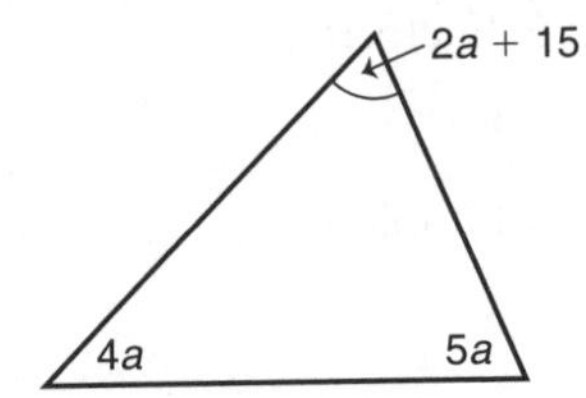

16 A rectangle has sides as shown in the diagram.
Find x and y and the area of the rectangle.

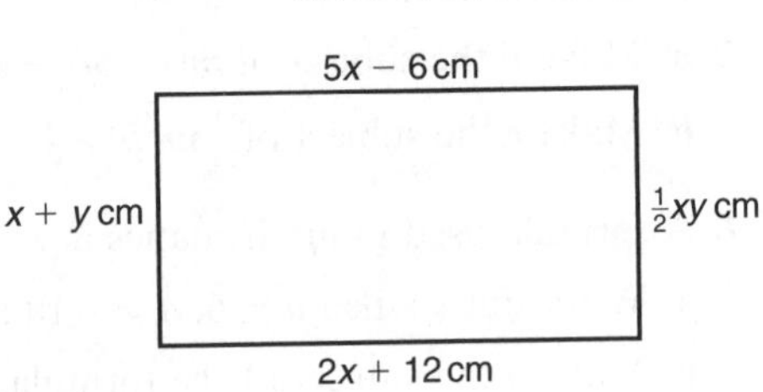

17 y is directly proportional to x. If $y = 12$ when $x = 3$, find

a the formula for y in terms of x

b y when $x = 2$

c x when $y = 8$.

18 y is directly proportional to x^2. If $y = 12$ when $x = 2$, find

a the formula for y in terms of x

b y when $x = 4$

c x when $y = 27$.

19 y is inversely proportional to x. If $y = 8$ when $x = 3$, find

a the formula for y in terms of x

b y when $x = 2$

c x when $y = 6$.

20 y is inversely proportional to $\sqrt{x}$. If $y = 18$ when $x = 4$, find

a the formula for y in terms of x

b y when $x = 9$

c x when $y = 9$.

21 Solve the simultaneous equations $x + y = 8$ and $2x - y = 1$.

22 Solve the simultaneous equations $y = 5x - 2$ and $y = 4x + 3$.

23 A farmyard has only horses and ducks in it. Altogether there are 17 heads and 58 legs in the farmyard.

a Let x be the number of horses and y be the number of ducks. Form two equations involving x and y.

b Solve the equations to find how many ducks there are.

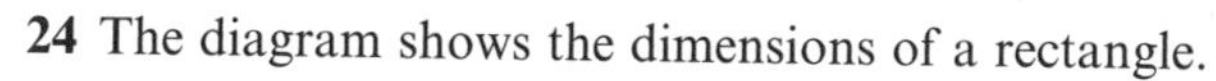

24 The diagram shows the dimensions of a rectangle.

a Form two equations in x and y.

b Solve your equations to find x and y.

c Calculate the area of the rectangle.

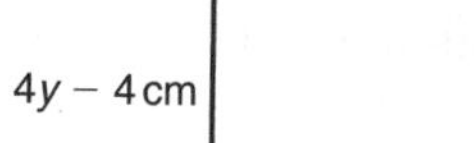

25 Solve the simultaneous equations $y = x + 1$ and $y = x^2 + 2x - 11$.

26 Solve the equation $2x^2 + 3x - 1 = 0$, giving your answers to 3 s.f.

27 Solve the simultaneous equations $x^2 + y^2 = 5$ and $y = x + 1$.

28 The length of a rectangle is 4 cm more than the width. The area of the rectangle is $32\,\text{cm}^2$.

a Show that $x^2 + 4x - 32 = 0$.

b Solve the equation $x^2 + 4x - 32 = 0$.

c Hence find the perimeter of the rectangle.

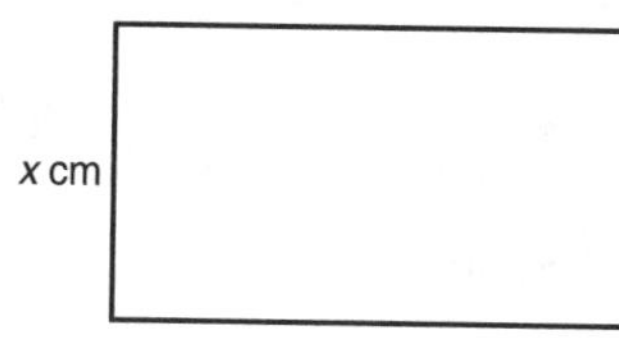

29 The two rectangles shown in the diagram have the same area. All dimensions are in cm. Find the value of x and the area of each rectangle.

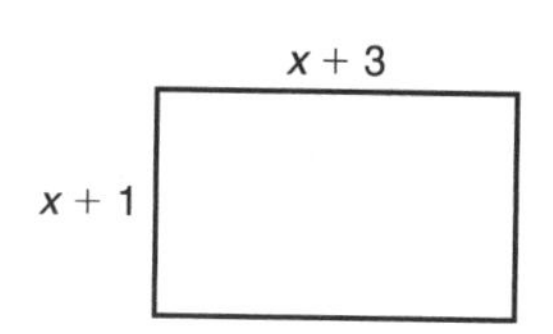

30 Solve the following inequalities, showing the answers on a number line.

a $4x - 3 > 1$ **b** $2(x + 2) \leqslant 5(x - 1)$

31 n is an integer such that $-6 \leqslant 3n < 6$. Write down the possible values of n.

32 Solve these quadratic inequalities.

a $x^2 < 9$ **b** $x^2 + x - 2 \leqslant 0$

33 $f(x) = 3x - 2$

a Find (i) $f(2)$ (ii) $f^{-1}(x)$ (iii) $ff(x)$

b If $f(a) = 7$, find the value of a.

c Solve the equation $f(x) = f^{-1}(x)$.

34 $f(x) = x + 3$ and $g(x) = \dfrac{2}{x}$

a Find (i) $f(2)$ (ii) $fg(2)$ (iii) $gf(2)$

b Which value of x must be excluded from the domain of $gf(x)$?

c Solve the equation $gf(x) = 1$.

35 $f(x) = x - 1$ and $g(x) = x^2$

a Are there any values of x that should be excluded from the domain of f?

b What is the range of (i) f (ii) g?

c Find (i) $fg(2)$ (ii) $fg(x)$

d Solve the equation $gf(x) = f^{-1}(x)$.

Exercise 96★ (Revision)

1 Simplify the following.

a $\dfrac{a}{4} + \dfrac{2a}{3}$ **b** $\dfrac{a}{4} \div \dfrac{2a}{3}$ **c** $\dfrac{a-b}{a} \times \dfrac{ab}{a-b}$ **d** $\dfrac{x+2}{5} - \dfrac{x-3}{3}$

2 Add the fractions.

a $\dfrac{1}{x+2} + \dfrac{4}{x-3}$ **b** $x + \dfrac{x(x+y)}{x-y}$

3 Simplify:

a $\dfrac{x^2 + x}{x}$ **b** $\dfrac{x^2 + x}{x+1}$

4 a Factorise $x^2 - x - 72$.

b Simplify $\dfrac{x^2 - 81}{x^2 - x - 72}$.

5 a Factorise $3x^2 + 32x - 11$.

b Simplify $\dfrac{3x^2 + 32x - 11}{x^2 - 121}$.

6 Add the fractions.

a $2 + \dfrac{x-1}{x^2 - 2x - 3}$ **b** $\dfrac{x}{x+1} + \dfrac{x^2 + x - 12}{x^2 - 2x - 3}$

7 A formula used in engineering is $I = \frac{1}{3}M(a^2 + b^2)$.

a Find I to 3 s.f. when $M = 2$, $a = 3.5$ and $b = 5.4$.

b Make a the subject of the formula.

8 A formula used in mathematics is $S = \frac{n}{2}(2a + (n - 1)d)$.

a Calculate S when $n = 54$, $a = 3$ and $d = 2$.

b Make d the subject of the formula.

9 A formula used in science is $\frac{1}{f} = \frac{1}{u} + \frac{1}{v}$.

a Calculate f when $u = 3$ and $v = 5$.

b Make u the subject of the formula.

c Calculate u when $f = 4$ and $v = 6$.

10 Solve the equation $\frac{x + 5}{15} - \frac{x - 5}{10} = 1 + \frac{2x}{15}$.

11 When the same number, x, is added to the top and bottom of the fraction $\frac{123}{456}$ the answer is $\frac{1}{2}$. Form an equation in x and solve it to find the number.

12 A triangle has two angles as shown in the diagram.

a Find the third angle in terms of x.

There are three different isosceles triangles with these three angles.

b Find the three different values of x which make the triangle isosceles.

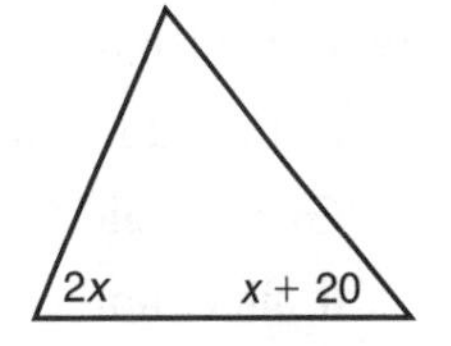

13 A ladder leaning against a vertical wall just reaches a window 5 m above the ground. Let x be the distance of the foot of the ladder from the wall and let l be the length of the ladder.

a Write down an equation connecting x and l.

b Chris thinks the ladder is unsafe, so she pulls the base of the ladder out a further 2 metres. The ladder now only reaches 4 m up the wall. Write down another equation involving x and l.

c Solve your equations to find the length of the ladder.

14 The speed of a skier sliding down a mountain, v m/s, is proportional to the square root of the distance, d m, she has moved from her starting point.
Given that $v = 20$ when $d = 100$, find

a the formula for v in terms of d

b the speed of the skier when she is 49 m from her starting point

c the distance travelled when the skier's speed is 10 m/s.

15 y is inversely proportional to the square of x. Given that $y = 10$ when $x = 2$, find

a the formula for y in terms of x

b the value of y when $x = \frac{1}{4}$

c the value of x when $y = \frac{1}{4}$.

16 The temperature of a drink, T °C, is inversely proportional to the square root of the time, m minutes, after it has been poured (for $m \geqslant 2$).
After 4 minutes the temperature is 60 °C.

a Find the formula for T in terms of m.

b Find the temperature after 9 minutes.

c Find how long it takes for the drink to cool to 50 °C.

17 Solve the simultaneous equations $y = 3x + 8$ and $5x + 2y + 6 = 0$.

18 Solve the simultaneous equations $6x - 5y = 7$ and $4x + 3y = 11$.

19 In the triangle shown, AC = CB and AD = AB.

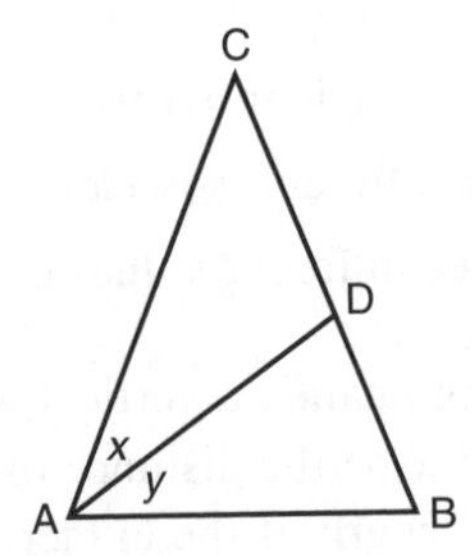

a Find an equation connecting x and y.

Given angle $C = 40°$

b find another equation connecting x and y

c solve these equations to find x and y.

20 Max sold 14 tickets for a concert. He sold x tickets at \$12 each and y tickets at \$18 each. He collected \$204.

a Write down two equations connecting x and y.

b Solve your equations to find how many of each kind of ticket he sold.

21 Anne and Brian have some CDs. If Anne were to give Brian a CD, they would both have an equal number of CDs. If Brian were to give Anne a CD, then Anne would have twice as many CDs as Brian.

a Let x and y be the number of CDs Anne and Brian have respectively. Form a pair of simultaneous equations in x and y.

b Solve these equations to find how many CDs they each have.

22 Mandy is investigating whether it is possible to have a right-angled triangle where the sides are three consecutive whole numbers. She lets x, $x + 1$ and $x + 2$ stand for the lengths of the sides.

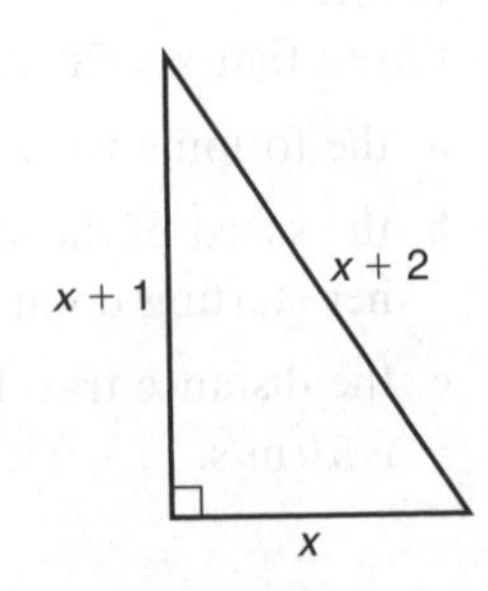

a Form and simplify an equation for x.

b Solve the equation and show that only one triangle is possible.

23 The time, t minutes, left for recording on a video tape is given by $t = 11r^2 - 16$, where r cm is the radius of the unused tape.

a Mary wants to record an episode of 'Coastwatch' which lasts three-quarters of an hour. She finds a tape with 2.5 cm radius of unused tape on it. Is there enough blank tape?

b Franz wants to put marks on the casing of a tape to indicate 1 hour, 2 hours and 3 hours recording time left. Where should he put the marks?

24 The area of the triangle shown is $24\,\text{cm}^2$.

a Show that $x^2 + 6x - 16 = 0$.

b Solve the equation $x^2 + 6x - 16 = 0$.

c Hence find the height of the triangle.

$x + 4$ cm

$2x + 4$ cm

25 The trapezium shown has an area of $21\,\text{cm}^2$.

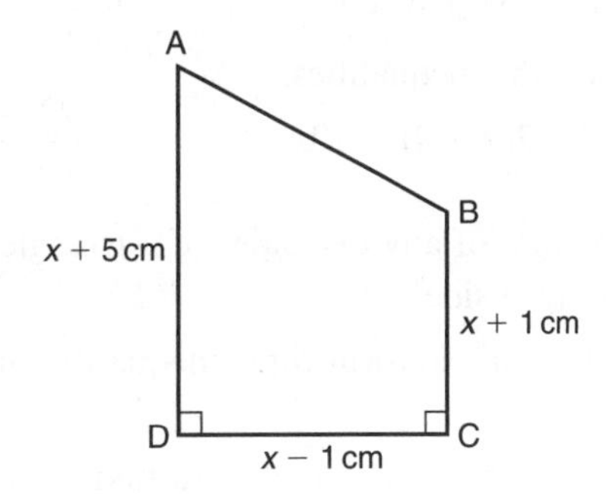

a Show that $x^2 + 2x - 24 = 0$.

b Solve the equation $x^2 + 2x - 24 = 0$.

c Hence find the length AB.

26 Solve the simultaneous equations $5x + 9y = 1$ and $x^2 - 9y = 13$.

27 Solve the simultaneous equations $x^2 - xy + y^2 = 7$ and $2x - y = 5$.

28 Shane takes off from the top of a skateboard ramp. The ramp can be modelled by $y = 3 - \frac{x}{2}$, while Shane's path can be modelled by $y = 1 + 2x - \frac{1}{2}x^2$, with units in metres. Find, by calculation, the co-ordinates of where Shane takes off and lands.

29 **a** Find where the line $3x + 4y = 25$ intersects the circle $x^2 + y^2 = 25$.

b What does your answer tell you about the line and the circle?

30 Kayleigh is designing a pendant. The outline of the pendant is given by $4x^2 + y^2 = 4$, with units in centimetres. She wants to divide up the pendant with two straight lines as shown in the diagram. Kayleigh finds that the lines $y = 2x - 1$ and $y = -2x - 1$ are suitable.

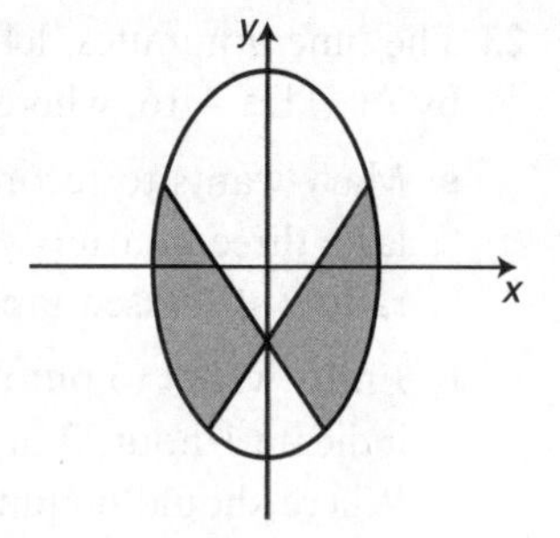

a Kayleigh needs to find where the line $y = 2x - 1$ intersects the outline of the pendant. Solve the simultaneous equations $y = 2x - 1$ and $4x^2 + y^2 = 4$ to find these points.

b Where does the other line intersect the outline?

31 n is an integer such that $-4 < 5n + 3 \leqslant 13$. Write down the possible values of n.

32 Solve the inequalities.

a $x^2 - 16 > 0$ **b** $x^2 - 16x < 0$

33 Solve the inequalities.

a $x^2 - 4x + 3 \leqslant 0$ **b** $2 + x - x^2 < 0$

34 Solve the inequalities.

a $4 - 2(x + 4) < 10x$ **b** $\dfrac{22 + x}{4} \geqslant 3x$

35 The sum of any two sides of a triangle must be greater than the third side.

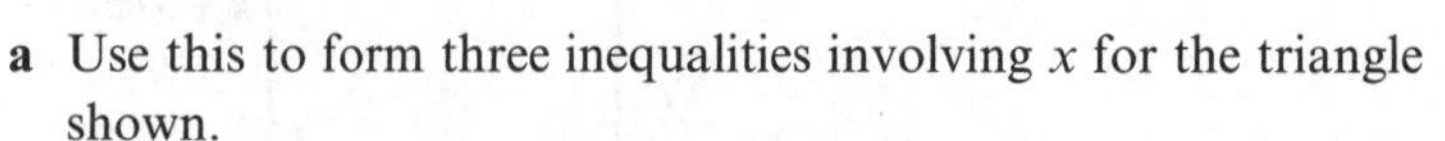

a Use this to form three inequalities involving x for the triangle shown.

b Solve these inequalities to find the least and greatest values of x.

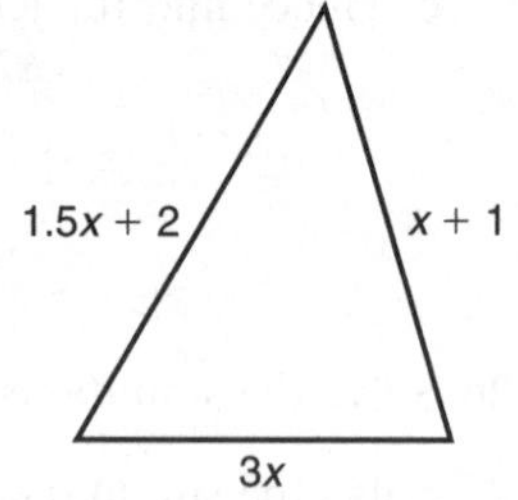

36 $f(x) = \dfrac{2x + 2}{2x + 5}$

a Find $f(2)$.

b Solve the equation $f(x) = 0$.

c Find the two values of x so that $f(x) = x$.

d What value of x must be excluded from the domain of f?

37 $f(x) = \dfrac{x}{x + 2}$

a What value of x must be excluded from the domain of f?

b Find $f^{-1}(x)$.

c Solve the equation $f(x) = f^{-1}(x)$.

38 $f(x) = \sqrt{2 + x}$ and $g(x) = x^2$

a What is the domain of f?

b Find (i) $gf(x)$ (ii) $fg(x)$

c What is the domain of $fg(x)$?

d Solve the equation $g(x) = fg(x)$.

Graphs & sequences 5 (Revision)

Straight-line graphs

Graphs of the form $y = mx + c$

The equation of any straight-line graph can always be written in the form $y = mx + c$, where m is the **gradient** and c is the **y-intercept**.

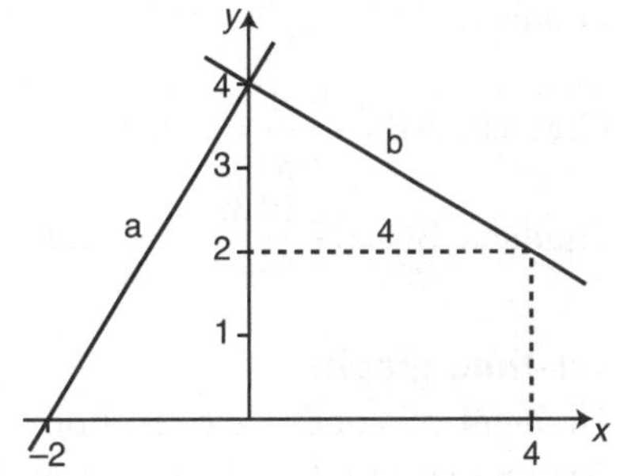

Line a has a gradient of $\frac{4}{2} = 2$ (so $m = 2$), and a y-intercept of $+4$.

Therefore its equation is $y = 2x + 4$.

Line b has a gradient of $-\frac{2}{4} = -\frac{1}{2}$ $\left(\text{so } m = -\frac{1}{2}\right)$, and a y-intercept of $+4$.

Therefore its equation is $y = -\frac{1}{2}x + 4$.

Graphs of the form $ax + by = c$

The graph of $3x + 4y = 12$ can be arranged to give $y = -\frac{3}{4}x + 3$, showing that it is a straight line with gradient $-\frac{3}{4}$ and y-intercept at 3. To sketch the graph of $3x + 4y = 12$, substitute $x = 0$, to get the y-intercept of 3, and substitute $y = 0$ to get the x-intercept of 4.

Other graphs

These graphs are often used to model real-life situations.

Quadratic graphs $y = ax^2 + bx + c$

Parabolas

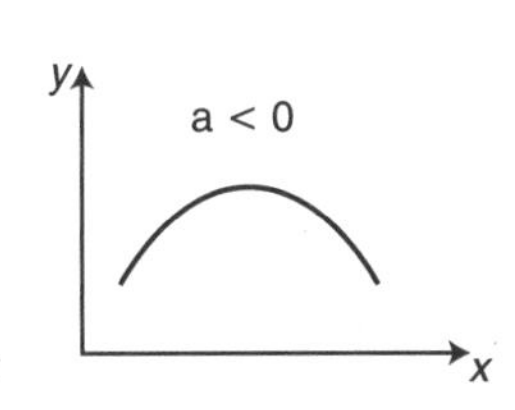

Solution of $0 = ax^2 + bx + c$

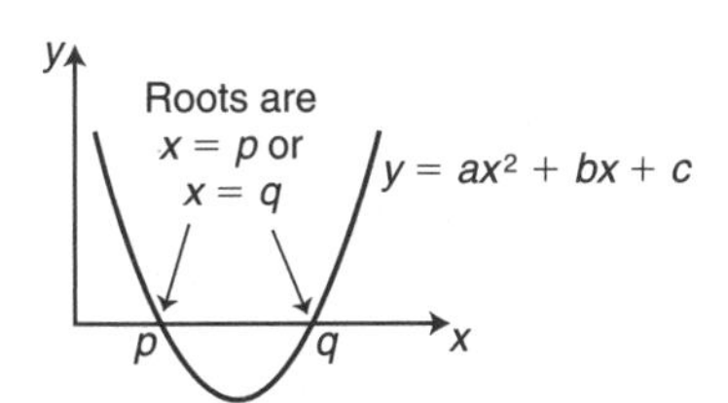

Cubic graphs $y = ax^3 + bx^2 + cx + d$

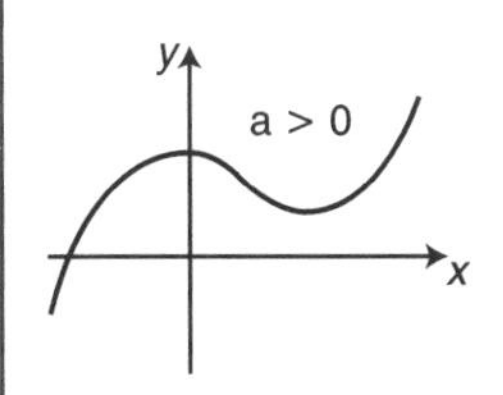

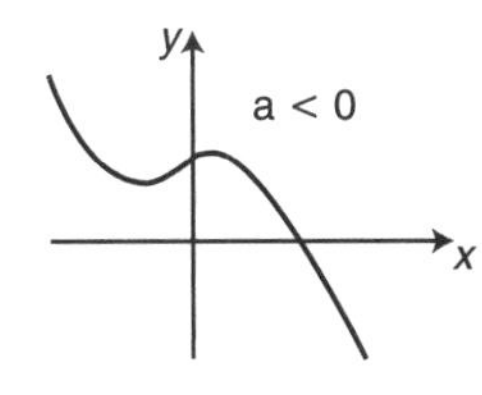

Reciprocal graphs $y = \frac{a}{x}$

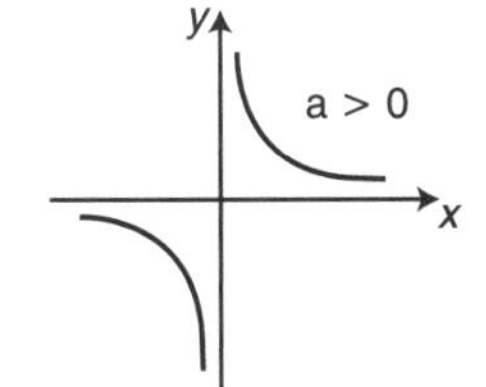

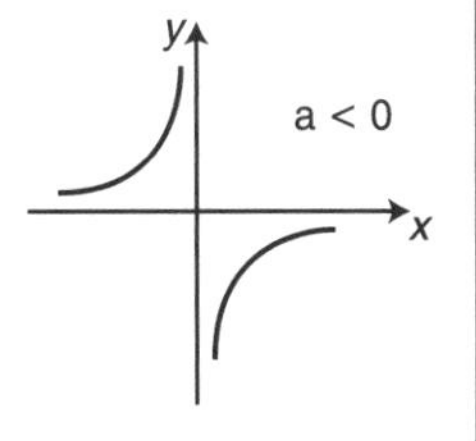

Distance/speed–time graphs

Distance–time graphs

Gradient of slope = speed

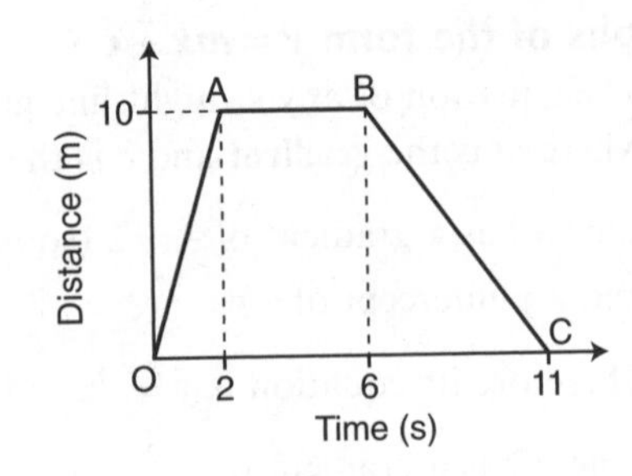

Velocity

Gradient OA $= \dfrac{10\,\text{m}}{2\,\text{s}} = 5\,\text{m/s}$

Gradient AB $= \dfrac{0\,\text{m}}{4\,\text{s}} = 0\,\text{m/s}$

Gradient BC $= \dfrac{-10\,\text{m}}{5\,\text{s}} = -2\,\text{m/s}$

Speed–time graphs

Gradient of slope = acceleration
Area under graph = distance travelled

Acceleration

Gradient OA $= \dfrac{10\,\text{m/s}}{2\,\text{s}} = 5\,\text{m/s}^2$ (speeding up)

Gradient AB $= \dfrac{0\,\text{m}}{3\,\text{s}} = 0\,\text{m/s}^2$ (constant speed)

Gradient BC $= \dfrac{-10\,\text{m/s}}{4\,\text{s}} = -2.5\,\text{m/s}^2$ (slowing down)

Average speed

Average speed $= \dfrac{\text{distance travelled}}{\text{time}} = \dfrac{\frac{1}{2} \times (3+9) \times 10}{9} = 6\dfrac{2}{3}\,\text{m/s}$

Simultaneous equations

To solve simultaneous equations graphically

- Draw the graphs for both equations on one set of axes.
- The solution is at the intersection points of the graphs.
- If the graphs don't intersect, there is no solution.
- If the graphs are the same, there are an infinite number of solutions.

Solve the simultaneous equations

$$x + y = 6,\ 2x - y = 0$$

Solution is $x = 2$, $y = 4$.

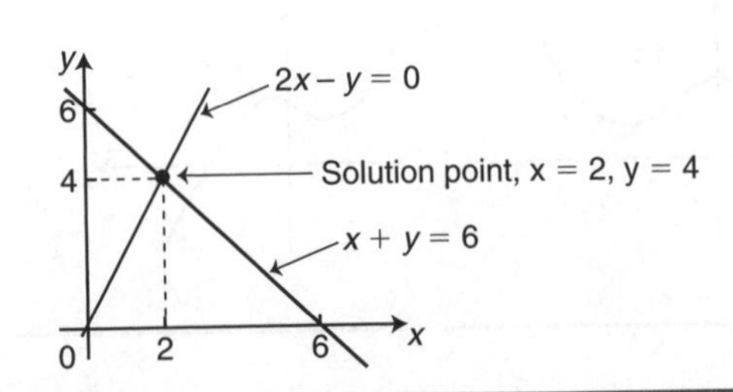

Simultaneous inequalities

Inequalities can be shown graphically by shading regions to identify solutions in unshaded regions.

Solve the inequalities $x \geqslant 0$, $y \geqslant 0$, $x + y \leqslant 3$ and $y < 2$ by drawing suitable lines (solid or broken) and shading *unwanted* regions.

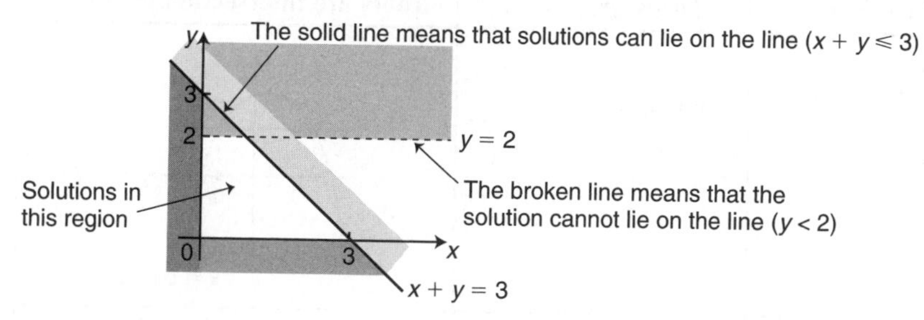

Sequences

To continue a sequence

Find the difference between each term.

Sequence	2		7		12		17		22	...
Differences		5		5		5		5		

The difference is 5, so the next term is 27.

To find the rule if the first row of differences is constant

If the differences are equal to a, the formula for the nth term will be $an \pm b$ where b is another constant.

Sequence	−5		−2		1		4 ...
Differences		**+3**		**+3**		**+3**	

Therefore the rule is $+\mathbf{3}n + b$. When $n = 1$, the formula must give the first term as −5. Thus $3 \times 1 + b = -5$, giving $b = -8$. Therefore the rule for the nth term (t_n) is $t_n = 3n - 8$.

Solution of equations

- To solve simultaneous equations graphically, draw both graphs on one set of axes. The co-ordinates of the intersection points are the solutions of the simultaneous equations.

To find roots of ...	Draw graphs ...	Roots are intersection points ...
$d = ax^2 + bx + c$	$y = ax^2 + bx + c$ $y = d$	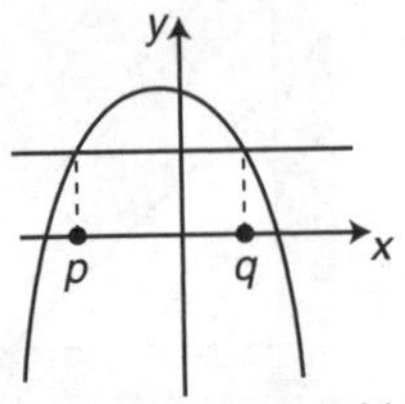 $x = p$ or q
$x + d = ax^2 + bx + c$	$y = ax^2 + bx + c$ $y = x + d$	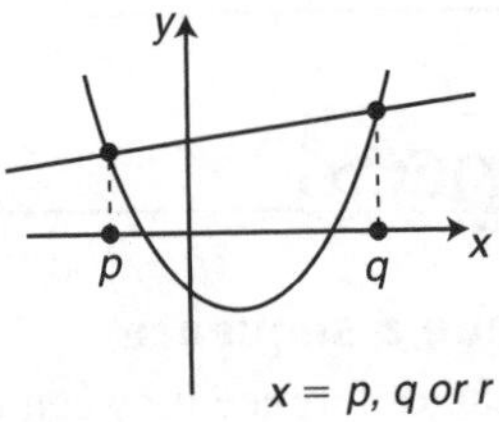 $x = p, q$ or r
$cx + d = ax^3 + bx^2$	$y = ax^3 + bx^2$ $y = cx + d$	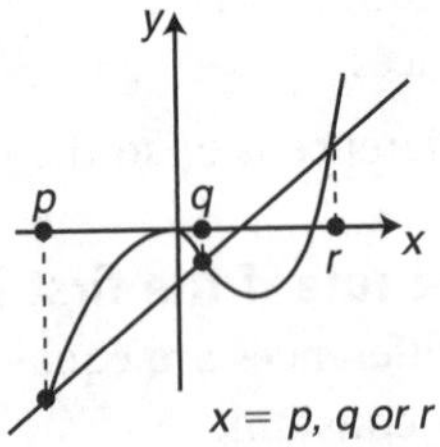 $x = p, q$ or r

Gradients of curves

Estimating gradients of curves

The gradient of a curve at a point P can be *estimated* by drawing the best-fitting tangent to the curve by eye, then finding the gradient of this line, using

$$\text{Gradient} = \frac{\text{'rise'}}{\text{'run'}}$$

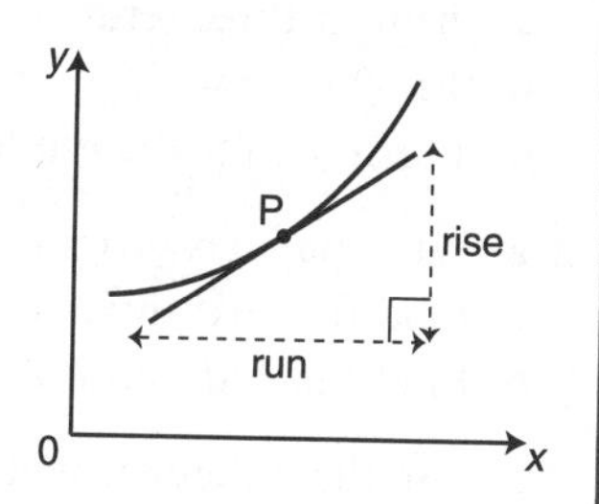

Calculating gradients of curves

The gradient of a curve at any point can be found by differentiating.

If $y = ax^n$ then $\frac{dy}{dx} = nax^{n-1}$, where a is a constant.

Stationary points occur where $\frac{dy}{dx} = 0$.

Minimum points occur where $\frac{dy}{dx} = 0$ and the curve is ∪-shaped near this point. The gradient around this point changes from − to +.

Maximum points occur where $\frac{dy}{dx} = 0$ and the curve is ∩-shaped near this point. The gradient around this point changes from + to −.

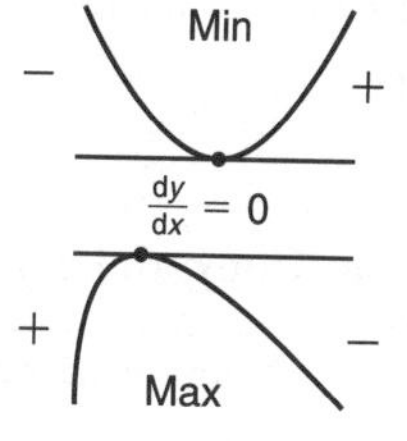

Motion

If a particle moves in a straight line such that it is s from a fixed point O after time t:

Displacement, $s = f(t)$

Velocity, $v = \frac{ds}{dt}$

Acceleration, $a = \frac{dv}{dt}$

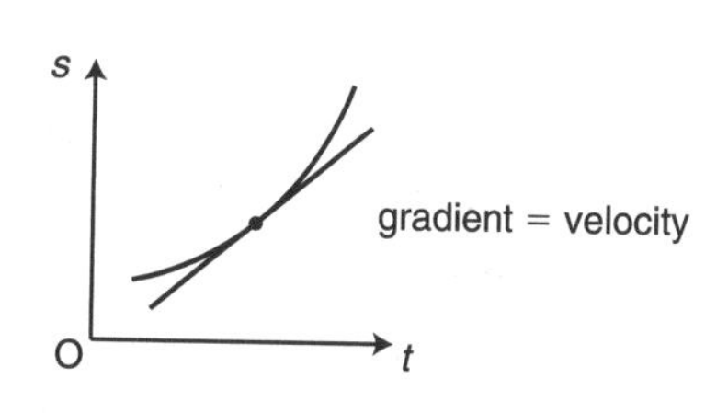

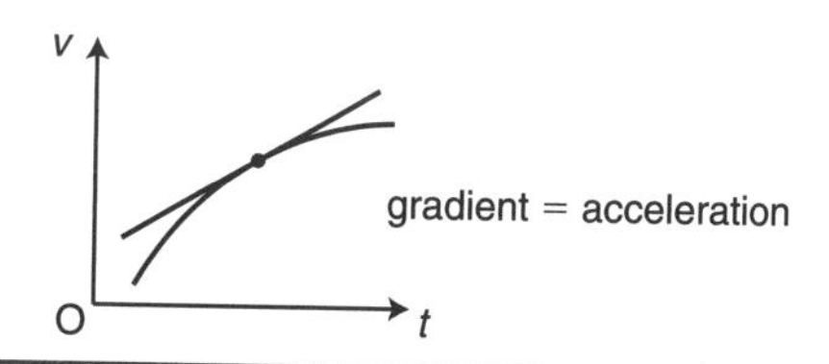

Exercise 97 (Revision)

1 For the sequence 3, 9, 15, 21, ..., find

a the next three terms

b the 50th term

c a formula for the nth term

2 a What name is given to the sequence 1, 4, 9, 16, ... ?

b Find the next three terms of this sequence.

c Find the 20th term of this sequence.

3 a Use the difference method to find the next three terms of the sequence 1, 2, 4, 7, 11, ...

b Find the terms a and b in the sequence ..., a, 2, 9, 16, 23, b, ...

4 a How many terms are in the sequence 7, 10, 13, 16, ..., 94?

b The formula for the nth term of the sequence 4, 10, 18, 28, ... is $n^2 + an$. Find the value of a.

5 Here is a sequence of blue tiles surrounding white tiles.

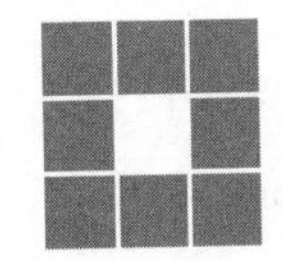
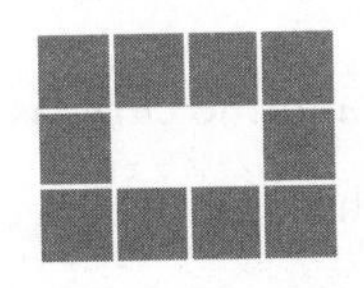
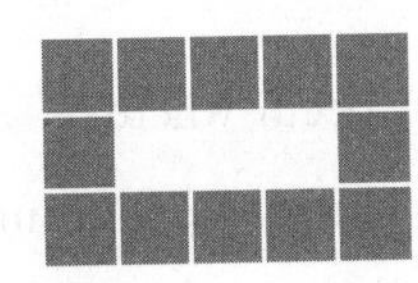

a Copy and complete this table.

Number of white tiles (w)	1	2	3	4	5
Number of blue tiles (b)	8				

b Find a formula giving b in terms of w.

c How many blue tiles are needed to surround the pattern with 25 white tiles?

d A pattern has 126 blue tiles. How many white tiles are there?

6 An exercise has 100 questions.

a A mathematics teacher asks one pupil to do each question numbered $5n - 2$, where n is an integer. How many questions does the pupil do?

b Another pupil is told to do questions 2, 6, 10, 14, ... Find a formula to generate this sequence. How many questions does this pupil do?

7 Garden trellis is made from strips of wood joined together with bolts. It comes in different sizes as shown in the diagrams. (Black dots represent the bolts.)

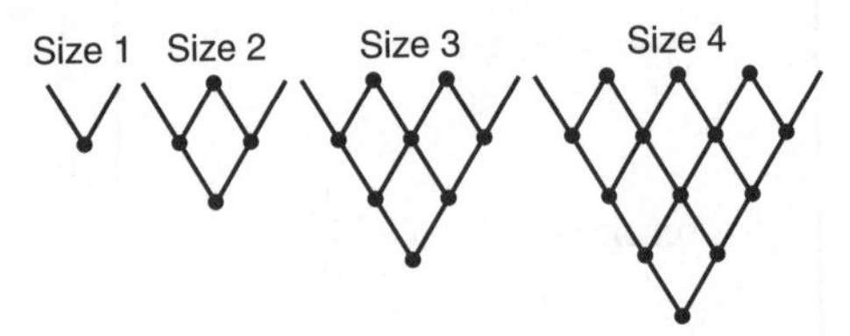

a Copy and complete this table.

Size	1	2	3	4	5	6
No. of bolts	1	4				

b Explain how the table could be extended.

c What size of trellis uses 89 bolts?

8 The diagram shows four graphs

a $y = 3$ **b** $x = 3$ **c** $y = \frac{1}{2} - 2x$ **d** $y = \frac{1}{2}x - 2$

Which graph is which?

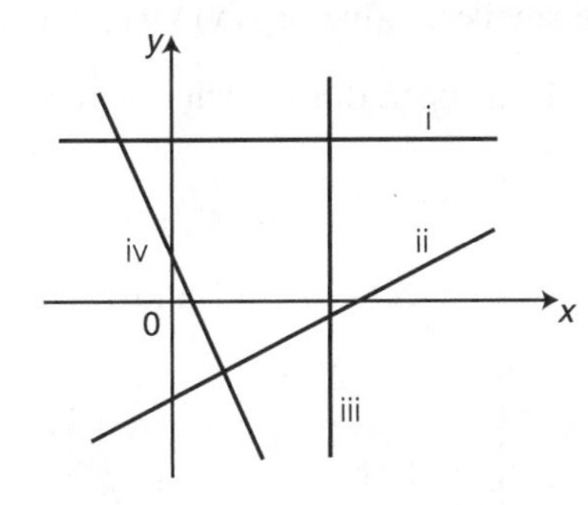

9 The three vertices of a triangle are A(1, 1), B(3, 2) and C(−1, 5).

a Find the gradients of AB, BC and AC.

b Which of AB, BC and AC are perpendicular?

c Find the lengths of AB and BC, giving your answers as square roots.

d Hence find the exact area of the triangle.

10 a Draw on one set of axes the graphs of:

$x + y = 7$,

$3x - y = 9$ for $0 \leqslant x \leqslant 5$.

b Use these graphs to solve the simultaneous equations $x + y = 7$ and $3x - y = 9$.

11 a Draw on one set of axes the graphs of $x + y = 6$, $x = 2$ and $y = 2$.

b By shading the unwanted regions, find the region satisfying $x + y \leqslant 6$, $x > 2$ and $y \geqslant 2$.

12 a Draw on one set of axes the two graphs $y = 2x - 1$ and $y = 2 - \frac{1}{2}x$ for $-1 \leqslant x \leqslant 3$.

b Use your graph to solve the simultaneous equations $y = 2x - 1$ and $y = 2 - \frac{1}{2}x$.

c By shading the *unwanted* regions, find the region satisfying $y \leqslant 2x - 1$ and $y \geqslant 2 - \frac{1}{2}x$.

13 Solve these inequalities.

a $3x - 1 \leqslant 2x + 7$ **b** $2(x + 1) > 3(x - 2)$

14 Zoë cycles at a constant speed from her house at 09:00, arriving at Greta's house, 10 km away, at 09:30. She stays there for one hour before cycling off further in exactly the same direction to visit Harry, who lives 14 km away from her home, arriving there at 11:00. She stays there for one hour also, before she returns home, cycling at 10 km/h.

a Draw a distance–time graph representing Zoë's travels.

b Use your graph to find how fast Zoë cycles between visiting Greta and Harry.

c At what time does Zoë return home?

d What is her average speed while on the move?

15 a Draw the graph of $y = x^2 - 5x + 3$ for $0 \leqslant x \leqslant 5$.

b Use your graph to solve the equation $0 = x^2 - 5x + 3$.

16 a Draw an accurate graph of $y = x^2 + x - 1$ for $-3 \leqslant x \leqslant 2$.

b Use your graph to solve the equation $x^2 + x - 1 = 0$.

c Use your graph to solve the equation $x^2 + x = 2$.

d Use your graph to find the smallest value of $f(x)$ where $f(x) = x^2 + x - 1$.

17 If the graph of $y = x^2 + 2x - 1$ has been drawn, what is the equation of the line that should be drawn to solve these?

a $x^2 + 2x = 2$ **b** $x^2 + x = 3$

18 This diagram shows the graphs of $y = x^2 - 3x + 1$ and $y = x - 1$.

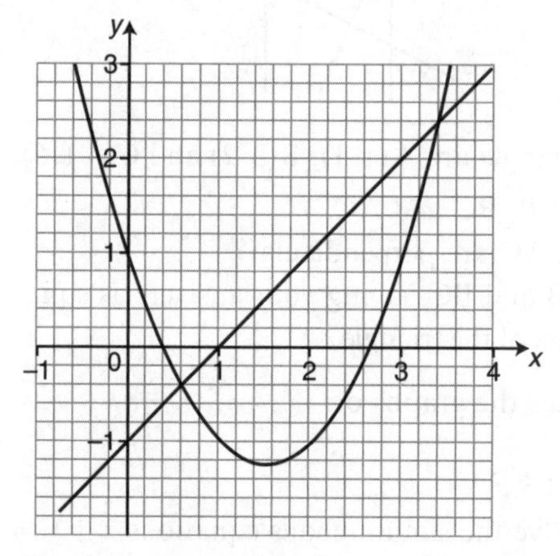

a What equation in x is solved by the intersection points of these graphs?

b Use the diagram to solve the equation in part **a**.

19 a Copy and complete the table for the graph $y = x^3 + x^2 + x + 3$.

x	−3	−2	−1	0	1	2	3
x^3	−27			0			
x^2	9			0			
x	−3			0			
+3	3			3			
y	−18			3			

b Draw the graph of $y = x^3 + x^2 + x + 3$ for $-3 \leqslant x \leqslant 3$.

c Draw the tangent to the curve at $x = 1$ and calculate an estimate of its gradient.

20 a Copy and complete the table for the graph $y = \frac{4}{x}$.

x	1	2	3	4	5
y	4				

b Draw the graph of $y = \frac{4}{x}$ for $1 \leqslant x \leqslant 5$.

21 Find $\frac{dy}{dx}$ for the following:

a $y = x^2$ **b** $y = 3x$ **c** $y = x^3 + 3x^2$

22 Find the gradient of the following curves at $x = 2$.

a $y = (x + 1)(x + 2)$ **b** $y = x(3x + x^2)$ **c** $y = (x - 2)^2$

23 **a** Find $\frac{dy}{dx}$ for the curve $y = x^2 - 4x + 1$.

b Find the co-ordinates of the point where the gradient of the curve is zero.

c State whether this point is a maximum or a minimum.

24 A car moves in a straight line such that its distance from fixed point O, x m, t seconds later is given by $x = 2t^2 + 5t$.

a Find the velocity at $t = 2$.

b Find the acceleration at $t = 2$.

Exercise 97★ (Revision)

1 For the sequence 201, 197, 193, 189, . . . , find

a the next three terms

b the 50th term

c a formula for the nth term

2 **a** Use the difference method to find the next three terms of the sequence 7, 12, 20, 31, 45, ...

b Find the terms a and b of the sequence . . . a, 21, 39, 55, 69, 81, b, . . .

3 **a** How many terms are in the sequence 100, 96, 92, 88, . . . , 24?

b The formula for the nth term of the sequence 3, 10, 21, 36, . . . is $an^2 + bn$. Find the values of a and b.

4 The formula for the sum of n terms of a certain series is $5n^2 - 3n$.

a What are the first and second terms of the series?

b How many terms are needed to make a sum of 938?

5 Andy is investigating bees' honeycombs for his biology project. He wants to know the connection between the number of cells and the number of walls for different arrangements of the cells.

He first considers cells in a line.

a Copy and complete this table, and find a formula for w in terms of c.

No. of cells (c)	No. of walls (w)
1	6
2	
3	
4	
5	26

He next considers cells arranged in triangles.

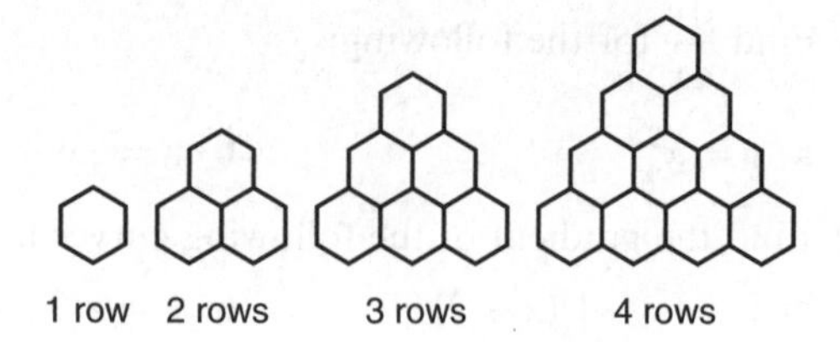

b Copy and complete this table.

No. of rows (r)	1	2	3	4	5
No. of cells (c)	1				15
No. of walls (w)	6			42	
$r + c$	2				

c What formula connects w and $r + c$?

d Andy's teacher tells him that $c = \frac{1}{2}(r^2 + r)$. Use this and your answer to part **a** to find a formula connecting w and r.

e How many walls are in a triangular honeycomb with 10 rows?

6 a Match these equations to the graphs.

$2y = x + 4$ $\quad$ $y = x$ $\quad$ $y + 2x = -2$ $\quad$ $y + 2x = 4$

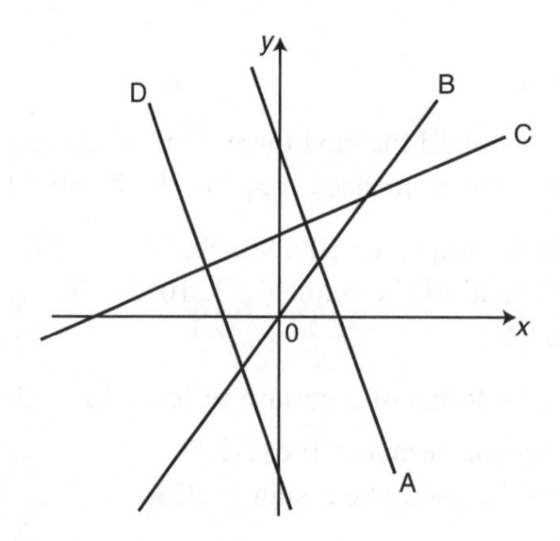

b Explain why A and D are parallel to each other.

7 a The equation of line L_1 is $y = mx + c$.
State what m and c represent.

b Find the equation of line L_2 which is parallel to L_1 and passes through point (4, 1).

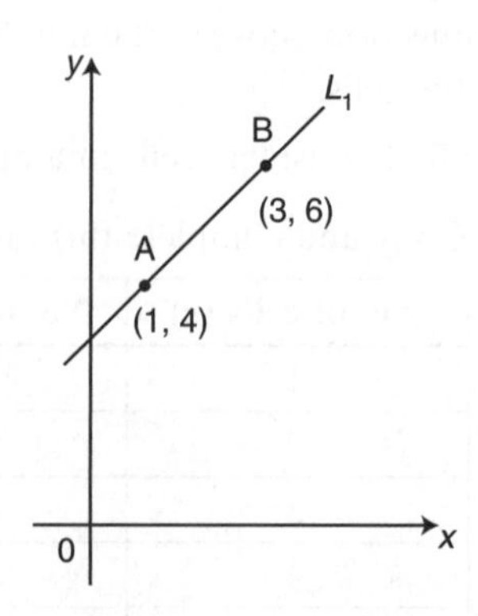

8 Lines L_1 and L_2 intersect at P(1, 4). The co-ordinates of P are the solutions to which two simultaneous equations?

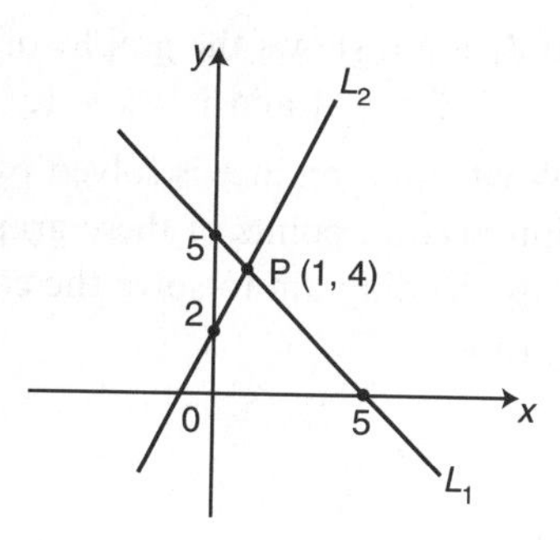

9 The position of a pebble thrown from the top of a cliff, relative to the axes shown, is given by $y = 3x - 0.2x^2$, all units being in metres.

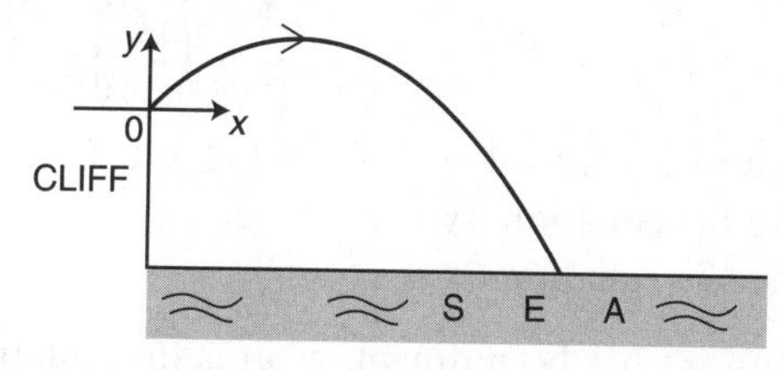

a Copy and complete this table and use the values to draw the graph for $0 \leqslant x \leqslant 20$.

x	0	4	8	12	16	20
y	0		11.2		−3.2	

b Use your graph to find the maximum height of the pebble above the sea, given that the cliff is 20 m high.

c What is the horizontal distance from the cliff when the pebble is level with the cliff-top?

10 This diagram shows four graphs:

a $y = 4 - x^2$ b $y = (x - 2)^2 - 1$ c $3y = 6 - x$ d $y = (x + 1)^3$

Which graph is which?

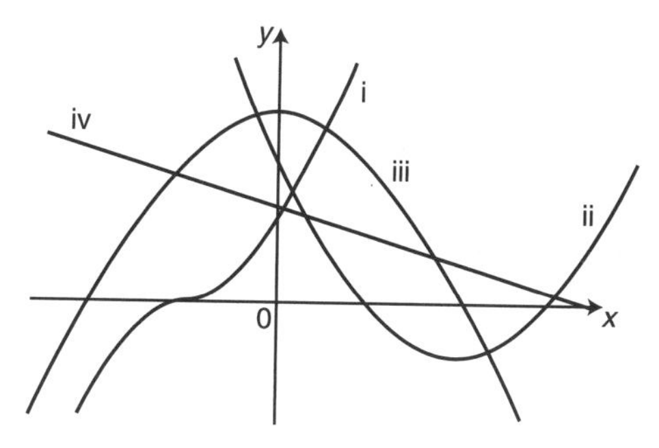

11 *Answer this question on graph paper.*

a Construct a table of values of y for $-4 \leqslant x \leqslant 2$ where $y = x^3 + x^2$.

b Plot the graph of $y = x^3 + x^2$.

c Use your graph to solve the equation $1 = x^3 + x^2$.

d By drawing a suitable line on your graph, solve the equation $2x = x^3 + x^2$.

12 This diagram shows the graphs of $y = x^3 - 3x^2 + 1$ and $y = x - 1$.

a What equation in x is solved by the intersection points of these graphs?

b Use the diagram to solve the equation in part **a**.

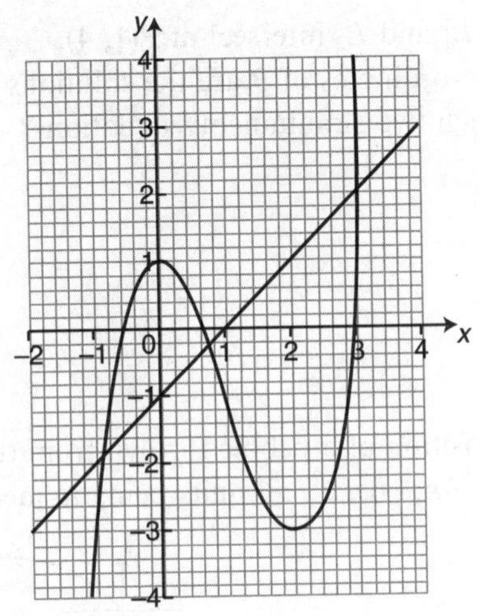

13 **a** Draw an accurate graph of $y = 4x - 1 - x^2$ for $-1 \leqslant x \leqslant 5$.

b Use your graph to solve the equation $4x - x^2 = -1$.

c Use your graph to solve the equation $5x - x^2 = 3$.

14 If the graph of $y = 4 + 2x - x^2$ has been drawn, what is the equation of the line that should be drawn to solve these?

a $1 + 2x - x^2 = 0$ **b** $x^2 - 4x - 2 = 0$

15 Jamal runs a 100 m race in 14.2 s. This is the speed–time graph for Jamal's race.

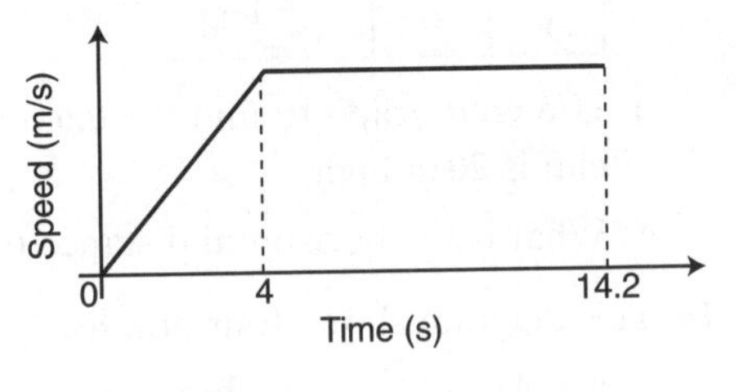

a Find his average speed for the race.

b What is the greatest speed reached by him in the race?

c Find his acceleration over the first four seconds.

d Sidd also runs in this race, accelerating constantly from the start for 5 s until he reaches a top speed of 8 m/s, which he maintains until the finish. Does Sidd beat Jamal?

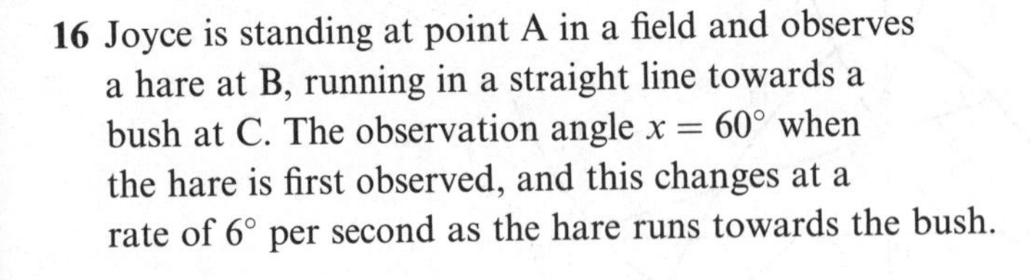

16 Joyce is standing at point A in a field and observes a hare at B, running in a straight line towards a bush at C. The observation angle $x = 60°$ when the hare is first observed, and this changes at a rate of 6° per second as the hare runs towards the bush.

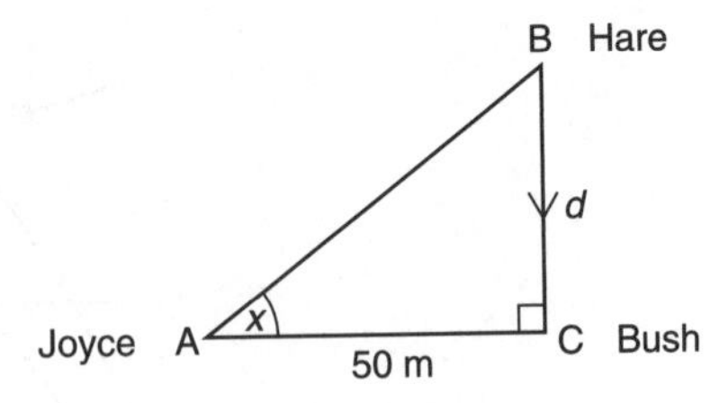

a If d is the distance of the hare from the bush, copy and complete this table.

***t* (s)**	0	2	4	6	8	10
***d* (m)**	86.6				10.6	

b Draw a graph of d(m) against time (s) and comment on it.

c Use the graph to *estimate* the gradient of the graph at $t = 6$ s and state what this represents.

17 A speed–time graph is shown for a journey of a car between two sets of traffic lights.

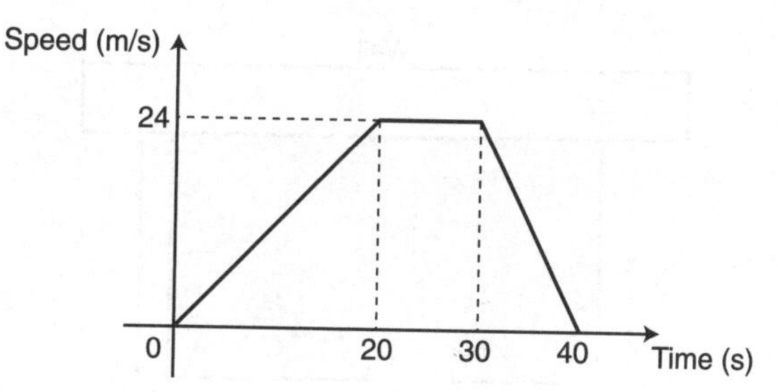

Find:

a the acceleration over the first 20 s

b the retardation over the final 10 s

c the mean speed for the whole journey.

18 Lemon Mobile Phones want to market a new phone card containing both peak units, costing 50 cents per unit, and off-peak units, costing 25 cents per unit. Let x be the number of peak units on a card and y the number of off-peak units on a card.

a There is to be a maximum of 30 units on a card. Express this as an inequality in x and y.

b The maximum cost of a card is to be \$10. Form another inequality in x and y and show it simplifies to $2x + y \leqslant 40$.

c Using both x and y axes from 0 to 40, show both inequalities on one graph and shade the unwanted regions.

d If units are only available in multiples of 10, mark the possible combinations on your graph. Remember there must be some of each type of unit on the card. What combination of units would you recommend?

19 Find $\frac{dy}{dx}$ for the following:

a $y = 2x + \frac{3}{x}$ **b** $y = (2x + 1)^2$ **c** $y = \frac{2}{x^2}$

20 Find and classify the stationary points on the curve $y = x + \frac{9}{x}$.

21 A farmer wishes to enclose a rectangular field with one side formed by a stone wall.

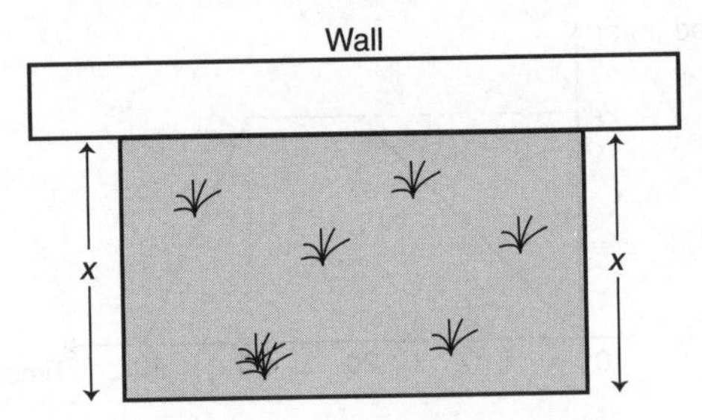

If the total length of the fence is 150 m

a show that the area of the enclosed field, $A\,\text{m}^2$, is $A = x(150 - 2x)$.

b Find $\frac{dA}{dx}$ and use it to calculate the maximum area and the value of x at which this occurs.

22 A bicycle moves in a straight line such that its distance from fixed point O, x m, t seconds later is given by $x = \frac{2}{3}t^3 - \frac{9}{2}t^2 + 4t$. Find

a when the bicycle is stationary

b the acceleration of the bicycle at $t = 3$.

Shape and space 5 (Revision)

Compass constructions

60° angle (equilateral triangle)
Draw arc from A to intersect AB at P. Keeping the same radius, draw arc from P to intersect arc at Q.
$\angle BAQ = 60°$

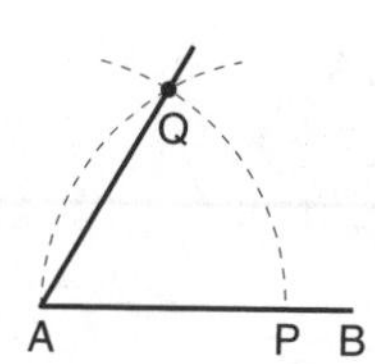

Bisecting an angle
Draw arc from A to intersect lines at P and Q. Keeping the same radius, draw arcs from P and Q to intersect at R. Draw AR.
$\angle PAR = \angle QAR$.

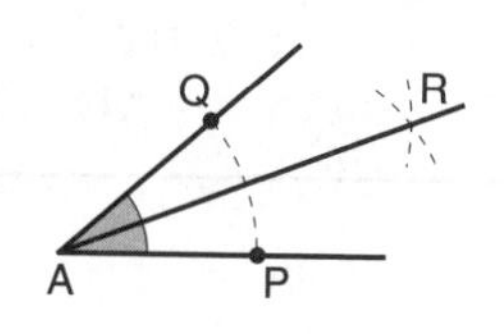

Perpendicular bisector of a line
Draw arc from A, with radius more than $\frac{1}{2}$ AB above and below line. Keeping the same radius, draw arcs from B to intersect those from A above and below the line. These points are P and Q. Draw line PQ, the perpendicular bisector of AB.

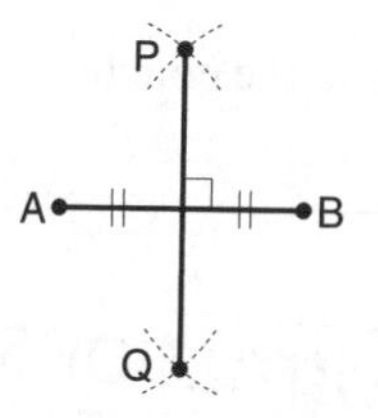

Similar triangles

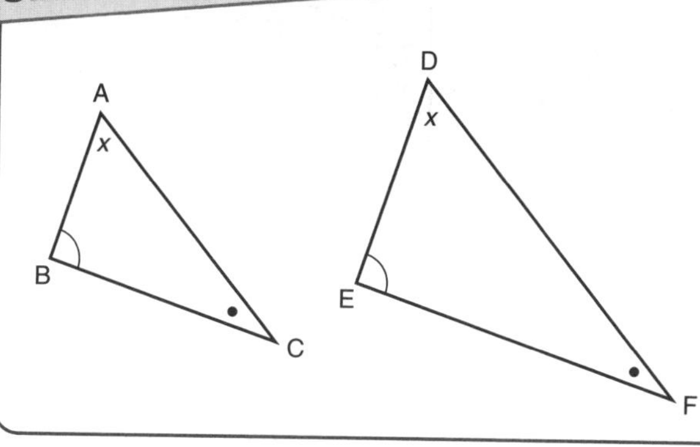

Triangle ABC is similar to triangle DEF, as corresponding angles are equal.

$$\Rightarrow \frac{AB}{DE} = \frac{AC}{DF} = \frac{BC}{EF}$$

Congruence

For triangles to be congruent (the same size and shape), the corresponding sides and angles must be: **SSS, SAS, AAS** or **ASA** (in that order) or **RHS** (right angle, hypotenuse, side).

Use congruent triangles to prove that AD = BC.

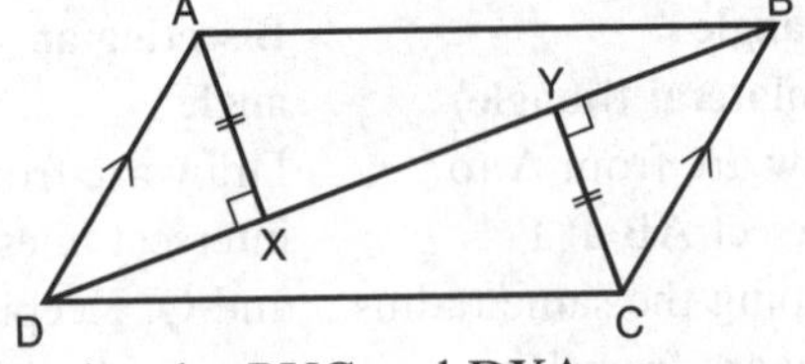

- State the triangles.
- State the three conditions for congruency and give reasons in brackets
- Write the conclusion

In triangles BYC and DXA

1 AX = CY (Given)
2 ∠DBC = ∠BDA (Alternative angles)
3 ∠BYC = ∠DXA (Given)

∴ △BYC = △DXA (SAA)
∴ AD = BC

Transformations

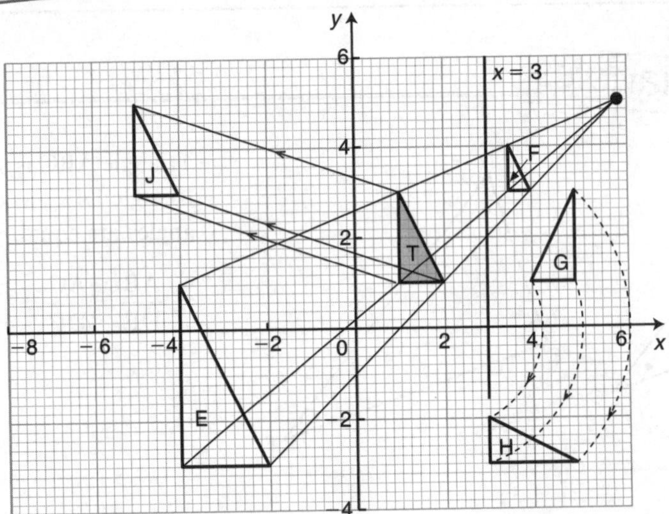

- △G is a **reflection** of △T in the line $x = 3$.
- △H is a **rotation** of △G 90° clockwise (−90°), centre (2, 0).
- △J is a **translation** of △T by the vector $\begin{pmatrix} -6 \\ +2 \end{pmatrix}$.
- △E is an **enlargement** of △T by a scale factor 2, centre (6, 5). The area of △E is four times the area of △T.
- △F is an **enlargement** of △T by a scale factor $\frac{1}{2}$, centre (6, 5). The area of △F is one-quarter of the area of △T.

Circle theorems

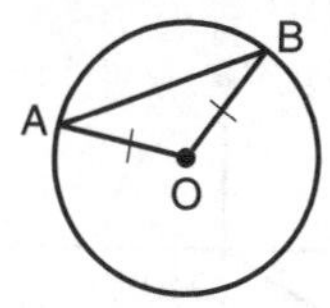

Triangle OAB is isosceles.

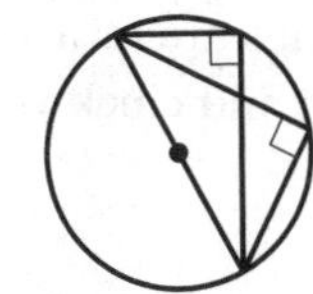

Angles off the diameter in a semicircle = 90°.

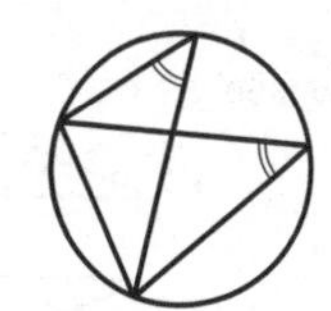

Angles in same segment off a chord are equal.

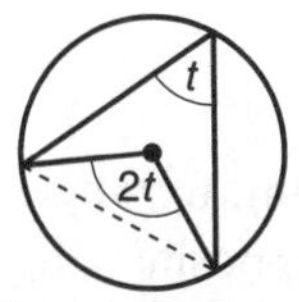

Angle at centre is twice angle formed at circumference in same segment off a chord.

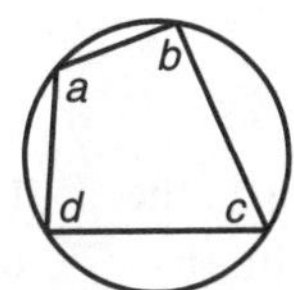

$a + c = 180°$
$b + d = 180°$

Opposite angles in a cyclic quadrilateral sum to 180°.

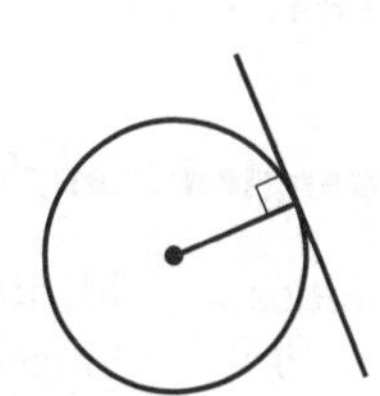

Angles off a tangent to the radius = 90°.

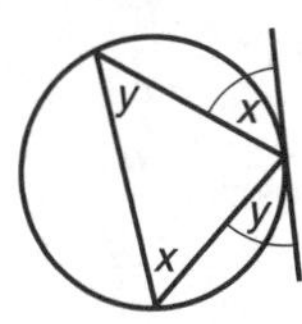

Alternate Segment Theorem: angle between chord and tangent = angle in alternate segment.

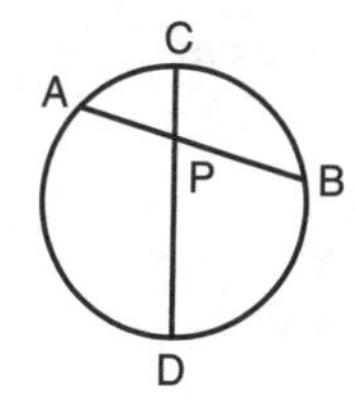

Chords intersecting inside the circle.

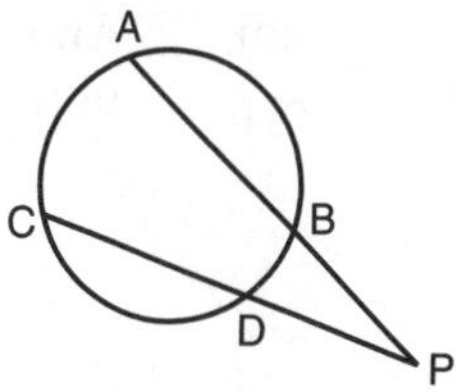

Chords intersecting outside the circle.

In both cases:
$AP \times PB = CP \times PD$.

Trigonometry

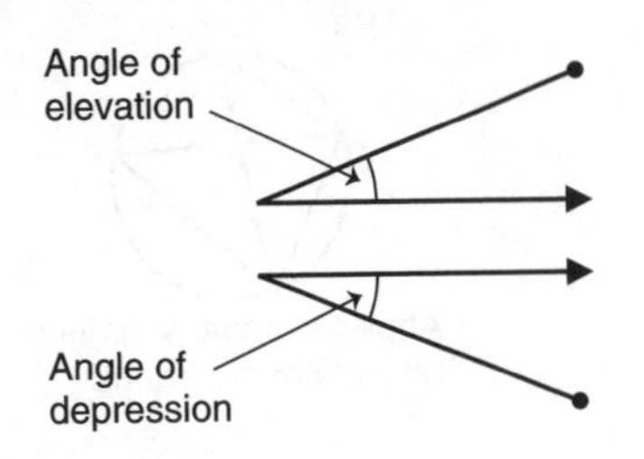

Bearings are measured from North and clockwise.

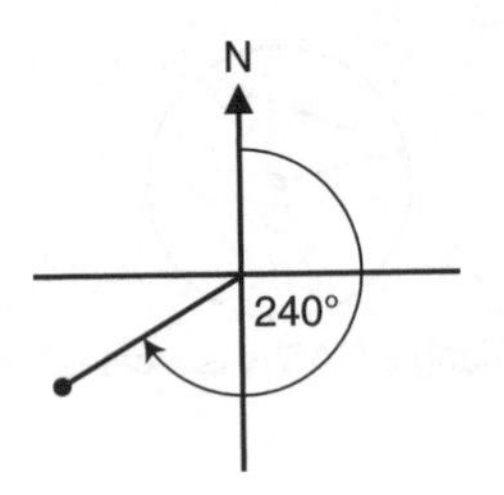

Right-angled triangles

Trig. ratios:

$\tan x = \dfrac{\text{opp}}{\text{adj}}$

$\sin x = \dfrac{\text{opp}}{\text{hyp}}$

$\cos x = \dfrac{\text{adj}}{\text{hyp}}$

Identify the **hypotenuse**. This is the longest side: the side opposite the right angle.
Then, the **opposite** side is opposite the angle.
And, the **adjacent** side is adjacent to the angle.

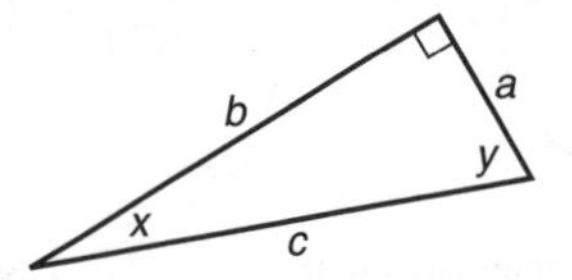

For angle x,
opposite $= a$,
adjacent $= b$

$\tan x = \dfrac{a}{b}$

$\sin x = \dfrac{a}{c}$

$\cos x = \dfrac{b}{c}$

But, for angle y,
opposite $= b$,
adjacent $= a$

$\tan y = \dfrac{b}{a}$

$\sin y = \dfrac{b}{c}$

$\cos y = \dfrac{a}{c}$

Pythagoras' Theorem: $a^2 + b^2 = c^2$

Using trig. ratios to find a length

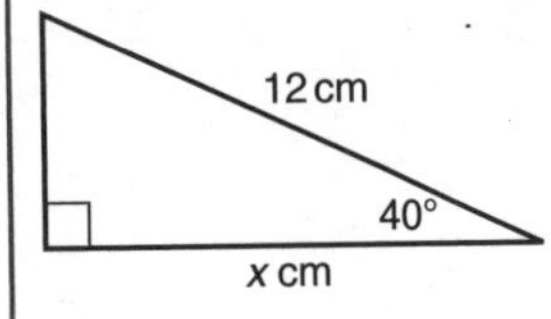

Identify the ratio:

$\dfrac{x}{12} = \cos 40°$

$x = 12 \cos 40°$

$x = 9.19\,\text{cm}$ (3 s.f.)

Using trig. ratios to find an angle

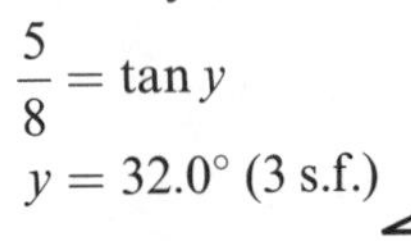

Identify the ratio:

$\dfrac{5}{8} = \tan y$

$y = 32.0°$ (3 s.f.)

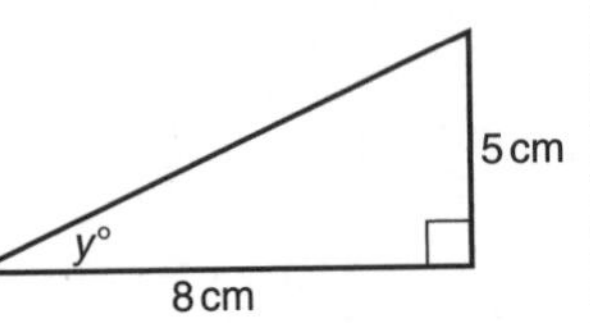

Using Pythagoras to find the hypotenuse

$z^2 = 7^2 + 11^2$

$= 170$

$z = 13.0\,\text{cm}$ (3 s.f.)

z
11 cm
7 cm

Using Pythagoras to find a shorter side

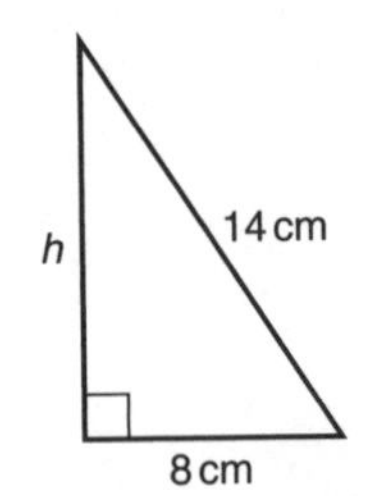

$h^2 + 8^2 = 14^2$

$h^2 = 14^2 - 8^2$

$h^2 = 132$

$h = 11.5\,\text{cm}$ (3 s.f.)

Non-right-angled triangles

Cosine rule

$a^2 = b^2 + c^2 - 2bc \cos A$

Label the vertices with capital letters, then, length a is opposite A, ...

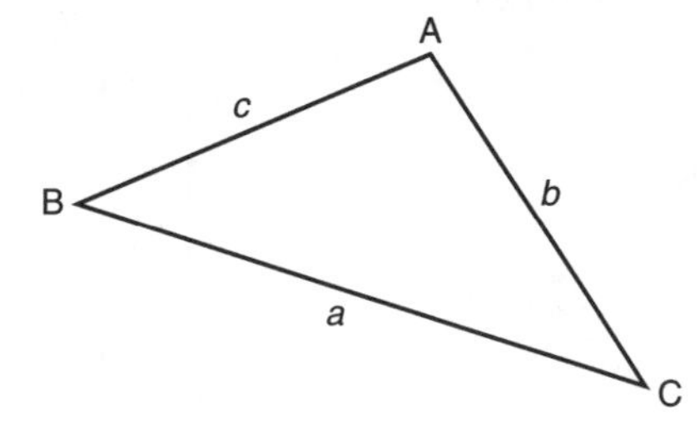

Sine rule

$$\frac{a}{\sin A} = \frac{b}{\sin B} = \frac{c}{\sin C}$$

or $$\frac{\sin A}{a} = \frac{\sin B}{b} = \frac{\sin C}{c}$$

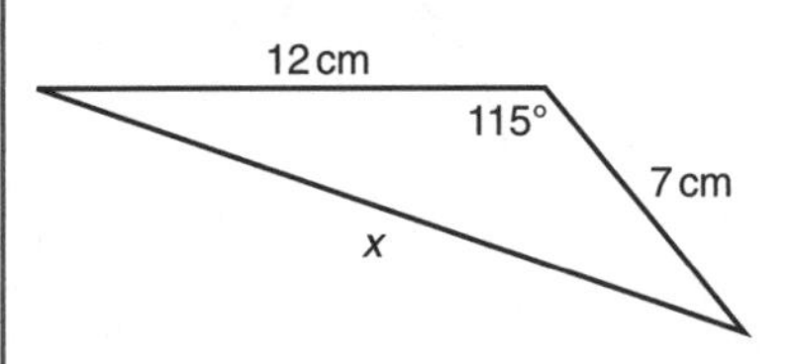

$x^2 = 12^2 + 7^2 - (2 \times 12 \times 7 \times \cos 115°)$

$x^2 = 263.99$

$x = 16.2\,\text{cm (3 s.f.)}$

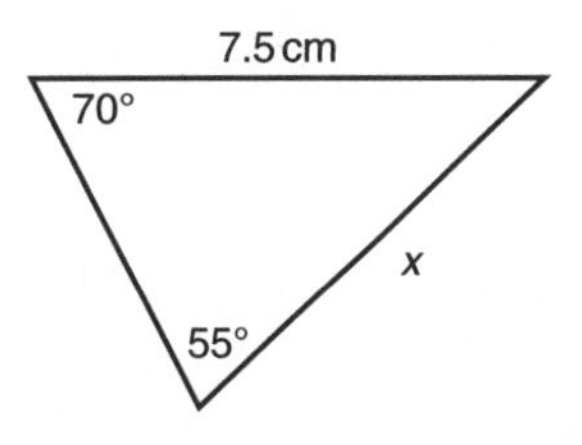

$$\frac{x}{\sin 70°} = \frac{7.5}{\sin 55°}$$

$$x = \frac{7.5 \times \sin 70°}{\sin 55°}$$

$x = 8.60\,\text{cm (3 s.f.)}$

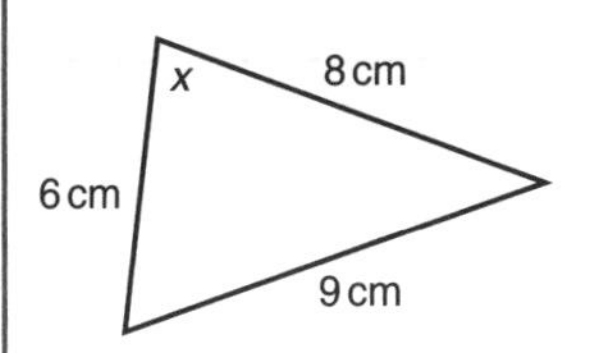

$9^2 = 6^2 + 8^2 - (2 \times 6 \times 8 \times \cos x)$

$96 \cos x = 36 + 64 - 81$

$$\cos x = \frac{19}{96}$$

$x = 75.6°\ \text{(3 s.f.)}$

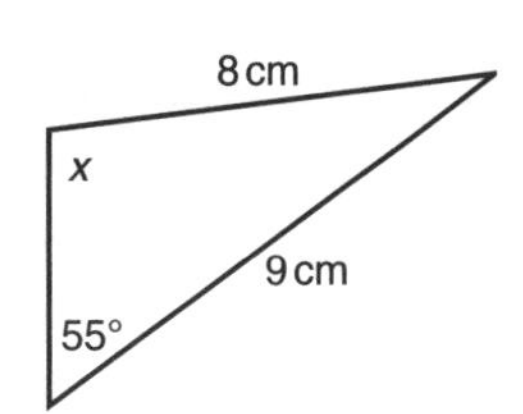

$$\frac{\sin x}{9} = \frac{\sin 55°}{8}$$

$$\sin x = \frac{9 \times \sin 55°}{8}$$

$\sin x = 0.92155$

$x = 67.2°$ or $(180° - 67.2°) = 112.8°$ (3 s.f.)

Graphs of sin θ, cos θ, tan θ

- These graphs are for $0° \leqslant \theta \leqslant 180°$.

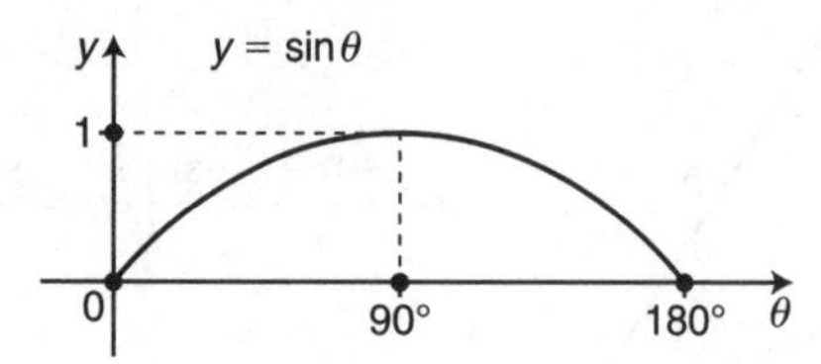

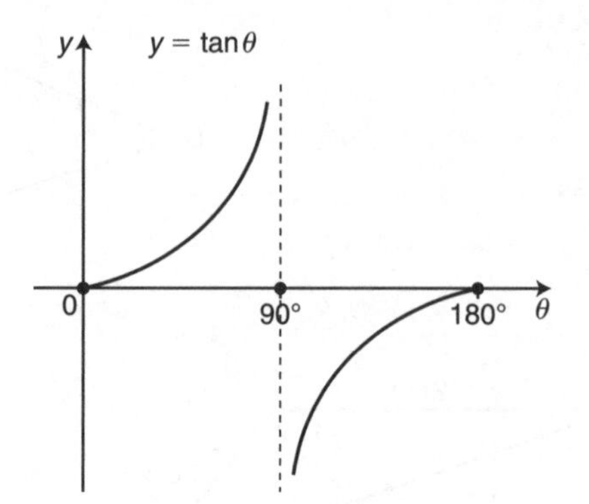

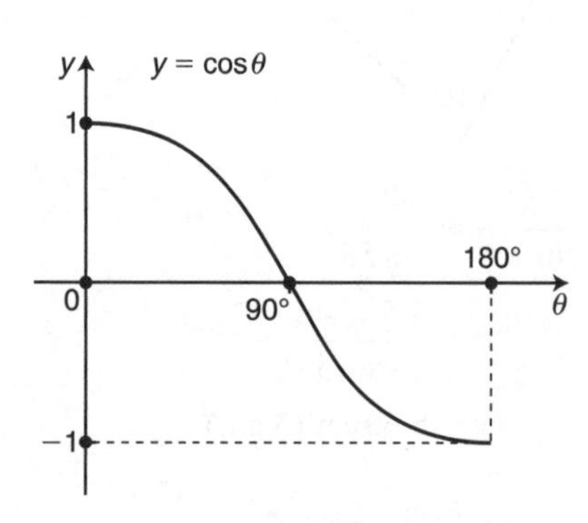

Area of a triangle

- For triangle ABC, area $= \frac{1}{2} \times a \times b \times \sin C$

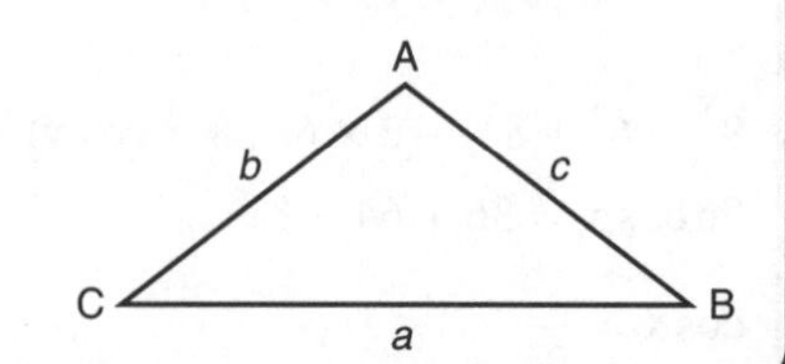

3D trigonometry

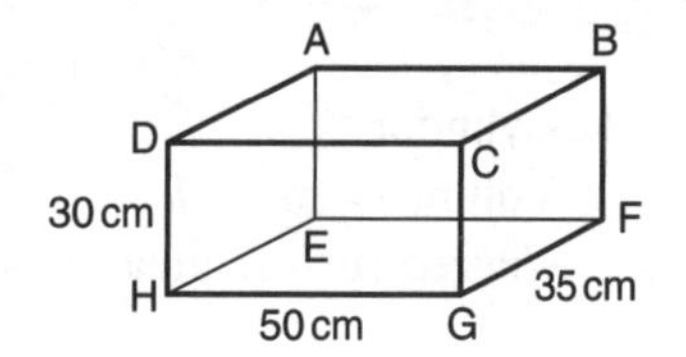

The diagram shows a box in the shape of a cuboid. The base is a rectangle; length = 50 cm, width = 35 cm, height = 30 cm.

a Calculate the length of the longest rod that can be placed in the box.

b When the rod is in the box, calculate the angle between the rod and the floor of the box.

Extract and draw right-angled triangles:

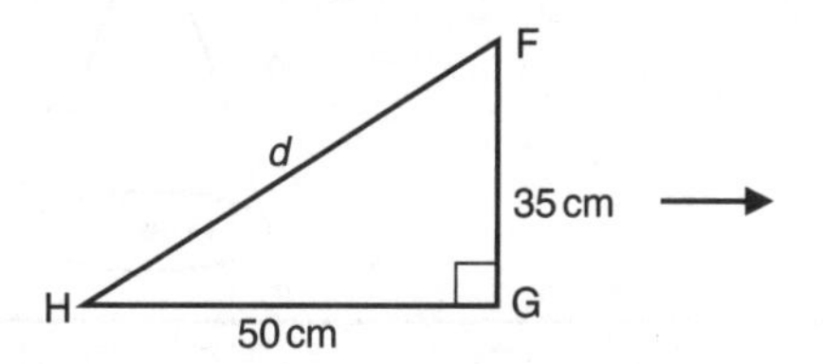

$d^2 = 50^2 + 35^2 = 3725$

$d = \sqrt{3725} = 61.033$ cm (3 d.p.)

Use the exact value of d in the next calculation!

a $l^2 = d^2 + 30^2 = 3725 + 30^2 = 4625$

$l = 68.0$ cm (3 s.f.)

b $\sin x = \frac{30}{l} = \frac{30}{\sqrt{4625}} \Rightarrow x = 26.2°$ (3 s.f.)

Arcs, sectors and segments

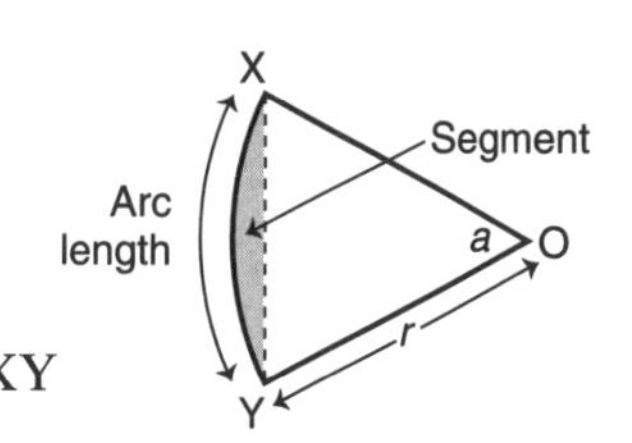

- Arc length XY $= \frac{a}{360} \times 2\pi r$
- Sector area OXY $= \frac{a}{360} \times \pi r^2$
- Segment area = area of sector OXY − area of triangle OXY

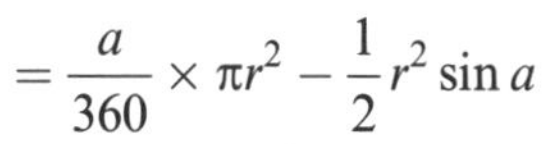

$= \frac{a}{360} \times \pi r^2 - \frac{1}{2} r^2 \sin a$

Surface areas and volumes of solids

- Cylinder
  Volume $= \pi r^2 \times h$
  Curved surface area $= 2\pi rh$

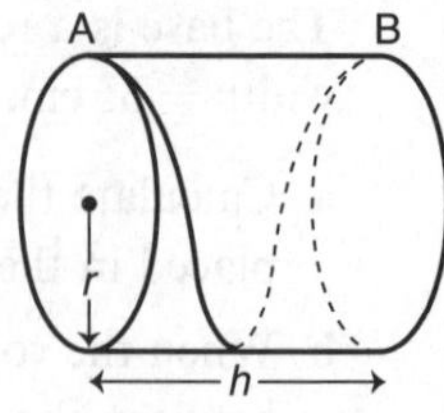

- Cone
  Volume $\frac{1}{3}\pi r^2 \times h$
  Curved surface area $= \pi rl$

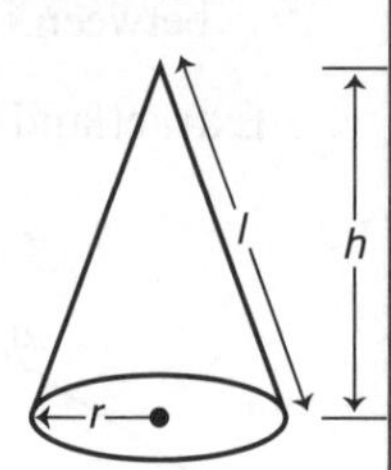

- Sphere
  Volume $\frac{4}{3}\pi r^3$
  Surface area $= 4\pi r^2$

Similar solids and shapes

When applying a scale factor, k, of enlargement of more than 1

- to an area, the larger area = smaller area × (scale factor)^2. So $A = ak^2$
- to a volume, the larger volume = smaller volume × (scale factor)^3. So $V = vk^3$

Vectors

Vector notation

Vectors have magnitude and direction and can be written as bold letters: **v**, **u**, . . ., with capitals covered by an arrow: $\overrightarrow{OP}$, $\overrightarrow{OQ}$, . . . or on co-ordinate axes as column vectors $\begin{pmatrix} 3 \\ -1 \end{pmatrix}$, $\begin{pmatrix} 0 \\ 6 \end{pmatrix}$, . . .

Multiplication of a vector by a scalar

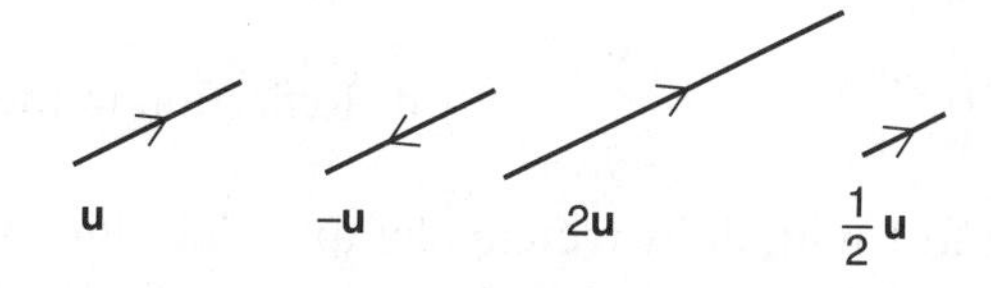

Vector geometry

- $\overrightarrow{AC} = \overrightarrow{AB} + \overrightarrow{BC} = 2\mathbf{a} + \mathbf{b}$

  $\therefore \quad \overrightarrow{AD} = \overrightarrow{AC} + \overrightarrow{CD}$

  $\qquad = 2\mathbf{a} + \mathbf{b} + (-3\mathbf{a}) = \mathbf{b} - \mathbf{a}$

  AB is parallel to DC. $\therefore$ ABCD is a trapezium

  Ratio of AB:DC = 2:3. $\therefore$ 2DC = 3AB

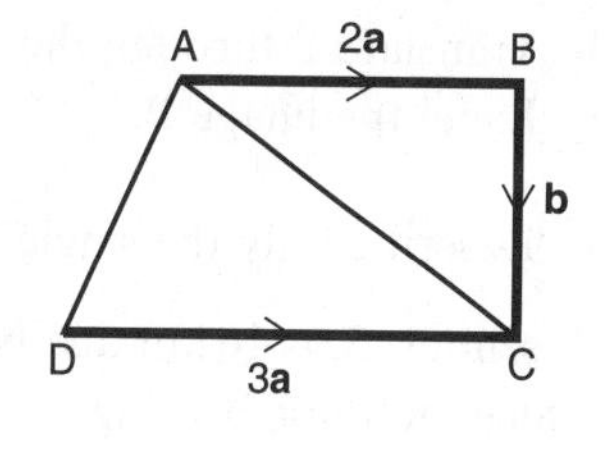

- 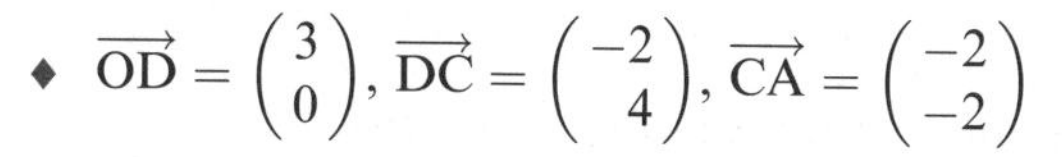

  $\overrightarrow{OD} = \begin{pmatrix} 3 \\ 0 \end{pmatrix}$, $\overrightarrow{DC} = \begin{pmatrix} -2 \\ 4 \end{pmatrix}$, $\overrightarrow{CA} = \begin{pmatrix} -2 \\ -2 \end{pmatrix}$

  $\therefore \quad \overrightarrow{OA} = \overrightarrow{OD} + \overrightarrow{DC} + \overrightarrow{CA}$

  $$= \begin{pmatrix} 3 \\ 0 \end{pmatrix} + \begin{pmatrix} -2 \\ 4 \end{pmatrix} + \begin{pmatrix} -2 \\ -2 \end{pmatrix}$$

  $$= \begin{pmatrix} -1 \\ 2 \end{pmatrix}$$

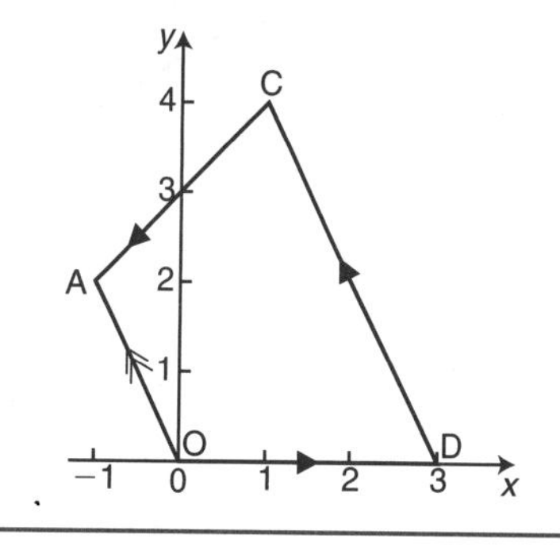

Exercise 98 (Revision)

1 Coventry is 150 km from London on a bearing of 315°. Felixstowe is 120 km from London on a bearing of 060°.

a Using a scale of 1 cm:10 km, construct, using ruler and compasses only, a triangle showing the positions of London, Coventry and Felixstowe.

b From your drawing, estimate the distance between Coventry and Felixstowe.

2 A point P has co-ordinates (3, 5). State the new position of P after each of these.

a Reflection in the x-axis

b Rotation about O, 90° clockwise

c Translation along $\begin{pmatrix} -4 \\ 4 \end{pmatrix}$

d Reflection in the line $x = 5$

3 Using a scale of 1 cm to 1 unit, draw rectangular axes, labelling both axes from −6 to 6. On your diagram, draw and label the triangle P with vertices (2, 2), (2, 4) and (3, 4).

a Rotate P, −90° about the point (−1, 2). Label the image A.

b Translate P through the vector $\begin{pmatrix} -6 \\ 1 \end{pmatrix}$.
Label the image B.

c Describe fully the single transformation that maps A to B.

4 A school, S, is 10 km due North of the church, C. A village, V, is 5 km due west of C. Calculate these bearings.

a S from V

b V from S

5 Find lengths of sides p and q and sizes of angles r and s in these triangles.

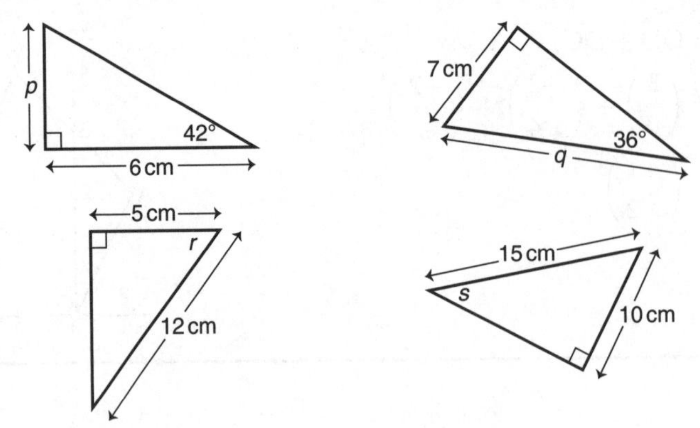

6 Mira is investigating whether it is possible to have a right-angled triangle where the sides are three consecutive even numbers.
She lets x, $x + 2$ and $x + 4$ stand for the lengths of the sides.

a Form and simplify an equation for x.

b Solve the equation and show that only one triangle is possible.

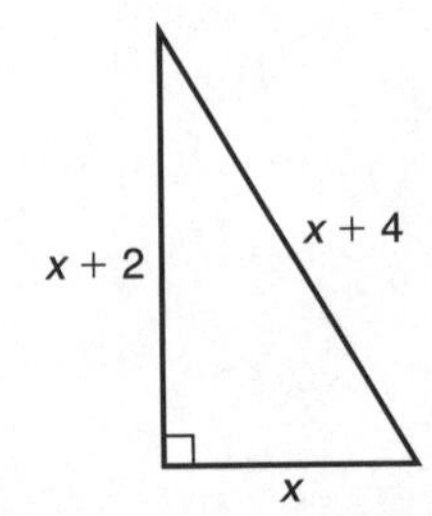

7 State whether the pairs of triangles are congruent. If they *are* congruent, give reasons.

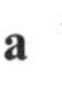
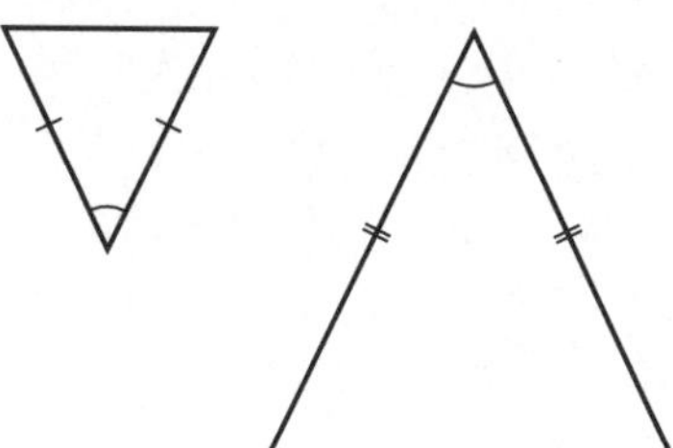

b

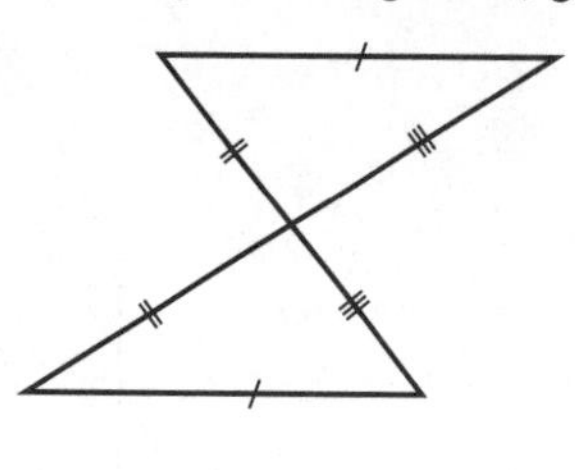

8 Find the sides x and y.

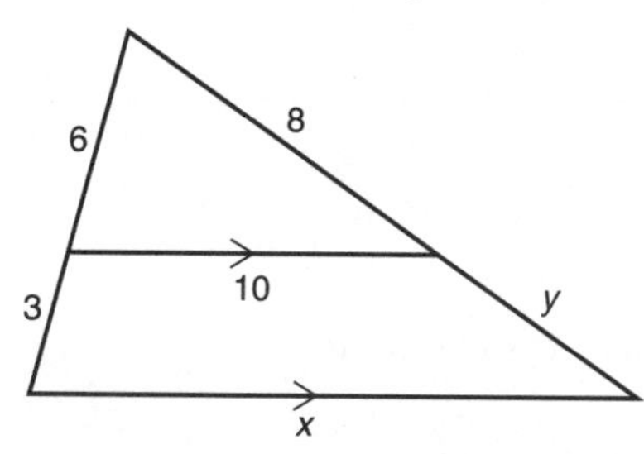

9 A triangle has two angles as shown in this diagram.

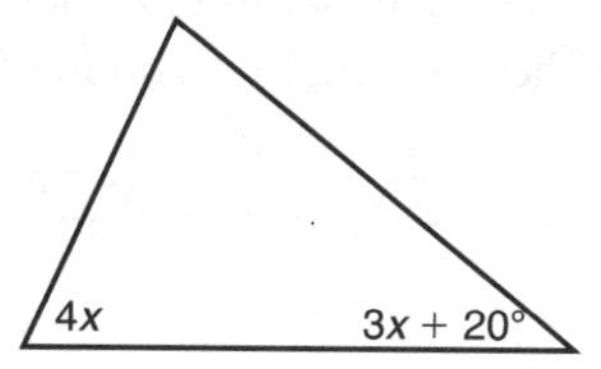

a Find the third angle in terms of x.

b There are three different isosceles triangles with these angles. Write down three equations giving these angles.

c Solve the equations to find the three different values of x which make the triangle isosceles.

10 a For a regular octagon, work out the size of the exterior angle.

b Calculate the sum of the interior angles of a regular octagon.

c Explain why regular octagons will *not* tessellate (meet around a point without any gaps or overlaps).

11 A cylindrical can of beans has a height of 15 cm and base radius of 4 cm. It is put into a cylindrical saucepan of radius 10 cm already fairly full with water, and as a result it is completely covered.

a By how much does the water level rise?

b If it was stood on its end, and the water was originally 7 cm deep, by how much would the water level rise now?

12 This figure shows a frustum with a circular base.

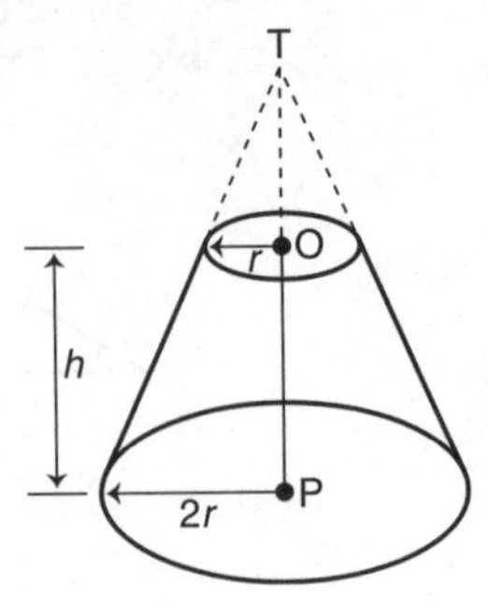

a Use similar triangles to show that OT = h.

b Show that the volume V of the frustum is $V = \frac{7\pi r^2 h}{3}$

c Write the formula in part **b** in the form $V = kr^2h$, where k is correct to 4 significant figures.

d The volume of another frustum, where $r = h$, is $35\,\text{cm}^3$. Find the value of r.

13 A floating toy is in the shape of a circular pyramid of height 20 cm. The top section protruding above the water is of height 4 cm and volume $50\,\text{cm}^3$.

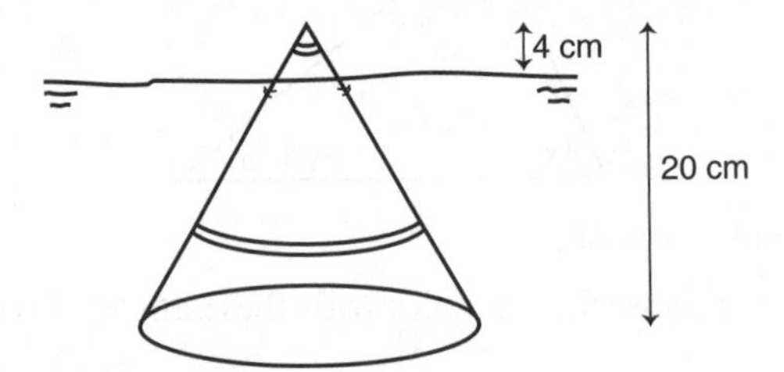

Find the volume of the cone beneath the water level.

14 Chords AD and BC intersect at Q.
AQ = 6 cm, AD = 10 cm and BQ = 3 cm.
Calculate the length of BC.

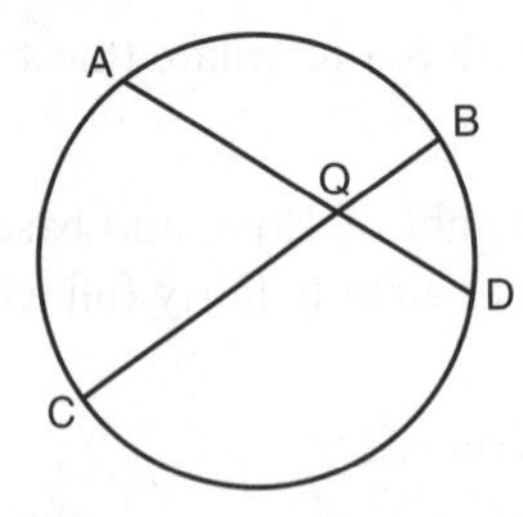

UNIT 5 ◆ Shape and space

15 Find angles x and y, giving reasons for each step of your calculation.

a

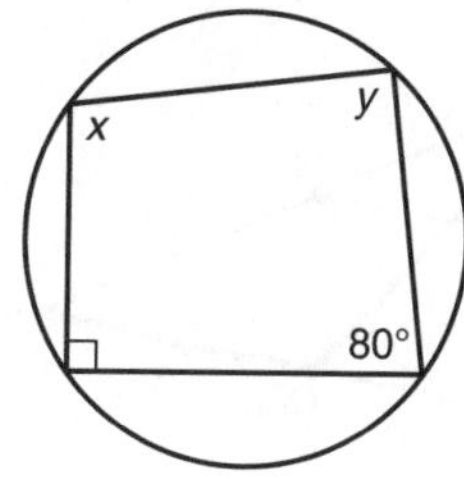

b

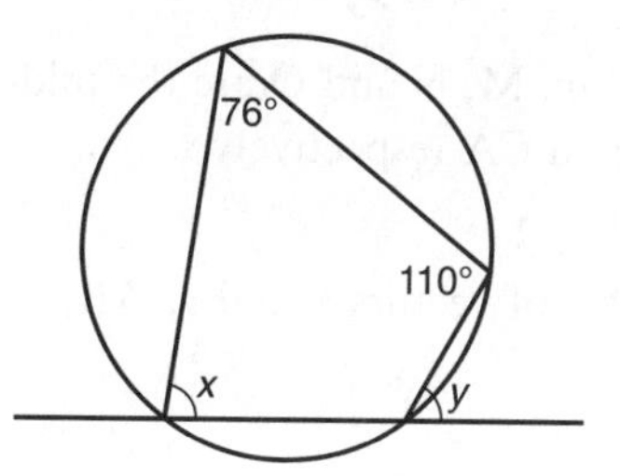

c

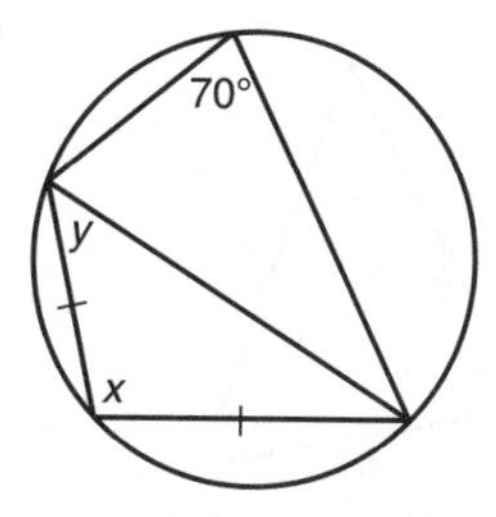

d

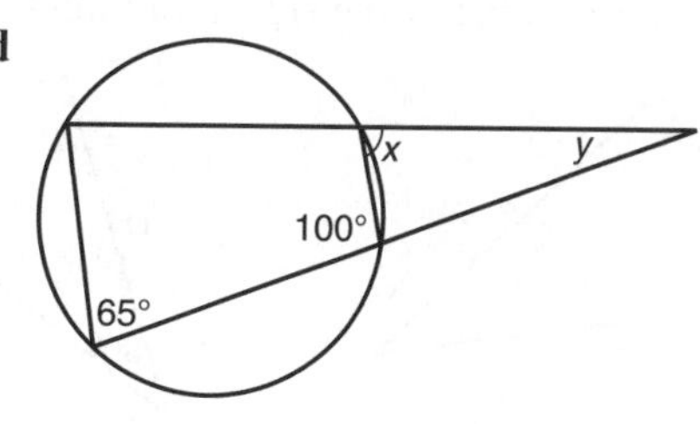

16 a Find the values of x, y and z.

b Find a triangle similar to △ABF.

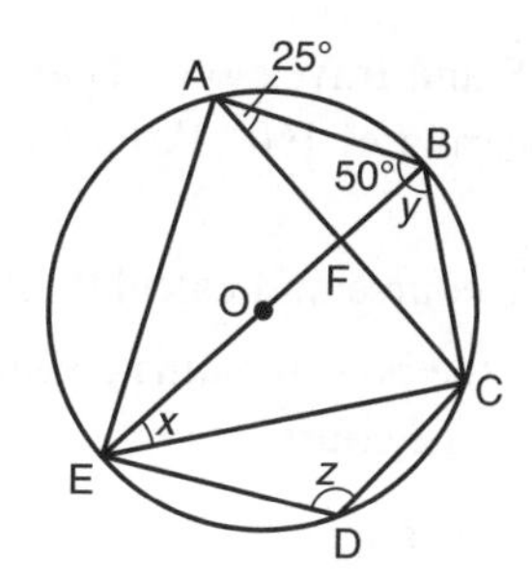

17 In the diagram O is the centre of the circle and the lines PT and QT are tangents to the circle.

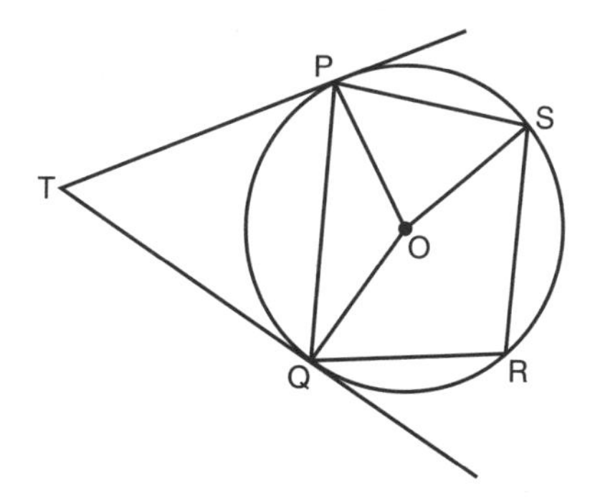

If angle PTQ = 48° and angle PSO = 42°, find the size of the following angles, giving brief reasons for your answers.

a angle QPO **b** angle QPS **c** angle QRS

18 $\mathbf{u} = \begin{pmatrix} 2 \\ 3 \end{pmatrix}$ and $\mathbf{v} = \begin{pmatrix} -1 \\ 5 \end{pmatrix}$. Find $2\mathbf{v}$, $\mathbf{u} + \mathbf{v}$, $\mathbf{u} - \mathbf{v}$ and the length of $2\mathbf{u} - 3\mathbf{v}$.

19 In the diagram, M, N and O are the mid-points of AB, BC and CA respectively.

$\overrightarrow{OA} = \mathbf{x}$, $\overrightarrow{OB} = \mathbf{y}$.

Find, in terms of vectors $\mathbf{x}$ and $\mathbf{y}$, $\overrightarrow{AB}$, $\overrightarrow{OM}$, $\overrightarrow{ON}$ and $\overrightarrow{MN}$.

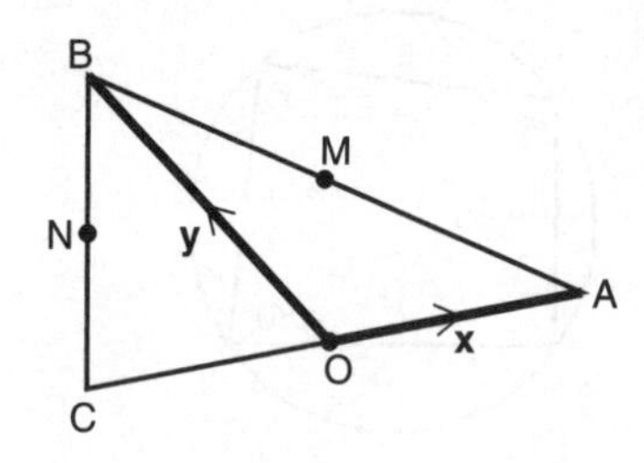

20 Calculate the value of the side x and angles y and z.

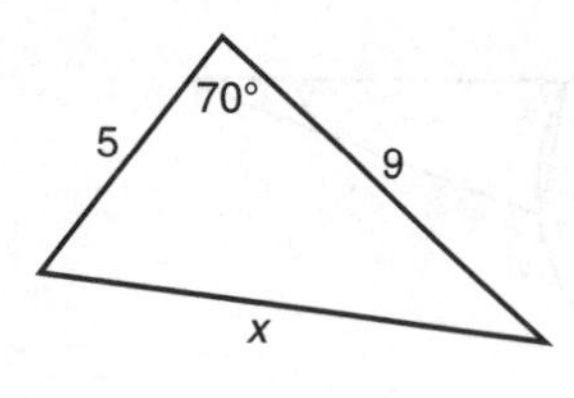

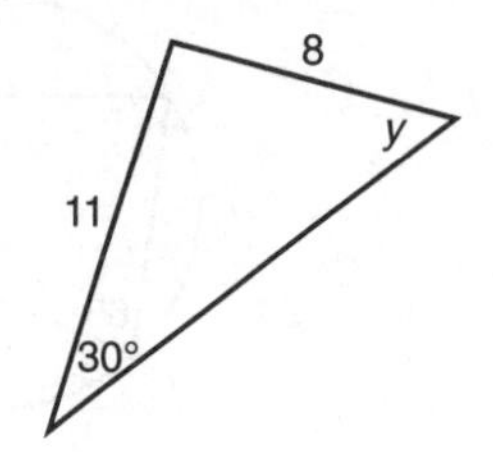

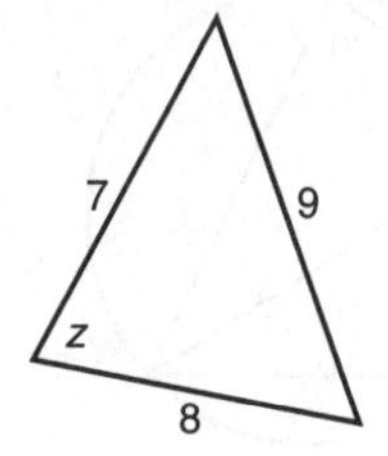

21 A speed boat starts a race at S and travels on a bearing of 060°. After 5 minutes, it rounds buoy A and continues on a bearing of 160°. After 6 km, it rounds buoy B and heads back to the start on a bearing of 315°.

a Draw a neat diagram of the course and calculate the distances SA and BS.

b Assuming that the speed of the boat remains constant throughout, find the time, in minutes, it took to complete the course.

22 PQR and PRS are right-angled triangles.

∠RPQ = 30°

∠PSR = 50°

PR = 10 m

Calculate

a length PQ

b length QR

c length RS

d area PQRS.

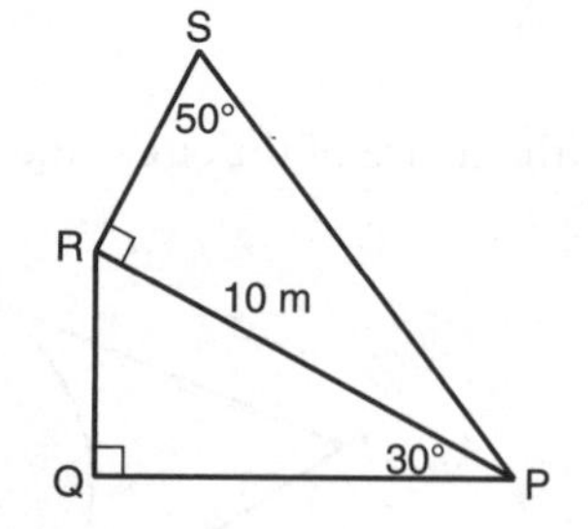

23 A circular Big Wheel funfair ride of radius 20 m rotates in a clockwise direction. The lowest point is 10 m above the ground and the highest point on the wheel is P. The wheel rotates at 2° per second.

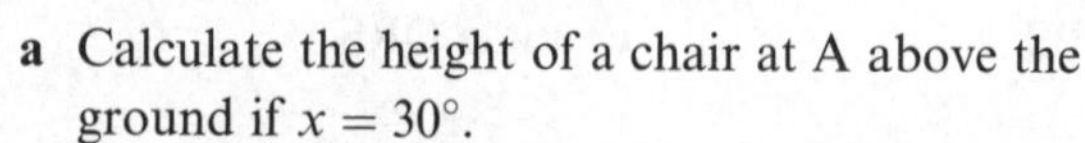

a Calculate the height of a chair at A above the ground if $x = 30°$.

b Find the height of the chair 20 s after reaching point A, above the ground.

c What is the height of the chair 130 s after reaching point A, below point P?

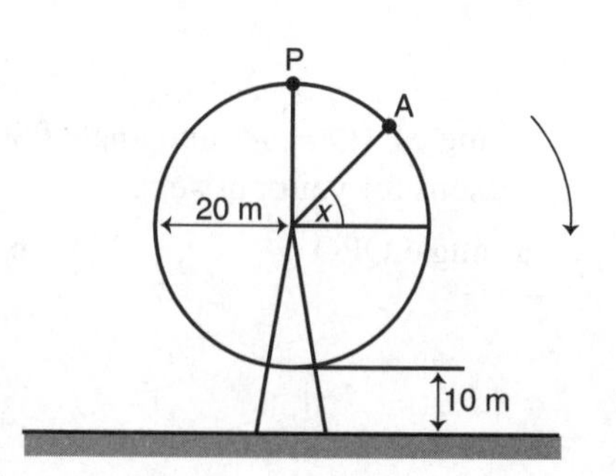

24 ABCDEF is a triangular prism.
The base ABCD is a square of length 18 cm.
The face ADEF is perpendicular to the base.
The height DE = 10 cm.

Calculate

a length BD **b** length BE

c angle DBE.

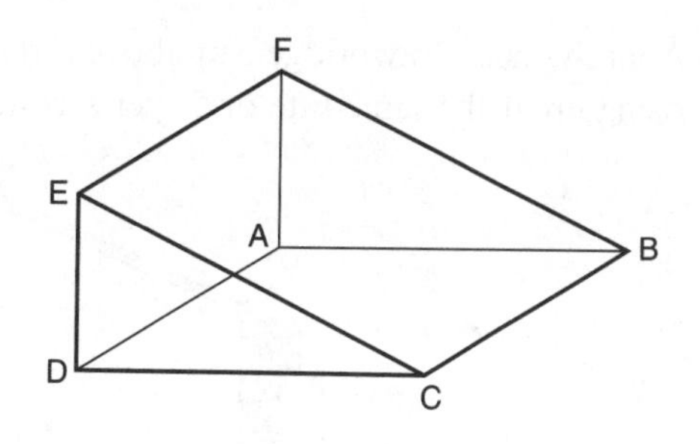

Exercise 98★ (Revision)

1 PQR and PRS are right-angled triangles.
$\angle RPQ = 40°$.
$\angle PSR = 60°$.
PR = 10 m.

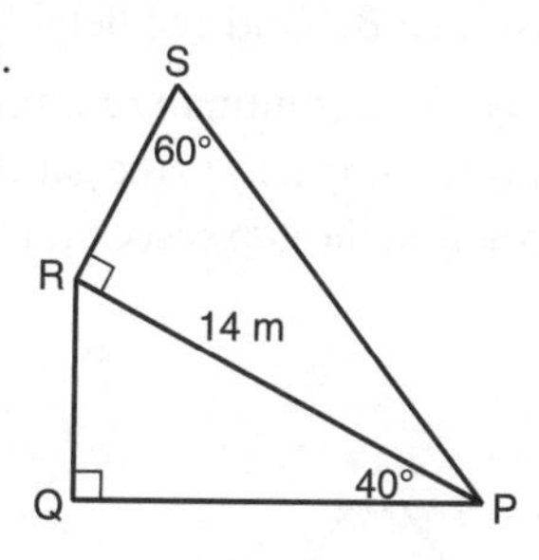

Calculate

a length PQ **b** length QR **c** length RS **d** area PQRS

2 Find the angle x and length y in each of these triangles.

a

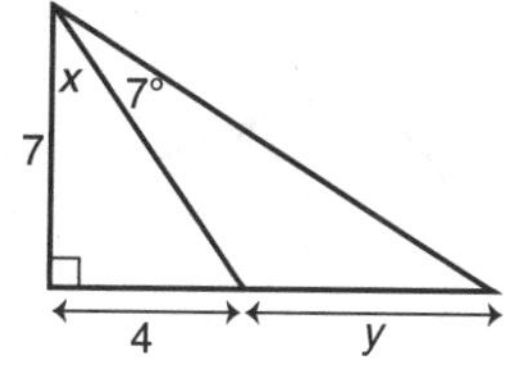

b

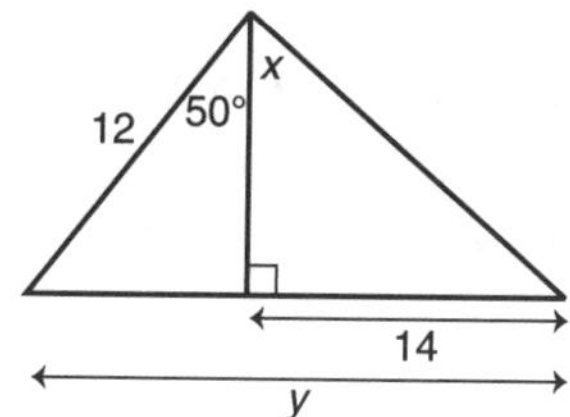

c

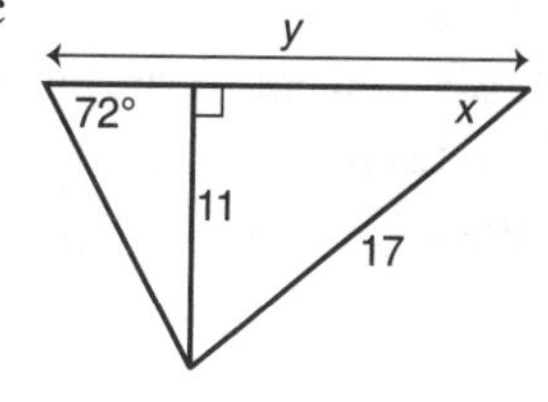

3 A pulley P is vertically above B and PB = 24 m. C is a point level with B and 18 m from it. A 50 m rope has one end pegged at C and the other end at Q, hanging over the pulley as shown.

a Find the length CP.

b How far is Q above B?

c The peg C is moved to a new position 14 m further from B and still on the same level. By how much does Q rise?

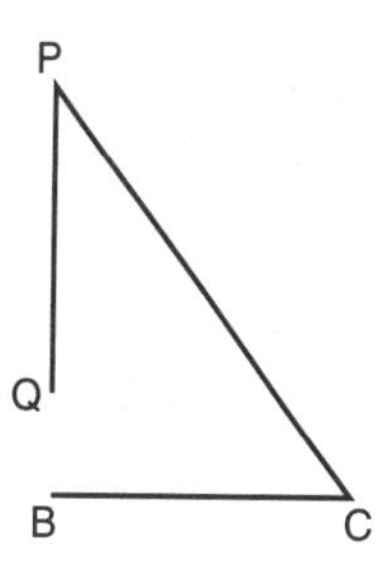

4 A mechanical drawbridge 5 m above a river consists of two equal spans of length 15 m both rising up at the same rate of 5° per second.

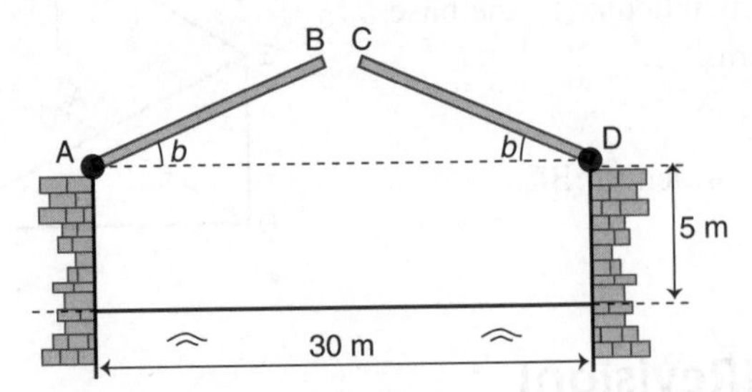

a When $b = 20°$, find distance BC and the height of B above the river.

b At the moment when $b = 20°$, a stuntman runs up the slope AB starting from A at a constant speed of 5 m/s attempting to jump across the gap. He can clear a horizontal distance of 5 m. Does he make the jump across the gap successfully?

5 Find the sides x and y.

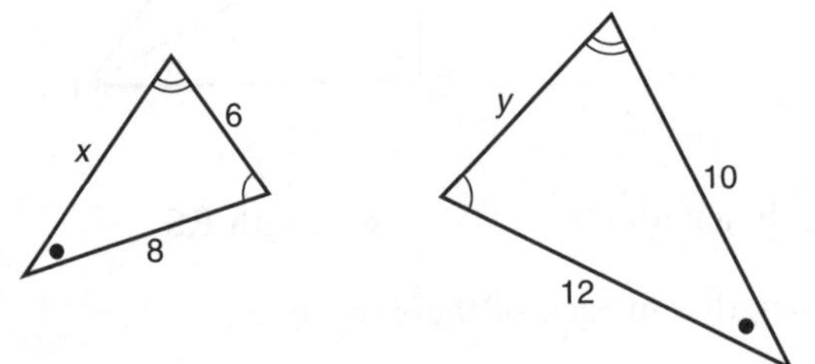

6 AC is a tangent to the circle with centre at O. AB = 5 cm and BC = 12 cm.

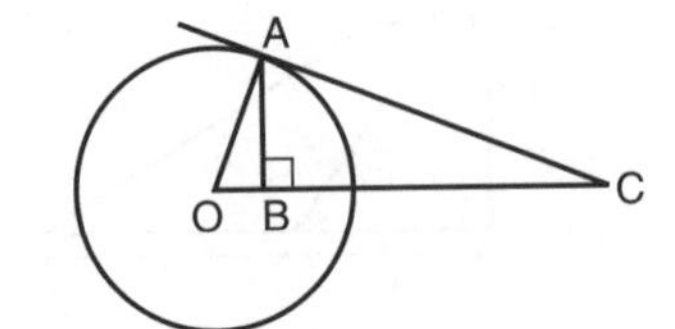

a Write down two triangles that are similar to △ABC.

b Calculate the length of OB.

c Calculate the length of AC.

7 a Draw x and y axes from −4 to 8 and on these axes draw and label the triangle P with vertices (1, 1), (1, 3) and (2, 1).

Four transformations are defined:

A Reflection in $y = -1$

B Reflection in $y + x = 0$

C Enlargement, centre 0, scale factor 2

D Translation along $\begin{pmatrix} -3 \\ -1 \end{pmatrix}$

b Draw on the same axes, triangles Q, R, S and T where:

(i) A(P) = Q (ii) B(Q) = R (iii) C(P) = S (iv) D(S) = T

c Find the angle and centre of rotation that takes P onto R.

d Find the scale factor and centre of enlargement that takes P onto T.

8 This diagram shows a hollow metal pipe.

a Show that the volume V of metal is $V = \pi h(R - r)(R + r)$.

b Find the volume of metal in a pipe where $R = 12.5$ cm, $r = 8.5$ cm and $h = 20$ cm. Take $\pi = 3\frac{1}{7}$.

c Another pipe has a cross-sectional area of $28\,\text{cm}^2$ and volume of $126\,\text{cm}^3$. Find h.

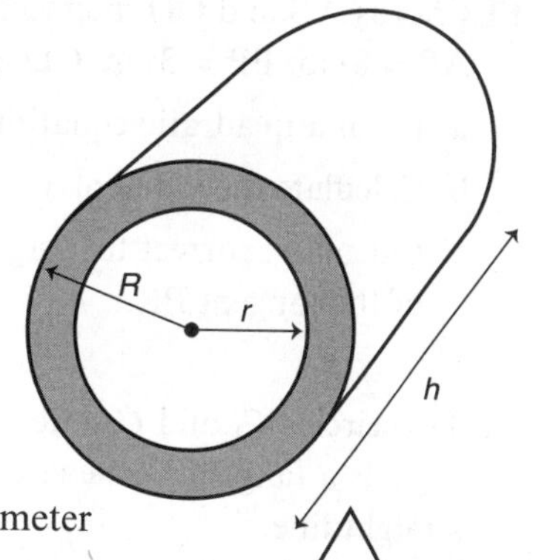

9 This figure shows three identical spherical billiard balls of diameter 6 cm inside an equilateral triangle.

Work out the perimeter of the frame, leaving your answer in surd form.

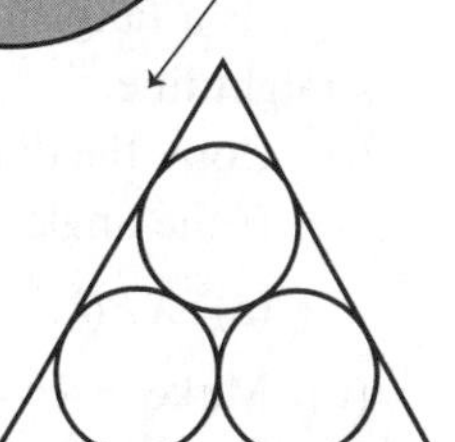

10 A sector with an angle of 144° is cut out of a circle of radius 10 cm. The remainder is turned into a cone by bringing two radii together.

a Work out the circumference of the base of the cone in terms of π.

b Find the base radius of the cone.

c Calculate the height of the cone.

d If the volume of the cone is the same as a sphere, find the radius of the sphere, leaving your answer as a cube root. (Vol. sphere $= \frac{4}{3}\pi r^3$)

11 If PA and PB are tangents, PA is parallel to BC and angle APB is 78°, calculate the angles x, y and z, being careful to state reasons for each step of your working.

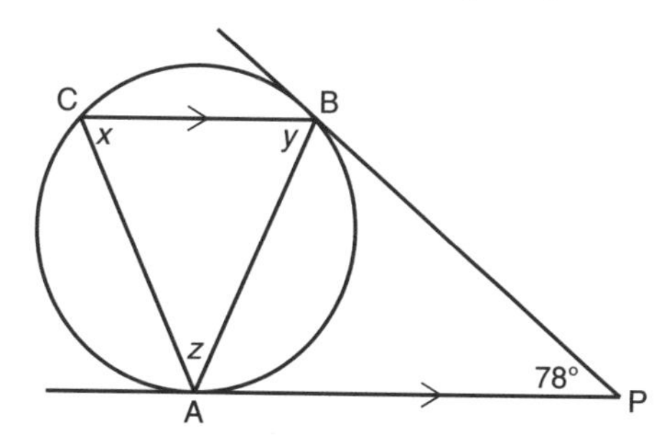

12 In the figure, the perimeter is $4r$.

a Find the arc length AB, in terms of r.

b Show that the angle s is 114.6°, correct to 4 significant figures.

c Find, in terms of r, the shaded sector area.

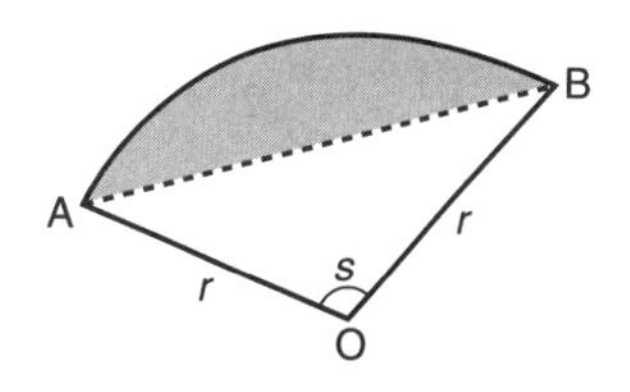

13 Chords AB and CD intersect at P.
AP = 8 cm, BP = 3 cm, CD = 2 cm and DP = x cm.

a Form a quadratic equation in x.

b Calculate the value of x.

c Calculate, correct to 3 significant figures, the length of the tangent PR.

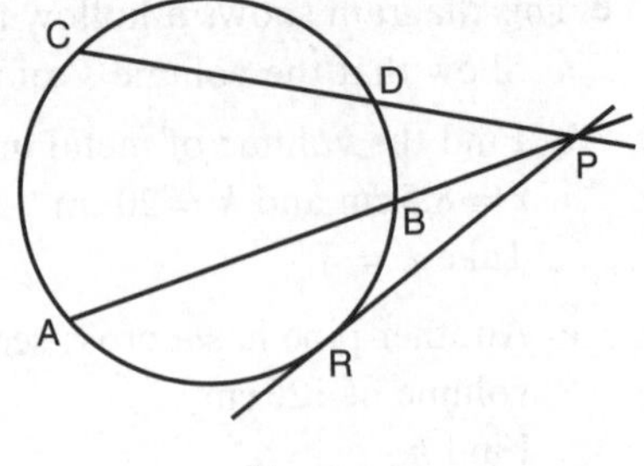

14 Two circles C_1 and C_2 intersect at P and Q. The line AQT is a tangent to the circle C_1 at Q. XPB is a straight line.

(i) Copy the diagram and join PQ.
If the angle $TQX = 100°$, calculate the angles

(a) XPQ (b) QAB

(ii) Make a second copy of the given diagram and prove that, whatever the size of angle TQX, XQ is parallel to AB.

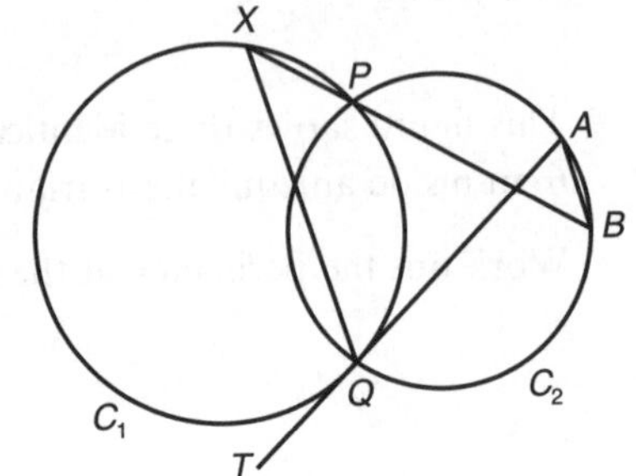

15 Chords AB and CD intersect at P.
AB = 8 cm, BP = 6 cm and DP = 7 cm.
Calculate the length of CD.

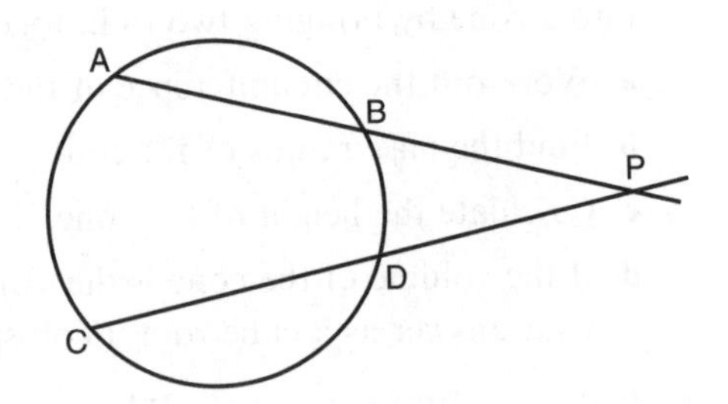

16 You are given these vectors:

$$\mathbf{x} = \begin{pmatrix} 1 \\ 1 \end{pmatrix} \quad \mathbf{y} = \begin{pmatrix} 2 \\ -3 \end{pmatrix} \quad \mathbf{z} = \begin{pmatrix} -5 \\ 5 \end{pmatrix}$$

a If $\mathbf{x} + \mathbf{y} + \mathbf{z} = \mathbf{w}$, find the length of $\mathbf{w}$.

b If $p\mathbf{x} + q\mathbf{y} + \mathbf{z} = \mathbf{0}$, find values of p and q.

17 P, Q, R and S are the mid-points of lines OB, OA, AC and CB respectively.
$\overrightarrow{OA} = \mathbf{a}$, $\overrightarrow{OB} = \mathbf{b}$ and $\overrightarrow{AC} = \mathbf{c}$.

a Write down in terms of **a**, **b** and **c** expressions for

(i) $\overrightarrow{OP}$ and $\overrightarrow{OQ}$

(ii) $\overrightarrow{PQ}$

(iii) $\overrightarrow{OR}$ and $\overrightarrow{OS}$

(iv) $\overrightarrow{SR}$

b Explain why PQRS is a parallelogram.

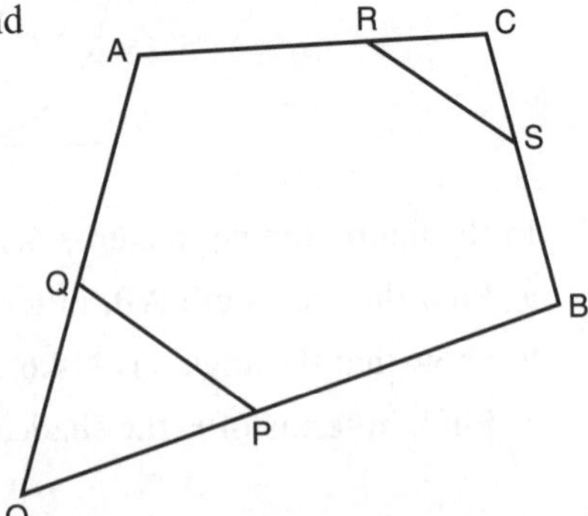

18 A regular tetrahedron of side 10 cm is shown. P is directly below O on horizontal base ABC and Q is the mid-point of AB.

a Find the length of AP, PQ and OQ.

b What angle does OA make with ABC?

c What angle does OAB make with ABC?

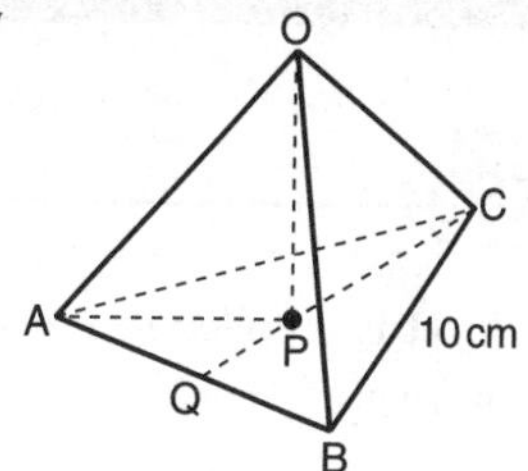

19 Mel is a bird-watcher who spots a crow flying directly towards her at a constant height of 25 m above her eye level.

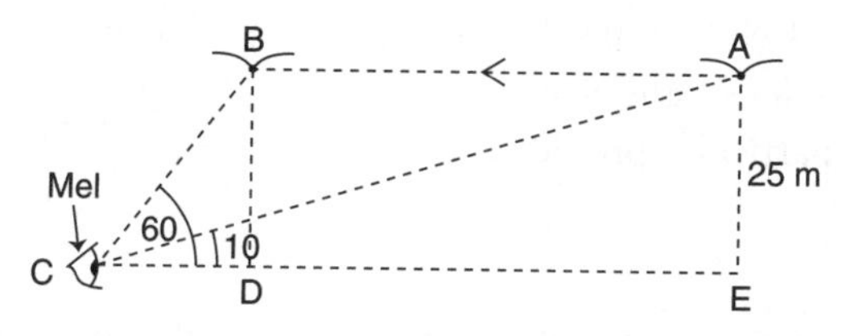

The bird is initially at A, at an angle of elevation of 10°. Ten seconds later, it is at B, where the angle of elevation is 60°.

a Calculate the distance CD

b Calculate the distance CE

c Calculate the crow's speed.

20 A Swing-Ball tennis game consists of a tennis ball attached to a 1m string, one end of which is fixed to a vertical pole of height h metres.

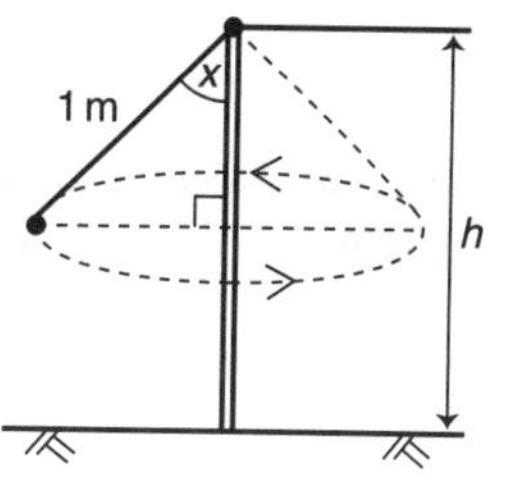

$\sin 30° = \cos 60° = \frac{1}{2}$

$\sin 60° = \cos 30° = \frac{\sqrt{3}}{2}$

a Given that, when angle $x = 60°$, the height of the ball above the ground is 1.5 m, find h.

b When $x = 60°$, the ball takes $\sqrt{3}$ seconds to perform one revolution of its circular path. Find the speed of the ball in terms of π.

c When $x = 30°$, the ball takes 2 seconds to perform one revolution of its circular path. Find the percentage change in the ball's speed from that at $x = 60°$.

Handling data 5 (Revision)

Sets

A **set** is a collection of objects, described by a list or a rule. $A = \{1, 3, 5\}$

Each object is an **element** or a **member** of the set. $1 \in A,\ 2 \notin A$

Sets are **equal** if they have exactly the same elements. $B = \{5, 3, 1\},\ B = A$

The **number of elements** of set A is given by $n(A)$. $n(A) = 3$

The **empty set** is the set with no members. $\{\ \}$ or $\emptyset$

The **universal set** contains all the elements being discussed in a particular problem. $\mathscr{E}$

B is a **subset** of A if every member of B is a member of A. $B \subset A$

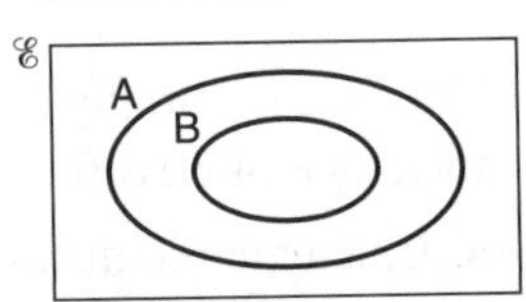

The **complement** of set A is the set of all elements not in A. A'

The **intersection** of A and B is the set of elements which are in both A and B. $A \cap B$

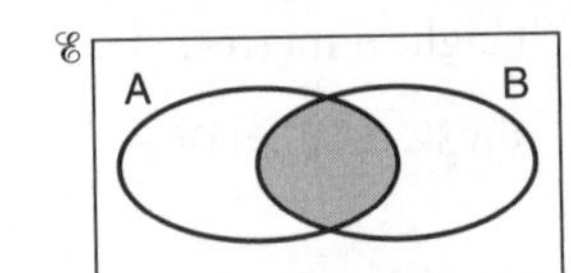

The **union** of A and B is the set of elements which are in A or B or both. $A \cup B$

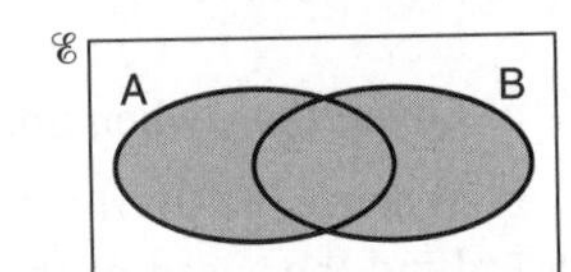

The Venn diagram shows the combinations of languages studied by the pupils in a class at a school.

1 pupil does not study English, French or German.
2 pupils study English, French and German.
3 pupils study English and French, but not German.
4 pupils study English and German, but not French.
5 pupils study French and German, but not English.
6 pupils study German only.
7 pupils study French only.
8 pupils study English only.

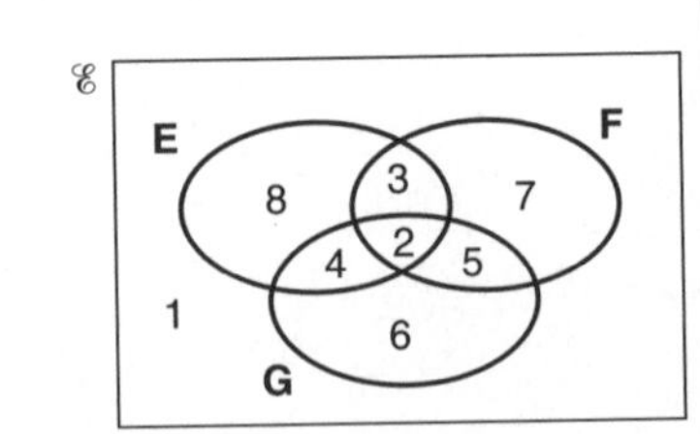

There are $(2 + 3 + 4 + 8 =)17$ pupils studying English. There are 36 pupils in the class.

Probability

Probability

- The **probability** of an event E happening, p(E)

$$p(E) = \frac{\text{number of desired outcomes}}{\text{total number of possible outcomes}}$$

(Impossible) $0 \leqslant p(E) \leqslant 1$ (Certain)

- The probability of an event not happening is $p(\bar{E})$.

$p(E) + p(\bar{E}) = 1$

Relative frequency

Experimental probability is measured by

$$\textbf{relative frequency} = \frac{\text{number of successes}}{\text{total number of trials}}$$

Independent events

If A and B are **independent** events

$$p(A \text{ and } B) = p(A) \times p(B)$$

$$p(A \text{ or } B) = p(A) + p(B)$$

To calculate the probability of an event occurring, it is necessary to consider *all* the ways in which that event can happen.

A bag contains four white beads and one black bead. Three beads are removed from the bag individually, being replaced before each selection. Find the probability that from the three beads there are at least two black beads.

Let B and W be the number of black and white beads respectively.

$p(B \geqslant 2) + p(B < 2) = 1$

so $\quad p(B \geqslant 2) = 1 - p(B < 2)$

$= 1 - p(B \leqslant 1)$

$= 1 - p(B = 0 \text{ or } B = 1)$

$= 1 - [p(B = 0) + p(B = 1)]$

$$p(B = 0) = p(W = 3) = \left(\frac{4}{5}\right)^3 = \frac{64}{125}$$

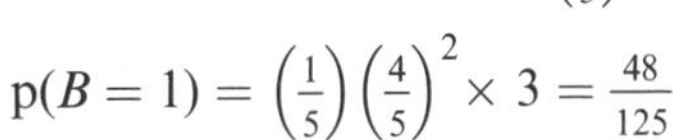

$$p(B = 1) = \left(\frac{1}{5}\right)\left(\frac{4}{5}\right)^2 \times 3 = \frac{48}{125}$$

$$p(B \geqslant 2) = 1 - \left(\frac{64}{125} + \frac{48}{125}\right)$$

so $\quad p(B \geqslant 2) = \frac{13}{125}$

UNIT 5 ◆ Handling data

Conditional events

Two (or more) events might not be **independent**.

An event might be **conditional** on the outcome of another event.

For example: A bag contains three white balls and two black balls. Two balls are to be picked out, at random, and without replacement.

The chances of the second pick will depend on the outcome of the first pick.

$$\text{p(second ball is black)} = \begin{cases} 0.25 \text{ if the first ball was black} \\ 0.50 \text{ if the first ball was white} \end{cases}$$

Probability diagrams

Two dice are thrown. Y equals the difference between the scores.

- The table displays the set of possible outcomes.
- There are 36 equally likely outcomes.
- There are 10 outcomes when Y = 1.
- $p(Y = 1) = \frac{10}{36} = \frac{5}{8}$

First dice (columns); Second dice (rows)

Y	1	2	3	4	5	6
1	0	1	2	3	4	5
2	1	0	1	2	3	4
3	2	1	0	1	2	3
4	3	2	1	0	1	2
5	4	3	2	1	0	1
6	5	4	3	2	1	0

A bag contains two red balls, three green balls and four white balls. Two balls are taken out of a bag without replacement.

- The tree diagram displays the set of possible outcomes and their probabilities.
- p(both the same colour)

$= \frac{2}{72} + \frac{6}{72} + \frac{12}{72} = \frac{20}{72} = \frac{5}{18}$

$\frac{2}{9}$ R: $\frac{1}{8}$ R — RR $\frac{2}{72}$; $\frac{3}{8}$ G — RG $\frac{6}{72}$; $\frac{4}{8}$ W — RW $\frac{8}{72}$

$\frac{3}{9}$ G: $\frac{2}{8}$ R — GR $\frac{6}{72}$; $\frac{2}{8}$ G — GG $\frac{6}{72}$; $\frac{4}{8}$ W — GW $\frac{12}{72}$

$\frac{4}{9}$ W: $\frac{2}{8}$ R — WR $\frac{8}{72}$; $\frac{3}{8}$ G — WG $\frac{12}{72}$; $\frac{3}{8}$ W — WW $\frac{12}{72}$

G = {event that Gurdeep passes an exam} and H = {event that Hassan passes an exam}.

- The Venn diagram shows the probabilities of events:

  $p(G) = \frac{7}{10}$; $p(H) = \frac{8}{10}$

  $p(G \text{ and } H) = \text{p(they both pass)} = \frac{6}{10}$

  $p(G) \times p(H) = \frac{56}{100}$

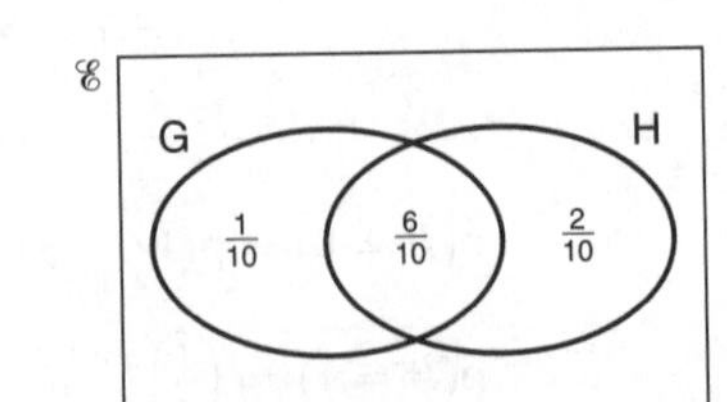

- Events G and H are not independent.

  Perhaps they revise together!

Handling data

Discrete data are data that can be listed and counted, for example the number of peas in a pod, or the number of pages in a book.

Continuous data are data that cannot be listed and counted, for example time, mass and length.

Averages

Example set of data: 23, 5, 7, 8, 10, 7

$$\textbf{mean} = \frac{\text{total of all values}}{\text{total number of values}} = \frac{23+5+7+8+10+7}{6} = 10$$

mode = value that occurs most often = 7

median = value in the middle when data is arranged in ascending order = 7.5
(The median of 5, 7, 7, 8, 10, 23 is 7.5, because it is in the middle of 7 and 8.)

Displaying data

Coloured marbles in a bag are distributed as shown in the table.

Colour	Frequency	Pie-chart angle
Red	1	$\frac{1}{10} \times 360° = 36°$
Yellow	3	$\frac{3}{10} \times 360° = 108°$
Blue	2	$\frac{2}{10} \times 360° = 72°$
Green	4	$\frac{4}{10} \times 360° = 144°$
	$\sum = 10$	

Marble colour	
Red	●
Yellow	● ● ●
Blue	● ●
Green	● ● ● ●

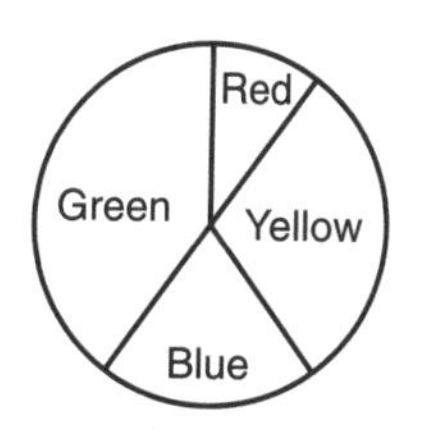

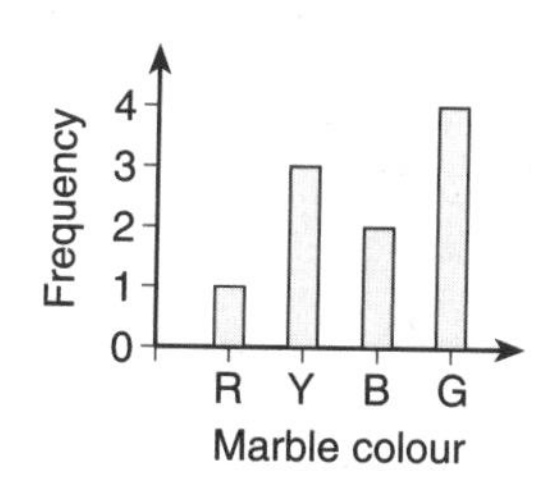

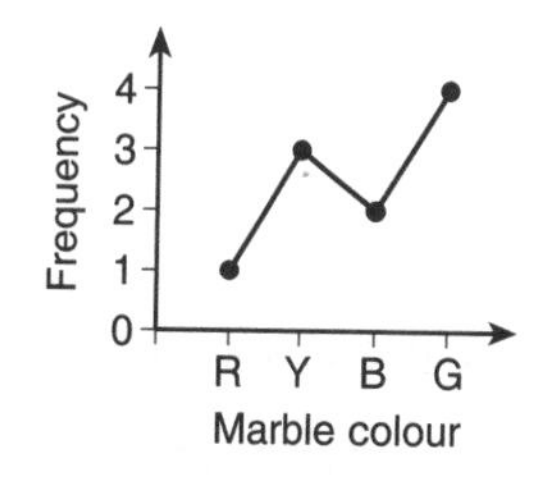

Grouped data: Estimate of mean $= \frac{\sum f \times t}{\sum f} = \frac{\text{sum of (frequency} \times \text{mid-point)}}{\text{sum of frequencies}}$

Time (s)	Mid-point t	Frequency f	$f \times t$
20–25	22.5	5	$5 \times 22.5 = 112.5$
25–35	30.0	15	$15 \times 30.0 = 450$
35–40	37.5	10	$10 \times 37.5 = 375$
		$\sum f = 30$	$\sum ft = 937.5$

An estimate of the mean is $937.5 \div 30 = 31.25$ s.

Histograms

The horizontal axis is a continuous number line.
The **area** of each bar represents the frequency.
The **frequency density** is given by

$$\text{frequency density (height of bar)} = \frac{\text{frequency}}{\text{width of group}}$$

Group time (sec)	Frequency	Width	Frequency density (frequency ÷ bar width)
15–20	12	5	$12 \div 5 = 2.4$
20–28	16	8	$16 \div 8 = 2$
28–36	8	8	$8 \div 8 = 1$
36–40	5	4	$5 \div 4 = 1.25$

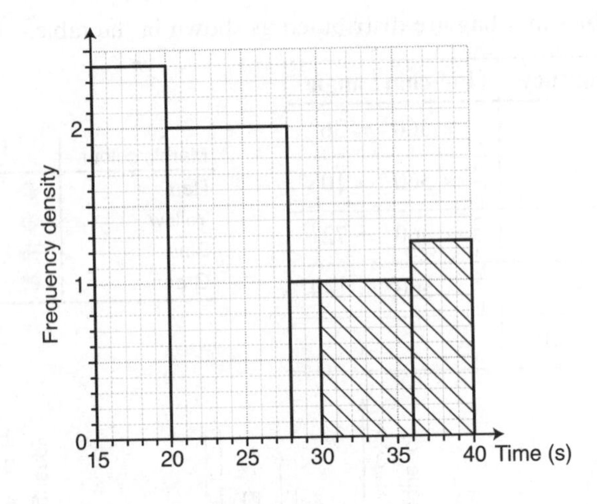

The percentage > 30 $= \dfrac{6 \times 1 + 4 \times 1.25}{41} \times 100\% = 27\%$

Cumulative frequency

Cumulative frequency is always plotted on the vertical axis against the **end-points** of each group. Points are joined by a smooth curve or straight line segments. This cumulative frequency curve can be used to *estimate* quartiles.

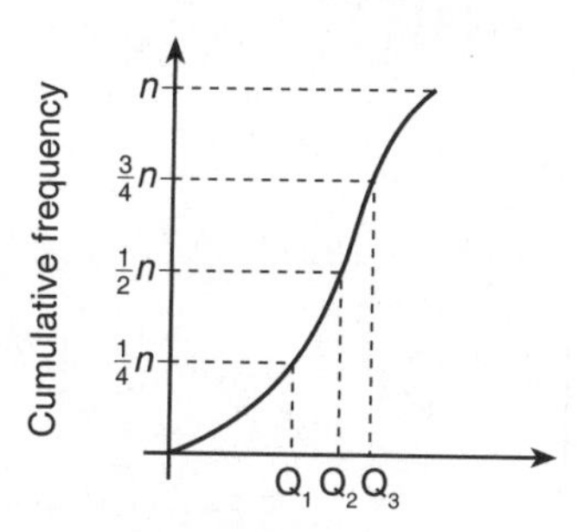

Quartiles	Cumulative frequency	Percentile (%)
Lower quartile Q_1	$\frac{1}{4}n$	25
Median Q_2	$\frac{1}{2}n$	50
Upper quartile Q_3	$\frac{3}{4}n$	75

Measures of spread

Range = largest value − smallest value

Interquartile range = upper quartile − lower quartile = $Q_3 - Q_1$

Exercise 99 (Revision)

1 A = $\{a, c, e, g, i, k\}$ and B = $\{d, e, f, g\}$.

a List the members of the set.

(i) $A \cap B$ (ii) $A \cup B$

b Show sets A and B in a Venn diagram.

2 The universal set, $\mathscr{E} = \{21, 22, 23, \ldots, 30\}$.

X = {multiples of 3}

Y = {prime numbers}

Z = {odd numbers}

Copy the Venn diagram and write the members of $\mathscr{E}$ in the appropriate regions.

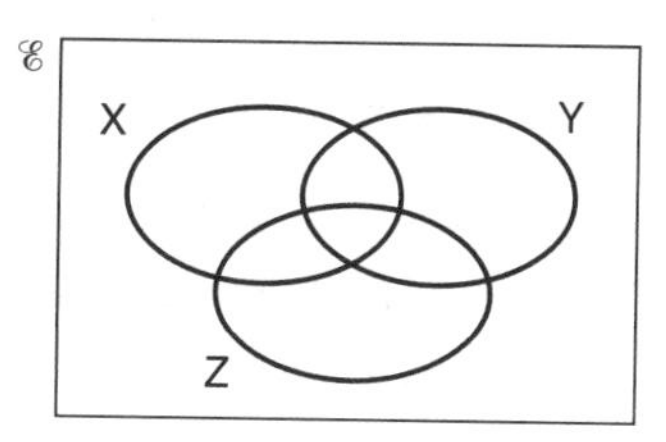

3 Copy the Venn diagram and shade the region representing $(P \cup R) \cap Q'$.

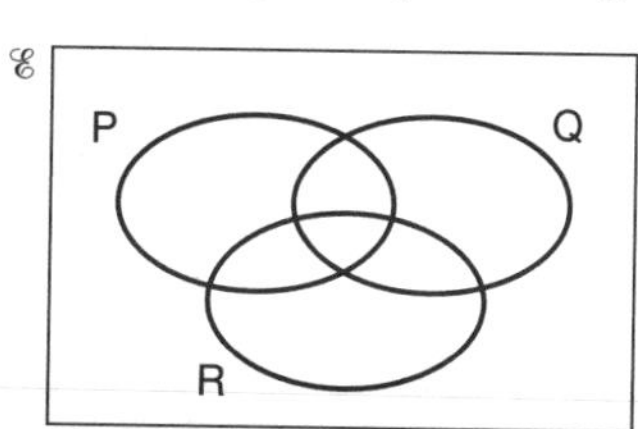

4 In a class of 30 girls, 18 play netball, 12 do gymnastics and 4 don't do games at all.

a How many girls do both gymnastics and netball?
b How many girls play netball, but not gymnastics?

5 From a pack of 52 playing cards

H = {Hearts}
R = {Red cards}
P = {Picture cards}

Copy the Venn diagram and label the three sets, H, R and P.

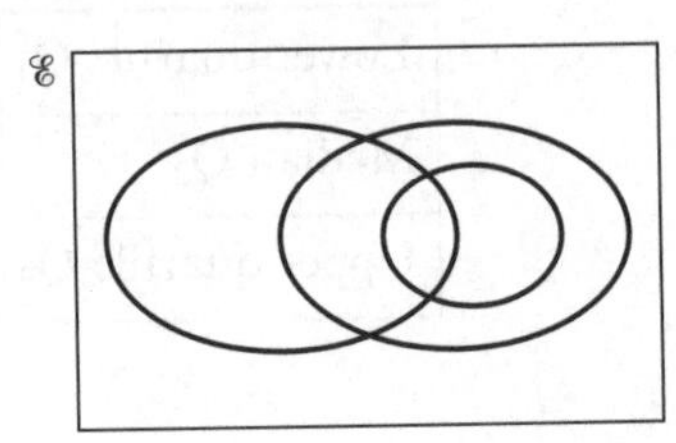

6 When selecting a card from a normal pack of 52 cards, the events 'select a spade' and 'select a red card' are mutually exclusive. Explain what *mutually exclusive* means.

7 A pupil is selected at random from the group shown.

Work out these probabilities.

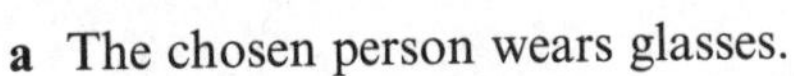

a The chosen person wears glasses.
b He/she does not wear glasses.
c It is a male pupil.

8 The probability that the penalty taker for Midchester Rovers scores at each penalty is 0.8. Ron has two penalties during the season. Calculate these probabilities.

a Ron scores two goals.
b Ron scores only one goal.

9 Two normal six-sided dice are thrown. X is the difference between the scores.

a Copy and fill in this table to show all the possible values of X.
b Calculate $p(X = 1)$
c Calculate $p(X > 2)$

X	1	2	3	4	5	6
1						
2						
3						
4						
5						
6						

10 Lesley and Nicky both have their driving test on the same day. Their probabilities of passing are $\frac{3}{5}$ and $\frac{3}{4}$ respectively.

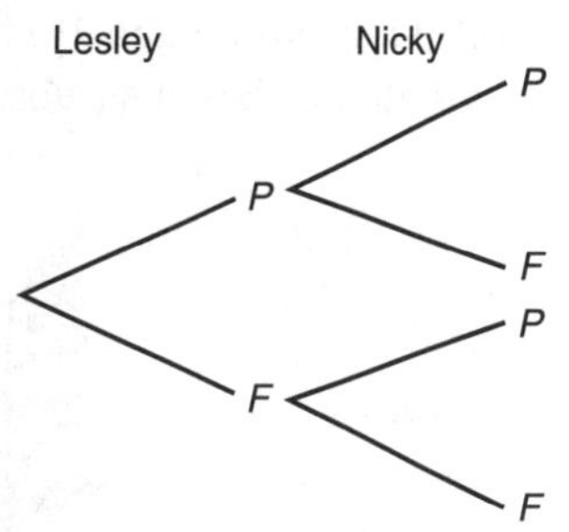

a Copy and complete this tree diagram.

Work out the probabilities of these events.

b Both pass their tests.

c At least one of them passes.

11 A packet of fruit gums contains two yellow, three red, four green and six black sweets placed randomly in the pack. If you take the top two sweets, what is the probability of these events?

a They are both black ones.

b They are the same colour.

c They are different colours.

12 The ages of the players at a football club are given in this table.

17	23	25	30	24	18
18	36	20	20	27	18
24	34	32	32	22	20
26	25	27	23	24	21
21	19	33	29	25	25

a Work out the median age of the players.

b Calculate the mean age of the players.

13 Find the mean, median and mode of these scores from a class test.

9 7 6 5 8 7 6 7 4 5

14 a Construct a frequency table of these scores thrown by the dice in a game of ludo.

1	4	5	6	3	2	3	5	1	3	6	4	3	6	2
3	3	3	5	6	4	5	4	5	2	1	4	1	2	6

b Calculate the mean score.

15 The total number of points scored in a series of basketball matches is given in the frequency table.

a How many matches are included in the survey?

b Calculate an estimate for the mean number of points scored per match.

c Estimate the probability that the number of points scored in a match is between 125 and 155.

Points, x	Frequency, f
121–130	7
131–140	11
141–150	15
151–160	13
161–170	9
171–180	5

For Questions 16 and 17, refer to this frequency table, which shows the distribution of heights of a batch of laurel shrubs grown at a garden centre.

Height (cm)	f
20–40	15
40–50	18
50–55	22
55–60	19
60–70	18
70–90	8

16 Use the garden centre data in the table for this question.

a Calculate the frequency density for each group.

b Draw a histogram of the data.

c Calculate, to one decimal place, an estimate of the mean height of a shrub.

17 Use the garden centre data in the table for this question.

a Calculate the cumulative frequencies.

b Draw the cumulative frequency graph.

c From your graph, work out the median, quartiles and interquartile range.

18 A garage recorded the mileage of a sample of 150 cars at their 12-month service. The results are displayed in this cumulative frequency diagram.

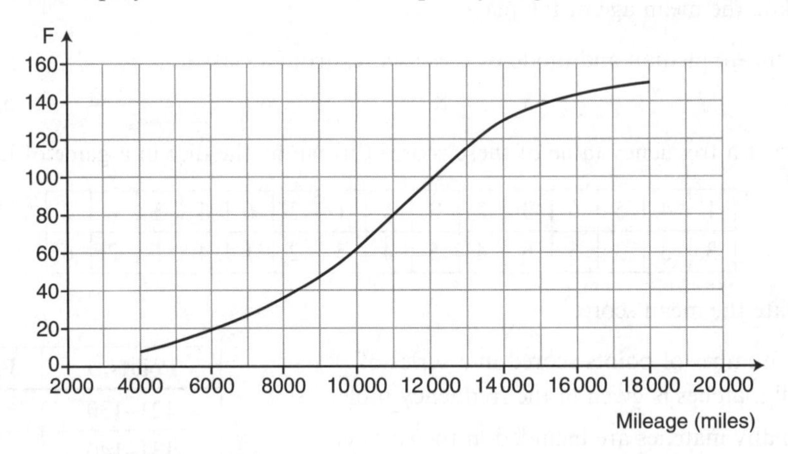

a Write down the median, quartiles and interquartile range of the mileages.

b Calculate the percentage of cars with less than 14 000 miles on the clock.

19 This cumulative frequency diagram shows the distribution of adult female heights of two South American tribes.

a Find the median and interquartile range of the height of each tribe.

b Referring to your statistics, comment briefly on the distribution of the heights of the two tribes.

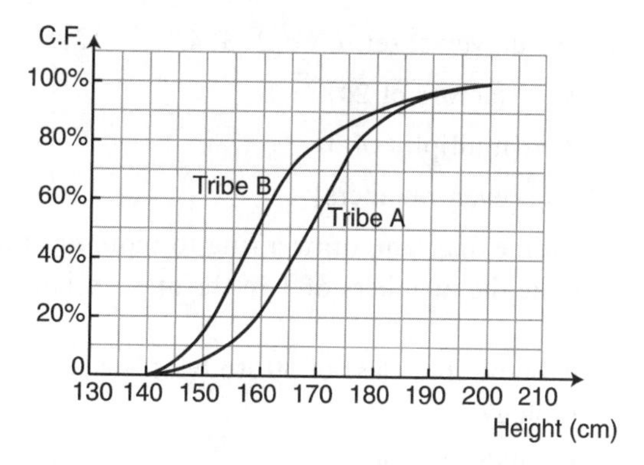

20 Ramon recorded the lengths of time, in seconds, that it took some pupils to solve a puzzle. The results are shown in the unfinished table and histogram.

Length (mins)	Frequency
$60 < h \leqslant 80$	
$80 < h \leqslant 90$	
$90 < h \leqslant 100$	50
$100 < h \leqslant 120$	50
$120 < h \leqslant 150$	45

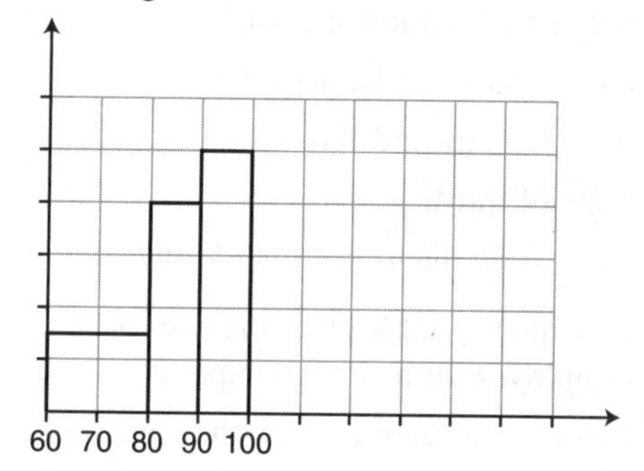

a Use the information in the histogram to complete the table.

b Use the information in the table to complete the histogram.

c Find an estimate for the median length of time taken. You must make your method clear.

Exercise 99★ (Revision)

1 Copy the Venn diagram and shade the region representing $(A' \cup B') \cap C$.

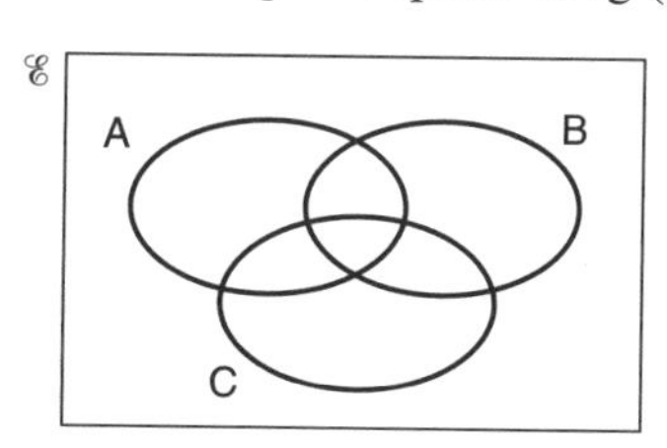

2 $A = \{\alpha, \beta, \chi, \delta, \varepsilon, \phi\}$ and $B = \{\chi, \phi, \eta, \lambda, o, \theta\}$

a List the members of the set.

(i) $A \cap B'$ (ii) $A' \cap B$

b Show sets A and B in a Venn diagram.

3 The universal set, ℰ = {2, 3, 4, …, 12}.

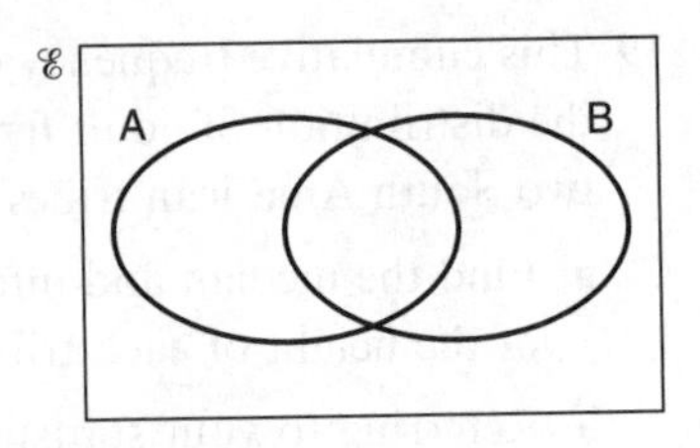

A = {factors of 24}

B = {multiples of 3}

C = {even numbers}

On the diagram, draw a ring to represent the set C, and write the members of ℰ in the appropriate regions.

4 All boys in a class of 30 study at least one of the three sciences: Physics, Chemistry and Biology.

14 study Biology.

15 study Chemistry.

6 study Physics and Chemistry.

7 study Biology and Chemistry.

8 study Biology and Physics.

5 study all three.

Use the Venn diagram to work out how many boys study Physics.

5 This frequency table gives the ages of the employees at a small company.

a Calculate an estimate for the mean age.

b Estimate the median age.

Group	Frequency
20–30	12
30–40	9
40–50	6
50–60	3

6 This bar chart shows the size of the e-mails in a phone-in mailbox.

a Construct the frequency table.

b Calculate the mean, median and mode of the size of e-mails.

c Display the results in a pie chart.

7 There are two bags. The red bag contains seven black balls and three white balls, and the green bag contains four black balls and six white balls. A ball is chosen at random from each bag. Draw a tree diagram and calculate the probability of these events.

a Selecting two white balls

b Selecting a ball of each colour

8 A black dice and a red dice are thrown together and their scores are added. What is the probability that the sum is

a a multiple of 3?

b more than 9?

c a prime number?

9 Three pairs of socks are lying separately in a drawer. Amir grabs two socks without looking. Calculate the probability of these events.

a He selects a matching pair

b He selects a pair of odd socks

10 Two pupils from a class of 25 are to be chosen at random to escort a VIP guest around the school. The probability that two girls are chosen is 0.4. Calculate the probability that two boys are chosen.

11 A bag contains three red balls, four green balls and two black balls. Two balls are selected, *without replacement*.

a Draw and fill in a probability tree diagram.

b Thus work out the probability of choosing two of the same colour.

12 A cook has eight eggs in the fridge; two of them are 'bad'. She breaks three of the eggs into a bowl to make an omelette.

a Complete the probability tree diagram.

b Work out the probability that she breaks at least one bad egg.

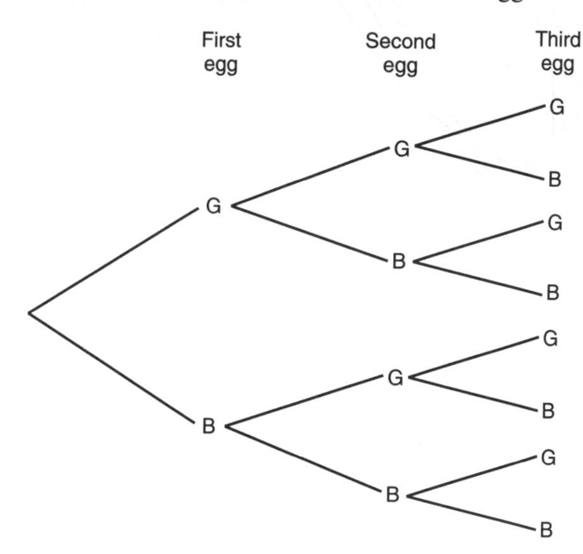

13 A bag contains 16 marbles. Some are white; the remainder are black. Two marbles are taken out consecutively *without* replacement. The probability that two white marbles are picked is $\frac{1}{12}$.

a How many white marbles are in the bag?

b What is the probability that at least one of the two marbles will be white?

14 5000 fourteen-year-old children completed a numeracy test as part of a government survey into educational standards. The results are given in this table.

Marks, x	Frequency, f
0–10	75
10–25	304
25–40	503
40–55	1105
55–70	1469
70–85	994
85–100	550

a Draw a cumulative frequency diagram for the results.

Use your diagram to estimate these.

b The median mark and the interquartile range

c The percentage who scored at least half-marks

15 Researchers investigated a new health diet by setting a fitness test (scored out of 100) to 75 volunteers before and then after a three-week period on the diet. The cumulative frequency diagram displays the results.

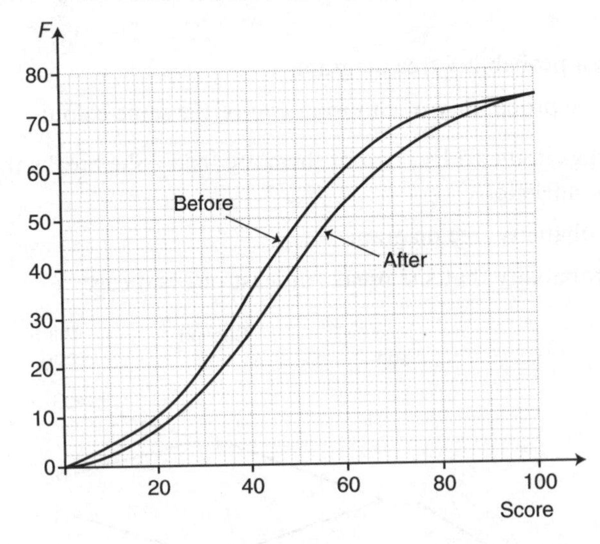

a Write down the median, quartiles and interquartile range for each set of results.

b Compare the results and comment on the effectiveness of the diet.

16 The mass of each potato from a 20 kg sack gave the distribution given in the frequency table.

Mass (g)	f
60–80	15
80–100	31
100–120	53
120–140	40
140–160	24
160–180	9

a Calculate the mean mass of a potato.

b Draw the cumulative frequency curve for these data.

c Work out the median, quartiles and interquartile range from your diagram.

17 Motor-cycle despatch riders in London were sent questionnaires about their work. This histogram shows the age distribution of those who responded.

a Construct a frequency table.

b Use your frequency table to calculate the total number of responses.

c Estimate the mean age of the despatch riders who replied.

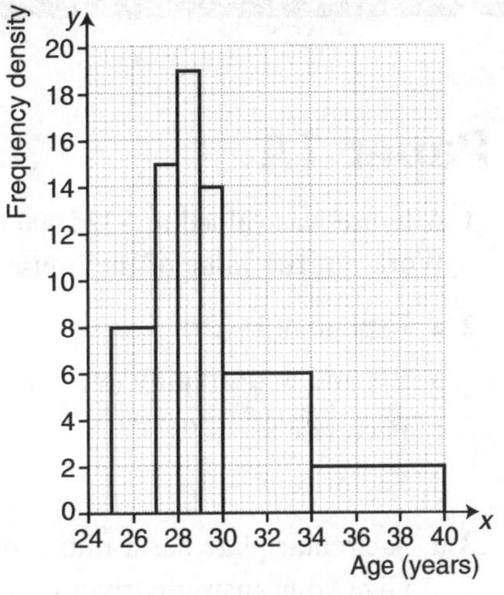

18 This frequency table gives the weights of a catch of cod landed from one boat at Aberdeen one morning.

A histogram is drawn of this distribution and the tallest bar is 7 cm high. Work out the height of the bar for these groups.

a $1.5 \leqslant w < 1.75$

b $2 \leqslant w < 2.5$

c $0 \leqslant w < 1$

Weight, w (kg)	Frequency, f
$0 \leqslant w < 1$	16
$1 \leqslant w < 1.5$	24
$1.5 \leqslant w < 1.75$	25
$1.75 \leqslant w < 2$	35
$2 \leqslant w < 2.5$	29
$2.5 \leqslant w < 4$	21

19 This frequency table and the histogram show the match times of the 64 first-round games in the Ladies' Singles at a tennis tournament.

Time (min)	Frequency
30–45	
45–60	
60–70	
70–80	12
80–90	10
90–120	6
120–150	3

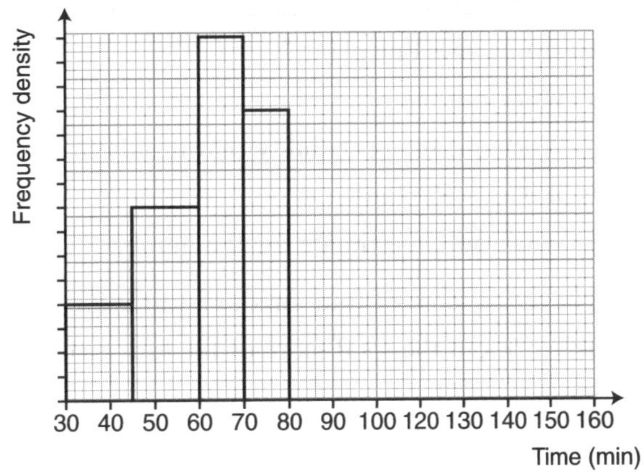

a Copy and complete the frequency table.

b Copy and complete the histogram and write in the scale on the frequency density axis.

c Extend the frequency table and thus calculate an estimate of the mean length of a game.

Examination papers

Paper 1A

1 A house was valued at \$325 000 in January 2003. During the year, it rose in value by 6%. Work out the value of the house in January 2004. **(3)**

2 **a** Expand $4(3 - 2t)$. **(1)**

b Expand and simplify $(x - 4)(x + 2)$. **(2)**

c Factorise $8r - 20s$. **(1)**

d Factorise $p^2 - 3p$ **(1)**

3 **a** A circular plate has a radius of 8.6 cm. Work out the area of the plate. Give your answer correct to 3 significant figures. **(2)**

b Work out the area of the shape in the diagram. **(4)**

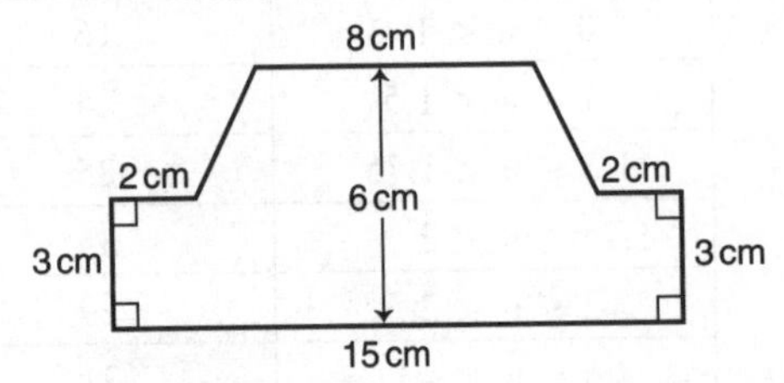

4 A rare plant will have one, two, three, four or five flowers each year. The probabilities are given in the table.

Flowers	1	2	3	4	5
Probability	0.08	0.13	0.35	0.28	

Harshil grows one plant.

a Work out the probability that it has five flowers. **(2)**

b Work out the probability that it only has one or two flowers. **(2)**

Ishil plants 60 plants.

c Work out an estimate for the number of plants with three flowers. **(2)**

5 The mean score of four girls in a test is 73.

a Work out the total score of the four girls. **(1)**

Pasha joins the group and the mean score is then 69.

b Work out Pasha's score. **(3)**

6 Two points, P and Q, are plotted on a centimetre grid. P has co-ordinates (3, 4) and Q has co-ordinates (7, 10). R is the mid-point of PQ.

a Work out the co-ordinates of R. **(2)**

b Use Pythagoras' Theorem to work out the length of PQ. **(4)**

Give your answer correct to 3 significant figures.

7 A = {p, q, r, s, t} and B = {p, q, u, v}.

a List the members of the set.

(i) A ∩ B (ii) A ∪ B **(2)**

b Show sets A and B in a Venn diagram. **(1)**

8 The grouped frequency table gives information about the heights of 90 athletes at a running club.

Height (h cm)	Frequency
$160 < h \leqslant 165$	8
$165 < h \leqslant 170$	18
$170 < h \leqslant 175$	27
$175 < h \leqslant 180$	22
$180 < h \leqslant 185$	10
$185 < h \leqslant 190$	5

a Work out the percentage of athletes who are taller than 180 cm. **(2)**

b Work out an estimate for the mean height of the athletes. **(4)**

Height (h cm)	Cumulative frequency (F)
$160 < h \leqslant 165$	
$160 < h \leqslant 170$	
$160 < h \leqslant 175$	
$160 < h \leqslant 180$	
$160 < h \leqslant 185$	
$160 < h \leqslant 190$	

c Complete the cumulative frequency table. **(1)**

d Draw a cumulative frequency graph for your table. **(2)**

e Use your graph to find an estimate for the median height of the athletes.
Show your method clearly. **(2)**

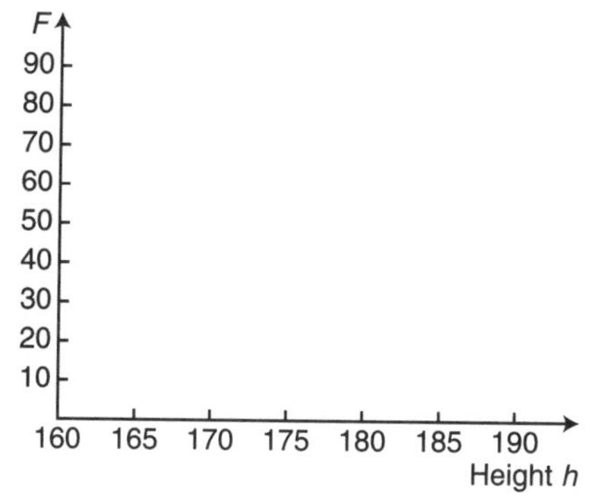

9 Calculate the perimeter of the sector shown in the diagram.
Give your answer correct to 3 significant figures. **(4)**

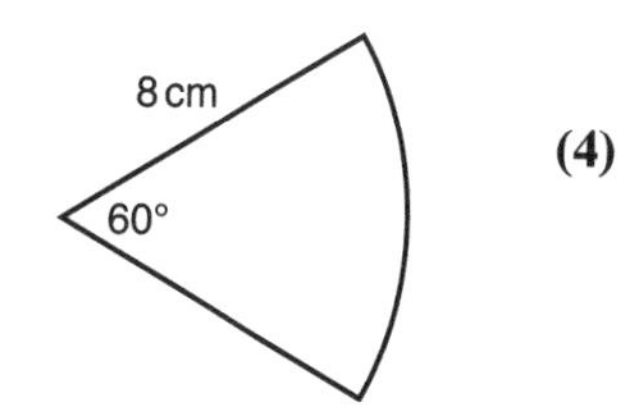

10 Solve the equation $6(x - 3) = x + 5$. **(3)**

11 a Find the gradient of the line with equation $2x - 5y = 18$. **(3)**

b Work out the co-ordinates of the point of intersection of the lines with equations $2x - 5y = 18$ and $8x + 6y = 7$. **(4)**

12 Convert the recurring decimal $0.\dot{1}\dot{2}$ to a fraction in its lowest form. **(4)**

13 a Describe fully the single transformation which maps flag P to flag Q. **(3)**

b Give the co-ordinates of the image of the top of flag P after it has been rotated by $+90°$ about the origin. **(2)**

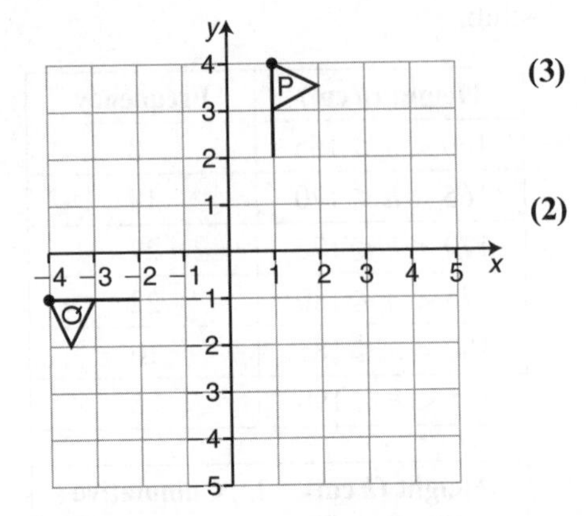

14 The lengths of the sides of a rectangle are 4.53 cm and 6.75 cm, correct to 3 significant figures.

a Work out the upper bound for the area of the rectangle. **(2)**

b Give the area of the rectangle to an appropriate degree of accuracy. You must show working to explain how you obtained your answer. **(2)**

15 Express the algebraic fraction $\frac{x^2 - 25}{2x^2 + 7x - 15}$ as simply as possible. **(3)**

16 a Solve the inequality $15 - 4x < 7$. **(3)**

b Represent your solution to part **a** on a number line. **(1)**

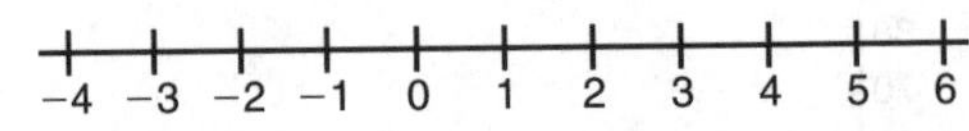

17 The force of attraction between two objects, F newtons, is inversely proportional to the square of their distance apart, d km. When $d = 5$, $F = 10$.

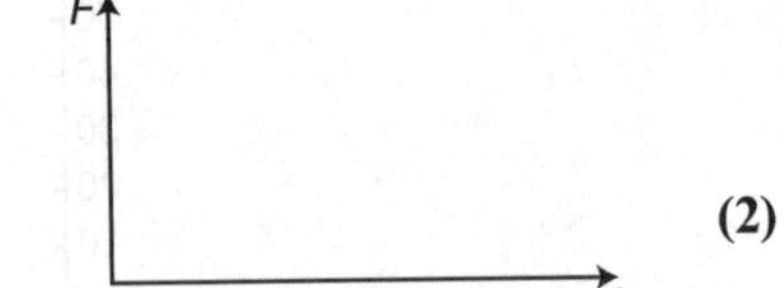

a Express F in terms of d. **(2)**

b Copy the axes and sketch the graph of F against d on the axes. **(2)**

c Calculate the force of attraction between the two objects when they are 12 km apart. **(1)**

18 The diagram shows a solid cone.

The radius of the base is 4.5 cm and the vertical height is 10.8 cm.

Giving your answers correct to 3 significant figures:

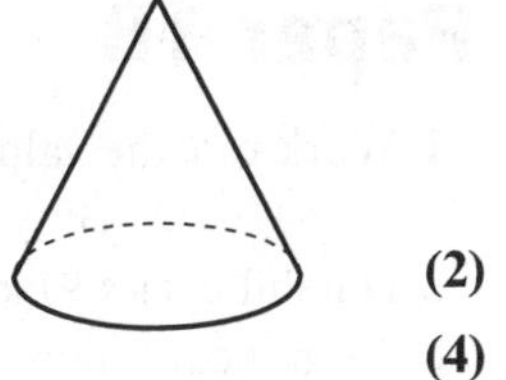

a Calculate the volume of the cone. **(2)**

b Calculate the total surface area of the cone. **(4)**

19 $f: x \mapsto 3x + 1$

$g: x \mapsto x^2$

a Find the value of

(i) $f(2)$

(ii) $gf(3)$ **(2)**

b Express the inverse function f^{-1} in the form $f^{-1}: \mapsto \ldots$ **(2)**

c Express the composite function fg in the form $fg: x \mapsto \ldots$ **(1)**

d Solve the equation $fg(x) = 4$. **(1)**

20

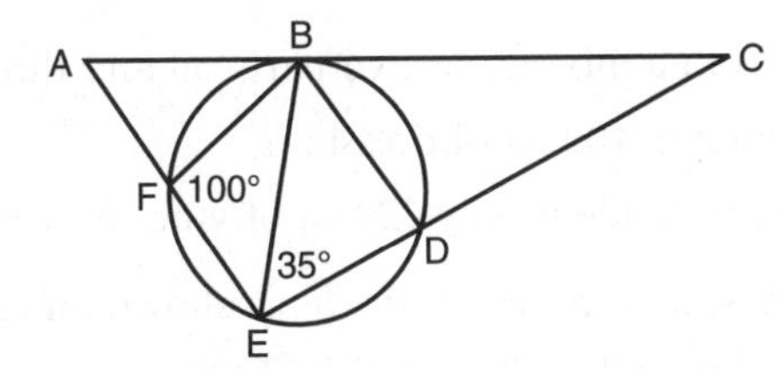

B, D, E and F are points on the circumference of a circle. ABC is a tangent to the circle at B. CDE is a straight line. Angle BFE = 100°. Angle BED = 35°.

Work out the size of angle ACE. You must give a reason for each step of your working. **(5)**

Paper 1B

1 Work out the value of $\frac{5.1 + 3.3}{5.1 - 2.7}$. **(2)**

2 Harshil cycles 91 km in 4 hours 15 minutes. Work out his average speed in km/h, correct to 3 significant figures. **(3)**

3 To approximately convert temperature from the Celsius scale (°C) to temperature in the Fahrenheit scale (°F), you

Double the temperature and add 30

The temperature of an object is F (in °F) and C (in °C).

Write down a formula for F in terms of C. **(2)**

4 a Express 360 as the product of its prime factors. **(2)**

b Work out the highest common factor of 135 and 360. **(2)**

5 A salad dressing is made from a mixture of two parts oil and three parts vinegar.

a How much oil is contained in 100 ml of dressing? **(2)**

b How much dressing can be made if only 120 ml of vinegar is available? **(1)**

6 Ramon recorded the lengths, in minutes, of the films shown on television in one week. The results are shown in the unfinished table and histogram.

Length (min)	Frequency
$60 < h \leqslant 80$	
$80 < h \leqslant 90$	
$90 < h \leqslant 100$	52
$100 < h \leqslant 120$	60
$120 < h \leqslant 150$	24

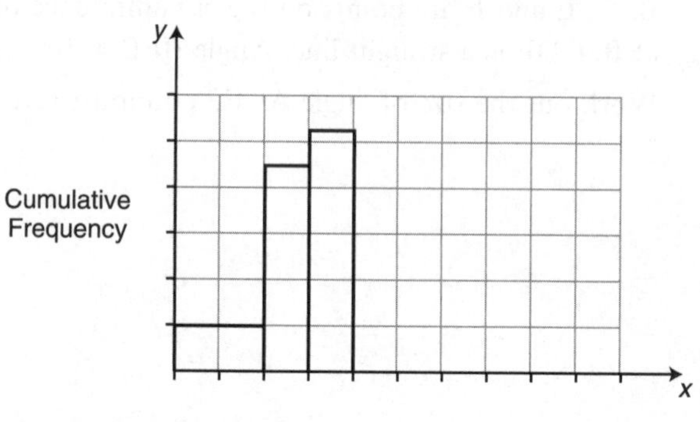

a Use the information in the histogram to complete the table. **(2)**

b Use the information in the table to complete the histogram. **(2)**

c Find an estimate for the median length of a film. You must make your method clear. **(2)**

7 Work out $3\frac{1}{7} \times 1\frac{2}{5}$. Give your answer as a mixed fraction in its simplest form. **(3)**

8 $T = 2\pi\sqrt{\dfrac{l}{g}}$

Yoosuf uses this formula to calculate T from values of l and g.

The values of $l = 42.6$ and $g = 9.8$ have been recorded correct to 1 decimal place.

a Using $\pi = 3$, work out an *estimate* of T. **(2)**

b Write down the upper bound for the value of l. **(1)**

c Using the calculator value of π, calculate the upper bound for the value of T, correct to 2 decimal places. **(3)**

9 a Simplify $a \times a \times b \times b \times b$. **(1)** **b** Simplify $c^2 \times d^3$. **(1)**

c Simplify $\dfrac{e^6}{e^2}$. **(1)** **d** Simplify $\dfrac{f^3 \times f^4}{f^5}$. **(1)**

10 Calculate the size of angle ABC. Give your answer correct to 1 decimal place. **(3)**

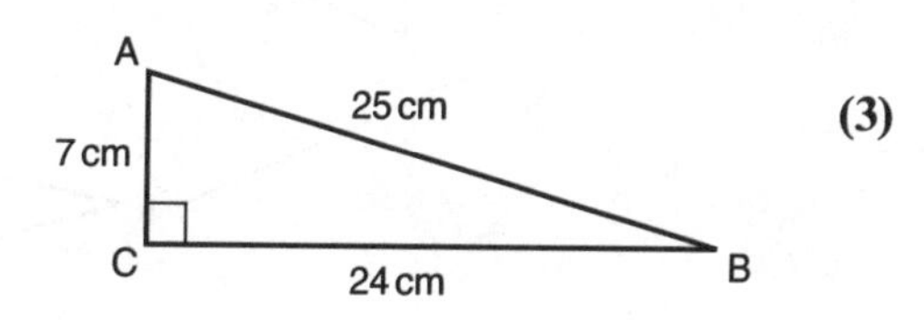

11 The two cylinders are mathematically similar.

a The diameter of the large cylinder is 12 cm. Calculate the radius of the small one. **(2)**

b If it takes 5 ml of paint to cover the small cylinder, how much paint would be needed to cover the large one? **(2)**

20 cm

30 cm

c Both cylinders are made of the same material. The small one weighs 120 g. How much does the large one weigh? **(2)**

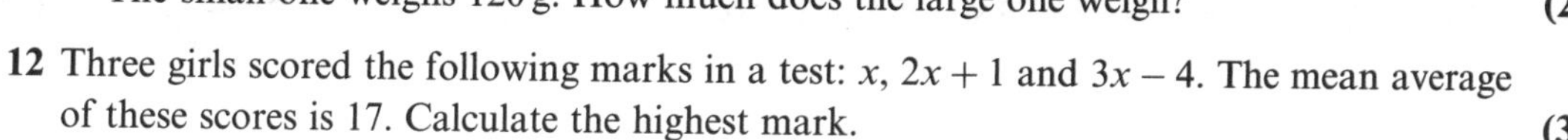

12 Three girls scored the following marks in a test: x, $2x + 1$ and $3x - 4$. The mean average of these scores is 17. Calculate the highest mark. **(3)**

13 Shade the region representing $(P \cup R') \cap Q$. **(3)**

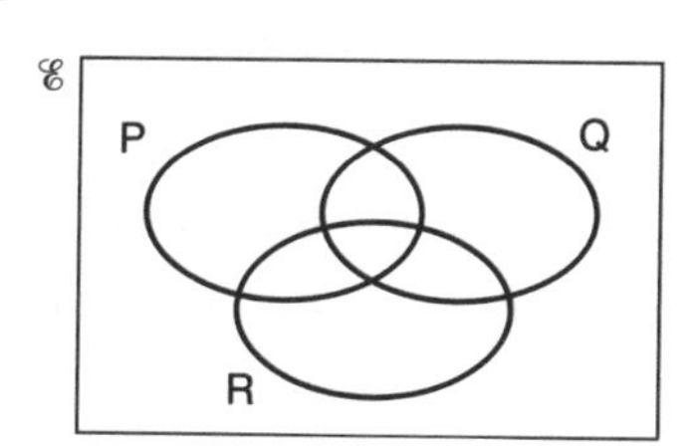

14 Avogadro's number, $N = 602\,000\,000\,000\,000\,000\,000\,000$.

a Write N in standard form. **(1)**

b Calculate the value of $(N \times 50)$, and write your answer in standard form. **(2)**

15 Chi-Loon plays three games of backgammon against his teacher. The probability that his teacher wins any game is $\frac{3}{4}$. There are no draws.

a What is the probability that Chi-Loon wins the first match? **(1)**

b Copy and complete the probability tree diagram for Chi-Loon. **(3)**

c Calculate the probability that Chi-Loon wins all three games. **(2)**

d Calculate the probability that Chi-Loon wins more games than he loses. **(3)**

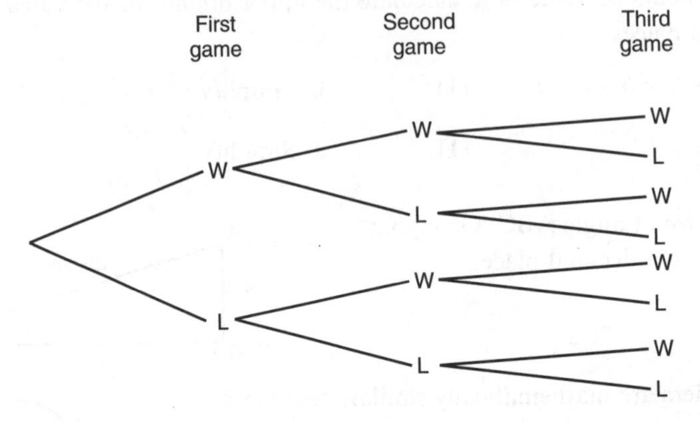

16 A, B, E and D are four points on the circumference of a circle. The chords BD and AE intersect at C.

AB = 5.0 cm, AC = 3.6 cm, CE = 6.5 cm and BC = 4.8 cm.

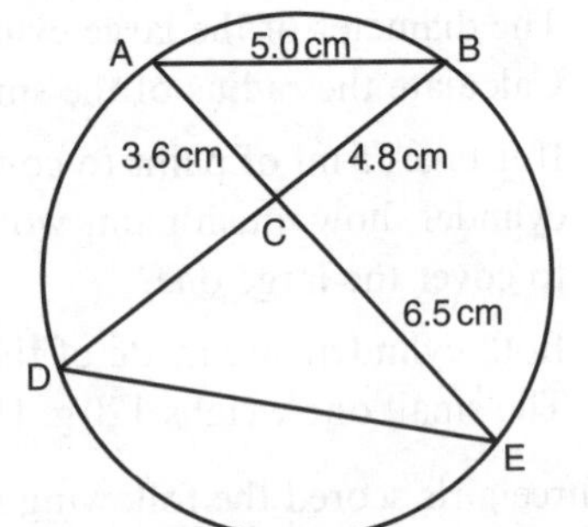

a Calculate the length of CD. **(3)**

b Calculate the size of angle ACB.
Give your answer correct to 3 significant figures. **(3)**

17 A new car loses 18% of its value in the first year. If it is worth \$26 240 after one year, what did it cost when it was new? **(3)**

18 A body is moving in a straight line which passes through a fixed point O.

The displacement, s metres, of the body from O at time t seconds is given by $s = t^3 - 2t^2 + 6t$.

a Find an expression for the velocity, v m/s, at time t seconds. **(2)**

b Find the acceleration after 2 seconds. **(2)**

19 a Expand and simplify $(3a - 4b)(4a + 3b)$ **(2)**

b Simplify $(3cd^3)^2$. **(2)**

c Simplify $(f^{-6}g^4)^{-\frac{1}{2}}$. **(2)**

d Simplify $(2h^3)^{-2}$. **(2)**

20 a Copy and complete the table of values for $y = x^3 - 3x - 1$. **(2)**

x	-2	$-\frac{3}{2}$	-1	$-\frac{1}{2}$	0	$\frac{1}{2}$	1	$\frac{3}{2}$	2
y		$\frac{1}{8}$				$-2\frac{3}{8}$			

b On a copy of the grid, draw the graph of $y = x^3 - 3x - 1$. **(2)**

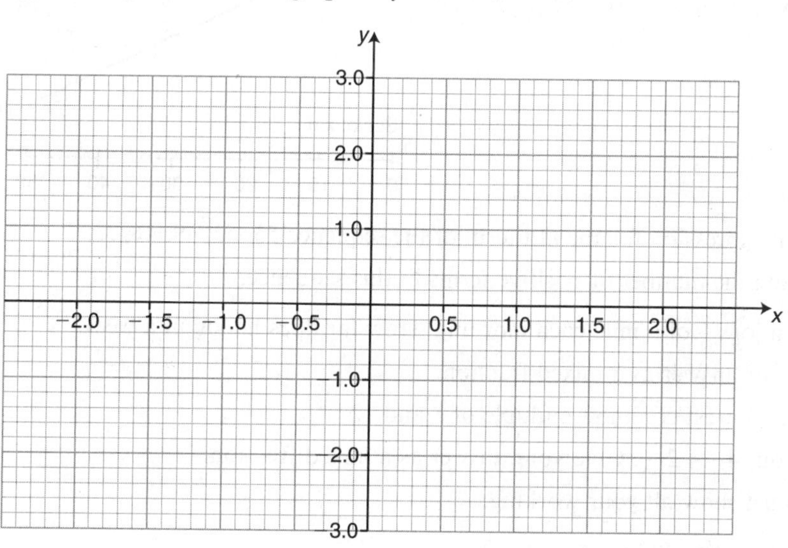

c Use the graph to estimate, correct to 1 decimal place, the solutions of $x^3 - 3x - 1 = 0$. **(2)**

d Use the graph to estimate, correct to 1 decimal place, the solutions of $x^3 - 3x + 1 = 0$. **(2)**

21 Solve the simultaneous equations: $x + 2y = 2$

$$x^2 + 2y^2 = 18$$ **(5)**

22 B is the mid-point of OY.

$\overrightarrow{OB} = \mathbf{b}$ and $\overrightarrow{OA} = \mathbf{a}$.

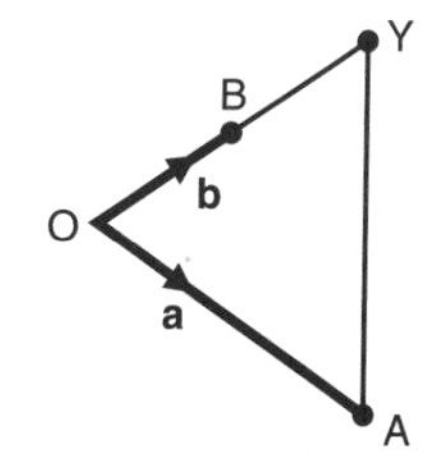

a Write down, in terms of **a** and **b**, expressions for

(i) $\overrightarrow{OY}$

(ii) $\overrightarrow{AY}$ **(2)**

X is the mid-point of OA.

b Write down, in terms of **a** and **b**, an expression for $\overrightarrow{XB}$. **(2)**

c Thus show that XB is parallel to AY. **(2)**

Paper 2A

1 The diagram shows the position of towns A and B on a map.

a Find the bearing of B from A. **(2)**

b Work out the distance of B from A. **(2)**

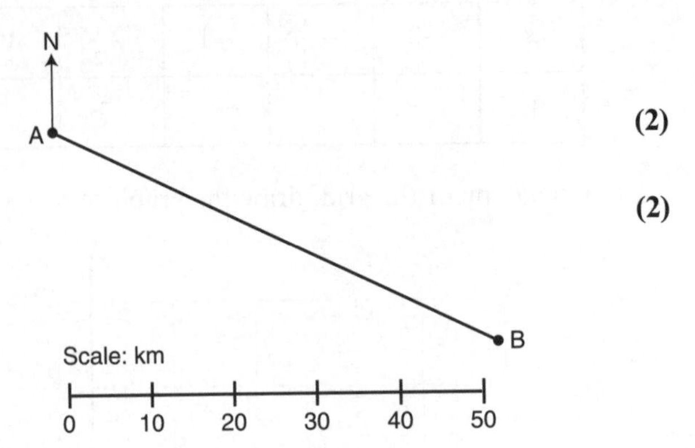

2 a Find an expression for the nth term of the sequence 55, 52, 49, 46, … **(2)**

b Calculate the number of positive terms in this sequence. **(2)**

3 a Hussein jogs $\frac{7}{8}$ of a mile each morning. How far does he jog in a week? Give your answer as a mixed fraction. **(2)**

b Find the lowest common multiple of 8 and 12. **(2)**

c Work out $4\frac{5}{8} - 2\frac{11}{12}$. Give your answer as a mixed fraction. **(2)**
You must show all your working.

4 Solve the equation $2(3x - 4) = x + 4$. **(3)**

5 a On a copy of the grid, draw the line $2y = x$. **(2)**

b On a copy of the grid, show clearly the region defined by the inequalities:

$2y \leqslant x$

$x < 4$

$y \geqslant 0$ **(3)**

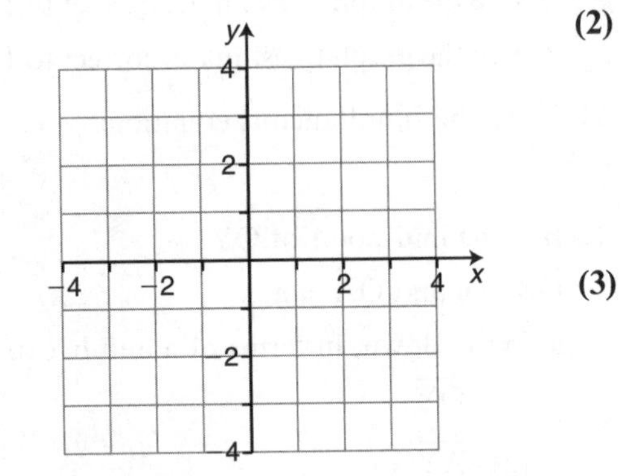

6 a A can of gasoline costs \$1.20, of which 87 cents is purchase tax. Calculate the purchase tax as a percentage of the total cost. **(2)**

b The purchase tax on a can of oil is 70%. If the purchase tax amounts to 91 cents, calculate the full price. **(2)**

7 The diagram shows a circle, centre O.
PQ = 12 cm.

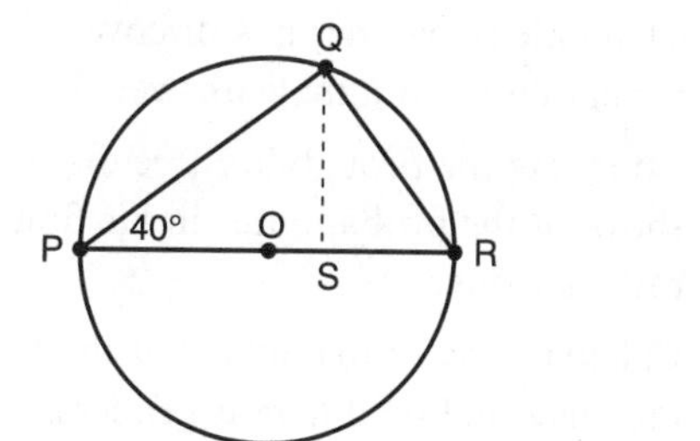

a Explain why angle PQR is a right angle, and write down the value of angle QRS. **(2)**

b Calculate the length QS. **(2)**

c Calculate the length QR. **(2)**

8 The table shows information about the litters born to a group of dogs at an animal farm.

Litter size	Number
3	4
4	5
5	6
6	9
7	6

a Write down the mode of these litters. **(1)**

b Write down the median litter size. **(2)**

c Calculate the mean number of puppies per litter. **(3)**

Another dog gives birth to a litter of four puppies.

d State, without further calculations, whether the mean will increase, decrease or stay the same. **(1)**

e Explain your answer to part **d**. **(1)**

9 The straight line L passes through the points (0, 5) and (3, −4).

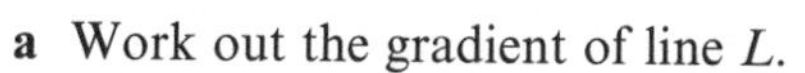

a Work out the gradient of line L. **(2)**

b Write down the equation of L. **(2)**

c Write down the equation of the line through the point (−3, 4) that is parallel to L. **(1)**

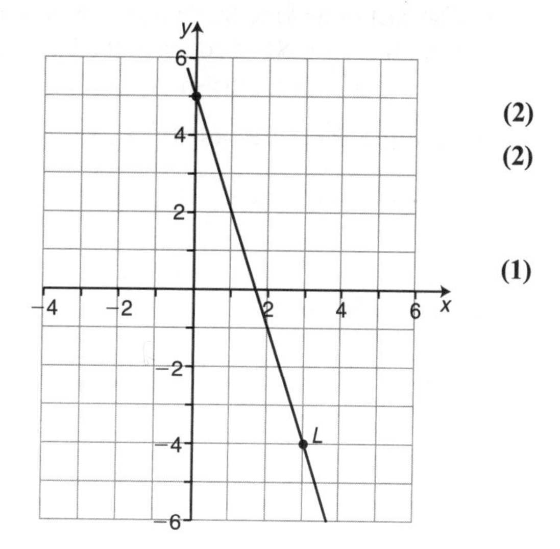

10 The Roman Colosseum was completed in AD 80. It had tiers of seating for 5.0×10^4 spectators and about 80 entrances for the crowds. The population within the city walls of Rome at that time was approximately 5.0×10^5, with three times more living outside the walls.

Giving your answers to 2 significant figures, and in standard form:

a What was the approximate population of Rome plus those living outside the walls? **(2)**

b Calculate the number of spectators per entrance for a capacity crowd. **(2)**

11 A telephone researcher has discovered that the probability that a male answers his call is $\frac{5}{8}$.

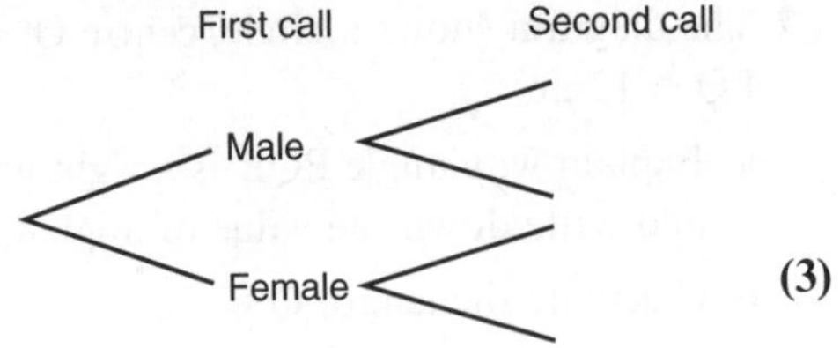

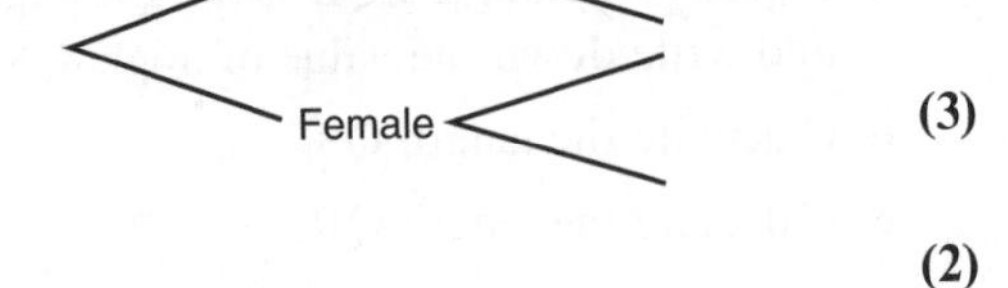

a Complete the probability tree diagram to show all the probabilities in his first two calls one day. **(3)**

b Calculate the probability that the two calls are answered by different genders. **(2)**

12 $f(x) = \dfrac{1}{x+1}$ and $g(x) = 2x$.

a Solve the equation $f^{-1}(x) = 3$. **(2)**

b Solve the equation $fg(x) = gf(x)$. **(3)**

13 The large and the junior footballs are mathematically similar. The large football has volume $= 400\,\text{cm}^3$. The height of the large football is 1.5 times the height of the junior football.

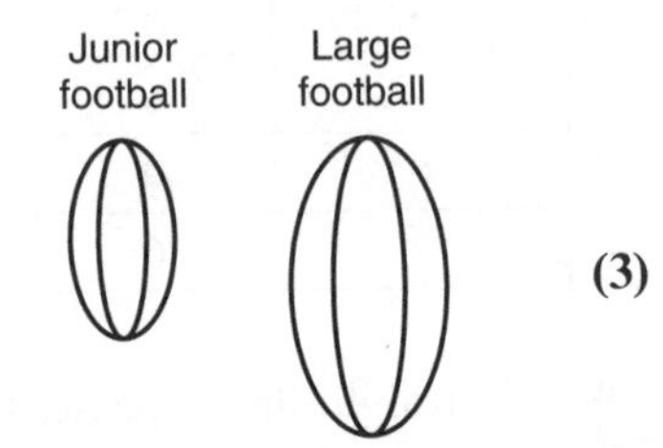

a Calculate the volume of the junior football. **(3)**

b The cost of making the ball is proportional to the surface area. If it costs \$1.35 to make the large football, calculate the cost of making the small one. **(3)**

14 a Factorise the expression $3x^2 + 2x - 5$. **(2)**

b Simplify fully $\dfrac{x^2 - 5x + 6}{x^2 - 4}$. **(3)**

15 Apply the cosine rule to form an expression for $\cos\theta$ in terms of x. **(3)**

16 Show that $\sqrt{12} + \sqrt{27} = \sqrt{75}$. **(3)**

17 There are seven girls and five boys in a class. The teacher selects two at random, *without replacement*, to represent the class on the school council.

a Calculate the probability that he selects two boys. **(2)**

b Calculate the probability that he selects at least one girl. **(2)**

On another occasion he is selecting, at random and *without replacement*, a larger group.

c Calculate the probability that a girl is not chosen in the first three picks. **(2)**

18 Solve the equation $5x^2 + 11x - 6 = 0$.

Give the answers correct to 3 significant figures. **(4)**

19 The diagram shows a solid cylinder of length 14 cm and radius 3.5 cm.

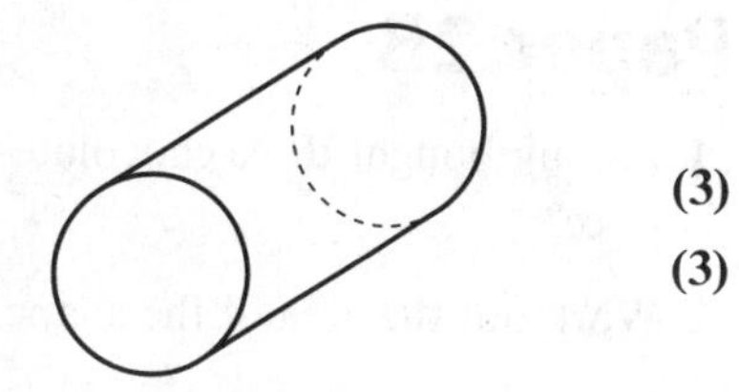

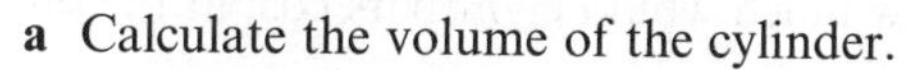

a Calculate the volume of the cylinder. **(3)**

b Calculate the total surface area of the cylinder. **(3)**

Give your answers correct to 3 significant figures.

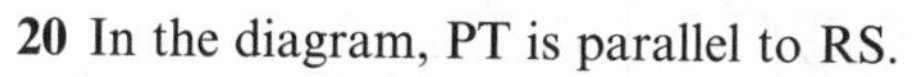

20 In the diagram, PT is parallel to RS.

PQ = 5.6 cm, QS = 4.2 cm, RS = 5.4 cm and RT = 11.2 cm.

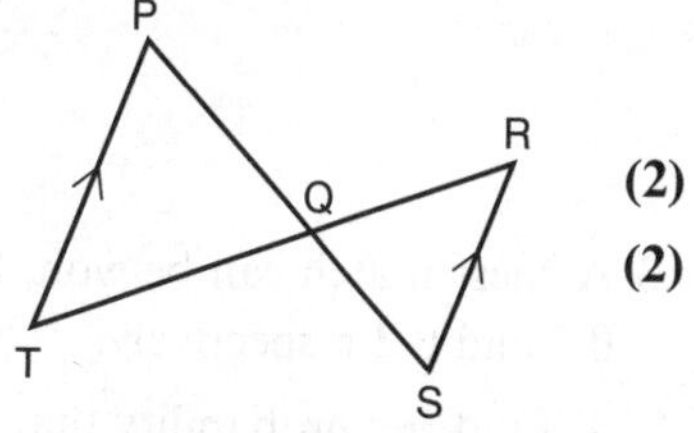

a Calculate the length of PT. **(2)**

b Calculate the length of QR. **(2)**

21 Karim occasionally cycles the 24 km journey to school. When he hurries and increases his normal average speed, v, by 6 km/h his journey time is cut by 40 minutes. How long does it normally take him to cycle to school? **(4)**

Paper 2B

1 Farouk bought three chocolate bars for \$2.52. How much would eight bars cost at the same price? **(2)**

2 Work out the area of the shape. **(4)**

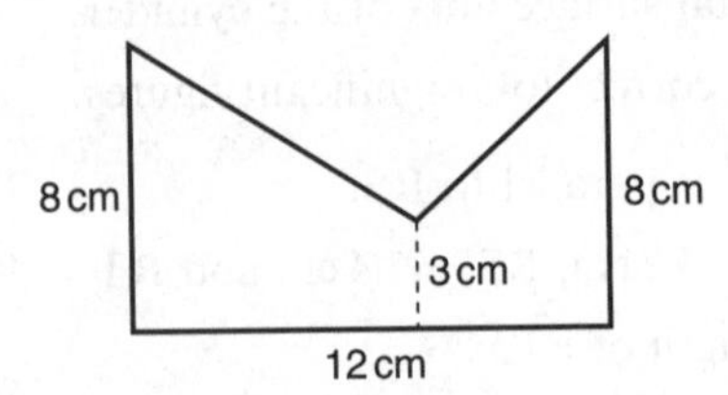

3 A chess match can be won, lost or drawn. The probabilities that Andreiy wins or loses are 0.5 and 0.2 respectively.

a Find the probability that Andreiy draws a match. **(1)**

b During a season Andreiy lost 56 matches. Work out an estimate for the number of matches that he won. **(2)**

4 Calculate the value of $(\sqrt{15} + \frac{1}{7})^3$. Write down all the figures on your calculator display. **(2)**

5 a Expand $p(p - 3)$. **(1)**

b Expand and simplify $2(3q - 6) + 3(q + 4)$. **(2)**

6 The first four terms of a sequence are 1, 2, 5, 9. The rule for this sequence is

Next term = sum of the previous two terms + 2

a Work out the next two terms of the sequence. **(2)**

Consecutive terms of this sequence are 519 and 841.

b Calculate the value of the term before these two. **(3)**

7 The two figures in the diagram are mathematically similar.

a Express tan x as a fraction in its lowest form. **(2)**

b Calculate the height, h, of the smaller triangle. **(2)**

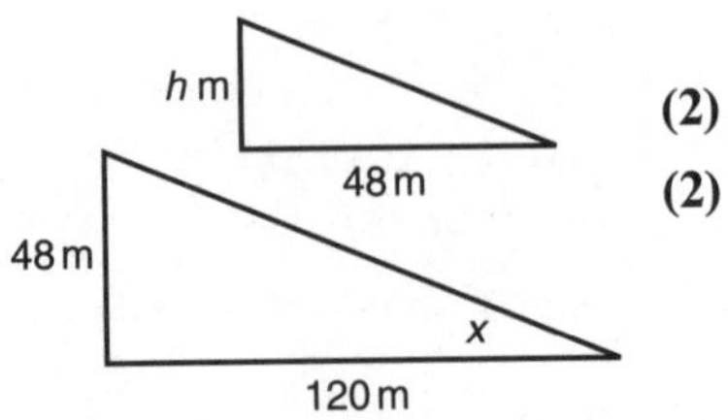

Examination papers

8 The table shows the weights of 40 people who join a slimming club.

Weight (kg)	Frequency
$70 \leqslant w < 75$	4
$75 \leqslant w < 80$	11
$80 \leqslant w < 85$	14
$85 \leqslant w < 90$	6
$90 \leqslant w < 95$	5

a Write down the modal group. **(1)**

b Work out an estimate that a slimmer chosen at random weighs less than 80 kg. **(2)**

c Complete the cumulative frequency table and draw the cumulative frequency graph. **(3)**

Weight (kg)	Cumulative frequency (F)
$70 \leqslant w < 75$	
$70 \leqslant w < 80$	
$70 \leqslant w < 85$	
$70 \leqslant w < 90$	
$70 \leqslant w < 95$	

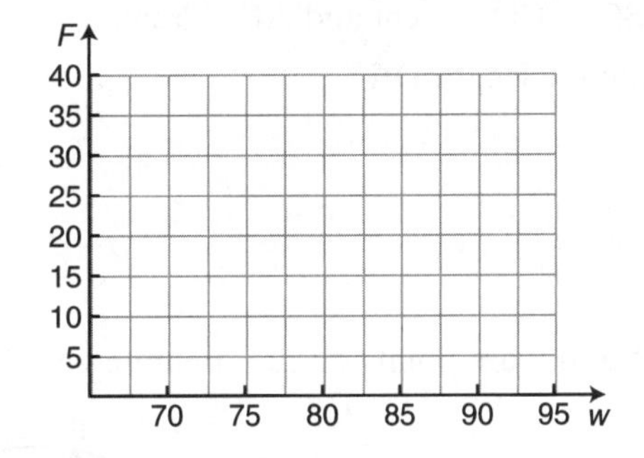

d Use the graph to estimate the interquartile range of the data. Show your method clearly. **(2)**

9 Simplify, leaving your answer in index form:

a $3^5 \times 3^4$ **(1)**

b $3^8 \div 3^4$ **(1)**

c Solve the equation $4^{y-1} = 1$. **(1)**

10 Solve the simultaneous equations: $3x + 4y = 3$ **(4)**

$$2x - 5y = 25$$

11 The universal set, $\mathscr{E} = \{2, 3, 4, \ldots, 10\}$.
A = {factors of 12}
B = {multiples of 3}

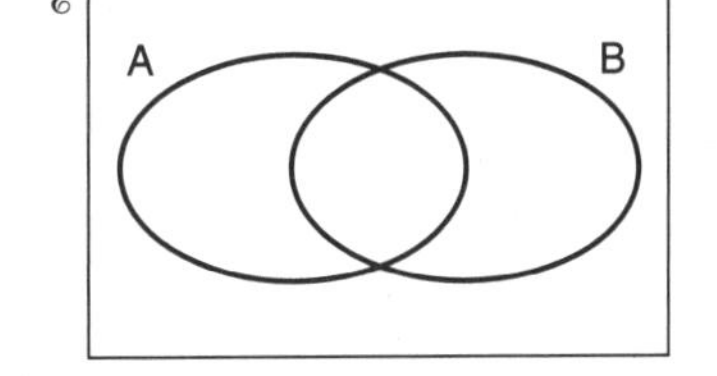

a On the diagram, shade the region that represents the set $A' \cap B$. **(2)**

C = {prime numbers}

b On a copy of the diagram, draw a ring to represent the set C, and write the members of $\mathscr{E}$ in the appropriate regions. **(2)**

12 a Find the lowest common multiple of 72 and 96. **(2)**

b Calculate the exact value of $\frac{7}{96} - \frac{5}{72}$. **(2)**

13 Make u the subject of the formula $s = ut + \frac{1}{2}at^2$. **(3)**

14 A quantity of perfume is to be sold in a cylindrical bottle. The height of the bottle, h, is inversely proportional to the radius, r. When the radius is 5 cm, the height is 8 cm.

a Find a formula for h in terms of r. **(3)**

b Calculate the radius, correct to 3 significant figures, when the height is 6 cm. **(2)**

15 Arthur buys a car for \$1500 and sells it for \$1875. Calculate his percentage profit. **(3)**

16 Angles EAB, EBC and ECD are 90°.
AB = BC = CD = 2 cm and AE = 5 cm.
Calculate the length DE. **(5)**

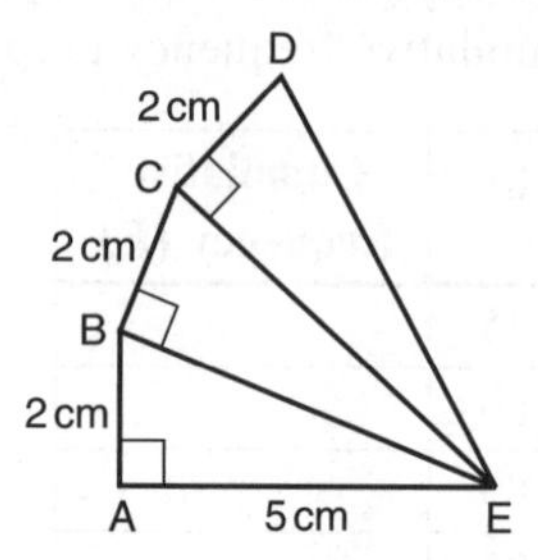

17 P, Q, R and S are points on the circumference of a circle. O is the centre of the circle.

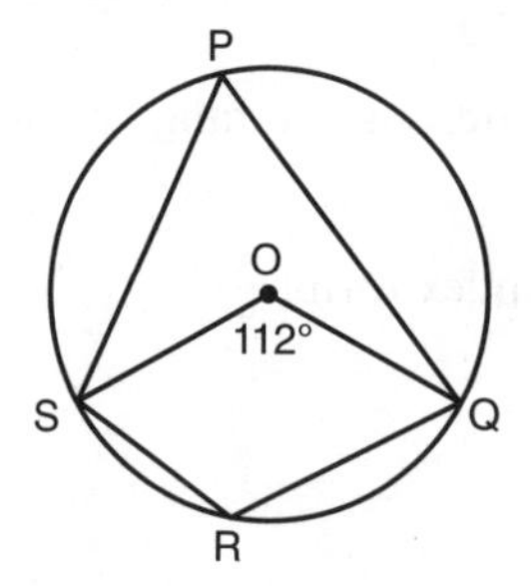

a Find the size of angle SPQ. **(1)**

b Give a reason for your answer. **(1)**

c Find the size of angle SRQ. **(1)**

d Give a reason for your answer. **(1)**

e If a tangent is drawn to touch the circle at S, calculate the acute angle between the tangent and the line SQ. **(2)**

18 A curve has equation $y = x^3 + 6x^2 - 16$. Find

a $\dfrac{dy}{dx}$ **(2)**

b the co-ordinates of the turning points. **(4)**

c Identify each turning point as a maximum or minimum, and justify your answer. **(2)**

19 a Complete the table of values for $y = x^2 + \frac{1}{x}$. **(2)**

x	0.2	0.4	0.6	0.8	1.0	1.2	1.4	1.6	1.8	2.0
y			2.03			2.27	2.67		3.80	

b On a copy of the grid, draw the graph of $y = x^2 + \frac{1}{x}$ for $0.2 \leqslant x \leqslant 2.0$. **(2)**

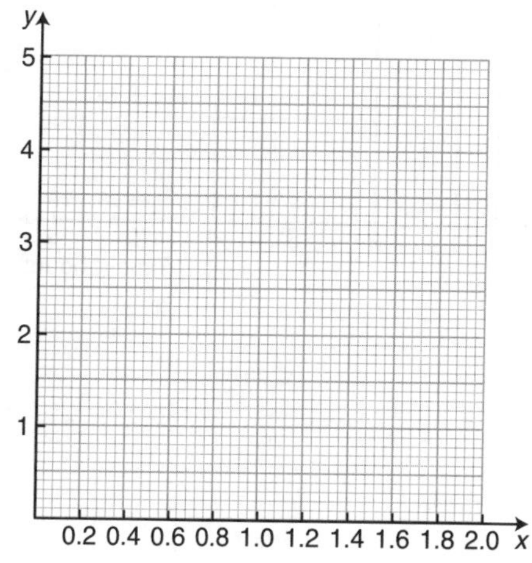

c Use your graph to find estimates, correct to 2 decimal places, for the solutions to the equation $x^2 + \frac{1}{x} = 3$. **(2)**

d By drawing a suitable line on the graph, find estimates, correct to 2 decimal places, for the solutions to the equation $x^2 - 2x + \frac{1}{x} = 0$. **(2)**

20 A circle of radius 10 cm is drawn through the vertices of an equilateral triangle.

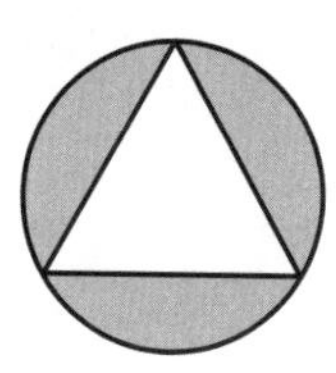

a Calculate the length of the side of the triangle. **(3)**

b Calculate the area enclosed between the equilateral triangle and the circle. **(4)**

21 The winning time in a school 100 m race was given as 14.6 seconds.

The length of the track was measured to the nearest metre, and the time was measured to 1 decimal place.

a Calculate the upper bound for the winning time. **(2)**

b Calculate the upper bound for the average speed of the winner. **(3)**

22 The unfinished table and histogram give information about the survival times of a group of gladiators in the Colosseum.

Time (t hours)	Frequency
$0 \leqslant t < 0.5$	10
$0.5 \leqslant t < 1$	
$1 \leqslant t < 2$	13
$2 \leqslant t < 4$	
$4 \leqslant t < 6$	8

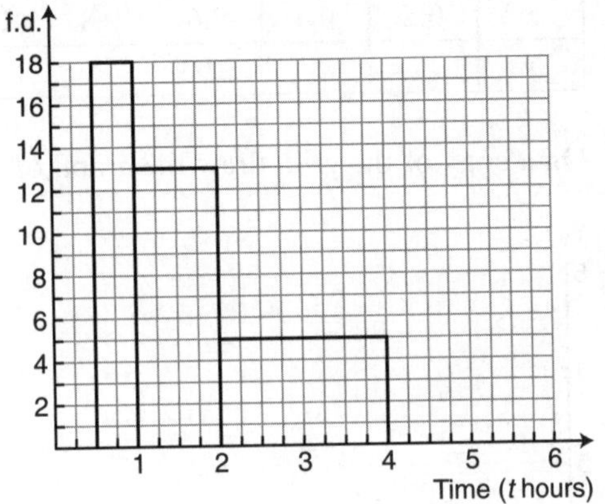

a Use the histogram to complete the table. **(2)**

b Use the table to complete the histogram. **(2)**

Skills practice 1

Number

Work out these.

1 0.67×100 **2** 0.7×100 **3** $5.8 \div 0.1$ **4** $93 \div 0.1$ **5** $34.6 + 3.46$ **6** $1.56 + 0.96$

7 $0.7 - 2.3$ **8** $0.5 - 1.85$ **9** 6.59×7 **10** 4.91×8 **11** $3.43 \div 7$ **12** $7.36 \div 8$

13 3×2^0 **14** 2×3^0 **15** $(3 \times 2)^3$ **16** $(3 + 2)^3$ **17** $10^3 \div 10^4$ **18** $10^2 \div 10^4$

19 $\sqrt{169}$ **20** $\sqrt{144}$ **21** $3\frac{3}{7} + 2\frac{4}{21}$ **22** $4\frac{5}{9} + 3\frac{1}{3}$ **23** $3\frac{3}{7} \div 4\frac{2}{3}$ **24** $2\frac{1}{7} \div 2\frac{1}{2}$

Give your answers to these in standard form.

25 $6 \times (4.5 \times 10^4)$ **26** $7 \times (3.8 \times 10^6)$

Correct these to 3 significant figures.

27 0.046 86 **28** 0.035 79

What is the reciprocal of each of these?

29 6 **30** 8

Simplify and write each of these as a single fraction.

31 $\dfrac{1.5}{7.5}$ **32** $\dfrac{1.2}{2.8}$ **33** $6 \times \dfrac{0.5}{3.5}$ **34** $7 \times \dfrac{0.2}{3.6}$

Algebra

Simplify these.

1 $4x^2 + 2x^2$ **2** $3p^3 + 7p^3$ **3** $4x^2 \times 2x^2$ **4** $5x^2 \times 3x^2$

5 $\dfrac{4a^2}{2a}$ **6** $\dfrac{6a^2}{3a}$ **7** $x(x^2 - 2) - x$ **8** $p(p^3 + 3) + p$

Factorise these.

9 $4ab^2 - 6a^2b$ **10** $8a^3b + 4ab^2$ **11** $x^2 - x - 2$ **12** $x^2 - 5x + 6$

Solve each of these for x.

13 $\dfrac{92}{x} = 4$ **14** $\dfrac{112}{x} = 7$ **15** $8.7 - x = 7.8$ **16** $12.6 - x = 9.5$

17 $3x - 5 = 8 - x$ **18** $5x - 7 = 7 - 2x$ **19** $3x^2 + 5 = 17$ **20** $5x^2 - 10 = 35$

Substitute $a = 3$, $b = -2$ and $c = 1$ to find the value of each of these.

21 $a^2 - b + c$ **22** $a^3 - 2b - c$ **23** $\dfrac{a}{b^2} + 2b$ **24** $\dfrac{2a}{b^2} + b$

25 $(a - b)^2$ **26** $(b - a)^2$

Skills practice 2

Number

Work out these.

1 $6.7 \div 100$ **2** $81 \div 1000$ **3** 45.3×0.01 **4** 99.8×0.01

5 $67.8 - 123$ **6** $34.9 - 87.5$ **7** $3^5 + 2^4$ **8** $2^5 + 3^4$

9 7.54×8 **10** 5.29×7 **11** $(0.6)^2$ **12** $(0.2)^3$

13 $125^{\frac{1}{3}}$ **14** $216^{\frac{1}{3}}$ **15** $585 \div 13$ **16** $1394 \div 17$

17 $54 \div 2^{-2}$ **18** $64 \div 4^{-2}$ **19** $0.39 + 10^{-3}$ **20** $0.87 + 10^{-2}$

21 $3 + 5 \times 6$ **22** $8 + 6 \times 3$ **23** $\sqrt{3600}$ **24** $\sqrt{4900}$

Estimate these.

25 767.9×0.0582 **26** $79.92 \div 0.386$ **27** $\dfrac{\sqrt{7946}}{92.8}$ **28** $\dfrac{\sqrt{389.2}}{5.2}$

Algebra

Simplify these.

1 $\dfrac{6a^3b^2}{3a^2b^3}$ **2** $\dfrac{8a^3b^3}{2a^2b^2}$ **3** $\dfrac{xy^2}{z^2} \times \dfrac{yz}{x}$ **4** $\dfrac{x^2y^2}{z^2} \times \dfrac{y^2z}{x^2}$

5 $\dfrac{3x + x^2}{2x}$ **6** $\dfrac{2a + a^2}{2a}$ **7** $\dfrac{xy^2}{6} \div \dfrac{x^2y}{15}$ **8** $\dfrac{x - y}{2} - \dfrac{y - x}{3}$

Factorise these.

9 $x^2 - 5x + 6$ **10** $x^2 - x - 6$

11 $6a^3b^2 - 27ab^3$ **12** $4x^2 - 9y^4$

Make x the subject of these.

13 $ax + b = c$ **14** $c + cx = a$ **15** $a(x - b) = c$

16 $d = c(x + d)$ **17** $ax = by - cx$ **18** $\left(b + \dfrac{a}{x}\right)^2 = c$

Substitute $a = -1$, $b = -2$ to find the value of these.

19 ab^3 **20** ba^3 **21** $(a - b)^2$ **22** $(b - a)^2$

23 $a - b^3$ **24** $b - a^3$ **25** $(ab)^2$ **26** $(ba)^2$

Skills practice 3

Number

Work out these.

1 2.67×0.7 **2** 4.97×0.3 **3** $5.8 \div 0.2$ **4** $93 \div 0.3$

5 35.6×1.4 **6** 48.7×1.3 **7** $40.8 \div 12$ **8** $68.4 \div 12$

9 $45 - 0.01$ **10** $6.3 - 0.001$ **11** $3^5 - 3^4$ **12** $4^4 - 4^3$

13 $\sqrt{1.44}$ **14** $\sqrt{0.81}$ **15** $10^3 \div 10^4$ **16** $10^2 \div 10^4$

17 $3^{-2} \times 6^3$ **18** $6^{-2} \times 3^3$ **19** $2^{\frac{1}{3}} \times 2^{\frac{1}{3}} \times 2^{\frac{1}{3}}$ **20** $3^{\frac{1}{2}} \times 3^{\frac{1}{2}}$

21 $163.2 \div 17$ **22** $101.4 \div 13$ **23** $6^4 \div 3^4$ **24** $4^4 \div 2^4$

Give your answer to these in standard form.

25 $(2.4 \times 10^5) \times (1.5 \times 10^{-3})$ **26** $(4.5 \times 10^5) \times (1.4 \times 10^{-7})$

27 $(2.4 \times 10^4) \div (1.6 \times 10^{-3})$ **28** $(3.6 \times 10^{-5}) \div (1.5 \times 10^3)$

Algebra

Simplify these.

1 $3x^2 - 6x^2 + 2x^2$ **2** $x^2 - 4x^2 + 3x^2$ **3** $4xy^2 \times 2x^2y$

4 $3yx^2 \times 4xy^2$ **5** $\dfrac{4ba^2}{2ab^2}$ **6** $\dfrac{6xy^3}{3x^2y}$

7 $x - (x^2 - 2) + x^2$ **8** $p - (p^3 + 3) + 3$ **9** $\dfrac{a}{3} + \dfrac{a}{2}$

10 $\dfrac{a}{2} + \dfrac{a}{6}$ **11** $\dfrac{x}{2y} \times \dfrac{y^2}{3}$ **12** $\dfrac{a^2}{2b} \times \dfrac{2b^2}{a}$

Solve these for x.

13 $\dfrac{19}{x} = 3.8$ **14** $\dfrac{17}{x} = 3.4$ **15** $8.7 - 2x = 7.8$

16 $12.6 - 2x = 9.5$ **17** $3(x - 5) = 2(8 - x)$ **18** $5(x - 6) = 2(3 - 2x)$

19 $\dfrac{2x^2 - 4}{2} = 23$ **20** $\dfrac{3x^2 + 12}{4} = 15$ **21** $0 = x^2 - 5x - 14$

22 $0 = x^2 - 6x + 9$

Substitute $a = 3.2$ and $b = -0.8$ to find the value of these.

23 $\dfrac{a^2}{a} - b$ **24** $\dfrac{b^2}{b} + a$ **25** $(a - b)^3$ **26** $(b - a)^3$

Skills practice 4

Number

Calculate these.

1 $3.5 - 0.1$ **2** 7.4×0.2 **3** $7.4 \div 0.2$ **4** 0.3^2 **5** $\sqrt{0.81}$

6 $(\sqrt{3})^2$ **7** $10^4 \div 10^{-2}$ **8** 5% of \$64 **9** $(8^{\frac{1}{3}})^2$ **10** $99^2 - 1^2$

Estimate these to 1 significant figure.

11 5121×0.022

12 $7243 \div 0.035$

13 $(6.02 \times 10^{-4}) \times (1.12 \times 10^6)$

14 $(9.81 \times 10^3) \div (4.89 \times 10^{-3})$

15 $\sqrt{11.3 \times 9.81}$

16 $\dfrac{7.8 \times 9.4}{36.1 + 0.2}$

Give your answer to these in standard form correct to 3 s.f.

17 $(3.5 \times 10^8) + (7.2 \times 10^7)$

18 $(3.5 \times 10^8) \div (7.2 \times 10^7)$

19 $(3.5 \times 10^8) \times (7.2 \times 10^7)$

20 $(3.5 \times 10^8) - (7.2 \times 10^7)$

Algebra

Simplify these.

1 $a^5 \times a^3$ **2** $a^5 \div a^3$ **3** $\dfrac{9a^2b^3}{3a^3b^2}$ **4** $(4x^2y^3)^2$

5 $x(x + 2) - x(x - 2)$ **6** $(x + 3)(x + 2)$ **7** $(x - 3)(x - 2)$ **8** $(x + 3)(x - 3)$

9 $\dfrac{x}{2} + \dfrac{x}{3}$ **10** $\dfrac{x+1}{3} - \dfrac{x-1}{4}$ **11** $\dfrac{x^2 + x}{x + 1}$ **12** $\dfrac{x - 1}{x^2 - 1}$

Solve for x.

13 $\dfrac{1.4}{x} = 0.28$ **14** $2.7 - 3.2x = 12.3$ **15** $5.2 + 2.3x = 7.3 - 1.2x$

16 $x^2 - 25 = 0$ **17** $x^2 = 25x$ **18** $x^2 + 2x - 15 = 0$

19 $x^2 - 2x = 8$ **20** $2x^2 - 5x - 3 = 0$ **21** $\dfrac{x}{4} - 1 = \dfrac{x}{3}$

22 $\dfrac{x+2}{4} = \dfrac{x-2}{5}$

Make x the subject of:

23 $2(x + a) = 5(x - b)$ **24** $\dfrac{4}{3}\pi x^3 = p$ **25** $E = q\sqrt{\dfrac{x}{\pi}}$ **26** $P = a\sqrt{\dfrac{b}{x}}$

Substitute $a = 1.2$ and $b = -4.8$ to find the value of these.

27 $a^2 + b^2$ **28** $\dfrac{2a + 3b}{4a}$ **29** $\dfrac{a + b}{a - b}$ **30** $\dfrac{-b + \sqrt{b^2 - 4a}}{2a}$

CHALLENGES

1 Calculate $1\frac{1}{2} \times 1\frac{1}{3} \times 1\frac{1}{4} \times 1\frac{1}{5} \times \ldots 1\frac{1}{2005}$.

2 How many squares are there on a chessboard?

3 I shuffle a pack of 52 cards and deal the cards out. What is the probability that the third card I deal is a heart?

4 What is the angle between the hands of a clock at 14:18?

5 If a hot-air balloon flew right round the Equator at a constant height of 1 km, how much further than the circumference of the Earth will it travel?

6 A, B, C and D are four different digits. If ABCD $\times$ 9 = DCBA, find ABCD.

7 Find the exact value of $666\,666\,666^2 - 333\,333\,333^2$.

8 If $x - y = 3$ and $x^2 - y^2 = 63$, what is $x^2 + y^2$?

9 A teenager wrote his own age after that of his father. From this four-digit number he subtracted the difference of their ages to get 4289. Find the sum of the father's and the son's ages.

10 A rock climber climbed up a rock face at a uniform rate. At 9 o'clock he was one-sixth of the way up, and at 11 o'clock he was three-quarters of the way up. What fraction of the total rock face had he completed at 10 o'clock?

11 The two circles are concentric. The line shown is 20 cm long and is a tangent to the inner circle. Find the shaded area.

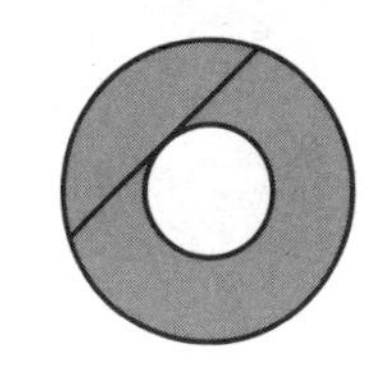

12 The diagram shows a sailing boat with two parallel masts of height 12 m and 8 m. Use similar triangles to find the height h m above the deck where the two supporting wires cross.

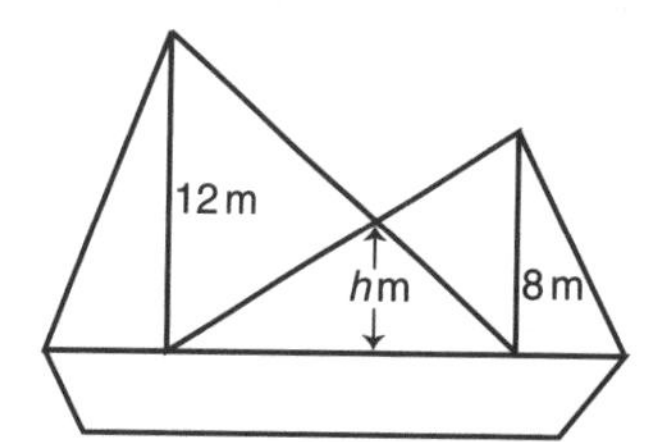

13 Make the following shape from a single sheet of A4 paper using only a pair of scissors. No gluing or sticking is allowed.

(A standard A4 paper is 21 cm wide and 29.7 cm long.)

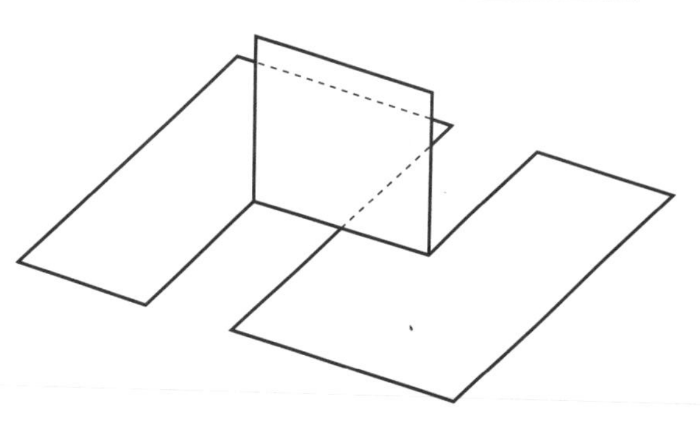

14 The diagram shows a cuboid of side 4 cm. M and N are the mid-points of the sides AB and AC respectively. Using surds, find the exact area of the triangle DMN.

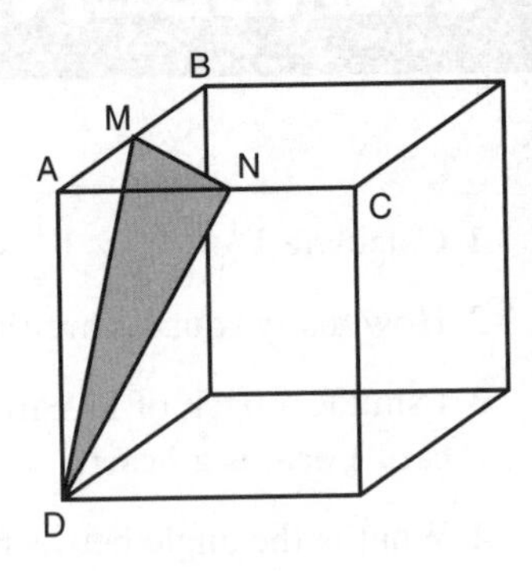

15 What is wrong with the following proof?

Let $a = b$
$\Rightarrow a^2 = ab$
$\Rightarrow a^2 - b^2 = ab - b^2$
$\Rightarrow (a + b)(a - b) = b(a - b)$
$\Rightarrow a + b = b$
Substituting $a = b = 1$ gives $2 = 1$

16 A two-digit number is increased by 36 when the digits are reversed. The sum of the digits is 10. Find the original number.

17 A sheet of paper 8 cm by 6 cm is folded so that one corner is over a diagonally opposite corner. What is the length of the crease?

18 *Remove* five matches to leave three of the original squares.

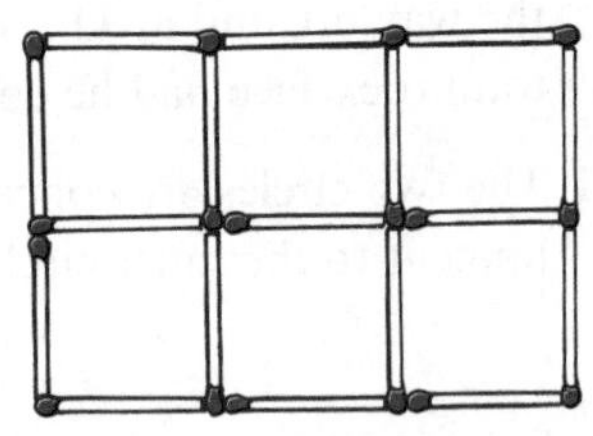

19 The amounts of porridge put out for the three bears were in the ratio 10 : 7 : 4. If father bear's bowl contained $\frac{1}{4}$ litre more than mother bear's bowl, how much did baby bear get? (Baby bear got the least of the three.)

20 The diagram shows a square PQRS inside a right-angled isosceles triangle ABC.
PQ = QR = 2 cm.
Find the area of the triangle ABC.

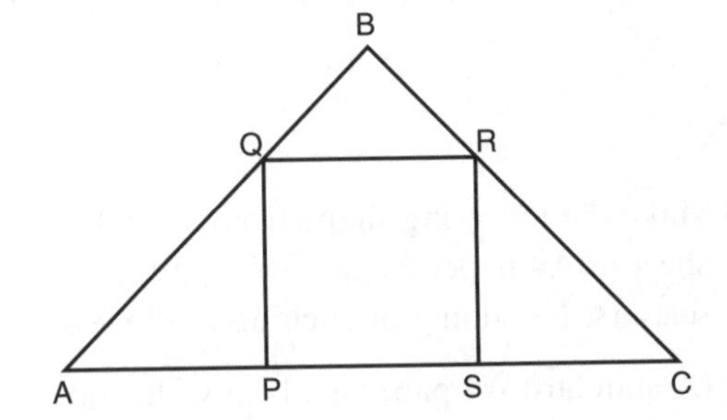

21 Gertie the goat is tied *outside* a 6 m by 10 m rectangular fenced enclosure by a 6 m long rope which is tied to the middle of the shorter side. (See diagram.)

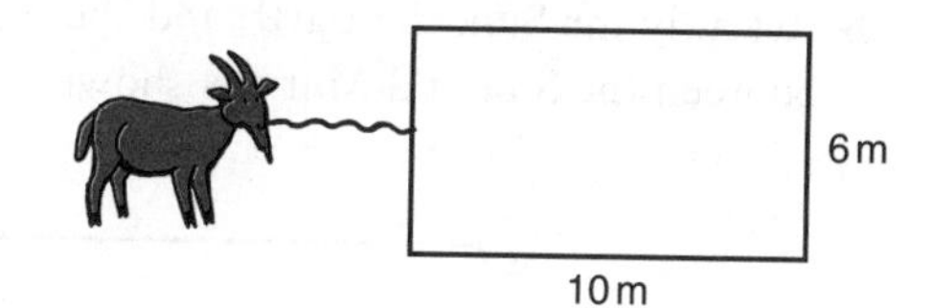

a What area can Gertie graze?
(*Note*: The rope cannot go over the enclosure: it must go round the corners.)

b Gertie thinks the grass is greener inside the enclosure, so she jumps over the fence. What area inside the enclosure can she graze now?

22 *Change* the positions of two matches to reduce the number of unit squares from five to four.
'Loose ends', i.e. matches not used as sides of unit squares, are not allowed.

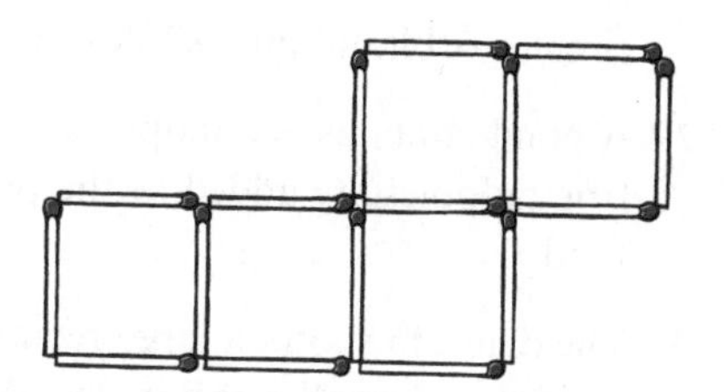

23 I eat 'Choc-o-bars' at the rate of 40 per month. The 'Amazing New Choc-o-bar' is introduced and contains 20% less chocolate than the old one. How many extra bars must I purchase each month to keep my chocolate consumption the same?

24 After throwing a dart n times at a dartboard, my percentage of hits is p%. With my next throw I make a hit and my percentage of hits is now $(p+1)$%. Prove that $n+p=99$.

25 Two concentric circles have radii of 10 cm and 26 cm. What is the length of the longest straight line that can be drawn between them?

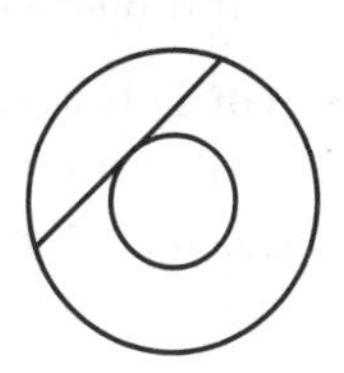

26 a Assuming a water molecule is a cube of side 10^{-9} m, find the volume of a water molecule in m^3.

b The volume of a cup of tea is $200\,cm^3$. How many molecules of water are there in the cup of tea?

c If all these molecules are placed end to end in a straight line, how many times would the line go around the Earth? Take the circumference of the Earth to be 40 000 km.

27 A tetromino is a two-dimensional shape consisting of four squares joined together along common edges.
One tetromino is shown in the diagram.
Two tetrominoes are the same if one is simply a rotation of the other in its plane.

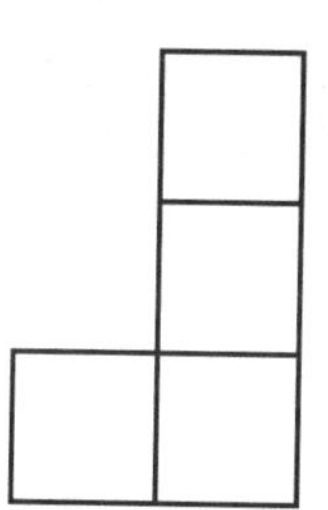

a Draw diagrams of the six other possible tetrominoes.

b Is it possible to fit these seven tetrominoes into a 7×4 rectangle so that none of them overlap?
Give reasons. (You might find it helpful to shade the tetrominoes as if they had been cut out from a chessboard.)

28 Recently the Sun, the Earth and the planet Mars were all in a straight line with the Earth between the Sun and Mars as shown in the diagram.

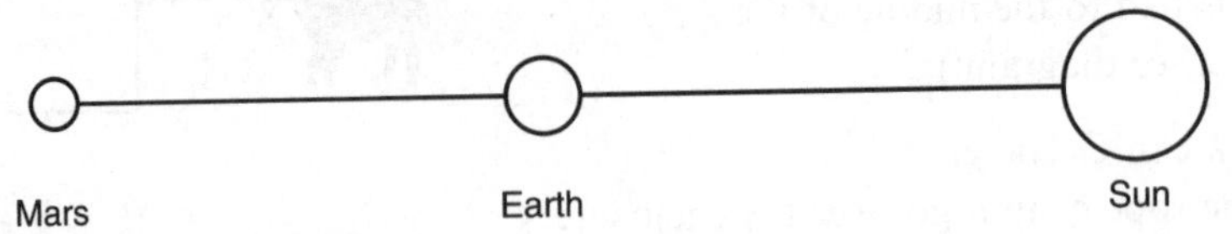

How many days will it take before the Sun, Earth and Mars are in a straight line again with the Earth between the Sun and Mars? Assume that the Earth travels at constant speed in a circular orbit around the Sun in 365 days and that Mars also travels at constant speed in a circular orbit, taking 687 days to complete one orbit.

29 A pond contains 300 tadpoles. f are frog tadpoles and the rest are toad tadpoles. If 100 more frog tadpoles are added to the pond, the probability of catching a frog tadpole is doubled. Find f.

30 The diagram shows a large rectangle composed of 13 identical smaller rectangles. Both the length and breadth of each of these smaller rectangles are a whole number of centimetres.

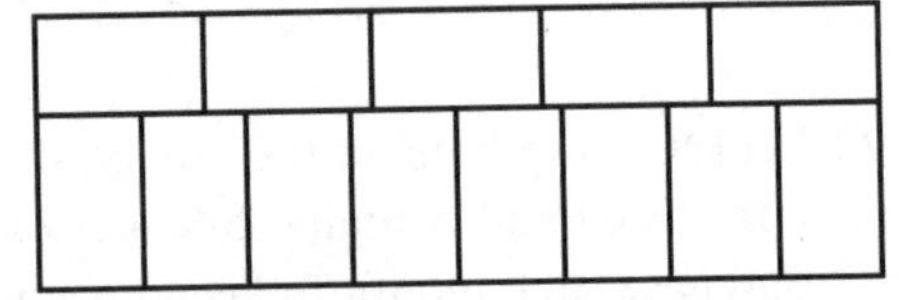

a What is the *smallest* possible value for the area of the large rectangle?

b If there are m rectangles in the bottom row and n rectangles in the top row, find a formula that gives a *possible* area of the large rectangle (not necessarily the smallest area).

31 ABCD is a square of side 6 cm.
AFB and DEC are equilateral triangles.
Calculate the distance EF exactly.

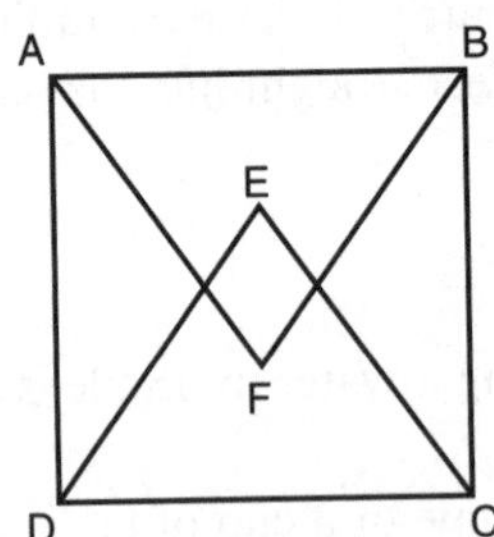

32 AB is a diameter of a circle of radius 1 cm.
Two circular arcs of equal radius are drawn with centres A and B meeting on the circumference of the circle as shown.
Calculate the shaded area exactly.

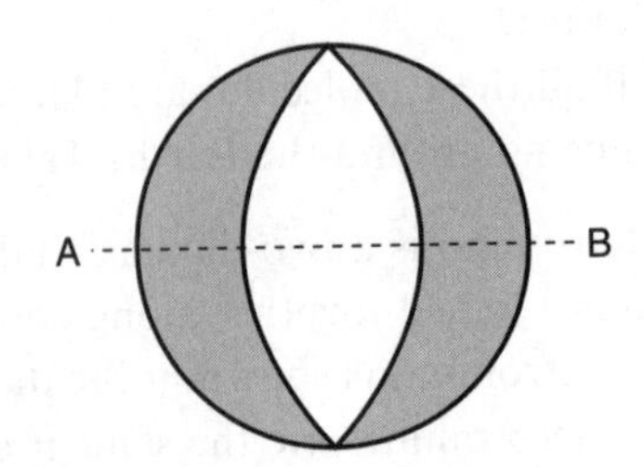

33 The pursuit of the true value of π has fascinated many people for many years. Investigate the approximations shown here and, where possible, rank them in order of accuracy.

Approximation of π	**Source**
3	*Bible* 1 Kings 7:23
$3\frac{1}{8}$	Babylon 2000 BCE; found on a clay tablet in 1936
$256 \div 81$	Egypt 2000 BCE; found on the 'Rhind Papyrus'
$22 \div 7$	Syracuse 250 BCE; Archimedes
$377 \div 120$	Greece 140 BCE; Hipparchus
$355 \div 113$	China AD 450; Tsu Chung-Chih
$\sqrt{10}$	India AD 625; Brahmagupta
$864 \div 275$	Italy AD 1225; Fibonacci
$\frac{2 \times 2 \times 2 \times 4 \times 4 \times 6 \times 6 \times 8 \times 8 \cdots}{1 \times 3 \times 3 \times 5 \times 5 \times 7 \times 7 \times 9 \cdots}$	England AD 1655; Wallis
$\sqrt[4]{9^2 + \frac{19^2}{22}}$	India AD 1910; Ramanujan

CHALLENGES

World population

Our planet is a rocky sphere, with a diameter of **12 800 km** and a mass of **6.6×10^{21}** tonnes, spinning through space on its path round the Sun. Viewed from space the Earth is a mainly white and blue planet, covered with clouds, water and some land. It is a watery planet, as **70%** of the surface is covered with water.

The dominant species is humans, primates with an average mass of **50 kg**. Humans are only able or willing to live on **12%** of the available land area, as the rest is considered either too cold, too dry or too steep.

Until the modern era, world population grew slowly. It is estimated that the population grew at about **0.05%** per year, reaching about **300 million** in **AD 1**. During the following **16 centuries** the growth rate fluctuated, partly because of diseases such as the Black Death. The world's population broke through the one billion threshold in **1804**. The **second billion** took another **123 years**, but the **third billion** only took another **33 years**.

The world population growth rate peaked at about **2%** per year in **1960**. The population of humans grew to about **6 billion** by the year **2000**, and was increasing at around **1.5%** every year. At this rate the number of human beings will double by about the year **2050**.

The recent global population explosion is not only the consequence of increased birth rates but also the result of an unprecedented decrease in death rate (now down to two people every second) because of significant advances in public health and medicine.

However, the United Nations estimates that world population will stabilise at **12 billion** in the year **2100**, assuming that effective family planning will result in a universally low birth rate. Global average births are now **2.7** per woman, down from **5** in **1950**. Education plays a key role, as almost half of the **6 billion** population are under the age of **25**.

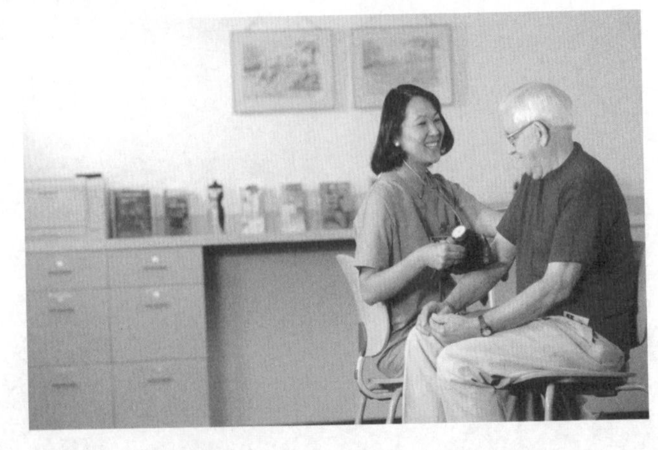

Exercise

1 What year did the world population first reach

a two billion **b** three billion?

2 a What was the increase in world population from the year 2000 to 2001?

b What was the average increase in the population every second during the year 2000?

c Approximately how many people were born every second in the year 2000?

3 What percentage of the total surface area of the Earth is suitable for humans to live on?

4 The surface area, S, of a sphere is given by $S = 4\pi r^2$.

Work out the area available for humans to live on, giving your answer in km^2.

5 Work out the average number of humans per km^2 of habitable land in the year 2000.

6 a What factor do you multiply by to increase a quantity by 1.5%?

b Work out the world population in

(i) 2001 (ii) 2002

7 Show that the world population, T years after 2000, is given by $6 \times 10^9 \times (1.015)^T$, assuming a growth rate of 1.5%.

8 Fill in the table below and then draw a graph showing the world population from the years 2000 to 2100. Use your graph to estimate when the world population doubles to 12 billion.

Year	2000	2020	2040	2060	2080	2100
Population						

9 Work out the number of humans per km^2 of habitable land in

a 2100 **b** 2300

10 What would be the population of the world if there were an average of 1 human for every $25\ \text{m}^2$ of habitable land, and when will this occur?

11 Work out the world population in the year 4100.

12 Work out the mass of humans in the year 4100. Compare this to the mass of the Earth. What conclusions can you draw?

Fact finders

Mount Vesuvius

There have been many volcanic eruptions and earthquakes of catastrophic proportions in the world's history, many of which have been far more powerful than Italy's Mount Vesuvius.

In **1815**, **92 000** people died owing to the Tambora volcano's eruption in Indonesia, and in **1976**, **242 000** people perished when an earthquake shook Tangshan in eastern China.

However, the eruption of Vesuvius in **AD 79** has become infamous because of the extraordinary documentation of the event and its impact on the towns of Pompei and Herculaneum. The instability of the area is due to the northward movement of the African tectonic plate at **3 cm per year**. The eruption lasted three days, during which time the sky was dark with dust and ash ejected **17 km** into the air, covering nearby villages and towns in just a few hours. Pompei (population **20 000**) was covered with **7 m** of hot ash, while Herculaneum (population **5000**) was hit by a rapid mud-flow **25 m** deep, which captured people and animals as they tried to escape. During the eruption, **4 km^3** of magma and **10^6 m^3** of lapilli (fine ash) were spewed out in **24 hours**, causing **3600** lives to be lost.

Many artefacts were perfectly preserved in Herculaneum, owing to the protection of the deep mud-flow. These are kept in the Naples Museum, where they have inspired poets, philosophers and scientists over the years.

The present-day summit of **1281 m** would have been dwarfed by the height of the volcano in the year **AD 700** of **3000 m**. Since **1631** there have been major eruptions from Vesuvius in **1760**, **1794**, **1858**, **1861**, **1872**, **1906**, **1929**, **1933** and **1944**.

The top of the crater today is an enormous circular chasm **600 m** in diameter and **200 m** deep. Walking around the summit takes about **one hour** and vapour can still be seen escaping from fumaroles (cracks), some of which have temperatures of **500 °C**, indicating that Vesuvius is still active. If the volcano were to erupt to the same strength as in **AD 79**, it would require the evacuation of about **600 000** people. Vesuvius is modelled as a right-circular pyramid whose top has been blown off. Its cross-section is as shown.

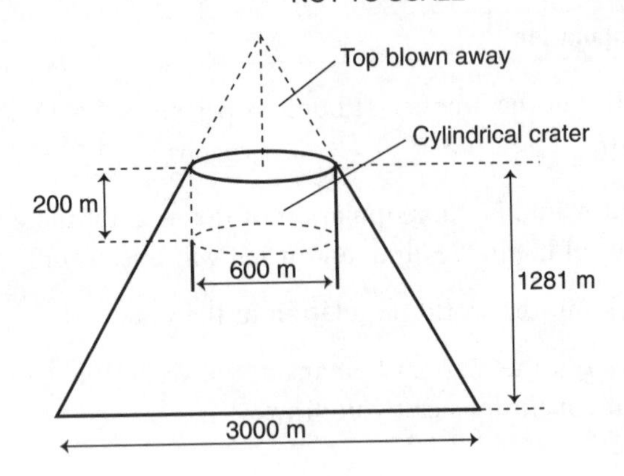

Exercise

1 How many years elapsed between the infamous volcanic destruction of Pompei and the Tambora eruption?

2 What percentage of the population of Pompei and Herculaneum perished because of the volcanic activity?

3 How many years will the African tectonic plate take to shift one mile northwards? (1 mile $\approx$ 1600 m)

4 What is the percentage decrease in the height of Vesuvius from that in the year AD 700?

5 Given that C degrees Celsius is related to F degrees Fahrenheit by the formula $C = 5(F - 32)/9$, find the temperature around some of the fumaroles in the crater of Vesuvius today in degrees Fahrenheit.

6 Calculate the speed a tourist travels, in m/s, walking around the top of the crater of Vesuvius.

7 Given that the approximate areas of Herculaneum and Pompei are half a square mile and four square miles respectively, calculate an estimate in m^3 in standard form to three significant figures of these:

a the volume of hot mud-flow that filled Herculaneum

b the volume of hot ash that fell on Pompei.

8 a Since 1631, what is the average number of years between major volcanic activities in Vesuvius?

b Using these data, when would you reasonably expect the next eruption of Vesuvius?

9 Calculate in m^3/s, in standard form correct to 3 significant figures, the rate at which magma was thrown out from Vesuvius in AD 79.

10 a Given that the average density of magma from Vesuvius in AD 79 was approximately 4500 kg/m^3, calculate the mass of magma spilled out in tonnes per second.

b Compare your answer in **a** to an average family car of mass 1 tonne. (1 tonne = 1000 kg)

11 Calculate an estimate of the volume of Mount Vesuvius in m^3 in standard index form correct to 3 significant figures. (Volume of cone $= \frac{1}{3}\pi r^2 h$)

12 Calculate an estimate of the area of the slopes of Mount Vesuvius in km^2 correct to 3 significant figures. (Curved surface area of cone $= \pi r l$)

The human body

The human body is an amazing collection of organs, nerves, muscles and bones that is more complex and robust than any machine ever invented. You were born with **300** bones and when you get to be an adult you will have **206**. Your skull comprises **29** bones, with **100 000** hairs on its scalp and with about **70** hairs being lost per day.

There are **45 miles** of nerves in the skin of an average person, and in each square cm of your skin there are **96 cm** of blood vessels, **200** nerve cells and **16** sweat glands. About **80%** of the body is made up of water. About **50 000** of the cells in your body will die and be renewed in the time it has taken for you to read this sentence, and **15 million** blood cells are destroyed per second.

Human blood travels **60 000 miles per day** on its journey around the body, and in the same period a human breathes **23 040** times, with each breath drawing in about **4.2 litres** of air.

Nerve impulses to and from the brain travel at about **288 km/h,** whilst a sneeze leaves the head at **160 km/h** and a person blinks about **6 250 000** times a year.

A facial frown uses **43** muscles whilst a smile only uses **17** muscles, and every **200 000** frowns creates a single wrinkle.

In a lifetime of **70 years**, a person:

- walks a distance equivalent to **5** times around the equator
- experiences about **3000 million** heartbeats and in this time the heart pumps about **216 million litres** of blood
- produces **45 000 litres** of saliva
- drinks **72 000 litres** of water
- could grow a beard of **9 m** length if they are male.

We are truly remarkable creatures.

Fact finders

Exercise

Give all answers to 3 significant figures where appropriate.

1 What percentage of your bones as an adult are in your skull?

2 How many muscles will have been used to create 100 wrinkles?

3 In a lifetime of 70 years, how many litres of water are drunk per hour?

4 How many blinks will have taken place in a classroom of 20 children in a one-hour lesson?

5 Assume a male can grow facial hair from the age of 18 years. Find the speed of growth of beard in mm per second, expressing your answer in standard form.

6 Calculate the average heart rate in beats per minute of a human.

7 Given that the radius of the Earth is 6370 km, calculate the average speed a human walks in a lifetime in m/s.

8 A glass of water is $250\,\text{cm}^3$. How many glasses does a human drink per day on average?

9 Calculate the volume of air breathed in a 70-year lifetime by a human in cubic millimetres in standard form.

Compare this to a rectangular classroom of dimension $10\,\text{m} \times 8\,\text{m} \times 3\,\text{m}$.

10 A Maths teacher, Mr Quixote, is drawn to a scale of 1 : 60.

a If he stubs his toe, how many seconds will it take before he feels any pain?

b Mr Quixote has an unusual reaction to stubbing his toe, in that it triggers a sneeze once his brain registers the incident. Unfortunately Mr Quixote has a cold.

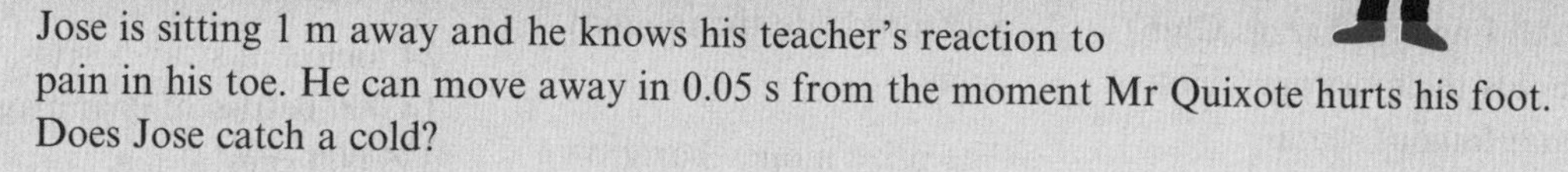

Jose is sitting 1 m away and he knows his teacher's reaction to pain in his toe. He can move away in 0.05 s from the moment Mr Quixote hurts his foot. Does Jose catch a cold?

Wimbledon

Each year at Wimbledon, the 'All England Tennis Club' holds the most prestigious and famous tennis tournament in the world. The first championships were held in July **1877**. Since then many changes have taken place to accommodate the modern game.

The prize money has grown dramatically over the years: see the table below.

Year	Men's Singles Winner	Women's Singles Winner
1968	$2850	$1070
1990	$328 570	$295 710
2004	$860 710	$800 000

Greg Rusedski has one of the world's fastest serves, which was measured at **238 km/h**. In **1999** he served **68** aces that made up **79%** of all of his points won that year at the tournament.

The Centre Court seats **13 813** people, with the entire club having a capacity of **34 500** spectators.

The tournament is broadcast on TV around the world over the two weeks, with the following total hours:

Europe **2509 hours,**
USA **360 hours,**
Latin America **720 hours,**
Africa Middle East **628 hours,**
Asia Pacific **1449 hours**.

68% of this coverage was shown live.

The crowds consume over the two weeks:

24 tonnes of strawberries,
14 000 bottles of champagne,
335 000 cups of tea,
140 000 ice creams and
12 000 kg of smoked salmon.

Approximately **1750 dozen** tennis balls are used over the **650** matches. They have to be stored at a precise temperature of **68 °F**, and **200** ball boys and girls are used.

Exercise

Give all answers to 3 significant figures where appropriate.

1 How many years ago was the first Wimbledon championships?

2 Given that the formula to convert temperature in Fahrenheit (°F) to Celsius (°C) is:

$$C = \frac{5F - 160}{9}$$

find the temperature that the tennis balls have to be stored at in °C.

3 Find the percentage increase in prize money from 1990 to 2004 for the winners of the:

a men's singles **b** women's singles.

4 How many balls are used on average per match in the championships?

5 Given that the length of a tennis court is 23.78 m, estimate how long Greg Rusedski's opponent would have had to react to his world record service.

6 How many points did Rusedski win at Wimbledon in 1999?

7 Find the percentage difference in prize money between the winners of the men's and women's singles title in:

a 1968 **b** 1990 **c** 2004

8 How many hours of Wimbledon are broadcast live on TV per day worldwide?

9 Assuming that the club is full for the entire two weeks, find the number of kilograms of strawberries consumed

a per person per day **b** per day at the Centre Court.

10 Assume that the club is full for the entire two weeks and the costs for food and drink are:

Strawberries	$25 per kg
Bottle of champagne	$40
Cup of tea	$1.50
Ice cream	$5
Salmon per kg	$25

Calculate the mean amount spent by a spectator at the championships.

Answers to activities and exercises

Unit 1

Number 1

ACTIVITY 1 (page 1)

Time (h)	1	2	4	6	7	8
Speed (km/h)	160	80	40	26.7	22.9	20

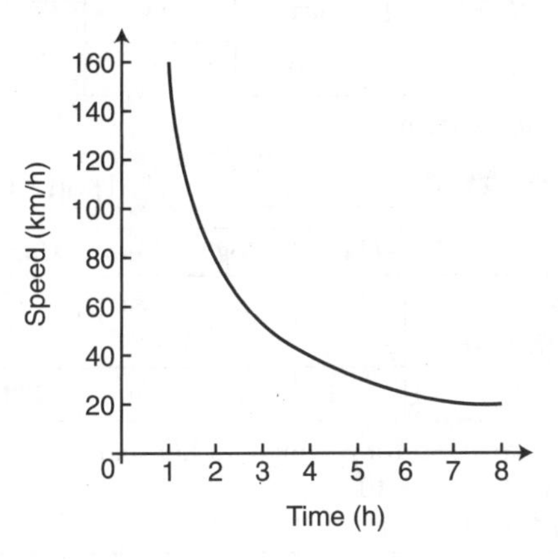

EXERCISE 1 (page 2)

1 a 4 days **b** 2 days **c** $2\frac{2}{3}$ days

3 a 8 men **b** 16 men **c** 32 men

5 a 60 years **b** 15 years **c** 1200 years

7 a 12 hours **b** 72 km/hour

EXERCISE 1★ (page 3)

1 a

Number of light bulbs (N)	Power of each bulb (P)
6	**500**
5	600
2	**1500**
30	100
N	P

b $3000 = NP$

3 a $12\,\text{g/cm}^3$ **b** $1.5\,\text{cm}^3$

5

Number of men	Number of tunnels	Time in years
100 000	4	**4**
100 000	**2**	2
20 000	8	**40**
400 000	2	0.5

7 a 5 **b** 1000 **c** 2.26×10^8 tonnes

9 a 16 tonnes (2 s.f.)
b (i) 4 minutes (ii) 30 seconds (iii) 70 seconds (iv) 8 seconds

EXERCISE 2 (page 6)

1 0.375 **3** 0.08 **5** 0.28125
7 $0.\dot{2}$ **9** $0.\dot{1}\dot{8}$ **11** $0.2\dot{6}$
13 $0.3\dot{8}$ **15** $\frac{9}{16}$ **17** $\frac{3}{20}, \frac{5}{64}$
19 $\frac{1}{3}$ **21** $\frac{5}{9}$ **23** $\frac{7}{9}$
25 $\frac{7}{90}$ **27** $\frac{1}{30}$ **29** $\frac{1}{18}$

EXERCISE 2★ (page 6)

1 $0.4\dot{6}$ **3** $0.04\dot{6}$ **5** $2.\dot{3}\dot{0}$ **7** $0.3\dot{0}\dot{1}$
9 $\frac{11}{16}, \frac{7}{40}, \frac{3}{15}$ **11** $\frac{19}{20}, \frac{3}{25}, \frac{5}{64}$ **13** $\frac{24}{99}$
15 $\frac{10}{33}$ **17** $9\frac{19}{990}$ **19** $\frac{3}{110}$ **21** $\frac{412}{999}$
23 $\frac{128}{333}$ **25** $\frac{11}{90}$ **27** $\frac{28}{495}$ **29** $0.\dot{0}3\dot{7}$

REVISION EXERCISE 3 (page 7)

1 a 2 hours **b** 6 mowers
3 a 0.2 **b** 0.125 **c** 0.05
5 a $\frac{2}{9}$ **b** $\frac{7}{90}$ **c** $\frac{23}{99}$

REVISION EXERCISE 3★ (page 7)

1 a 16 km/litre **b** 40 litres
3 $0.\dot{1}5384\dot{6}$
5 a $\frac{2}{9}$ **b** $\frac{1}{90}$ **c** $\frac{67}{99}$ **d** $3\frac{1}{22}$

Algebra 1

ACTIVITY 2 (page 9)

Variables	Related? yes (Y) or no (N)
Area of a circle (A) and its radius (r)	Y
Circumference of a circle (C) and its diameter (d)	Y
Volume of water in a tank (V) and its weight (w)	Y
Distance travelled (D) at constant speed and time taken (t)	Y
Number of pages in a book (N) and its thickness (t)	Y
Mathematical ability (M) and a person's height (h)	N
Wave height in the sea (W) and wind speed (s)	Y
Grill temperature (T) and time to toast bread (t)	Y

EXERCISE 4 (page 11)

1 a $y = 5x$ b 30 c 5
3 a $y = 2x$ b 10 cm c 7.5 kg
5 1950 sales

EXERCISE 4★ (page 11)

1 a $v = 9.8t$ b 49 m/s c 2.5 s
3 a $d = 150$ m b 1500 km c 266.7 g
5 a $h = 3y/2$ b 0.75 m c 4 months

EXERCISE 5 (page 14)

1 a $y = 4x^2$ b 144 c 4
3 a $v = 2w^3$ b 54 c 4
5 a $y = 5t^2$ b 45 m c $\sqrt{20}\,\text{s} \simeq 4.47$ s

EXERCISE 5★ (page 15)

1

g	2	4	**6**
f	12	**48**	108

3 a $R = \left(\frac{5}{256}\right)s^2$ b 113 km/hour
5 $x = 10\sqrt{2}$

ACTIVITY 3 (page 16)

$t^2 \simeq 3.95 \times 10^{-20} d^3$

Planet	d (million km)	t (earth days) (2 s.f.)
Mercury	57.9	**88**
Jupiter	**778**	4300
Venus	108	220
Mars	228	680
Saturn	1430	11 000
Uranus	2870	31 000
Neptune	4500	60 000
Pluto	5950	91 000

Ask pupils to compare these with the actual values.

EXERCISE 6 (page 17)

1 a $y = \frac{12}{x}$ b $y = 6$ c $x = 4$
3 a $m = \frac{36}{n^2}$ b $m = 9$ c $n = 6$
5 a $I = 4 \times \frac{10^5}{d^2}$ b 0.1 candle-power

EXERCISE 6★ (page 18)

1

b	2	5	**10**
a	50	**8**	2

3 a $R = \frac{2}{r^2}$ b $\frac{2}{9}$ ohm

5 a

Day	N	t
Mon	400	25
Tues	**447**	20
Wed	500	**16**

b 407 approx.

ACTIVITY 4 (page 19)

	Hare	Dog	Man	Horse
Pulse (beats/min)	**200**	135	**83**	65
Mass (kg)	3	**12**	70	**200**

Total heartbeats for a human of life-span 75 years $\simeq 3.27 \times 10^9$
According to theory:

	Hare	Dog	Man	Horse
Life-span (years)	31.1	46.1	75	95.8

Theory clearly not correct

REVISION EXERCISE 7 (page 20)

1 a $y = 6x$ b $y = 42$ c $x = 11$
3 a $c = \frac{3}{4}a^2$ b \$675 c 28.3 m^2

REVISION EXERCISE 7★ (page 20)

1 a $y^2 = 50z^3$ b 56.6 c 5.85
3 1500 m

EXERCISE 8 (page 22)

1, 3

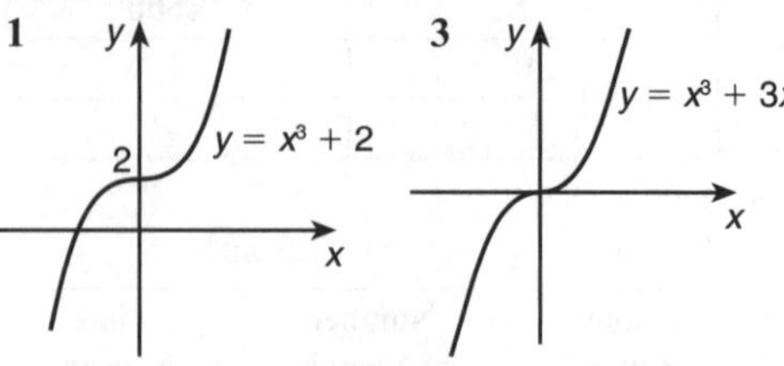

5

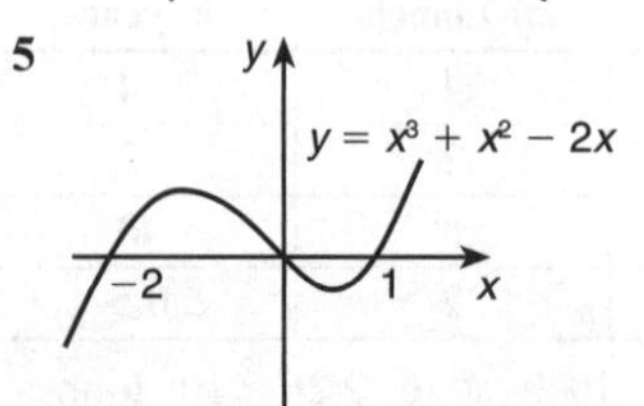

7 a $V = x^2(x-1) = x^3 - x^2$

b 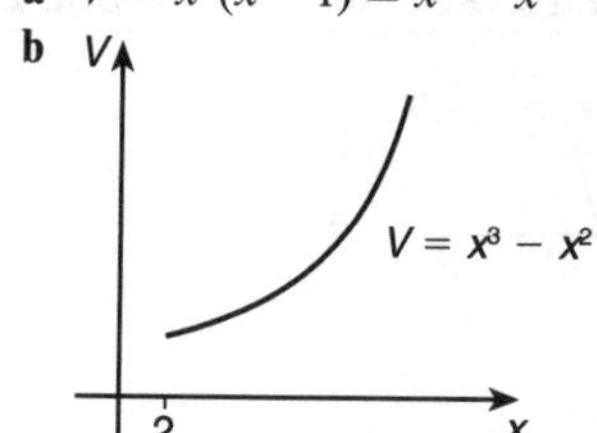

c $48\,m^2$

d $4.6\,m \times 4.6\,m \times 3.6\,m$

EXERCISE 8★ (page 22)

1 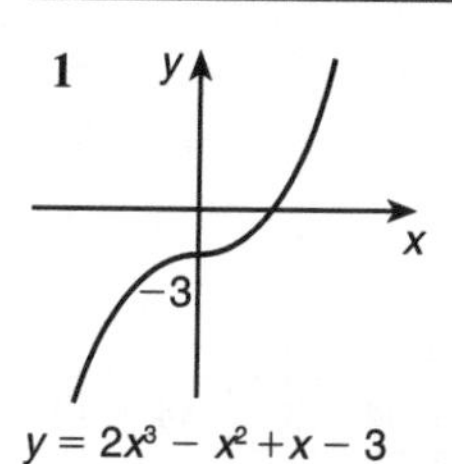

3 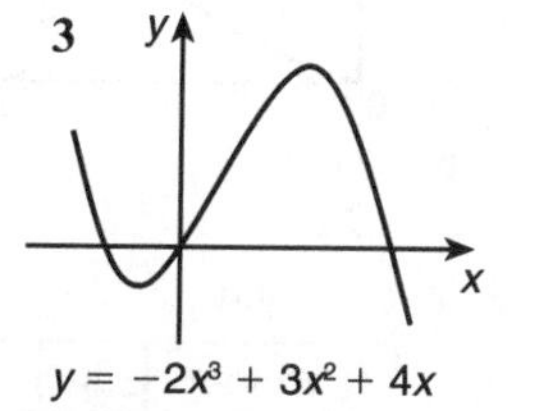

5 a

t	0	1	2	3	4	5
v	0	26	46	54	44	10

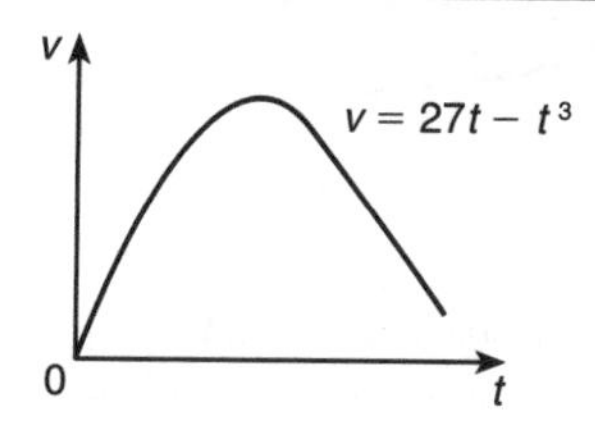

b $v_{max} = 54$ m/s and occurs at $t = 3$ s

c $v \geqslant 30$ m/s when $1.2 \leqslant t \leqslant 4.5$, so for about 3.3 s

7 a $A = 100\pi = 2\pi r^2 + 2\pi rh$

$\Rightarrow 100\pi - 2\pi r^2 = 2\pi rh$

$\Rightarrow \frac{50}{r} - r = h$

$\Rightarrow V = \pi r^2 h = \pi r^2\left(\frac{50}{r} - r\right) = 50\pi r - \pi r^3$

b

r	0	1	2	3	4	5	6	7
V	0	153.9	289.0	386.4	427.3	392.7	263.9	22.0

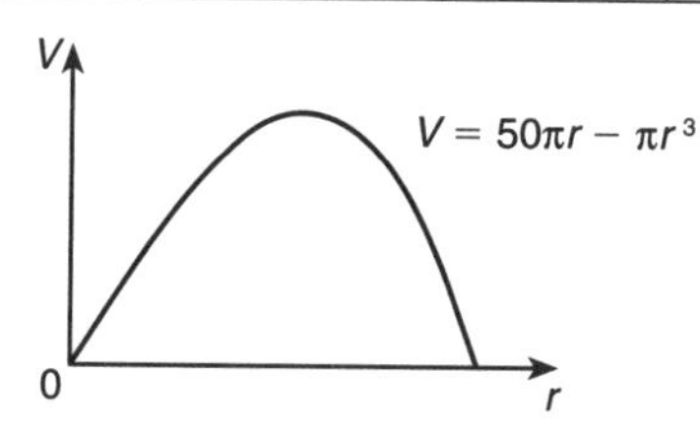

c $V_{max} = 427.3\,cm^3$

d $d = 8.16$ cm, $h = 8.2$ cm

ACTIVITY 5 (page 24)

Year interval	Fox numbers	Rabbit numbers	Reason
A–B	Decreasing	Increasing	Fewer foxes to eat rabbits
B–C	Increasing	Increasing	More rabbits attract more foxes into the forest
C–D	Increasing	Decreasing	More foxes to eat rabbits, so rabbit numbers decrease
D–A	Decreasing	Decreasing	Fewer rabbits to be eaten by foxes, so fox numbers decrease

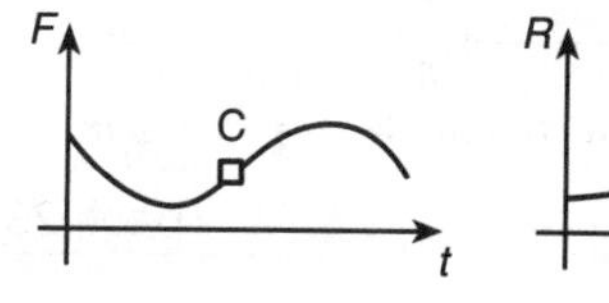

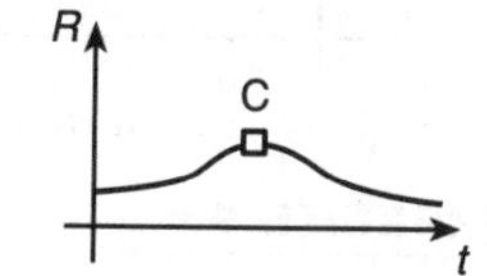

ACTIVITY 6 (page 24)

$y = \frac{3}{x}$

x	−3	−2	−1	0	1	2	3
y	−1	−1.5	−3	∅	3	1.5	1

$y = -\frac{3}{x}$

x	−3	−2	−1	0	1	2	3
y	1	1.5	3	∅	−3	−1.5	−1

As x approaches 0, y approaches ∞, so at $x = 0$, y is not defined, as denoted by ∅.

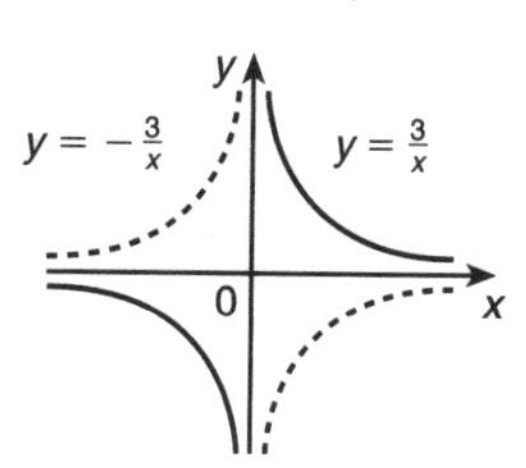

EXERCISE 9 (page 25)

1 3

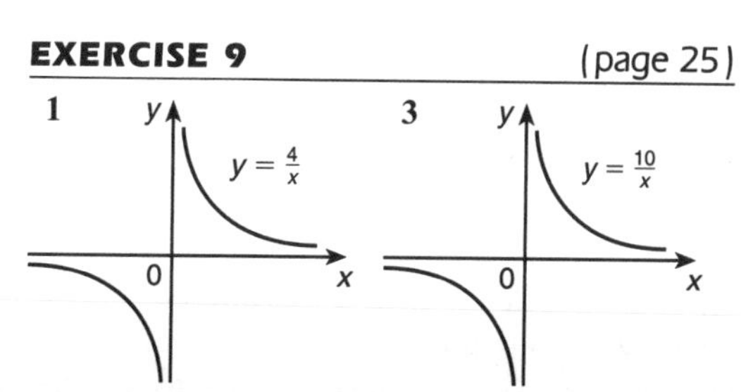

5 a

t (months)	1	2	3	4	5	6
y	2000	1000	667	500	400	333

b

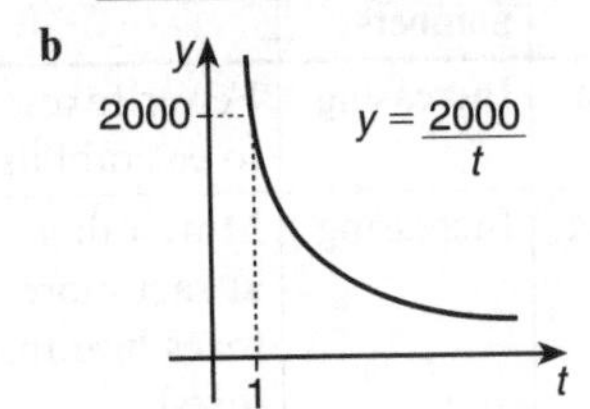

c 3.3 months **d** 2.7 months approx

7 a $k = 400$

m (mins)	5	6	7	8	9	10
t(°c)	80	67	57	50	44	40

b

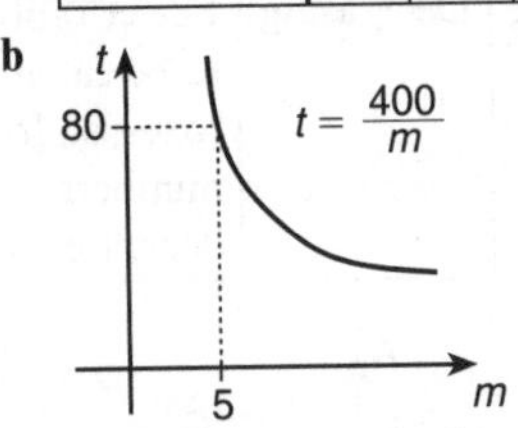

c 53 °C **d** 6 mins 40 s **e** $5.3 \leqslant m \leqslant 8$

EXERCISE 9★ (page 26)

1 **3**

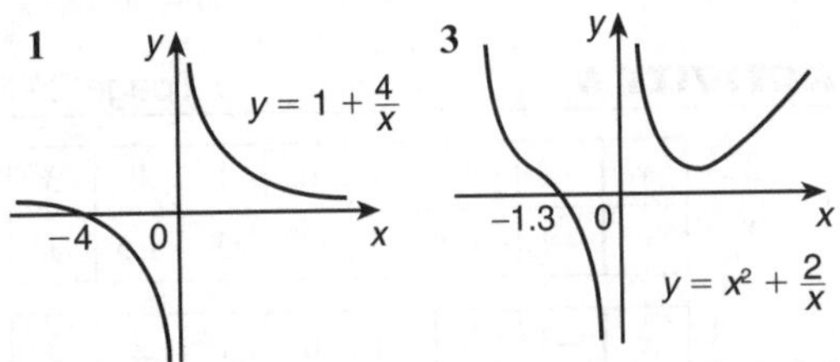

5 a

x(°)	30	35	40	45	50	55	60
d(m)	3.3	7.9	10	10.6	10	8.6	6.7

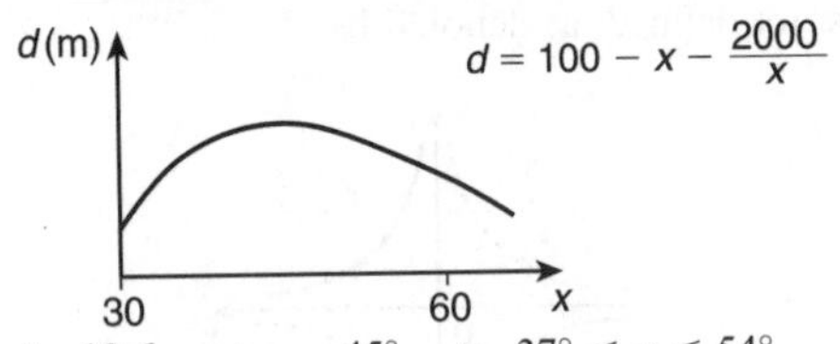

b 10.6 m at $x = 45°$ **c** $37° \leqslant x \leqslant 54°$

7 a

r (m)	1	2	3	4	5
A(m²)	106	75	90	126	177

b

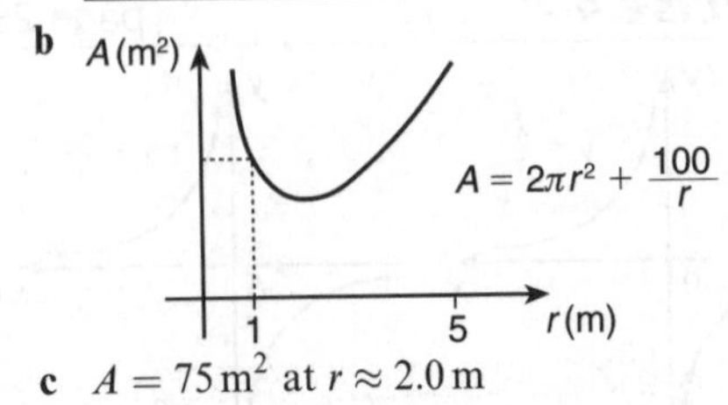

c $A = 75\,\text{m}^2$ at $r \approx 2.0$ m

REVISION EXERCISE 10 (page 28)

1

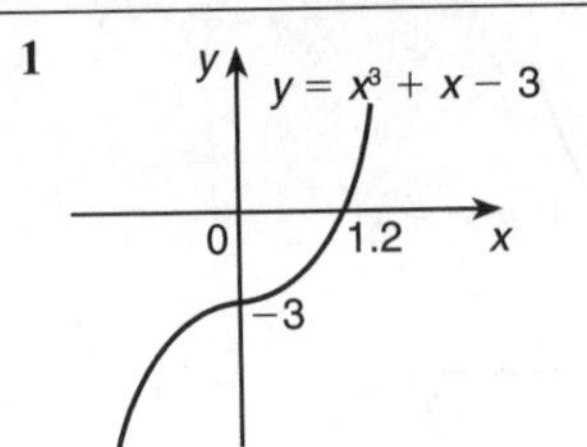

3 a

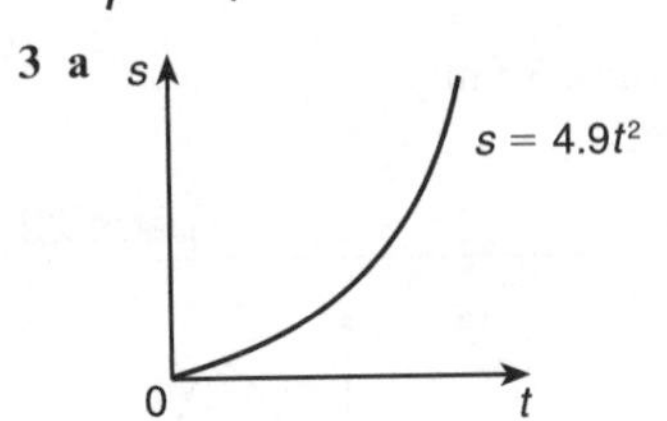

b 30.6 m

c 3.2 s

5 a $k = 2800$

t (weeks)	30	32	34	36	38	40
x (kg)	93	88	82	78	74	70

b

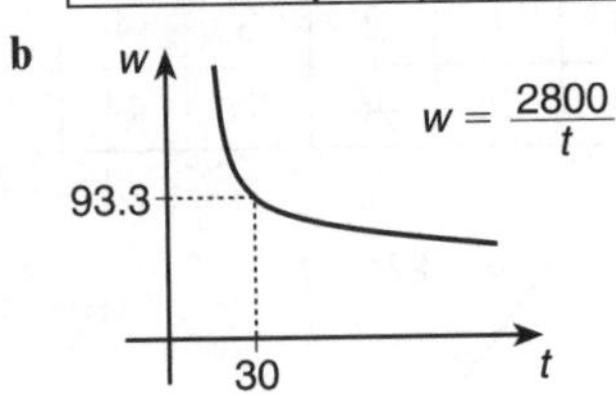

c 35 weeks

d Clearly after 500 weeks, Nick cannot weigh 5.6 kg. So there is a domain over which the equation fits the situation being modelled.

REVISION EXERCISE 10★ (page 29)

1

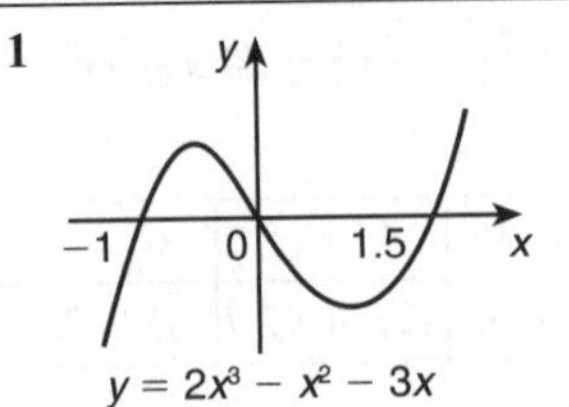

3 a

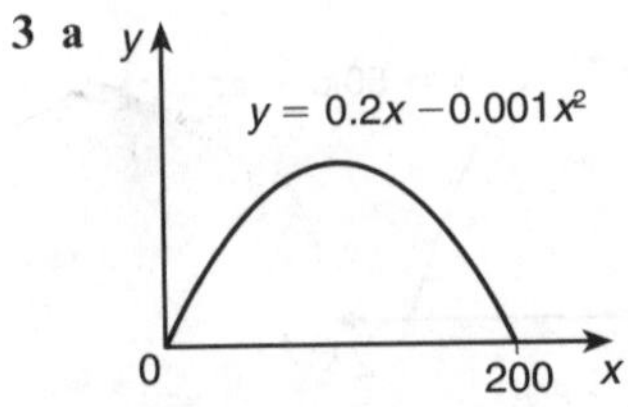

b 10 m

c $29 \leqslant x \leqslant 170$

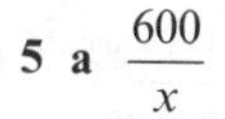

5 a $\frac{600}{x}$

c

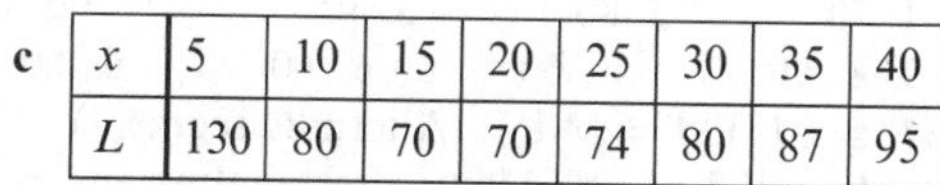

x	5	10	15	20	25	30	35	40
L	130	80	70	70	74	80	87	95

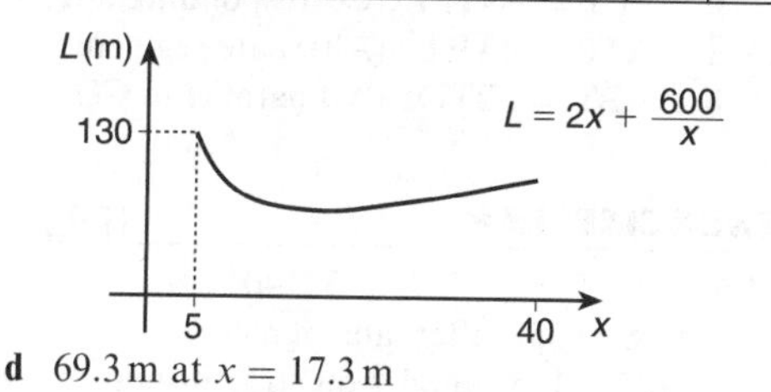

d 69.3 m at $x = 17.3$ m

e $11.6 < x < 25.9$

Shape and space 1

ACTIVITY 7 (page 30)

When BC > 10 cm a unique solution results for any length BC. When BC = 5 cm a right-angled triangle results. When BC < 5 cm there is no solution. When 5 cm < BC < 10 cm the ambiguous case results, and there are two triangles possible for each length BC.

This shows that **ASS** is not necessarily a condition for congruency, but another condition for congruency is **RHS** (right angle, hypotenuse, side). The fourth condition for congruency is when two angles and a side are given *in that order* (**AAS**).

EXERCISE 11 (page 32)

1 Yes, SAS
3 No
5 Yes, RHS
7 AB = XY, BC = YZ, ∠CAB = ∠ZXY
9 1 BE = DE (Given)
2 AE = CE **(Given)**
3 ∠AEB = ∠CED **(Vert. opp)**
∴ △ABE ≡ △CDE **(SAS)**
∴ ∠EDC = ∠**EBA**
∠ECD = ∠**EAB**
and AB = **DC**

EXERCISE 11★ (page 34)

1 1 WX = YZ (**Opp. sides of parallelogram**)
2 WZ = YX (**Opp. sides of parallelogram**)
3 WY is common
∴ △WXY ≡ △YZW (SSS)
∴ ∠WXY = ∠**YZW**
∠XWY = ∠**ZYW**
and ∠XYW = ∠**ZWY**

3 △ADE ≡ △BCE (SAS)
∴ AD = BC

5 △APY ≡ △BPZ (RHS)
∴ ∠AYP = ∠BZP
∴ △XYZ is isosceles
∴ XY = XZ

7 △BMO ≡ △BNO (SSS)
∴ ∠MBO = ∠NBO

9 △s EBF, FCG, GDH, HAE are congruent (SAS)
∴ EFGH is a square.

11 △POX ≡ △POY (RHS)
∴ PX = PY

13 In triangles ABX and CDX: XB = XD (△XBD is isosceles), AX = CX (△ACX is isosceles), AB = CD (given).
△AXB ≡ △CXD (SSS).

ACTIVITY 8 (page 36)

- Make sure that all pupils know how to use the compass.
- The base line AB should be set up before the lesson. It is better to use two piles of sand to mark the base line rather than posts, which can be moved. Make sure AB lines up with magnetic North. The hedge, or some other suitable boundary, should be roughly parallel to the base line.
- Measure the length of the hedge and give it to the pupils *after* they have made their scale drawings.
- The whole practical should be written up as a report.
- Pupils will find it easier to make the scale drawing if magnetic North is first drawn along a line on lined paper.

INVESTIGATE (page 37)

a If the base line is too short, the two bearings from A and B to, say, Y will intersect at an acute angle, thus producing a larger error than if they had intersected at a larger angle. Similarly, if AB is many times longer than XY, accuracy is affected. Diagrams should be drawn, by the pupils, to illustrate this inaccuracy.

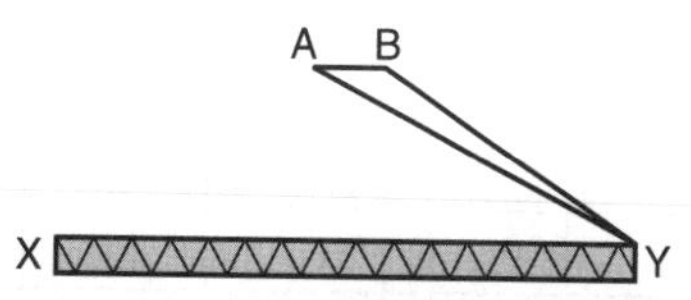

b If the angle between the base line and one end of the hedge is small, again, the two bearings will intersect at an acute angle, thus producing a larger error than if they had intersected at a larger angle. Diagrams should be drawn, by the pupils, to illustrate this inaccuracy.

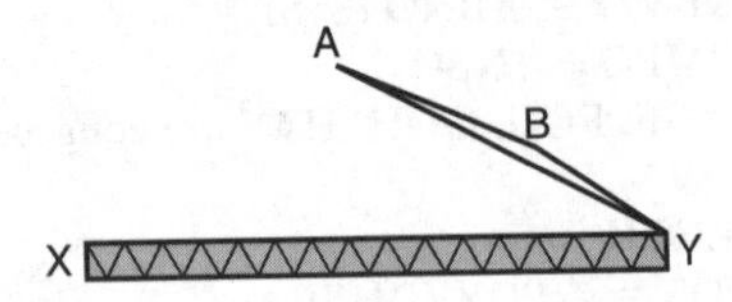

ACTIVITY 9 (page 38)

$\angle PQO = y$ and $\angle PRO = x$ (isosceles triangles)

$\therefore\ \angle QOT = 2y$ and $\angle ROT = 2x$.

$\therefore\ \angle QOR = 2y + 2x = 2(x + y)$

$= 2 \times \angle QPR$

EXERCISE 12 (page 40)

1 100° 3 45° 5 280° 7 60° 9 60°

11 $\angle ADB$ and $\angle BCA$ are angles in the same segment

13 LKMN is an isosceles trapezium

EXERCISE 12★ (page 41)

1 140° 3 115° 5 54°

7 119° 9 $3x$

11 $\angle ADB = x°$ ($\triangle ABD$ is isosceles),

$\angle BDC = (180 - 4x)°$ ($\triangle BCD$ is isosceles)

$\therefore\ \angle ADC = (180 - 3x)°$

$\therefore\ \angle ADC + \angle ABC = 180°$, so quadrilateral is concyclic

13 $\angle BEC = \angle CDB$ (angles in the same segment)

$\therefore\ \angle CEA = \angle BDA$

15 Let $\angle QYZ = y°$ and $\angle YZQ = x°$. So $\angle ZWX = y°$ and $\angle PWZ = 180 - y°$

In $\triangle YQZ$: $x + y + 20 = 180$

$\therefore\ x + y = 160$ (i)

In $\triangle PWZ$: $180 - y + x + 30 = 180$

$\therefore\ y - x = 30$ (ii)

From (i) and (ii) $x = 65$ and $y = 95$

$\therefore$ angles of the quadrilateral are 65°, 85°, 95°, 115°

ACTIVITY 10 (page 43)

	$Q\hat{P}B$	$O\hat{P}B$	$O\hat{B}P$	$B\hat{O}P$	$B\hat{A}P$
C_1	50°	40°	40°	100°	50°
C_2	62°	28°	28°	124°	62°
C_3	$x°$	$(90 - x)°$	$(90 - x)°$	$2x°$	$x°$

EXERCISE 13 (page 44)

1 70° 3 80° 5 30° 7 100°

9 **a** 90° **b** 60° **c** 60° **d** 60°

11 **a** $\angle NTM = \angle$**NPT** (Alternate segment)

b $\angle PLT = \angle$**NTM** (Corresponding angles)

13 **a** $\angle ATC = \angle$**ABT** (Alternate segment)

b $\angle ABT = \angle BTD$ **(AB parallel to CD)**

EXERCISE 13★ (page 46)

1 65° 3 140°

5 $\angle ATE = 55°$ (alternate segment),

$\angle TBC = 125°$ (angles on straight line),

$\angle BTC = 35°$ (angle sum of triangle),

$\angle ATB = 90°$ (angles on straight line)

$\therefore$ AB is a diameter

7 **a** 56° **b** 68°

9 **a** 55° **b** 35°

c $\angle ADE = 70°$ (angle sum of triangle),

$\angle ACD = 35°$ (angle sum of triangle),

$\therefore\ \triangle ACD$ is isosceles

11 20°

13 **a** Triangles ACG and ABF are right-angled.

b Angles ACG and ABF are equal and in the same segment of the chord FG.

15 **a** $\angle EOC = 2x°$ (angle at centre twice angle at circumference)

$\angle CAE = (180 - 2x)°$ (opposite angles of a cyclic quadrilateral)

$\angle AEB = x°$ (angle sum of triangle AEB)

$\therefore$ triangle ABC is isosceles

b $\angle ECB = x°$ (base angles of isosceles triangle) $\therefore\ \angle BEC = (180 - 2x)° = \angle CAE$

Since $\angle BAE$ and $\angle BEC$ are equal and angles in the alternate segment, BE must be the tangent to the larger circle at E.

EXERCISE 14 (page 50)

1 22.5 cm 3 10.5 cm 5 18 cm

7 4 cm 9 12 cm

EXERCISE 14★ (page 52)

1 3 cm 3 $x^2 - 22x + 120 = 0$, $x = 10$ or 12 cm

5 8 cm 7 24 cm 9 12 cm

REVISION EXERCISE 15 (page 53)

1 In triangles ABE and CDE

DE = **EB** (Given)

$\angle BAE = \angle$**ECD** (Alternate)

$\angle AEB = \angle CED$ **(Vertically opposite)**

$\therefore\ \triangle ABE \equiv \triangle CDE$ **(AAS)**

$\therefore\ \angle ABE = \angle$**CDE**

AB = **CD**

AE = **EC**

3 20° **5** 65° **7** 13.5 cm

9 a 80° **b** 100° **c** 50° **d** 50°

REVISION EXERCISE 15★ (page 55)

1 ZY = ZX, WX = WY (Constructions), ZW is common ∴ △ZXW ≡ △ZYW (SSS)

3 a 55°

b 35°

c ∠TDC = 35° (Angles in the same segment) ∴ ∠EDC = 90° and EC is the diameter

5 $x = 3$ cm, $y = 12$ cm

7 a 40°

b 50°

c ∠ZXT = ∠WVT (Angle in alternate segment) ∴ XZ is parallel to WV (Alternate angles)

Sets

EXERCISE 16 (page 58)

1 6

3 22

5 a 10 **b** 8 **c** 5

EXERCISE 16★ (page 59)

1 23

3 100

5 $8 \leqslant x \leqslant 14$, $0 \leqslant y \leqslant 6$

EXERCISE 17 (page 60)

1

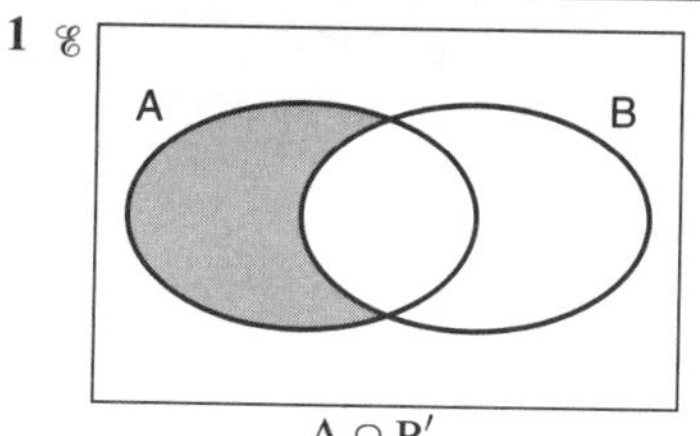

A ∩ B′

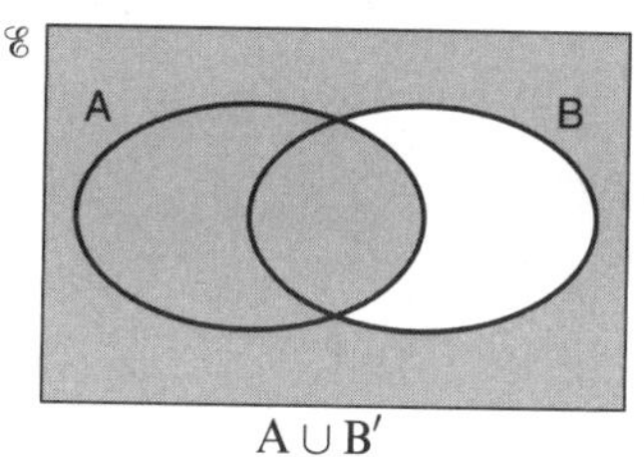

A ∪ B′

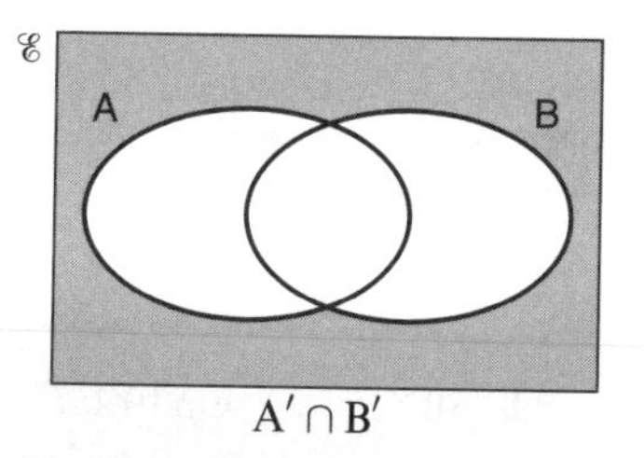

A′ ∩ B′

A′ ∪ B′

3

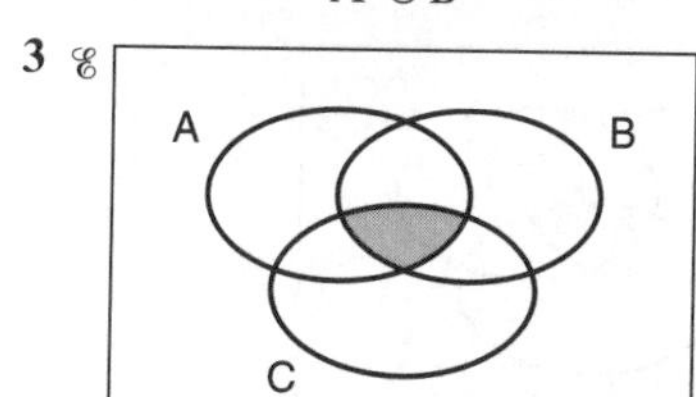

A ∩ B ∩ C

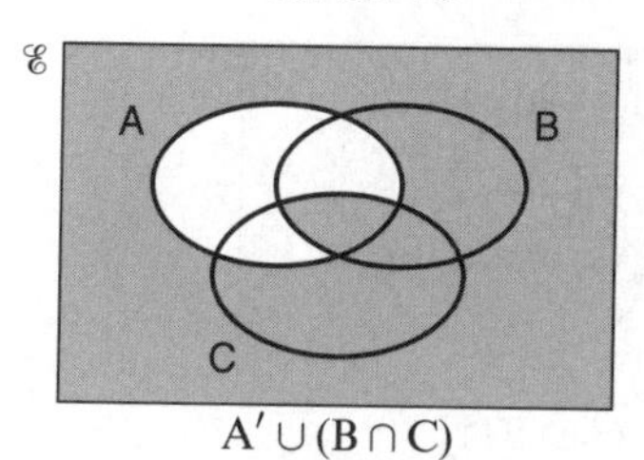

A′ ∪ (B ∩ C)

5 A ∩ B′, A′ ∩ B′, A′ ∪ B′

EXERCISE 17★ (page 61)

1

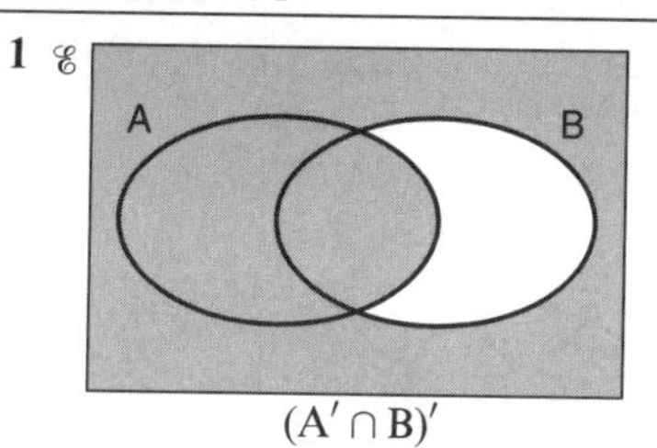

(A′ ∩ B)′

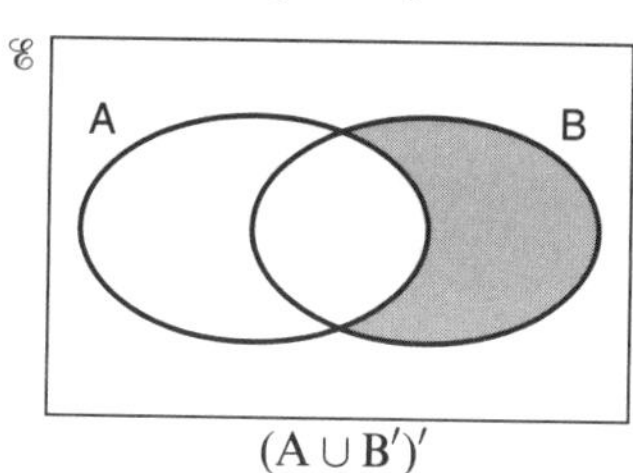

(A ∪ B′)′

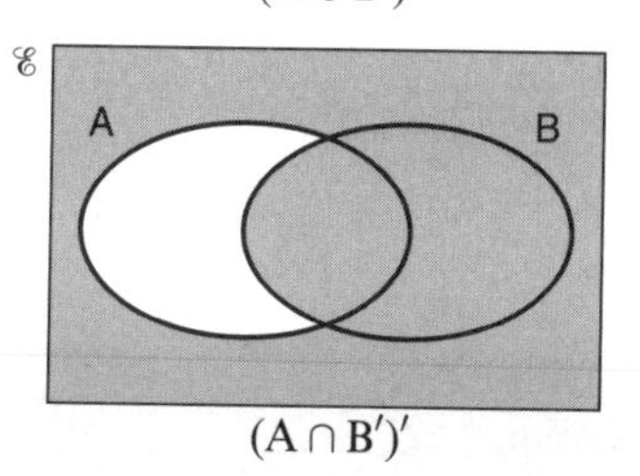

(A ∩ B′)′

ℰ A B

$(A \cup B)'$

3 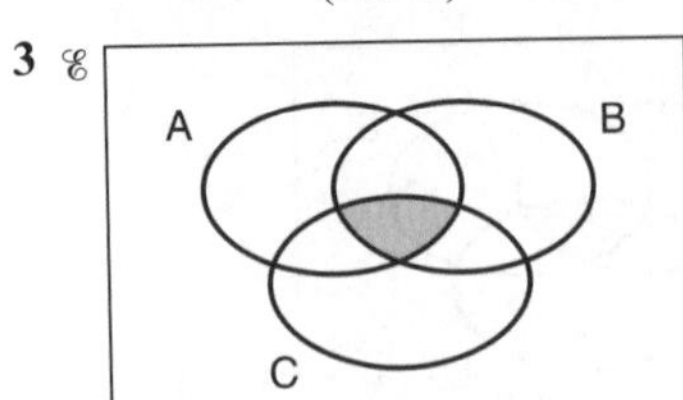

$A \cap B \cap C$

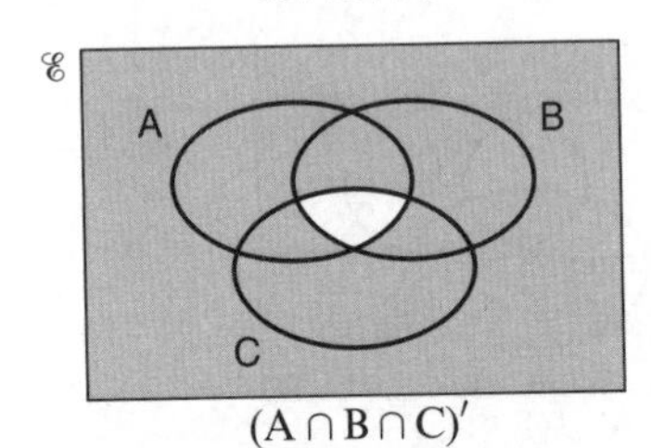

$(A \cap B \cap C)'$

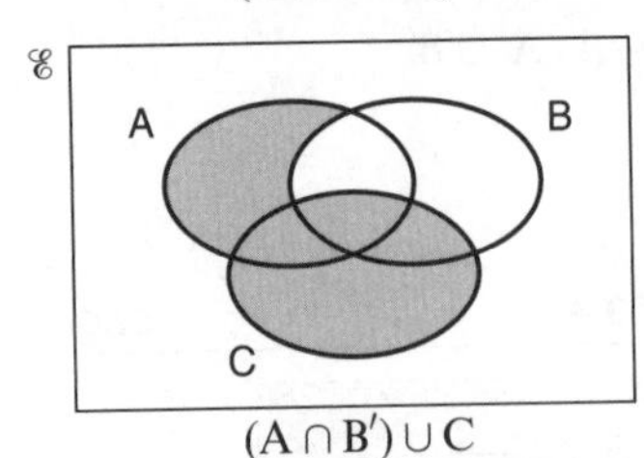

$(A \cap B') \cup C$

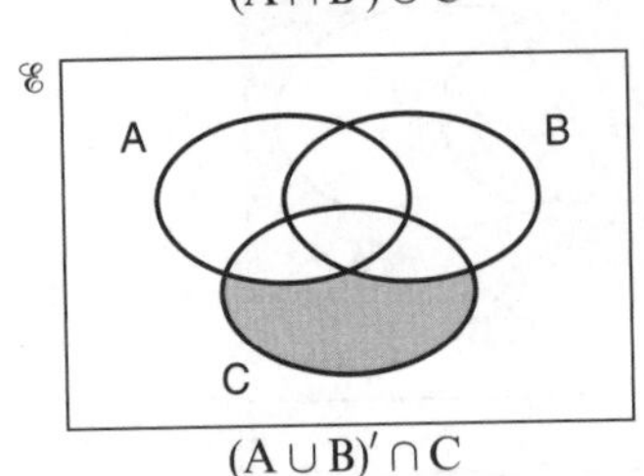

$(A \cup B)' \cap C$

5 $B \cap (A \cup C)'$, $B' \cap C$, $(B' \cap C) \cup (A \cap B \cap C')$

EXERCISE 18 (page 63)

1 a {Tuesday, Thursday}
b {Red, Amber, Green}
c {1, 2, 3, 4, 5, 6}
d {−1, 0, 1, 2, 3, 4, 5, 6}
3 a $\{x : x < 7, x \in \mathbb{N}\}$
b $\{x : x > 4, x \in \mathbb{N}\}$
c $\{x : 2 \leqslant x \leqslant 11, x \in \mathbb{N}\}$
d $\{x : -3 < x < 3, x \in \mathbb{N}\}$
e $\{x : x \text{ is odd}, x \in \mathbb{N}\}$
f $\{x : x \text{ is prime}\}$

EXERCISE 18★ (page 63)

1 a {2, 4, 6, 8, 10, 12}
b {3, 7, 11, 15, 19, 23}
c {2, 4, 6}
d {Integers between 1 and 12 inclusive}
3 a $\varnothing$
b $\left\{1, \frac{1}{2}, \frac{1}{4}, \frac{1}{8}, \frac{1}{16}\right\}$
c {2}
d {−3, 2}

5 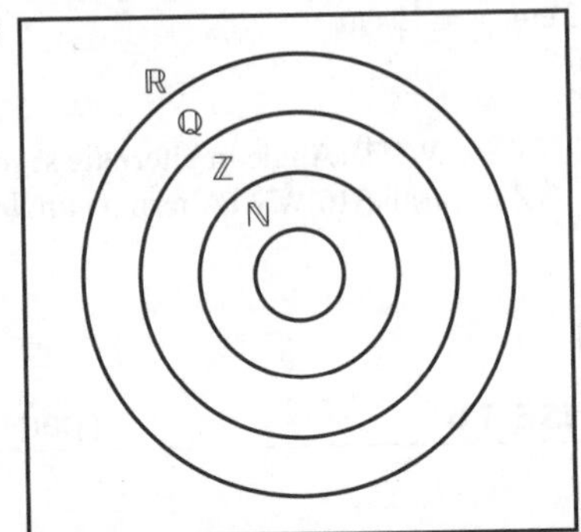

REVISION EXERCISE 19 (page 64)

1 a ℰ A = 20, B, 7, 10

b 17 **c** 30
3 a 17% **b** 52% **c** 31%
5 $A' \cup B'$
7 a $\{x : x \text{ is even}, x \in \mathbb{N}\}$
b $\{x : x \text{ is a factor of } 24, x \in \mathbb{N}\}$
c $\{x : -1 \leqslant x \leqslant 4, x \in \mathbb{N}\}$

REVISION EXERCISE 19★ (page 65)

1 34
3 2
5 a $(A \cup B') \cap C$
b $A \cup B \cup C'$
7 a $\{x : x > -5, x \in \mathbb{N}\}$
b $\{x : 4 < x < 12, x \in \mathbb{N}\}$
c $\{x : x \text{ is a multiple of } 3, x \in \mathbb{N}\}$ or $\{x : x = 3y, y \in \mathbb{N}\}$

Examination practice 1

1 a 1.2 mpg **b** 1920 gallons
3 $\frac{9}{40}, \frac{7}{8}, \frac{7}{128}$
5 a $p = \frac{1}{5}q^2$ **b** 80 **c** ± 10

7 a $x = 4\sqrt[3]{y}$ **b** $y = 512$

9 a

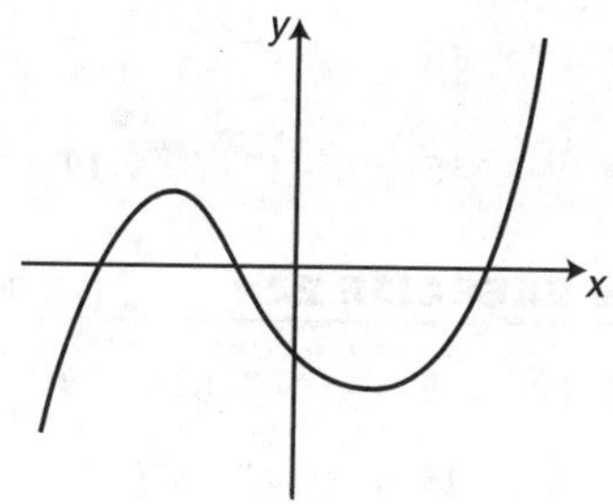

b $x = -3, -1, 1$

11 a 80° **b** 74°

13 a BP = 5.4 cm **b** CQ = 4 cm

15 a

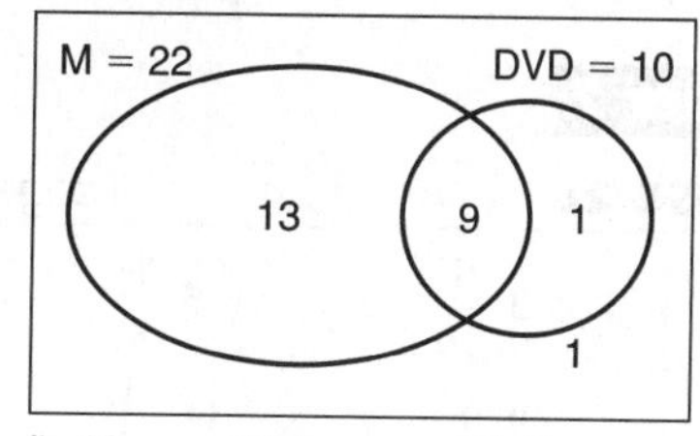

b i) 13 ii) 1 iii) 9

Answers

Unit 2

Number 2

EXERCISE 20 (page 73)

1 $\frac{1}{9}$ 3 $\frac{1}{10}$ 5 $\frac{1}{8}$
7 $\frac{1}{8}$ 9 9 11 $\frac{1}{64}$
13 3 15 2 17 $\frac{1}{16}$
19 $\frac{1}{9}$ 21 $\frac{1}{16}$ 23 $\frac{1}{2}$
25 16 27 625 29 2.84
31 0.167 33 0.0123 35 46 700
37 0.0370 39 0.111 41 64
43 a 45 c^{-3} 47 e
49 a^{-2}

EXERCISE 20★ (page 73)

1 $\frac{1}{64}$ 3 5 5 1 7 512
9 12.5 11 4 13 10 15 8
17 $\frac{1}{2}$ 19 $\frac{1}{36}$ 21 0.364 23 1.92
25 0.001 37 27 a^2 29 $2c^{-4}$ 31 $3a^{-2}$
33 $12a^{-1}$ 35 a 37 c^{-1} 39 2
41 2 43 −3 45 $k = 2\frac{1}{3}$ 47 a^{-3}
49 b 51 $2a^{-2}$ 53 $5a^{-2}$ 55 $\frac{1}{3}$
57 $-9a^6$ 59 $\frac{b}{a}$

EXERCISE 21 (page 75)

1 $\frac{1}{3}$ 3 $\frac{1}{4}$ 5 $\frac{1}{9}$ 7 $\frac{1}{4}$
9 27 11 $\frac{1}{27}$ 13 0.224 15 0.376
17 36.5 19 0.0540 21 b^{-1} 23 d
25 f^{-1}

EXERCISE 21★ (page 75)

1 $\frac{1}{6}$ 3 6 5 16 7 $\frac{1}{16}$
9 1.29 11 0.894 13 2.85 15 0.351
17 a^{-2} 19 c^{-2} 21 e 23 9
25 $\frac{1}{36}$ 27 0.952 29 $x = 3, y = -2$

REVISION EXERCISE 22 (page 76)

1 $\frac{1}{16}$ 3 4 5 5 7 $\frac{1}{64}$ 9 $\frac{1}{11}$
11 $\frac{1}{81}$ 13 $\frac{1}{6}$ 15 a^2 17 d^{-2} 19 $b^{2\frac{1}{2}}$

REVISION EXERCISE 22★ (page 76)

1 $\frac{1}{32}$ 3 1 5 4 7 64 9 $\frac{1}{9}$
11 4 13 $3c$ 15 a^2 17 $2a^{-1}$ 19 $-9a^6$

Algebra 2

EXERCISE 23 (page 78)

1 $x = -2$ or $x = -1$ 3 $x = -3$ or $x = 2$
5 $x = -5$ or $x = -2$ 7 $x = -3$ or $x = 5$
9 $x = 3$ 11 $x = -6$ or $x = 2$
13 $x = -1$ or $x = 0$ 15 $x = 0$ or $x = 4$
17 $x = \pm 2$ 19 $x = -6$ or $x = 6$

EXERCISE 23★ (page 78)

1 $x = -5$ or $x = -1$ 3 $x = -1$ or $x = 4$
5 $x = -8$ or $x = -7$ 7 $x = -5$ or $x = 9$
9 $x = 7$ 11 $x = -5$ or $x = 8$
13 $x = 0$ or $x = 13$ 15 $x = 0$ or $x = -17$
17 $x = \pm 9$ 19 $x = -11$ or $x = 11$

EXERCISE 24 (page 79)

1 $x = \pm\frac{7}{2}$ 3 $x = \pm\frac{9}{4}$
5 $x = -2$ or $x = 0$ 7 $x = 0$ or $x = 1$
9 $x = 2$ or $x = 3$ 11 $x = 0.5$ or $x = 2$
13 $x = -1.5$ or $x = -1$ 15 $x = -2$ or $x = -1$
17 $x = -3$ or $x = 3$ 19 $x = 0$ or $x = 2$
21 $x = -2$ or $x = -\frac{1}{3}$ 23 $x = -\frac{1}{3}$ or $x = 2$
25 $x = -2$ or $x = 3$ 27 $x = -2$ or $x = -\frac{2}{3}$
29 $x = 4$ or $\frac{2}{3}$

EXERCISE 24★ (page 79)

1 $x = \pm\frac{5}{7}$ 3 $x = \pm\frac{8}{3}$
5 $x = -2$ or $x = 0$ 7 $x = 0$ or $x = \frac{3}{2}$
9 $x = 1$ or $x = 2$ 11 $x = 1.5$ or $x = 2$
13 $x = -9$ or $x = -\frac{4}{3}$ 15 $x = -\frac{1}{3}$ or $x = 1.5$
17 $x = -\frac{1}{2}$ or $x = -\frac{1}{4}$ 19 $x = \frac{2}{5}$ or $x = 5$
21 $x = 0.8$ or $x = 1.5$ 23 $x = -\frac{4}{3}$ or $x = 7$
25 $x = -4$ or $x = 4$ 27 $x = 0$ or $x = 3$
29 $x = -5$ (repeated root)
31 $x = 0.25$ or $x = 7$ 33 $x = \frac{2}{3}$ or $x = 1.5$
35 $x = \frac{5}{3}$ (repeated root)

EXERCISE 25 (page 81)

1 $x = -3.45$ or $x = 1.45$
3 $x = -1.65$ or $x = 3.65$
5 $x = -5.46$ or $x = 1.46$
7 $x = 1.84$ or $x = 8.16$
9 $x = -14.2$ or $x = 0.211$
11 $x = -1.53$ or $x = 21.5$
13 $x = -2.90$ or $x = 6.90$
15 $x = -7.04$ or $x = 17.0$
17 $x = -3.56$ or $x = 0.561$
19 $x = -0.541$ or $x = 5.54$
21 $x = -3.37$ or $x = 2.37$
23 $x = -1.83$ or $x = 3.83$
25 $x = -1.58$ or $x = -0.423$
27 $x = -1.13$ or $x = 0.883$
29 $x = -2.79$ or $x = 1.79$
31 $x = 0.172$ or $x = 5.83$
33 $x = 1.30$ or $x = -0.876$
35 $x = 0.310$ or $x = 1.29$

EXERCISE 25★ (page 81)

1 $x = 0.171$ or $x = 5.83$
3 $x = 0.190$ or $x = 15.8$
5 $x = -7.58$ or $x = 1.58$
7 $x = -12.7$ or $x = 0.315$
9 $x = 1.59$ or $x = 4.41$
11 $x = -\frac{1}{3}$ or $x = 2$
13 $x = -1.69$ or $x = 7.69$
15 $x = -0.414$ or $x = 2.41$
17 $x = 0.258$ or $x = 7.74$
19 $x = -1$ or $x = 3.5$
21 $x = -1.86$ or $x = -0.537$
23 $x = -2.77$ or $x = 0.271$
25 $x = -0.522$ or $x = 1.09$
27 $x = 0.137$ or $x = 1.46$
29 $x = -0.257$ or $x = 2.59$
31 $x = 3.11$ or $x = -1.61$
33 $x = 0.105$ or $x = 5.37$
35 $x = -2.16$ or $x = 1.16$

ACTIVITY 11 (page 82)

- $y = x^2 + 8x + 15$ cuts the x-axis at two points, when $x = -5$ and $x = -3$, so two solutions.
  $b^2 - 4ac = 4$; two solutions are $x = -5$ and $x = -3$
  $y = x^2 + 8x + 16$ cuts the x-axis at one point, when $x = -4$, so one solution.
  $b^2 - 4ac = 0$; one solution is $x = -4$.
  $y = x^2 + 8x + 17$ does not cut the x-axis, so no solution.
  $b^2 - 4ac = -4$, no solution.
- $b^2 - 4ac > 0$ means two solutions.
  $b^2 - 4ac = 0$ means one solution.
  $b^2 - 4ac < 0$ means no solutions.

INVESTIGATE (page 82)

$k < 16$

EXERCISE 26 (page 82)

1 1   3 0   5 0   7 2   9 0

EXERCISE 26★ (page 83)

1 2   3 2   5 1   7 2   9 2

EXERCISE 27 (page 85)

1 2 or 5
3 12, 8 or −8, −12
5 2.61
7 3.22
9 2.32

EXERCISE 27★ (page 86)

1 4
3 3.21
5 8, 9 or −9, −8
7 2.32 by 4.43
9 **a** 1414
b 446 terms too small, 447 terms too big.
11 6 days

EXERCISE 28 (page 88)

1 $-4 < x < 4$
3 $x \leqslant 5$ or $x \geqslant 5$
5 $-9 \leqslant x \leqslant 9$
7 $-5 < x < 5$
9 $x < -4$ or $x > 4$
11 $-3 \leqslant x \leqslant 1$
13 $x < -4$ or $x > -3$
15 $-1 < x < \frac{1}{2}$
17 $x \leqslant 5$ or $x \geqslant -2$
19 $-5 < x < 3$

EXERCISE 28★ (page 89)

1 $-5 < x < 5$
3 $x \leqslant -4$ or $x \geqslant 4$
5 $-5 < x < 5$
7 $x < 3$ or $x > 7$
9 $-6 < x < 2$
11 $-7 < x < -3$
13 $x < -4$ or $x > 3$
15 $x < -4$ or $x > 2$
17 $-\frac{1}{2} \leqslant x \leqslant 2$
19 $x \leqslant -3$ or $x \geqslant \frac{1}{6}$
21 $x \leqslant 0$ or $x \geqslant \frac{1}{4}$
23 $-8 < x < 2$
25 $\frac{1}{2} < x < 1$
27 $6 <$ width < 8

REVISION EXERCISE 29 (page 89)

1 **a** $x = \pm 5$ **b** $x = -4$ or $x = 0$
3 **a** $x = -1.24$ or $x = 3.24$
b $x = 0.232$ or $x = 1.43$
5 1.70

REVISION EXERCISE 29★ (page 90)

1 **a** $x = \pm 4.47 (\pm\sqrt{20})$ **b** $x = 0$ or $x = 9$
3 **a** $x = -0.573$ or $x = 2.91$
b $x = -4.46$ or $x = 0.459$
5 width 5, length 6
7 **a** $4 < x < 8$
b $x \leqslant -2.32$ or $x \geqslant 4.32$

Graphs 2

EXERCISE 30 (page 92)

1 $x = -2.2$ or $x = 2.2$
3 $x = -1$ or $x = 2$
5 $x = -3.8$ or $x = 1.8$
7 $x = 0.6$ or $x = 3.4$
9 $x = -2.9$ or $x = 3.4$
11 No solutions

EXERCISE 30★ (page 92)

1 $x = -1.3$ or $x = 2.3$
3 $x = -2.6$ or $x = -0.4$
5 $x = 2$
7 $x = -2.7$ or $x = 2.2$
9 $x = -2.8$ or $x = 3.2$
11 No solutions

EXERCISE 31 (page 95)

1 a $x = 0$ or $x = 3$
b $x = -0.56$ or $x = 3.56$
c $x = 0.38$ or $x = 2.62$
d $x = -0.24$ or $x = 4.24$
e $x = -0.79$ or $x = 3.79$
f $x = 0.21$ or $x = 4.79$
3 a $x = 1$ or $x = 3$
b $x = -0.65$ or $x = 4.65$
c $x = 0.70$ or $x = 4.30$
d $x = -0.56$ or $x = 3.56$
5 a $2x^2 + 2x - 1 = 0$ b $x^2 + 5x - 5 = 0$
7 a $y = 2x + 2$ b $y = x$
c $y = -3x - 3$
9 $(2.71, 3.5)$, no

EXERCISE 31★ (page 96)

1 a $x = 5$ or $x = 0$
b $x = 4.30$ or $x = 0.70$
c $x = 3.73$ or $x = 0.27$
d $x = 0.76$ or $x = 5.24$
3 a $x = -1.78$ or $x = 0.28$
b $x = -2.35$ or $x = 0.85$
c $x = -2.28$ or $x = -0.22$
5 a $6x^2 - 7x - 2 = 0$ b $5x^2 - 7x - 4 = 0$
7 a $y = x + 2$ b $y = -2x - 1$
9 a Yes b 17.9 m, so legal

EXERCISE 32 (page 98)

1 a $x = -1.73$, $x = 0$ or $x = 1.73$
b $x = -1.53$, $x = -0.35$ or $x = 1.9$
c $x = -1.62$, $x = 0.62$ or $x = 1$
3 a $x = -0.53$, $x = 0.65$ or $x = 2.88$
b $x = -1$ or $x = 2$
c $x = -1.11$, $x = 1.25$ or $x = 2.86$
5 $x = -1$ or $x = 3$

EXERCISE 32★ (page 98)

1 a $x = 1.78$
b $x = -3.30$, $x = -1.05$, $x = 1.05$ or $x = 3.30$
c $x = -2.84$, $x = -1.46$ or $x = 0.96$
3 $y = 3x + 8$
5 $x = -1.25$, $x = 0.45$ or $x = 1.80$

INVESTIGATE (page 98)

a $k < -16$ or $k > 16$
b $k = -16$ or $k = 16$
c $-16 < k < 16$

ACTIVITY 12 (page 99)

◆

x	0	2	4
$2x$	0	4	8
$-\frac{1}{4}x^2$	0	-1	-4
$y = 2x - \frac{1}{4}x^2$	0	3	4

x	6	8	10
$2x$	12	16	20
$-\frac{1}{4}x^2$	-9	-16	-25
$y = 2x - \frac{1}{4}x^2$	3	0	-5

x	0	2	4	6	8	10
$\frac{1}{4}x$	0	$\frac{1}{2}$	1	$1\frac{1}{2}$	2	$2\frac{1}{2}$
$y = \frac{1}{4}x - 1$	-1	$-\frac{1}{2}$	0	$\frac{1}{2}$	1	$1\frac{1}{2}$

◆

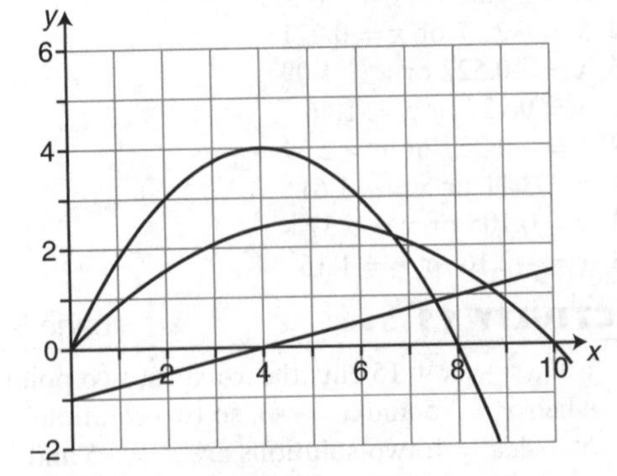

◆ Mary is not successful.
◆ When angle changes, Peter is made wet.

EXERCISE 33 (page 100)

1 $x = -1.7$, $y = 5$; $x = 1.7$, $y = 5$
3 $x = -3$, $y = -5$; $x = 1$, $y = 3$
5 $x = 2$, $y = 7$; $x = -1$, $y = -2$
7 $x = 2$, $y = 2$; $x = 4$, $y = 6$
9 $x = -1.46$ and $y = 1.37$; or
$x = 1.93$ and $y = 0.52$

11 $x = -1.56$ and $y = -2.56$; or $x = 2.56$ and $y = 1.56$
13 $x = -2$ and $y = 0$; or $x = -0.71$ and $y = 0.65$; or $x = 0.71$ and $y = 1.35$
15 $x = 23.7$ and $y = 21.8$

EXERCISE 33★ (page 101)

1 $x = 1, y = 3; x = -2, y = 3$
3 $x = 2, y = -3; x = -3, y = 7$
5 $x = -2, y = 8; x = 2.5, y = 3.5$
7 $x = -0.5, y = -3; x = 0.4, y = -1.2$
9 $x = -1.93$ and $y = -0.86$; or $x = 0.73$ and $y = 4.46$
11 $x = -0.48$ and $y = 3.96$; or $x = 1.31$ and $y = 0.38$; or $x = 3.17$ and $y = 3.34$
13 $x = -1.23$ and $y = -4.14$; or $x = 1.63$ and $y = 10.1$
15 $x = 44$ and $y = 22$; length is 49.2 m

REVISION EXERCISE 34 (page 102)

1 (137, −50)
3 a $y = 4$ b $y = x + 1$ c $y = 2x - 1$
5 $xy = 30$, $x + y = 12$; $x = 8.45$, $y = 3.55$

REVISION EXERCISE 34★ (page 103)

1 $x^2 + y^2 = 144$; (−7, 9.75), (7, 9.75)
3 a $y = 1$ b $y = x$ c $y = 3x - 1$
5 $x + y = 5$, $x^2 + y^2 = 16$
$x = 1.18$, $y = 3.82$ or vice versa

Shape and space 2

EXERCISE 35 (page 105)

	km	m	cm	mm
1	**5**	5000	5×10^5	5×10^6
3	**3**	3000	3×10^5	3×10^6
5	2	**2000**	2×10^5	2×10^6
7	0.05	50	**5000**	5×10^4
9	1	10^3	10^5	$\mathbf{10^6}$
11	15	$\mathbf{1.5 \times 10^4}$	1.5×10^6	1.5×10^7

EXERCISE 35★ (page 105)

	km	m	cm	mm
1	**500**	5×10^5	5×10^7	5×10^8
3	$\mathbf{2.5 \times 10^4}$	2.5×10^7	2.5×10^9	2.5×10^{10}
5	5×10^3	$\mathbf{5 \times 10^6}$	5×10^8	5×10^9
7	5×10^{-4}	0.5	**50**	500
9	9×10^3	9×10^6	9×10^8	$\mathbf{9 \times 10^9}$
11	4×10^{-6}	4×10^{-3}	0.4	**4**

13 a 2×10^{14} b 1×10^6

EXERCISE 36 (page 106)

	km^2	m^2	cm^2	mm^2
1	**2**	2×10^6	2×10^{10}	2×10^{12}
3	**1**	10^6	10^{10}	10^{12}
5	5×10^{-5}	**50**	5×10^5	5×10^7
7	6×10^{-4}	6×10^2	$\mathbf{6 \times 10^6}$	6×10^8
9	10	10^7	10^{11}	$\mathbf{10^{13}}$
11	4×10^{-2}	$\mathbf{4 \times 10^4}$	4×10^8	4×10^{10}

EXERCISE 36★ (page 107)

	km^2	m^2	cm^2	mm^2
1	**80**	8×10^7	8×10^{11}	8×10^{13}
3	6×10^{-3}	**6000**	6×10^7	6×10^9
5	6	6×10^6	$\mathbf{6 \times 10^{10}}$	6×10^{12}
7	2×10^9	2×10^{15}	2×10^{19}	$\mathbf{2 \times 10^{21}}$
9	$\mathbf{7 \times 10^{-2}}$	7×10^4	7×10^8	7×10^{10}
11	4×10^{-10}	4×10^{-4}	**4**	4×10^2

EXERCISE 37 (page 108)

	km^3	m^3	cm^3	mm^3
1	**1**	10^9	10^{15}	10^{18}
3	**4**	4×10^9	4×10^{15}	4×10^{18}
5	8×10^{-9}	**8**	8×10^6	8×10^9
7	4×10^{-12}	4×10^{-3}	$\mathbf{4 \times 10^3}$	4×10^6
9	10^{-3}	10^6	10^{12}	$\mathbf{10^{15}}$
11	7×10^{-7}	$\mathbf{7 \times 10^2}$	7×10^8	7×10^{11}

13 10^3 15 10

EXERCISE 37★ (page 109)

	km^3	m^3	cm^3	mm^3
1	**70**	7×10^{10}	7×10^{16}	7×10^{19}
3	6×10^{-7}	**600**	6×10^8	6×10^{11}
5	3×10^{-7}	3×10^2	$\mathbf{3 \times 10^8}$	3×10^{11}
7	5×10^7	5×10^{16}	5×10^{22}	$\mathbf{5 \times 10^{25}}$
9	$\mathbf{4 \times 10^{-6}}$	4×10^3	4×10^9	4×10^{12}
11	3×10^{-18}	3×10^{-9}	3×10^{-3}	**3**

13 5.12×10^5 15 10^3 17 10^{45}

EXERCISE 38 (page 111)

1 18.8 cm, 28.3 cm^2
3 33.6 cm, 58.9 cm^2
5 22.3 cm, 30.3 cm^2
7 50.8 cm, 117 cm^2
9 46.8 cm, 39.5 cm^2
11 37.7 cm, 37.7 cm^2

	Radius in cm	Circumference in cm	Area in cm^2
13	0.955	**6**	2.86
15	2.11	13.3	**14**
17	8.28	**52**	215
19	5.17	32.5	**84**

21 2.07 km **23** 6.03 m

EXERCISE 38★ (page 113)

1 20.6 cm, 25.1 cm^2
3 37.7 cm, 92.5 cm^2
5 43.7 cm, 99.0 cm^2
7 66.8 cm, 175 cm^2
9 37.7 cm, 56.5 cm^2
11 29.7 cm, 63.3 cm^2
13 $r = 3.19$ cm, $P = 11.4$ cm
15 16.0 m
17 2 cm
19 a 40 100 km **b** 464 m/s
21 6.28 km
23 $r = 1.79$ cm, $A = 7.53$ cm^2

EXERCISE 39 (page 116)

1 8.62 cm **3** 25.6 cm
5 38.4 cm **7** 63.6 cm
9 34.4° **11** 115°
13 14.3 cm **15** 10.6 cm

EXERCISE 39★ (page 117)

1 10.95 cm **3** 38.3 cm
5 25.1° **7** 121°
9 13.4 cm **11** 117 cm
13 33.0 mm **15** 15.5 cm
17 4.94 cm

EXERCISE 40 (page 120)

1 12.6 cm^2 **3** 61.4 cm^2
5 170 cm^2 **7** 11.9 cm^2
9 76.4° **11** 129°
13 5.86 cm **15** 8.50 cm

EXERCISE 40★ (page 121)

1 15.8 cm^2 **3** 625 cm^2
5 53.3° **7** 103°
9 4.88 cm **11** 19.7 cm
13 11.5 cm **15** 5.08 cm
17 1.45 cm^2 **19** 6.14 cm

EXERCISE 41 (page 125)

1 120 cm^3 **3** 48 cm^3, 108 cm^2
5 1.57 m^3, 7.85 m^2 **7** 800 m^3
9 0.785 m^2 **11** $16\frac{2}{3}$ cm

EXERCISE 41★ (page 127)

1 3000 cm^3
3 1.2×10^5 cm^3, 1.84×10^4 cm^2
5 229 cm^3, 257 cm^2
7 1.18 cm^3, 9.42 cm^2
9 405 cm^3, 426 cm^2
11 29.5 m

EXERCISE 42 (page 131)

1 2.57×10^6 m^3 **3** 9820 cm^3, 2410 cm^2
5 2090 cm^3, 628 cm^2 **7** 25.1 cm^3
9 396 m^3, 311 m^2 **11** 12 cm, 1810 cm^2
13 61 cm

EXERCISE 42★ (page 133)

1 3.15×10^4 cm^3 **3** 39.3 cm^3
5 2150 cm^3, 971 cm^2 **7** 2.92×10^5 m^3
9 5.12×10^8 km^2 **11** 0.417 cm
13 12 cm

EXERCISE 43 (page 137)

1 16 cm^2
3 a Angles are the same **b** 8.55 cm^2
5 213 cm^2 **7** 84.4 cm^2 **9** 6 cm
11 3 cm **13** 3 cm **15** 13.3 cm

EXERCISE 43★ (page 140)

1 675 cm^2 **3** 45 cm^2
5 7.5 cm **7** 10 cm
9 100 cm^2 **11** 44%
13 19% **15** 75 cm^2

EXERCISE 44 (page 144)

1 800 cm^3 **3** 675 cm^3
5 14.1 cm^3 **7** 25.3 cm^3
9 15.1 cm **11** 9.64 cm
13 5.06 cm **15** 27.8 cm

EXERCISE 44★ (page 147)

1 135 cm^2 **3** 31.25 cm^2
5 25.0 cm **7** 11.9 cm
9 72.8% **11** $33.75
13 40 m^2, 1500 cm^3
15 a 25 cm **b** 48 g
17 a 2000 quills **b** 22.5 m
c 810 cm^2 **d** 240 g

REVISION EXERCISE 45 (page 149)

1 3×10^6 mm
3 10^6 cm^3
5 22.3 cm^2, 21.6 cm
7 132 m^3
9 27 litres

REVISION EXERCISE 45★ (page 150)

1 4 km
3 8×10^3 m^3
5 6.98 cm^2, 11.0 cm
7 335 cm^3, 289 cm^2
9 a $6.75 b 12 cm

Handling data 2

EXERCISE 46 (page 154)

1 a $\frac{1}{36}$ b $\frac{25}{36}$ c $\frac{5}{36}$ d $\frac{5}{18}$
3 a $\frac{4}{25}$ b $\frac{9}{25}$ c $\frac{6}{25}$ d $\frac{12}{25}$
5 a $\frac{4}{9}$ b $\frac{4}{9}$ c $\frac{1}{9}$
7 a $\frac{1}{169}$ b $\frac{1}{4}$ c $\frac{30}{169}$ d $\frac{1}{8}$

EXERCISE 46★ (page 155)

1 a $\frac{4}{15}$ b $\frac{8}{15}$ c $\frac{3}{5}$
3 a $\frac{1}{9}$ b $\frac{4}{9}$ c $\frac{5}{9}$
5 a $\frac{1}{4}$ b $\frac{1}{2}$ c $\frac{15}{32}$ d $\frac{1}{16}$
7 a $\frac{1}{5}$ b $\frac{13}{35}$ c $\frac{4}{5}$

ACTIVITY 13 (page 157)

p(X to B) $= \frac{1}{2}$; p(A to B) $= \frac{1}{2}$

$p(A) = \frac{1}{2}$; $p(B) = \frac{1}{2} + \left(\frac{1}{2} \times \frac{1}{2}\right) = \frac{3}{4}$

A: 30; B: 45; B not via A: 30

$p(A) = \frac{1}{2}$

$p(B) = \frac{1}{2} + \left(\frac{1}{2} \times \frac{1}{3}\right) = \frac{2}{3}$

$p(C) = \left(\frac{1}{2} \times \frac{1}{3} \times \frac{1}{2}\right) + \left(\frac{1}{2} \times \frac{1}{3} \times \frac{1}{2}\right) + \left(\frac{1}{2} \times \frac{1}{2}\right) = \frac{5}{12}$

$p(D) = \frac{1}{2} \times \frac{1}{3} = \frac{1}{6}$

A: 30; B: 40; C: 25; D: 10

Compare the theoretical figures to your observed and find percentage differences. Would more vehicles improve the goodness of fit? If a computer program is used, take a much larger sample and compare.

EXERCISE 47★ (page 158)

1 a $\frac{1}{9}$ b $\frac{4}{9}$
3 a (i) $\frac{43}{63}$ (ii) $\frac{20}{63}$ b $\frac{2}{7}$

c Let X be the no. of beads added to the box.

$$p(W_2) = \frac{2}{7} \times \frac{(2+X)}{(7+X)} + \frac{5}{7} \times \frac{2}{(7+X)}$$
$$= \frac{2}{(7(7+X))} \times [(2+X)+5] = \frac{2}{7}$$

Therefore true!

5 a 0.0034 b 0.0006 c 0.0532
7 a $\frac{1}{8}$ b $\frac{8}{15}$ c $\frac{13}{60}$
9 a $p(H_1) = \frac{1}{4}$

b $p(H_2) = p(HH) + p(\overline{H}H)$
$$= \frac{1}{4} \times \frac{12}{51} + \frac{3}{4} \times \frac{13}{51} = \frac{1}{4}$$

c $p(H_3) = p(HHH) + p(H\overline{H}H) + p(\overline{H}HH) + p(\overline{H}\overline{H}H)$
$$= \frac{1}{4} \times \frac{12}{51} \times \frac{11}{50} + \frac{1}{4} \times \frac{39}{51} \times \frac{12}{50} + \frac{3}{4} \times \frac{13}{51} \times \frac{12}{50} + \frac{3}{4} \times \frac{38}{51} \times \frac{13}{50}$$
$$= \frac{1}{4}$$

INVESTIGATE (page 160)

Clearly there will always be someone left standing, so this is a ludicrous example of testing for telepathy despite the fact that the person 'selected' has a chance of $\frac{1}{1024}$ of being chosen.

ACTIVITY 14 (page 161)

Rank in order of safest first:
Motor racing, Smoking, Influenza, Drinking, Run over by a vehicle, Football = Rock Climbing, Tornadoes = Floods, Earthquakes, Lightning, Bites of venomous creatures, Falling Aircraft, Meteorite.

Points for discussion:
Does the number of participants affect this table?
How do you think the 'meteorite' statistic was evaluated?
Does the ranking come out as you expected? It might be interesting to compare your table to your guessed ranking for the activities.

REVISION EXERCISE 48 (page 161)

1 a $\frac{9}{25}$ b $\frac{12}{25}$
3 a $\frac{9}{25}$ b $\frac{12}{25}$
5 a $\frac{1}{36}$ b $\frac{5}{18}$ c $\frac{4}{9}$

REVISION EXERCISE 48★ (page 162)

1 a $\frac{1}{6}$ **b** $\frac{5}{18}$ **c** $\frac{13}{18}$

3 a $\frac{1}{32}$ **b** $\frac{5}{16}$ **c** $\frac{3}{16}$

5 a 1 : 3 : 5 **b** $\frac{4}{45}$

c (i) $\frac{16}{2025}$ **(ii)** $\frac{164}{2025}$ **(iii)** $\frac{344}{2025}$

Examination practice 2

1 a x^7 **b** x^3 **c** x^{10} **d** x^{-3}

3 a b **b** c^{-3} **c** b^{-2} **d** c **e** 1

5 6.79 cm

9 a $x = 3\,\text{cm}$ **b** $y = 4\,\text{cm}$ **c** $16\,\text{cm}^2$

11 a $50.3\,\text{cm}^2$ **b** $4021\,\text{cm}^3$ **c** $1474\,\text{cm}^2$

13 a $\frac{2}{15}$ **b** $\frac{8}{15}$ **c** $\frac{16}{15}$ **d** $\frac{64}{225}$ **e** $\frac{139}{225}$

Unit 3

Number 3

EXERCISE 49 (page 175)

1 \$346, \$17 992
3 \$9.92
5 a \$108.93 **b** \$105.80 **c** \$61.76
7 a \$15.65 **b** \$1.83 **c** \$225
9 \$9000, \$3416.67
11 \$24 250, \$5895.83

EXERCISE 49★ (page 176)

1 a \$148.75 **b** \$7.875 per tonne
3 a \$19.57 **b** \$52.17
5 a \$101 000 **b** \$1942.31 **c** 32.7%
7 a \$12 000, \$4416.67 **b** 30.7%
9 \$20 000
11 b 70.2% **c** 235.8%

EXERCISE 50 (page 180)

1 a \$20 **b** \$575.48
3 12.68% **5** 19.56% **7** 0.80% **9** 0.95%
11 \$5
13 a \$289.33 **b** \$14 091 **c** \$3291

EXERCISE 50★ (page 180)

1 a \$15.40 **b** \$245.46
3 15.39% **5** 23.14% **7** 1.39% **9** 1.71%
11 \$9.74
13 a 8.25% **b** \$278.95

ACTIVITY 16 (page 182)

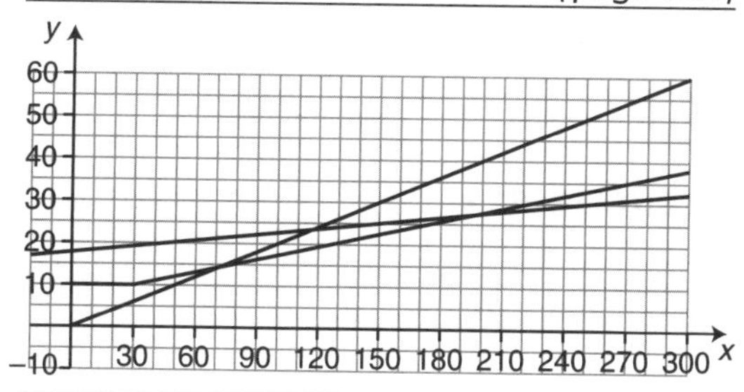

Pay-as-you-go	$0 \leqslant t \leqslant 70$
Speakeasy	$70 \leqslant t \leqslant 220$
Chatterbox	$220 \leqslant t$

EXERCISE 51 (page 184)

1 \$154.80 **3** HK\$1198.56
5 €159.09 **7** €3.74
9 €3.26 **11** €3.27
13 €6.27

EXERCISE 51★ (page 185)

1 VB200 590.65
3 €95.11
5 6.54%
7 a €16.54 **b** €23.08
c 16.54%, 11.5%
9 a 1.161 **b** 1.277
11 a €61 538 **b** \$73 846 **c** \$92 307
13 a \$58.05, \$60 **b** \$122.55, \$120
c €71.67

REVISION EXERCISE 52 (page 186)

1 \$14 508
3 a \$8600 **b** \$24 700
5 14.03%
7 a \$35.40 **b** \$1029.60

REVISION EXERCISE 52★ (page 187)

1 \$8549.46
3 \$30 000
5 a €218.53 **b** 13.73

Algebra 3

EXERCISE 53 (page 192)

1 $(-2, 4), (3, 9)$
3 $(-1, 1), (4, 16)$
5 $x = 1, y = 2$ or $x = 2, y = 3$
7 $x = -3.45, y = -7.90$ or $x = 1.45, y = 1.90$
9 $x = -0.2, y = 1.4$ or $x = 1, y = -1$
11 $x = -2, y = -1.5$ or $x = 3, y = 1$
13 $x = -2.87, y = 4.87$ or $x = 0.871, y = 1.13$
15 $x = 1, y = -1$
17 $x = 2.17, y = 0.172$ or $x = 7.83, y = 5.83$
19 a $y = 3$ **b** $x = 29.85$ and $y = 3$
c 29.85 cm

EXERCISE 53★ (page 193)

1 $x = -1.54, y = -4.17$ or $x = 4.54, y = 20.2$
3 $x = 1, y = 0$ or $x = 7, y = -12$
5 $x = -2, y = 2$ or $x = -1, y = 3$
7 $x = \frac{2}{3}, y = \frac{1}{3}$ or $x = \frac{1}{3}, y = \frac{2}{3}$
9 $(6, -6)$; tangent
11 a $y = 1.5$
b A$(-1.68, 1.5)$, B$(1.68, 1.5)$, AB = 3.35 cm
13 a $y = -3$
b $(-2.65, -3), (2.65, -3)$, diameter is 5.30 cm
15 (7.53, 0.88), No
17 $(2.67, -1.78)$
19 a A(0, 0), B(4, 8) and C(8, 6)
b 90°
c Diameter
21 $x = 5, y = 3$

EXERCISE 54 (page 197)

1 a 5 b −5 c 2 d 1
3 a 4 b −11 c $-\frac{1}{2}$ d −2
5 a 8 b 3 c $1\frac{1}{4}$ d 0
7 a 9 b −26 c $1\frac{1}{8}$ d 1
9 a 2 b 0 c 10
11 a $\frac{1}{5}$ b −1 c $\frac{1}{1+2a}$
13 a $1\frac{1}{2}$ b $2\frac{1}{2}$ c $2-\frac{1}{y}$
15 3
17 3
19 3

EXERCISE 54★ (page 197)

1 a 12 b 7 c 8 d $8-2p$
3 a 11 b $2\frac{1}{4}$
c 3 d p^2-2p+3
5 a 0 b $1\frac{1}{4}$ c 0 d $p(p+2)$
7 a −6 b $2\frac{1}{8}$ c 2 d p^3+2
9 a $-\frac{1}{3}$ b $\frac{1}{5}$ c $-\frac{1}{195}$
11 a $\sqrt{8}$ b 0 c $2\sqrt{a^2+a}$
13 a −4 b $\frac{2}{3}$ c $\frac{9y+2}{3y-4}$
15 1
17 −2, 3
19 $\frac{1}{2}$

EXERCISE 55 (page 198)

1 a $-2x+1$ b $2x+5$ c $2x+3$
3 a $4x+1$ b $8x-3$ c $8x-6$
5 a $3+x$ b $3+3x$ c $3x-9$
7 a x^2-1 b x^2 c $2-x^2$
9 a $9x^2-3x$ b $3x^2-3x$ c x^2+x
11 1
13 1
15 $-\frac{1}{2}$

EXERCISE 55★ (page 199)

1 a $2+x$ b $2+2x$ c $2x-4$
3 a x^2+4x+5 b x^2+3 c x^2+1
5 a $2x^2+x$ b $x-2x^2$ c $8x^2-2x$
7 a $3-9x^2$ b $9-3x^2$ c $9-3x^2$
9 a x^2 b x^2 c $\frac{1}{x^2}$
11 ±3
13 −6, −1
15 4
17 2

EXERCISE 56 (page 202)

1 a $\{-2, 1, 4, 7\}$
b All real numbers
3 a $\{2, 0, 6, 20\}$
b $\{y: y \geqslant 0$, y a real number$\}$
5 a $\{17, 5, 1\}$
b $\{y: y \geqslant 1$, y a real number$\}$
7 a $\{-11, -4, -3, 5\}$
b $\{y: y \geqslant -3$, y a real number$\}$
9 a $\{6, 4, 3, 2\}$
b $\{y: 0 < y \leqslant 12$, y a real number$\}$
11 a $\{2, \sqrt{3}, \sqrt{2}, 1\}$
b $\{y: y \geqslant 3$, y a real number$\}$

EXERCISE 56★ (page 202)

1 a $\{8, 5, 2, -1\}$
b All real numbers
3 a $\{0, 8, 24\}$
b $\{y: y \geqslant 0$, y a real number$\}$
5 a $\{27, 11, 3\}$
b $\{y: y \geqslant 2$, y a real number$\}$
7 a $\{-10, 0, 10, 68\}$
b $\{y: y \geqslant 2$, y a real number$\}$
9 a $\left\{1, \frac{1}{2}, \frac{1}{3}, \frac{1}{4}\right\}$
b $\{y: 0 < y \leqslant 1$, y a real number$\}$
11 a $\left\{-2, -\frac{1}{2}, 0, \frac{1}{4}\right\}$
b $\{y: -2 \leqslant y < 1$, y a real number$\}$

EXERCISE 57 (page 203)

1 −1
3 $\{x: x < 2$, x a real number$\}$
5 0
7 None
9 $\{x: -2 < x < 2$, x a real number$\}$

EXERCISE 57★ (page 203)

1 $\frac{1}{2}$
3 $\{x: x > 9$, x a real number$\}$
5 −1
7 ±1
9 $\{x: x \leqslant -2$, x a real number$\}$

EXERCISE 58 (page 205)

1 $fg(3) = 6$, $gf(3) = 6$
3 $fg(1) = 9$, $gf(1) = 3$
5 $fg(4) = 5$, $gf(4) = \frac{4}{5}$
7 a $x-1$ b $x-1$ c $x-8$ d $x+6$
9 a $2x+4$ b $2x+2$ c $4x$ d $x+4$

11 a $(x+2)^2$ b x^2+2 c x^4 d $x+4$
13 a x b x c $x-12$ d $x+12$
15 a 7 b 6
17 a $\{x: x \neq -3, x \text{ a real number}\}$
b All real numbers
19 a $\{x: x \geqslant \frac{1}{2}, x \text{ a real number}\}$
b $\{x: x \geqslant 0, x \text{ a real number}\}$

EXERCISE 58★ (page 206)

1 $fg(-3) = 19, gf(-3) = 8$
3 $fg(2) = 82, gf(2) = 36$
5 $fg(-3) = -4\frac{1}{2}, gf(-3) = -\frac{3}{7}$
7 a $x-2$ b $x-4$ c $\dfrac{x-12}{4}$ d $4x$
9 a $2(x-2)^2$ b $2x^2-2$ c $8x^4$ d $x-4$
11 a $\dfrac{1}{x}$ b $\dfrac{1}{x-2}+2$
c $\dfrac{x-2}{5-2x}$ d $x+4$
13 a $4\sqrt{\dfrac{x}{4}+4}$ b $\sqrt{x+4}$
c $16x$
d $\sqrt{\dfrac{1}{4}\sqrt{\left(\dfrac{x}{4}+4\right)}+4}$
15 a $\frac{5}{4}$ b $-\frac{1}{2}$
17 a $\{x: x \neq 5, x \text{ a real number}\}$
b $\{x: x \neq \pm 2, x \text{ a real number}\}$
19 a $\{x: x \geqslant 1, x \text{ a real number}\}$
b $\{x: x \geqslant -2, x \text{ a real number}\}$

EXERCISE 59 (page 208)

1 7 3 4 5 $\dfrac{x-4}{6}$ 7 $27-3x$
9 $\dfrac{x}{3}+6$ 11 $\dfrac{1}{3}\left(\dfrac{1}{x}-4\right)$ 13 $\dfrac{3}{4-x}$
15 $\sqrt{x-7}$
17 a 4 b $\frac{5}{2}$ c 1
19 $x = -5$

EXERCISE 59★ (page 209)

1 17 3 3.2 5 $\dfrac{4}{3}-\dfrac{x}{24}$
7 $2-\frac{3}{2x}$ 9 $\dfrac{7}{4-x}$ 11 $\sqrt{x^2-7}$
13 $\sqrt{\dfrac{x-16}{2}}$ 15 $\dfrac{4x-3}{x+2}$
17 a 4 b 7 c 0
19 $x = 2$
21 $x = 1$ or $x = 2$

REVISION EXERCISE 60 (page 209)

1 a 13 b -2 c 7
3 a $5x-1$ b $5x+3$
5 $-2, 3$
7 a all x b $x \geqslant 1$ c $x \geqslant 0$ d all x
9 a $\dfrac{1}{2}\left(\dfrac{x}{4}-3\right)$ b $7-x$
c $\dfrac{1}{x}-3$ d $\sqrt{x-4}$
11 $(-3, 9), (4, 16)$
13 $(-5, -3), (2, 4)$
15 a AB, $x = 1.7$
b A(1.7, 1.05), B(1.7, −1.05) length = 2.1 m

REVISION EXERCISE 60★ (page 210)

1 a 4 b 3 c 0
3 a $4-2x$ b $7-2x$
5 $-4, 2$
7 a $x \geqslant 3$ b $x \geqslant 0$
c $x \geqslant 0$ d all real numbers
9 a $\dfrac{1}{2}\left(1-\dfrac{x}{4}\right)$ b $4-\dfrac{3}{2-x}$
c $\dfrac{x^2+3}{2}$ d $2+\sqrt{x}$
11 $(-0.5, 0.5), (3, 18)$
13 $(-4, -9), (1.5, 2)$
15 $(15, 1.66), (15, -1.66)$

Graphs 3

EXERCISE 61 (page 214)

1 a 2 b 3 c -1 d -2 e $x = 1$
3 a -0.25 b -1 c -0.44
d -4 e $x = \pm 0.71$
5 a 4 b 2 c -4 d $x = 3$

EXERCISE 61★ (page 215)

1 a 2 b 4 c -1 d -3 e $x = 1.5$
3 a 1 b -0.37 or 1.37
c -1.3, 0.17 or 1.13 d -1.13, -0.17, 1.3
5 b

x co-ordinate	−4	−3	−2	−1	0	1	2	3	4
Gradient	−8	−6	−4	−2	0	2	4	6	8

c Straight line gradient 2 passing through the origin

EXERCISE 62 (page 218)

1 a i 1 m/s ii 0 m/s iii 2 m/s
b 0–20 secs gradually increased speed then slowed down to a stop.

20–30 secs stationary.
30–40 secs speed increasing.
40–50 secs travelling at constant speed of 2 m/s.
50–60 secs slowing down to a stop.

3 b i $-9.6\,°\text{C/min}$
ii $-6.7\,°\text{C/min}$
iii $-5.3\,°\text{C/min}$

5 b i $0.5\,\text{m/sec}^2$
ii $1.5\,\text{m/sec}^2$
iii $2.5\,\text{m/sec}^2$

EXERCISE 62★ (page 220)

1 a

t (min)	0	20	40	60	80	100	120
N	10	40	160	640	2560	10 240	40 960

b i 0.7 **ii** 44 **iii** 710

3 a

t (min)	0	10	20	30	40	50	60	70	80	90
V (cm^3)	2000	1700	1445	1228	1044	887	754	641	545	463

b i -27.6 **ii** -8.86
c $t = 0$, $-32.5\,\text{cm}^3/\text{min}$

5 a i 1.67 **ii** -1.67 **iii** 0
b Max at $t = 0, 4, 8, 12$ at ± 2.36 m/hr

REVISION EXERCISE 63 (page 222)

1 b $-3, 0, 5$
3 b 2.6 mm/s, 6.3 mm/s
Height increases at an increasing rate

REVISION EXERCISE 63★ (page 223)

1 a

x	0	1	2	3	4
y	1	3	9	27	81

b 3.3, 9.9

3 a 2
b $y = 2x + 1$

Shape and space 3

ACTIVITY 19 (page 224)

Vectors: acceleration, a pass in hockey, velocity, rotation of 180°, force, 10 km on a bearing of 075°
Scalars: volume, area, temperature, price, length, density

ACTIVITY 20 (page 225)

◆ b: $\begin{pmatrix} 2 \\ -2 \end{pmatrix}$; 2.8; 135°

c: $\begin{pmatrix} 4 \\ -2 \end{pmatrix}$; 4.5; 116.6°

d: $\begin{pmatrix} 6 \\ 1 \end{pmatrix}$; 6.1; 080.5°

e: $\begin{pmatrix} -2 \\ 1 \end{pmatrix}$; 2.2; 296.6°

◆ $\overrightarrow{OH} = \begin{pmatrix} 8 \\ 0 \end{pmatrix}$; $\overrightarrow{OH} = \mathbf{a} + \mathbf{b} = \mathbf{c} + \mathbf{d} + \mathbf{e}$

EXERCISE 64 (page 227)

1 a $\overrightarrow{XY} = \mathbf{x}$ **b** $\overrightarrow{EO} = 4\mathbf{y}$
c $\overrightarrow{WC} = -8\mathbf{y}$ **d** $\overrightarrow{TP} = -4\mathbf{x}$

3 a $\overrightarrow{HJ}$ **b** $\overrightarrow{HN}$ **c** $\overrightarrow{HL}$ **d** $\overrightarrow{HO}$

5 a $\overrightarrow{DC} = \mathbf{x}$ **b** $\overrightarrow{DB} = \mathbf{x} + \mathbf{y}$
c $\overrightarrow{BC} = -\mathbf{y}$ **d** $\overrightarrow{AC} = \mathbf{x} - \mathbf{y}$

7 a $\overrightarrow{DC} = \mathbf{x}$ **b** $\overrightarrow{AC} = \mathbf{x} + \mathbf{y}$
c $\overrightarrow{BD} = \mathbf{y} - \mathbf{x}$ **d** $\overrightarrow{AE} = \frac{1}{2}(\mathbf{x} + \mathbf{y})$

9 a $\overrightarrow{AB} = \mathbf{x} - \mathbf{y}$ **b** $\overrightarrow{AD} = 3\mathbf{x}$
c $\overrightarrow{CF} = 2\mathbf{y} - 3\mathbf{x}$ **d** $\overrightarrow{CA} = \mathbf{y} - 3\mathbf{x}$

11 a $2\mathbf{x} + 4\mathbf{y}$ **b** $4\mathbf{y} - 2\mathbf{x}$
c $2\mathbf{y} - 2\mathbf{x}$ **d** $3\mathbf{x} - 4\mathbf{y}$

EXERCISE 64★ (page 230)

1 a $\overrightarrow{AB} = \mathbf{y} - \mathbf{x}$ **b** $\overrightarrow{AM} = \frac{1}{2}(\mathbf{y} - \mathbf{x})$
c $\overrightarrow{OM} = \frac{1}{2}(\mathbf{x} + \mathbf{y})$

3 a $\overrightarrow{AB} = \mathbf{y} - \mathbf{x}$; $\overrightarrow{OD} = 2\mathbf{x}$; $\overrightarrow{DC} = 2\mathbf{y} - 2\mathbf{x}$
b DC = 2AB and they are parallel lines

5 a $\overrightarrow{AB} = \mathbf{y} - \mathbf{x}$; $\overrightarrow{OC} = -2\mathbf{x}$; $\overrightarrow{OD} = -2\mathbf{y}$;
$\overrightarrow{DC} = 2\mathbf{y} - 2\mathbf{x}$
b DC = 2AB and they are parallel lines

7 $\overrightarrow{AB} = \mathbf{y} - \mathbf{x}$; $\overrightarrow{BC} = \mathbf{y} - 2\mathbf{x}$; $\overrightarrow{AD} = 2\mathbf{y} - 4\mathbf{x}$;
$\overrightarrow{BD} = \mathbf{y} - 3\mathbf{x}$

9 a $\overrightarrow{MA} = \frac{3}{5}\mathbf{x}$; $\overrightarrow{AB} = \mathbf{y} - \mathbf{x}$; $\overrightarrow{AN} = \frac{3}{5}(\mathbf{y} - \mathbf{x})$;
$\overrightarrow{MN} = \frac{3}{5}\mathbf{y}$
b OB and MN are parallel; $\overrightarrow{MN} = \frac{3}{5}\overrightarrow{OB}$

EXERCISE 65 (page 233)

1 $\mathbf{p} + \mathbf{q} = \begin{pmatrix} 6 \\ 8 \end{pmatrix}$; $\mathbf{p} - \mathbf{q} = \begin{pmatrix} -2 \\ -2 \end{pmatrix}$;

$2\mathbf{p} + 3\mathbf{q} = \begin{pmatrix} 16 \\ 21 \end{pmatrix}$

3 $\mathbf{p} + \mathbf{q} = \begin{pmatrix} 4 \\ 6 \end{pmatrix}$; $\mathbf{p} - \mathbf{q} = \begin{pmatrix} -2 \\ -2 \end{pmatrix}$;

$2\mathbf{p} + 5\mathbf{q} = \begin{pmatrix} 17 \\ 24 \end{pmatrix}$

5 $\mathbf{v}+\mathbf{w}=\begin{pmatrix}4\\5\end{pmatrix}, \sqrt{41}$

$2\mathbf{v}-\mathbf{w}=\begin{pmatrix}5\\-2\end{pmatrix}, \sqrt{29}$

$\mathbf{v}-2\mathbf{w}=\begin{pmatrix}1\\-7\end{pmatrix}, \sqrt{50}$

7 a Chloe $\begin{pmatrix}5\\7\end{pmatrix}$; Leo $\begin{pmatrix}4\\5\end{pmatrix}$; Max $\begin{pmatrix}3\\2\end{pmatrix}$

b Chloe: $\sqrt{74}$ km, 2.9 km/hour
Leo: $\sqrt{41}$ km, 2.1 km/hour
Max: $\sqrt{13}$ km, 1.2 km/hour

EXERCISE 65★ (page 234)

1 $\mathbf{p}+\mathbf{q}=\begin{pmatrix}5\\0\end{pmatrix}$, 5, 090°

$\mathbf{p}-\mathbf{q}=\begin{pmatrix}-1\\2\end{pmatrix}, \sqrt{5}$, 333°

$2\mathbf{p}-3\mathbf{q}=\begin{pmatrix}-5\\5\end{pmatrix}, \sqrt{50}$, 315°

3 $m=-1, n=-2$

5 a $\begin{pmatrix}10.4\\6\end{pmatrix}$ km b $\begin{pmatrix}13\\-7.5\end{pmatrix}$ km

7 a $m=1, n=3$ b $p=3, q=10$

ACTIVITY 21 (page 236)

Time	$t=0$	$t=1$	$t=2$	$t=3$	$t=4$
$\mathbf{r}$	$\begin{pmatrix}12\\5\end{pmatrix}$	$\begin{pmatrix}9\\9\end{pmatrix}$	$\begin{pmatrix}6\\13\end{pmatrix}$	$\begin{pmatrix}3\\17\end{pmatrix}$	$\begin{pmatrix}0\\21\end{pmatrix}$

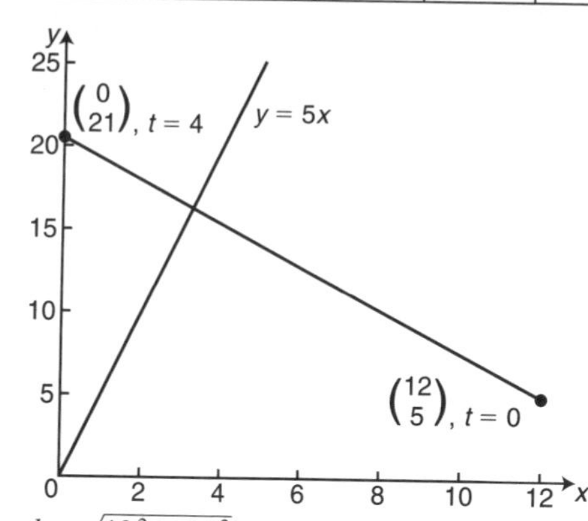

▸ $d=\sqrt{12^2+16^2}=20$ km

$V=\dfrac{20\text{ km}}{4\text{ min}}=300$ km/hour, 323.1°

▸ About 12:03

▸ $x=12-3t$

$y=5+4t$

$y=5x$

so

$5+4t=5(12-3t) \therefore t=\frac{55}{19}$

Boundary is crossed at 12:02:54

REVISION EXERCISE 66 (page 236)

1 $\mathbf{p}+\mathbf{q}=\begin{pmatrix}1\\5\end{pmatrix}; \sqrt{26}$

$\mathbf{p}-\mathbf{q}=\begin{pmatrix}5\\3\end{pmatrix}; \sqrt{34}$

$3\mathbf{p}-2\mathbf{q}=\begin{pmatrix}13\\10\end{pmatrix}; \sqrt{269}$

3 a $\overrightarrow{AB}=3\mathbf{y}+\mathbf{x}$ b $\overrightarrow{AC}=2\mathbf{y}+2\mathbf{x}$

c $\overrightarrow{CB}=-\mathbf{x}+\mathbf{y}$

5 a $\overrightarrow{AB}=\mathbf{y}-\mathbf{x}$ b $\overrightarrow{FB}=2\mathbf{y}-\mathbf{x}$

c $\overrightarrow{FD}=\mathbf{y}-2\mathbf{x}$

7 a $\begin{pmatrix}4\\1\end{pmatrix}$ b $\begin{pmatrix}6\\-4\end{pmatrix}$

c $\sqrt{29}$ d $v=1, w=2$

REVISION EXERCISE 66★ (page 238)

1 $m=3, n=1$

3 a $\overrightarrow{XM}\begin{pmatrix}-1\\3\end{pmatrix}$

$\overrightarrow{XZ}=\begin{pmatrix}-10\\6\end{pmatrix}$

b $v\begin{pmatrix}7\\3\end{pmatrix}$

c $\begin{pmatrix}8\\0\end{pmatrix}+w\begin{pmatrix}-10\\6\end{pmatrix}$

d $v=\dfrac{2}{3}, w=\dfrac{1}{3}$

5 a

t	$t=0$	$t=1$	$t=2$	$t=3$
$\mathbf{r}$	$\begin{pmatrix}12\\7\end{pmatrix}$	$\begin{pmatrix}1\\10\end{pmatrix}$	$\begin{pmatrix}-10\\13\end{pmatrix}$	$\begin{pmatrix}-21\\16\end{pmatrix}$

c 41 000 km/hour, 285.3°

Handling data 3

EXERCISE 67 (page 243)

1 f.d.: 0.20, 0.40, 0.90, 0.60, 0.45, 0.05

3 a f.d.: 3.5, 7, 10, 24, 38, 16, 9

b 6.5–7 kg c 55%

5 a f.d.: 3.5, 9.5, 12, 13.6, 10.4, 2.5

c $\bar{x}=26.8$ years

EXERCISE 67★ (page 245)

1 a f.d.: 0.04, 0.07, 0.087, 0.113, 0.024, 0.012

b 51.4% c $\bar{x}=368.5$

3 a f.d.: 10, 13, 15, 15, 13, 7, 7

b $\bar{x}=9.77$ yrs

c 6.5 cm, 7.5 cm, 7.5 cm, 6.5 cm, 3.5 cm, 3.5 cm

5 a 522 customers

b \$1566

c $\bar{x} = 52.5$ customers per hour. Not a useful statistic

d 10.00–12.00, 1 staff; 12.00–14.00, 4 staff; 14.00–18.00, 2 staff; 18.00–20.00, 3 staff

7 a 6, 8

b f.d.: 36, 17, 6, 1

c $\bar{x} = 97.7$ min

REVISION EXERCISE 68 (page 248)

1 a f.d.: 2.8, 8, 14.32, 20.64, 6, 1.56

b 9 cm

c 2.34 cm

REVISION EXERCISE 68★ (page 248)

1 a f.d.: 4, 0.5, 0.267, 1.2, 1.5, 0.7, 0.267, 0.133

b 2105 min = 35 hours 5 min

c 30 min 30 secs

d 67 min 54 secs

e \$11.84

Examination practice 3

1 \$4900

3 (4.73, 6.46), (1.27, −0.466)

5 a $-\frac{1}{2}$

b $x = 1$

c $f^{-1}: x \to \frac{1}{x} + 1$

d $x = 0$

7 a i 5 **ii** −3

b $x = -0.39$ or $x = 1.72$

9 a

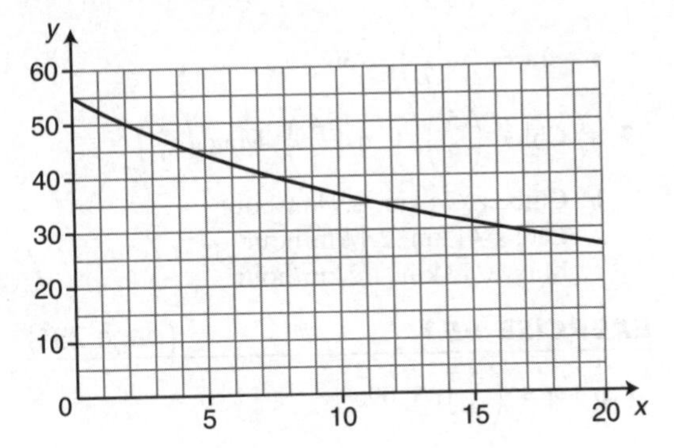

b $2 \leqslant m \leqslant 11.4$

c −1.5

11 a $2n$ **b** $2m$ **c** $2m - 2n$

d NM is parallel to AB and half the length

13 a $\begin{pmatrix} 10 \\ 25 \end{pmatrix}$ **b** 26.9 **c** 68.2° **d** $\begin{pmatrix} 25 \\ 27.5 \end{pmatrix}$

15 a freq 20, 70, 58, 60, 44, 16

b $\frac{2x}{5}\ \frac{y}{2}$

Unit 4

Number 4

EXERCISE 69 (page 259)

1 $\frac{57}{100}$
3 $\frac{47}{99}$
5 $\frac{7}{1}$
7 Irrational
9 $\frac{3}{1}$
11 $\frac{1}{1}$
13 Irrational
15 $\frac{0}{1}$
17 e.g. 2.5
19 e.g. $\sqrt{53}$
21 $\frac{2}{\pi}$

EXERCISE 69★ (page 260)

1 Irrational
3 $\frac{2}{5}$
5 $\frac{3}{5}$
7 $\frac{3}{2}$
9 Irrational
11 $\frac{1}{1}$
13 Irrational
15 e.g. 3.5
17 e.g. $\sqrt{7}$
19 e.g. $\sqrt{2} \times \sqrt{8}$
21 $\frac{9}{\pi}$
23 a e.g. $3:4:5$
b e.g. $\sqrt{2}:\sqrt{2}:2$
c e.g. $1:2:\sqrt{5}$
d e.g. $1:\sqrt{3}:2$

EXERCISE 70 (page 261)

1 $6\sqrt{5}$ 3 $4\sqrt{3}$ 5 32
7 20 9 8 11 105
13 $2\sqrt{2}$ 15 4 17 8

EXERCISE 70★ (page 261)

1 $5\sqrt{11}$ 3 $4\sqrt{7}$ 5 99
7 96 9 $56\sqrt{7}$ 11 $120\sqrt{2}$
13 9 15 4 17 6

EXERCISE 71 (page 263)

1 $2\sqrt{3}$ 3 $3\sqrt{2}$ 5 $4\sqrt{3}$ 7 $3\sqrt{5}$
9 $3\sqrt{3}$ 11 $6\sqrt{2}$ 13 $\sqrt{50}$ 15 $\sqrt{27}$
17 $\sqrt{54}$ 19 $\frac{1}{2}$ 21 $\frac{2}{5}$ 23 $\frac{2}{3}$
25 12, $10\sqrt{2}$, $\sqrt{26}$

EXERCISE 71★ (page 263)

1 $2\sqrt{7}$ 3 $3\sqrt{11}$ 5 $4\sqrt{5}$ 7 $3\sqrt{13}$
9 $5\sqrt{3}$ 11 $5\sqrt{3}$ 13 $\sqrt{75}$ 15 $\sqrt{80}$
17 $\sqrt{63}$ 19 $\frac{1}{6}$ 21 $\frac{9}{10}$ 23 $\frac{7}{13}$
25 8, $8\sqrt{2}$

ACTIVITY 22 (page 264)

a i $\sqrt{2}$ ii 45°
iii

$\sin 45° = \frac{1}{\sqrt{2}}$	$\cos 45° = \frac{1}{\sqrt{2}}$	$\tan 45° = 1$

b i $\sqrt{3}$ ii 60°, 30°
iii

$\sin 30° = \frac{1}{2}$	$\cos 30° = \frac{\sqrt{3}}{2}$	$\tan 30° = \frac{1}{\sqrt{3}}$
$\sin 60° = \frac{\sqrt{3}}{2}$	$\cos 60° = \frac{1}{2}$	$\tan 60° = \sqrt{3}$

EXERCISE 72 (page 265)

1 $3 + 2\sqrt{2}$ 3 $4 - 2\sqrt{3}$
5 $21 + 12\sqrt{3}$ 7 $27 - 18\sqrt{2}$
9 -4 11 $5 + 2\sqrt{6}$
13 $1 + \sqrt{2} - \sqrt{5} - \sqrt{10}$

EXERCISE 72★ (page 265)

1 $9 + 4\sqrt{5}$
3 $18 - 8\sqrt{2}$
5 13
7 $41 + 24\sqrt{2}$
9 $10 - 2\sqrt{21}$
11 2
13 $15 + 10\sqrt{2} - 6\sqrt{7} - 4\sqrt{14}$
15 $4\sqrt{3}$, 2, $2\sqrt{2}$

EXERCISE 73 (page 266)

1 $\frac{\sqrt{3}}{3}$ 3 $\frac{\sqrt{5}}{5}$ 5 $\sqrt{3}$ 7 $2\sqrt{2}$
9 $\frac{2\sqrt{3}}{3}$ 11 $\frac{3\sqrt{2}}{4}$ 13 $\frac{2+\sqrt{2}}{2}$

ACTIVITY 23 (page 266)

a By long division b $\frac{\sqrt{2}}{2}$
c 0.707 107 d $\sqrt{2} + 1$
e $\frac{\sqrt{2} - 1}{3}$

EXERCISE 73★ (page 267)

1 $\frac{\sqrt{13}}{13}$ 3 $\sqrt{a}$ 5 $2\sqrt{3}-1$
7 $\frac{\sqrt{3}}{3}$ 9 $2+\sqrt{5}$ 11 $\frac{-(1+\sqrt{5})}{4}$
13 $\frac{5-4\sqrt{2}}{7}$

REVISION EXERCISE 74 (page 267)

1 $0.\dot{3}$ and $\sqrt{25}$ 3 e.g. $\sqrt{11}$
5 $5\sqrt{3}$ 7 18
9 $2\sqrt{2}$ 11 $3\sqrt{3}$
13 $6+2\sqrt{5}$ 15 -1
17 $2\sqrt{3}$ 19 $1+\sqrt{3}$

REVISION EXERCISE 74★ (page 268)

1 $(\sqrt{3})^2$ and $0.\dot{2}\dot{3}$ 3 e.g. $\sqrt{37}$
5 $8\sqrt{5}$ 7 75
9 $4\sqrt{3}$ 11 $19\sqrt{6}$
13 8 15 $\sqrt{14}-5$
17 $2\sqrt{6}$ 19 2
21 $5\sqrt{3}$, 18

Algebra 4

EXERCISE 75 (page 269)

1 $\frac{3}{2}$ 3 $\frac{x}{y}$ 5 $x+2$ 7 $\frac{1}{x+2}$
9 $\frac{x+y}{x-y}$ 11 $\frac{x+1}{x-3}$ 13 $\frac{x+3}{x+2}$

EXERCISE 75★* (page 270)

1 $\frac{3}{5}$ 3 $\frac{y}{x}$ 5 $\frac{x-4}{x+3}$ 7 $\frac{x+3}{x+4}$
9 $\frac{r-3}{r+1}$ 11 $\frac{a-b}{a+b}$ 13 $\frac{2(x+8)}{y}$

EXERCISE 76 (page 271)

1 $\frac{5x+3}{6}$ 3 $\frac{x-2}{4}$ 5 $\frac{8x+12}{15}$ 7 $\frac{5x+3}{4}$
9 $\frac{1-4x}{5}$ 11 $\frac{7x+2}{12}$ 13 $\frac{3x-17}{10}$ 15 $\frac{3x+2}{6}$
17 $\frac{39-2x}{12}$ 19 $\frac{23x-11}{10}$

EXERCISE 76★ (page 272)

1 $\frac{9x+13}{10}$ 3 $\frac{1-2x}{12}$
5 $\frac{16-3x}{14}$ 7 $\frac{6x-2}{35}$
9 $\frac{x-230}{15}$ 11 $\frac{x-6}{72}$
13 $\frac{47x-22}{60}$ 15 $\frac{5x-8}{6}$
17 $\frac{23x+7}{18}$ 19 $\frac{59-78x}{10}$

EXERCISE 77 (page 274)

1 $\frac{5}{6x}$ 3 $\frac{1}{4x}$
5 $\frac{x-4}{2(x-2)}$ 7 $\frac{2x}{(x-1)(x+1)}$
9 $\frac{x+8}{(x-1)(x+2)}$ 11 $\frac{x^2+x+2}{x(x+2)}$
13 $\frac{2x+3}{(x+2)(x+1)}$

EXERCISE 77★ (page 274)

1 $\frac{17}{15x}$ 3 $\frac{x-1}{(x+3)(x+2)}$
5 $\frac{x^2+x+1}{1+x}$ 7 $\frac{1}{x+1}$
9 $\frac{x+2}{x+1}-\frac{x-1}{x-2}$ 11 $\frac{1}{x+1}$
13 $\frac{3x+4}{(x+1)^2}$ 15 $\frac{8}{(x-3)(x+1)}$

EXERCISE 78 (page 276)

1 21 3 2
5 $\frac{1}{2}$ 7 3
9 -8 11 $-\frac{2}{3}$
13 $\frac{1}{2}$ 15 230
17 -6 19 0

EXERCISE 78★ (page 276)

1 $\frac{2}{5}$ 3 15 5 $\frac{1}{3}$
7 0 9 15 11 7
13 1 15 $\frac{7}{5}$ 17 $-\frac{8}{3}$
19 $-\frac{5}{13}$ 21 6 km

EXERCISE 79 (page 279)

1 $-7, 2$
3 $1, 2$
5 6
7 $-\frac{2}{3}$
9 $-4, 5$
11 $0, 6$
13 $-1, 4$
15 $-\frac{1}{3}, 2$

EXERCISE 79★ (page 279)

1 $-3.5, 1$
3 $-4, 1$
5 4
7 $-6, 6$
9 $-\frac{7}{3}, 2$
11 $-0.768, 0.434$
13 $-2, 6$
15 60
17 -2.5
19 $-\frac{2}{3}, 5$

REVISION EXERCISE 80 (page 280)

1 3
3 $\frac{x+3}{x-3}$
5 $\frac{x-3}{6}$
7 $\frac{3x-21}{20}$
9 $\frac{-3}{(x+1)(x-2)}$
11 $\frac{x^2+2x-4}{x(x-2)}$
13 $\frac{8}{3}$
15 -7
17 $-2, 1$
19 $-8, 2$

REVISION EXERCISE 80★ (page 281)

1 $\frac{2}{3}$
3 $\frac{x+4}{x-7}$
5 $\frac{x-11}{12}$
7 $\frac{5x+7}{12}$
9 $\frac{3x^2+7x+24}{(x+3)(3x-1)}$
11 $\frac{-1}{(x-4)(x+1)}$
13 -1
15 $\frac{15}{61}$
17 $\frac{1}{3}$ or 2
19 -0.464 or 6.46

Graphs 4

EXERCISE 81 (page 284)

1 $\frac{dy}{dx} = 3x^2$
3 $\frac{dy}{dx} = 5x^4$
5 $\frac{dy}{dx} = 8x^7$
7 $\frac{dy}{dx} = 10x^9$
9 $\frac{dy}{dx} = -x^{-2}$
11 $\frac{dy}{dx} = -x^{-2}$
13 4
15 32
17 448

EXERCISE 81★ (page 285)

1 $\frac{dy}{dx} = 12x^{11}$
3 $\frac{dy}{dx} = 1$
5 $\frac{dy}{dx} = -3x^{-4}$
7 $\frac{dy}{dx} = -6x^{-7}$
9 $\frac{dy}{dx} = \frac{1}{2}x^{-\frac{1}{2}}$
11 $\frac{dy}{dx} = \frac{1}{2}x^{-\frac{1}{2}}$
13 $-\frac{3}{16}$
15 $-\frac{1}{4}$
17 $\frac{1}{4}$
19 0

EXERCISE 82 (page 286)

1 $\frac{dy}{dx} = 8x$
3 $\frac{dy}{dx} = 6x^2$
5 $\frac{dy}{dx} = 15x^4$
7 $\frac{dy}{dx} = -6x^{-3}$
9 $\frac{dy}{dx} = 2x^{-2}$
11 $\frac{dy}{dx} = 5x^4 + 2x$
13 $\frac{dy}{dx} = 6x^2 + 4$
15 $\frac{dy}{dx} = 9x^2 + 8x^3$
17 $\frac{dy}{dx} = 4x$
19 $\frac{dy}{dx} = -x^{-2}$
21 $\frac{dy}{dx} = -6x^{-4}$
23 $\frac{dy}{dx} = 2x - 2x^{-3}$
25 $\frac{dy}{dx} = 3x^2 + 3x^{-4}$

EXERCISE 82★ (page 287)

1 $\frac{dy}{dx} = 3x^2 + 4x$
3 $\frac{dy}{dx} = 2x + 4$
5 $\frac{dy}{dx} = 2x - 1$
7 $\frac{dy}{dx} = 4x + 7$
9 $\frac{dy}{dx} = 2x$
11 $\frac{dy}{dx} = 2x + 6$
13 $\frac{dy}{dx} = 8x - 4$
15 $\frac{dy}{dx} = 8x - 2x^{-3}$
17 $\frac{dy}{dx} = -6x^{-4}$
19 $\frac{dy}{dx} = 8x^3 + 2x^{-3}$
21 $\frac{dy}{dx} = 5$
23 $(3, -9)$

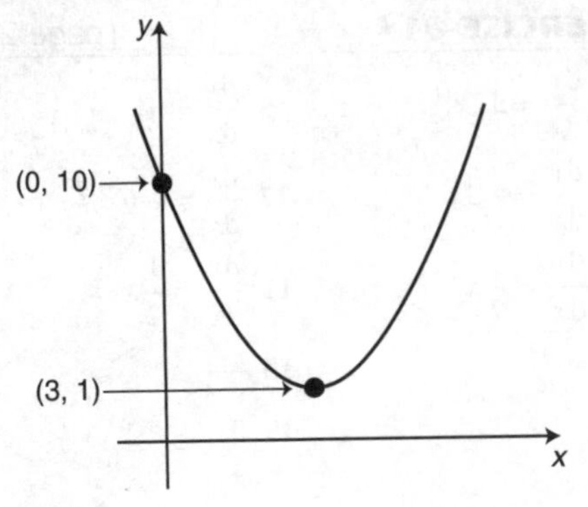

25 (3, 1)

EXERCISE 83 (page 290)

1 a $2x+4$ **b** $(-2, 6)$
3 a $x=3$ **b** $y=24$
5 a 15 **b** 6 **c** $y=6x+9$
7 a -1 **b** 10 **c** $y=10x+19$
9 a $8t+8$
b i 16° per min **ii** 48° per min
11 a $3x^2-12$
b $x=-2, 2$
c $(-2, 21), (2, -11)$
d Maximum: $(-2, 21)$, minimum: $(2, -11)$; Curve-shape
13 a $6-2x$ **b** $(1, 20)$ **c** $y-4x=16$
15 a $6-2x$ **b** 20
17 a $-2-2x$ **b** 9
19 a $4-x^{-2}$ **b** $x=-\frac{1}{2}, \frac{1}{2}$
c $\left(-\frac{1}{2}, -4\right), \left(\frac{1}{2}, 4\right)$

EXERCISE 83★ (page 293)

1 a $6x-7$ **b** $y=5x-7$
3 a $-1, -3$ **b** $2x+4$ **c** $-2, 2$
d $y=-2x-6$ and $y=2x+2$
5 Minimum at $(-3, 54)$, maximum at $(3, 54)$
7 a 11 150 **b** 305 per year
9 a $80-40t$ **b** $t=2$
c 350° **d** 120° per hour
11 a $6x^2-2x-4$ **b** $-\frac{2}{3}$
c $(-\frac{2}{3}, 11\frac{17}{27}), (1, 7)$
d Maximum: $(-\frac{2}{3}, 11\frac{17}{27})$, minimum: $(1, 7)$; Curve-shape
13 a $2x-16x^{-2}$ **b** $x=2$ **c** $(2, 12)$
15 $(-\frac{1}{3}, 10\frac{5}{27}), (1, 9)$. Both y co-ordinates are positive.
17 a $3-27t^{-2}$ **b** 18 °C **c** 1.92 °C/month
19 40

EXERCISE 84 (page 297)

1 $10t$
3 a $40+10t$ **b** 70 m/s
5 32
7 a $v=3t^2+8t-5$ **b** $a=6t+8$
c $v=6$ m/s; $a=14$ m/s^2
9 a $2t+10$ **b** 14 m/s

EXERCISE 84★ (page 298)

1 a $v=8t+\frac{2}{t^2}$ **b** $a=8-\frac{4}{t^3}$
3 a $20+10t$ **b** 30 m/s, 40 m/s, 50 m/s
c 10 m/s^2
5 a $40-10t$ **b** 30 m/s, 20 m/s, 10 m/s, 0 m/s
c 80 m
7 a $\frac{ds}{dt}=-2+\frac{18}{t^2}$; $\frac{dv}{dt}=\frac{-36}{t^3}$
b 3 secs
c 28 m
9 a 8 secs **b** 3 secs **c** 170 m

REVISION EXERCISE 85 (page 300)

1 $\frac{dy}{dx}=5x^4+6x^2$
3 a $y=4$; $\frac{dy}{dx}=3$ **b** $y=3x-5$
5 a $3x^2+6x-9$ **b** $x=1, -3$
c Maximium at $(-3, 32)$; Minimum at $(1, 0)$

REVISION EXERCISE 85★ (page 300)

1 $\frac{dy}{dx}=8x^3-6x$
3 0
5 a $\frac{dy}{dx}=6x^2+6x-36$
b $(-3, 81), (2, -44)$
7 a $\frac{ds}{dt}=3t^2+5$; $\frac{dv}{dt}=6t$
b 18 m, 42 m
c 24 m/s

Shape and space 4

ACTIVITY 24 (page 302)

- Sine ratio

x	36°	72°	108°	144°	180°
$\sin x$ (calc)	0.588	0.951	0.951	0.588	0

$\sin x$	0	0.2	0.5	0.8	1.0
x (calc)	0°	11.5° 169°	30° 150°	53.1° 127°	90°

◆ Cosine ratio

x	36°	72°	108°	144°	180°
$\cos x$ (calc)	0.809	0.309	−0.309	−0.809	−1

$\cos x$	0	0.2	0.5	0.8	1.0
x (calc)	90°	78.5°	60°	36.9°	180°

◆ Tangent ratio

x	0°	60°	85°	95°	180°
$\tan x$ (calc)	0	1.73	11.4	−11.4	0

$\tan x$	0	0.5	5	10	−10
x (calc)	0	26.6°	78.7°	84.3°	−84.3°

EXERCISE 86 (page 304)

1 $x = 17.5°, 163°$
3 $x = 72.5°$
5 $x = 71.6°$
7 $x = 107°$
9 $x = 198°, 343°$
11 $x = 108°, 288°$

EXERCISE 86★ (page 304)

1 $x = 27.2°, 153°$
3 $x = -153°, -27.2°, 207°, 333°$
5 $x = -75.7°, 75.7°, 284°$
7 $x = -112°, 112°, 248°$
9 $x = -112°, 67.9°, 248°$
11 $x = -55.5°, 125°, 305°$

EXERCISE 87 (page 307)

1 $x = 5.94$
3 MN = 39.0 cm
5 AC = 37.8 cm
7 $x = 37.3°$
9 ∠ABC = 38.8°
11 ∠ACB = 62.2°

EXERCISE 87★ (page 309)

1 $x = 29.7$
3 ∠LMN = 67.4°, 113°
5 EF = 10.4 cm, ∠DEF = 47.5°, ∠FDE = 79.0°
7 13 km
9 BC = 261 m
11 PR = 115 m; 112 m

EXERCISE 88 (page 311)

1 $a = 7.26$
3 AB = 39.1 cm
5 RT = 24.2 cm
7 $X = 73.4°$
9 ∠ABC = 92.9°

EXERCISE 88★ (page 312)

1 $x = 9.34$
3 ∠XYZ = 95.5°
5 ∠BAC = 81.8°
7 QR = 4.18 cm, ∠PQR = 39.2°, ∠QRP = 62.8°
9 11.6 km

EXERCISE 89 (page 315)

1 a 9.64 b 38.9°
3 a 54.9° b 88.0° c 37.1°
5 a 7.88 b 6.13
7 a 79.1° b 7.77
9 a 16.8 km b 168°

EXERCISE 89★ (page 316)

1 247 km, 280°
3 a 50.4° b 7.01 m c 48.4°
5 ∠BXA = 75.9°
7 CS = 2.64 km; 040.2°
9 a $x = 9.23$ cm b BCD = 38.7°
11 ACB = 32.0°

EXERCISE 90 (page 318)

1 7.39 cm^2
3 36.2 cm^2
5 121 cm^2

EXERCISE 90★ (page 319)

1 173 cm^2
3 16.5 cm
5 53.5 cm^2

EXERCISE 91 (page 321)

1 a 11.7 cm b 14.2 cm c 34.4°
3 a 14.1 cm b 17.3 cm c 35.4°
5 a 4.47 m b 4.58 m c 29.2° d 12.6°
7 a 43.3 cm b 68.7 cm c 81.2 cm

EXERCISE 91★ (page 324)

1 a 16.2 cm b 67.9° c 55.3 cm^2
3 a 30.3° b 31.6° c 68.9°
5 a 15 m b 47.7° c €91 300
7 46.5 m

REVISION EXERCISE 92 (page 326)

1 a 22.9° b 22.9
3 a 50°, 60°, 70° b AB = 4.91 km
5 a AC = 42.4 cm b 33.9 cm
c 68.0° d 58.0°

REVISION EXERCISE 92★ (page 327)

1 BH = 506 m
5 a AC = 70.7 cm b 98.7 cm
c 27.9° d 216 000 cm^3

Handling data 4

EXERCISE 93 (page 332)

1 a 0.655 b 0.345
3 a $\frac{1}{36}$ b $\frac{5}{18}$
5 a $\frac{2}{9}$ b $\frac{5}{9}$ c $\frac{1}{9}$

EXERCISE 93★ (page 333)

1 a $\frac{1}{11}$ b $\frac{1}{3}$ c $\frac{3}{11}$
d $\frac{9}{55}$

3 a $\frac{1}{16}$ b $\frac{1}{4}$ c $\frac{15}{16}$

5 a $p(RR) = \frac{3}{5} \times \frac{5}{9} = \frac{1}{3}$

b

Outcome		Probability
Bag X	Bag Y	
4R + 4W	5R + 5W	$\frac{1}{3}$
5R + 3W	4R + 6W	$\frac{8}{15}$
6R + 2W	3R + 7W	$\frac{2}{15}$

c (i) $p(i) = p(WY \to X) = \frac{1}{3} \times \frac{1}{2} = \frac{1}{6}$
(ii) $p(ii) = p(RY \to X) = \frac{2}{15} \times \frac{3}{10} = \frac{1}{25}$
(iii) $p(iii) = p(RY \to X \text{ or } WY \to X)$
$= \frac{8}{15} \times \frac{4}{10} + \frac{2}{15} \times \frac{7}{10} = \frac{23}{75}$
(iv) $p(iv) = p(RY \to X \text{ or } WY \to X)$
$= \frac{1}{3} \times \frac{1}{2} + \frac{8}{15} \times \frac{6}{10} = \frac{73}{150}$

ACTIVITY 26 (page 335)

It is important to tell the respondents that they should keep their die score secret and tell the truth to all the questions. Clearly, the greater the sample, the more likely the chance of the final results reflecting those of the parent population.

Suggestions for possible questions:
Do you like mathematics?
Have you ever played truant from school?
Have you ever cheated in a test at school?

INVESTIGATE (page 336)

Obviously the answers to both situations could be guessed by knowing the probabilities. These investigations are to try to prove the results practically. Also, an IT application should be used to simulate a larger sample than can be generated manually and to find the sum of an infinite series!

For one die, practically, simply use

$$\bar{x} = \frac{\Sigma fx}{\Sigma f}$$

Theoretically, the frequencies that arise are directly related to their associated probabilities. So, given that f is equivalent to the probability of x occurring, p, we can write down

$$\bar{x} = \frac{\Sigma px}{\Sigma p} = \Sigma px \quad \text{as } \Sigma p = 1$$

$$= \frac{1}{6} \times 1 + \frac{5}{6} \times \frac{1}{6} \times 2 + \frac{5}{6} \times \frac{5}{6} \times \frac{1}{6} \times 3 + \cdots$$

This can be investigated using a spreadsheet. The theoretical proof is an infinite geometric progression, which requires subtle manipulation! The result is $\bar{x} = 6$.

For two dice, practically, this data collection will prove too tedious. A computer simulation would prove a better use of time.

Again, using the same logic as above:

$$\bar{x} = \frac{\Sigma px}{\Sigma p} = \Sigma px \quad \text{as } \Sigma p = 1$$

$$= \frac{1}{36} \times 1 + \frac{35}{36} \times \frac{1}{36} \times 2$$
$$+ \frac{35}{36} \times \frac{35}{36} \times \frac{1}{36} \times 3 + \cdots$$

This can be investigated using a spreadsheet. The theoretical proof is again an infinite geometric progression, which requires subtle manipulation! The result is $\bar{x} = 36$.

REVISION EXERCISE 94 (page 337)

1 a (i) $\frac{4}{9}$ (ii) $\frac{4}{9}$ b $\frac{7}{27}$

3 a $\frac{6}{25}$ b $\frac{19}{25}$ c $\frac{12}{43}$
d $\frac{31}{43}$ e 0.320

5 a 0.9 b 0.1 c 0.35

REVISION EXERCISE 94★ (page 338)

1 a $\frac{48}{125}$ b $\frac{12}{125}$ c $\frac{61}{125}$

3 a $\frac{19}{45}$
b (i) $\frac{2}{15}$ (ii) $\frac{8}{105}$

5 b (i) $\frac{9}{16}$ (ii) $\frac{7}{16}$ c $\frac{1}{4}$

Examination practice 4

1 a R b I c R d R e R

3 a $5\sqrt{2}$ b 18 c 12 d $2\sqrt{2}$

5 a $21 + 12\sqrt{3}$ b -3

7 $2\sqrt{3}$

9 a $x + 3$ b $\frac{x+2}{x-2}$ c $\frac{x-1}{x+2}$

11 a $\frac{1}{2x}$ b $\frac{5}{(x-2)(x+3)}$
c $\frac{-1}{(x-4)(x+1)}$

13 a -3 or 2 b $\frac{1}{3}$ or 2 c -2.24 or 6.24

15 a $\frac{dy}{dx} = x^2 + 2x - 7$ **b** $(-4, 17\frac{2}{3})$, $(2, -12\frac{1}{3})$

17 a $v = 40 - 10t$ **b** $t = 4$ seconds

c 80 m

19 a 47.1° **b** 131 cm^2

21 61.1°

23 a $\frac{8}{15}$ **b** $\frac{1}{12}$ **c** $\frac{1}{15}$ **d** $\frac{49}{360}$

Unit 5

Number 5

REVISION EXERCISE 95 (page 355)

1 a $\frac{7}{12}$ b $1\frac{1}{3}$ c 24 d 12
3 14.235, 14.25, 14.3, 14.532
5 a $504 = 2^3 \times 3^2 \times 7$
b HCF = 3, LCM = 210
7 $\frac{181}{333}$
9 49.3
11 a 37.625 b 37.62 c 1.398 d 1.40
13 a 5×10^4 b 5×10^1
15 3.05×10^{-3} mm
17 a 16 b $\frac{1}{3}$ c $\frac{1}{4}$ d 9
19 a $4\sqrt{3}$ b 1 c 12
21 a $48 000 b 12.5%
23 a 9.39 billion people b 1.40%
25 3.2 kg
27 45%
29 15%
31 a 329 blocks b 820 tonnes c 12.5 years
33 a 8565 gallons/second
b 39 m^2/second
35 6429 dominoes
37 $408
39 a $6.17 b $4.98

REVISION EXERCISE 95★ (page 358)

1 $\frac{2}{3} = \frac{24}{26}$, $\frac{13}{18} = \frac{26}{36}$, $\frac{3}{4} = \frac{27}{36}$, $\frac{7}{9} = \frac{28}{36}$
3 a $180 = 2^2 \times 3^2 \times 5$, $84 = 2^2 \times 3 \times 7$
b LCM = 1260, HCF = 12
5 10 million grains
7 a 8×10^{-3} b 9×10^{-3}
9 3:14
11 a 6.4 b 2.8
13 a $a^4, a = 2$ b $a^{-\frac{1}{3}}, a = 8$ c $2a^{-2}, a = \frac{1}{4}$
15 a $\frac{5\sqrt{2}}{2}$ b $3(2 - \sqrt{3})$
17 e.g. $\sqrt{10}$ or $\sqrt{13}$
19 a 0.029 b 0.409%
21 $11 \times 11 = 121$, $111 \times 111 = 12321$, etc.
Therefore after cancelling:
fraction $= (111\,111\,111)^2$
$= 12\,345\,678\,987\,654\,321$
23 30 tonnes
25 a 33.3% b (i) 38.5 °C (ii) 37 °C
27 To reduce 56 kg by 10%, 0.9 kg
29 20%
31 'Open' after 16 years, 'closed' after 100 years
33 a 16.4 cm/day b 4.17×10^{10} m^3
c 110 m
35 5.50 g/cm^3
37 112 mph
39 a 1:64 b 64 spheres
c 1:16 d 4:1

Algebra 5

REVISION EXERCISE 96 (page 373)

1 a x^{10} b x^2 c y^2 d $4x^6y^4$
3 a $z^2 + 3z$ b $14 - x$
5 a $(x + 2)(x - 2)$ b $(x + 1)(x + 2)$
7 a $v - \frac{1}{m}$ b $\sqrt{\frac{12}{\pi h}}$
9 a 3.14 b $g\left(\frac{T}{2\pi}\right)^2$
11 a 3 b 2 c 2
13 10
15 15
17 a $y = 4x$ b 8 c 2
19 a $y = \frac{24}{x}$ b 12 c 4
21 (3, 5)
23 a $x + y = 17$, $4x + 2y = 58$ b 5
25 (3, 4), (−4, −3)
27 (1, 2), (−2, −1)
29 $x = 5$, area 48
31 −2, −1, 0, 1
33 a (i) 4 (ii) $\frac{x + 2}{3}$ (iii) $9x - 8$
b 3 c 1
35 a No b (i) all reals (ii) $y \geqslant 0$
c (i) 3 (ii) $x^2 - 1$ d 0, 3

REVISION EXERCISE 96★ (page 376)

1 a $\frac{11a}{12}$ b $\frac{3}{8}$ c b d $\frac{21 - 2x}{15}$
3 a $x + 1$ b x
5 a $(3x - 1)(x + 11)$ b $\frac{3x - 1}{x - 11}$
7 a 27.6 b $\sqrt{\frac{3I}{M} - b^2}$
9 a $\frac{15}{8}$ b $\frac{fv}{v - f}$ c 12
11 $\frac{x + 123}{x + 456} = \frac{1}{2}$, 210
13 a $x^2 + 25 = l^2$
b $(x + 2)^2 + 16 = l^2$
c 5.15 m to 3 s.f.
15 a $y = \frac{40}{x^2}$ b 640 c 12.6
17 (−2, 2)

19 a $2x + 3y = 180$ **b** $2x + 2y = 140$
c $x = 30, y = 40$
21 a $x - 1 = y + 1, x + 1 = 2(y - 1)$
b $x = 7, y = 5$
23 a 2.5 cm equals 52.75 minutes, so yes
b 2.63 cm , 3.52 cm, 4.22 cm
25 b $-6, 4$ **c** 5
27 (3, 1), (2, −1)
29 a (3, 4) **b** tangent
31 −1, 0, 1, 2
33 a $1 \leqslant x \leqslant 3$ **b** $x < -1$ or $x > 2$
35 a $2.5x + 3 < 3x, 4x + 1 > 1.5x + 2,$
$4.5x + 2 > x + 1$
b $\frac{2}{5} < x < 6$
37 a -2 **b** $\frac{2x}{1 - x}$ **c** −1 or 0

Graphs and sequences 5

REVISION EXERCISE 97 (page 386)

1 a 27, 33, 39 **b** 297 **c** $6n - 3$
3 a 16, 22, 29 **b** $a = 5, b = 30$
5 a

Number of white tiles (w)	1	2	3	4	5
Number of blue tiles (b)	8	**10**	**12**	**14**	**16**

b $b = 2w + 6$ **c** 56 blue tiles
d 60 white tiles
7 a 1, 4, 8, 13, 19, 26
b Add one more than before
c Size 12
9 a AB, $\frac{1}{2}$: BC, $\frac{-3}{4}$: AC, −2
b AB perpendicular to AC
c $AB = \sqrt{5}$, $BC = \sqrt{25}$
d Area = 5 units squared

11

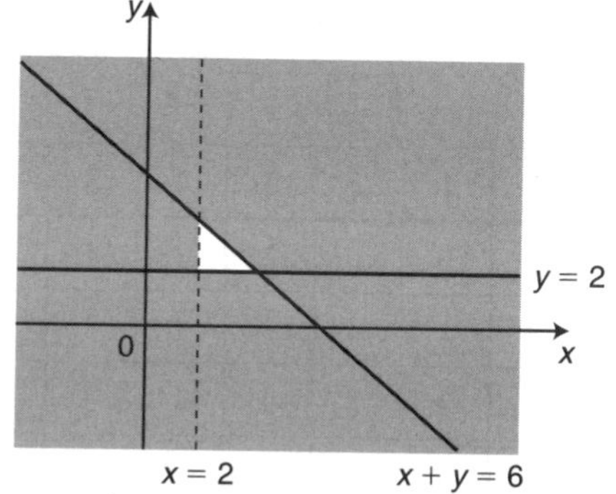

13 a $x \leqslant 8$ **b** $x < 8$

15 a

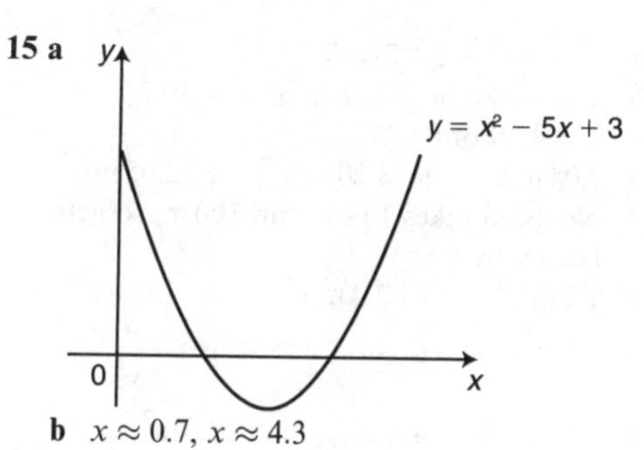

b $x \approx 0.7, x \approx 4.3$
17 a $y = 1$ **b** $y = x + 2$
19 a

x	−2	−1	1	2	3
y	−3	2	6	17	42

b

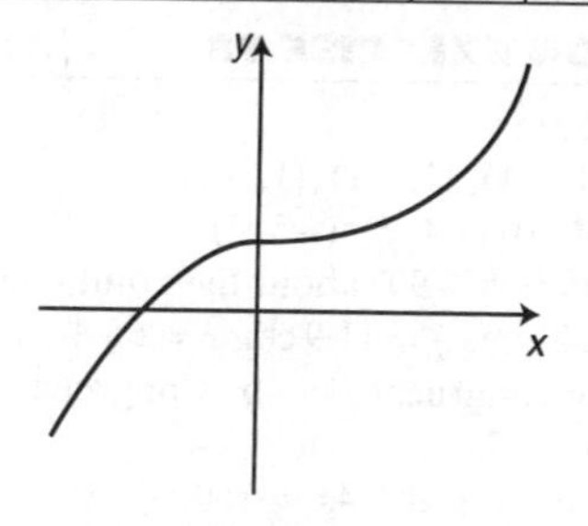

c 6
21 a $2x$ **b** 3 **c** $3x^2 + 6x$
23 a $2x - 4$ **b** $x = 2, y = -3$
c (2, −3) is a minimum

REVISION EXERCISE 97★ (page 389)

1 a 185, 181, 177 **b** 5 **c** $-4n + 205$
3 a 20 terms **b** $a = 2, b = 1$
5 a $w = 5c + 1$

No. of cells (c)	1	2	3	4	5
No. of walls (w)	6	**11**	**16**	**21**	**26**

b

No. of rows (r)	1	2	3	4	5
No. of cells (c)	1	**3**	**6**	**10**	15
No. of walls (w)	6	**15**	**27**	42	**60**
$r + c$	2	**5**	**9**	**14**	**20**

c $w = 3(r + c)$
d $w = \frac{1}{2}(3r^2 + 9r)$
e 195 walls
7 a $y = x + 3$ **b** $y = x - 3$
9 a

x	0	4	8	12	16	20
y	0	8.8	11.2	7.2	−3.2	−20

b 31.3 m **c** 15 m
11 c $x = 0.74$
d $x = -2, x = 0,$ or $x = 1$

13 a $x^3 - 3x^2 - x + 2 = 0$
b $x = -0.86$ or $x = 0.75$ or $x = 3.12$
c $x = 4.30$ or 0.697
15 a $7.04\,\text{m/s}$ **b** $8.20\,\text{m/s}$ **c** $2.05\,\text{m/s}^2$
d No. Sidd takes 15 s to run 100 m, so gets beaten by 0.8 s.
17 a $1.2\,\text{m/s}^2$ **b** $2.4\,\text{m/s}^2$ **c** $15\,\text{m/s}$
19 a $2 - \frac{3}{x^2}$ **b** $8x + 4$ **c** $-\frac{4}{x^3}$
21 b $\frac{dA}{dx} = 150 - 4x$, $A_{max} = 2812.5\,\text{m}^2$, $x = 37.5\,\text{m}$

Shape and space 5

REVISION EXERCISE 98 (page 404)

1 b 215 km
3 a (−1, −1), (1, −1), (1, −2)
b (−4, 3), (−4, 5), (−3, 5)
c Rotation +90° about the point (−3, −2)
5 $p = 5.40\,\text{cm}$, $q = 11.9\,\text{cm}$, $r = 65.4°$, $s = 41.8°$
7 a Not congruent **b** Congruent, SSS
9 a $160° - 7x$
b $4x = 3x + 20°$, $4x = 160° - 7x$
$160° - 7x = 3x + 20°$
c $x + 20°$, 32°, 35°
11 a 2.4 cm **b** $1\frac{1}{3}$ cm
13 $6200\,\text{cm}^3$
15 a $x = 100°$, $y = 90°$ **b** $x = 70°$, $y = 76°$
c $x = 110°$, $y = 35°$ **d** $x = 65°$, $y = 35°$
17 a ∠TPQ = 66° (Triangle TPQ is isosceles)
∠OPT = 90° (TP is tangent)
∴ ∠QPO = 24°
b ∠OPS = 42° (Isosceles triangle)
∴ ∠QPS = 66°
c ∠QRS = 114° (Opposite angles of a cyclic quadrilateral)
19 $\overrightarrow{AB} = \mathbf{y} - \mathbf{x}$; $\overrightarrow{OM} = \frac{1}{2}(\mathbf{x} + \mathbf{y})$; $\overrightarrow{ON} = \frac{1}{2}(\mathbf{y} - \mathbf{x})$
$\overrightarrow{MN} = -\mathbf{x}$ (MN is parallel to AC)
21 a SA = 2.63 km, BS = 6.12 km
b 28.1 minutes
23 a 40 m **b** 26.5 m **c** 4.68 m

REVISION EXERCISE 98★ (page 409)

1 a PQ = 10.7 m
b QR = 9 m
c RS = 7 m
d Area PQRS = $97.2\,\text{m}^2$
3 a CP = 30 m **b** QB = 4 m **c** 10 m
5 $x = 6\frac{2}{3}$, $y = 9$
7 a $\sqrt{13}$ **b** $p = 1$, $q = 2$
c (−1, −1) 90 degrees clockwise
d Centre (3, 1), SF 2

9 Perimeter = $18(1 + \sqrt{3})$ cm
11 $x = y = 51°$, $z = 78°$
13 a $x + 2x - 24 = 0$
b $x = 4\,\text{cm}$ **c** 4.90 cm
15 5 cm
17 a (i) $\frac{1}{2}\mathbf{b}, \frac{1}{2}\mathbf{a}$, (ii) $\frac{1}{2}\mathbf{a} - \frac{1}{2}\mathbf{b}$
(iii) $\mathbf{a} + \frac{1}{2}\mathbf{c}, \frac{1}{2}\mathbf{a} + \frac{1}{2}\mathbf{b} + \frac{1}{2}\mathbf{c}$
(iv) $\frac{1}{2}\mathbf{a} - \frac{1}{2}\mathbf{b}$
b PQ = SR; PQ // SR
19 a CD = 14.4 m
b CE = 142 m
c 12.7 m/s

Handling data 5

REVISION EXERCISE 99 (page 419)

1 a (i) $A \cap B = \{e, g\}$
(ii) $A \cup B = \{a, c, d, e, f, g, i, k\}$
b
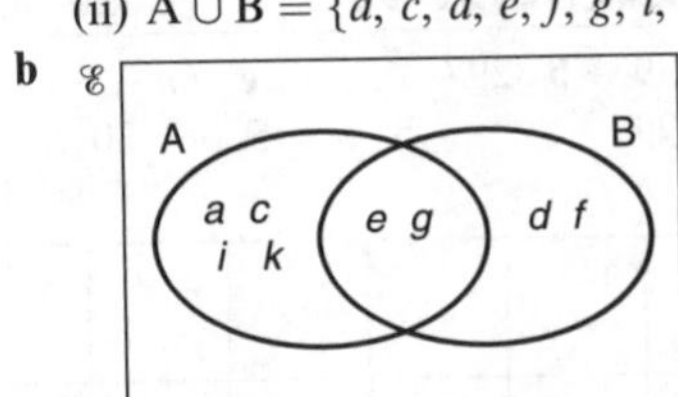

3
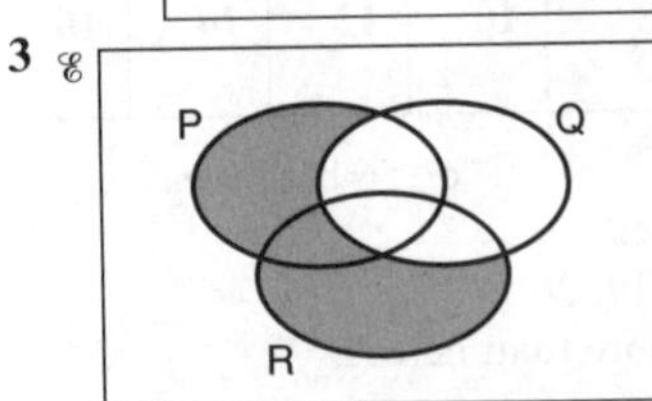

5
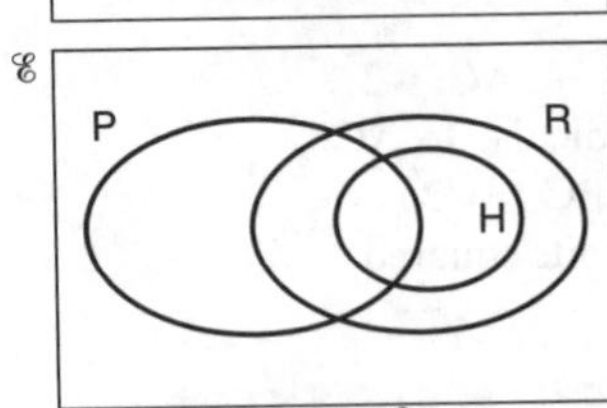

7 a $\frac{1}{3}$ **b** $\frac{2}{3}$ **c** $\frac{1}{4}$

9 a

X	1	2	3	4	5	6
1	0	1	2	3	4	5
2	1	0	1	2	3	4
3	2	1	0	1	2	3
4	3	2	1	0	1	2
5	4	3	2	1	0	1
6	5	4	3	2	1	0

b $p(X = 1) = \frac{5}{18}$

c $p(X > 2) = \frac{1}{3}$

11 a $\frac{1}{7}$ **b** $\frac{5}{21}$ **c** $\frac{16}{21}$

13 Mean = 6.4, median = 6.5, mode = 7

15 a 60 matches **b** 149 points **c** 0.6

17 a (20, 0), (40, 0), (50, 33), (55, 55), (60, 74), (70, 92), (90, 100)

b

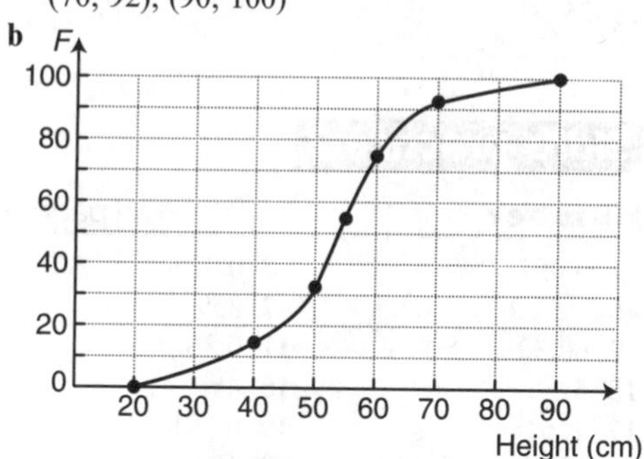

c 53 cm, 47 cm, 61 cm, 14 cm

19 a $m = 159$, IQR = 14; $m = 168$, IQR = 14

b Tribe B are, on average, 9 cm taller.

REVISION EXERCISE 99★ (page 423)

1

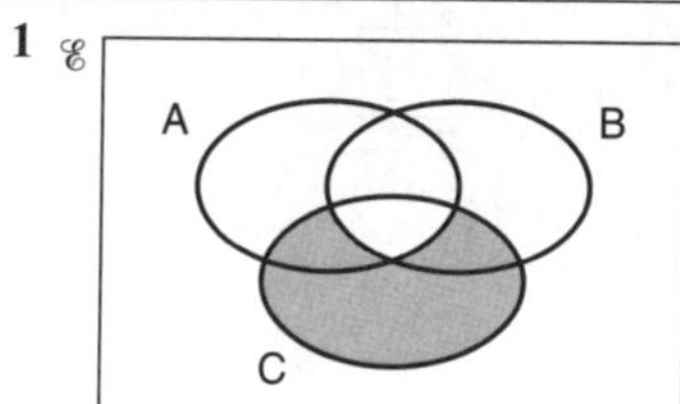

3

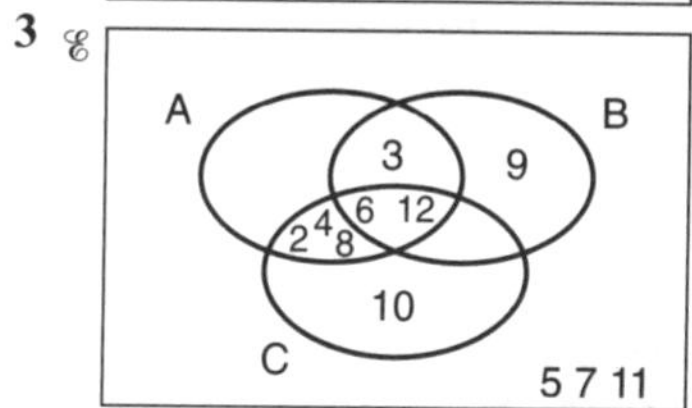

5 a 34.6 **b** 33.5

7 a $\frac{1}{3}$ **b** $\frac{1}{6}$

9 a $\frac{1}{5}$ **b** $\frac{4}{5}$

11 b $\frac{5}{18}$

13 a $(x - 5)(x + 4) = 0$, so $x = 5$

b $\frac{13}{24}$

15 a Before: 41; 29, 57; 28

After: 48; 33, 63; 30

c Diet improves fitness

17 a $f = 16, 15, 19, 14, 24, 12$

b 100 responses **c** 29.95 years

19 a $f = 6, 12, 15, 12, 10, 6, 3$

b f.d. = 0.4, 0.8, 1.5, 1.2, 1, 0.2, 0.1

c 72.1 min

Examination papers

Paper 1A (page 428)

1 \$344 500

3 a 232 cm^2 **b** 73.5 cm^2

5 a 292 **b** 53

7 a (i) p, q (ii) {p, q, r, s, t, u, v}

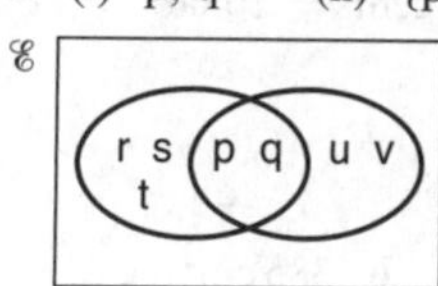

9 24.4 cm

11 a $\frac{2}{5}$ **b** (2.75, −2.5)

13 a Reflection in line $x + y = 0$

b (−4, 1)

15 $\frac{x - 5}{2x - 3}$

17 a $F = 250 \times \frac{1}{d^2}$ **c** 1.74 N

19 a (i) 7 (ii) 100

b $x \mapsto \frac{x - 1}{3}$

c $x \mapsto 3x^2 + 1$

d $x = \pm 1$

Paper 1B (page 432)

1 3.5

3 $F = 2C + 30 = 2(C + 15)$

5 a 40 ml **b** 200 ml

7 $4\frac{2}{5}$

9 a a^2b^3 **b** c^2d^3 **c** e^4 **d** f^2

11 a 4 cm **b** 11.25 ml **c** 405 g

13

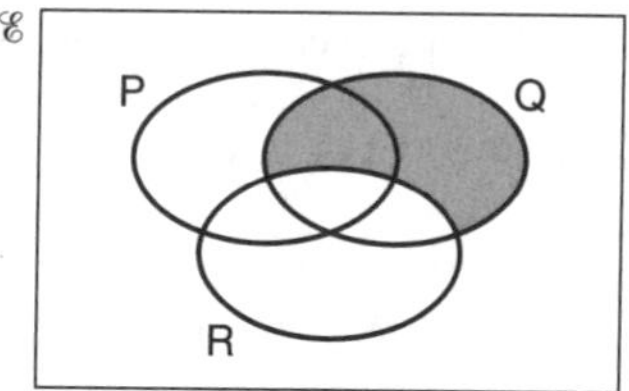

15 a $\frac{1}{4}$ **c** $\frac{1}{64}$ **d** $\frac{5}{32}$

17 \$32 000

19 a $12a^2 - 7ab - 12ab^2$ **b** $9c^2d^6$

c f^3g^{-2} **d** $\frac{1}{4h^6}$

21 $(4, -1), \left(\frac{8}{3}, \frac{-7}{3}\right)$

Paper 2A (page 436)

1 a 115° **b** 59 km

3 a $6\frac{1}{8}$ **b** 24 **c** $1\frac{17}{24}$

7 a Angles in a semi-circle; 50°
b 7.7 cm
c 10.1 cm

9 a −3 **b** $y + 3x - 5 = 0$
c $y + 3x + 5 = 0$

11 b $\frac{15}{32}$

13 a $119\,\text{cm}^3$ **b** 60 cents = \$0.60

15 $\cos\theta = \dfrac{164 - x^2}{160}$

17 a $\frac{5}{33}$ **b** $\frac{28}{33}$ **c** $\frac{1}{22}$

19 a $539\,\text{cm}^3$ **b** $385\,\text{cm}^2$

21 2 hours

Paper 2B

(page 440)

1 \$6.72
3 a 0.3 **b** 140
5 a $p^2 - 3p$ **b** $9q$
7 a $\frac{2}{5}$ **b** 19.2 m
9 a 3^9 **b** 3^4 **c** 1
11

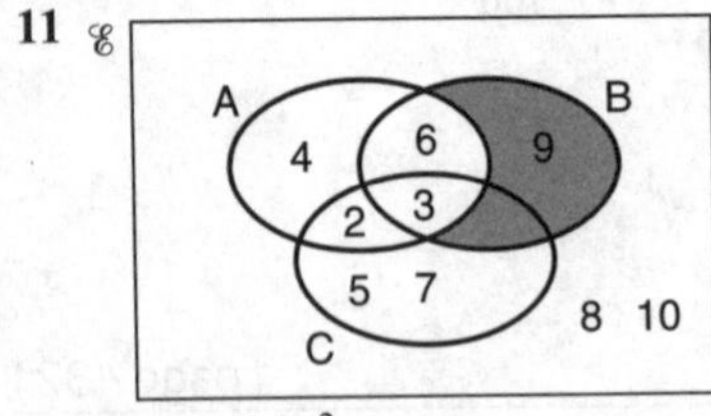

13 $u = \dfrac{2s - at^2}{2t}$
15 25%
17 a 56°
b Angle at centre is twice angle at circumference
c 124°
d Opposite angles of a cyclic quadrilateral
e 56°
19 a 5.04, 2.66, 1.89, 2, 3.185, 4.5
c 0.35, 1.53
d 1, 1.62
21 a 14.65 secs **b** 6.91 m/s

Skills practice 1

Number

(page 445)

1 67 **3** 58 **5** 38.06
7 −1.6 **9** 46.13 **11** 0.49
13 3 **15** 216 **17** 10^{-1}
19 13 **21** $5\frac{13}{21}$ **23** $\frac{36}{49}$
25 2.7×10^5 **27** 0.0469 **29** $\frac{1}{6}$
31 $\frac{1}{5}$ **33** $\frac{6}{7}$

Algebra

(page 445)

1 $6x^2$ **3** $8x^4$
5 $2a$ **7** $x^3 - 3x$
9 $2ab(2b - 3a)$ **11** $(x - 2)(x + 1)$
13 $x = 23$ **15** $x = 0.9$
17 $x = 3.25$ **19** $x = \pm 2$
21 12 **23** $-3\frac{1}{4}$
25 25

Skills practice 2

Number

(page 446)

1 0.067 **3** 0.453
5 −55.2 **7** 259
9 60.32 **11** 0.36
13 5 **15** 45
17 216 **19** 0.391
21 33 **23** 60
25 50 **27** 1

Algebra

(page 446)

1 $\dfrac{2a}{b}$ **3** $\dfrac{y^3}{z}$
5 $\dfrac{3 + x}{2}$ **7** $\dfrac{5y}{2x}$
9 $(x - 2)(x - 3)$ **11** $3ab^2(2a^2 - 9b)$
13 $x = \dfrac{c - b}{a}$ **15** $x = \dfrac{c}{a} + b$
17 $x = \dfrac{by}{a + c}$ **19** 8
21 1 **23** 7
25 4

Skills practice 3

Number

(page 447)

1 1.869 **3** 29 **5** 49.84
7 3.4 **9** 44.99 **11** 162
13 1.2 **15** 10^{-1} **17** 24
19 2 **21** 9.6 **23** 16
25 3.6×10^2 **27** 1.5×10^7

Algebra

(page 447)

1 $-x^2$ **3** $8x^3y^3$ **5** $\dfrac{2a}{b}$
7 $x + 2$ **9** $\dfrac{5a}{6}$ **11** $\dfrac{xy}{6}$
13 5 **15** 0.45 **17** 6.2
19 ± 5 **21** 7, −2 **23** 4
25 64

Skills practice 4

Number (page 448)

1 3.4
3 37
5 0.9
7 10^6
9 4
11 100
13 600
15 10
17 4.22×10^8
19 2.52×10^{16}

Algebra (page 448)

1 a^8
3 $\dfrac{3b}{a}$
5 $4x$
7 $x^2 - 5x + 6$
9 $\dfrac{5x}{6}$
11 x
13 5
15 0.6
17 0, 25
19 −2, 4
21 −12
23 $\dfrac{2a + 5b}{3}$
25 $\pi\left(\dfrac{E}{q}\right)^2$
27 24.5
29 −0.6

CHALLENGES (page 449)

1 1003
3 $\frac{1}{4}$
5 2π km
7 333 333 332 666 666 667
9 59 (43 & 16)
11 100π
13

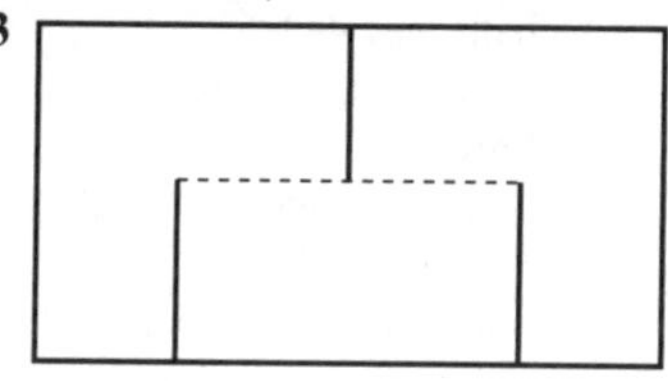

15 Dividing by zero $(a - b)$ in the third line
17 7.5 cm
19 $\frac{1}{3}$ litre
21 **a** $70.7\,\text{m}^2$ **b** $34.4\,\text{m}^2$
23 10
25 24 cm
27 **a**

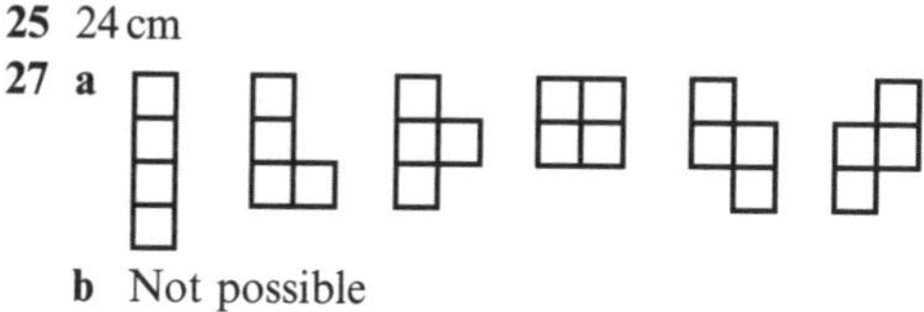

b Not possible
29 60
31 $6(\sqrt{3} - 1)$
33 Investigate this how you wish. A typical method might be to use percentage errors from $\pi = 3.141\,592\,654$ (to 10 s.f.).

Rank	Percentage error (2 s.f.)
1 Ramanujan	-3.3×10^{-8}
2 Tsu Chung-Chih	$+8.5 \times 10^{-6}$
3 Hipparchus	$+2.4 \times 10^{-3}$
4 Fibonacci	$+7.2 \times 10^{-3}$
5 Archimedes	+0.040
6 Babylon	−0.53
7 Egypt	+0.60
8 Brahmagupta	+0.66
9 *Bible*	−4.5
10 Wallis (as printed!)	−5.4

Investigate Wallis' expression further.

Fact finder: World population (page 455)

1 **a** 1927 **b** 1960
3 3.6%
5 324
9 **a** 1438 **b** 28 235
11 2.27×10^{23}

Fact finder: Mount Vesuvius (page 457)

1 1736 years
3 ≈53 333 years
5 932 °F
7 **a** $3.20 \times 10^7\,\text{m}^3$ **b** $7.25 \times 10^7\,\text{m}^3$
9 $4.63 \times 10^4\,\text{m}^3/\text{s}$
11 $3.69 \times 10^9\,\text{m}^3$

Fact finder: The human body (page 459)

1 14.1%
3 0.117 litres per hour
5 5.49×10^{-6} mm/sec
7 81.5 beats/min
9 $2.47 \times 10^{15}\,\text{mm}^3$, 10 300 × volume of the classroom.

Fact finder: Wimbledon (page 461)

1 129 in 2006
3 **a** 162% **b** 171%
5 0.360 secs
7 **a** +166% **b** +11.1% **c** +7.59%
9 **a** 0.0497 kg **b** 686 kg

Index

D

E

F

G

H

I

K

L

M

N

O

P

Q

R

S

T

U

V

W